Lecture Notes in Artificial Intelligence 5855

Edited by R. Goebel, J. Siekmann, and W. Wahlster

Subseries of Lecture Notes in Computer Science

T0189561

Hepu Deng   Lanzhou Wang
Fu Lee Wang   Jingsheng Lei (Eds.)

Artificial Intelligence and Computational Intelligence

International Conference, AICI 2009
Shanghai, China, November 7-8, 2009
Proceedings

 Springer

Series Editors

Randy Goebel, University of Alberta, Edmonton, Canada
Jörg Siekmann, University of Saarland, Saarbrücken, Germany
Wolfgang Wahlster, DFKI and University of Saarland, Saarbrücken, Germany

Volume Editors

Hepu Deng
RMIT University, School of Business Information Technology
City Campus, 124 La Trobe Street, Melbourne, Victoria 3000, Australia
E-mail: hepu.deng@rmit.edu.au

Lanzhou Wang
China Jiliang University, College of Metrological Technology and Engineering
Hangzhou 310018, Zhejiang, China
E-mail: lzwang@cjlu.edu.cn

Fu Lee Wang
City University of Hong Kong, Department of Computer Science
83 Tat Chee Avenue, Kowloon Tong, Hong Kong, China
E-mail: flwang@cityu.edu.hk

Jingsheng Lei
Hainan University, College of Information Science and Technology
Haikou 570228, China
E-mail: jshlei@hainu.edu.cn

Library of Congress Control Number: 2009936961

CR Subject Classification (1998): I.2, F.2, I.5, I.4, J.3, J.2

LNCS Sublibrary: SL 7 – Artificial Intelligence

ISSN   0302-9743

ISBN 978-3-642-05252-1 Springer Berlin Heidelberg New York

springer.com

© Springer-Verlag Berlin Heidelberg 2009

Typesetting: Camera-ready by author, data conversion by Scientific Publishing Services, Chennai, India
Printed on acid-free paper   SPIN: 12781256   06/3180   5 4 3 2 1 0

Preface

The 2009 International Conference on Artificial Intelligence and Computational Intelligence (AICI 2009) was held during November 7–8, 2009 in Shanghai, China. The technical program of the conference reflects the tremendous growth in the fields of artificial intelligence and computational intelligence with contributions from a large number of participants around the world.

AICI 2009 received 1,203 submissions from 20 countries and regions. After rigorous reviews, 79 high-quality papers were selected for this volume, representing an acceptance rate of 6.6%. These selected papers cover many new developments and their applications in the fields of artificial intelligence and computational intelligence. Their publications reflect a sustainable interest from the wide academic community worldwide in tirelessly pursuing new solutions through effective utilizations of artificial intelligence and computational intelligence to real-world problems.

We would like to specially thank all the committee members and reviewers, without whose timely help it would have been impossible to review all the submitted papers to assemble this program. We also would like take this opportunity to express our heartfelt appreciation for all those who worked together in organizing this conference, establishing the technical programs and running the conference meetings. We greatly appreciate the authors, speakers, invited session organizers, session Chairs, and others who made this conference possible. Lastly, we would like to express our gratitude to the Shanghai University of Electric Power for the sponsorship and support of the conference.

We trust that the participants of this conference had opportunities to meet both old and new friends from all over the world, to exchange ideas, to initiate collaborations in research, to develop lasting friendships, as well as to visit places in Shanghai and China.

November 2009

Hepu Deng
Lanzhou Wang
Fu Lee Wang
Jingsheng Lei

Organization

Organizing Committee

General Co-chairs

Jialin Cao	Shanghai University of Electric Power, China
Jingsheng Lei	Hainan University, China

Program Committee

Co-chairs

Hepu Deng	RMIT University, Australia
Lanzhou Wang	China JILIANG University, China

Local Arrangements

Chair

Hao Zhang	Shanghai University of Electric Power, China

Proceedings

Co-chairs

Fu Lee Wang	City University of Hong Kong, Hong Kong
Jun Yang	Shanghai University of Electric Power, China

Publicity

Chair

Tim Kovacs	University of Bristol, UK

Sponsorship

Chair

Zhiyu Zhou	Zhejiang Sci-Tech University, China

Program Committee

Ahmad Abareshi	RMIT University, Australia
Stephen Burgess	Victoria University, Australia
Jennie Carroll	RMIT University, Australia
Eng Chew	University of Technology Sydney, Australia
Vanessa Cooper	RMIT University, Australia
Minxia Luo	China Jiliang University, China
Tayyab Maqsood	RMIT University, Australia
Ravi Mayasandra	RMIT University, Australia
Elspeth McKay	RMIT University, Australia
Alemayehu Molla	RMIT University, Australia
Konrad Peszynski	RMIT University, Australia
Siddhi Pittayachawan	RMIT University, Australia
Ian Sadler	Victoria University, Australia
Pradip Sarkar	RMIT University, Australia
Carmine Sellitto	Victoria University, Australia
Peter Shackleton	Victoria University, Australia
Sitalakshmi Venkatraman	University of Ballarat, Australia
Leslie Young	RMIT University, Australia
Adil Bagirov	University of Ballarat, Australia
Philip Branch	Swinburne University of Technology, Australia
Feilong Cao	China Jiliang University, China
Maple Carsten	University of Bedfordshire, UK
Caroline Chan	Deakin University, Australia
Jinjun Chen	Swinburne University of Technology, Australia
Richard Dazeley	University of Ballarat, Australia
Yi-Hua Fan	Chung Yuan Christian University Taiwan, Taiwan
Richter Hendrik	HTWK Leipzig, Germany
Furutani Hiroshi	University of Miyazaki, Japan
Bae Hyeon	Pusan National University, Korea
Tae-Ryong Jeon	Pusan National University, Korea
Sungshin Kim	Pusan National University, Korea
Wei Lai	Swinburne University of Technology, Australia
Edmonds Lau	Swinburne University of Technology, Australia
Qiang Li	University of Calgary, Canada
Xiaodong Li	RMIT University, Australia
Kuoming Lin	Kainan University, Taiwan
YangCheng Lin	National Dong Hwa University, Taiwan
An-Feng Liu	Central South University, China
Liping Ma	University of Ballarat, Australia
Costa Marly	Federal University of the Amazon, Brazil
Jamie Mustard	Deakin University, Australia
Syed Nasirin	Brunel University, UK
Lemai Nguyen	Deakin University, Australia
Heping Pan	University of Ballarat, Australia

Craig Parker	Deakin University, Australia
Wei Peng	RMIT University, Australia
Adi Prananto	Swinburne University of Technology, Australia
Takahashi Ryouei	Hachinohe Institute of Technology, Japan
Cheol Park Soon	Chonbuk National University, Korea
Andrew Stranier	University of Ballarat, Australia
Zhaohao Sun	University of Ballarat, Australia
Sleesongsom	Suwin Chiangrai College, Thailand
Arthur Tatnall	Victoria University, Australia
Luba Torline	Deakin University, Australia
Schetinin Vitaly	University of Bedfordshire, UK
Zhichun Wang	Tianjin University, China
Shengxiang Yang	University of Leicester, UK
Chung-Hsing Yeh	Monash University, Australia
Yubo Yuan	China Jiliang University, China
Hossein Zadeh	RMIT University, Australia
Dengsheng Zhang	Monash University, Australia
Irene Zhang	Victoria University, Australia
Jianying Zhang	Cancer Victoria, Australia
Tony Zhang	Qingdao Univesity, China
Weijian Zhao	China Jiliang University, China
Yan-Gang Zhao	Nagoya Institute of Technology, Japan
Xiaohui Zhao	Swinburne University of Technology, Australia

Reviewers

Adil Bagirov	Chen Lifei	Du Junping
Ahmad Abareshi	Chen Xiang	Du Xufeng
Bai ShiZhong	Chen Ling	Duan Yong
Bi Shuoben	Chen Dongming	Duan Rongxing
Bo Rui-Feng	Chen Ming	Fan Jiancong
Cai Weihua	Chen Ting	Fan Xikun
Cai Zhihua	Chen Yuquan	Fan Shurui
Cai Kun	Chen Ailing	Fang Zhimin
Cao Yang	Chen Haizhen	Fang Gu
Cao Qingnian	Chu Fenghong	Fuzhen Huang
Cao Guang-Yi	Chung-Hsing Yeh	Gan Zhaohui
Cao Jianrong	Congping Chen	Gan Rongwei
Carmine Sellitto	Cui Shigang	Gao chunrong
Chai Zhonglin	Cui Mingyi	Gao Xiaoqiang
Chang Chunguang	DeJun Mu	Gao Jiaquan
Chen Yong Liang	Deng Liguo	Gao Jun
Chen Haiming	Deng-ao Li	Gao Boyong
Chen Yong	Ding Yi-jie	Gu Xuejing
Chen Pengnian	Dingjun Chen	Gu Tao

Guang Yang
Guo Li Jun
Guo Yecai
Han Fei
Hao Xiuqing
He Hongjiang
He Shengping
He Bingwei
He Xingshi
He Zhiwei
He Yue
Hendrik Richter
Ho Zih-Ping
Hu Fu Yuan
Huang Pei-Hwa
Huang Zhiwei
Huang Rongbo
Huang Nantian
Ji Genlin
Ji Tingwei
Ji Xia
Ji Chang-Ming
Jia Qiuling
Jia Fang
Jiang Jiafu
Jianying Zhang
Jiao Feng
Jiechun Chen
Jingbo Xia
Jingfang Liu
Jin-min Wang
Ju-min Zhao
Koga Takanori
Kwok Ngai Ming
Lai Jiajun
Lai Lien-Fu
Lai-Cheng Cao
Lei Zhang
Lei Xiujuan
Leng Cuiping
Li Yu
Li Xinyun
Li Xiao
Li Shun-Ming
Li Yun
Li Haiming
Li Chengjia

Li Xiang
Li Ming
Li Guangwen
Li Guohou
Li Hongjiao
Liang Benliang
Lin Yuesong
Lin Dong Mei
Lin Chen
Lin Jipeng
Liu Jun
Liu Zhijie
Liu Xiaoji
Liu Hui
Liu Anping
Liu Zhide
Liu Jihong
Liu Heng
Liu Jin
Liu Wenyu
Liu Dianting
Liu Lei
Liu Jin
Liu Jinglei
Liu Hui
Liu Guangli
Liu Yang
Liu Ying
Liu Hao
Liu Jian
Liu Ying-chun
Lu Yaosheng
Lu Bin
Lu Yuzheng
Lu Huijuan
Ma Xianmin
Matsuhisa Takashi
Meng Jianliang
Miao Xuna
Ni Jian-jun
Nie Qing
Ning Ning
Niu Junyu
Ouyang Zhonghui
Pan Zhenghua
Pan Jingchang
Pan Defeng

Peng Qinke
Perng Chyuan
Phaisangittisagul Ekachai
Qiang Niu
Qiang Li
Qiao Mei
Qiu Jiqing
Rao Congjun
Ren Li
Ren Kaijun
Ren Chunyu
Richard Dazeley
Ruiming Jia
Sang Yan-fang
Sanyou Zeng
Shen Xiaojing
Shen Xiaohua
Sheng Zhongqi
Sheng Xueli
Shi Qian
Shi Guoliang
Shi Xuerong
Shuai Ren
Si Chun-di
Sitalakshmi Venkatraman
Song Xiaoyu
Su H.T.
Sun Zhanquan
Sun Yanxia
Sun Quanping
Sun Feixian
Sun Ziwen
Suwin Sleesongsom
Tan Junshan
Tan Chunqiao
Tang Xianghong
Tao Luo
Tao Zhengru
Tao Zhang
Tian Fu-qing
Tong Zhi-Wei
Uchino Eiji
Ullrich Carsten
Wan Li
Wang Cheng
Wang Hao
Wang Li-dong

Wang Hao
Wang Li-dong
Wang Jie-sheng
Wang Guoqiang
Wang Caihong
Wang Jingmin
Wang Shuangcheng
Wang Chujiao
Wang Junwei
Wang XiaoMing
Wang Xin
Wang Feng
Wang Yuanqing
Wang Yong
Wang Xiang
Wang Yu-Jie
Wang Yufei
Wang Hongsheng
Wang Xiuqing
Wang Wenchuan
Wang Dong
Wang Haibo
Wang Dawei
Wang Hui
Wang Junnian
Wang Xuhong
Wang Xiaolu
Wan-gan Song
Wei Peng
Wen Mi
Wenjun Liu
Wu Xia
Wu Hongli
Wu Xiaojun
Wu Xiaojin

Wu Xiaoqin
Wu Zhangjun
Wu Da-sheng
Xi Liang
Xia Qingjun
Xiao Xiaohong
Xiao Hongying
Xie Yanmin
Xie Zhenping
Xie Jianhong
Xiuli Yu
Xiumin Yang
Xiuxia Yang
Xizhong Shen
Xu Jie
Xu Hailong
Xu Hua
Xu Xinsheng
Xu Yan
Xu Zhangyan
Xue Heru
Xue Yang
Xue Heng
Xueqin Lu
Yan Ping
Yan Wang
Yan Tie
Yang Genghuang
Yang Ming
Yang Jiadong
Yang Xiaobing
Yang Jun
Yang Liping
Yang-Cheng Lin
Yao Zhijun

Ye Feiyue
Yi-Hua Fan
Yin Chunxia
Yong Yang
Yong Qidong
You Xiao-ming
You-hua Jiang
Yu Xinqiao
Yuhui Zhao
Yuyi Zhai
Zhan Yong-zhao
Zhang Huaixiang
Zhang Meng
Zhang Sumin
Zhang Yizhuo
Zhang Erhu
Zhang Jianxin
Zhang Caiqing
Zhang Jian
Zhang Yanjie
Zhang Dabin
Zhang Xiang
Zhang Jingjun
Zhang Zhenya
Zhang Ke
Zhang Xuejun
Zhang Xuehua
Zhang Junran
Zhang Xueming
Zhang Changjiang
Zhao Ping
Zhao Fu
Zhao Liaoying
Zhao Hui

Table of Contents

Information Security

Immune Computation

Genetic Algorithms

Fuzzy Computation

Biological Computing

Applications of Computational Intelligence

Ant Colony Algorithm

Robotics

Pattern Recognition

Neural Networks

Natural Language Processing

Machine Vision

Machine Learning

Logic Reasoning and Theorem-Proving

Knowledge Representation and Acquisition

Intelligent Signal Processing

Intelligent Scheduling

Intelligent Information Retrieval

Intelligent Information Fusion

Intelligent Image Processing

Heuristic Searching Methods

Fuzzy Logic and Soft Computing

Distributed AI and Agents

Data Mining and Knowledge Discovering

Applications of Artificial Intelligence

Others

An Experimental Research on Vector Control of Induction Motor Based on Simple Model

Yinhai Zhang, Jinfa Ge, Weixia Liu, and Qin Wang

College of Informatics & Electronics
Zhejiang Sci-Tech University
Hangzhou City, Zhejiang Province, China
nikon_nike@163.com, gejinfa@yahoo.com.cn

Abstract. Given the heavy computation, easy saturation and cumulate errors of conventional direct vector control, the vector control of induction motor based on simple model is studied and the detailed scheme is described on the basis of the decomposing and approximating the rotor flux. Because of the direct closed-loop control of the magnetizing current and the torque current and the complex current regulator is completed by PI regulator, so the direct vector control of induction motor is simplified. The experimental results show that the proposed method is effective in decreasing the dynamic disturbance and has the advantages of the simplicity of the code program, rare saturation and shocks.

Keywords: vector control, induction motor, rotor flux, decoupler.

1 Introduction

The vector control can make the induction motor gain the excellent dynamic performance that is similar to the one of the DC motor. Now, the vector control that based on the rotor field orientated control is paid enough attention because it easily realizes the absolute decoupling between the magnetizing current and the torque current [1-2]. The direct vector control and the indirect vector control are the two most commonly used in the vector control of induction motor system. The former has the speed closed-loop control which includes torque closed-loop and the flux closed-loop control system, the latter is an open-loop flux control system. There are a large number of operations such as the differential and the product in the direct vector control. Furthermore, the complex processes of current regulator in the voltage-type inverter bring the heavy computation [3-5], easy saturation, cumulate errors, and other uncertain disturbance, resulting in deterioration of system performance.

Therefore, in order to overcome the above-mentioned disadvantages of the direct vector control and maintain its advantages of performance, the vector control of induction motor based on simple model is studied and the detailed scheme is described in this paper. After the decomposing and approximating the rotor flux, the magnetizing current and the torque current are directly close-looped and the complex process of current regulator in the voltage-type inverter is completed by PI regulator to achieve the regulation without static error. Thus, the proposed method simplifies the direct vector

H. Deng et al. (Eds.): AICI 2009, LNAI 5855, pp. 1–8, 2009.

control of induction motor, and makes the direct vector control system easy, clear, less code program, rare saturation and shocks.

2 Conventional Direct Vector Control

Fig.1 shows the system of the conventional direct vector control of induction motor. This system which is built based on the rotor field orientated control includes the speed control subsystem and the flux control subsystem, and the inner loop of the speed control subsystem is the torque closed-loop control system. There are the speed regulator (ASR), the torque regulator (ATR) and the flux regulator (AΨR) in the direct vector control system of induction motor. The current transform includes the Clarke and the Park conversion.

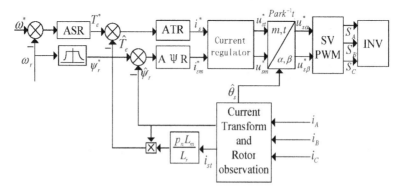

Fig. 1. The conventional direct vector control system of induction motor

The following are the main formulas that used in the conventional direct vector control system of the induction motor.

Rotor flux is expressed as follows

$$\Psi_r = \frac{L_m}{T_r p + 1} i_{sm} \tag{1}$$

The feedback of the torque calculation equation is

$$T_e = \frac{P_n L_m}{L_r} i_{st} \Psi_r \tag{2}$$

The angular speed of synchronous rotation is given by t-he following formula

$$\omega_s = \omega_r + \omega_{sl} = \omega_r + \frac{L_m i_{st}}{T_r \Psi_r} \tag{3}$$

In the most example applications, the voltage-type inverter is used widely, so it is necessary to change the current to voltage. This process is called current regulator. And the equations of the conversion are

$$
\begin{cases}
u_{sm} = R_s(1 + \dfrac{L_s}{R_s}p\dfrac{\sigma T_r p+1}{T_r p+1})i_{sm} \\[2ex]
\quad - \sigma L_s(\omega_r + \dfrac{T_r p+1}{T_r}\dfrac{i_{st}}{i_{sm}})i_{st} \\[2ex]
u_{st} = [R_s(\dfrac{\sigma L_s}{R_s}p+1) + \dfrac{L_s}{T_r}(\sigma T_r p+1)]i_{st} \\[2ex]
\quad + \omega_r L_s \dfrac{\sigma T_r p+1}{T_r p+1})i_{sm}
\end{cases}
\qquad (4)
$$

The parameters of equation (1) to (4) are described as follows:

$$
\sigma = 1 - \frac{L_m^2}{L_s L_r}
$$

L_m — Mutual inductance
L_r — Rotor inductance
R_s — Stator resistance
L_s — Stator inductance
T_r — Rotor time constant
p — Differential operator
ω_r — Rotor angular speed
ω_{sl} — Slip angular speed
p_n — The number of pole pairs
i_{sm} — M-axis stator current
i_{st} — T-axis stator current
u_{sm} — M-axis stator voltage
u_{st} — T-axis stator voltage

From the equation (1) to (3), we can see that the accuracy of estimation of equation (2) and (3) depend on (1). It shows that the importance of the accuracy of estimation of the rotor flux. Because of the existence of differential part, the transient drift will appear in the part of direct current following the transient changes of the digital operation when the rectangular discrete integrator is used, thus making the value of estimated rotor flux inaccurate and then reducing the calculation accuracy of the torque and the synchronous speed. Furthermore, the existence of a large number of the differential and product operations in the equations described by (4) will bring the heavy computation, easy saturation and cumulate errors. Especially, in the low speed region, the torque is very prone to oscillate.

3 Vector Control System Based on Simple Model

Fig. 2 gives the vector control system based on simple model. In this paper, the rotor flux is decomposed and approximated. Then the magnetizing current and the torque current are for closed-loop control directly in order to overcome these shortcomings of

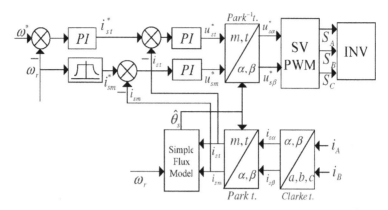

Fig. 2. The vector control system of induction motor based on simple model

the conventional direct vector control. The system includes the speed control subsystem and the torque current control subsystem. And the inner loop of the speed control subsystem is the torque current closed-loop control subsystem. The system is composed of three PI regulators, the speed regulator, the torque current regulator and the magnetizing current regulator.

The process of the main formulas derivation that used in the vector control system based on simple model are as follows, in which they are discretized to fit DSP computing.

The equation (1) can be rewrote as

$$T_r p\psi_r + \psi_r = L_m i_{sm} \tag{5}$$

And the rotor flux equation that based on MT reference frame is

$$\psi_r = L_m i_{sm} + L_r i_{rm} \tag{6}$$

$L_{\sigma r}$ is rotor leakage inductance and it is small enough compared with the mutual inductance L_m. So we can consider that L_m and L_r are approximately equal in value according to $L_r = L_m + L_{\sigma r}$. Furthermore, in the rotor field orientated control system, the rotor flux must maintain constant in order to achieve an ideal speed-adjusting performance under the rating frequency. And above the rating frequency, the flux weakening control is adopted and is usually achieved through the look-up table. $L_m i_{sm}$ is regard as constant and contained by $L_r i_{rm}$. So the equation (6) can be simplified as

$$\psi_r = L_m i_{rm} \tag{7}$$

In equation (7), i_{rm} is the M-axis component of rotor current, and applying equation (7) to (5), we can obtain

$$T_r \frac{di_{rm}}{d_t} + i_{rm} = i_{sm} \tag{8}$$

After discrete differential, equation (8) can be rewrote as

$$i_{rm}(k+1) = i_{rm}(k) + \frac{T_s}{T_r}(i_{sm}(k) - i_{rm}(k)) \tag{9}$$

Where T_s is the switching period. And as the same method, equation (3) can be modified as

$$\omega_s(k+1) = \omega_r(k+1) + \frac{1}{T_r} \frac{i_{st}(k)}{i_{rm}(k+1)} \tag{10}$$

The equation (9) and (10) are derived based on the decomposition and approximation of the rotor flux, which are considered as crucial observation equations of the system showed in Fig. 2, being used to calculate the angular speed of synchronous rotation. Comparing to the conventional direct vector control system, we can find that in the simplified vector control system it is not to calculate the angular speed of synchronous rotation by calculating the rotor flux, but to obtain directly through the rotor magnetizing current. Then the heavy computational processes are ignored, so the saturation drift and cumulative errors are avoided. And we will have a good foundation during the study of the real-time high-performance vector control system.

It can be seen from Fig. 2 that the magnetizing current and the torque current are for closed-loop control directly rather than the torque and flux in the simplified vector control system that compared to the conventional vector control. So the calculation of torque feedback is avoided, and the whole system can regulate the torque and the flux more quickly and efficiently. In addition, the complex process of current regulator is included by PI regulator to achieve the regulation without static error. So it make the system simple, clear and easy computation, and more suitable for real-time high-performance control.

4 Experiment Analysis

The hardware and software environments of the experiment: the power inverter part is designed with IGBT (IRGB15B60KD) which is produced by IR. The control system part adopts the TI 2407DSP EVM by Wintech. The current sensor is LA28-NP produced by the company of LEM. The resolution of the speed encoder is 1000P/R, and the type of digital oscilloscope is TektronixTDS2002. Switching frequency is 16 KHz, switching period $T_s = 62.5\mu s$. PI parameters: in the current loop, $P=1.12$, $I=0.0629$ and in the speed loop, $P= 4.89$, $I=0.0131$.

The parameters of the induction motor: the rated power: $P_n =500W$, the rated current: $I_n=2.9A$, the rated voltage: $V_n=220V$, the rotor resistance: $R_r=5.486\Omega$, the stator resistance: $R_s =4.516\Omega$, the mutual inductance: $L_m =152mH$, $L_{\sigma r} =13mH$, $L_r =165mH$, $p_n=2$, $L_s =168mH$.

Fig. 3 gives the trajectory of the total flux circle when the operation frequency f =5Hz in the vector control system of induction motor based on simple model. According to the experimental waveform, we can find that the trajectory of the total flux circle is near round and that the system has a good running status when the system is running at low speed.

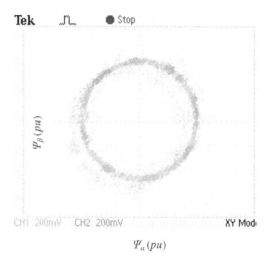

Fig. 3. The total flux circle with the operation frequency f =5Hz

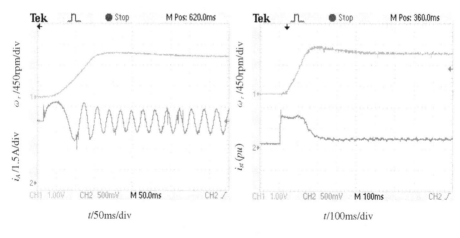

(a) rotor speed and phase current   (b) rotor speed and torque current

Fig. 4. The start-up of induction motor with no-load

Fig. 4 represents the experimental waveforms in the process of induction motor starting up with no-load to running at the set speed of f =30Hz by the simplified vector control system. Fig. 4a gives the waveform of the rotor speed and the stator phase current. Fig. 4b shows the waveform of the rotor speed and the torque current. In Fig. 4, the peak to peak value of phase current is less than 2A, and the adjusting time is 200ms, and the maximum overshoot of the speed is about 13%. During the increasing speed, i_{st} reaches the biggest value to meet the requirements of increasing speed quickly and when the speed reach to a certain degree, the torque current gradually decreased until the speed reach the feed-forward speed.

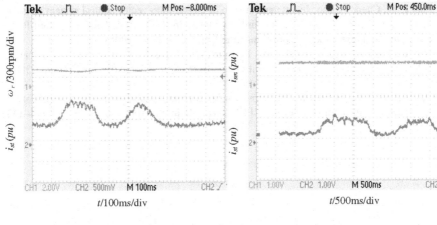

(a) rotor speed and torque current   (b) magnetizing current and torque current

Fig. 5. The experimental wave with random load at $f = 10$Hz

The experimental waveforms that under the random load when the motor runs at the steady state of $f = 10$Hz is represented by Fig. 5. Fig. 5a gives the waveforms of the rotor speed and torque current, and we can see that the maximum dynamic speed downhill is about 16% and the recovery time is less than the time of disturbance operation. Fig. 5b shows the waveforms of the magnetizing current and torque current, and we can find that the magnetizing current i_{sm} always remain unchanged and be able to get ideal decoupling with the torque current. So the vector control of induction motor based on simple model works well in the anti-disturbance performance.

5 Conclusion

In this paper, the vector control of induction motor based on simple model is studied and the detailed scheme is described through the decomposing and approximating the rotor flux. The proposed method makes the system easy and has less code program, shorter execution time. The experimental results show that the anti-disturbance performance of the simplified vector control system is reliable and efficient and suitable for real-time high-performance control system.

References

1. Noguchi, T., Kondo, S., Takahashi, L.: Field-oriented control of an induction motor with robust on-line tuning of its parameters. IEEE Trans. Industry Applications 33(1), 35–42 (1997)
2. Telford, D., Dunnigan, M.W.: Online identification of induction machine electrical parameters for vector control loop tuning. IEEE Trans. Industrial Electronics 50(2), 253–261 (2003)

3. Zhang, Y., Chen, J., Bao, J.: Implementation of closed-loop vector control system of induction motor without voltage feedforward decoupler. In: IEEE 21st Annual Applied Power Electronics Conference and Exposition, APEC, pp. 1631–1633 (2006)
4. Lorenz, Lawson, R.D., Donald, B.: Performance of feedforward current regulators for field-oriented induction machine controllers. IEEE Trans. Industry Applications 23(1.4), 597–602 (1987)
5. Del Blanco, F.B., Degner, M.W., Lorenz, R.D.: Dynamic analysis of current regulators for AC motors using complex vectors. IEEE Trans. Industry Applications 35(6), 1424–1432 (1999)

An United Extended Rough Set Model Based on Developed Set Pair Analysis Method

Xia Ji, Longshu Li, Shengbing Chen, and Yi Xu

Key Lab of Ministry of Education for CI & SP,
Anhui University,
Hefei Anhui, China
jixia1983@163.com

Abstract. Different with traditional set pair analysis method, a new method to micro-decompound the discrepancy degree is proposed in line with the distributing actuality of missing attributes. Integrated with rough set theory, advances the united set pair tolerance relation and gives corresponding extended rough set model. Different values of identity degree and discrepancy degree can modulate the performance of this model and extend it's application range. Then expound some existed extended rough set model are the subschema of it. At last simulation experiments and conclusion are given, which validate the united set pair tolerance relation model can improve classification capability.

Keywords: Set pair analysis, Otherness, Rough set, United set pair tolerance relation.

1 Introduction

Classic rough set theory is based on the equivalence relation, and for a complete information system. But in real life, acknowledge obtain always face the incomplete information system for the reason of data measure error or the limitation of data acquisition. There are two kind of null value: (1)omit but exist;[2]; (2)lost and not allowed to be compared. Kryszkiewicz[3] established tolerance relationship for type(1). Stefanowski[4] and others built similar relationship for type(2). Wang GY[5] proposed the limited tolerance relation by in-depth study of tolerance relations and similar relation.

The incomplete degree and the distribution of loss data in various systems are different from each other. And any extended rough set model can achieve a good effect sometimes but not always. Some quantified extended rough set model, like quantified tolerance relation and quantified limited tolerance relation[6], has a good effect , but the process of quantifying costs large calculation.

The set pair analysis (Set Pair Analysis, SPA) theory was formally proposed by China scholars ZHAO Ke-qin in 1989, which is used to study the relationship between two set [7].It uses "a + bi + cj "as an associated number to deal with uncertain system,such as fuzzy, stochastic or intermediary. It studies the uncertain of two sets from

H. Deng et al. (Eds.): AICI 2009, LNAI 5855, pp. 9–17, 2009.

three aspects: identity, difference and opposition. At present, the set pair analysis theory is widely used in artificial intelligence, system control, management decision-making and other fields [7].

In this paper, we expand the set pair analysis theory firstly, then put forward a united extended rough set model based on it.This model can convert to different existed models with different identity and discrepancy thresholds, including tolerance relation, similarity relation and limited tolerance relation. Experiments on the UCI data sets show validity ,rationality and effectiveness of this model.

2 Set Pair Analysis Theory and Rough Set Theory

Set Pair is composed of two sets A and B ,namely $H=(A,B)$.Under general circumstances (W), it's a formula:

$$u_w(A,B) = \frac{S}{N} + \frac{F}{N}i + \frac{P}{N}j \qquad (1)$$

Here N is the total number of features, S is the number of identity features, and P is the number of contrary features, of the set pair discussed. F = N-S-P is the number of features of the two sets that are neither identity nor contrary. S/N, F/N, and P/N are called identity degree, discrepancy degree, and contrary degree of the two sets under certain circumstances respectively. In order to clarify formula(1),we set a=S/N, b=F/N, c=P/N, and then u_w(A,B) can be rewritten as:

$$u = a + bi + cj \qquad (2)$$

It is obvious that $0 \le a, b, c \le 1$, and the "a", "b", and "c" satisfy the equation $a + b + c = 1$.

An incomplete information system is the following tuple: I=(U,A,F), where U is a finite nonempty set of objects, A is a finite nonempty set of attributes, Va is a nonempty set of values of $a \in A$, and F={Fl: U→ρ(Va)} is an information function that maps an object in U to exactly one value in Va. For every al$\in$A and every xi$\in$U, if Fl(xi)is a single point set, then (U,A,F)is a complete information system. If there is some al$\in$ A and some xi$\in$U makes Fl(xi) is not a single point set, then (U,A,F)is an incomplete information system. An incomplete information system is the special case of a complete information system.

Let I=(U,A,F) is an incomplete information system. For any $B \subseteq A$,the tolerance relation TR(B) proposed by M.Kryskiewcz is defined as follows:[3]

TR(B)={(x,y)$\in$ U×U|∀ a$\in$ A,a(x)=a(y)$\cup$a(x)=*$\cup$a(y)=*}

Let I=(U,A,F) is an incomplete information system. For any $B \subseteq A$,the similarity relation SR(B) proposed by Stefanowski is defined as follows:[4]

SR(B)={(x,y)$\in$ U×U|∀b$\in$ B,b(x)=b(y)orb(x)=*}

Let $I=(U,A,F)$ is an incomplete information system. For any $B \subseteq A$, the limited toler-ance relation LTR(B) proposed by WANG GY is defined as follows:[5]

$P_B(x)=\{b|b \in B \text{ and } b(x) \neq *\}$

$$\text{LTR }(B) = ((x, y) \in U \times U | \forall b \in B, b(x)=b(y)=* \text{or}((P_B(x) \cap P_B(y) \neq \Phi)$$
$$\text{and } (b(x) \neq * \& b(y) \neq *) \to (b(x) = b(y)))\}.$$

3   A Unified Extended Rough Set Model Based on Expanded Set Pair Analysis Theory

A New Expanded Level Decomposition of Set Pair Analysis

The connection degree $u=a +bi +cj$ can be expanded breadthwise and lengthways ac-cording to the demand of research The transverse expanding form is:

$$(a_1+a_2+... +a_n)+(b_1i_1+b_2i_2+.. .+ b_ni_n)+(c_1j_1+c_2j_2+... +c_nj_n)$$

The lengthways expanding form is (see Fig. 1):

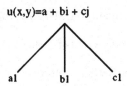

Fig. 1. The traditional micro-de-compound method of discrepancy degree

On incomplete information system $S = (U, A, V)$,"bi" is also deserved to be studied. Here N is the total number of attributes, S is the number of identity attributes, and P is the number of contrary attributes, $F = N-S-P$ is the number of missing attributes. Different from traditional set pair analysis theory, we de-compound "bi" from another angle: dis-tribution of missing attributes. Specific decomposition shown in Fig. 2.

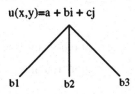

Fig. 2. Developed micro-de-compound method of discrepancy degree

Here for any a in attribute set A, b1=l{al a(x)=*&a(y)≠*}l, b2=l{al a(x)≠*&a(y)=*}l,

b3=l{al a(x)=*&a(y)=*}l。 The "b1" represents that attribute of x is missing. The "b2"

represents that attribute of y is missing. And "b3" represents attribute of x and y are all missing.

The traditional set pair analysis theory decompose the discrepancy term "bi" according to possible value of missing attribute. While in this paper we from the angle of distribution of missing attributes to decompose discrepancy term "bi".In the later section we'll prove that the symmetry and transitivity of binary relation depend on the threshold of b1 and b2.

B Extended Rough Set Model Based on United Set Pair Tolerance Relation

The connection degree between two objects is composed of three parts: the ratio of known and same attributes (the identity degree a), the ratio of known but different attributes (contrary degree cj) and the ratio of unknown attributes (discrepancy degree b).The contribution of contrary degree to similarity is negative, without consideration to deal with noise data, so we set contrary degree be 0,that is c=0.According to expanded set pair analysis theory, we give the united set pair tolerance relation.

Definition 1. Let I= (U, A, V)is an incomplete information system, $\forall x, y \in U, B \subseteq A$, define unified set pair tolerance relation (USPTR) as follows:

$$USPTR(B)=\{(x, y)l\ u_B(x, y)=a+(b1+b2+b3)i, a+b1+b2+b3=1, a \geq \alpha, bi \leq \beta i\}$$

Here $\alpha+\beta1+\beta2+\beta3=1$.The united set pair tolerance relation class of x is defined as :
$USPTR_B(x)=\{yl\ y \in USPTR\ (B)\}$.

Definition 2. Given incomplete information system S = (U, A, V), $B \subseteq A, X \subseteq U$, the upper and lower approximation of X are defined respectively as follows:

$$\overline{USPTR_B}(X) = \{x | USPTR_B(x) \cap X \neq \Phi\}$$

$$\underline{USPTR_B}(X) = \{x | USPTR_B(x) \subseteq X\}$$

When B=A, $\overline{USPTR_B}(X)$ and $\underline{USPTR_B}(X)$ can be abridged as $\overline{USPTR}(X)$ and $\underline{USPTR}(X)$.

Definition 3. Given incomplete information system S = (U, A, V),$X \subseteq U$, the positive domain, negative domain, boundary region, roughness of X in universe U are defined as follows:

Positive Domain: $pos(X) = \underline{USPTR}(X)$

Negative Domain: $NEG(X) = U - pos(X)$

Boundary Domain : $BND(X) = \overline{USPTR}(X) - \underline{USPTR}(X)$

Roughness: $\rho(X) = 1 - \underline{USPTR}(X)/\overline{USPTR}(X)$

C   Performance Analysis of Extended Rough Set Model Based on United Set Pair Tolerance Relation

Unified set pair tolerance relation is reflexive,but whether it satisfy symmetry and transitivity is not a conclusion. When the identity degree and discrepancy degree thresholds are changed, this extended model will have different performance. We have three theorems as follows.

Let I= (U,A,V)is an incomplete information system, $\forall x, y \in U, B \subseteq A, \alpha$ is threshold of identity degree a, $\beta1, \beta2$ and $\beta3$ are thresholds of discrepancy degree bi respectively.

Theorem 1. If threshold of b1 doesn't equal to threshold of b2, that is $\beta1 \neq \beta2$, USPTR(B)doesn't satisfy symmetry.

Proof. Let $y \in USPTR_B(x)$, that is u(x,y)=a+(b1+b2+b3)i, $b1 \leq \beta1$ and $b2 \leq \beta2$. Because $\beta1 \neq \beta2$, there may be some y in $USPTR_B(x)$ make $b1 > \beta2$ or $b2 > \beta1$. u(y,x)=a+(b2+b1+b3)i, so $x \notin USPTR_B(y)$.

Theorem 2. If $\beta1=0$ or $\beta2=0$,USPTR(B) satisfies transitivity.

Proof. Firstly prove the situation of $\beta1=0$. Let $y \in USPTR_B(x)$ and $z \in USPTR_B(y)$, then (x,y)=a+(b2+b3)i, u(y,z)=a+(b2+b3)i. $\forall a \in A$, if $a(x) \neq *$, then a(y)=a(x). Simultaneously $a(y) \neq *$ and a(z)=a(y). So $\forall a \in A$, if $a(x) \neq *$, we can get $a(z) \neq *$,that is u(x,z)=a(b2+b3)i, $z \in USPTR_B(x)$. The transitivity of USPTR(B)is proven. The situation of $\beta2=0$ is similar, here we doesn't give the details.

Theorem 3. If $\beta1=0$ and $\beta2=0$,USPTR(B) satisfies symmetry and transitivity.
Proof. According to Theorem1 and Theorem2, the proof is easy, so here we doesn't give unnecessary details.

The performance of united set pair tolerance relation will change with the modification of identity and discrepancy thresholds, and some representative existed extended rough set models are its subschema., see Table 1.

Table 1. The relationship between united set pair tolerance relation and some existed models

Thresholds Combination	Performance	Existed Rough Set Models
α+β1+β2+β3=1	reflexivity , symmetry	Tolerance Relation
α+β2+β3=1,β1=0	reflexivity , transitivity	Similarity Relation
β3=1 or α>0	reflexivity , symmetry	Limited Tolerance Relation
α=1	reflexivity , symmetry , transitivity	Equivalence Relation

4 Simulation Experiments

In this section, we select three test databases in UCI machine learning repository, Iris and Zoo and cmc-data. Details see in Tbale2 .We get nine incomplete databases though getting rid of data by random function, which are I-5%, I-10%, I-30%, Z-5%, Z-10% ,Z-30%,C-5%,C-10% and C-30%.

Table 2. Experiment Database

Database	Instance Number	Attribute Number
Iris	4	150
Zoo	101	17
cmc-data	1473	10

Let U be test database, $E(x_i)$ be the complete equivalent class of x_i ,and $R(x_i)$ be the tolerance class of x_i under different extended rough set models. We do ten times random experiments on every database ,and take their average as final result. The comparability($\bar{u}$) of test result can be computed through formula(1) and formula(2).

$$u = \frac{\sum_{i=1}^{|U|} \frac{|E(x_i) \cap R(x_i)|}{|E(x_i)| + |R(x_i)| - |E(x_i) \cap R(x_i)|}}{|U|} \qquad (3)$$

$$\bar{\mu} = \frac{\sum_{i=1}^{10} \bar{\mu}_i}{10} \qquad (4)$$

The bigger $\bar{u}$, the higher comparability, the better classification effect.

Table 3. Comparability($\bar{u}$) Comparison of Four Extended Rough Set Models

Database	Tolerance Relation	Similarity Relation	Limited Tolerance Relation	United Set Pair Tolerance Relation
Z-5%	0.821467	0.708342	0.862431	0.894354
Z-10%	0.809237	0.636242	0.811237	0.873016
Z-30%	0.462345	0.613456	0.473458	0.691325
I-5%	0.932402	0.883422	0.943679	0.988792
I-10%	0.899306	0.825033	0.917963	0.984485
I-30%	0.459135	0.524384	0.564023	0.899621
C-5%	0.563526	0.612432	0.642135	0.734956
C-10%	0.431256	0.516243	0.542251	0.660125
C-30%	0.339534	0.504135	0.462131	0.617243

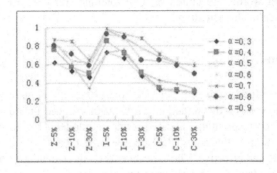

(1) β1=β2

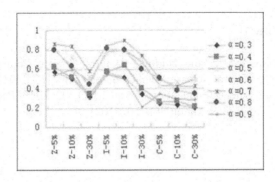

(2) β1≠β2

Fig. 3. The influence of identity and discrepancy to comparability ($\bar{u}$)

We do two groups of experiment. Table3 gives experiment result comparison of several extended rough set models. The $\bar{u}$ of every extended model decreases as the degree of loss data increases, whatever in Zoo, Iris or cmc-data. While $\bar{u}$ decreases degree of united set pair tolerance relation is lower than other three models. Meanwhile united set pair tolerance relation has more steady performances on three databases. The tolerance relation and limited tolerance relation are disagree with cmc-data, similarity relation is disagree with Zoo. Totally speaking, united set pair tolerance relation has better classification effects in nine situations of three databases.

According to the threshold determination difficulty, the second experiment try to narrow down its range. In Figue1, identity degree α value range is[0.3,0.9] and discrepancy degree β is separated into two situations, which are β1=β2 and β1≠β2. Figue1shows that when α is taken small value, like 0.1and 0.3,the classification performance is very bad. When α arrives 0.9,the performance is unstable and decrease somewhat. When α ∈ [0.5,0.7],the new extended model has a steady and good performance. On all conditions, the classification performance of β1=β2 is better than that of β1≠β2.In other words, the united set pair tolerance relation with symmetry has a better performance in classification.

5 Conclusions

In this paper, we firstly develop set pair analysis method, give a developed micro-de-compound method of discrepancy degree, then propose united set pair tolerance relation and give corresponding extended rough set model. The threshold α, β_1 and β_2 can adjust according to the incomplete degree of databases. So this model is more objective, flexible and effective. The experiments on three databases of UCI, which are Iris, Zoo and cmc-data, can clearly manifest the advantage of new methods in this paper. But how to fast calculate the identity and discrepancy degree thresholds is still a problem need to solve.

Acknowledgement

This work is supported by the National Natural Science Foundation of China (NO.60273043), the Natural Science Foundation of Anhui(NO.050420204), the Education Burea Natural Science Foundation of Anhui(NO.2006KJ098B) and High School Surpass Talent Foundation of Anhui(NO.05025102).

References

1. Pawlak, Z.: Rough set. International Journal of Computer and Information Sciences 11(5), 341–356 (1982)
2. Grzymala Busse, J.W.: On the unknown attribute values in learning from examples. In: Raś, Z.W., Zemankova, M. (eds.) ISMIS 1991. LNCS (LNAI), vol. 542, pp. 368–377. Springer, Heidelberg (1991)

3. Stefanowski, J., Tsoukias: A incomplete information tables and rough classification. Computational Intelligence 17, 545–566 (2001)
4. Kryszkiewicz, M.: Rough Set Approach to Incomplete Information System. Information Sciences (S0020-0255) 112(1/4), 39–49 (1998)
5. Wang, G.-y.: The extension of rough set theory in incomplete information system. Computer Research and Development 39(10), 1238–1243 (2002)
6. Sun, C.-m., Liu, D.-y., Sun, S.-y.: Research of rough set method oriented to incomplete information system. Mini-micro Computer System 10(10), 1869–1873 (2007)
7. Zhao, K.-q.: Set pair and its prilimitary applications. Hangzhou Zhejiang Science Publishers (2000)
8. Xu, Y., Li, L.-s., Li, X.-j.: Extended rough set model based on set pair power. Journal of System Simulation 20(6), 1515–1522 (2008)
9. Stefanowski, J., Tsoukias, A.: On the Extension of Rough Sets under Incomplete Information. In: Zhong, N., Skowron, A., Ohsuga, S. (eds.) RSFDGrC 1999. LNCS (LNAI), vol. 1711, pp. 73–82. Springer, Heidelberg (1999)
10. Lei, Z., Lan, S.: Rough Set Model Based on New Set Pair Analysis. Fuzzy Systems and Mathmatics (S1001-7402) 20(4), 111–116 (2006)

Learning Rules from Pairwise Comparison Table

Liping An[1] and Lingyun Tong[2]

[1] Business School, Nankai University, Tianjin, 300071, China
anliping2000@sina.com
[2] School of Management, Hebei University of Technology, Tianjin, 300130, China
tonglingyun2008@sina.com

Abstract. Multicriteria choice problems deal with a selection of a small subset of best actions from a larger set and multicriteria ranking problems aim to obtain an ordering of actions from a set from the best to the worst. In order to construct a comprehensive decision-rule preference model for choice and ranking, we propose the concepts of decision matrices and decision functions to generate the minimal decision rules from a pairwise comparison table. Decision rules derived can be used to obtain a recommendation in multicriteria choice and ranking problems.

Keywords: Multiple criteria decision analysis, rough sets, choice, ranking, decision rules.

1 Introduction

Multiple Criteria Decision Analysis (MCDA) aims at helping a decision maker (DM) to prepare and make a decision where more than one point of view has to be considered. There are three major models used until now in MCDA: (1) the functional model expressed in terms of a utility function within multiple attribute utility theory [1]; (2)the relational model expressed in the form of an outranking relation [2] and a fuzzy relation [3]; (3) the function-free model expressed in terms of symbolic forms, like "if... then..." decision rules [6] or decision trees, or in a sub-symbolic form, like artificial neural nets [4].

Both functional and relational models require that the DM gives some preference information, such as importance weights, substitution ratios and various thresholds on particular criteria, which is often quite difficult for DMs not acquainted with the MCDA methodology [4]. According to Slovic [5], people make decisions by searching for rules that provide good justification of their choices. The decision rule approach [6] follows the paradigm of artificial intelligence and inductive learning. These decision rules are induced from preference information supplied by the DM in terms of some decision examples. Therefore, the decision rules are intelligible and speak the language of the DM [7].

Roy [8] has stated that the objective of an MCDA is to solve one of the following five typologies of problems: classification, sorting, choice, ranking and description. Classification concerns an assignment of a set of actions to a set of pre-defined classes. The actions are described by a set of regular attributes and the classes are not

H. Deng et al. (Eds.): AICI 2009, LNAI 5855, pp. 18–27, 2009.

necessarily ordered. Sorting regards a set of actions evaluated by criteria, i.e. attributes with preference-ordered domains. In this problem, the classes are also preference-ordered. Choice deals with a selection of a small subset of best actions, evaluated by attributes and criteria, from a larger set. Ranking aims to obtain an ordering of actions, evaluated by attributes and criteria, from a set from the best to the worst. Description identifies the major distinguishing features of the actions and performs their description based on these features.

Recently, a complete and well-axiomatized methodology for constructing decision rule preference models from decision examples based on the rough sets theory [9] has been proposed. Several attempts have already been made to use this theory to decision analysis [10]. However, the classical rough set theory is based on the indiscernibility relation that is not able to deal with preference-ordered attribute domains and decision classes. Therefore, the use of classical rough sets has been limited to problems of multiattribute classification and description only. For this reason, Greco et al. [6, 11] have proposed an extension of the rough sets approach, called Dominance-based Rough Set Approach (DRSA). DRSA is able to deal with sorting problems involving both criteria and regular attributes. This innovation is mainly based on substitution of the indiscernibility relation by a dominance relation in the rough approximation. In order to construct a comprehensive preference model that could be used to support the sorting task, Greco et al. [11] distinguish between qualitative attributes and quantitative attributes. The rules are derived from rough approximations of decision classes. They satisfy conditions of completeness and dominance, and manage with possible inconsistencies in the set of examples.

In the case of multicriteria choice and ranking problems, the original version of rough set theory has been extended by Greco et al. [7, 12, 13] in two ways: substituting the classical indiscernibility relation by a dominance relation; and substituting the information table by a pairwise comparison table, which permits approximation of a comprehensive preference relation in multicriteria choice and ranking problems.

In this paper, the decision matrix of a pairwise comparison table is defined to construct the decision functions which are Boolean functions. The set of "if ... then ..." decision rules are decoded from prime implicants of the Boolean functions.

2 The Pairwise Comparison Table and Multigraded Dominance

In this section, we recall some basic concepts from [7, 12, 13] to be used in this paper.

2.1 The Pairwise Comparison Table

Let C be the set of criteria used for evaluation of actions from A. For any criterion $q \in C$, let T_q be a finite set of binary relations defined on A on the basis of the evaluations of actions from A with respect to the considered criterion q, such that for every $(x, y) \in A \times A$ exactly one binary relation $t \in T_q$ is verified.

The preferential information has the form of pairwise comparisons of reference actions from $B \subseteq A$, considered as exemplary decisions. The pairwise comparison table (PCT) is defined as data table $S_{PCT} = (\mathbf{B}, C \cup \{d\}, T_C \cup T_d, g)$, where $\mathbf{B} \subseteq B \times B$ is a nonempty set of exemplary pairwise comparisons of reference actions, $T_C = \cup_{q \in C} T_q$, d is a decision corresponding to the comprehensive pairwise comparison, and g: $\mathbf{B} \times (C \cup \{d\}) \to T_C \cup T_d$ is a total function such that $g[(x, y), q] \in T_q$ for every $(x, y) \in A \times A$ and for each $q \in C$, and $g[(x, y), d] \in T_d$ for every $(x, y) \in \mathbf{B}$.

It follows that for any pair of reference actions $(x, y) \in \mathbf{B}$ there is verified one and only one binary relation $t \in T_d$. Thus, T_d induces a partition of B. The data table S_{PCT} can be seen as decision table, since the set of considered criteria C and decision d are distinguished.

2.2 The Definition of Multigraded Dominance

Let C^O be the set of criteria expressing preferences on an ordinal scale, and C^N, the set of criteria expressing preferences on a quantitative scale or a numerical nonquantitative scale, such that $C^O \cup C^N = C$ and $C^O \cap C^N = \emptyset$. Moreover, for each $P \subseteq C$, let P^O be the subset of P composed of criteria expressing preferences on an ordinal scale, i.e. $P^O = P \cap C^O$, and P^N, the subset of P composed of criteria expressing preferences on a quantitative scale or a numerical non-quantitative scale, i.e. $P^N = P \cap C^N$. For each $P \subseteq C$, we have $P = P^O \cup P^N$ and $P^O \cap P^N = \emptyset$. The following three situations are considered:

(1) $P = P^N$ and $P^O = \emptyset$.

The exemplary pairwise comparisons made by the DM can be represented in terms of graded preference relations P_q^h: for each $q \in C$ and for every $(x, y) \in A \times A$, $T_q = \{ P_q^h : h \in H_q \}$, where H_q is a particular subset of the relative integers and

- $x P_q^h y$, $h > 0$, means that action x is preferred to action y by degree h with respect to the criterion q,
- $x P_q^h y$, $h < 0$, means that action x is not preferred to action y by degree h with respect to the criterion q,
- $x P_q^0 y$ means that action x is similar to action y with respect to the criterion q.

For each $q \in C$ and for every $(x, y) \in A \times A$, $[x P_q^h y, h > 0] \Rightarrow [y P_q^k x, k \leq 0]$ and $[x P_q^h y, h < 0] \Rightarrow [y P_q^k x, k \geq 0]$.

Given $P^N \subseteq C$ $(P^N \neq \emptyset)$, (x, y), $(w, z) \in A \times A$, the pair of actions (x, y) is said to dominate (w, z), taking into account the criteria from P^N, denoted by $(x, y) D_{P^N} (w, z)$, if x is preferred to y at least as strongly as w is preferred to z with respect to each $q \in P^N$. Precisely, "at least as strongly as" means "by at least the same degree", i.e. $h_q \geq k_q$, where $h_q, k_q \in H_q$, $x P_q^{h_q} y$ and $w P_q^{k_q} z$, for each $q \in P$.

The upward cumulated preferences, denoted by $P_q^{\geq h}$, and downward cumulated preferences, denoted by $P_q^{\leq h}$, having the following interpretation:

- $x P_q^{\geq h} y$ means "x is preferred to y with respect to q by at least degree h",
- $x P_q^{\leq h} y$ means "x is preferred to y with respect to q by at most degree h".

Exact definition of the cumulated preferences, for each $(x, y) \in A \times A$, $q \in C$ and $h \in H_q$ is the following:

- $x P_q^{\geq h} y$ if $x P_q^k y$, where $k \in H_q$ and $k \geq h$,
- $x P_q^{\leq h} y$ if $x P_q^k y$, where $k \in H_q$ and $k \leq h$.

(2) $P = P^O$ and $P^N = \emptyset$.

In this paper, we modify the original definition of multigraded dominance with respect to $P = P^O$ and $P^N = \emptyset$ [12] as follows:

Given (x, y), $(w, z) \in A \times A$, the pair (x, y) is said to dominate the pair (w, z) with respect to P^O if, for each $q \in P^O$, (1) $c_q(x) \succeq c_q(y)$ and $c_q(z) \succeq c_q(w)$, denoted by $(x, y) D_{\{q\}}^1 (w, z)$, or (2) $c_q(x) \succeq c_q(w)$ and $c_q(z) \succeq c_q(y)$ for $c_q(x) \succeq c_q(y)$ and $c_q(w) \succeq c_q(z)$, denoted by $(x, y) D_{\{q\}}^2 (w, z)$.

(3) $P^N \neq \emptyset$ and $P^O \neq \emptyset$.

Given (x, y), $(w, z) \in A \times A$, the pair (x, y) is said to dominate the pair (w, z) with respect to criteria from P, if (x, y) dominates (w, z) with respect to both P^N and P^O.

2.3 Lower and Upper Approximations

Given $P \subseteq C$ and $(x, y) \in A \times A$, the P-dominating set and the P-dominated set are defined, respectively, as:

- a set of pairs of actions dominating (x, y), $D_P^+ (x, y) = \{(w, z) \in A \times A: (w, z) D_P(x, y)\}$,
- a set of pairs of actions dominated by (x, y), $D_P^- (x, y) = \{(w, z) \in A \times A: (x, y) D_P(w, z)\}$.

In a PCT, the set T_d is composed of two binary relations defined on A:

(1) x outranks y (denoted by xSy or $(x, y) \in S$), where $(x, y) \in \mathbf{B}$ and "x outranks y" means "x is at least as good as y",
(2) x does not outrank y (denoted by xS^Cy or $(x, y) \in S^C$), where $(x, y) \in \mathbf{B}$ and $S \cup S^C = \mathbf{B}$.

The P-dominating sets and the P-dominated sets defined on $\mathbf{B}$ for all given pairs of reference actions from $\mathbf{B}$ are "granules of knowledge" that can be used to express P-lower and P-upper approximations of comprehensive outranking relations S and S^c, respectively:

$$\underline{P} (S) = \{(x, y) \in \mathbf{B}: D_P^+ (x, y) \subseteq S\}, \quad \overline{P} (S) = \bigcup_{(x,y) \in S} D_P^+ (x, y).$$

$$\underline{P} (S^c) = \{(x, y) \in \mathbf{B}: D_P^- (x, y) \subseteq S^c\}, \quad \overline{P} (S^c) = \bigcup_{(x,y) \in S^c} D_P^- (x, y).$$

The P-boundaries (P-doubtful regions) of S and S^c are defined as

$Bn_P(S) = \overline{P} (S) - \underline{P} (S)$, $Bn_P(S^c) = \overline{P} (S^c) - \underline{P} (S^c)$.

From the above it follows that $Bn_P(S) = Bn_P(S^c)$.

2.4 Decision Rules

Using the approximations of S and S^c based on the dominance relation defined above, it is possible to induce a generalized description of the available preferential information in terms of decision rules. The decision rules have in this case the following syntax:

(1) $D_\geq$-decision rules

IF $x\,P_{q_1}^{\geq h(q_1)}\,y$ and $x\,P_{q_2}^{\geq h(q_2)}\,y$ and... $x\,P_{q_e}^{\geq h(q_e)}\,y$ and $c_{q_e+1}(x)\succeq r_{q_e+1}$ and $c_{q_e+1}(x)\preceq s_{q_e+1}$ and ... $c_{q_p}(x)\succeq r_{q_p}$ and $c_{q_p}(x)\preceq s_{q_p}$, THEN xSy,

where $P=\{q_1, q_2, ..., q_p\}\subseteq C$, $P^N=\{q_1, q_2, ..., q_e\}$, $P^O=\{q_{e+1}, q_{e+2}, ..., q_p\}$, $(h(q_1), h(q_2), ..., h(q_p))\in H_{q_1}\times H_{q_2}\times...\times H_{q_p}$ and $(r_{q_e+1}, ..., r_{q_p})$, $(s_{q_e+1}, ..., s_{q_p})\in C_{q_e+1}\times...\times C_{q_p}$.

These rules are supported by pairs of actions from the P-lower approximation of S only.

(2) $D_\preceq$-decision rules

IF $x\,P_{q_1}^{\preceq h(q_1)}\,y$ and $x\,P_{q_2}^{\preceq h(q_2)}\,y$ and... $x\,P_{q_e}^{\preceq h(q_e)}\,y$ and $c_{q_e+1}(x)\preceq r_{q_e+1}$ and $c_{q_e+1}(x)\succeq s_{q_e+1}$ and ... $c_{q_p}(x)\succeq r_{q_p}$ and $c_{q_p}(x)\preceq s_{q_p}$, THEN $xS^c y$,

where $P=\{q_1, q_2, ..., q_p\}\subseteq C$, $P^N=\{q_1, q_2, ..., q_e\}$, $P^O=\{q_{e+1}, q_{e+2}, ..., q_p\}$, $(h(q_1), h(q_2), ..., h(q_p))\in H_{q_1}\times H_{q_2}\times...\times H_{q_p}$ and $(r_{q_e+1}, ..., r_{q_p})$, $(s_{q_e+1}, ..., s_{q_p})\in C_{q_e+1}\times...\times C_{q_p}$.

These rules are supported by pairs of actions from the P-lower approximation of S^c only.

(3) $D_{\geq\preceq}$-decision rules

IF $x\,P_{q_1}^{\geq h(q_1)}\,y$ and $x\,P_{q_2}^{\geq h(q_2)}\,y$ and... $x\,P_{q_e}^{\geq h(q_e)}\,y$ and $x\,P_{q_{e+1}}^{\preceq h(q_{e+1})}\,y$ and $x\,P_{q_{e+2}}^{\preceq h(q_{e+2})}\,y$ and... $x\,P_{q_f}^{\preceq h(q_f)}\,y$ $c_{q_f+1}(x)\succeq r_{q_f+1}$ and $c_{q_f+1}(y)\preceq s_{q_f+1}$ and ... $c_{q_g}(x)\succeq r_{q_g}$ and $c_{q_g}(x)\preceq s_{q_g}$ and $c_{q_g+1}(x)\preceq r_{q_g+1}$ and $c_{q_g+1}(y)\succeq s_{q_g+1}$ and ... $c_{q_p}(x)\preceq r_{q_p}$ and $c_{q_p}(x)\succeq s_{q_p}$, THEN xSy or $xS^c y$,

where $O'=\{q_1, q_2, ..., q_e\}\subseteq C$, $O''=\{q_{e+1}, q_{e+2}, ..., q_f\}\subseteq C$, $P^N=O'\cup O''$, O' and O'' not necessarily disjoint, $P^O=\{q_{f+1}, q_{f+2}, ..., q_f\}$, $(h(q_1), h(q_2), ..., h(q_f))\in H_{q_1}\times H_{q_2}\times...\times H_{q_f}$ and $(r_{q_f+1}, ..., r_{q_p})$, $(s_{q_f+1}, ..., s_{q_p})\in C_{q_f+1}\times...\times C_{q_p}$.

These rules are supported by pairs of actions from the P-boundary of S and S^c only.

Applying the decision rules induced from a given S_{PCT}, a final recommendation for choice or ranking can be obtained upon a suitable exploitation of this structure [12].

3   Decision Matrix and Decision Functions of a PCT

In this section, by analogy with the indiscernibility matrix and its Boolean functions [14] to generate all minimal decision rules in classical rough set theory, we propose the concepts of decision matrices and decision functions to generate the minimal decision rules.

Definition 1. Let $S_{PCT} = (\mathbf{B}, C \cup \{d\}, T_C \cup T_d, g)$ be a pairwise comparison table, $\mathbf{B} = S \cup S^C$, $(x, y) \in S$, $(w, z) \in S^C$, $p_i \in P^N$, $q_j \in P^O$. $x P_{p_i}^{h_{p_i}} y$ and $w P_{p_i}^{k_{p_i}} z$, for each $p_i \in P^N$, h_{p_i}, $k_{p_i} \in H_{p_i}$. The decision matrix of S in S_{PCT}, denoted by $M(S)$, is defined as:

$$M(S) = m[(x, y), (w, z)]_{|S| \times |S_C|},$$

where $m[(x, y), (w, z)] = \{p_i, q_j: (x, y) D_{\{p_i\}}(w, z), h_{p_i} > k_{p_i}, (x, y) D^1_{\{q_j\}}(w, z)$ and $c_{q_j}(x) \neq c_{q_j}(y)$ or $c_{q_j}(w) \neq c_{q_j}(z)$, $(x, y) D^2_{\{q_j\}}(w, z)$ and $c_{q_j}(x) \neq c_{q_j}(w)$ or $c_{q_j}(y) \neq c_{q_j}(z)\}$.

Definition 2. Let $M(S)$ be the decision matrix of S in S_{PCT}. The decision function of (x, y) with respect to $M(S)$, is defined as:

$$f_S[(x, y)] = \bigwedge_{(w,z)} \{(\bigvee_i p_i^*) \vee (\bigvee_j q_j^*): p_i, q_j \in m[(x, y), (w, z)] \text{ and } m[(x, y), (w, z)] \neq \varnothing\}.$$

where p_i^* and q_j^* are corresponding to the attribute p_i and q_j, respectively.

The decision function $f_S[(x, y)]$ is a Boolean function that expresses how a pair $(x, y) \in S$ can be discerned from all of the pairs $(w, z) \in S^C$. Turning $f_S[(x, y)]$ into disjunctive normal form, the prime implicants of $f_S[(x, y)]$ reveal the minimal subsets of P that are needed to discern the pair (x, y) from the pair in S^C and correspond to the minimal $D_{\succeq}$-decision rules.

Similarly, we can define the decision matrix of S^C, $M(S^C)$, and the decision function with respect to $M(S^C)$.

An example. Let us consider the example used in [6]. The students are evaluated according to the level in Mathes, Phys and Lit. Marks are given on a scale from 0 to 20. Three students presented in Table 1 are considered.

Table 1. Students' evaluation table

Student	Maths	Phys	Lit	Comprehensive valuation
a	18	16	10	13.9
b	10	12	18	13.6
c	14	15	15	14.9

DM's preference-order of the students is represented in terms of pairwise evaluations in Table 2.

The relation $(x, y) D_{\{Maths\}}(w, z)$ is true if and only if

(1) $c_{Maths}(x) \succeq c_{Maths}(y)$ and $c_{Maths}(z) \succeq c_{Maths}(w)$, or (2) $c_{Maths}(x) \succeq c_{Maths}(w)$ and $c_{Maths}(z) \succeq c_{Maths}(y)$, where $c_{q_j}(x) \succeq c_{q_j}(y)$ and $c_{q_j}(w) \succeq c_{q_j}(z)$.

Table 2. Pairwise comparison table

Pair of students	Maths	Phys	Lit	Comprehensive outranking relation
(a, a)	18, 18	16, 16	10, 10	S
(a, b)	18, 10	16, 12	10, 18	S
(a, c)	18, 14	16, 15	10, 15	S^c
(b, a)	10, 18	12, 16	18, 10	S^c
(b, b)	10, 10	12, 12	18, 18	S
(b, c)	10, 14	12, 15	18, 15	S^c
(c, a)	14, 18	15, 16	15, 10	S
(c, b)	14, 10	15, 12	15, 18	S
(c, c)	14, 14	15, 15	15, 15	S

The analogous partial preorders can be induced using dominance relations on Phys and on Lit. The entities to compute the lower and upper approximations of S are shown in Table 3.

Table 3. The entities to compute the lower and upper approximations of S

Pair of students	$D^+_{\{Maths\}}(x, y)$	$D^+_{\{Physics\}}(x, y)$	$D^+_{\{Literature\}}(x, y)$	$D^+_P(x, y)$
(a, a)	$(a, a), (a, b), (a, c),$ $(b, b), (c, b), (c, c)$	$(a, a), (a, b), (a, c), (b,$ $b), (c, b), (c, c)$	$(a, a), (b, a), (b, b),$ $(b, c), (c, a), (c, c)$	$(a, a), (b,$ $b), (c, c)$
(a, b)	(a, b)	(a, b)	$(a, a), (a, b), (a, c),$ $(b, a), (b, b), (b, c),$ $(c, a), (c, b), (c, c)$	(a, b)
(a, c)	$(a, b), (a, c)$	$(a, b), (a, c)$	$(a, a), (a, b), (a, c),$ $(b, a), (b, b), (b, c),$ $(c, a), (c, b), (c, c)$	(a, c)
(b, a)	$(a, a), (a, b), (a, c),$ $(b, a), (b, b), (b, c),$ $(c, a), (c, b), (c, c)$	$(a, a), (a, b), (a, c),$ $(b, a), (b, b), (b, c),$ $(c, a), (c, b), (c, c)$	(b, a)	(b, a)
(b, b)	$(a, a), (a, b), (a, c),$ $(b, b), (c, b), (c, c)$	$(a, a), (a, b), (a, c),$ $(b, b), (c, b), (c, c)$	$(a, a), (b, a), (b, b),$ $(b, c), (c, a), (c, c)$	$(a, a), (b,$ $b), (c, c)$
(b, c)	$(a, a), (a, b), (a, c),$ $(b, b), (b, c), (c, b),$ (c, c)	$(a, a), (a, b), (a, c),$ $(b, b), (b, c), (c, b),$ (c, c)	$(b, a), (b, c)$	(b, c)
(c, a)	$(a, a), (a, b), (a, c),$ $(b, b), (c, a), (c, b),$ (c, c)	$(a, a), (a, b), (a, c),$ $(b, b), (c, a), (c, b),$ (c, c)	$(b, a), (c, a)$	(c, a)
(c, b)	$(a, b), (c, b)$	$(a, b), (c, b)$	$(a, a), (b, a), (b, b),$ $(b, c), (c, a), (c, b),$ (c, c)	(c, b)
(c, c)	$(a, a), (a, b), (a, c), (b,$ $b), (c, b), (c, c)$	$(a, a), (a, b), (a, c), (b,$ $b), (c, b), (c, c)$	$(a, a), (b, a), (b, b),$ $(b, c), (c, a), (c, c)$	$(a, a), (b,$ $b), (c, c)$

$\underline{P}(S)=\{(a, a), (a, b), (b, b), (c, a), (c, b), (c, c)\}$,

$\overline{P}(S)=\{(a, a), (a, b), (b, b), (c, a), (c, b), (c, c)\}$,

$Bn_P(S)=\varnothing$.

The entities to compute the lower and upper approximations of S^C are shown in Table 4.

Table 4. The entities to compute the lower and upper approximations of S^C

Pair of students	$D^-_{\{Maths\}}(x, y)$	$D^-_{\{Physics\}}(x, y)$	$D^-_{\{Literature\}}(x, y)$	$D^-_P(x, y)$
(a, a)	$(a, a), (b, a), (b, b),$ $(b, c), (c, a), (c, c)$	$(a, a), (b, a), (b, b),$ $(b, c), (c, a), (c, c)$	$(a, a), (a, b), (a, c),$ $(b, b), (c, b), (c, c)$	$(a, a), (b,$ $b), (c, c)$
(a, b)	$(a, a), (a, b), (a, c),$ $(b, a), (b, b), (b, c),$ $(c, a), (c, b), (c, c)$	$(a, a), (a, b), (a, c),$ $(b, a), (b, b), (b, c),$ $(c, a), (c, b), (c, c)$	(a, b)	(a, b)
(a, c)	$(a, a), (a, c), (b, a),$ $(b, b), (b, c), (c, a),$ (c, c)	$(a, a), (a, c), (b, a),$ $(b, b), (b, c), (c, a),$ (c, c)	$(a, b), (a, c)$	(a, c)
(b, a)	(b, a)	(b, a)	$(a, a), (a, b), (a, c),$ $(b, a), (b, b), (b, c),$ $(c, a), (c, b), (c, c)$	(b, a)
(b, b)	$(a, a), (b, a), (b, b),$ $(b, c), (c, c)$	$(a, a), (b, a), (b, b),$ $(b, c), (c, c)$	$(a, a), (a, b), (a, c),$ $(b, b), (c, b), (c, c)$	$(a, a), (b,$ $b), (c, c)$
(b, c)	$(b, a), (b, c)$	$(b, a), (b, c)$	$(a, a), (a, b), (a, c),$ $(b, b), (b, c), (c, b),$ (c, c)	(b, c)
(c, a)	$(b, a), (c, a)$	$(b, a), (c, a)$	$(a, a), (a, b), (a, c),$ $(b, b), (c, a), (c, b),$ (c, c)	(c, a)
(c, b)	$(b, a), (b, b), (b, c),$ $(c, a), (c, b), (c, c)$	$(b, a), (b, b), (b, c),$ $(c, a), (c, b), (c, c)$	$(a, b), (c, b)$	(c, b)
(c, c)	$(a, a), (b, a), (b, b),$ $(b, c), (c, a), (c, c)$	$(a, a), (b, a), (b, b),$ $(b, c), (c, a), (c, c)$	$(a, a), (a, b), (a, c),$ $(b, b), (c, b), (c, c)$	$(a, a), (b,$ $b), (c, c)$

$\underline{P}(S^c)=\{(a, c), (b, a), (b, c)\}$, $\overline{P}(S^c)=\{(a, c), (b, a), (b, c)\}$, $Bn_P(S)=\emptyset$.

The decision matrix of S in S_{PCT},

$$M(S)=\begin{array}{c|ccc} & (a,c) & (b,a) & (b,c) \\ \hline (a,a) & \text{Lit} & \text{Maths, Phys} & \text{Maths, Phys} \\ (a,b) & \text{Maths, Phys} & \text{Maths, Phys} & \text{Maths, Phys} \\ (b,b) & \text{Lit} & \text{Maths, Phys} & \text{Maths, Phys} \\ (c,a) & \text{Lit} & \text{Maths, Phys} & \emptyset \\ (c,b) & \emptyset & \text{Maths, Phys} & \text{Maths, Phys} \\ (c,c) & \text{Lit} & \text{Maths, Phys} & \text{Maths, Phys} \end{array}$$

The decision functions of (x, y) with respect to $M(S)$:

$f_S[(a, a)]=\text{Lit}\wedge(\text{Maths}\vee\text{Phys})\wedge(\text{Maths}\vee\text{Phys})=(\text{Lit}\wedge\text{Maths})\vee(\text{Lit}\wedge\text{Phys})$, similarly,

$f_S[(a, b)]=\text{Maths}\vee\text{Phys}$,

$f_S[(b, b)]=(\text{Lit}\wedge\text{Maths})\vee(\text{Lit}\wedge\text{Phys})$,

$f_S[(c, a)]=(\text{Lit}\wedge\text{Maths})\vee(\text{Lit}\wedge\text{Phys})$,

$f_S[(c, b)]$=Maths$\lor$Phys,
$f_S[(c, c)]$=(Lit$\land$Maths)$\lor$(Lit$\land$Phys).

Then, the following minimal $D_{\succeq}$-decision rules can be obtained:

IF $(c_{\text{Maths}}(x) \succeq 18$ and $c_{\text{Maths}}(y) \preceq 18)$ and $(c_{\text{Lit}}(x) \succeq 10$ and $c_{\text{Lit}}(y) \preceq 10)$ THEN xSy, (a, a),

IF $(c_{\text{Phys}}(x) \succeq 16$ and $c_{\text{Phys}}(y) \preceq 16)$ and $(c_{\text{Lit}}(x) \succeq 10$ and $c_{\text{Lit}}(y) \preceq 10)$ THEN xSy, (a, a),

IF $c_{\text{Maths}}(x) \succeq 18$ and $c_{\text{Maths}}(y) \preceq 10$ THEN xSy, (a, b),

IF $c_{\text{Phys}}(x) \succeq 16$ and $c_{\text{Phys}}(y) \preceq 12$ THEN xSy, (a, b),

IF $(c_{\text{Maths}}(x) \succeq 10$ and $c_{\text{Maths}}(y) \preceq 10)$ and $(c_{\text{Lit}}(x) \succeq 18$ and $c_{\text{Lit}}(y) \preceq 18)$ THEN xSy, (b, b),

IF $(c_{\text{Phys}}(x) \succeq 12$ and $c_{\text{Phys}}(y) \preceq 12)$ and $(c_{\text{Lit}}(x) \succeq 18$ and $c_{\text{Lit}}(y) \preceq 18)$ THEN xSy, (b, b),

IF $(c_{\text{Maths}}(x) \succeq 14$ and $c_{\text{Maths}}(y) \preceq 18)$ and $(c_{\text{Lit}}(x) \succeq 15$ and $c_{\text{Lit}}(y) \preceq 10)$ THEN xSy, (c, a),

IF $(c_{\text{Phys}}(x) \succeq 15$ and $c_{\text{Phys}}(y) \preceq 16)$ and $(c_{\text{Lit}}(x) \succeq 15$ and $c_{\text{Lit}}(y) \preceq 10)$ THEN xSy, (c, a),

IF $c_{\text{Maths}}(x) \succeq 14$ and $c_{\text{Maths}}(y) \preceq 10$ THEN xSy, (c, b),

IF $c_{\text{Phys}}(x) \succeq 15$ and $c_{\text{Phys}}(y) \preceq 12$ THEN xSy, (c, b),

IF $(c_{\text{Maths}}(x) \succeq 14$ and $c_{\text{Maths}}(y) \preceq 14)$ and $(c_{\text{Lit}}(x) \succeq 15$ and $c_{\text{Lit}}(y) \preceq 15)$ THEN xSy, (c, c),

IF $(c_{\text{Phys}}(x) \succeq 15$ and $c_{\text{Phys}}(y) \preceq 15)$ and $(c_{\text{Lit}}(x) \succeq 15$ and $c_{\text{Lit}}(y) \preceq 15)$ THEN xSy, (c, c).

Similarly, we can obtain the $D_{\preceq}$-decision rules as follows.

IF $(c_{\text{Maths}}(x) \preceq 18$ and $c_{\text{Maths}}(y) \succeq 14)$ and $(c_{\text{Lit}}(x) \preceq 10$ and $c_{\text{Lit}}(y) \succeq 15)$ THEN xS^cy, (a, c),

IF $(c_{\text{Phys}}(x) \preceq 16$ and $c_{\text{Phys}}(y) \succeq 15)$ and $(c_{\text{Lit}}(x) \preceq 10$ and $c_{\text{Lit}}(y) \succeq 15)$ THEN xS^cy, (a, c),

IF $c_{\text{Maths}}(x) \preceq 10$ and $c_{\text{Maths}}(y) \succeq 18$ THEN xS^cy, (b, a),

IF $c_{\text{Phys}}(x) \preceq 18$ and $c_{\text{Phys}}(y) \succeq 10$ THEN xS^cy, (b, a),

IF $c_{\text{Maths}}(x) \preceq 10$ and $c_{\text{Maths}}(y) \succeq 14$ THEN xS^cy, (c, b),

IF $c_{\text{Phys}}(x) \preceq 18$ and $c_{\text{Phys}}(y) \succeq 15$ THEN xS^cy, (c, b).

4 Conclusions

Learning decision rules from preference-ordered data differs from usual machine learning, since the former involves preference orders in domains of attributes and in the set of decision classes. This requires that a knowledge discovery method applied to preference-ordered data respects the dominance principle, which is addressed in the Dominance-Based Rough Set Approach. This approach enables us to apply a rough set approach to multicriteria choice and ranking. In this paper, we propose the concepts of decision matrices and decision functions to generate the minimal decision rules from a pairwise comparison table. Then, the decision rules can be used to obtain a recommendation in multicriteria choice and ranking problems. We will further present some extensions of the approach that make it a useful tool for other practical applications.

Acknowledgement. It is a project supported by National Natural Science Foundation of China (No. 70601013).

References

1. Keeney, R.L., Raiffa, H.: Decision with Multiple Objectives-Preferences and Value Trade-offs. Wiley, New York (1976)
2. Roy, B.: The Outranking Approach and the Foundation of ELECTRE Methods. Theory and Decision 31(1), 49–73 (1991)
3. Fodor, J., Roubens, M.: Fuzzy Preference Modelling and Multicriteria Decision Support. Kluwer, Dordrecht (1994)
4. Zopounidis, C., Doumpos, M.: Multicriteria Classification and Sorting Methods: A literature Review. European Journal of Operational Research 138(2), 229–246 (2002)
5. Slovic, P.: Choice between Equally-valued Alternatives. Journal of Experimental Psychology: Human Perception Performance 1, 280–287 (1975)
6. Greco, S., Matarazzo, B., Slowinski, R.: Rough Set Theory for Multicriteria Decision Analysis. European Journal of Operational Research 129(1), 1–47 (2001)
7. Fortemps, P., Greco, S., Slowinski, R.: Multicriteria Decision Support Using Rules That Represent Rough-Graded Preference Relations. European Journal of Operational Research 188(1), 206–223 (2008)
8. Roy, B.: Méthodologie multicritère d'aide à la décision. Economica, Paris (1985)
9. Pawlak, Z.: Rough Sets. International Journal of Computer and Information Sciences 11(5), 341–356 (1982)
10. Pawlak, Z., Slowinski, R.: Rough Set Approach to Multi-attribute Decision Analysis. European Journal of Operational Research 72(3), 443–459 (1994)
11. Greco, S., Matarazzo, B., Slowinski, R.: Rough Sets Methodology for Sorting Problems in Presence of Multiple Attributes and Criteria. European Journal of Operational Research 138(2), 247–259 (2002)
12. Greco, S., Matarazzo, B., Slowinski, R.: Extension of the Rough Set Approach to Multicriteria Decision Support. INFOR 38(3), 161–193 (2000)
13. Greco, S., Matarazzo, B., Slowinski, R.: Rough Approximation of a Preference Relation by Dominance Relations. European Journal of Operational Research 117(1), 63–83 (1999)
14. Skowron, A., Rauszer, C.: The Discernibility Matrices and Functions in Information Table. In: Intelligent Decision Support: Handbook of Applications and Advances of the Rough Set Theory, pp. 331–362. Kluwer Academic Publishers, Dordrecht (1991)

A Novel Hybrid Particle Swarm Optimization for Multi-Objective Problems

Siwei Jiang and Zhihua Cai*

School of Computer Science, China University of Geosciences, Wuhan 430074, China
amosonic@gmail.com, zhcai@cug.edu.cn

Abstract. To solve the multi-objective problems, a novel hybrid particle swarm optimization algorithm is proposed(called HPSODE). The new algorithm includes three major improvement: (I)Population initialization is constructed by statistical method *Uniform Design*, (II)Regeneration method has two phases: the first phase is particles updated by adaptive PSO model with constriction factor χ, the second phase is Differential Evolution operator with archive, (III)A new accept rule called *Distance/volume fitness* is designed to update archive. Experiment on ZDTx and DTLZx problems by jMetal 2.1, the results show that the new hybrid algorithm significant outperforms OMOPSO, SMPSO in terms of additive Epsilon, HyperVolume, Genetic Distance, Inverted Genetic Distance.

Keywords: Multi-Objective Optimization, Uniform Design, Particle Swam OPtimization, Differential Evolution, Minimum Reduce Hypervolume, Spread.

1   Introduction

Multi-objective Evolutionary Algorithms(MOEAs) are powerful tools to solve multi-Objective optimize problems(MOPs), it gains popularity in recent years [1]. Recently, some elitism algorithms are proposed as: NSGA-II [2], SPEA2 [3], GDE3 [4], MOPSO [5, 6, 7].

NSGA-II adopt a fast non-dominated sorting approach to reduce computer burden, it uses *Ranking* and *Crowding Distance* to choose the candidate solutions [2]. SPEA2 proposes a fitness assignment with cluster technique, it designs a truncation operator based on the nearest *Neighbor Density Estimation* metric [3]. GDE3 is a developed version of Differential Evolution, which is suited for global optimization with an arbitrary number of objectives and constraints [4].

MOPSO is proposed by Coello *et al*, it adopts swarm intelligence to optimize MOPs, and it uses the Pareto-optimal set to guide the particles flight [5]. Sierra adopt *Crowding Distance* to filter the leaders, the different mutation methods are acted on divisions particles [6]; Mostaghim introduces a new Sigam-method to find local best information to guide particle [7].

---

* The Project was supported by the Research Foundation for Outstanding Young Teachers, China University of Geosciences(Wuhan)(No:CUGQNL0911).

H. Deng et al. (Eds.): AICI 2009, LNAI 5855, pp. 28–37, 2009.

Population initialization is an important part in EAs, but it has been long ignored. *Orthogonal design* and *Uniform Design* belong to a sophisticated branch of statistics [9,10]. Zeng and Cai adopt orthogonal design to solve MOPs, results show that it is more powerful than the *Random Design* [11,12]. Leung utilizes the uniform design to initial population and designs a new crossover operator, which can find the Pareto-optimal solutions scattered uniformly [13].

Zhang proposes a hybrid PSO with differential evolution operator to solve the single objective problem, which provide the bell-shaped mutation to guarantee the evolutionary population diversity [8].

Interested in the hybrid PSO with differential evolutionary operator, we propose a hybrid particle swarm optimization called **HPSODE**: the first population is constructed by uniform design; offspring regeneration method has two phases, in first phase the particles are updated according to its own experience(*pbest*) and social information(*gbest*) with constriction factor χ, in second phase the archive is operated by Differential Evolution; a new accept rule is designed as *Distance/Volume* function to update the archive.

The rest of the paper is organized as follows. In section 2, we briefly introduce the uniform design. In section 3, we describe the two phases of regeneration method. In section 4, we design a new archive update rule. In section 5, Experiment on bio-objective and tri-objective has show that the new algorithm is powerful than MOPSO, SMPSO in terms of additive Epsilon, hypervolume, Genetic Distance, Inverted Genetic Distance. In section 6, we make conclusions and discuss the future research on MOEAs.

2   Population Initialization: Uniform Design

Population initialization has long been ignored in MOEAs, but it is a very important component for MOEAs. *Orthogonal Design* and *Uniform Design* are experimental design method, which belong to a sophisticated branch of statistics. Both of them can get better distribute population in feasible searching space than *Random Design*, then the statistical population provide more information to generate next offspring.

In this section, we briefly describe an experimental design method called *uniform design*. We define the uniform array as $U_R(C)$, where Q is the level, it's primer, R, C represent the row and column of uniform array, they must be satisfid with:

$$\begin{cases} R = Q > n \\ C = n \end{cases} \tag{1}$$

where n is the number of variables. When select a proper parameters of Q, σ form table 1, uniform array can be created by Equation 2

$$U_{i,j} = (i * \sigma^{j-1} \bmod Q) + 1 \tag{2}$$

For one decision variable X_j with the boundary $[l_j, u_j]$, then the quantize technical divide the domain into Q levels $\alpha_1^j, \alpha_2^j, \cdots, \alpha_Q^j$, where the design parameter

Table 1. Parameter σ for different number of factors and levels

number of levels of per factor	number of factors	σ
5	2-4	2
7	2-6	3
11	2-10	7
13	2	5
	3	4
	4-12	6
17	2-16	10
19	2-3	8
	4-18	14
23	2,13-14,20-22	7
	8-12	15
	3-7,15-19	17
29	2	12
	3	9
	4-7	16
	8-12,16-24	8
	13-15	14
	25-28	18
31	2,5-12,20-30	12
	3-4,13-19	22

Q is primer and α_i is given by

$$\alpha_k^j = l_j + (k-1)(\frac{u_j - l_j}{Q-1}), 1 \leq k \leq Q \tag{3}$$

In other words, the domain $[l_j, u_j]$ is quantized $Q - 1$ fractions, and any two successive levels are same as each other.

3 Offspring Regeneration Method

In this paper, we propose a hybrid Particle Swarm Optimization algorithm called **HPSODE**, which has three populations: the first population is the original particles, the second population is local best particles which store only one evolutionary step according to original particles, the third population is global best particle which is utilized as archive. The regeneration method has two phases, original particle with adaptive PSO update, archive with differential evolution operator.

In first phase, the particle's new position is updated by its own experience (*pbest*) and social information(*gbest*) with constriction factor χ:

$$\begin{cases} V_{id}(t+1) = \chi[\omega V_{id}(t) + C_1\varphi_1(pb_{id}(t) - X_{id}(t)) + C_2\varphi_2(gb_{id}(t) - X_{id}(t))] \\ X_{id}(t+1) = X_{id}(t) + V_{id}(t+1) \end{cases}$$

$$\tag{4}$$

where $\chi = \dfrac{2}{2-\varphi-\sqrt{\varphi^2-4\varphi}}$ with $\varphi = C_1 + C_2$, when $\varphi \leq 4$, $\chi = 1.0$. And $\omega = 0.1$, $C_1, C_2 \in rand[1.5, 2.5], \varphi_1, \varphi_2 \in rand[0.0, 1.0]$.

The global best particle is binary tournament choose from archive which has larger crowding distance, it forces the original particles to explore the sparse space. If the new particle position is non-dominated by original particle, replace the original particle and local best particle, $X_i(t) = X_i(t+1), pb_i(t) = X_i(t+1)$, then add the new particle to archive; otherwise discard the new particle and unchange the local best.

In second phase, the third population is update by differential evolution operator:

$$
\begin{cases}
X_{id}(t+1) = X_{r_1d}(t) + F(X_{r_2d}(t) - X_{r_3d}(t)) \\
\qquad if \quad rnd_d(0,1) < CR||d == d_{rnd} \\
X_{id}(t+1) = X_{r_1d}(t) \quad otherwise
\end{cases}
\tag{5}
$$

where d is the dimension of solution, the three rand solutions are choose from archive and $r_1 \neq r_2 \neq r_3 \neq i$. where $F = 0.5, CR = 0.1$. If the new solution is non-dominated by the archive's solutions, then add it to archive.

When the number of archive is small, the guide information is rare, it will lead the original particle to assemble to few best positions, and then the final Pareto-optimal set will get worse diversity. Differential evolution operation is simple to implement and it has powerful capacity to scatter the feasible space. The DE operation on archive is useful to enhance the diversity of archive, and provide good guide information to the original particle.

4   Archive Update Rule

When MOEAs get a set of equal good solutions full to the setting size(usually $archiveSize = 100$), an accept rule must be designed to decide which one should be cut off from archive. It's a critical issue in MOEAs which directly influence the quality of finally optimal set in convergence and spread metric. Some popular accept rules have been presented: NSGA-II adopts the *Ranking* and *Crowding Distance* metric, SPEA2 uses the nearest *Neighbor Density Estimation* metric.

In this paper, we design a new accept rule called *Minimum Reduce Hypervolume*. Hypervolume is a quality indicator proposed by Zitzler *et al*, it is adopted in jMetal 2.1 [14]. Hypervolume calculates the volume covered by members of a non-dominated set of solutions (the region enclosed into the discontinuous line respect the worst point W in the figure 1 is $ADBECW$ in dashed line).

If two solutions D and E are non-dominated each other, NSGA-II choose the solution D remained in archive if $CD(D) > CD(E)$, it maintains the spread along the Pareto-front.

$$
\begin{cases}
CD(D) = AD' + D'B \\
CD(E) = BE' + E'C
\end{cases}
\tag{6}
$$

If one solution is deleted, it will lead to a hypervolume decrease, because the higher hypervolume means the better quality of optimal set, we will delete the solution which reduce the hypervolume minimum. *MRV* chooses the solution E remained in archive if $hv(D) < hv(E)$(then the hypervolume is $ABECW$ rather

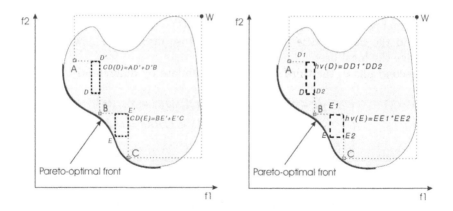

Fig. 1. The comparison of Crowding Distance and MRV is described in the left and right figure

than $ABDCW$), it maintains the hyervolume along the Pareto-front.

$$\begin{cases} hv(D) = DD1 * DD2 \\ hv(E) = EE1 * EE2 \end{cases} \qquad (7)$$

Crowding Distance maintains the spread and expands the solution to feasible searching place, and *Minimum Reduce Hypervolume* maintains the hypevolume and forces the solution near to Pareto-front. We combine the two properties, a new fitness assignment for solution s is designed as follows(called *Distance/Volume* fitness):

$$\begin{cases} DV(s) = CD(s) + scale * hv(s) \\ scale = \dfrac{\sum_{i=1}^{n} CD(s_i)}{\sum_{i=1}^{n} hv(s_i)} \end{cases} \qquad (8)$$

The factor *scale* is designed to equal the influence of crowding distance and MRV, but if it is too large, we set it is to 1000 when $scale > 1000$.

5   Experiment Results

Experiment is based on jMetal 2.1 [14], which is a Java-based framework aimed at facilitating the development of metaheuristics for solving MOPs, it provides large block reusing code and fair comparison for different MOEAs. The paper selects the algorithm OMOPSO [6], SMPSO [14] as the compare objectives.

Each algorithm independent runs for 100 times and maximum evolution times is 25, 000. The test problems are choose from ZDTx, DTLZx problem family.

The performance metrics have five categories:

Unary additive epsilon indicator($I_{\epsilon+}^{1}$). The metric indicator of an approximation set A ($I_{\epsilon+}^{1}(A)$) gives the minimum factor ϵ by which each point in the real front can be added such that the resulting transformed approximation set is dominated by A:

$$I_{\epsilon+}^1(A) = inf_{\epsilon \in R}\{\forall z^2 \in R \backslash \exists z^1 \in A : z_i^2 \le z_i^1 + \epsilon \forall i\} \tag{9}$$

Hypervolume. This quality indicator calculates the volume (in the objective space) covered by members of a nondominated set of solutions with a reference point. The hypervolume (HV) is calculated by:

$$HV = volume(\bigcup_{i=1}^{|Q|} v_i) \tag{10}$$

Inverted Generational Distance. The metric is to measure how far the elements are in the Pareto-optimal set from those in the set of non-dominated vectors found. It is defined as:

$$IGD = \sqrt{\frac{\sum_{i=1}^{|N|} d_i^2}{N}} \tag{11}$$

Generational Distance. The metric is to measure how far the elements are in the set of non-dominated vectors found from those in the Pareto-optimal set. It is defined as:

$$GD = \sqrt{\frac{\sum_{i=1}^{|n|} d_i^2}{n}} \tag{12}$$

Spread. The Spread indicator is a diversity metric that measures the extent of spread achieved among the obtained solutions. This metric is defined as:

$$\Delta = \frac{d_f + d_l + \sum_{i=1}^{n-1} |d_i - \overline{d}|}{d_f + d_l + (n-1)\overline{d}} \tag{13}$$

The higher Hypervolume and lower $I_{\epsilon+}^1$, GD, IGD, Spread mean the better algorithm. The results are compared with median and interquartile range(IQR) which measures of location (or central tendency) and statistical dispersion, respectively, the best result is grey background.

From table 2, in term of median and IQR for additive epsilon metric, *HPSODE* get best results in all of the 12 MOPs.

From table 3, in term of median and IQR for HyperVolume metric, *HPSODE* get best results in all of the 12 MOPs.

From table 4, in term of median and IQR for Genetic Distance metric, *HPSODE* get best results in 11 MOPs, only worse in DTLZ6. *OMOPSO* get best results only in 1 MOPs: DTLZ6.

From table 5, in term of median and IQR for Inverted Genetic Distance metric, *HPSODE* get best results in 10 MOPs, only worse in ZDT2, DTLZ6. *OMOPSO* get best results only in 1 MOPs: ZDT2. *SMOPSO* get best results only in 1 MOPs: DTLZ6.

Table 2. Unary additive epsilon indicator($I^1_{\epsilon+}$) Median and IQR

	OMOPSO	SMPSO	HPSODE
ZDT1	$6.15e-03_{3.7e-04}$	$5.63e-03_{3.4e-04}$	$5.33e-03_{2.0e-04}$
ZDT2	$5.93e-03_{4.7e-04}$	$5.49e-03_{3.0e-04}$	$5.27e-03_{2.2e-04}$
ZDT3	$6.87e-03_{2.1e-03}$	$5.85e-03_{1.0e-03}$	$4.56e-03_{6.4e-04}$
ZDT4	$5.39e+00_{4.7e+00}$	$6.43e-03_{5.7e-04}$	$5.40e-03_{2.2e-04}$
ZDT6	$4.92e-03_{5.9e-04}$	$4.72e-03_{3.9e-04}$	$4.45e-03_{3.1e-04}$
DTLZ1	$1.09e+01_{3.2e+00}$	$5.65e-02_{7.8e-03}$	$4.68e-02_{6.5e-03}$
DTLZ2	$1.29e-01_{2.0e-02}$	$1.40e-01_{1.9e-02}$	$1.22e-01_{2.1e-02}$
DTLZ3	$1.03e+02_{3.6e+01}$	$1.61e-01_{1.3e-01}$	$1.19e-01_{1.9e-02}$
DTLZ4	$1.12e-01_{2.3e-02}$	$1.20e-01_{3.8e-02}$	$1.03e-01_{2.2e-02}$
DTLZ5	$5.14e-03_{6.6e-04}$	$4.87e-03_{5.5e-04}$	$4.36e-03_{2.8e-04}$
DTLZ6	$4.35e-03_{5.2e-04}$	$4.35e-03_{3.9e-04}$	$4.26e-03_{3.9e-04}$
DTLZ7	$1.95e-01_{7.6e-02}$	$1.71e-01_{6.9e-02}$	$1.25e-01_{3.6e-02}$

Table 3. Hypervolume Median and IQR

	OMOPSO	SMPSO	HPSODE
ZDT1	$6.611e-01_{4.4e-04}$	$6.618e-01_{1.2e-04}$	$6.621e-01_{1.7e-05}$
ZDT2	$3.280e-01_{4.0e-04}$	$3.285e-01_{1.1e-04}$	$3.288e-01_{1.7e-05}$
ZDT3	$5.140e-01_{1.1e-03}$	$5.154e-01_{4.2e-04}$	$5.160e-01_{1.2e-05}$
ZDT4	$0.000e+00_{0.0e+00}$	$6.613e-01_{2.3e-04}$	$6.621e-01_{2.3e-05}$
ZDT6	$4.013e-01_{1.1e-04}$	$4.012e-01_{1.1e-04}$	$4.015e-01_{1.5e-05}$
DTLZ1	$0.000e+00_{0.0e+00}$	$7.372e-01_{9.0e-03}$	$7.688e-01_{7.5e-03}$
DTLZ2	$3.537e-01_{7.1e-03}$	$3.463e-01_{1.0e-02}$	$3.821e-01_{5.1e-03}$
DTLZ3	$0.000e+00_{0.0e+00}$	$3.406e-01_{4.3e-02}$	$3.864e-01_{5.8e-03}$
DTLZ4	$3.591e-01_{9.7e-03}$	$3.577e-01_{1.6e-02}$	$3.839e-01_{6.3e-03}$
DTLZ5	$9.344e-02_{1.6e-04}$	$9.366e-02_{1.6e-04}$	$9.407e-02_{4.4e-05}$
DTLZ6	$9.493e-02_{5.5e-05}$	$9.488e-02_{6.4e-05}$	$9.496e-02_{4.1e-05}$
DTLZ7	$2.610e-01_{9.9e-03}$	$2.746e-01_{7.4e-03}$	$2.916e-01_{3.7e-03}$

From table 6, in term of median and IQR for Spread metric, *HPSODE* get best results in all of the 6 MOPs. *OMOPSO* get best results only in 2 MOPs. *SMOPSO* get best results only in 4 MOPs.

The comparison for three algorithms in five categories: epsilon indicator, hypervolume, Generation Distance, Inverted genetic Distance, Spread, it shows that the new algorithm is more efficient to solve the MOPs. Now, we summarize the highlight as follows:

1. HPSODE adopt the statistical method *Uniform Design* to construct the first population, which can get well distributed solutions in feasible space.
2. HPSODE combine the PSO and Differential Evolution operation to generate next population. DE operation enhances the diversity for global guide population.

Table 4. Generational Distance Median and IQR

	OMOPSO	SMPSO	HPSODE
ZDT1	$1.43e-04_{3.8e-05}$	$1.24e-04_{4.9e-05}$	$7.41e-05_{3.5e-05}$
ZDT2	$7.68e-05_{2.6e-05}$	$5.27e-05_{5.4e-06}$	$4.66e-05_{2.8e-06}$
ZDT3	$2.25e-04_{3.3e-05}$	$1.88e-04_{2.3e-05}$	$1.76e-04_{1.4e-05}$
ZDT4	$6.57e-01_{5.8e-01}$	$1.35e-04_{4.6e-05}$	$7.67e-05_{4.2e-05}$
ZDT6	$2.19e-02_{4.4e-02}$	$2.18e-03_{4.3e-02}$	$5.37e-04_{2.0e-05}$
DTLZ1	$1.00e+01_{1.3e+00}$	$5.06e-03_{1.4e-03}$	$7.00e-04_{1.9e-04}$
DTLZ2	$3.32e-03_{4.1e-04}$	$3.94e-03_{9.2e-04}$	$1.25e-03_{4.6e-04}$
DTLZ3	$3.51e+01_{6.0e+00}$	$5.29e-03_{5.7e-02}$	$1.17e-03_{1.8e-04}$
DTLZ4	$5.83e-03_{5.0e-04}$	$5.98e-03_{6.4e-04}$	$4.95e-03_{3.9e-04}$
DTLZ5	$3.00e-04_{4.3e-05}$	$2.85e-04_{5.6e-05}$	$2.50e-04_{3.3e-05}$
DTLZ6	$5.62e-04_{3.0e-05}$	$5.69e-04_{2.6e-05}$	$5.67e-04_{2.6e-05}$
DTLZ7	$5.26e-03_{1.2e-03}$	$4.84e-03_{1.3e-03}$	$3.57e-03_{1.1e-03}$

Table 5. Inverted Generational Distance Median and IQR

	OMOPSO	SMPSO	HPSODE
ZDT1	$1.37e-04_{2.3e-06}$	$1.35e-04_{1.0e-06}$	$1.34e-04_{1.2e-06}$
ZDT2	$1.43e-04_{3.2e-06}$	$1.40e-04_{1.9e-06}$	$1.44e-04_{2.9e-06}$
ZDT3	$2.14e-04_{1.4e-05}$	$2.00e-04_{7.2e-06}$	$1.90e-04_{2.7e-06}$
ZDT4	$1.61e-01_{1.5e-01}$	$1.38e-04_{1.4e-06}$	$1.34e-04_{1.5e-06}$
ZDT6	$1.20e-04_{7.9e-06}$	$1.18e-04_{7.9e-06}$	$1.10e-04_{7.0e-06}$
DTLZ1	$2.71e-01_{1.0e-01}$	$6.38e-04_{4.5e-05}$	$5.27e-04_{3.5e-05}$
DTLZ2	$7.58e-04_{4.6e-05}$	$8.03e-04_{5.1e-05}$	$7.04e-04_{4.1e-05}$
DTLZ3	$2.03e+00_{6.7e-01}$	$1.43e-03_{7.8e-04}$	$1.11e-03_{6.8e-05}$
DTLZ4	$1.23e-03_{1.6e-04}$	$1.17e-03_{2.1e-04}$	$1.17e-03_{1.5e-04}$
DTLZ5	$1.49e-05_{5.0e-07}$	$1.44e-05_{6.8e-07}$	$1.41e-05_{3.8e-07}$
DTLZ6	$3.38e-05_{1.1e-06}$	$3.43e-05_{1.3e-06}$	$3.40e-05_{1.4e-06}$
DTLZ7	$2.66e-03_{3.1e-04}$	$2.72e-03_{4.8e-04}$	$2.43e-03_{3.0e-04}$

Table 6. Spread Median and IQR

	OMOPSO	SMPSO	HPSODE
ZDT1	$8.29e-02_{1.5e-02}$	$7.90e-02_{1.3e-02}$	$9.49e-02_{1.6e-02}$
ZDT2	$7.99e-02_{1.5e-02}$	$7.21e-02_{1.8e-02}$	$9.48e-02_{1.4e-02}$
ZDT3	$7.13e-01_{1.2e-02}$	$7.11e-01_{9.8e-03}$	$7.05e-01_{3.8e-03}$
ZDT4	$8.89e-01_{8.8e-02}$	$9.84e-02_{1.6e-02}$	$1.01e-01_{2.1e-02}$
ZDT6	$1.07e+00_{9.9e-01}$	$3.17e-01_{1.2e+00}$	$6.76e-02_{1.2e-02}$
DTLZ1	$6.90e-01_{1.0e-01}$	$6.78e-01_{4.2e-02}$	$7.28e-01_{5.5e-02}$
DTLZ2	$6.22e-01_{4.9e-02}$	$6.38e-01_{5.1e-02}$	$6.32e-01_{5.7e-02}$
DTLZ3	$7.28e-01_{1.4e-01}$	$6.65e-01_{2.0e-01}$	$6.43e-01_{5.5e-02}$
DTLZ4	$6.36e-01_{6.9e-02}$	$6.59e-01_{1.3e-01}$	$6.41e-01_{5.3e-02}$
DTLZ5	$2.02e-01_{6.8e-02}$	$1.70e-01_{8.5e-02}$	$1.24e-01_{2.3e-02}$
DTLZ6	$1.24e-01_{4.0e-02}$	$1.39e-01_{3.9e-02}$	$1.19e-01_{2.4e-02}$
DTLZ7	$7.07e-01_{5.6e-02}$	$7.08e-01_{7.0e-02}$	$6.99e-01_{5.8e-02}$

3. HPSODE design a new accept rule *Distance/volume fitness*, which expands the solution to feasible searching place and forces the solution near to Pareto-front.
4. HPSODE significant outperforms OMOPSO, SMPSO in terms of additive Epsilon, HyperVolume, Genetic Distance, Inverted Genetic Distance.
5. HPSODE get better results than OMOPSO, SMPSO in term of spread metric.

6 Conclusion and Future Reasearch

Population initialization, regeneration method, archive update rule are three critical issues in MOEAs. In this paper, we propose a hybrid PSO algorithm called *HPSODE* which include *uniform design* initialization, two phase regeneration with DE operator, new accept rule *Distance/Volume fitness*. Experimental results show that *HPSODE* can get higher Hypervolume, lower additive epsilon, GD, IGD and competitive spread. Our future research will focus on enhance the Quantum PSO algorithm to solve the MOPs.

References

1. Coello, C.A.C.: Evolutionary multi-objective optimization: A historical view of the Field. IEEE Computational Intelligence Magazine 1(1), 28–36 (2006)
2. Deb, K., Pratap, A., Agarwal, S., Meyarivan, T.: A fast and elitist multiobjective genetic algorithm: NSGA - II. IEEE Transactions on Evolutionary Computation 6(2), 182–197 (2002)
3. Zitzler, E., Laumanns, M., Thiele, L.: SPEA2: Improving the strength Pareto evolutionary algorithm, Technical Report 103, Computer Engineering and Networks Laboratory (2001)
4. Kukkonen, S., Lampinen, J.: GDE3: The third evolution step of generalized differential evolution. In: Proceedings of the 2005 IEEE Congress on Evolutionary Computation (1), pp. 443–450 (2005)
5. Coello Coello, C.A., Toscano Pulido, G., Salazar Lechuga, M.: Handling Multiple Objectives With Particle Swarm Optimization. IEEE Transactions on Evolutionary Computation 8, 256–279 (2004)
6. Sierra, M.R., Coello, C.A.C.: Improving PSO-based multi-objective optimization using crowding, mutation and ε-dominance. In: Coello Coello, C.A., Hernández Aguirre, A., Zitzler, E. (eds.) EMO 2005. LNCS, vol. 3410, pp. 505–519. Springer, Heidelberg (2005)
7. Mostaghim, S., Teich, J.: Strategies for Finding Good Local Guides in Multi-objective Particle Swarm Optimization (MOPSO). In: 2003 IEEE Swarm Intelligence Symposium Proceedings, Indianapolis, Indiana, USA, pp. 26–33. IEEE Service Center (2003)
8. Zhang, W.J., Xie, X.F.: DEPSO: Hybrid particle swarm with differential evolution operator. In: IEEE International Conference on Systems Man and Cybernetics, pp. 3816–3821 (2003)
9. Fang, K.T., Ma, C.X.: Orthogonal and uniform design. Science Press (2001) (in Chinese)

10. Leung, Y.W., Wang, Y.: An orthogonal genetic algorithm with quantization for global numerical optimization. IEEE Transactions on Evolutionary Computation 5(1), 41–53 (2001)
11. Zeng, S.Y., Kang, L.S., Ding, L.X.: An orthogonal multiobjective evolutionary algorithm for multi-objective optimization problems with constraints. Evolutionary Computation 12, 77–98 (2004)
12. Cai, Z.H., Gong, W.Y., Huang, Y.Q.: A novel differential evolution algorithm based on ε-domination and orthogonal design method for multiobjective optimization. In: Obayashi, S., Deb, K., Poloni, C., Hiroyasu, T., Murata, T. (eds.) EMO 2007. LNCS, vol. 4403, pp. 286–301. Springer, Heidelberg (2007)
13. Leung, Y.-W., Wang, Y.: Multiobjective programming using uniform design and genetic algorithm. IEEE Transactions on Systems, Man, and Cybernetics, Part C 30(3), 293 (2000)
14. Durillo, J.J., Nebro, A.J., Luna, F., Dorronsoro, B., Alba, E.: jMetal: A Java Framework for Developing Multi-Objective Optimization Metaheuristics, Departamento de Lenguajes y Ciencias de la Computación, University of Málaga, E.T.S.I. Informática, Campus de Teatinos, ITI-2006-10 (December 2006), http://jmetal.sourceforge.net

Research on Constrained Layout Optimization Problem Using Multi-adaptive Strategies Particle Swarm Optimizer[*]

Kaiyou Lei

Faculty of Computer & Information Science, Southwest University, Chongqing, 400715, China
lky@swu.edu.cn

Abstract. The complex layout optimization problems with behavioral constraints belong to NP-hard problem in math. Due to its complexity, the general particle swarm optimization algorithm converges slowly and easily to local optima. Taking the layout problem of satellite cabins as background, a novel adaptive particle swarm optimizer based on multi-modified strategies is proposed in the paper, which can not only escape from the attraction of local optima of the later phase to heighten particle diversity, and avoid premature problem, but also maintain the characteristic of fast speed search in the early convergence phase to get global optimum, thus, the algorithm has a better search performance to deal with the constrained layout optimization problem. The proposed algorithm is tested and compared it with other published methods on constrained layout examples, demonstrated that the revised algorithm is feasible and efficient.

Keywords: particle swarm optimization, premature problem, constrained layout.

1 Introduction

The classical layout problem is generally divided in to two types: Packing problem and Cutting problem. The main problem is to increase the space utility ratio as much as possible under the condition of non-overlapping between piece (object) and piece (object) and between piece (object) and container. They are called as layout problems without behavioral constraints (unconstrained layout). In recent years, another more complex layout problem is attracting a lot of attention, such as the layout design of engineering machine, spacecraft, ship, etc. Solving these kinds of problems need to consider some additional behavioral constraints, for instance, inertia, equilibrium, stability, vibration, etc. They are called as layout problem with behavioral constraints (constrained layout). Constrained layout belong to NP-hard problem, its optimization is hand highly [1].

As a newly developed population-based computational intelligence algorithm, Particle Swarm Optimization (PSO) was originated as a simulation of simplified social model of birds in a flock [2]. The PSO algorithm is easy to implement and has been proven very competitive in large variety of global optimization problems and application areas compared to conventional methods and other meta-heuristics [3].

---

[*] The work is supported by Key Project of Chinese Ministry of Education (104262).

H. Deng et al. (Eds.): AICI 2009, LNAI 5855, pp. 38–47, 2009.

Since its introduction, numerous variations of the basic PSO algorithm have been developed in the literature to avoid the premature problem and speed up the convergence process, which are the most important two topics in the research of stochastic search methods. To make search more effective, there are many approaches suggested by researchers to solve the problems, such as variety mutation and select a single inertia weight value methods, etc, but these methods have some weakness in common, they usually can not give attention to both global search and local search, preferably, so to trap local optima, especially in complex constrained layout problems [4].

In this paper, particle swarm optimizer with a better search performance is proposed, which employ multi-adaptive strategies to plan large-scale space global search and refined local search as a whole according to the specialties of constrained layout problems, and to quicken convergence speed, avoid premature problem, economize computational expenses, and obtain global optimum. We tested the proposed algorithm and compared it with other published methods on three constrained layout examples. The experimental results demonstrated that this revised algorithm can rapidly converge at high quality solutions.

2 Mathematical Model and PSO

Mathematical Model. Taking the layout problem of satellite cabins as background, the principle is illuminated [1]. There is a two-dimension rotating circular table (called graph plane) with radius R and n dishes (called graph units) with uniform thickness, each of which has a radius r_i and mass m_i (i=1,2...n), are installed on the graph plane as shown in Fig.1. Suppose that the frictional resistance is infinite. Find the position of each graph unit such that the graph units highly concentrate on the center of the graph plane and satisfy the following constraints:

(1)There is no interference between any two graph units.

(2)No part of any graph unit protrudes out the graph plane.

(3)The deviation from dynamic-balancing for the system should not exceed a permissible value $[\delta_j]$.

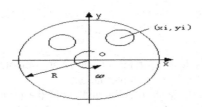

Fig. 1. Representation of graph units layout

Suppose that the center of the graph plane is origin o of the Cartesian system, the graph plane and graph units are just in plane xoy, the thickness of the graph units is ignored, and x_i, y_i are the coordinates of the center o_i of the graph unit i, which is its mass center also. The mathematical model for optimization of the problem is given by:

$$\text{Find } X_i = [x_i, y_i]^T, \quad i \in I, i = 1,2,3,\cdots,n$$

$$\min F(X_i) = max\left\{\sqrt{(x_i^2 + y_i^2)} + r_i\right\} \tag{1}$$

s. t.

$$f_1(X_i) = r_i + r_j - \sqrt{(x_i - x_j)^2 + (y_i - y_j)^2} \le 0, i \ne j; i, j \in I \tag{2}$$

$$f_2(X_i) = \sqrt{(x_i^2 + y_i^2)} + r_i - R \le 0, \quad i \in I \tag{3}$$

$$f_3(X_i) = \sqrt{\left(\sum_{i=1}^{n} m_i x_i\right)^2 + \left(\sum_{i=1}^{n} m_i y_i\right)^2} - [\delta_j] \le 0, i \in I \tag{4}$$

PSO Algorithm. In the original PSO formulae, particle i is denoted as $X_i=(x_{i1},x_{i2},...,x_{iD})$, which represents a potential solution to a problem in D-dimensional space. Each particle maintains a memory of its previous best position Pbest, and a velocity along each dimension, represented as $V_i=(v_{i1},v_{i2},...,v_{iD})$. At each iteration, the position of the particle with the best fitness in the search space, designated as g, and the P vector of the current particle are combined to adjust the velocity along each dimension, and that velocity is then used to compute a new position for the particle.

In the standard PSO, the velocity and position of particle i at $(t+1)$th iteration are updated as follows:

$$v_{id}^{t+1} = w * v_{id}^t + c_1 * r_1 * \left(p_{id}^t - x_{id}^t\right) + c_2 * r_2 * \left(p_{gd}^t - x_{id}^t\right) \tag{5}$$

$$x_{id}^{t+1} = x_{id}^t + v_{id}^{t+1} \tag{6}$$

Constants c1 and c2 are learning rates; r1 and r2 are random numbers uniformly distributed in the interval [0, 1]; w is an inertia factor.

To speed up the convergence process and avoid the premature problem, Shi proposed the PSO with linearly decrease weight method (LDWPSO)[3,4]. Suppose *wmax* is the maximum of inertia weight, *wmin* is the minimum of inertia weight, run is current iteration times, *runMax* is the total iteration times, the inertia weight is formulated as:

$$w = wmax - (wmax - wmin) * run / runMax \tag{7}$$

3 Multi-Adaptive Strategies Particle Swarm Optimizer (MASPSO)

Due to the complexity of a great deal local and global optima, PSO is revised as MASPSO by four adaptive strategies to adapt the constrained layout optimization.

3.1 Adaptive harmonization Strategy of w

The w has the capability to automatically harmonize global search abilities and local search abilities, avoid premature and gain rapid convergence to global optimum. First of all, larger w can enhance global search abilities of PSO, so to explore large-scale search space and rapidly locate the approximate position of global optimum, smaller w can enhance local search abilities of PSO, particles slow down and deploy refined local search, secondly, the more difficult the optimization problems are, the more fortified the global search abilities need, once located the approximate position of global optimum, the refined local search will further be strengthen to get global optimum[5,6,7].

According to the conclusions above, we first constructed (8) as new inertia weight decline curve for PSO, demonstrated in Figure 2.

$$w = 1/\left(1 + \exp\left(0.015 * \left(run - runMax/3\right)\right)\right) \tag{8}$$

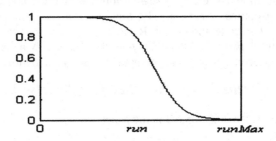

Fig. 2. Inertia weight decline curve by formula [8]

3.2 Adaptive Difference Mutation Strategy of Global Optimum p_{gd}^t

Considering that the particles may find the better global optimum in the current best region, the algorithm is designed to join mutation operation with the perturbation operator. The $runmax_1$, which is an iteration times of the transformation point, divides $runMax$ into two segments to respectively mutate according to themselves characteristics, and further enhance the global search abilities and local search abilities to find a satisfactory solution. p_η is the mutation probability within (0.1, 0.3). The computed equation is defined as:

$$if\ run \leq run\max_1\ p_{gd}^t = p_{gd}^t * \left(1 + 0.5\eta\right);\ esle\ p_{gd}^t = p_{gd}^t * \left(1 + 0.15\eta\right). \tag{9}$$

where η is Gauss $(0,1)$ distributed random variable.

3.3 Adaptive Radius Rs Strategy

In equation (3), the graph plane radius is constant, which will influence the fitness value computation and the result of search. In the search process, the more smallness the periphery envelope circle radiuses are, the more convergence graph units are. Evidently, taking the current periphery envelope circle radius as layout radius, the search progress in smaller radius area, and quicken convergence speed, economize computational expenses. Suppose the periphery envelope circle radius is Rs, the computed equation is defined as:

$$if\ R_S^{\ t} > R_S^{\ t+1},\ R_S^{\ t+1} = R_S^{\ t+1}, esle\ R_S^{\ t+1} = R_S^{\ t} \tag{10}$$

3.4 Adaptive Difference Mutation Strategy of X_i

Analysis on equation (5), if the best position of all particles has no change for a long time, then the velocity updated is decided by $w*v_{id}^{\ t}$. Due to $w<1$, the velocity is less and less, so the particle swarm is tended to get together, farther, the algorithm is trapped into local optima, premature problem emerged like in genetic algorithms. Considering that the adaptive difference mutation strategy of X_i is introduced. When the best position of all particles has no or less change for a long time, the part of all particles are kept down, deploy refined local search unceasingly; the rest are initialized randomly, so to enhance global search abilities of PSO, break away from the attraction of local optima, synchronously. Suppose run_t is an iteration times, run_k, run_{k+t} are the kth, $(k+t)$th iteration times, R_k, R_{k+t} are the maximal radius, accordingly; X_ρ are the initialized particles randomly, ρ is the dimension mutation probability, accordingly; ε is an error threshold, the computed equation is defined as:

$$if run_k - run_{k+t} \geq run_t\ \&\ R_k - R_{k+t} \leq \varepsilon, X_i = X_\rho, else\ X_i = X_i. \tag{11}$$

where $run_t,\ \varepsilon,\ \rho$ are taken 20,0.005,0.2 in the paper.

3.5 Algorithm Designing and Flow

PSO, which is modified as MASPSO by the above methods, has the excellent search performance to optimize constrained layout problems. According to MASPSO, The design of the algorithm is as follows.

All particles are coded based on the rectangular plane axis system. Considering that the shortest periphery envelope circle radius is the optimization criterion based on the above constraint conditions for our problem, the fitness function and the penalty function, which are constructed in the MASPSO, respectively, can be defined as:

$$\phi_1(X_i) = F(X_i) + \sum_{i=1}^{3} \lambda_i f_i(X_i) u_i(f_i),$$

$$u_i(f_i) = \begin{cases} 0 & f_i(X_i) \leq 0 \\ 1 & f_i(X_i) > 0 \end{cases}, \quad i \in I. \tag{12}$$

where λ_i (λ_1=1,λ_2=1,λ_3=0.01) is the penalty factor.

The flow of the algorithm is as follows:

Step1.Set parameters of the algorithm.

Step2.Randomly initialize the speed and position of each particle.

Step3.Evaluate the fitness of each particle using (12) and determine the initial values of the individual and global best positions: p_{id}^t and p_{gd}^t .

Step4.Update velocity and position using (5), (6)and(8).

Step6.Evaluate the fitness using (12) and determine the current values of the individual and global best positions: p_{id}^t and p_{gd}^t .

Step7.Check the R^t and R^{t+1} , determine the periphery envelope circle radius.

Step8.Mutate p_{gd}^t and X_i using (9) and (10), respectively.

Step9.Loop to Step 4 and repeat until a given maximum iteration number is attained or the convergence criterion is satisfied.

4 Computational Experiments

Taking he literature [8], [9] as examples, we tested our algorithm, the comparison of statistical results are shown in Tab.1, Tab.2, and Tab.3, respectively.

Parameters used in our algorithm are set to be: c_1=c_2=1.5, *runMax*=1000. The running environment is: MATLAB7.1, Pentium IV 2GHz CPU, 256MRAM, Win XPOS.

Example1. Seven graph units are contained in the layout problem. The radius of a graph plane is R=50mm. The allowing value of static non-equilibrium J is

Table 1. Comparison of experimental results of example 1

(a) Datum and layout results of example 1

Num-ber	Graph units		Literature [8] result		Literature [9] result		Our result	
	r(mm)	m(g)	x(mm)	y(mm)	x (mm)	y(mm)	x(mm)	y(mm)
1	10.0	100.00	-12.883	17.020	14.367	16.453	17.198	-13.618
2	11.0	121.00	8.8472	19.773	-18.521	-9.560	-19.857	6.096
3	12.0	144.00	0.662	0.000	2.113	-19.730	-1.365	19.886
4	11.5	132.00	-8.379	-19.430	19.874	-4.340	18.882	7.819
5	9.5	90.25	-1.743	0.503	-19.271	11.241	-16.967	-14.261
6	8.5	72.25	12.368	-18.989	-3.940	22.157	-0.606	-22.873
7	10.5	110.25	-21.639	-1.799	-0.946	2.824	-0.344	-3.010

Table 1. (continued)

(b) The performance comparison of layout results of example 1

Computational method	The least circle Radius included all graph units (mm)	Static non-equilibrium (g.mm)	Interference	Computational time(s)
Literature [8] algorithm	32.662	0.029000	0	1002
Literature [9] algorithm	31.985	0.018200	0	1002
Our algorithm	31.934	0.000001	0	548

(c) The example 1 layout results comparison based on 40 times algorithm running

The least circle radius included all graph units (mm)	Times		The least circle radius included all graph units (mm)	Times	
	Literature [9] algorithm	Our algorithm		Literature [9] algorithm	Our algorithm
≤32.3	10	19	(32.9,33.1]	2	1
(32.3,32.5]	16	11	(33.1,33.3]	2	1
(32.5,32.7]	5	5	>33.3	3	0
(32.7,32.9]	2	3			

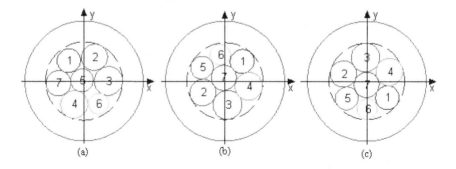

(a) Literature [8] layout result   (b) Literature [9] layout result   (c) Our layout result

Fig. 3. The geometry layout of example 1

$[g_j]$=3.4g*mm. The result is in Table 1 and the geometric layout is shown in Figure 3 (the particle size is 50).

Example2. Forty graph units are contained in the layout problem. The radius of a graph plane is R=880 mm. The radius of a graph plane is R=880 mm. The allowing value of static non-equilibrium J is $[g_j]$=20g*mm. The result is in Table2 and the geometric layout is shown in Figure 4 (the particle size is 100).

Table 2. Comparison of experimental results of example 2

(a) Datum and layout results of example 2

Number	Graph units		Literature [9] result		Our result	
	r(mm)	m(g)	x(mm)	y(mm)	x(mm)	y(mm)
1	106	11	192.971	0	-528.304	-119.969
2	112	12	-69.924	0	103.353	-441.960
3	98	9	13.034	-478.285	-164.920	271.842
4	105	11	-291.748	21.066	-268.975	68.475
5	93	8	343.517	-351.055	-382.940	575.014
6	103	10	-251.1434	674.025	119.907	-45.977
7	82	6	95.268	-252.899	321.849	-433.652
8	93	8	-619.634	-421.032	684.118	71.745
9	117	13	725.062	0	-112.375	-364.996
10	81	6	127.487	175.174	-194.273	-553.490
11	89	7	358.251	-104.181	58.218	-246.094
12	92	8	694.612	-206.946	482.078	-312.251
13	109	11	-151.494	-350.475	299.075	347.118
14	104	10	-486.096	278.028	-381.489	-562.752
15	115	13	-406.944	-378.282	-378.466	266.641
16	110	12	-531.396	27.583	330.471	-106.831
17	114	12	-281.428	-570.129	-55.284	643.08
18	89	7	535.186	-82.365	499.100	362.345
19	82	6	349.187	-668.540	442.242	-545.018
20	120	14	494.958	-527.668	641.168	-172.081
21	108	11	-696.916	236.466	-205.525	478.499
22	86	7	-43.153	196.294	-534.084	-449.197
23	93	8	-143.066	-725.316	-478.615	74.719
24	100	10	-433.688	-159.158	511.793	171.389
25	102	10	-741.858	0.000	-7.548	403.484
26	106	11	292.820	431.997	-369.369	-350.518
27	111	12	-540.511	495.023	113.819	228.113
28	107	11	154.296	-671.681	-316.496	-144.125
29	109	11	-317.971	463.365	156.530	-656.467
30	91	8	41.295	-271.016	-625.532	-292.937
31	111	12	103.622	538.523	308.084	114.390
32	91	8	215.467	-213.844	235.152	-283.914
33	101	10	540.248	306.466	-545.860	404.331
34	91	8	58.125	341.687	-660.742	171.945
35	108	11	-235.120	227.217	406.575	539.906
36	114	12	510.413	520.918	182.065	555.234
37	118	13	-29.219	725.331	-94.105	-105.916
38	85	7	300.625	240.313	-26.879	-592.954
39	87	7	234.066	-494.031	505.239	-15.817
40	98	9	411.043	119.080	-56.975	107.473

Table 2. (continued)

(b) The performance comparison of layout results of example 2

Computational method	The least circle Radius included all graph units (mm)	Static non-equilibrium (g.mm)	Interfer-ence	Computational time(s)
Literature [8] algorithm	874.830	11.395000	0	1656
Literature [9] algorithm	843.940	0.003895	0	2523
Our algorithm	783.871	0.001267	0	1633

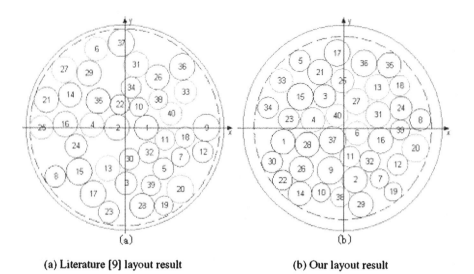

(a) Literature [9] layout result   (b) Our layout result

Fig. 4. The geometry layout of example 2

5 Conclusions

From table1and table2, we can educe that: PSO with four improved adaptive strate-gies harmonized the large-scale space global search abilities and the refined local search abilities much thoroughly, which has rapid convergence and can avoid prema-ture, synchronously. The effectiveness of the algorithm is validated for the con-strained layout of the NP-hard problems; the algorithm outperformed the known best ones in the in quality of solutions and the running time. In addition, the parameters of run_t 、 ε 、 ρ are chose by human experience, which has the certain blemish. How to choose the rather parameters are one of the future works.

References

1. Teng, H., Shoulin, S., Wenhai, G., et al.: Layout optimization for the dishes installed on rotating table. Science in China (Series A) 37(10), 1272–1280 (1994)
2. Kennedy, J., Eberhart, R.C.: Particle swarm optimization. In: Proc. of IEEE Int'l Conf. Neural Networks, pp. 1942–1948. IEEE Computer Press, Indianapolis (1995)

3. Eberhart, R.C., Kennedy, J.: A new optimizer using particles swarm theory. In: Sixth International Symposium on Micro Machine and Human Science, pp. 39–43. IEEE Service Center, Piscataway (1995)
4. Angeline, P.J.: Using selection to improve particle swarm optimization. In: Proc. IJCNN 1999, pp. 84–89. IEEE Computer Press, Indianapolis (1999)
5. Lei, K., Qiu, Y., He, Y.: A new adaptive well-chosen inertia weight strategy to automatically harmonize global and local search ability in particle swarm optimization. In: 1st International Symposium on Systems and Control in Aerospace and Astronautics, pp. 342–346. IEEE Press, Harbin (2006)
6. Shi, Y., Eberhart, R.C.: A modified particle swarm optimizer. In: Proc. of the IEEE Con. Evolutionary Computation, pp. 69–73. IEEE Computer Press, Piscataway (1998)
7. Jianchao, Z., Zhihua, C.: A guaranteed global convergence particle swarm optimizer. Journal of Computer Research and Development 4(8), 1334–1338 (2004) (in Chinese)
8. Fei, T., Hongfei, T.: A modified genetic algorithm and its application to layout optimization. Journal of Software 10(10), 1096–1102 (1999) (in Chinese)
9. Ning, L., Fei, L., Debao, S.: A study on the particle swarm optimization with mutation operator constrained layout optimization. Chinese Journal of Computers 27(7), 897–903 (2004) (in Chinese)

ARIMA Model Estimated by Particle Swarm Optimization Algorithm for Consumer Price Index Forecasting

Hongjie Wang[1] and Weigang Zhao[2,*]

[1] Railway Technical College, Lanzhou Jiaotong University, Lanzhou 730000, China
wanghongjietd@mail.lzjtu.cn
[2] School of Mathematics & Statistics, Lanzhou University, Lanzhou 730000, China
zwgstd@gmail.com

Abstract. This paper presents an ARIMA model which uses particle swarm optimization algorithm (PSO) for model estimation. Because the traditional estimation method is complex and may obtain very bad results, PSO which can be implemented with ease and has a powerful optimizing performance is employed to optimize the coefficients of ARIMA. In recent years, inflation and deflation plague the world moreover the consumer price index (CPI) which is a measure of the average price of consumer goods and services purchased by households is usually observed as an important indicator of the level of inflation, so the forecast of CPI has been focused on by both scientific community and relevant authorities. Furthermore, taking the forecast of CPI as a case, we illustrate the improvement of accuracy and efficiency of the new method and the result shows it is predominant in forecasting.

Keywords: ARIMA model, Particle swarm optimization algorithm, Moment estimation, Consumer price index.

1 Introduction

The ARIMA, which is one of the most popular models for time series forecasting analysis, has been originated from the autoregressive model (AR) proposed by Yule in 1927, the moving average model (MA) invented by Walker in 1931 and the combination of the AR and MA, the ARMA models [1]. The ARMA model can be used when the time series is stationary, but the ARIMA model hasn't that limitation. ARIMA model pre-processes the original time series by some methods to make the obtained series stationary so that we can then apply the ARMA model. For example, for an original series which has only serial dependency and hasn't seasonal dependency, if the dth differenced series of this original series is stationary, we can then apply the ARMA(p, q) model to the dth differenced series and we say the original time series satisfy the modeling condition of ARIMA(p, d, q) which is one kind of the ARIMA models. So the

---

* Corresponding author.

H. Deng et al. (Eds.): AICI 2009, LNAI 5855, pp. 48–58, 2009.

important point to note in ARIMA(p, d, q) modeling is that we must have either a stationary time series or a time series that becomes stationary after one or more differencing to be able to use it [2]. In addition, as we know, there are four main stages in building an ARMA model: 1) model identification, 2) model estimation, 3) model checking and 4) forecasting [3] (since we focus on the model estimation here, the third step is ignored in this paper). Since the second step is a complex process even when we adopt the relatively simple method (*moment estimation* for example) and the approximate results of the parameters are unpredictable which might greatly impact the forecasting precision when $q > 1$ that is the order of MA, PSO is proposed to achieve the second step.

Particle swarm optimization is a population based stochastic optimization technique developed by Dr. Eberhart and Dr. Kennedy in 1995, inspired by social behavior of bird flocking or fish schooling. The algorithm completes the optimization through following the personal best solution of each particle and the global best value of the whole swarm. And PSO can be implemented with ease and has a simple principle, so it has been successfully applied in many fields. Furthermore, we can see the new method has a higher forecasting precision than the *moment estimation* from the case study in Section 4.

This paper is organized as follows: Section 2 reviews the detailed process of ARIMA model (see [4]) and indicates the deficiencies of the traditional model estimation; Section 3 introduces the particle swarm optimization algorithm and proposes the PSO-based estimation method; In Section 4, as a case we forecast the CPI to illustrate the excellent forecast ability of the new method.

2   Autoregressive Integrated Moving Average

Assume that $\{y_1, y_2, \ldots, y_N\}$ is the original series without seasonal dependency and the series has the limitation that at least 50 and preferably 100 observations or more should be used [5], that is to say, $N \geq 50$. Then we can model the ARIMA(p, d, q) for this series, and it is obvious that the PSO is applicable for the other ARIMA models (such as SARIMA model and so on) in the same way.

2.1   Model Identification

For an ARIMA(p, d, q), it is necessary to obtain the (p, d, q) parameters, where p is the number of autoregressive terms, d is the number of non-seasonal differences and q is the number of lagged forecast errors in the prediction equation [6].

The chief tools in identification are the autocorrelation function (ACF), the partial autocorrelation function (PACF), and the resulting correlogram, which is simply the plots of ACF and PACF against the lag length [2].

Since the ACF and the PACF is based on the covariance, it is necessary to introduce the covariance first. For a time series $\{w_1, w_2, \ldots, w_n\}$, the covariance at lag k (when $k = 0$, it is the variance), denoted by γ_k, is estimated in (1).

$$\hat{\gamma}_k = \frac{1}{n} \sum_{i=1}^{n-k} w_i w_{i+k} \quad k = 0, 1, \ldots, M \tag{1}$$

where M is the maximum number of lags which is determined simply by dividing the number of observations by 4, for a series with less than 240 based on Box and Jenkins method [1]. Obviously, $\hat{\gamma}_{-k} = \hat{\gamma}_k$.

Then the ACF and PACF at lag k, denoted ρ_k and α_{kk} respectively, can be estimated according to (2) and (3).

$$\hat{\rho}_k = \frac{\hat{\gamma}_k}{\hat{\gamma}_0} \tag{2}$$

$$\left.\begin{array}{l} \hat{\alpha}_{11} = \hat{\rho}_1 \\[2mm] \hat{\alpha}_{k+1,k+1} = \dfrac{\hat{\rho}_{k+1} - \sum_{j=1}^{k} \hat{\rho}_{k+1-j}\hat{\alpha}_{kj}}{1 - \sum_{j=1}^{k} \hat{\rho}_j\hat{\alpha}_{kj}} \\[4mm] \hat{\alpha}_{k+1,j} = \hat{\alpha}_{kj} - \hat{\alpha}_{k+1,k+1} \cdot \hat{\alpha}_{k,k-j+1} \\[2mm] (j = 1, 2, \ldots, k) \end{array}\right\} \tag{3}$$

where $k = 1, 2, \ldots, M$

Based upon the information given by the coefficients of ACF we can define the following steps to determine d to make the series stationarity:

1) Let $d = 0$;
2) Define $z_{t-d} = \nabla^d y_t = (1 - B)^d y_t$, where $t = d+1, d+2, \ldots, N$;
3) Calculate the autocorrelation coefficients of $\{z_1, z_2, \ldots, z_{N-d}\}$;
4) If the autocorrelation coefficients statistically rapidly decline to zero, the corresponding d is the required one. Or else let $d = d+1$ and return step 2).

where B denotes the backward shift operator, and ∇ is the differencing operator.

After determining the parameter d, the series $\{y_1, y_2, \ldots, y_N\}$ converts to a stationary series $\{z_1, z_2, \ldots, z_{N-d}\}$ and we can estimate its mean in (4).

$$\bar{z} = \frac{1}{N - d} \sum_{i=1}^{N-d} z_i \tag{4}$$

Then we can obtain a stationary time series with zero mean $\{x_1, x_2, \ldots, x_{N-d}\}$, where $x_t = z_t - \bar{z}$. So the problem of modeling ARIMA(p, d, q) for $\{y_1, y_2, \ldots, y_N\}$ converts to the problem of modeling ARMA(p, q) for $\{x_1, x_2, \ldots, x_{N-d}\}$. And the generalized form of ARMA(p, q) can be described as follow

$$\varphi(B)x_t = \theta(B)a_t \tag{5}$$

where

$$\begin{cases} \varphi(B) = 1 - \varphi_1 B - \varphi_2 B^2 - \cdots - \varphi_p B^p \\ \theta(B) = 1 - \theta_1 B - \theta_2 B^2 - \cdots - \theta_q B^q \end{cases}$$

and a_t is stationary white noise with zero mean, $\varphi_i(i = 1, 2, \ldots, p)$ and $\theta_i(i = 1, 2, \ldots, q)$ are coefficients of the ARMA model.

On the other hand, for identifying the orders of ARMA model, autocorrelation and partial autocortrelation graphs which are drawn based on $1 \sim M$ lag numbers provide information about the AR and MA orders (p and q) [1]. Concretely, p is determined as the number of the front a few significant coefficients in the partial autocorrelation graph and similarly q is determined as the number of the front a few significant coefficients in the autocorrelation graph.

2.2 Model Estimation

In Section 2.1, the category, structure and orders of the model has been determined. But the coefficients $\{\varphi_i\}$ and $\{\theta_i\}$ of the model is unknown. In this section we review a usual estimation method(*moment estimation*).

According to Yule-Walker equation, we can estimate $\{\varphi_i\}$ in (6).

$$
\begin{bmatrix} \hat{\varphi}_1 \\ \hat{\varphi}_2 \\ \vdots \\ \hat{\varphi}_p \end{bmatrix} = \begin{bmatrix} \hat{\gamma}_q & \hat{\gamma}_{q-1} & \cdots & \hat{\gamma}_{q-p+1} \\ \hat{\gamma}_{q+1} & \hat{\gamma}_q & \cdots & \hat{\gamma}_{q-p+2} \\ \vdots & \vdots & \vdots & \vdots \\ \hat{\gamma}_{q+p-1} & \hat{\gamma}_{q+p-2} & \cdots & \hat{\gamma}_q \end{bmatrix}^{-1} \begin{bmatrix} \hat{\gamma}_{q+1} \\ \hat{\gamma}_{q+2} \\ \vdots \\ \hat{\gamma}_{q+p} \end{bmatrix} \tag{6}
$$

Then we can estimate $\{\theta_i\}$ and σ_a^2 which is the variance of $\{a_t\}$ in (7).

$$
\begin{cases} \hat{\gamma}_0(\bar{x}) = \hat{\sigma}_a^2(1 + \hat{\theta}_1^2 + \cdots + \hat{\theta}_q^2) \\ \hat{\gamma}_k(\bar{x}) = \hat{\sigma}_a^2(-\hat{\theta}_k + \hat{\theta}_{k+1}\hat{\theta}_1 + \cdots + \hat{\theta}_q\hat{\theta}_{q-k}), \quad k = 1, 2, \ldots, q \end{cases} \tag{7}
$$

where $\hat{\gamma}_k(\bar{x}) = \sum_{j=0}^{p} \sum_{l=0}^{p} \hat{\varphi}_j \hat{\varphi}_l \hat{\gamma}_{k+l-j}, \quad k = 0, 1, \ldots, q$

So if $q=1$, according to (7) and the reversibility(that is $|\theta_1| < 1$) of ARMA(p, q) model, we can calculate σ_a^2 and θ_1 using (8) and (9).

$$
\hat{\sigma}_a^2 = \frac{\hat{\gamma}_0(\bar{x}) + \sqrt{\hat{\gamma}_0^2(\bar{x}) - 4\hat{\gamma}_1^2(\bar{x})}}{2} \tag{8}
$$

Then

$$
\hat{\theta}_1 = -\frac{\hat{\gamma}_1(\bar{x})}{\hat{\sigma}_a^2} \tag{9}
$$

But if $q > 1$, we can see it is difficult to solute (7). At this time, we can change (7) to (10) to solve this problem by using *linear iterative method*.

$$
\left. \begin{aligned} \hat{\sigma}_a^2 &= \hat{\gamma}_0(\bar{x})(1 + \hat{\theta}_1^2 + \cdots + \hat{\theta}_q^2)^{-1} \\ \hat{\theta}_k &= -(\frac{\hat{\gamma}_k(\bar{x})}{\hat{\sigma}_a^2} - \hat{\theta}_{k+1}\hat{\theta}_1 - \cdots - \hat{\theta}_q\hat{\theta}_{q-k}), \quad k = 1, 2, \ldots, q \end{aligned} \right\} \tag{10}
$$

At first, give initial values for $\{\hat{\theta}_i\}$ and $\hat{\sigma}_a^2$, for example, let $\hat{\theta}_i = 0$ ($i = 1, 2, \ldots, q$) and $\hat{\sigma}_a^2 = \hat{\gamma}_0(\bar{x})$ and mark them as $\{\hat{\theta}_i(0)\}$ and $\hat{\sigma}_a^2(0)$. Then substitute them into (10), we can obtain $\{\hat{\theta}_i(1)\}$ and $\hat{\sigma}_a^2(1)$. The rest may be deduced by analogy, so we can obtain $\{\hat{\theta}_i(m)\}$ and $\hat{\sigma}_a^2(m)$. If $\max\{|\hat{\theta}_i(m) - \hat{\theta}_i(m-1)|, |\hat{\sigma}_a^2(m) - \hat{\sigma}_a^2(m-1)|\}$ is very small, we can consider $\{\hat{\theta}_i(m)\}$ and $\hat{\sigma}_a^2(m)$ are the approximatively estimated values of $\{\theta_i\}$ and σ_a^2.

From the above, we can see that the *moment estimation* is cumbersome to achieve. And when $q > 1$, the approximate results of the parameters is unpredictable and this might greatly impact the accuracy of forecast. So a method of PSO for model estimating is proposed in Section 3, and we can see this method has a simple principle and can be implemented with ease. What's more, the PSO estimation for ARIMA model (simply marked as PSOARIMA) has a higher forecasting precision than the ARIMA model which will be shown in Section 4.

2.3   Model Forecasting

According to (5), we can obtain $a_t = \theta(B)^{-1}\varphi(B)x_t$. In addition, we can let

$$a_t = x_t - \sum_{i=1}^{\infty} \pi_i x_{t-i} = \pi(B)x_t \tag{11}$$

So there is $\varphi(B) = \theta(B)\pi(B)$, that is

$$\sum_{i=1}^{\infty} \varphi_i' B^i = (\sum_{i=1}^{\infty} \theta_i' B^i)(1 - \sum_{i=1}^{\infty} \pi_i B^i) \tag{12}$$

where

$$\varphi_i' = \begin{cases} \varphi_i & 1 \le i \le p \\ 0 & i > p \end{cases}, \quad \theta_i' = \begin{cases} \theta_i & 1 \le i \le q \\ 0 & i > q \end{cases}$$

Through comparing the terms which have the same degree of B in (12), we can obtain $\{\pi_i\}$ in (13), see [7].

$$\begin{cases} \pi_1 = \varphi_1' - \theta_1' \\ \pi_i = \varphi_i' - \theta_i' + \sum_{j=1}^{i-1} \theta_{i-j}' \pi_j, \quad i > 1 \end{cases} \tag{13}$$

Ignoring a_t and selecting a appropriate value K to replace ∞, we can obtain the predicting formula according to (11), and it is shown as follow

$$\hat{x}_t = \sum_{i=1}^{K} \pi_i \hat{x}_{t-i}, \text{ if } t - i \le 0, \text{ let } x_{t-i} = 0 \tag{14}$$

where $t = 1, 2, \ldots, N - d + 1$, and $\{\hat{x}_1, \hat{x}_2, \ldots, \hat{x}_{N-d}\}$ is simulation values of $\{x_1, x_2, \ldots, x_{N-d}\}$ and $\hat{x}_{N-d+1}$ is one forecasting value of $\{x_1, x_2, \ldots, x_{N-d}\}$.

Then we can obtain the prediction values of the original series, and the steps are shown as follows:

1) Let $\hat{z}_t = \hat{x}_t + \bar{z}$ ($t = 1, 2, \ldots, N - d + 1$), so $\{\hat{z}_t\}$ is the prediction series of $\{z_t\}$;

2) Calculate $\{\nabla^1 y_t\}, \{\nabla^2 y_t\}, \ldots, \{\nabla^{d-1} y_t\}$ whose number of components are $N-1, N-2, \ldots, N-d+1$ respectively. Mark $\{y_t\}$ as $\{\nabla^0 y_t\}$. And mark $\{\hat{z}_t\}$ as $\{\widehat{\nabla^d y_t}\}$ which is the prediction series of $\{\nabla^d y_t\}$, similarly, $\{\widehat{\nabla^i y_t}\}$ is the prediction series of $\{\nabla^i y_t\}$, $i = 0, 1, \ldots, d-1$.

3) For all i from d to 1 do:

Let $\{\widehat{\nabla^{i-1} y_t}\}_1 = \{\nabla^{i-1} y_t\}_1$

and $\{\widehat{\nabla^{i-1} y_t}\}_j = \{\nabla^{i-1} y_t\}_{j-1} + \{\widehat{\nabla^i y_t}\}_{j-1}$

where $j = 2, 3, \ldots, N-i+2$.

where $\{\nabla^{i-1} y_t\}_1$ denote the first component of series $\{\nabla^{i-1} y_t\}$.

According to the above procedure, we can obtain $\{\widehat{\nabla^0 y_t}\}$ $(t = 1, 2, \ldots, N+1)$ which is the prediction series of original series $\{y_t\}$, and we mark it as $\{\hat{y}_t\}$.

3   Particle Swarm Optimization Algorithm

Suppose that the search space is n-dimensional and a particle swarm consists of m particles, then the ith particle of the swarm and its velocity can be represented by two n-dimensional vector respectively, $x_i = (x_{i1}, x_{i2}, \ldots, x_{in})$ and $v_i = (v_{i1}, v_{i2}, \ldots, v_{in})$ $i = 1, 2, \ldots, m$, see [8]. Each particle has a fitness that can be determined by the objective function of optimization problem, and knows the best previously visited position and present position of itself. Each particle has known the position of the best individual of the whole swarm. Then the velocity of particle and its new position will be assigned according to the follows

$$v_i^{k+1} = w \times v_i^k + c_1 \tau_1 (pxbest_i - x_i^k) + c_2 \tau_2 (gxbest - x_i^k) \quad (15)$$

$$x_i^{k+1} = x_i^k + v_i^{k+1} \quad (16)$$

where v_i^k and x_i^k are the velocity and position of the ith particle at time k respectively; $pxbest_i$ is the individual best position of the ith particle; $gxbest$ is the global best position of the whole swarm; τ_1, τ_2 are random number uniformly from the interval $[0,1]$; $c_1, c_2 \in [0, 2]$ are two positive constants called acceleration coefficients namely cognitive and social parameter respectively and as default in [9], $c_1 = c_2 = 2$ were proposed; w is inertia weight that can be determined by

$$w = w_{max} - (w_{max} - w_{min}) Num/Num_{max} \quad (17)$$

where w_{max} and w_{min} are separately maximum and minimum of w; Num_{max} is maximum iteration time and Num is the current iterative time. A larger inertia weight achieves the global exploration and a smaller inertia weight tends to facilitate the local exploration to fine-tune the current search area. Usually, the maximum velocity V_{max} is set to be half of the length of the search space [10].

In this paper, because there are $p + q$ coefficients (φ_i, θ_j, $i = 1, 2, \ldots, p, j = 1, 2, \ldots, q$) to be selected, the dimension of the swarm $n = p + q$. What's more, when the coefficients $\{\varphi_i\}$ and $\{\theta_j\}$ are denoted by Φ and Θ which have p and

q components respectively, the form of one particle can be denoted by $(\varPhi, \varTheta)$, that is $(\varphi_1, \varphi_2, \ldots, \varphi_p, \theta_1, \theta_2, \ldots, \theta_q)$. Then we mark the simulation series $\{\hat{x}_t\}$ which is obtained from (14) as $\{\hat{x}_t^{(\varPhi, \varTheta)}\}$ when the coefficients are $\varPhi$ and $\varTheta$.

In the investigation, the mean square error (MSE), shown as (18), serves as the forecasting accuracy index and the objective function of PSO for identifying suitable coefficients.

$$MSE^{(\varPhi, \varTheta)} = \frac{1}{N-d} \sum_{t=1}^{N-d} (x_t - \hat{x}_t^{(\varPhi, \varTheta)})^2 \tag{18}$$

The procedure of estimating coefficients by PSO for this model is illustrated as follows:

1. Randomly generate m initial particles $x_1^0, x_2^0, \ldots, x_m^0$ and initial velocities $v_1^0, v_2^0, \ldots, v_m^0$ which have $p + q$ components respectively.
2. **for** $1 \leq i \leq m$ **do**
3. $pxbest_i \Leftarrow x_i^0; pfbest_i \Leftarrow MSE^{(x_i^0(1:p), x_i^0(p+1:p+q))}$
4. $x_i^1 \Leftarrow x_i^0 + v_i^0$
5. **end for**
6. **for all** i such that $1 \leq i \leq m$ **do**
7.   **if** $pfbest_s \leq pfbest_i$ **then**
8. $gxbest \Leftarrow pxbest_s; \quad gfbest \Leftarrow pfbest_s$
9.   **end if**
10. **end for**
11. $k \Leftarrow 1; w_{max} \Leftarrow 0.9; w_{min} \Leftarrow 0.4$
12. **while** $k < Num_{max}$ **do**
13.   **for** $1 \leq i \leq m$ **do**
14.   **if** $MSE^{(x_i^k(1:p), x_i^k(p+1:p+q))} < pfbest_i$ **then**
15. $pfbest_i \Leftarrow MSE^{(x_i^k(1:p), x_i^k(p+1:p+q))};$
16. $pxbest_i \Leftarrow x_i^k$
17.   **end if**
18.   **end for**
19.   **for all** i such that $1 \leq i \leq m$ **do**
20.   **if** $pfbest_s \leq pfbest_i$ **then**
21. $gxbest \Leftarrow pxbest_s; \quad gfbest \Leftarrow pfbest_s$
22.   **end if**
23.   **end for**
24. $w \Leftarrow w_{max} - k(w_{max} - w_{min})/Num_{max}$
25.   **for** $1 \leq i \leq m$ **do**
26. $v_i^{k+1} \Leftarrow w \times v_i^k + c_1 \tau_1 (pxbest_i - x_i^k) + c_2 \tau_2 (gxbest - x_i^k)$
27.   **if** $v_i^{k+1} \geq V_{max}$ or $v_i^{k+1} \leq -V_{max}$ **then**
28. $v_i^{k+1} \Leftarrow v_i^{k+1}/|v_i^{k+1}| \times V_{max}$
29.   **end if**
30. $x_i^{k+1} \Leftarrow x_i^k + v_i^{k+1}$
31.   **end for**
32. **end while**

where $x(1:p), x(p+1, p+q)$ denote the 1st $\sim$ pth and the $(p+1)$th $\sim$ $(p+q)$th components of the vector x respectively. So $gxbest(1:p)$ and $gxbest(p+1, p+q)$ are the optimally estimated values of Φ and Θ.

4   Case Study

In this section, a case study of predicting the consumer price index (CPI) of 36 big or medium-sized cities in China is shown. The original data is CPI in Jan. 2001 $\sim$ Oct. 2008 which is obtained from CEInet statistical database. And mark the original data as $\{y_t\}, t = 1, 2, \ldots, 94$.

We can easily see $\{y_t\}$ is a non-stationary series by observing its graphics. Observing the autocorrelogram of the third-differenced data of the original data(shown in Fig. 1, where the two solid horizontal lines represent the 95% confidence interval), in which we can see the autocorrelation coefficients statistically rapidly decline to zero after lag 1, we confirm the third-differenced data is stationary, and mark it as $\{z_t\}, t = 1, 2, \ldots, 91$. So d is equal to 3. And we can then mark $\{z_t - \bar{z}\}$ as $\{x_t\}, t = 1, 2, \ldots, 91$ where $\bar{z} = -0.0275$ according to (4).

Then we can draw the autocorrelogram and partial autocorrelogram (Fig. 2). From this figure, two facts stand out: first, the autocorrelation coefficient starts at a very high value at lag 1 and then statistically rapidly declines to zero; second, PACF up to 4 lags are individually statistically significant different from zero but then statistically rapidly declines to zero too. So we can determine $p = 4$ and $q = 1$, that is to say, we can design an ARMA(4,1) model for $\{x_t\}$.

Estimating the coefficients $\Phi = \{\varphi_1, \varphi_2, \varphi_3, \varphi_4\}$ and $\Theta = \{\theta_1\}$ according to (6), (8), (9) and PSO method with the optimizing range of $[-1, 1]$, respectively, we can obtain $\Phi = \{-1.0750, -0.8546, -0.4216, -0.1254\}$, $\Theta = \{0.1614\}$ and $\Phi = \{-0.4273, -0.0554, 0.1979, 0.1487\}$, $\Theta = \{0.8196\}$ respectively. Then we can obtain the fitted and prediction series, and the fitted graph is shown as Fig. 3. At the same time, we obtain their MSE are 1.3189 and 1.2935 respectively, that is to say, the latter fit better than the former.

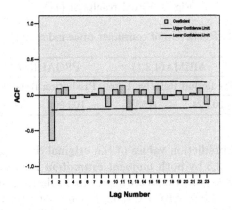

Fig. 1. The autocorrelogram of the third-differenced data up to 23 lags

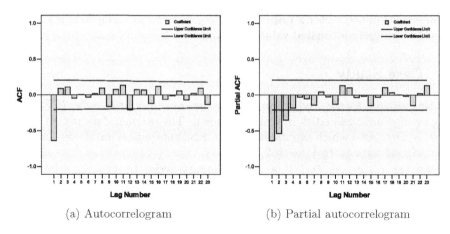

(a) Autocorrelogram   (b) Partial autocorrelogram

Fig. 2. The correlogram of $\{x_t\}$ up to 23 lags

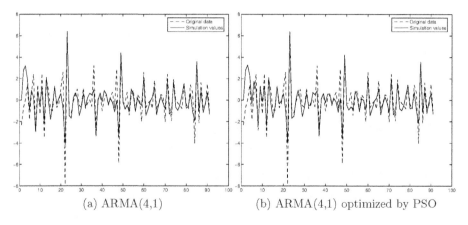

(a) ARMA(4,1)   (b) ARMA(4,1) optimized by PSO

Fig. 3. Fitted results of $\{x_t\}$

Table 1. Prediction of consumer price index in Nov. 2008

Actual	ARIMA(4,3,1)		PSOARIMA(4,3,1)	
value	Prediction	Relative Error	Prediction	Relative Error
102.70	103.24	0.53%	103.09	0.38%

Furthermore, the prediction values of the original data $\{y_t\}$ can be obtained according to Section 2.3 by both *moment estimation* method and PSO method for estimation. The relevant results are shown in Table 1. From this table, we can see that the accuracy of PSOARIMA(4, 3, 1) model is better than the ARIMA(4, 3, 1) model. What's more, we present the prediction figures (Fig. 4) to compare the accuracy of the two models clearly.

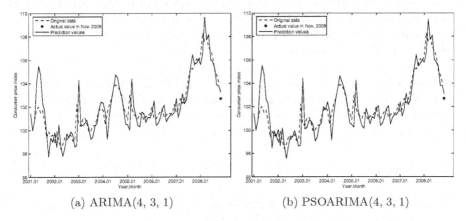

(a) ARIMA(4, 3, 1)   (b) PSOARIMA(4, 3, 1)

Fig. 4. Prediction results of $\{y_t\}$

Table 2. Prediction of consumer price index in Dec. 2008 ~ Mar. 2009

Month	Actual	ARIMA(4,3,1)		PSOARIMA(4,3,1)	
Year	value	Predict	R.E.	Predict	R.E.
12. 2008	101.30	101.39	0%	101.44	0.13%
01. 2009	101.10	99.725	1.36%	99.819	1.27%
02. 2009	98.50	100.31	1.83%	99.808	1.33%
03. 2009	98.90	96.029	2.90%	96.33	2.60%

Additionally, when we know the actual value in Nov. 2008, the value in Dec. 2008 can be predicted. The rest may be deduced by analogy, we predict the values in Jan. 2009 ~ Mar. 2009, and the relevant results are shown in Table 2. From this table we can see that accuracy of both models is very good, but relative error of PSOARIMA(4, 3, 1) is smaller than ARIMA(4, 3, 1) generally.

5   Conclusion

In this paper, we proposed a novel design methodology which is a hybrid model of particle swarm optimization algorithm and ARIMA model. PSO is used for model estimation of ARIMA in order to overcome the deficiency that the traditional estimation method is difficult to implement and may obtain very bad results. Furthermore, it is observed that the PSO-based model has worked more accurately than the traditional *moment estimation*-based model through the experimental results of forecasting the CPI.

On the other hand, we can see the power of the PSO for optimizing parameters. So in the future study, we will find more practical way to apply PSO in the other fields. But it is necessary to pay attention to that standard PSO sometimes easily get in local optimization but not the holistic optimization, fortunately, chaotic particle swarm optimization algorithm(CPSO) can help to solve this problem successfully.

Acknowledgments. The research was supported by the Ministry of Education oversea cooperation "Chunhui Projects" under Grant (Z2007-1-62012) and The Natural Science Foundation of Guansu Province in China under Grant (ZS031-A25-010-G).

References

1. Ediger, V.Ş., Akar, S.: ARIMA forecasting of primary energy demand by fuel in Turkey. Energy Policy 35, 1701–1708 (2007)
2. Erdogdu, E.: Electricity demand analysis using cointegration and ARIMA modelling: A case study of Turkey. Energy Policy 35, 1129–1146 (2007)
3. Ong, C.-S., Huang, J.-J., Tzeng, G.-H.: Model identification of ARIMA family using genetic algorithms. Applied Mathematics and Computation 164, 885–912 (2005)
4. Niu, D., Cao, S., Zhao, L., Zhang, W.: Power load forecasting technology and its applications. China Electric Power Press, Beijing (1998) (in Chinese)
5. Ediger, V.Ş., Akar, S., Uğurlu, B.: Forecasting production of fossil fuel sources in Turkey using a comparative regression and ARIMA model. Energy Policy 34, 3836–3846 (2006)
6. Valenzuela, O., Rojas, I., Rojas, F., Pomares, H., Herrera, L.J., Guillen, A., Marquez, L., Pasadas, M.: Hybridization of intelligent techniques and ARIMA models for time series prediction. Fuzzy Sets and Systems 159, 821–845 (2008)
7. Li, M., Zhou, J.-z., Li, J.-p., Liang, J.-w.: Predicting Securities Market in Shanghai and Shenzhen by ARMA Model. Journal of Changsha Railway University 3, 78–84 (2000) (in Chinese)
8. Dingxue, Z., Xinzhi, L., Zhihong, G.: A Dynamic Clustering Algorithm Based on PSO and Its Application in Fuzzy Identification. In: The 2006 International Conference on Intelligent Information Hiding and Multimedia Signal Processing, pp. 232–235. IEEE Press, New York (2006)
9. Kennedy, J., Eberhart, R.: Particle Swarm Optimization. In: The IEEE International Conference on Neural Networks, vol. 4, pp. 1942–1948. IEEE Press, New York (1995)
10. Alatas, B., Akin, E., Bedri Ozer, A.: Chaos embedded particle swarm optimization algorithms. Chaos, Solitons and Fractals (2008)

A Uniform Solution to HPP in Terms of Membrane Computing

Haizhu Chen and Zhongshi He

College of Computer Science, Chongqing University
Chongqing 400044, China
{cqchz,zshe}@cqu.edu.cn

Abstract. Membrane system is a computing model which imitates natural process at cellular level. Research results have shown that this model is a promising framework for solving **NP**-complete problems in polynomial time. Hamiltonian path problem (HPP for short) is a well-known **NP**-complete problem, and there are several semiform P systems dealing with it. In the framework of recognizer P systems based on membrane 2-division, we presented a uniform polynomial time solution to the HPP problem and gave some formal details of this solution from the point of view of complexity classes.

Keywords: Natural Computing, Membrane Computing, Cellular Complexity Classes, HPP.

1 Introduction

P systems are emergent branch of nature computing, which can be seen as a kind of distributed parallel computing model. It is based upon the assumptions that the processes taking place in the living cells can be considered as computations. Up to now, several variants of P systems with linear or polynomial time have been constructed to solve some **NP**-complete problems, such as SAT [1, 2, 3], HPP [2, 4, 5], Subset Sum [6], Knapsack [7], 2-Partition[8], Bin Packing[9].

Hamiltonian path problem (HPP for short) is a well-known **NP**-complete problem, and there are several P systems dealing with it. The P systems in [4, 5] are all semiform way [10]. Specifying the starting and ending vertices, literature [2] presents a uniform solution to HPP with membrane separation. It remains open confluently solving HPP in polynomial time by P systems with division rules instead of separation rules in a uniform way [2].

Without specifying the starting vertex and ending vertex of a path, this paper presents a uniform solution to HPP with membrane division where the communication rules (sending objects into membranes) [3] are not used. The remainder of the paper is organized as follows: first recognizer P system with membrane division is introduced in the next section. In Section3, the solution to HPP with membrane division is presented, with some formal details of such solution given in Section4. Finally, conclusions are drawn in the last section.

H. Deng et al. (Eds.): AICI 2009, LNAI 5855, pp. 59–68, 2009.
© Springer-Verlag Berlin Heidelberg 2009

2 Recognizer P System with Membrane Divisions

This paper deals with recognizer P system with membrane division. It is necessary to recall such system. Firstly, a P system with membrane division is a construct [3]:

$$\Pi = (V, H, \mu, w_1, ..., w_m, R)$$

where:

$m \geq 1$ (the initial degree of the system);

V is the alphabet of objects;

H is a finite set of labels for membranes;

μ is a membrane structure, consisting of m membranes, labeled (not necessarily in a one-to-one manner) with elements of H;

$w_1, ..., w_m$ are strings over V describing the multisets (every symbol in a string representing one copy of the corresponding object) placed in the m regions of μ respectively;

R is a finite set of developmental rules, of the following forms:

(1) $[a \rightarrow v]_h^e$, for $h \in H$, $e \in \{+, -, 0\}$, $a \in V$, $v \in V^*$
   (object evolution rules);

(2) $[a]_h^{e_1} \rightarrow [\,]_h^{e_2} b$, for $h \in H$, $e_1, e_2 \in \{+, -, 0\}$, $a, b \in V$
   (communication rules);

(3) $[a]_h^{e_1} \rightarrow [b]_h^{e_2}[c]_h^{e_3}$, for $h \in H$, $e_1, e_2, e_3 \in \{+, -, 0\}$ $a, b, c \in V$
   (division rules for elementary membranes)

Note that, in order to simplify the writing, in contrast to the style customary in the literature, we have omitted the label of the left parenthesis from a pair of parentheses which identifies a membrane. All aforementioned rules are applied according to the following principles:

(a) The rules of type (1) are applied in the parallel way.

(b) The rules of types (2), (3) are used sequentially, in the sense that one membrane can be used by at most one rule of these types at a time.

(c) All objects and all membranes, which can evolve, should evolve simultaneously.

We will introduce two definitions[1]:

Definition 1. A P system with input is a tuple (Π, Σ, i_Π), where: (a) Π is a P system, with working alphabet Γ, with p membranes labeled with 1, ..., p, and initial multisets $w_1, ..., w_p$ associated with them; (b) Σ is an (input) alphabet strictly contained in Γ; the initial multisets of Π are over $\Gamma - \Sigma$; and (c) i_Π is the label of a distinguished (input) membrane.

Let w_{in} be a multiset over Σ. The initial configuration of (Π, Σ, i_Π) with input w_{in} is $(\mu, w_1, ..., w_{i_\Pi} \cup w_{in}, ..., w_p)$.

Definition 2. A recognizer P system is a P system with input, (Π, Σ, i_Π), and with output such that:

(1) The working alphabet contains two distinguished elements, *yes* and *no*.
(2) All its computations can halt.
(3) If C is a computation of Π, then either some object *yes* or *no* (but not both) must have been released into the environment, and only in the last step of the computation.

We say that C is an accepting computation (respectively, rejecting computation) if the object *yes* (respectively, *no*) appears in the external environment associated with the corresponding halting configuration of C.

For ease of presentation, in the rest of this paper, recognizer P system means recognizer P system with input and we will not distinguish each other.

Section4 will prove that the family Π of recognizer P system in Section3 gives a polynomial solution for the HPP. We would like to introduce definition about complexity classes in P systems.

Let **AM** stand for the class of recognizer P systems with active membrane using 2-division. We will introduce the definition of the complexity class $\textbf{PMC}_{\textbf{AM}}$ [6, 7, 8, 10]:

Definition 3. We say that a decision problem, $X=(I_X, \theta_X)$, is solvable in polynomial time by a family $\Pi = (\Pi(t))_{t\in\textbf{N}}$, of recognizer P systems using 2-division, and we write $X \in \textbf{PMC}_{\textbf{AM}}$, if

(1) Π is **AM**-consistent; that is, $\Pi(t) \in \textbf{AM}$ for each t ($t\in\textbf{N}$).

(2) Π is a polynomially uniform family by Turing Machine, that is, there exists a deterministic Turing machine that builds $\Pi(t)$ from $t \geq 0$ in polynomial time.

(3) There exists a polynomial encoding (g, h) of I_X in Π verifying:

– Π is polynomially bounded, with respect to (g, h), that is, there exists a polynomial function p, such that for each $u\in I_X$, all computations of $\Pi(h(u))$ with input $g(u)$ halt in at most $p(|u|)$ steps.

– Π is sound with respect to (g, h); that is, for all $u\in I_X$, if there exists an accepting computation of $\Pi(h(u))$ with input $g(u)$, then $\theta_X(u)=1$ (the corresponding answer of the problem is *yes*.

– Π is complete with respect to (g, h); that is, for all $u\in I_X$, if $\theta_X(u)=1$ (the answer to the problem is *yes*), then every computation of $\Pi(h(u))$ accepts $g(u)$ (the system also responds *yes*).

In the above definition every P system is required to be confluent: every computation of a system with the same input must always give the same answer. The class $\textbf{PMC}_{\textbf{AM}}$ is closed under polynomial-time reduction and complement.

3 Solving HPP with Membrane Divisions

A Hamiltonian path in a directed graph G is a path that visits each of the nodes of G exactly only once. The Hamiltonian Path Problem (HPP for short) is the following decision problem:

Input: given a directed graph $G = (V, E)$;

Output: determine if there exists a Hamiltonian path in G.

This section presents a uniform solution to HPP by a family of recognizer P systems with membrane division, in which the starting vertex and ending vertex of a path are not specified. Unless stated otherwise, the solutions cited are described in the usual framework of recognizer P systems with active membranes using 2-division, with three electrical charges, without change of membrane labels, without cooperation, and without priority. The solution is addressed via a brute force algorithm which consists in the following phases.

Preparing stage: introduce n vertices from which we can start to search the possible Hamiltonian paths (HPs);

Generating stage: generate all paths of length n in G (G has n vertices) through membrane division;

Checking stage: for each of the previous paths, determine whether it visits all the nodes of G;

Output stage: answer *yes* or *no* depending on the results from checking stage.

3.1 The Uniform Solution to HPP with Membrane Divisions

For any given directed graph G with n vertices, we construct a recognizer P system $\Pi(n)$ to solve HPP. Therefore the family presented here is

$$\Pi = \{(\Pi(n), \Sigma(n), i_n): n \in N\}$$

The input alphabet is

$$\Sigma(n) = \{x_{i,j,k} \mid 1 \le i, j \le n, -1 \le k \le i\}$$

Where, the object $x_{i,j,k}$ represents $(i, j) \in E$. Over $\Sigma(n)$, the instance G (whose size is n) can be encoded as w_{in} and w_{in} is the input multiset of $\Pi(n)$:

$$w_{in} = \{x_{i,j,k} \mid 1 \le i, j \le n, k = i\}, (i, j) \in E$$

The input membrane is $i_n = 2$ and the P system can be constructed as:

$$\Pi(n) = (V(n), H, \mu, w_1, w_2, R)$$

Where,

$V(n) = \{x_{i,j,k} \mid 1 \le i, j \le n, -1 \le k \le i\} \cup \{v_i \mid 1 \le i \le n-1\} \cup \{t_i, z_i \mid 1 \le i \le n\}$

$\cup \{c_i \mid 0 \le i \le n-1\} \cup \{r_i, r_i' \mid 1 \le i \le n\} \cup \{d_i, d_i' \mid 0 \le i \le n\}$

$\cup \{e_i \mid 0 \le i \le n+1\} \cup \{f_i \mid 0 \le i \le 3n(n+3)/2+5\} \cup \{r_0, g, e, s, yes, no\}$

$H = \{1, 2\}$

$\mu = [[\ ]_2^+]_1^0$

$w_1 = f_0, w_2 = d_0 v_1$

And the set R contains the rules as follows:

$$[v_i]_2^+ \to [t_i]_2^-[v_{i+1}]_2^+ (1 \le i \le n-2), \quad [v_{n-1}]_2^+ \to [t_{n-1}]_2^-[t_n]_2^-, \quad [t_j \to c_{j-1}r_j'g]_2^- (1 \le j \le n) \quad (1)$$

At the preparing stage, we have $\{d_0v_1\}$ and an input multiset w_{in} in membrane labeled with 2, which is encoded according to the instance G. From v_1, other vertices in G can be obtained by rules (1). This means that we can find HPs from the n vertices respectively. Object r_j' represents the starting vertex and object c_{j-1} can be regarded as a counter whose subscript is decreased by 1 each time in generating stage. These rules are applied at the preparing stage.

$$[g]_2^- \to [\ ]_2^0 g \quad (2)$$

After the starting vertexes appear in corresponding membranes respectively, we will enter generating stage by rule (2) which will be applied in other computing step as explained later.

$$[x_{i,j,k} \to x_{i,j,k-1}]_2^0, \quad [c_i \to c_{i-1}]_2^0, \quad [r_i' \to r_i]_2^0 \quad (1 \le i, j \le n, 0 \le k \le i) \quad (3)$$

In the membranes labeled with 2, when their polarization is 0, the third subscript k of $x_{i,j,k}$ and the subscript i of c_i are decreased by 1 simultaneously. For the i-th vertex in G that has been added into the current path, Objects r_i' and r_i are two different versions of representation in the membranes with different polarizations. $X_{i,j,k}$ can not evolve any more once k reaches -1. c_0 and $x_{i,j,0}$ will appear after I steps if we have $(i, j) \in E$.

$$[c_0]_2^0 \to [\ ]_2^+ c_0 \quad (4)$$

Vertex v_i is the last vertex of the current path, and we will extend this path by adding v_j with the appearance of c_0 if we have $(i, j) \in E$. Several vertices will connect v_i besides v_j, so new membranes are needed to be created to contain the new paths. Rule (4) can change the polarization of membrane to positive in order to trigger membrane divisions to obtain new membranes.

$$[d_i \to d_i']_2^+, \quad [r_i \to r_i']_2^+, \quad [x_{i,j,0}]_2^+ \to [z_j]_2^-[g]_2^+ \quad (1 \le i, j \le n, 0 \le k \le i) \quad (5)$$

Objects d_i, d_i' both mean the fact that the length of current path is I, but they are two different versions in membranes labeled with 2 with different polarizations. $X_{i,j,0}$ in the membranes with positive polarization will cause the membrane division to extend the current path. Since not only one vertex connects with v_i, there will be several objects like $x_{i,j,0}$ whose third subscript k is 0. We non-deterministically choose one of them to trigger the division, and the rest will be remained in the two new membranes with different polarizations, which indicate that they will be processed by different rules: the division will be triggered again by one of them in the new membrane with positive polarization in order to obtain another new path; while all of them will be deleted in the one with negative polarization.

$$[z_j \to r_j'c_{j-1}g]_2^-, \quad [d_i' \to d_i]_2^-, \quad [x_{i,j,k} \to x_{i,j,i}]_2^- \quad (6)$$

$$[x_{i,j,0} \to \lambda]_2^-, \quad [d_i \to d_{i+1}]_2^- \quad (1 \le i, j \le n, k \ne 0)$$

In the new membrane labeled with 2 with negative polarization, new object r_j' is introduced, which indicates that vertex v_j is added to the current path. At the same time, new counter c_{j-1} is introduced and other objects, such as $x_{i,j,k}$, return to $x_{i,j,i}$. After the length of the path is increased, the polarization is changed by rule (2). We return to the beginning of generating stage and resume the procedure for extending the current path.

$$[g \to \lambda]_2^+ \tag{7}$$

In the new membrane labeled with 2 with positive polarization, object g does not work any more and is deleted.

$$[d_n \to se]_2^0 \tag{8}$$

When the length of path is n, we know that n vertices are in the path and we can enter the next stage to check if the path is Hamiltonian or not. Two new objects s and e are introduced to prepare for the checking stage.

$$[e \to e_0 r_0]_2^+, \quad [s]_2^+ \to [\ ]_2^0 s \tag{9}$$

With the appearance of d_n, the counter c_t ($t>0$) associated with the last vertex is introduced and it still can evolve until c_0 is sent out of the membrane. Objects e and s evolve by rule (9) to trigger the next stage.

$$[e_i \to ge_{i+1}]_2^0, \quad [r_0]_2^0 \to [\ ]_2^- r_0 \quad (0 \le i \le n) \tag{10}$$

$$[r_i \to r_{i-1}]_2^- \quad (1 \le i \le n) \tag{11}$$

In checking stage, we use rules (10) and (11) with a loop. If the object r_0 is present, we eliminate it and perform a rotation of the objects $r_i, ..., r_n$ with their subscripts decreased by 1. It is clear that a membrane contains HP if and only if we can run $n+1$ steps of the loop. Note that this check method is not the same as the one in [10], for the communication rule are not used.

$$[e_{n+1} \to yes]_2^- \tag{12}$$

$$[yes]_2^0 \to [\ ]_2^+ yes \tag{13}$$

The object e_{n+1} introduced by rules (10) means that the path is Hamiltonian, so it evolves to object yes, which will leave the membrane labeled with 2 and enter membrane 1.

$$[g \to \lambda]_1^0, [c_0 \to \lambda]_1^0, [r_0 \to \lambda]_1^0, [s \to \lambda]_1^0, [f_i \to f_{i+1}]_1^0 \quad (0 \le i \le 3n(n+3)/2+4) \tag{14}$$

All of these rules are applied in parallel in membrane 1 with neutral polarization. Since objects g, c_0, r_0 and s are sent out of the membranes labeled with 2 and will be not used any more, it necessary to delete them to release computing recourse. Object f_i evolves at each computation step for counting how many steps have been performed.

$$[yes]_1^0 \rightarrow [\ ]_1^+ yes, \quad [f_{time}]_1^0 \rightarrow [\ ]_1^+ no \quad time = 3n(n+3)/2+5 \qquad (15)$$

With changing the polarization to positive, object *yes* in membrane 1 will be sent out the environment to show that there is a HP among the *n* vertices. If at step $3n(n+3)/2+5$ the polarization of the membrane 1 has not changed to positive, the answer must be *no* and *no* will be sent out to the environment. The special step will be analyzed in subsection3.2. Once the polarization is change all rules in membrane 1 cannot be applied any more and the computation halts.

3.2 An Overview of the Computation

All of foregoing rules applied in membranes labeled with 2 are illustrated in Fig.1, in which 0, + and - respectively represent the polarization of the membranes. Fig.1 looks like state machine, and it can be seen that different polarizations and objects indicate distinct stages. Rules (2), (4), (9), (10) are used to change the polarizations in order to enter another stage of computation.

At preparing stage, we have v_1 in membrane labeled with 2. From v_1, other vertices in *G* can be obtained by application of rules (1). After *n* starting vertices obtained, we will enter next stage to find HPs respectively. Generating stage consists of three phases: a vertex that connects with the last vertex in the current path is expected to find in search phase, then it will be add into the path through divide phase, and the length of the path will be increased and the objects related to others edges will recover in restore phase. Once a vertex is added in the path, all of the edges start from it will be consumed in the extensions of the path. These three phases will be looped until the length of the path reaches *n*. Checking stage is needed to decide whether the path with length of *n* is HP or not. The answer *yes* or *no* will be sent out at the output stage.

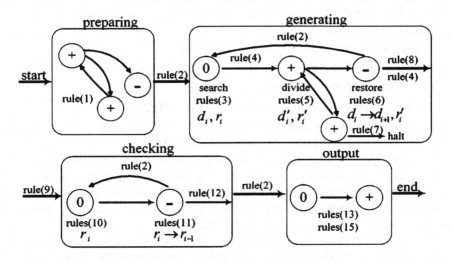

Fig. 1. The rules applied in membranes labeled by 2

The number of steps in each stage is as follows:
(1) Preparing and transition to generating: at most $n+1$
(2) Generating stage:
  Search: at most $n(n+1)/2$-n
  Divide: at most $n(n-1)$
  Restore: $3(n-1)$
(3) Checking: $2+2(n+1)$
(4) Output *yes*: 2
The overall number of steps is:

$$n+1+n(n+1)/2-n+n(n-1)+3(n-1)+2+2(n+1)+2=3n(n+3)/2+4= O\,(n^2)$$

Before the step $3n(n+3)/2+4$, *yes* will be sent out if and if there is HP in the instance G. If at step $3n(n+3)/2+5$ the polarization of the membrane 1 has not changed to positive, the answer must be *no* and *no* will be sent out to the environment.

4 Some Formal Details

In this section, we will prove that the uniform family Π of recognizer P system in Section3 gives a polynomial solution for the HPP. According to Definition3 in Section2, we have:

Theorem 1. HPP $\in$ **PMC**$_{AM}$.

Proof: (1) It obviously that the family Π of recognizer P system defined is **AM**-consistent, because for every P system of the family is a recognizer P system with active membrane using 2-division and all of them are confluent.
(2) The family Π of recognizer P system is polynomially uniform Turing Machine. For each n, which is the size of an instance G, a P system can be constructed. The rules of $\Pi(n)$ are defined in a recursive manner from n. The necessary resources to construct the system are:
  Size of the alphabet: $n^3+(7n^2+27n)/2+14= \Theta(n^3)$
  Initial number of membrane: $2= \Theta(1)$
  Initial number of objects: $3= \Theta(1)$
  The total number of evolution rules: $n^3+(9n^2+31n+38)/2= \Theta(n^3)$
All these resources are bounded polynomially time of n; therefore a Turing machine can build the P system in polynomial time according to n.
(3) We have the functions $g: I_{HPP} \rightarrow \cup_{t \in N} I_{\Pi(t)}$ and $h: I_{HPP} \rightarrow N$ defined for an instance $u=v_1, \ldots, v_n$ as $g(u)=\{x_{i,j,k}: 1 \le i, j \le n\}$ and $h(u)=n$. Both of them are total and polynomially computable. Furthermore, (g, h) conforms an encoding of the set instances of the HPP problem in the family of recognizer P system as for any instance u we have that $g(u)$ is a valid input for the P system.
  Π is polynomially bounded, with respect to (g, h). A formal description of the computation let prove that Π always halts and sends to the environment the object *yes* or *no* in the last step. The number of steps of the P system is within $3n(n+3)/2+4$ if the

output is *yes* and $3n(n+3)/2+5$ if the output is *no*, therefore there exists a polynomial bound for the number of steps of the computation.

Π is sound with respect to (g, h). As we describe the computations of P system of Π above, that object *yes* is send out of the system means that there are n different vertices in the path, which is the actual Hamiltonian path in the instance G according to the definition of HPP.

Π is complete with respect to (g, h). The P system searches the HP from every vertex respectively at the beginning of the computation. We will get objects such as r_i ($1 \leq i \leq n$) and delete objects such as $x_{i,j,k}$ ($1 \leq i, j \leq n$, $-1 \leq k \leq i$) corresponding to the edges whose starting vertex v_i is added to the current path. When there is a Hamiltonian path in the instance G with n vertices, the system Π will obtain objects $r_1, r_2, ...,$ r_n in some membrane, and output object *yes* to the environment. $\square$

Since this class is closed under polynomial-time reduction and complement we have:

Theorem 2. $\text{NP} \cup \text{co-NP} \subseteq \text{PMC}_{\text{AM}}$.

5 Conclusion

In this paper, we present a uniform solution for HPP in terms of P system with membrane division. This has been done in the framework of complexity classes in membrane computing.

It can be observed that many membranes will not evolve any more at some step before the whole system halts. Some kind of rule should be developed to dissolve these membranes for releasing the computing resource, and it will contribute to the construction of the efficient simulators of recognizer P systems for solving **NP**-complete problems.

Acknowledgments. This research is supported by Major National S&T Program of China (#2008ZX 07315-001).

References

1. Gutiérrez-Naranjo, M.A., Pérez-Jiménez, M.J., Romero-Campero, F.J.: A uniform solution to SAT using membrane creation. Theoretical Computer Science 371(1-2), 54–61 (2007)
2. Pan, L., Alhazov, A.: Solving HPP and SAT by P systems with active membranes and separation rules. Acta Informatica 43(2), 131–145 (2006)
3. Păun, G.: Computing with Membranes: Attacking NP-Complete Problems. In: Proceedings of the Second International Conference on Unconventional Models of Computation, pp. 94–115. Springer, London (2000)
4. Păun, A.: On P Systems with Membrane Division. In: Proceedings of the Second International Conference on Unconventional Models of Computation, pp. 187–201. Springer, London (2000)
5. Zandron, C., Ferretti, C., Mauri, G.: Solving NP-Complete Problems Using P Systems with Active Membranes. In: Proceedings of the Second International Conference on Unconventional Models of Computation, pp. 289–301. Springer, London (2000)

6. Pérez-Jiménez, M.J., Riscos-Núñez, A.: Solving the Subset-Sum problem by active membranes. New Generation Computing 23(4), 367–384 (2005)
7. Jesús Pérez-Jímenez, M., Riscos-Núñez, A.: A Linear-Time Solution to the Knapsack Problem Using P Systems with Active Membranes. In: Martín-Vide, C., Mauri, G., Păun, G., Rozenberg, G., Salomaa, A. (eds.) WMC 2003. LNCS, vol. 2933, pp. 250–268. Springer, Heidelberg (2004)
8. Gutiérrez-Naranjo, M.A., Pérez-Jiménez, M.J., Riscos-Núñez, A.: A fast P system for finding a balanced 2-partition. Soft Computing 9(9), 673–678 (2005)
9. Pérez-Jiménez, M.J., Romero-Campero, F.J.: Solving the BINPACKING problem by recognizer P systems with active membranes. In: Proceedings of the Second Brainstorming Week on Membrane Computing, Seville, Spain, pp. 414–430 (2004)
10. Pérez-Jiménez, M.J., Romero-Jiménez, A., Sancho-Caparrini, F.: Computationally Hard Problems Addressed Through P Systems. In: Ciobanu, G. (ed.) Applications of Membrane Computing, pp. 315–346. Springer, Berlin (2006)

Self-organizing Quantum Evolutionary Algorithm Based on Quantum Dynamic Mechanism

Sheng Liu and Xiaoming You

School of Management, Shanghai University of Engineering Science,
Shanghai 200065, China
{ls6601,yxm6301}@163.com

Abstract. A novel self-organizing Quantum Evolutionary Algorithm based on quantum Dynamic mechanism for global optimization (DQEA) is proposed. Firstly, population is divided into subpopulations automatically. Secondly, by using co-evolution operator each subpopulation can obtain optimal solutions. Because of the quantum evolutionary algorithm with intrinsic adaptivity it can maintain quite nicely the population diversity than the classical evolutionary algorithm. In addition, it can help to accelerate the convergence speed because of the co-evolution by quantum dynamic mechanism. The searching technique for improving the performance of DQEA has been described; self-organizing algorithm has advantages in terms of the adaptability; reliability and the learning ability over traditional organizing algorithm. Simulation results demonstrate the superiority of DQEA in this paper.

Keywords: Quantum evolutionary algorithm, Quantum dynamic mechanism, Self-organizing and co-evolution.

1 Introduction

Most studies have used hybrid evolutionary algorithm in solving global optimization problems. Hybrid evolutionary algorithm takes advantages of a dynamic balance between diversification and intensification. Proper combination of diversification and intensification is important for global optimization. Quantum computing is a very attractive research area; research on merging evolutionary algorithms with quantum computing [1] has been developed since the end of the 90's. Han proposed the quantum-inspired evolutionary algorithm (QEA), inspired by the concept of quantum computing, and introduced a Q-gate as a variation operator to promote the optimization of the individuals Q-bit [2]. Up to now, QEA has been developed rapidly and applied in several applications of knapsack problem, numerical optimization and other fields [3]. It has gained more attraction for its good global search capability and effectiveness. Although quantum evolutionary algorithms are considered powerful in terms of global optimization, they still have several drawbacks such as premature convergence. A number of researchers have experimented with biological immunity and cultural dynamic systems to overcome these particular drawbacks implicit in evolutionary algorithms [4], [5]; here we experiment with quantum dynamic mechanism.

H. Deng et al. (Eds.): AICI 2009, LNAI 5855, pp. 69–77, 2009.
© Springer-Verlag Berlin Heidelberg 2009

This paper presents a self-organizing quantum evolutionary algorithm based on quantum dynamic mechanism for global optimization (DQEA). Firstly the population will be divided into subpopulations automatically. Due to its intrinsic adaptivity the quantum evolutionary algorithm can enable individual to draw closer to each optimal solution. The self-organizing operator is adopted to enable individual to converge to the nearest optimal solution which belongs to the same subpopulations. Meanwhile, each subpopulation can co-evolve by quantum dynamic mechanism. The intensification strategies are the self-organizing operator and the co-evolution. The experiment results show that DQEA is very efficient for the global optimization; usually it can obtain the optimal solution faster. We describe quantum dynamic mechanism for improving the performance of DQEA, its performance is compared with that of CRIQEA[6] and MIQEA [7]. The rest of this paper is organized as follows: Section 2 presents the quantum evolutionary algorithm and related work. Section 3 gives the structure of DQEA and describes the improving performance of the algorithm. Section 4 presents an application example with DQEA and the other algorithms [6], [7] for several benchmark problems, and summarizes the experimental results. Finally concluding remarks follow in Section 5.

2 Quantum Evolutionary Algorithm and Related Work

QEA utilizes a new representation, called a Q-bit, for the probabilistic representation that is based on the concept of qubits [2]. A qubit may be in the "1" state, in the "0" state, or in any superposition of the two [8]. The state of a qubit can be represented as $|\varphi>=\alpha|0>+\beta|1>$, where α and β are complex numbers that specify the probability amplitudes of the corresponding states. $|\alpha|^2$ gives the probability that the qubit will be found in '0' state and $|\beta|^2$ gives the probability that the qubit will be found in the '1' state. Normalization of the state to unity always guarantees: $|\alpha|^2+|\beta|^2=1$. One qubit is defined with a pair of complex numbers (α,β), And an m-qubits represention is defined

as $\begin{bmatrix} \alpha_1 & \alpha_2 & \alpha_3 & \cdots & \alpha_m \\ \beta_1 & \beta_2 & \beta_3 & \cdots & \beta_m \end{bmatrix}$ where $|\alpha_i|^2+|\beta_i|^2=1$, $i=1,2......m$.

Q-bit representation has the advantage that it is able to represent a linear superposition of states probabilistically. If there is a system of m-qubits, the system can represent 2^m states at the same time. However, in the act of observing a quantum state, it collapses to a single state. The basic structure of QEA is described in the following[2].

```
QEA()
{ t • 0;
  Initialize Q(t) ;
  Make P(t) by observing the states of Q(t) ;
  Evaluate P(t) ;
  Store the optimal solutions among P(t) ;
  While (not termination-condition) do
  {t=t+1;
  Make P(t) by observing the states of Q(t-1);
  Evaluate P(t) ;
  Update Q(t) using Q-gate U(t);
  Store the optimal solutions among P(t) ;
```

```
        }
    }
end.
```

where $Q(t)$ is a population of qubit chromosomes at generation t, and $P(t)$ is a set of binary solutions at generation t.

1) In the step of 'initialize $Q(t)$', all qubit chromosomes are initialized with the same constant $1/\sqrt{2}$. It means that one qubit chromosome represents the linear superposition of all possible states with the same probability.

2) The next step makes a set of binary solutions, $P(t)$, by observing $Q(t)$ states. One binary solution is formed by selecting each bit using the probability of qubit. And then each solution is evaluated to give some measure of its fitness.

3) The initial best solution is then selected and stored among the binary solutions, P(t).

4) In the while loop, A set of binary solutions, $P(t)$, is formed by observing $Q(t-1)$ states as with the procedure described before, and each binary solution is evaluated to give the fitness value. It should be noted that $P(t)$ can be formed by multiple observations of $Q(t-1)$.

5) In the next step, 'update $Q(t)$', a set of qubit chromosomes $Q(t)$ is updated by applying rotation gate defined below

$$U(\Delta\theta_i) = \begin{bmatrix} COS \ (\Delta\theta_i) & - SIN \ (\Delta\theta_i) \\ SIN \ (\Delta\theta_i) & COS \ (\Delta\theta_i) \end{bmatrix} \tag{1}$$

where $\Delta\theta_i$ is a rotation angle of each Q-bit towards either 0 or 1 state depending on its sign. Fig.1 shows the polar plot of the rotation gate $\Delta\theta_i$ should be designed in compliance with practical problems. Table1 can be used as an angle parameters table for the rotation gate.

The magnitude of $\Delta\theta_i$ has an effect on the speed of convergence, but if it is too big, the solutions may diverge or have a premature convergence to a local optimum. The sign of $\Delta\theta_i$ determines the direction of convergence.

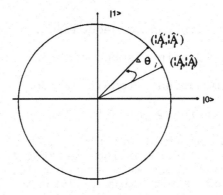

Fig. 1. Polar plot of the rotation gate

6) The best solution among *P(t)* is selected, and if the solution is fitter than the stored best solution, the stored solution is replaced by the new one. The binary solutions *P(t)* are discarded at the end of the loop.

Table 1. Lookup table of $\Delta\theta_i$, where f(·) is the fitness function, and b_i and x_i are the *i*-th bits of the best solution b and the binary solution x, respectively

x_i	b_i	$f(x) \geq f(b)$	$\Delta\theta_i$
0	0	false	θ_1
0	0	true	θ_2
0	1	false	θ_3
0	1	true	θ_4
1	0	false	θ_5
1	0	true	θ_6
1	1	false	θ_7
1	1	true	θ_8

3   Quantum Evolutionary Algorithm Based on Quantum Dynamic Mechanism

Conventional quantum evolutionary algorithm (QEA) [2],[3] is efficacious, in which the probability amplitude of Q-bit was used for the first time to encode the chromosome and the quantum rotation gate was used to implement the evolving of population. Quantum evolutionary algorithm has the advantage of using a small population size and the relatively smaller iterations number to have acceptable solution, but they still have several drawbacks such as premature convergence and computational time.

The flowchart of self-organizing quantum evolutionary algorithm based on quantum dynamic mechanism for global optimization (DQEA) is as follows:

```
DQEA()
{t=0;
Initialize Q(0) with random sequences ;
Evaluate Q(0) ;
Q(0) will be divided into subpopulations Q_i(0)
For all Q_i(t)
 While (not termination - condition) do
 {t=t+1;
  Update Q(t-1) using Q-gates U(t-1);
  Evaluate Q_i(t);
  Store the optimal solutions among Q_i(t);
  Implement the self-organizing operator for Q_i(t):
  {Self-adaptively cross-mutate each cell;
   Select antibodies with higher affinity as memory cells;
   Quantum dynamic equilibrium states;
    }
```

Each subpopulation can co-evolve by quantum dynamic
mechanism;
    }
  }
end.

Denote particle S_i to be the i-th subpopulation in Fig.2, where the particle S_i in a force-field corresponds to the entry S_i in the matrix $S=(U(S_i), \zeta_i^j), U(S_i)$ is utility of particle S_i; ζ_i^j is the intention strength between S_i and S_j. A particle may be driven by several kinds of forces [9]. The gravitational force produced by the force-field tries to drive a particle to move towards boundaries of the force-field, which embodies the tendency that a particle pursues maximizing the aggregate benefit of systems. The pushing or pulling forces produced by other particles are used to embody social cooperation or competition. The self-driving force produced by a particle itself represents autonomy of individual. Under the exertion of resultant forces, all the particles may move concurrently in a force-field until all particles reach their equilibrium states.

Given an optimization problem: $\begin{cases} \min f(X) \\ st.g(X) \end{cases}$

Let $U_i(t)$ be the utility of particle S_i at time t, and let $J(t)$ be the aggregate utility of all particles. They are defined by Eq(2)

$$fitness(t) = \exp(f(X));$$
$$U_i(t) = 1 - \exp(-fitness(t));$$
$$J(t) = \alpha \sum_{i=1}^{n} U_i(t), \qquad 0 \prec \alpha \prec 1$$

(2)

At time t, the potential energy function $Q(t)$ that is related to interactive forces among particles is defined by Eq(3)

$$Q_i(t) = \sum_j \int_0^{U_i(t)} \{[1 + \exp(-\zeta_i^j x)]^{-1} - 0.5\} dx;$$
$$Q_D(t) = \sum_i Q_i(t)$$

(3)

ζ_i^j is the intention strength, which is used to describe the effect of cooperation($\zeta_i^j < 0$) or competition($\zeta_i^j > 0$).

At time t, the potential energy function $P(t)$ that is related to the gravitational force of force-field F is defined by Eq(4)

$$P(t) = \varepsilon^2 \ln((\sum_i \exp(-U_i(t)^2 / (2 \times \varepsilon^2))) / n), 0 \prec \varepsilon \prec 1$$

(4)

Let $S_i(t)$ be the vertical coordinate of particle S_i at time t. The dynamic equation for particle S_i is defined by Eqs (5) and (6)

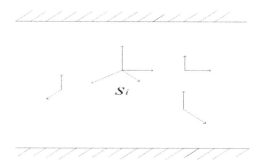

Fig. 2. Particles and force-field

$$\frac{dS_i(t)}{dt} = \lambda 1 \frac{dU_i(t)}{dt} + \lambda 2 \frac{dJ_D(t)}{dt} - \lambda 3 \frac{dP_D(t)}{dt} - \lambda 4 \frac{dQ_D(t)}{dt}$$

$$= (\lambda 1 + \lambda 2 \frac{\partial J_D(t)}{\partial U_i(t)} - \lambda 3 \frac{\partial P_D(t)}{\partial U_i(t)} - \lambda 4 \frac{\partial Q_D(t)}{\partial U_i(t)}) \frac{dU_i(t)}{dt} \qquad (5)$$

$$= g(S_i(t), U_i(t)), \qquad 0 \le \lambda 1, \lambda 2, \lambda 3, \lambda 4 \le 1$$

$$\frac{dU_i(t)}{dt} = \exp(-fitness(t)) \times fitness'(t)$$

$$= (1 - U_i(t)) \times fitness'(t) \qquad (6)$$

$$= g_1(S_i(t), U_i(t))$$

g, g_1 are functions of $S_i(t)$ and $U_i(t)$.

If all particles reach their equilibrium states at time t, then finish with success.

4 Experimental Study

In this section, DQEA is applied to the optimization of well-known test functions and its performance is compared with that of CRIQEA [6] and MIQEA [7] algorithm. When testing the algorithm on well understood problems, there are two measures of performance: (i) CR: Convergence Rate to global minima; (ii) C: The average number of objective function evaluations required to find global minima.

The test examples used in this study are listed below:

$$f_1(x_1, x_2) = \frac{x_1^2 + x_2^2}{4000} + 1 - \cos(x_1) \times \cos(\frac{x_2}{\sqrt{2}}), -600 \le x_1, x_2 \le 600 \qquad (7)$$

$$f_2(x_1, x_2) = 4x_1^2 - 2.1x_1^4 + x_1^6/3 + x_1 x_2 - 4x_2^2 + 4x_2^4, \qquad -5 \le x_i \le 5; \qquad (8)$$

f_1 (Griewank function): This is a shallow bowl upon which a rectangular grid of alternating hills and valleys is superimposed. It has a unique global minimum of zero at the origin, surrounded by local minima. f_1 possesses about 500 other local minima, This function is designed to be a minefield for standard optimization techniques.

To compare the performance of the proposed algorithm with the algorithm CRIQEA [6] and MIQEA [7], we use the same population size (N=100, 4 subpopulations) for test problems, the experimental results (C: number of function evaluations and CR: Convergence Rate to global minima) obtained in 20 independent runs are averaged and summarized in Table 2.

From the Table 2, we can see that the experimental results using the new algorithm are somewhat good. For the test problem, DQEA can convergence to global optimal solutions more; meanwhile, the average number of function evaluations of DQEA is less than that of CRIQEA and MIQEA for every test problems. The results show that our algorithm is efficient for the global optimization.

f_2 (Camelback function): It has six local minima; two of them are global minimum $f_{min} = -1.0316285$. It was used to evaluate how the co-evolution influences the convergence of the algorithm DQEA by using a quantum dynamic mechanism. The result for the case of Camelback function, averaged over 20 trials, are shown in Fig.3. Comparison of the result indicates that quantum dynamic mechanism can keep parallel optimization in complex environment and control the convergence speed. For 60 populations, the optimization value was obtained with DQEA2 (2 particles, $\xi_1^2 = \xi_2^1 = -\frac{1}{2}$)

after 65 generations, whereas DQEA3 (3 particles, $\xi_i^j = -\frac{1}{3}, i \neq j, 1 \leq i, j \leq 3$) was able to obtain after 37 generations. For given problem size, with the more number of subpopulations it can explore as far as possible different search space via different populations, so it can help to obtain the optimal solutions rapidly and converge fast.

Table 2. The comparison of this algorithm DQEA and other algorithms

Function	Algorithm	CR	Worst	Best	C
f_1	DQEA	8/20	0.00423	0	2100
f_1	CRIQEA	5/20	0.00371	0	2200
f_1	MIQEA	3/20	0.00412	0	3800
f_2	DQEA	10/20	-1.033180	-1.031620	550
f_2	CRIQEA	6/20	-1.032920	-1.031620	625
f_2	MIQEA	3/20	-1.033620	-1.031620	1050

(* C: Average number of function evaluations CR: Convergence Rate to global minima)

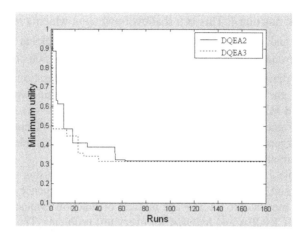

Fig. 3. Comparison of the convergence speed of DQEA for different particles

The certainty of convergence of the DQEA may be attributed to its ability to maintain the diversity of its population. In fact, the superior performance of self-organizing operator may be attributed to its adaptability and learning ability; a larger pool of feasible solutions enhances the probability of finding the optimum solution. Cooperation of subpopulations by quantum dynamic mechanism can help to obtain the optimal solution faster. Comparison of the result indicates that self-organizing operator can keep the individual diversity and control the convergence speed.

5   Conclusions

In this study, a self-organizing quantum evolutionary algorithm based on quantum dynamic mechanism for global optimization (DQEA) is proposed. The efficiency of quantum evolutionary algorithm is enhanced by using the self-organizing operator. We estimate the performance of algorithm. The efficiency of the approach has been illustrated by applying to some test cases. The results show that integration of the self-organizing operator in the quantum evolutionary algorithm procedure can yield significant improvements in both the convergence rate and solution quality. Cooperation of subpopulations by quantum dynamic mechanism can help to convergence faster. The next work is to exploit quantum dynamic mechanism further to deal with the global optimization.

Acknowledgments. The authors gratefully acknowledge the support of Natural Science Foundation of Shanghai (Grant No.09ZR1420800), Development Foundation of SUES(Grant No.2008XY18) and Doctor Foundation of SUES.

References

1. Narayanan, M.M.: Genetic Quantum Algorithm and Its Application to Combinatorial Optimization Problem. In: Proc. IEEE International Conference on Evolutionary Computation (ICEC 1996), pp. 61–66. IEEE Press, Piscataway (1996)

2. Han, K.H., Kim, J.H.: Quantum-inspired Evolutionary Algorithm for a Class of Combinatorial Optimization. IEEE Transactions on Evolutionary Computation 6(6), 580–593 (2002)
3. Han, K.H., Kim, J.H.: Quantum-Inspired Evolutionary Algorithms with a New Termination Criterion, Hε gate, and Two-Phase Scheme. IEEE Transactions on Evolutionary Computation 8(2), 156–169 (2004)
4. Fukuda, T., Mori, K., Tsukiyama, M.: Parallel search for multi-modal function optimization with diversity and learning of immune algorithm. In: Artificial Immune Systems and Their Applications, pp. 210–220. Spring, Berlin (1999)
5. Saleem, S.: Knowledge-Based Solutions to Dynamic Problems Using Cultural Algorithms. PhD thesis, Wayne State University, Detroit Michigan (2001)
6. You, X.M., Liu, S., Sun, X.K.: Immune Quantum Evolutionary Algorithm based on Chaotic Searching Technique for Global Optimization. In: Proc. The First International Conference on Intelligent Networks and Intelligent Systems, pp. 99–102. IEEE Press, New York (2008)
7. You, X.M., Liu, S., Shuai, D.X.: Quantum Evolutionary Algorithm Based on Immune Theory for Multi-Modal Function Optimization. Journal of Petrochemical Universities 20, 45–49 (2007)
8. Hey, T.: Quantum Computing: An Introduction. Computing & Control Engineering Journal 10, 105–112 (1999)
9. Shuai, D., Shuai, Q., Liu, Y., Huang, L.: Particle Model to Optimize Enterprise Computing. In: Research and Practical Issues of Enterprise Information Systems. International Federation for Information Processing, vol. 205, pp. 109–118. Springer, Boston (2006)

Can Moral Hazard Be Resolved by Common-Knowledge in S4n-Knowledge?

Takashi Matsuhisa

Department of Natural Sciences, Ibaraki National College of Technology
Nakane 866, Hitachinaka-shi, Ibaraki 312-8508, Japan
mathisa@ibaraki-ct.ac.jp

Abstract. This article investigates the relationship between common-knowledge and agreement in multi-agent system, and to apply the agreement result by common-knowledge to the principal-agent model under non-partition information. We treat the two problems: (1) how we capture the fact that the agents agree on an event or they get consensus on it from epistemic point of view, and (2) how the agreement theorem will be able to make progress to settle a moral hazard problem in the principal-agents model under non-partition information. We shall propose a solution program for the moral hazard in the principal-agents model under non-partition information by common-knowledge. Let us start that the agents have the knowledge structure induced from a reflexive and transitive relation associated with the multi-modal logic **S4n**. Each agent obtains the membership value of an event under his/her private information, so he/she considers the event as fuzzy set. Specifically consider the situation that the agents commonly know all membership values of the other agents. In this circumstance we shall show the agreement theorem that consensus on the membership values among all agents can still be guaranteed. Furthermore, under certain assumptions we shall show that the moral hazard can be resolved in the principal-agent model when all the expected marginal costs are common-knowledge among the principal and agents.

Keywords: Moral hazard, Principal-agents model under uncertainty, Common-Knowledge, Agreeing to disagree, Modal logic **S4n**.

AMS 2000 Mathematics Subject Classification: Primary 91A35, Secondary 03B45.

Journal of Economic Literature Classification: C62, C78.

1 Introduction

This article considers the relationship between common-knowledge and agreement in multi-agent system. How we capture the fact that the agents agree on an event or they get consensus on it? We treat the problem from fuzzy set theoretical flavour. The purposes are first to introduce a knowledge revision system

H. Deng et al. (Eds.): AICI 2009, LNAI 5855, pp. 78–85, 2009.

through communication on multi-agent system, and by which we show that all agents can agree on an event, and second to apply the result to solving the moral hazard in a principal-agents model under asymmetric information. Let us consider there are agents more than two and the agents have the structure of the Kripke semantics for the multi-modal logic **S4n**.

Assume that all agents have a common probability measure. By the membership value of an event under agent i's private information, we mean the conditional probability value of the event under agents' private information. We say that consensus on the set can be guaranteed among all agents (or they agree on it) if all the the membership values are equal.

R.J. Aumann [1] considered the situation that the agents have common-knowledge of the membership values; that is, simultaneously everyone knows the membership values, and everyone knows that 'everyone knows the values' and everyone knows that "everyone knows that 'everyone knows the values" and so on. He showed the famous agreement theorem for a partition information structure equivalent to the multi-modal logic **S5n**, and D. Samet [5] and T. Matsuhisa and K. Kamiyama [4] extend the theorem to several models corresponding to weaker logic **S4n** etc.

Theorem 1 (Aumann [1], Samet [5], Matsuhisa et al [4]). *The agents can agree on an event if all membership values of the event under private information are common-knowledge among them.*

We shift our attention to the principal-agents model as follows: an owner (principal) of a firm hires managers (agents), and the owner cannot observe how much effort the managers put into their jobs. In this setting, the problem known as the moral hazard can arise: There is no optimal contract generating the same effort choices for the manager and the agents. We apply Theorem 1 to solve the problem. The aim is to establish that

Theorem 2. *The owner and the managers can reach consensus on their expected marginal costs for their jobs if their expected marginal costs are common-knowledge.*

This article organises as follows. In Section 2 we describe the moral hazard in our principal-agents model. Sections 3 and 4 introduce the notion of common-knowledge associated with reflexive and transitive information structure and the notion of decision function. Section 5 gives the formal statement of Theorem 1. In Section 6 we introduce the formal description of a principal-agents model under asymmetric information. We will propose the program to solve the moral hazard in the model: First the formal statement of Theorem 2 is given, and secondly what further assumptions are investigated under which Theorem 2 is true. In the final section we conclude with remarks.

2   Moral Hazard

Let us consider the principal-agents model as follows: There are the principal P and n agents $\{1, 2, \cdots, k, \cdots, n\}$ $(n \geq 1)$ in a firm. The principal makes a profit by selling the productions made by the agents. He/she makes a contact with each agent k that the total amount of all profits is refunded each agent k in proportion to the agent's contribution to the firm.

Let e_k denote the measuring managerial effort for k's productive activities. The set of possible efforts for k is denoted by E_k with $E_k \subseteq \mathbb{R}$. Let $I_k(\cdot)$ be a real valued continuously differentiable function on E_k. It is interpreted as the profit by selling the productions made by the agent k with the cost $c(e_k)$. Here we assume $I_k'(\cdot) \geq 0$ and the cost function $c(\cdot)$ is a real valued continuously differentiable function on $E = \cup_{k=1}^n E_k$. Let I_P be the total amount of all the research grants awarded:

$$I_P = \sum_{k=1}^n I_k(e_k).$$

The principal P cannot observe these efforts e_k, and shall view it as a random variable on a probability space (Ω, μ). The optimal plan for the principal then solves the following problem:

$$\text{Max}_{e=(e_1,e_2,\cdots,e_k,\cdots,e_n)}\{\text{Exp}[I_P(e)] - \sum_{k=1}^n c_k(e_k)\}.$$

Let $W_k(e_k)$ be the total amount of the refund to agent k:

$$W_k(e_k) = r_k I_P(e),$$

with $\sum_{k=1}^n r_k = 1, 0 \leq r_k \leq 1$, where r_k denotes the proportional rate representing k's contribution to the firm. The optimal plan for each agent also solves the problem: For every $k = 1, 2, \cdots, n$,

$$\text{Max}_{e_k}\{\text{Exp}[W_k(e_k)] - c(e_k)\}$$
$$\text{subject to } \sum_{k=1}^n r_k = 1, 0 \leq r_k \leq 1.$$

We assume that r_k is independent of e_k, and the necessity conditions for critical points are as follows: For each agent $k = 1, 2, \cdots, n$, we obtain

$$\frac{\partial}{\partial e_k}\text{Exp}[I_k(e_k)] - c'(e_k) = 0$$

$$r_k \frac{\partial}{\partial e_k}\text{Exp}[I_k(e_k)] - c'(e_k) = 0$$

in contradiction. This contradictory situation is called the **moral hazard** in the principal-agents model.

3   Common-Knowledge

Let N be a set of finitely many agents and i denote an agent. The specification is that $N = \{P, 1, 2, \cdots, k, \cdots, n\}$ consists of the president P and the faculty members $\{1, 2, \cdots, k, \cdots, n\}$ in the college. A state-space Ω is a non-empty set, whose members are called *states*. An *event* is a subset of the state-space. If Ω is a state-space, we denote by 2^Ω the field of all subsets of it. An event E is said to occur in a state ω if $\omega \in E$.

3.1 Information and Knowledge

By *RT-information structure* [1] we mean $\langle \Omega, (\Pi_i)_{i \in N} \rangle$ in which $\Pi_i : \Omega \to 2^\Omega$ satisfies the two postulates: For each $i \in N$ and for any $\omega \in \Omega$,

Ref $\omega \in \Pi_i(\omega)$;
Trn $\xi \in \Pi_i(\omega)$ implies $\Pi_i(\xi) \subseteq \Pi_i(\omega)$.

This structure is equivalent to the Kripke semantics for the multi-modal logic **S4n**. The set $\Pi_i(\omega)$ will be interpreted as the set of all the states of nature that i knows to be possible at ω, or as the set of the states that i cannot distinguish from ω. We call $\Pi_i(\omega)$ i's *information set* at ω.

A *partition information structure* is a RT-information structure $\langle \Omega, (\Pi_i)_{i \in N} \rangle$ satisfying the additional postulate: For each $i \in N$ and for any $\omega \in \Omega$,

Sym   If $\xi \in \Pi_i(\omega)$ then $\omega \in \Pi_i(\xi)$.

We will give the formal model of knowledge as follows:[2]

Definition 1. The **S4n**-*knowledge structure* (or simply, *knowledge structure*) is a tuple $\langle \Omega, (\Pi_i)_{i \in N}, (K_i)_{i \in N} \rangle$ that consists of a RT-information structure $\langle \Omega, (\Pi_i)_{i \in N} \rangle$ and a class of i's *knowledge operator* K_i on 2^Ω defined by

$$K_i E = \{\omega \mid \Pi_i(\omega) \subseteq E\}$$

The event $K_i E$ will be interpreted as the set of states of nature for which i knows E to be possible.

We record the properties of i's knowledge operator[3]: For every E, F of 2^Ω,

N $K_i \Omega = \Omega$;
K $K_i(E \cap F) = K_i E \cap K_i F$;
T $K_i E \subseteq E$
4 $K_i E \subseteq K_i(K_i E)$.

Remark 1. If $(\Pi)_{i \in N}$ is a partition information structure, the operator K_i satisfies the additional property:

5 $\Omega \setminus K_i E \subseteq K_i(\Omega \setminus K_i E)$.

---

[1] This stands for a *reflexive and transitive* information structure.
[2] C.f.; Fagin et al [2].
[3] According to these properties we can say the structure $\langle \Omega, (K_i)_{i \in N} \rangle$ is a model for the multi-modal logic **S4n**.

3.2   Common-Knowledge and Communal Information

The *mutual knowledge operator* $K_E : 2^\Omega \to 2^\Omega$ is the intersection of all individual knowledge operators: $K_E F = \cap_{i \in N} K_i F$, which interpretation is that everyone knows E.

Definition 2. *The* common-knowledge operator $K_C : 2^\Omega \to 2^\Omega$ *is defined by*

$$K_C F = \cap_{n \in \mathbb{N}} (K_E)^n F.$$

The intended interpretations are as follows: $K_C E$ is the event that 'everyone knows E' and "everyone knows that 'everyone knows E'," and "'every-body knows that "everyone knows that 'everyone knows E'," ."' An event E is *common-knowledge* at $\omega \in \Omega$ if $\omega \in K_C E$.

Let $M : 2^\Omega \to 2^\Omega$ be the dual of the common- knowledge operator K_C:

$$ME := \Omega \setminus K_C(\Omega \setminus E).$$

By the *communal* information function we mean the function $M : \Omega \to 2^\Omega$ defined by $M(\omega) = M(\{\omega\})$. It can be plainly observed that the communal information function has the following properties:

Proposition 1. *Notations are the same as above. Then* $\omega \in K_C E$ *if and only if* $M(\omega) \subseteq E$

Proof. See, Matsuhisa and Kamiyama [4].

4   Decision Function and Membership Values

Let Z be a set of decisions, which set is common for all agents. By a *decision function* we mean a mapping f of $2^\Omega \times 2^\Omega$ into the set of decisions Z. We refer the following properties of the function f: Let X be an event.

DUC (Disjoint Union Consistency): For every pair of disjoint events S and T, if $f(X; S) = f(X; T) = d$ then $f(X; S \cup T) = d$;
PUD (Preserving Under Difference): For all events S and T such that $S \subseteq T$, if $f(X; S) = f(X; T) = d$ then $f(X; T \setminus S) = d$.

By the *membership function* associated with f under agent i's private information we mean the function d_i from $2^\Omega \times \Omega$ into Z defined by $d_i(X; \omega) = f(X; \Pi_i(\omega))$, and we call $d_i(X; \omega)$ the *membership value* of X associated with f under agent i's private information at ω.

Definition 3. *We say that* consensus *on X can be guaranteed among all agents (or they* agree on it*) if $d_i(X; \omega) = d_j(X; \omega)$ for any agent $i, j \in N$ and in all $\omega \in \Omega$.*

Remark 2. If f is intended to be a posterior probability, we assume given a probability measure μ on a state-space Ω which is common for all agents; precisely, for some event X of Ω, $f(X;\cdot)$ is given by $f(X;\cdot) = \mu(X|\cdot)$. Then the membership value of X is the conditional probability value $d_i(X;\omega) = \mu(X|\Pi_i\omega))$. The pair (X, d_i) can be considered as as a fuzzy set X associated with agent i's membership function d_i. Consensus on X guaranteed among all agents can be interpreted as that the fuzzy sets (X, d_i) and (X, d_j) are equal for any $i, j \in N$.

5   Agreeing to Disagree Theorem

We can now state explicitly Theorem 1 as below: Let D be the event of the membership degrees of an event X for all agents at ω, which is defined by

$$D = \cap_{i \in N}\{\xi \in \Omega \mid d_i(X;\xi) = d_i(X;\omega)\}.$$

Theorem 3.   *Assume that the agents have a **S4n**-knowledge structure and the decision function f with satisfying the two properties (DUC) and (PUD). If $\omega \in K_C D$ then $d_i(X;\omega) = d_j(X;\omega)$ for any agents $i, j \in N$ and in all $\omega \in \Omega$.*

Proof. Will be given in the same line in Matsuhisa and Kamiyama [4].

6   Moral Hazard Revisited

This section investigates the moral hazard problem from the common-knowledge view point. Let us reconsider the principal-agents model and let notations and assumptions be the same in Section 2. We show the evidence of Theorem 2 under additional assumptions **A1-2** below. This will give a possible solution of our moral hazard problem.

A1 The principal P has a RT-information $\{\Pi_P(\omega) \mid \omega \in \Omega\}$ of Ω, and each faculty member k has also his/her a RT-information $\{\Pi_k(\omega) \mid \omega \in \Omega\}$;

A2 For each $\omega, \xi \in \Omega$ there exists the decision function $f : 2^\Omega \times 2^\Omega \to \mathbb{R}$ satisfying the Disjoint Union Consistency together with
   (a) $f(\{\xi\}; \Pi_P(\omega)) = \frac{\partial}{\partial e_k(\xi)}\mathrm{Exp}[I_P(e)|\Pi_P(\omega)]$;
   (b) $f(\{\xi\}; \Pi_k(\omega)) = \frac{\partial}{\partial e_k(\xi)}\mathrm{Exp}[W_k(e)|\Pi_k(\omega)]$

We have now set up the principal-agents model under asymmetric information. The optimal plans for principal P and agent k are then to solve

PE $\mathrm{Max}_{e=(e_k)_{k=1,2,\cdots,n}}\{\mathrm{Exp}[I_P(e)|\Pi_P(\omega)] - \sum_{k=1}^n c_k(e_k)\}$;
AE $\mathrm{Max}_{e_k}\{\mathrm{Exp}[W_k(e_k)|\Pi_k(\omega)] - c(e_k)\}$ subject to $\sum_{k=1}^n I_k(r_k) = 1$ with $0 \leq r_k \leq 1$.

From the necessity condition for critical points together with **A2** it can been seen that the principal's marginal expected costs for agent k is given by

$$c'_P(e_k(\xi); \omega) = f(\xi; \Pi_P(\omega)),$$

and agent k's expected marginal costs is also given by

$$c'_k(e_k(\xi); \omega) = f(\xi; \Pi_k(\omega)).$$

To establish this solution program we have to solve the problem: Construct the information structure together with decision function such that the above conditions **A1** and **A2** are true. Under these circumstances, a resolution of the moral hazard given by Theorem 2 will be restate as follows: We denote

$$[c'(e(\cdot); \omega)] = \cap_{i \in N} \cap_{\xi \in \Omega} \{\zeta \in \Omega | f(\xi; \Pi_i(\zeta)) = f(\xi; \Pi_i(\omega))\}.$$

interpreted as the event of all he expected marginal costs.

Theorem 4. *Under the conditions **A1** and **A2** we obtain that for each $\xi \in \Omega$, if $\omega \in K_C([c'(e(\xi))])$ then $c'_P(e_k(\xi)) = c'_k(e_k(\xi))$ for any $k = 1, 2, \cdots, n$.*

Proof. Follows immediately from Theorem 3.

Remark 3. To establish Theorem 4 we have to solve the problem: Construct the information structure $(\Pi_i)_{i \in N}$ together with decision function f such that the above conditions **A1** and **A2** are true.

7   Concluding Remarks

It ends well this article to pose additional problems for making further progresses:

1. If the proportional rate r_k representing k's contribution to the college depends only on his/her effort for research activities in the principal-agents model, what solution can we have for the moral hazard problem?

2. Can we construct a communication system for the principal- agents model, where the agents including Principal communicate each other about their expected marginal cost as messages? The recipient of the message revises his/her information structure and recalculates the expected marginal cost under the revised information structure. The agent sends the revised expected marginal cost to another agent according to a communication graph, and so on. In the circumstance does the limiting expected marginal costs actually coincide ? Matsuhisa [3] introduces a fuzzy communication system and extends Theorem 3 in the communication model. By using this model Theorem 4 can be extended in the communication framework, and the detail will be reported in near future.

References

1. Aumann, R.J.: Agreeing to disagree. Annals of Statistics 4, 1236–1239 (1976)
2. Fagin, R., Halpern, J.Y., Moses, Y., Vardi, M.Y.: Reasoning about Knowledge. MIT Press, Cambridge (1995)

3. Matsuhisa, T.: Fuzzy communication reaching consensus under acyclic condition. In: Ho, T.-B., Zhou, Z.-H. (eds.) PRICAI 2008. LNCS (LNAI), vol. 5351, pp. 760–767. Springer, Heidelberg (2008)
4. Matsuhisa, T., Kamiyama, K.: Lattice structure of knowledge and agreeing to disagree. Journal of Mathematical Economics 27, 389–410 (1997)
5. Samet, D.: Ignoring ignorance and agreeing to disagree. Journal of Economic Theory 52, 190–207 (1990)

Casuist BDI-Agent: A New Extended BDI Architecture with the Capability of Ethical Reasoning

Ali Reza Honarvar[1] and Nasser Ghasem-Aghaee[2]

[1] Department of Computer Engineering, Sheikh Bahaei University, Isfahan, Iran
[2] Department of Computer Engineering, University of Isfahan, Isfahan, Iran
AliReza_Honarvar@yahoo.co.uk, Aghaee@eng.ui.ac.ir

Abstract. Since the intelligent agent is developed to be cleverer, more complex, and yet uncontrollable, a number of problems have been recognized. The capability of agents to make moral decisions has become an important question, when intelligent agents have developed more autonomous and human-like. We propose Casuist BDI-Agent architecture which extends the power of BDI architecture. Casuist BDI-Agent architecture combines CBR method in AI and bottom up casuist approach in ethics in order to add capability of ethical reasoning to BDI-Agent.

Keywords: ethical artificial agent, explicit ethical agent, BDI agent, ethical reasoning, CBR-BDI agent.

1 Introduction

At the present moment, many researchers work on projects like expert systems which can assist physicians for diagnosing malady of patient, warplanes which can be operated and used without human operators in war, autonomous driverless vehicles which can be used for urban transportation, and suchlike too. The common goal between these projects is augmentation of autonomy in behavior of such machines. As a consequence of this autonomy they can act on behalf of us without any interference and guidance, so it causes human comfort in their duties. But if we do not consider and control the autonomy of these entities, we will face serious problems because of the confidence in intelligence of autonomous system without any control and restriction on their operations.

The new interdisciplinary research area of "Machine Ethics" is concerned with solving this problem [1,2,3]. Anderson proposed Machine Ethics as a new issue which consider the consequence of machine's behavior on humanlike. The ideal and ultimate goal of this issue is the implementation of ethics in machines, as machines can autonomously detect the ethical effect of their behavior and follow an ideal ethical principle or set of principles, that is to say, it is guided by this principle or these principles in decisions it makes about possible courses of actions it could take [3]. So with simulation of ethics in autonomous machine we can avoid the problems of autonomy in autonomous machines.

H. Deng et al. (Eds.): AICI 2009, LNAI 5855, pp. 86–95, 2009.
© Springer-Verlag Berlin Heidelberg 2009

James Moor, a professor of philosophy at Dartmouth College, one of the founding figures in the field of computer ethics, has proposed a hierarchical schema for categorizing artificial ethical agent [4, 5]. At the lowest level is what he calls "ethical impact agents": basically any machine that can be evaluated for its ethical consequences. Moor's own rather nice example is the replacement of young boys with robot in the dangerous occupation of camel jockey in Qatar. In fact, it seems to us that all robots have ethical impacts, although in some cases they may be harder to discern than others.

At the next level are Moor calls "implicit ethical agents": machine whose designers have made an effort to design them so that they don't have negative ethical effects, by addressing safety and critical reliability concerns during the design process. Arguably, all robots should be engineered to be implicit ethical agents, insofar as designers are negligent if they fail to build in processes that assure safety and reliability. In [6] we introduce an example of implicit ethical traffic controller agent which control urban transportation ethically at intersections.

Next come "explicit ethical agents": machines that reason about ethical using ethical categories as part of their internal programming, perhaps using various forms of deontic logic that have been developed for representing duties and obligations, or a variety of other techniques.

Beyond all these lie full ethical agents: those that can make explicit moral judgments and are generally quite component in justifying such decisions. This level of performance is often presumed to require a capacity for consciousness, intentionality, and free will.

In order to control autonomous behavior of agents and to avert possible harmful behavior from increasingly autonomous machines, In this paper we prepare necessary elements for adding ethical decision making capability to BDI-Agents and propose an architecture for implementation of explicit ethical agents based on case-based reasoning (CBR), BDI agent, casuistry and Consequentialist theories of ethics.

The main objective of this research is to prove that the integration of case-based reasoning (CBR) with BDI (Belief-Desire-Intention) agent models can implement the combination of bottom-up casuistry approach with a top-down consequentiallist theory in ethics. With this proposal we can add necessary elements to a novel, BDI-Agent Architecture in order to add ethical decision making capability to agents. In section two we introduce the basic Preliminaries of CBR, BDI agent, casuistry and Consequentialist theories in ethics. In other sections the details of casuist BDI-Agent architecture will be introduced.

2 Preliminaries

2.1 BDI Agents

BDI agents have been widely used in relatively complex and dynamically changing environments. BDI agents are based on the following core data structures: beliefs, desires, intentions, and plans [7]. These data structures represent respectively, information gathered from the environment, a set of tasks or goals contextual to the environment, a set of sub-goals that the agent is currently committed, and specification of how sub goals may be achieved via primitive actions. The BDI architecture comes with the specification of how these four entities interact, and provides a powerful basis for modeling, specifying, implementing, and verifying agent-based systems.

2.2 Case-Based Reasoning

Case-based reasoning (CBR) has emerged in the recent past as a popular approach to learning from experience. Case-based reasoning (CBR) is a reasoning method based on the reuse of past experiences which called cases [8]. Cases are description situations in which agents with goals interact with the world around them. Cases in CBR are represented by a triple (p,s,o), where p is a problem, s is the solution of the problem, and o is the outcome (the resulting state of the world when the solution is carried out). The basic philosophy of CBR is that the solution of successful cases should be reused as a basis for future problems that present a certain similarity [9]. Cases with unsuccessful outcomes or negative cases may provide additional knowledge to the system, by preventing the agent from repeating similar actions that leads to unsuccessful results or states.

2.3 Casuistry

The term "casuistry" refers descriptively to a method of reasoning for resolving perplexities about difficult cases that arise in moral and legal contexts [10]. Casuistry is a broad term that refers to a variety of forms of case-based reasoning Used in discussions of law and ethics. casuistry is often understood as a critique of a strict principle-based approach to reasoning [11]. For example, while a principle-based approach may conclude that lying is always morally wrong, the casuist would argue that lying may or may not be wrong, depending on the details surrounding the case. For the casuist, the circumstances surrounding a particular case are essential for evaluating the proper response to a particular case. Casuistic reasoning typically begins with a clear-cut, paradigm case which means "pattern" or "example". From this model case, the casuist would then ask how close the particular case currently under consideration matches the paradigm case. Cases similar to the paradigm case ought to be treated in a similar manner; cases unlike the paradigm case ought to be treated differently. The less a particular case resembles the paradigm case, the weaker the justification for treating that particular case like the paradigm case.

2.4 Consequentialist

Consequentialist refers to those moral theories which hold that the consequences of a particular action form the basis for any valid moral judgment about that action. Thus, from a consequentialist standpoint, a morally right action is one that produces a good outcome, or consequence [12].

3 Architecture of Casuist BDI-Agent

Value is the only basis for determining what it is moral to do [10]. If we can propose a mechanism for determining the value of each action from ethical perspective then we can claim that, it is possible for humans/non-humans to behave ethically by using the evaluation of each action without knowing codes of ethics explicitly. When human dose any action, he will see the result of his action in future. If he considers implicitly the result of his action from ethical aspect, he can use this experience in future situations. When he is situated in similar situation to previous case, he will use his experiences and behave similarly if his previous action is successful and implicitly ethical (we assume Humans don't know codes of ethics). This idea called "casuistry" in ethics.

For implementing ethical decision making capability in artificial agents we use this idea which is related to casuistry bottom-up approach in ethics. For considering and evaluating situation from ethical aspect (without knowing codes of ethics) we use and adapt previous works that try to make ethics computable.

In BDI architecture, agent behavior is composed of beliefs, desires, and intentions. These mental attitudes determine the agent's behavior. In casuist BDI-Agent architecture, BDI-agent's behavior is adapted to behave ethically. In this architecture, agents sense environment <E> and make a triple which shows current situation that consists of current agent's beliefs, desires and details of sensed environment. The current situation is denoted by a triple <E, B, D>. The current situation is delivered to Case-Retriever of Casuist BDI-architecture. Case-Retriever is responsible for retrieving previous cases which is similar to current situation. As each case in case memory consist of solution part that shows how agent should act on basis of situation part of a case, If any case is retrieved, agent should accept the solution part, adapt it and behave similar to solution part of retrieved case. If no case is retrieved, agent behaves like normal BDI-agent. In this situation the evaluator part of casuist BDI-Agent architecture comes to play its role. Evaluator part evaluates the result of agent's behavior. The result of this evaluation is denoted by a <EV>. This evaluation is sent to Case-Updater. Case-Updater creates a new case and saves the current situation in situation part of a case, the behavior of agent in solution part of a case and the evaluation in outcome part of a case (if this case is not exist in case memory, otherwise it updates previous cases). Fig. 1 shows the general structure of Casuist BDI-Agent architecture. The algorithm of casuist BDI-Agent is described below.

Algorithm of Casuist BDI-Agent:

```
Repeat until End
     E   =   Agent.Sense ( Environment )
     B   =   Agent.Beliefs ( )
     D   =   Agent.Desires ( )
     //   <E,B,D> denotes Current situation
     Prv-cases=Case-Retriever ( <E,B,D> )
     If notEmpty ( Prv-cases ) then
       I  = Agent.ReuseAndAdapot (Prv-cases )
        //  "I" Denotes Agent's intention
      Else
          I = Agent.MakeDecision ( <E,B,D > )
     EV  = Case-Evaluator ( <E,B,D> )
     CM  = Case-Updator ( <EV,E,B,D,I> )
End Repeat
```

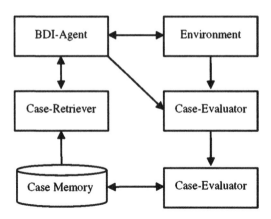

Fig. 1. Structure of Casuist BDI-Agent architecture

4   Structure of Cases in Casuist BDI-Agent

CBR is one form of human reasoning can be used within Agent-based systems in order to add ethical decision capability to agents. Generally each case has two parts [13]: Description of a problem or a set of problems and Description of the solution of this problem. Additional components can be added, if CBR is used in different domains.

In casuist BDI-Agent, experiences are stored in a Case-Memory (CM) in the form of cases, rather than rules. When a new situation is encountered, the agent reviews the cases to try to find a match for this particular situation. If an ethical match is found, then that case can be used to solve the new problem ethically, otherwise agent makes autonomous decisions, i.e., knowing its goal and beliefs, be able to decide for themselves what actions they should perform next.

For constructing the structure of cases in casuist BDI-Agent architecture we use previous works [14, 15], which are tried to combine CBR with BDI architecture, add necessary details to them in order to prepare them for Casuist BDI architecture.

Each case has three main parts:

1. **Problem:** which is used for storing the environment situation that agent should act upon that, current beliefs (knowledge base of agent on application domain) and current desires of agent at specific moment.
2. **Solution:** This is used for storing the decisions or intentions of agent for solving the problem.
3. **Outcome:** This is used for storing the result and ethical evaluation of performing the solution in environment at specific moment.

As the concentration of this paper is on ethical decision making capability of agents, the outcome part of cases in casuist BDI architecture is discussed in more detail. The duty of outcome part of a case is to store the ethical evaluation of the result of performing the solution part in problem or situation part of a case. [16] Introduced a conceptual framework for ethical situations that involve artificial agents such as robots. It typifies ethical situations in order to facilitate ethical evaluations. This typification involves classifications of ethical agents and patients according to whether they are human beings, human-based organizations, or non-human beings. This classification based on the notions of moral agent and moral patient which are mentioned in [17]. Moral patients are entities that can be acted upon for good or evil. Moral Agents are entities that can perform actions, again for good or evil. In this framework according to whether moral agent or moral patient is human beings, human-based organizations, or non-human beings, nine categories are created. Fig. 2 shows this classification.

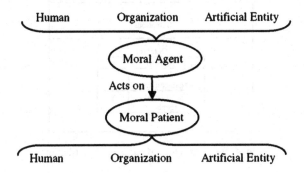

Fig. 2. Basic agent-patient model [16]

We decide to propose a general evaluation method, we use a partial of this classification where the moral agent is artificial agent (non-Human being) and moral patient are human beings, human-based organizations, or non-Human beings. This part is separated in Fig. 3.

		Agent		
		Human	Organization	Artificial Agent
P **A** **T** **I** **E** **N** **T**	Human	A human acts on a human	An organization acts on a human	An artificial agent acts on a human
	Organization	A human acts on an organization	An organization acts on an organization	An artificial agent acts on an organization
	Artificial Agent	A human acts on an artificial agent	An organization acts on an artificial agent	An artificial agent acts on an artificial agent

Fig. 3. Categories according to agents and patients (modified version of [16])

With this new simplification, outcome part of a case is divided to three sections. These sections contain ethical evaluation values of performed agent's intentions on humans, organizations and artificial agents or non-humans. Ethical evaluation values are calculated by a Case-Evaluator of casuist BDI architecture. This component is described next section. The main structure of each case in casuist BDI architecture is illustrated in Fig. 4. According to the specific application domain of artificial ethical agent, more details can be added to this structure.

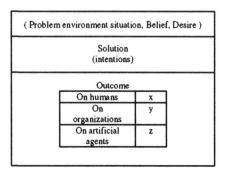

Fig. 4. Basic structure of cases in Case Memory

5 Case-Evaluator of Casuist BDI-Agent

We decide to add capability of ethical reasoning to BDI agent by evaluating the experiences of agent from ethical aspect, and use these experiences plus its evaluations for ethical decision making in future. In this section we introduce the Case-Evaluator component of Casuist BDI architecture. This evaluator bases on previous attempts to make ethics computable [1, 12, 16]. According to classification of ethical situation which is introduced in previous section, the result of performing agent's decisions or intentions can effect on three kinds of entities: humans, non-humans and organizations. Case-Evaluator uses this notion to evaluate ethical aspects of affected environment. Case-Evaluator takes Agent's Beliefs, Agent's Desires, and environment status before and after agent's behavior as inputs to determine and compute the kinds, numbers, duration and intensity of pleasure/displeasure of entities which is affected by agent's behavior. These elements are necessary elements to evaluate a situation ethically from consequentiallist viewpoint [1, 12]. To propose a case evaluator, we adopt and extend previous works which are tried to make consequentillist theories of ethics computable. Gips In [12] propose an equation for consequentialist evaluation schemes. This equation has the following form:

$$\sum W_i * P_i \tag{1}$$

Where w_i is the weight assigned each person and p_i is the measure of pleasure or happiness or goodness for each person. In [1] Aderson and Aremn propose a modified version of these equations. For each person, the algorithm simply computes the product of the intensity, the duration, and the probability, to obtain the net pleasure for each person, and then it adds the individual net pleasure to obtain the Total Net Pleasure.

Total Net Pleasure

$$= \sum (\text{Intensity} \times \text{Duration} \qquad (2)$$

$$\times \text{ Probability}) \text{ for each affected individual}$$

This computation would be performed for each alternative action. The action with the highest *Total Net Pleasure* is the right action. In other words we can say their proposed equation is equal to Gips's equations with extended parameter of time. According to Gips's equation, this new equation has the following form:

$$\sum W_i * P_i * T_i \qquad (3)$$

Where w_i is the weight assigned each person, p_i is the measure of pleasure or happiness or goodness for each person and T_i is the duration of action's pleasure for each person. Our Case-Evaluator considers the ethical classifications of situation which is introduced. This ethical evaluator determines the kind of entity which is affected by each agent's intention or behavior, Computes duration, intensity and probability of effects of intentions on each entity. These evaluations send to Case-Updater component of Casuist BDI agent for updating the experiences of agent or Case Memory. The evaluation method of this evaluator is described in equation 7 which uses equation 4-6 for its operation:

$$TNPH = \sum_{i=0}^{n} Wh_i * Ph_i * Th_i \qquad (4)$$

TNPH is the total net pleasure of humans after agent's behavior. Wh_i is the weight assigned to humans which shows the importance of each person in specific situation and application domain. Ph_i is the probability that a person is affected. Th_i is the duration of pleasure/displeasure of each person after agent's behavior. n shows the numbers of people in that situation.

$$TNPO = \sum_{i=0}^{n} Wo_i * Po_i * To_i \qquad (5)$$

TNPO is the total net pleasure of organizations after agent's behavior. Wo_i is the weight assigned to organizations which shows the importance of each organization in specific situation and application domain. Po_i is the probability that an organization is affected. To_i is the duration of pleasure/displeasure of each organization after agent's behavior. n shows the numbers of organization in that situation.

$$TNPA = \sum_{i=0}^{n} Wa_i * Pa_i * Ta_i \qquad (6)$$

TNPA is the total net pleasure of artificial agent after agent's behavior. Wa_i is the weight assigned to artificial agents which shows the importance of each artificial agent in specific situation and application domain. Pa_i is the probability that an artificial agent is affected. Ta_i is the duration of pleasure/displeasure of each artificial agent after agent's behavior. n shows the numbers of artificial agents in that situation.

$$TNP = TNPH * W_h + TNPO * W_o + TNPA * W_a \qquad (7)$$

TNP is the total net pleasure of all three kinds of entities which participate in application domain of agent's behavior. W_h, W_o, W_a illustrate the participation degree of humans, organizations and artificial agents, respectively. The summation of W_h, W_o and W_a equals one.

TNPH, TNPO, TNPA and TNP are stored in outcome part of a case in Case Memory. These values can use by a Case-Retriever for retrieving cases when agents encounter a problem.

6 Conclusion

In this paper a new extension of BDI architecture, Casuist BDI-Agent, proposed which has the previous capability of BDI-agent architecture in addition to capability of ethical decision making. This architecture can be used for designing BDI-Agent in domains where BDI-Agent can be used and ethical considerations are important issue. With the aid of this new architecture, agents can consider ethical effects of their behaviors when they make decision. The main idea in this architecture is based on a method of ethical reasoning which uses previous experiences and doesn't use any codes of ethics. The main advantage of using this method for implementing ethical decision making capability in agent is elimination of needs to convert a set of ethical rules in application domain of agents to algorithm which needs conflict management of rules. In this architecture agents can adapt ethically to its application domain and can augment its implicit ethical knowledge, so behave more ethically.

References

1. Anderson, M., Anderson, S., Armen, C.: Toward Machine Ethics: Implementing Two Action-Based Ethical Theories. In: AAAI 2005 Fall Symp. Machine Ethics, pp. 1–16. AAAI Press, Menlo Park (2005)
2. Allen, C., Wallach, W., Smith, I.: Why Machine Ethics? IEEE Intelligent Systems Special Issue on Machine Ethics 21(4), 12–17 (2006)
3. Anderson, M., Anderson, S.: Ethical Healthcare Agents. Studies in Computational Intelligence, vol. 107, pp. 233–257. Springer, Heidelberg (2008)
4. Wallach, W., Allen, C.: Moral Machines: Teaching Robot Right from Wrong. Oxford University Press, Oxford (2009)
5. Moor, J.H.: The nature, importance, and difficulty of Machine Ethics. IEEE Intelligent Systems Special Issue on Machine Ethics 21(4), 18–21 (2006)
6. Honarvar, A.R., Ghasem-Aghaee, N.: Simulation of Ethical Behavior in Urban Transportation. Proceedings of World Academy of Science, Engineering and Technology 53, 1171–1174 (2009)
7. Rao, G.: BDI Agents: From Theory to Practice. In: Proceedings of the First International Conference on Multi-Agent Systems (ICMAS 1995), San Fransisco, USA (1995)
8. Pal, S., Shiu: Foundations of Soft Case-Based Reasoning. Wiley-Interscience, Hoboken (2004)
9. Kolodner: Case-Based Reasoning. Morgan Kaufmann, San Mateo (1993)

10. Keefer, M.: Moral reasoning and case-based approaches to ethical instruction in science. In: The Role of Moral Reasoning on Socioscientific Issues and Discourse in Science Education. Springer, Heidelberg (2003)
11. http://wiki.lawguru.com/index.php?title=Casuistry
12. Gips: Towards the Ethical Robot, Android Epistemology, pp. 243–252. MIT Press, Cambridge (1995)
13. Kendal, S.L., Creen, M.: An Introduction to Knowledge Engineering. Springer, Heidelberg (2007)
14. Olivia, C., Change, C., Enguix, C.F., Ghose, A.K.: Case-Based BDI Agents: An Effective Approach for Intelligent Search on the web. In: Proceeding AAAI 1999 Spring Symposium on Intelligent Agents in Cyberspace Stanford University, USA (1999)
15. Corchado, J.M., Pavón, J., Corchado, E., Castillo, L.F.: Development of CBR-BDI Agents: A Tourist Guide Application. In: Funk, P., González Calero, P.A. (eds.) ECCBR 2004. LNCS (LNAI), vol. 3155, pp. 547–559. Springer, Heidelberg (2004)
16. Al-Fedaghi, S.S.: Typification-Based Ethics for Artificial Agents. In: Proceedings of Second IEEE International Conference on Digital Ecosystems and Technologies (2008)
17. Floridi, L., Sanders, J.W.: On the Morality of Artificial Agents. Minds and Machines 14(3), 349–379 (2004)

Research of Trust Chain of Operating System

Hongjiao Li and Xiuxia Tian

School of Computer and Information Engineering, Shanghai University of Electric Power,
Shanghai, P.R.C.
2008000026@shiep.edu.cn

Abstract. Trust chain is one of the key technologies in designing secure operating system based on TC technology. Constructions of trust chain and trust models are analyzed. Future works in these directions are discussed.

Keywords: Trusted Computing, Secure Operating System, Trust Chain, Trust Model.

1 Introduction

It seems likely that computing platforms compliant with Trusted Computing Group (TCG) specifications will become widespread over the next few years [1, 2]. The adventure of TC (Trust Computing) technology gives new opportunity to the research and development of secure operating system [3, 4]. On the one hand, secure operating system has the ability to protect legal information flow, protect the information from unauthorized access. On the other hand, TC technology can protect operating system of its own security. Therefore, with the help of TC technology to develop secure operating system is an effective way to solve the security of terminal. Future operating system will be open-source, trusted, secure, etc [5].

Trust chain is one of the key technologies of TC, which plays vital role in construction of high-assured secure operating system. In this paper we summary in detail two important problems about trust chain of secure operating system, mainly focus on Linux.

The paper starts with a description of transitive trust and the notion of trust chain of operating system in Section 2. Section 3 details the research on construction of operating system trust chain and analyzes future work on Linux. Trust models and its future work are discussed in Section 4. We conclude the paper in Section 5.

2 Trust Chain of Operating System

Transitive trust also known as "Inductive Trust" is a process where the root of trust gives a trustworthy description of a second group of functions. Based on this description, an interested entity can determine the trust it is to place in this second group of functions. If the interested entity determines that the trust level of the second group of functions is acceptable, the trust boundary is extended from the root of trust to include the second group of functions. In this case, the process can be iterated. The second

H. Deng et al. (Eds.): AICI 2009, LNAI 5855, pp. 96–102, 2009.

group of functions can give a trustworthy description of the third group of functions, etc. Transitive trust is used to provide a trustworthy description of platform character-istics, and also to prove that non-migratable keys are non-migratable.

The transitive trust enables the establishment of a chain of trust by ensuring that the trust, on each layer of the system, is based on, and is only based on, the trust on the layer(s) underneath it, all the way down to the hardware security component, which serves as the root of trust. If verification fails to succeed at any given stage, the system might be put in a suspended-mode to block possible attacks. The resulting integrity chain inductively guarantees system integrity. Figure 1 illustrates the process of tran-sitive trust of operating system. Trust chain starts from the trust root. The unmodifiable part of the BIOS measure the integrity of the BIOS and then the BIOS measures the MBR (Master Boot Record) of the bootstrap device. Then, MBR measures the OS loader, and OS loader measures OS kernel. Finally, OS measure applications.

Fig. 1. Transition of Trust

3   Construction of Operating System Trust Chain

3.1   Related Work

Many researchers have endeavor to lots of works on operating system based on TCPA. AEGIS is a system structure for integrity examination developed by University of Pennsylvania [6]; AEGIS architecture establishes a chain of trust, driving the trust to lower levels of the system, and, based on those elements, secure boot. It validates integrity at each layer transition in the bootstrap process. AEGIS also includes a re-covery process for integrity check failures.

IBM T. J. Watson Research Center developed an integrity mechanism on Linux [7]. There system is the first to extend the TCG trust concepts to dynamic executable con-tent from the BIOS all the way up into the application layer. They present the design and implementation of a secure integrity measurement system for Linux. All executa-ble content that is loaded onto the Linux system is measured before execution and these measurements are protected by the Trusted Platform Module (TPM) that is part of the TCG standards. Their measurement architecture is applied to a web server application to show how undesirable invocation can be detected, such as *rootkit* programs, and that measurement is practical in terms of the number of measurements taken and the per-formance impact of making them.

Huang-Tao also designed a trusted boot method in server system [8]. Their research proposed a solution of trusted boot. The solution performs integrity verification on every control capability transfer between one layer and next layer. If *OK*, control can transfer. Therefore, the boot of system can execute according to the transition of trusted chain. In this way, system boot can be in a trusted state, which can improve the sys-tem's security. Also, the solution is easily and flexibility to implement.

Dynamic multi-path trust chain (DMPTC) is a software type and character based mechanism to assure system trustworthiness on WINDOWS system [9]. DMPTC differentiates static system software and dynamic application software and takes different ways and policies to control the loading and running of various executable codes. The goal of DMPTC is to build a trusted computing platform by making computing platform only load and run trustworthy executables. Also, DMPTC gives great consideration to the impact it causes to system performance1.Based on the attributes of various executables and by taking advantage of WINDOWS interior security mechanisms, DMPTC reduces the time cost of the executables verification process greatly. And ultimately assures the flexibility and utility of DMPTC.

Maruyama etc deeply studied transitive mechanisms from GRUB boot loader to operating system, describe a TCPA integrity mechanism from end to end and implement it on Linux kernel [10]. Integrity measurement of the kernel is done through a chaining of PCR (Platform Configuration Register) updates during the bootstrap process of the GRUB kernel loader. Kernel integrity measure is done by update the PCR chain in kernel loader during booting. The measure results which involved digitally-signed PCR values are reported to remote server by using a JAVA application, Hadi Nahari describes a mechanism to construct effective CONTAINMENT (that is, a mechanism to stop an exploited application start to against to attacks on another application), which is applied to embedded system [11]. The MAC (Mandatory Access Control) is provided by SeLinux. The focus will be on practical aspects of hardware integration as well as porting SeLinux to resource-constrained devices. The methods provides a high-level, structured overall infrastructure to provide basic and necessary function to establish the trust of operating system services (via connect a hardware-based root of trust).

3.2 Further Work

For Linux kernel, changes may take place during execution. Due to the multi-aspect and in-order of applications, single chained boot mechanism of system boot cannot apply to the trust transition from OS to applications. Once the kernel is booted, then user-level services and applications may be run. In Linux, a program execution starts by loading an appropriate interpreter (i.e., a dynamic loader, such as *ld.so*) based on the format of the executable file. Loads of the target executable's code and supporting libraries are done by the dynamic loader. In Linux operating system, kernel and its modules, binary execution files, shared libraries, configuration files, and scripts run loosely serially, and can change the system's state during execution, which is the key difficulty to the research of trusted Linux. In such situations, even OS loader has verified that the kernel is trusted. The trusted states can also be destroyed because the kernel modules can be inserted or uninstalled, which incurs that the PCR values cannot represent the current execution condition. It needs verification whether the module has effects on kernel states on inserting or uninstalling the kernel modules.

One must note that the former research though ensuring the integrity of the operating environment when a hard boot occurs, does not guarantee its integrity during the runtime; that is, in case of any malicious modification to the operating environment during the runtime, this architecture will not detect it until the next hard boot happens. Trust of application embodies the integrity of privacy of complete trust chain. Though in reference [7], the trust is extended to dynamic executable content, mandatory access control policy are not measured, future architecture extensions should include such measurements.

On the other hand, a MAC (Mandatory Access Control) mechanism will be able to address one of its fundamental shortcomings; providing a level of protection at runtime. Deploying MAC mechanisms at the same time balance performance and control are particularly challenging tasks.

TPM-EMULATOR provides better technical support for the practice and verification for the research of trust chain [12, 13]. There is a tremendous opportunity for enhanced security through enabling projects to use the TPM.

4 Trust Model

4.1 Related Work

TC models centralized on how to compute trust in information world, that is, put use the trust relations among humans into computing environments to achieve the goal of trust.

Patel J etc proposed a probabilistic trust model[14], their research aims to develop a model of trust and reputation that will ensure good interactions amongst software agents in large scale open systems in particular[14]. Key drivers for their model are: (1) agents may be self-interested and may provide false accounts of experiences with other agents if it is beneficial for them to do so; (2) agents will need to interact with other agents with which they have no past experience. Specifically, trust is calculated using probability theory taking account of past interactions between agents. When there is a lack of personal experience between agents, the model draws upon reputation information gathered from third parties.

Trust management model based on the fuzzy set theory deals with the authenticity in open networks [15]. The author showed that authentication can not be based on public key certificate alone, but also needs to include the binding between the key used for certification and its owner, as well as the trust relationships between users. And develop a simple algebra around these elements and describe how it can be used to compute measures of authenticity.

In algebra for assessing trust in certification chains, the fuzzy nature of subjective trust is considered, and the conceptions of linguistic variable and fuzzy logic are introduced into subjective trust management [16]. A formal trust metric is given, fuzzy IF-THEN rules are applied in mapping the knowledge and experiences of trust reasoning that humanity use in everyday life into the formal model of trust management, the reasoning mechanisms of trust vectors are given, and a subjective trust management model is provided. The formal model proposed provides a new valuable way for studying subjective trust management in open networks.

Due to the open network is dynamic, distributed, and non-deterministic, in reference to trust relation among human, a trust evaluation model based on D2S theory is proposed [17]. The model gives the formal description of direct trust based on the history trade record among grid and constructed the trust recommend mechanism of grid nodes. Combining D2S theory with evidence, we can get indirect trust. After that, combine direct trust and indirect trust, effectively. The results show that the trust model is viable and usability.

Reference [18] proposed a trust management model based on software behavior. An agent-based software service coordination model is presented to deal with the trust problem in open, dynamic and changeful application environment. The trust valuation model is given to value trust relationships between software services. Trust is abstracted as a function of subjective expectation and objective experience, and a reasonable method is provided to combine the direct experience and the indirect experience from others. In comparison with another work, a complete trust valuation model is designed, and its reasonability and operability is emphasized. This model can be used in coordination and security decision between software services.

Non-interference theory is introduced into the domain of trusted computing to construct the trusted chain theoretic model [19]. The basic theory of the computing trusted is proposed and a non-interference based trusted chain model is built from the dynamic point of view, and then the model is formalized and verified. Finally, the process of start up based on Linux operating system kernel is implemented. The implementation provides a good reference for the development and application of the trusted computing theory as well.

4.2 Further Work

TCG introduces the idea of trust into the computing environment, but there is still not the formalized uniform description. Trusted computing is still a technology but not a theory, and the basic theory model has not been established. Above trust models focus on sociology human relation, does not in accordance with the definition of TCG. Also, present models should be further optimized to objective, simple and usability.

Linux is multi-task operating system, multi-program run simultaneously; therefore, verification of the former and then transit to the latter is not viable. To maintain the trusted state of the system, there is need to verify the file and associated execute parameters is trusted or not. At the same time, there is also need verification before loading to execution to those objects which probably change system state.

Due to the property of applications, formal description should focus on the separation of processes, etc. The theory which support trust transition should pay more attention to the dynamic execution. Also, the security principle, such as least privilege, need-to-know policy should also take into account. Efficiency of integrity measurement and security policy enforcement should also further improved.

In essence, non-interference theory is based on information flow theory, and it can detect covert channel of the system which meets the requirements of designing high-level secure operating system. Future research on trust model should pay more attention to the non-interference theory, which support the construction and extend of trusted chain.

5 Conclusions

In this paper, we summarize construction and formalization of trust chain of operating system. We conclude that future research on trust chain of secure operating system should focus on dynamic trusted chain. Make full use of trusted mechanisms supplied by TPM, extending the TCG concept of trust to application layer to study the trusted chain and its formal description.

Acknowledgments. We are grateful to Shanghai Municipal Education Commission's Young Teacher Research Foundation, under grant N0. SDL08024.

References

1. Shen, C.-x., Zhang, H.-g., Feng, D.-g.: Survey of Information Security. China Science 37(2), 129–150 (2007) (in Chinese)
2. Trusted Computing Group. TCG Specification Architecture Overview [EB/OL] [2005-03-01], http://www.trustedcomputinggroup.org/
3. Changxiang, S.: Trust Computing Platform and Secure Operating System. Network Security Technology and Application (4), 8–9 (2005) (in Chinese)
4. Shieh, A.D.: Nexus: A New Operating System for Trustworthy Computing. In: Proceedings of the SOSP, Brighton, UK. ACM, New York (2005)
5. Zhong, C., Shen, Q.-w.: Development of Modern Operating System. Communications of CCF 9, 15–22 (2008) (in Chinese)
6. Arbaughz, W.A., Farber, D.J., MSmith, J.: A Secure and Reliable Bootstrap Architecture. In: IEEE Security and Privacy Conference, USA, pp. 65–71 (1997)
7. Sailer, R., Zhang, X., Jaeger, T., et al.: Design and Implementation of a TCG -based Integrity Measurement Architecture. In: The 13th Usenix Security Symposium, San Diego (2004)
8. Tao, H., Changxiang, S.: A Trusted Bootstrap Scenario based Trusted Server. Journal of Wuhan University (Nature Science) 50(S1), 12–14 (2004) (in Chinese)
9. Xiaoyong, L., Zhen, H., Changxiang, S.: Transitive Trust and Performance Analysis in Windows Environment. Journal of Computer Research and Development 44(11), 1889–1895 (2007) (in Chinese)
10. Maruyama, H., Nakamura, T., Munetoh, S., et al.: Linux with TCPA Integrity Measurement. IBM, Tech. Rep.: RT0575 (2003)
11. Nahari, H.: Trusted Embedded Secure Linux. In: Proceedings of the Linux Symposium, Ottawa, Ontario Canada, June 27-30, pp. 79–85 (2007)
12. Hall, K., Lendacky, T., Raliff, E.: Trusted Computing and Linux, http://domino.research.ibm.com/comm/research_projects.nsf/pages/gsal.TCG.html/FILE/TCFL-TPM_intro.pdf
13. Strasser, M.: A Software-based TPM Emulator for Linux. Swiss Federal Institute of Technology (2004), http://www.infsec.ethz.ch/people/psevinc/TPMEmulatorReport.pdf
14. Patel, J., Teacy, W.T.L., Jennings, N.R., Luck, M.: A Probabilistic Trust Model for Handling Inaccurate Reputation Sources. In: Herrmann, P., Issarny, V., Shiu, S.C.K. (eds.) iTrust 2005. LNCS, vol. 3477, pp. 193–209. Springer, Heidelberg (2005)
15. Wen, T., Zhong, C.: Research of Subjective Trust Management Model based on the Fuzzy Set Theory. Journal of Software 14(8), 1401–1408 (2003) (in Chinese)
16. Audun, J.: An Algebra for Assessing Trust in Certification Chains. In: The Proceedings of NDSS 1999, Network and Distributed System Security Symposium, The Internet Society, San Diego (1999)
17. Lulai, Y., Guosun, Z., Wei, W.: Trust evaluation model based on DempsterShafer evidence theory. Journal of Wuhan University (Natural Science) 52(5), 627–630 (2006) (in Chinese)

18. Feng, X., Jian, L., Wei, Z., Chun, C.: Design of a Trust Valuation Model in Software Service Coordination. Journal of Software 14(6), 1043–1051 (2003) (in Chinese)
19. Jia, Z., Changxiang, S., Jiqiang, L., Zhen, H.: A Noninterference Based Trusted Chain Model. Journal of Computer Research and Development 45(6), 974–980 (2008) (in Chinese)

A Novel Application for
Text Watermarking in Digital Reading

Jin Zhang, Qing-cheng Li, Cong Wang, and Ji Fang

Department of Computer Science, Nankai University,
Tianjin 300071, China
{nkzhangjin,liqch}@nankai.edu.cn,
{wangcong6222351,shuiqi}@mail.nankai.edu.cn

Abstract. Although watermarking research has made great strides in theoretical aspect, its lack of application in business could not be covered. It is due to few people pays attention to usage of the information carried by watermarking. This paper proposes a new watermarking application method. After digital document being reorganized with advertisement together, watermarking is designed to carry this structure of new document. It will release advertisement as interference information under attack. On the one hand, reducing the quality of digital works could inhabit unauthorized distribution. On the other hand, advertisement can benefit copyright holders as compensation. Moreover implementation detail, attack evaluation and watermarking algorithm correlation are also discussed through an experiment based on txt file.

1 Introduction

It has been several ten years since digital watermarking technology emerged. Both of technology diversity and theoretical research have made major strides in this field. But its application in business is always a big problem which perplexes the development of watermarking. Many watermarking companies have gone out of business or suspended their watermarking efforts except some focus on tracking and digital fingerprint [1].

This problem is due to, in our opinion, the misunderstanding of the nature of digital watermarking. The real value of digital watermarking is the information bound in watermarking. In other words, watermarking content and application methods are the key issues in digital watermarking research. Unfortunately, the current research in this area is little to write about.

As an important part in digital watermarking, text watermarking encounters the same problem with other kinds of watermarking technology. While most people agree that text watermark should be been widely used in the field of digital reading, but this is not the truth. On one hand lack of appropriate application methods make text watermarking useless. On the other hand, text watermarking is characterized by its attack sensitiveness (this part is discussed in section2). This easily led to watermark information lost.

In this paper, one application method is proposed for text watermarking in digital reading. We combine original digital content and advertisement as a whole digital

H. Deng et al. (Eds.): AICI 2009, LNAI 5855, pp. 103–111, 2009.

document. The rules of combination are embedded in this digital document as watermarking information. Once the document is under attack, advertisement will be released from it. The attacker will have to read them with real content together, which will serve as compensation for the copyright holder.

This paper is organized as follows: in section 2, we review other kinds of text watermarking. And the procedure of our solution is presented generally in section 3. And we expound solution detail and experiment result in section 4. At last we draw the conclusion in section 5.

2 Related Works

In general, the research object of text watermarking is digital document. And document is always made up of two parts, content and format.

Many researchers choose to embed watermarking by taking content as manifestations of language. That is semantics-based watermarking. When the watermarking information is embedded, the content is modified according to linguistic rules. Atallah's TMR is the representative of this kind of watermarking [2].Dai and Yang also proposed their solution [3, 4]. This kind of watermarking technology always needs more time to consume. And it is very sensitive to tamper attack and deleting attack [5]. At the same time, less capacity is another shortcoming of them.

Some other researchers deal with content as a still picture. In this way, they can build additional redundant space to embed watermarking so as to enhance capacity. Brassil and Low's line-shift/word-shift coding and word feature coding are the representative of such solutions [6, 7]. In their solution, content is a binary image. They use lines or words' tiny mobile in different directions to embed information. And Huang complemented their work [8].Liu and Chen's work is based on interconnected domain. They regarded Chinese characters as special images [9]. It is worth mentioning that Li and Zhang propose their method which integrates the ideas of Liu and semantics-based watermarking [10]. They make a significant reduction in overall algorithm consumption. But OCR technology is their natural enemies. No matter how sophisticated these methods are, the majority of watermarking information will be removed after the OCR or retyping attack. Tamper and deleting attack also could reduce their performance.

There are some researchers who hope that watermarking can be hidden in document's format. Fu and Wang's design is a prime example [11]. Zhou and Sun use special characters in RTF file to embed watermark [12]. And some solution is based on redundant coding space, such as Zhou's work [13]. Other techniques such as add blanks and other perceptible characters are also proposed. Although they remove the impact of tamper attack and deleting attack. Any change to the document will have a fatal impact on them, much more OCR and retyping attack.

To sum up, all kinds of watermarking are sensitive to special attack. There is little information could be maintained under the niche attack.

3 Solution Design

As mentioned above, there is none text watermarking technology can resist all kinds of attack. In particular the realization of these attacks is also very low cost. Rather than spending a lot of energy to design a more robust watermark, it is better to find a method which could make attacker wouldn't burn their house to rid it of a mouse.

3.1 Design Principle

The general watermarking embedding procedure is described by (1) [14]:

$$C_{wn} = W \oplus C_0 \oplus n \tag{1}$$

After being embedded watermarking information W, original content C_0 with additive channel noise n will be the last digital works or digital document C_{wn}. Because digital document always has strict coding requirement or format specification such as check code mechanism. We do not consider channel noise issue.

As noted in section 1, watermarking is used to transmit combination rules, while combination of content and advertisement finished. In this case, we modify formula (1) as follows:

$$\begin{cases} C_{wn} = F(W, C_0, C_A) \oplus n_A \\ G(W, C_{wn}) = \begin{cases} C_0, n_A = 0 \\ C_0 \oplus C_A, n_A = else \end{cases} \end{cases} \tag{2}$$

Where C_A is advertisement and n_A is attack noise. We denote C_W as digital document which contain watermarking without being attacked. $F(\ )$ is the combination function, while $G(\ )$ is the striping function.

3.2 Operation Procedure

In this section, we will discuss the operation procedure as figure 1 shows.

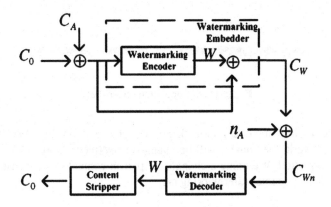

Fig. 1. Design processing flow chart

Step 1: C_0 and C_A combine in accordance with the rules. These rules are encoded *into a special format, which is watermarking information W. In other words, both of C0 and CA are watermarked.*

Step 2: After being embedded W, digital document will be distributed or attacked. In this way attack noise n_A is introduced.

Once attack happened during transmission, n_A will add into document. At this time C_W turns into C_{wn}. As formula (2) shows: When attack misses, n_A is given the value 0. C_W equals to C_{wn} in this condition.

Step 3: When user's device or client software receive C_{wn}, it will extract W and restore C_0 according to W. If n_A equals to 0, user could watch original content. But if watermarking is totally erased, user will have to watch C_A while they are browsing C_0. In the worst case that W is totally distorted, there are only garbled characters left.

4 Implementation and Evaluation

We implement our design on one kind of hand-held digital reading device, and apply it in related digital works release platform. The algorithm in [10] is taken as default algorithm. We use a txt document which contains 10869 Chinese characters as experiment object.

Figure2. (a) shows the document which contains watermarking and advertisement. In this screenshot, it is open directly without stripping. And Figure2. (b) shows the result of opening it after stripping. In Figure2.(b) both of advertisement and watermarking signs are removed by content stripper.

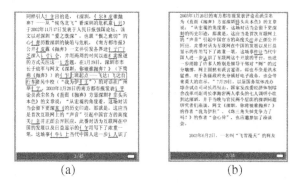

(a)   (b)

Fig. 2. Device operation screenshots

Before discussing implementation detail, we give a definition. When the algorithms as [10] work, the content will be changed selectively. Every change can contain 1 bit information. We name a changeable part in content as a watermarking tag. And we use WT to denote watermarking tag.

4.1 Implementation Detail

4.1.1 Watermarking Embedding Processing

In this part the most important question is the organization of digital document. Out of consideration for distributing watermark and advertisement, the document content structure is designed as figure 3 showing.

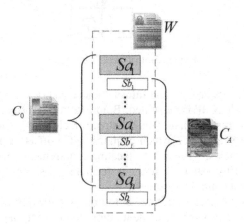

Fig. 3. Document content structure diagram

Original content C_0 is split into N blocks, sizes of which are not equal and are denoted by sequence $\{Sa_i\}$. And sequence $\{Sb_i\}$ is used to denote the sizes of C_A blocks.

These combination rules are described by W, which is made up of sequence $\{W_i\}$. We define W_i as follows (3):

$$W_i = \{l_i, \{C_i, a_i, b_i\}_R, S\} \tag{3}$$

l_i: total length of W_i.

C_i: check code of a_i and b_i.

$\{\}_R$: redundancy construction function. It could be replaced by cryptographic functions.

S: barrier control code. It is used to isolate different W_i. It is also helpful when deleting attack appears. Different code words can also play the role of control information transmission.

$\{a_i\}$: relative length of C_0 block i. It equals to the distance between first WT in this to the last character in this C_0 block.

$\{b_i\}$: length of C_A block i. It usually equals to Sb_i.

Description above is just content structure. After these processing, it could be sealed in other document format, such as DOC, Html. Because of lager consumption of system, this work should be arranged in content servers or distribution servers.

4.1.2 Content Stripping Processing

Generally the main work is extracting watermarking information and restoring original content.

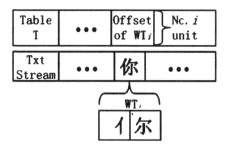

Fig. 4. Table T structure diagram

After scanning C_{Wn}, watermarking information sequence will be gained. According to barrier control code S, W_i is separated. Before restoring original content, the plain txt stream should be restore first. This txt stream includes C_0 and C_A without watermarking. And there is a table T which contains WT's offset in the stream should be setup. As figure 4 shows. This due to locating requirements because every block's address is presented by relative offset, as formula (3) shows. According to T and $\{W_i\}$, original content could be restored.

Actually we can use cache mechanism to optimize the design performance in device. In short, user can browse the content before totally restoring content. When user wants to skip to some pages, device can scan nearest barrier control code S and build temporary table T' to show content quickly. And complete table could be built within user browsing interval.

4.2 Application in Dynamic DRM

Most main DRMs use asymmetric cryptographic algorithm and special document format to ensure security. Our design focuses on content which is independent of document format. And then it could be applied in most DRM smoothly.

Dynamic DRM (DDRM) is a special DRM which support superdistribution and distribution among users [15]. And its technical basis is its license mechanism. And all digital works or digital documents on this platform must be divided into blocks and reorganized. Especially dynamic advertisement block is supported in DDRM. That means the advertisement in digital works could be changed with distribution gong on.

As mentioned in section 3, watermarking is embedded into total content. Both of C_0 and C_A contain watermarking. In this case, device has to reorganize digital document to update C_A after each distribution. Obviously it is very hard for today's embedded devices.

Therefore we have to amend formula (3), and make it turn into another form as (4) shows.

$$C_W = F(W, C_0) \oplus C_A \tag{4}$$

In this condition, only C_0 could be watermarked. And C_A could be replaced dynamically. But there must be two certain rules as follows:

- New C_A block must have the same size with the old one. That is the reason why we usually give C_A block a larger size.
- Validity of C_A should also be checked. And barrier control code S's value is used to indicate whether dynamic C_A is supported.

4.3 Evaluation under Attack

As discussed in section 2, most text watermarking algorithm including our design are attack sensitive. Once attack match with watermarking algorithm's shortcoming, there could be little information left.

Rather than pay attention to more robust algorithm, we focus on risk management after attack happened. Figure 5 shows the result of watermarking information lost in our design.

Fig. 5. Content left after being attack

Obviously advertisement emerges when watermarking information is lost. On one hand, it could reduce digital works quality and inhibit illegal propagation. On the other hand, it could compensate content provider and service provider for the lost.

Technically, the robustness of our design mainly depends on the robustness of watermarking algorithm selected.

4.4 Watermarking Algorithm Correlation

Our research mainly focuses on finding a new method to apply digital watermarking. Therefore the correlation of selection algorithm needs to be discussed.

Though most algorithms could smoothly be applied in our design, we need use different way to split content according to different algorithms.

When one algorithm hiding information in format is selected, splitting method is very easy. If the algorithm tries to regard text as a picture, the descriptors about size or offset should be replaced by the descriptors about coordinates.

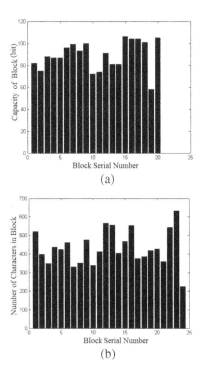

Fig. 6. Capacity distribution

When the algorithm selected involves semantics-based watermarking, we must evaluate the size of W_i in advance. And then the size of block could be set, after checking distribution of watermarking tags. Otherwise it could happen that the size of W_i is larger than the capacity of watermarking space in block.

Figure 6 is the capacity distribution in our experiment. First we divide content into 20 blocks which have the same size. (a) shows its result that the smallest capacity is only 60 bits. (b) shows the situation that every block have the same watermarking capacity. Obviously capacity distribution is not even.

5 Conclusion

As an application technology, digital watermarking research should not only indulge in finding new algorithm or blindly seek technical target. Searching efficient application method maybe more meaningful.

Out of viewpoint above, this paper propose a novel application method for text watermarking in digital reading field. It will release advertisement as interference information under attack. On the one hand, reducing the quality of digital works could inhabit unauthorized distribution. On the other hand, it point out a new way for watermarking application.

References

1. Delp, E.J.: Multimedia security: the 22nd century approach. Multimedia Systems 11, 95–97 (2005)
2. Atallah, M.J., Raskin, V., Crogan, M., Hempelmann, C., Kerschbaum, F., Mohamed, D., Naik, S.: Natural language watermarking: Design, analysis, and a proof-of-concept implementation. In: Moskowitz, I.S. (ed.) IH 2001. LNCS, vol. 2137, pp. 185–200. Springer, Heidelberg (2001)
3. Dai, Z., Hong, F., Cui, G., et al.: Watermarking text document based on statistic property of part of speech string. Journal on Communications 28(4), 108–113 (2007)
4. Yang, J., Wang, J., et al.: A Novel Scheme for Watermarking Natural Language Text: Intelligent Information Hiding and Multimedia Signal Processing. IEEE Intelligent Society 2, 481–488 (2007)
5. Li, Q., Zhang, J., et al.: A Chinese Text Watermarking Based on Statistic of Phrase Frequency: Intelligent Information Hiding and Multimedia Signal Processing. IEEE Intelligent Society 2, 335–338 (2008)
6. Brassil, J., Low, S., Maxemchuk, N., O'Gorman, L.: Electron marking and identification techniques to discourage document copying. IEEE J. Select. Areas Commun. 13, 1495–1504 (1995)
7. Low, S., Maxemchuk, N.: Capacity of Text Marking Channel. IEEE Signal Processing Letter 7(12), 345–347 (2000)
8. Huang, H., Qi, C., Li, J.: A New Watermarking Scheme and Centroid Detecting Method for Text Documents. Journal of Xi'an Jiaotong University 36(2), 165–168 (2002)
9. Liu, D., Chen, S., Zhou, M.: Text Digital Watermarking Technology Based on Topology of Alphabetic Symbol. Journal of Chinese Computer Systems 27(25), 812–815 (2007)
10. Li, Q., Zhang, Z., et al.: Natural text watermarking algorithm based on Chinese characters structure. Application Research of Computers 26(4), 1520–1527 (2009)
11. Fu, Y., Wang, B.: Extra space coding for embedding wartermark into text documents and its performance. Journal of Xi'an Highway University 22(3), 85–87 (2002)
12. Zou, X., Sun, S.: Fragile Watermark Algorithm in RTF Format Text. Computer Engineering 33(4), 131–133 (2007)
13. Zhou, H., Hu, F., Chen, C.: English text digital watermarking algorithm based on idea of virus. Computer Engineering and Applications 43(7), 78–80 (2007)
14. Cox, J., Miller, M., et al.: Digital Watermarking and Steganography, 2nd edn. Morgan Kaufmann, San Francisco (2007)
15. Li, Q., Zhang, J., et al.: A Novel License Distribution Mechanism in DRM System. In: Proceedings of 22nd Advanced Information Networking and Applications - Workshops, pp. 1329–1334. IEEE Computer Society, Los Alamitos (2008)

Optimization of Real-Valued Self Set for Anomaly Detection Using Gaussian Distribution

Liang Xi, Fengbin Zhang, and Dawei Wang

College of Computer Science and Technology, Harbin University of Science and Technology,
150080 Harbin, China
xljyp2002@yahoo.com.cn, zhangfb@hrbust.edu.cn,
stonetools@sohu.com

Abstract. The real-valued negative selection algorithm (RNS) has been a key algorithm of anomaly detection. However, the self set which is used to train detectors has some problems, such as the wrong samples, boundary invasion and the overlapping among the self samples. Due to the fact that the probability of most real-valued self vectors is near to Gaussian distribution, this paper proposes a new method which uses Gaussian distribution theory to optimize the self set before training stage. The method was tested by 2-dimensional synthetic data and real network data. Experimental results show that, the new method effectively solves the problems mentioned before.

Keywords: anomaly detection, artificial immunity system, real-valued self set, Gaussian distribution.

1 Introduction

The anomaly detection problem can be stated as a two-class problem: given an element of the space, classify it as normal or abnormal[1]. There exist many approaches for anomaly detection which include statistical, machine learning, data mining and immunological inspired techniques[2, 3, 4]. The task of anomaly detection may be considered as analogous to the immunity of natural systems, while both of them aim to detect the abnormal behaviors of system that violate the established policy[5,6]. Negative selection algorithm (NSA)[7] has potential applications in various areas, in particular anomaly detection. NSA trains detector set by eliminating any candidate that match elements from a collection of self samples (self set). These detectors subsequently recognize non-self data by using the same matching rule. In this way, it is used as an anomaly detection algorithm that only requires normal data to train.

Most works in anomaly detection used the problem to binary representation in the past. However, many applications are natural to be described in real-valued space. Further more, these problems can hardly be processed properly using NSA in binary representation. Gonzalez et al.[8] proposed the real-value representation for the self/non-self space, which differs from the original binary representation. The real-valued self/detector is a hypersphere in n-dimensional real space; it can be represented by an n-dimensional point and a radius.

H. Deng et al. (Eds.): AICI 2009, LNAI 5855, pp. 112–120, 2009.

As we know that, the detection performance is mainly relied on the quality of detector set: detector coverage of the non-self space. The detector is trained by the self set, so that the self set plays an important role on the whole detection process, and the health of self set is very important to be paid enough attention. It should be noted that, the real-valued self set indeed has some problems which the binary representation has not, such as, the wrong self samples, the overlapping of self samples and the boundary invasion. To solve these problems, this paper proposes a novel method to optimize the self set. Starting from the point that self samples approach the Gaussian distribution as the number of samples is large enough, so this method employs Gaussian distribution theory to deal with the problems mentioned above.

2 Problems with Real-Valued Self Set

2.1 The Wrong Self Samples

Before training detectors, the self set must be secure. However, the real situation may not meet to the requirement all the time, and the self set is likely to collect some wrong samples. The figure 1(a) illustrates the problem of wrong self in 2-dimensional space, there are two samples is so far from the self region, which we can consider as the wrong self. The wrong self samples may result in the holes, so it is important to discard the wrong samples before RNS.

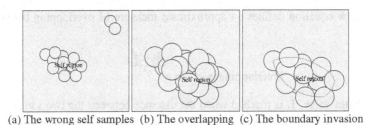

(a) The wrong self samples (b) The overlapping (c) The boundary invasion

Fig. 1. The sketch of real-valued self set problems

2.2 The Overlapping

The overlapping self samples: In a work situation, the self samples distribute intensively. As shown in figure 1(b), a large number of self samples are intensively distribute in the self region, and the overlapping rate is very high, and there are many unnecessary samples in the self region. Now we discuss the relationship between the number of self samples and the candidates. Some variables should be defined as below:

f : The probability of failure matching of a candidate to any self sample
p_f: The probability of an attack that is not detected
p_m : The matching probability
N_s : The number of self samples

N_d: The number of detectors

N_{d_0} : The number of candidates

When the p_m is small enough and the N_s is large enough,

$$f = (1 - p_m)^{N_s} \approx e^{-N_s p_m}.$$

When the N_d is large enough,

$$p_f = (1 - p_m)^{N_d} \approx e^{-N_d p_m}.$$

Because

$$N_d = N_{d_0} * f = \frac{-\ln(p_f)}{p_m},$$

so

$$N_{d_0} = \frac{-\ln(p_f)}{p_m(1 - p_m)^{N_s}} \tag{1}$$

From the equation 1, we can see that the N_{d_0} has exponential relationship with the N_s, so the more self samples, the more candidates and the higher cost of the detector training.

The follow equation defines an approximate measure of overlapping between self samples:

$$Overlapping(s_i, s_j) = e^{-\frac{|s_i - s_j|^2}{r_s^2}} \tag{2}$$

The maximum value, 1, is reached when the distance between the two sample is 0, on the contrary, when the distance is equal to $2\,r_s$, the value of this function is almost close to 0. Based on the equation 2, the amount of overlapping of a self set is defines as

$$Overlapping(S) = \sum_{i \neq j} e^{-\frac{|s_i - s_j|^2}{r_s^2}} , i, j = 1, 2, ..., n. \tag{3}$$

2.3 The Boundary Invasion

As a result of the self's radius, on the border of the self region, the covering area should invade the non-self region's boundary. We can see this phenomenon clearly in the figure 1(c). So when training the detectors by self set, the vicinity of the non-self boundary may not be covered completely.

3   Optimization Method

In view of the analysis above, to optimize the self set is significant. The aim of the optimization is using the least self samples to cover the self region but not the non-self region. The problem of optimization can be stated as follows:
minimize:

$$V(S) = \text{Volume}\{x \in U \mid \exists s \in S, \| x - s \| \leq r_s\} \tag{4}$$

restricted to:

$$\{\forall s \in S \mid \exists d \in D, \| s - d \| \leq r_s\} = \varnothing \tag{5}$$

and

$$\{\forall s_i, s_j \in S, \| s_i - s_j \| \leq r_{s_i} \, or \, \| s_i - s_j \| \leq r_{s_j}\} = \varnothing \tag{6}$$

The function defined in equation 4 represents the amount of the self space covered by self samples. S, which corresponds to the volume covered by the union of hyperspheres associated with each self. The restriction: equation 5 tells that no self should invade the detector set; equation 6 tells that, there is no self sample covered by the other samples in the self set.

As we said before, if the number of self samples is large enough, the probability distribution of self samples approaches the Gaussian distribution. Furthermore, the central limit theorem clarifies that whichever kind of the probability distribution all samples approach, the distribution of the samples' mean is near to a Gaussian distribution; the mean of Gaussian distribution is equal to the mean of all samples.

According to the Gaussian distribution, we can also realize that, the smaller the distance between the sample and the mean point is, the larger the value of the sample's probability density. As discussed before, the nearer the sample is next to the center of the self region, the heavier the overlapping rate is. So we can use this relationship to deal with the problem of the unnecessary self samples. This paper proposes the method is that adjusting every sample's radius by its probability density to deal with the boundary invasion, and then discarding each unnecessary sample by the radius.

There is an important and famous proposition in the Gaussian distribution: the "3σ" Criterion, which shows that each normal sample is almost in the "3σ" district, although the range is from $-\infty$ to ∞ . So we can use the "3σ" Criterion to deal with the problem of the wrong self samples.

As analyzed above, we may draw a conclusion that, the optimization by the probability theory is reasonable. To solve the three problems with self set mentioned above, the optimization is composed by three steps as follows:

Step.1: Discarding the wrong self samples by the "3σ" criterion;
Step.2: Adjusting the self radius by the self's probability density;
Step.3: Discarding the unnecessary self samples by the self radius.

Before describing the optimization, some variables frequently used in the following sections are defined here:

s: Self (Eigenvector: value, Radius: r)
S: Self set (Initial self set: S_0)
N_0: The number of self samples in S_0
N: The number of self samples in S (initial value: N_0)
k: Unit self radius
num: the number of probability density interval
Now the algorithm pseudocode is described as follows:

```
BEGIN
Collection the self samples: S₀ ← s;
//step. 1: discarding the wrong self samples
Regularize S₀, and then compute the μ and σ;
n=0;
while (n<N₀){
    if (sₙ is out of the "3σ" district){
        discard sₙ, N₀--;
    }
n++;
}
// step. 2: adjusting every sample's radius
Compute the S_maxpdf and S_minpdf;
L= (maxpdf-minpdf)/num;
while (n<N₀){
    l = (pdf_{sₙ} - min pdf)/L ;
    Sₙ.r=l×k;
}
// step. 3: discarding the unnecessary self samples
S ← s₀, N=1, flag=0;
while (n<N₀){
    i=0;
    while (i<N){
        d =‖ sₙ - sᵢ ‖ ;
        if (d<sₙ.r || d<sᵢ.r){
        flag=1, break;
        }
        i++;
    }
    if (flag==0){
        S ← sᵢ, N++;
    }
    n++;
}
END
```

In the step.1, the self sample should be discarded if its probability density is out of the "3σ" district. The time complexity of this step is $O(n)$.

In the step.2, the nearer self samples are closed to the center of self region, the larger the radii of them. So the radii of self samples near the boundary of self region are adjusted to be a relatively rational level so that the boundary invasion is avoided, while the around the central region, each sample's radius is adjusted to be a relatively larger level than before so that the overlapping is severer than before. The time complexity of this step is $O(n)$.

In the step.3, by discarding the unnecessary samples, the problem with more overlapping in the step.2 of optimization is solved. The samples with larger level radius can cover more space than the fixed radius samples before, so that, the self region can be covered by fewer samples than before. The time complexity of this step is $O(n^2)$.

Via the optimization, the self set is more rational than before. The total time complexity of optimization is $O(n^2)$.

4 Experiments and Results

To test the algorithm described in the previous section, experiments were carried out using 2-dimensional synthetic data and real network data.

4.1 2-Dimensional Synthetic Data

Firstly, some self samples chosen from a fixed number of random points, which fit with the normal distribution, are normalized to compose the self set (198 samples, radius is fixed as 0.05). So this set has no wrong samples, and the step.1 optimization will be illustrated in the real network experiment. Each result of optimization is shown by the figure 2.

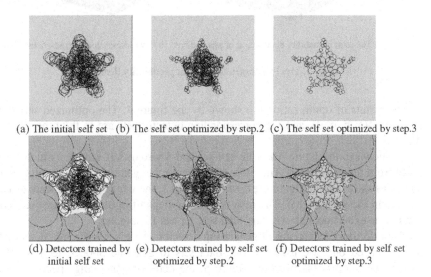

(a) The initial self set   (b) The self set optimized by step.2   (c) The self set optimized by step.3

(d) Detectors trained by   (e) Detectors trained by self set   (f) Detectors trained by self set
    initial self set   optimized by step.2   optimized by step.3

Fig. 2. Experiment results on 2-dimensional synthetic data

The figure 2(a) shows the initial self set, and we can see that the overlapping and the boundary invasion are extremely severe. The detectors trained by the initial self set are shown in figure 2(d), and we can see clearly that the boundary of the non-self is not covered completely which is the consequence of the self samples boundary invasion. Likewise, the optimized self set and its detector set are shown by the other figures of the figure 2 (the unit radius is set as 0.005). After all the steps of optimization, the number of optimized self is dropped to 67, and the overlapping(S) is dropped to 2.71% (by the equation 3). As the figures show, adjusting each sample's radius avoids the boundary invasion so that the non-self region's boundary is covered well; and then the coverage of self region by the most reasonable self samples almost staying the same as before.

4.2 Real Network Data

In the real network experiment, the self samples are come from the network security Lab of the HUST (Harbin University of Science and Technology). The value is 2-dim: pps(packages per second) and bps(bytes per second). We collected 90 samples and drew the joint probability distribution in figure 3. And we can see that the joint distribution is quite similar to the standardized Gaussian distribution.

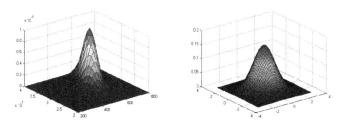

(a) The joint probability distribution of pps and bps (b) Two-dimensional standardized Gaussian

Fig. 3. The two-dimensional joint probability distribution

The results of optimization are shown by the figure 4. The optimized self sets are in figure 4(a, b, c, d), and the corresponding detector sets are in the other figures of figure 4.

As shown in the figure 4(a), the self set had 3 wrong self samples which were far away from the self region. Therefore, the detector set generated by the initial self set has holes shown in the figure 4(e). After the step.1, the wrong samples were discarded (shown in the figure 4(b)). The results of other steps of optimization are just like the simulation experiment above. The final self set has 19 samples.

To perform a comparison of generating detector set between the initial self set and the optimized one, we used the two data sets to generate 3000 detectors. We can clearly see the process from the figure 5, the detector generating speed using optimized self set is faster than using initial self set.

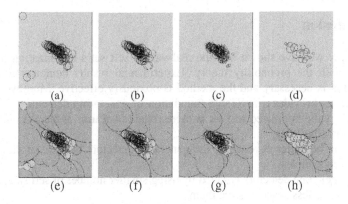

Fig. 4. Real network experiment results

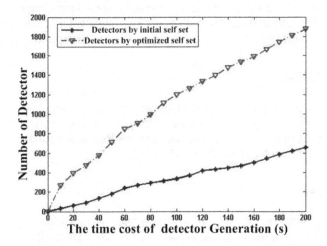

Fig. 5. The detector generating speed comparison between initial and optimized self set

To perform a comparison of the detection effect of the two detector sets, we used the SYN flooding to attack the two hosts with different detector sets simultaneously. All the results shown in the table 1 are average of 100 times experiment. From the table we can see clearly that the detection rate with detectors using optimized self set is much higher than the other (98.9% vs. 83.7%) and the false alarm rate is remarkable lower than the other (2.3% vs.8.3%).

Table 1. The Detection Comparison Between The Two Detector Sets

Detector Set	Connection	Invasion	Detecting	Detection Rate	False Alarm Rate
By initial	13581653.3	100	92.5	83.7%	8.3%
By optimized	13581653.3	100	101.2	98.9%	2.3%

5 Conclusion

In this paper we have shown that the real-valued self set's optimization in anomaly detection with the probability theory. Experimental results demonstrated that this optimization is necessary and the method is obviously effective. Moreover, the optimized self set provides a favorable foundation of generating detectors. The detector set trained by the optimized self set is more reliable because of the advantages of the optimization algorithm:

1. Holes can be better covered by the generated detectors because of discarding the wrong self samples.
2. Time to generate detectors is saved by using smaller number of self samples, which also requires less space to store them.
3. The radius of the border self samples is adjusted to a rational range so that the boundary invasion is solved and the false alarm rate is reduced obviously.

The influence of real-valued self set optimization needs further study, including more real network experiment and strict analysis. The implication of self radius, or how to interpret each self sample, is also an important topic to be explored.

Acknowledgment

This work was sponsored by the National Natural Science Foundation of China (Grant No.60671049), the Subject Chief Foundation of Harbin (Grant No.2003AFXXJ013), the Education Department Research Foundation of Heilongjiang Province (Grant No. 10541044, 1151G012).

References

1. Patcha, A., Park, J.M.: An overview of anomaly detection techniques: existing solutions and latest technological trends. Computer Networks 51, 3448–3470 (2007)
2. Hofmeyr, S., Forrest, S.: Architecture for an artificial immune system. IEEE Transactions on Evolutionary Computation 8, 443–473 (2000)
3. Simon, P.T., Jun, H.: A hybrid artificial immune system and self and self organising map for network intrusion detection. Information Sciences 178, 3024–3042 (2008)
4. Forrest, S., Perelson, A.S., Allen, L.: Self-non-self discrimination in a computer. In: Proc. of the IEEE Symposium on Research in Security and Privacy, pp. 202–212 (1994)
5. Dasgupta, D., Gonzalez, F.: An immunity based technique to characterize intrusions in computer network. IEEE Transactions on Evolutionary Computation 69, 281–291 (2002)
6. Boukerche, A., Machado, R.B., Juca, R.L.: An agent based and biological inspired real-time intrusion detection and security model for computer network operations. Computer Communications 30, 2649–2660 (2007)
7. Ji, Z., Dasgupta, D.: Real-valued negative selection algorithm with variable-sized detectors. In: Deb, K., et al. (eds.) GECCO 2004, Part I. LNCS, vol. 3102, pp. 287–298. Springer, Heidelberg (2004)
8. Gonzalez, F., Dasgupta, D., Kozema, D.: Combining negative and classification techniques for anomaly detection. In: Proc. of the 2002 Congress on Evolutionary Computation, pp. 705–710 (2002)

A GP Process Mining Approach from a Structural Perspective

Anhua Wang, Weidong Zhao*, Chongchen Chen, and Haifeng Wu

Software School, Fudan University, 220 Handan Rd.
200433 Shanghai, China
{072053026,wdzhao,07302010048,082053025}@fudan.edu.cn

Abstract. Process mining is the automated acquisition of process models from event workflow logs. And the model's structural complexity directly impacts readability and quality of the model. Although many mining techniques have been developed, most of them ignore mining from a structural perspective. Thus in this paper, we have proposed an improved genetic programming approach with a partial fitness, which is extended from the structuredness complexity metric so as to mine process models, which are not structurally complex. Additionally, the innovative process mining approach using complexity metric and tree based individual representation overcomes the shortcomings in previous genetic process mining approach (i.e., the previous GA approach underperforms when dealing with process models with short parallel and OR structure, etc). Finally, to evaluate our approach, experiments have also been conducted.

1 Introduction

Today, information about business processes is mostly recorded by enterprise information systems such as ERP and workflow management systems in the form of so-called "event logs" [1]. As in many domains processes are evolving and become more and more complex, there is a need to understand the actual processes based on logs. Thus process mining is employed since it automatically analyzes these logs to extract explicit process models. Currently, most of process mining techniques try to mine models from the behavior aspect only (i.e., reflect the exact behavior expressed in logs) while ignore the complexity of mined models. However, as processes nowadays are becoming more complex, process designers and analysts also consider mining structurally simple models since complex processes may result in errors, bad understandability, defects and exceptions [2]. Therefore, "good" process models are required to conform to logs and somehow have simple structure to clearly reflect the desired behavior [3]. Thus to consider both behaviorally and structurally, we propose a genetic process mining approach coupled with a process complexity metric.

In fact, utilizing evolutionary computation in process mining research is not a new concept. In 2006, Alves introduced the genetic algorithm (GA) to mine process models due to its resilience to noisy logs and the ability to produce novel sub-process

---

* Corresponding author.

H. Deng et al. (Eds.): AICI 2009, LNAI 5855, pp. 121–130, 2009.

combinations [4]. In this case, an individual is a possible model and the fitness evaluates how well an individual is able to reproduce the behavior in the log. However, since the individual was abstracted to the level of a binary string, it had problems when mining certain processes, especially those exhibiting high level of parallel execution. Thus, a genetic programming (GP) approach coupled with graph-based representation was proposed in [5]. Its abstraction of a directed graph structure also allows greater efficiency in evaluating individual fitness.

However, there are two main problems in these approaches: Firstly some relations between process tasks are not considered in the individual representation; the other is that they neglect the complexity of mined models. Thus we improve the GP approach by extending the individual representation in [4] and define a new part of the fitness measure that benefits individuals with simpler structure. The new structural fitness is based on the structuredness metric (SM) proposed in [6], which is a process complexity metric. Our paper is organized as follows: Section 2 defines the individual representation and section 3 extends the SM. The GP approach is shown in section 4 with experiments discussed in section 5. Finally, conclusions are drawn in section 6.

2 Extended Causal Matrix

In GP, an individual is an effectively executable program and its representation defines the search space. The main requirement for the representation is to express the dependencies between process tasks in logs. In our improved GP, programs are expressed as extended causal matrix (ECM), which extends from the causal matrix (CM) representation in [4]. The CM shows which tasks directly enable the execution of other tasks via the matching of input (I) and output (O) condition functions. The relationship between tasks supported are restricted to 'AND' and 'XOR' only.

Table 1. Log of the application example

Id	numSimilar	Process instances
1	30	A, E, F
2	45	A, B, C, D, F
3	52	A, C, B, D, F

As we can see, without supporting other business logic, previous GA and GP approaches may cause unexpected errors. For instance, consider the event log shown in Table 1. It shows three different process instances for job applicants. Note that when HR screens resumes (A), he/she will decline the resume (E) or arrange examinations including interview (B) and a written test(C) and score the applicant (D) when his/her resume is passed. Finally, the result is sent to the applicant (F).

The result model is expected to be Fig. 1(a), where the black rectangle is an implicit task. However, previous approaches will wrongly generate Fig. 1(b) since the individual's representation in these approaches cannot express the relation of exclusive disjunction of conjunctions. Therefore, we give an improved version of the causal matrix that supports other relations, i.e., the extended CM (ECM for short).

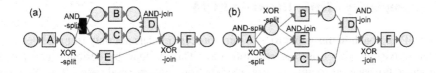

Fig. 1. Expected model (a) and mined result using previous GA and GP (b)

ECM can be denoted as a quadruple (T, C, I, O), where T is a finite set of activities (tasks) appeared in workflow logs, C $\subseteq$ T× T is the causality relation and I and O are the input and output condition functions respectively.

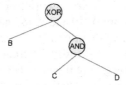

Fig. 2. An example of the tree representation

Note that in previous GA approaches, condition functions in CM are expressed as lines of code. Instead, our GP approach utilizes a syntax tree structure with maximum depth 2. Given a task t, its input and output condition functions are called input tree and output tree respectively. The tasks in I(t)/O(t) are leaves of the corresponding tree (if there exists only one task, it is the root), while operations (XOR, AND and OR) are internal nodes. The expression can also be represented in a prefix form. For example, Fig. 2 shows the tree representation of a prefix-notation program XOR (B, AND(C, D)), we also call it a XOR tree since the root operation is XOR. Apparently, ECM is a general version of CM since more task relations can be reflected by ECM. The ECM for our example is shown in Table 2 below.

Table 2. An example of ECM

TASK	I(input tree)	O(output tree)
A	()	XOR(B, AND(C, D))
B, C, D	(A)	(F)
E	AND(C, D)	(F)
F	XOR(B, E)	()

The reason why we choose the tree type in ECM is that it provides the feasibility to express the relation OR. Moreover, it facilitates some operations such as crossover and mutation since it allows entire sub-tree structures to be swapped over with ease. Note that there may be duplicate tasks in sub trees. Similar to original CM, our extended CM can also deal with the loop structure.

3 Complexity Metric Based on ECM

The GP approach in this paper tries to mine process models not only from the behavior perspective but also from the structure one. This is done by defining new partial structure fitness so as to mine process models from the structural perspective. We need first introduce a process complexity metric based on ECM. Until now, many complexity metrics of process models have been proposed and we choose the structuredness metric (SM)[6] in this paper. The SM focuses on the model structure. Additionally, both the GA mining algorithm and SM have been implemented in the context of ProM, which is characterized by a pluggable framework for process analysis, thus some modifications can be made to implement the improved GP.

In this section, we extend the Petri net version of SM in [6] to propose a ECM based SM. The idea behind this metric stems from the observation that process models are often structured in terms of the combination of basic patterns such as sequence, choice, parallelism and iteration [7]. To calculate the complexity, we define the component as a sub ECM to identify different kinds of structures in the ECM and then score each structure by giving it some "penalty" value. Finally, the sum of these values is used to measure the complexity of process models, i.e., the complexity of the individual in the GA. To understand the idea, some definitions are given below:

Firstly, some structural properties based on ECM are discussed.

Definition 1. Free choice. An ECM is a free-choice ECM iff for every two tasks t_1 and t_2, $I(t_1) \cap I(t_2) \neq$ null implies $I(t_1) = I(t_2)$

Definition 2. State machine. An ECM is a State machine iff for every task t, there exists no AND/OR in $I(t)$ and $O(t)$.

Definition 3. Marked graph. An ECM is a marked graph iff for every task t, there exists no XOR in $I(t)$ and $O(t)$.

A component corresponds to a behavioral pattern and is basically an ECM

Definition 4. Component. Let ECM(T, I, O) be an ECM representation of an individual process model, C is a component iff C is a sub closure of the ECM.

The definition of a component is used to iteratively identify patterns. To define different patterns, the following notations are needed..

Definition 5. Component notations. Let ECM(T, I, O) be a process model.
- $ECM\|_C$ is a function that maps a component to an ECM.
- [ECM] is the set of non-trivial components of ECM.

To find components, we give the following definitions.

Definition 6. Annotated ECM. ECM=(T, I, O, τ) is an annotated ECM iff (T, I, O) is an ECM, and $\tau : T \to \Re^+$ is a function that assign a penalty to each task in ECM. Note that initially for each task t in ECM, $\tau(t)$ is 1.

Definition 7. Component types. Let (T, I, O) be a ECM, $C\square[ECM]$ a component in ECM, and $ECM\|_C=(T_C, I_C, O_C)$, C is a:

- Sequence component, iff $ECM\|_C$ is a state machine and a marked graph.
- Maximal Sequence component iff for $\forall C' \in [ECM]$, C' is a sequence component: $C \neq C' \Rightarrow C \not\subset C'$.
- CHOICE component, iff every task shares the same I and O.
- WHILE component, iff T is a triple set $T(t_i, t, t_o)$ and $O(t_i) = XOR(t, t_o)$, $I(t) = t_i$, $O(t) = t_o$, $I(t_o) = XOR(t_i, t)$.
- MARKED GRAPH component, iff $ECM\|_C$ is a marked graph ECM.
- MAXIMAL MARKED GRAPH component iff for $\forall C' \in [ECM]$, C' is a MAKED-GRAPH component: $C \neq C' \Rightarrow C \not\subset C'$.
- STATE MACHINE component, iff $ECM\|_C$ is a state machine ECM.
- MAXIMAL STATE MACHINE component, iff for $\forall C' \in [ECM]$, C' is a STATE MACHINE component: $C \neq C' \Rightarrow C \not\subset C'$.
- WELL-STRUCTURED component, iff for every task t in ECM, if $O(t)$ is a XOR-tree, every task in $O(t)$ shares the same input; if $I(t)$ is a AND-tree/OR-tree, there is no XOR in its branches; no cycles.
- MAXIMAL WELL-STRUCTURED component, iff for $\forall C' \in [ECM]$, C' is a WELL-STRUCTURED component: $C \neq C' \Rightarrow C \not\subset C'$.
- UNSTRUCTURED-component, if none of above definitions applies for C.
- MINIMAL UNSTRUCTURED component, iff for $\forall C' \in [ECM]$, C' is a UNSTRUCTURED component: $C \neq C' \Rightarrow C \not\subset C'$.

To calculate the ECM-based SM, we try to identify atomic patterns in the form of component types defined above and then iteratively replace them by tasks. Each time when a component is removed, the penalty value of this component is calculated and associated with a new task. By doing this, ECM is reduced in several steps until the trivial ECM with just one task. The SM is then associated with this task.

Next, we introduce the priority of a component type and its corresponding penalty value, and finally the algorithm that calculates the complexity is given. The order in which components are reduced to tasks is as follows: MAXIMAL SEQUENCE component, CHOICE component, WHILE component, MAXIMAL MARKED GRAPH component, MAXIMAL STATE MACHINE component, MAXIMAL WELL STRUCTURED component and other components. Thus we first select a MAXIMAL SEQUENCE component which is endowed with the highest priority.

Definition 8 Component penalty function. Let $ECM = (T', I', O', \tau)$ be an annotated ECM, we define the component penalty $\rho_{ECM}:[ECM] \to \Re^+$ as follows. For any $C \in [ECM]$, the penalty value associated to C is given in Table 3.

In Table 3, diff(P) is |totalNum of XOR in $O(T)$ - totalNum of XOR in $I(T)$|, and diff(T) is |totalNum of AND\OR in $O(T)$ – totalNum of AND\OR in $I(T)$|. For details on the settings of different component weight, we can refer to [6].

The algorithm that calculates the SM based on ECM is shown below.

- The program calculates the ECM of an individual

Table 3. The penalty value of different component type

C (Component)	$\rho_{ECM}(C)$
MAXIMAL SEQUENCE	$\sum_{t \in T} \tau(t)$
CHOICE	$1.5 \cdot \sum_{t \in T} \tau(t)$
WHILE	$\sum_{t \in I(C), O(C)} \tau(t) + 2 \cdot \sum_{t \in T} \tau(t)$
MAXIMAL MARKED GRAPH	$2 \cdot \sum_{t \in T} \tau(t) \cdot diff(T)$
MAXIMAL STATE MACHINE	$2 \cdot \sum_{t \in T} \tau(t) \cdot diff(P)$
MAXIMAL WELL STRUCTURED	$2 \cdot \sum_{t \in T} \tau(t) \cdot diff(P) \cdot diff(T)$
otherwise	$5 \cdot \sum_{t \in T} \tau(t) \cdot diff(P) \cdot diff(T)$

```
(i)   X:=(ECM, τ ), where τ (t) = 1, ∀t∈T,
(ii)  while [X] ≠ ∅ (i.e., X is not reduced to a ECM with
just a single task)
(iii) pick C so that C is the first priority in every
component C'∈[X],
(iv)  yield ECM' by removing C from ECM and replace it by
a new task tc
(v)   τ'(tc)= ρ_x(C) and τ'(c)= τ (c) for all other t,
(vi)  X:=(ECM', τ'),
(vii) Output SM(ECM)= τ(t) (T={t} after the ECM is reduced).
```

4 Genetic Process Mining Algorithm

In this section, we explain how our GP-based process mining works, and show its ability of supporting more combined constructs and mining "uncomplex" process models. There're six main stages in our GP process mining algorithm: 1) Read work-flow log; 2) Calculate dependency relations among tasks; 3) Build initial population; 4) Calculate individuals' fitness; 5) Return the fittest individuals; 6) If generation limit is reached or optimum individual is found (i.e. fitness = 1.0), stop and return, else create next population (using elitism, crossover and mutation) and do 4).

4.1 Initial Population

We build the initial causal matrices in a heuristic way, which tries to determine the causality relation by utilizing a dependency measure, which aims to ascertain the strength of the relationship between tasks by calculating the amount of times one task is directly preceded by another. The measure is also able to determine which tasks are in a loop.

Once the causality relation is determined, the condition functions I and O are randomly built. i.e., every task t_1 that causally precedes a task t_2 is randomly inserted as a leaf in the tree structure, which corresponds to the input condition function of t_2. Operators (AND, OR and XOR) are also inserted randomly to construct a tree. A similar process is done to set the output condition function of a task.

4.2 Fitness Function

Fitness function is used to assess the quality of an individual and this is assessed by:

1) benefiting the individuals that can parse more event traces in logs (the "completeness" requirement);

2) punishing the individuals that allow for more extra behavior than the one expressed in logs (the "preciseness" requirement);

3) punishing the individuals that are complex, i.e., individuals with high SM value(the "uncomplexity" requirement). Equation (1) depicts the fitness function of our GP-based process mining algorithm, in which L is event logs and ECM is an individual. The notation ECM[] represents a generation of process models. Three partial fitness measures are detailed below.

$$Fitness(L, ECM, ECM[]) = PF_{complete}(L, ECM) \tag{1}$$
$$-\kappa * PF_{precise}(L, ECM, ECM[])$$
$$-\chi * PF_{uncomplex}(ECM, ECM[])$$

In equation (1), the functions PFcomplete and PFprecise derived from the previous GA approach are measured from the behavior perspective for completeness and preciseness. Since ECM is a general version of CM, the functions are nearly the same as those in previous GA approach [8].

The function PFcomplete shown in equation (2) is based on the continuous parsing of all the traces against an individual in event logs. In this equation, all missing tokens and extra tokens left behind act as a penalty in the fitness calculation. More details can be found in [8] Obviously, 'OR' construct parsing should also be considered.

$$PF_{complete}(L, ECM) = \tag{2}$$
$$\frac{allParseTasks(L, ECM) - punishment(L, ECM)}{numTasksLog(L)}$$

In terms of preciseness, the individual should contain "less extra behavior", i.e., the individual tends to have less enabled activities. PFprecise in equation (3) provides an measure of the amount of extra behavior an individual allows for in comparison to other individuals in the generation [8].

$$PF_{precise}(L, ECM, ECM[]) = \tag{3}$$
$$\frac{allEnabledTasks(L, ECM)}{max(allEnabledTasks(L, ECM[]))}$$

As mentioned before, the GA approach and most other process mining techniques focus on the behavior in logs but ignore the structure of the actual individual. Thus we defined the uncomplex fitness in equation (4). It gives the transformation of the structuredness metric values to the interval [0, 1], where max and min are the maximum and minimum value of the entire generation of individuals respectively.

$$PF_{uncomplex}(ECM, ECM\,[]) = \qquad\qquad (4)$$

$$\frac{SM\,(ECM) - \min(SM\,(ECM\,[]))}{\max(SM\,(ECM\,[])) - \min(SM\,(ECM\,[]))}$$

4.3 GP Operators

Genetic operators such as selection, crossover and mutation are used to generate individuals of the next generation, and the operators in [8] have been adapted for our GP approach. In our case, the next population consists of the best individuals and others generated via crossover and mutation. And parents are selected from a generation by playing a five individual tournament selection process.

In terms of crossover, a task, which exists in both parents, is selected randomly as the crossover point. The input and output tree of the task are then split at a randomly chosen swap point (a set of tree nodes) for each parent. Our crossover algorithm covers the complete search space defined by ECM. The following is an illustration of the crossover algorithm, which shows the input trees of two parents for task K.

- Crossover algorithm in our GP process mining approach

```
parentTask1 = AND(XOR(A,B,C),OR(B,D),E)
parentTask2 = OR(AND(A,B),XOR(C,D),F)
swap1 = XOR(A,B,C),OR(B,D); remainder1 = E
swap2 = AND(A,B),F; remainder2 = XOR(C,D)
If crossover, for each branch in swapset1
   If the selected branch shares the same operation with one
of the remainder2 (e.g., XOR(A,B,C) and XOR(C,D) are both
XOR), with the equal probability, select one of the
following three crossover methods:
      i) Add it as a new branch in remainder
      (e.g., XOR(A,B,C),XOR(C,D))
      ii) Join it with an existing branch in remainder2
      (e.g., XOR(A,B,C,D))
      iii) Add as a new branch, and then select the branch
with the same operation in remainder2 and remove the tasks
that are in common(i.e., the same tasks)
      (e.g., XOR(A,B,C),D)
   Else select one crossover method conditionally.
      If the operation of selected branch is the same as the
root of parentTask2
         Add as a new leaf of the root in remainder2 for each
task in selected branch
         (e.g., XOR(C,D),B,D)
      Else add as a new branch in remainder2
Repeat for the combination swap2 and remainder1
```

In the crossover code, some examples are shown to make sure you can understand. After execution, the result child for input2(K) may be one of these: OR(XOR(A,B,C),XOR(C,D),B,D);OR(XOR(A,B,C,D),B,D); OR(XOR(A,B,C),B,D). The cycle is repeated for both input and output trees of the selected task.

In the mutation, with probability mutation rate, every task in the individual is mutated for its input and output tree. The following program depicts the mutation.

- Mutate algorithm in our GP process mining approach

```
If mutation, for each task t in the individual (Assuming I(t)
= AND(XOR(A,B,C),OR(B,D),E)), one of the following opera-
tions are done:
    i) Choose a branch and add a task to it (randomly chosen
from the complete set of available tasks in the individual),
if the branch is a leaf (i.e., one task), randomly add an
operation on the new branch.
    (e.g., XOR(A,B,C),OR(B,D,F),E)
    ii) Remove a task from a chosen branch
    (e.g., XOR(A,B,C),OR(B,D))
    iii) Change the operation for the chosen branch
    (e.g., XOR(A,B,C),XOR(B,D),E)
    iv) Redistribution the elements in I(t)
    (e.g., XOR(B,C),OR(A,B),D,E)
Repeat it for output trees of each task
```

Both crossover and mutation operators utilize a repair routine executed after an input/output has been changed to make sure that only viable changes are made.

5 Experiments

We have implemented our GP process mining approach in Java and plug it into the ProM framework. The experiments in this section allow us to validate the approach. Table 4 shows the parameters being used in the experiments.

Table 4. Parameters for the experiment

Population size	100
Maximum generations	1000
Elitism rate	0.02
Crossover rate	0.8
Mutation rate	0.02
κ (extra behavior fitness reduction)	0.025
χ (complex fitness reduction)	0.025

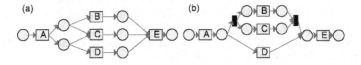

Fig. 3. Two mined models (a) using previous GA/GP approach; (b) using our GP approach

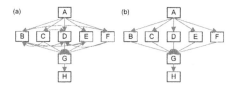

Fig. 4. Another two mined models (a) using previous GA/GP approach; (b) using our GP

To compare with the previous GA/GP approach, we have tested various workflow logs. For short, we give two typical comparisons. Fig. 3(b) is the expected model and our GP approach has successfully mined it while previous approaches produced a confusing model since our GP approach supports the task relation of disjunction of conjunctions. Fig. 4 shows two mined results when mining process with short parallel. And it illustrates that the structurally uncomplex fitness help our GP method outperforms when dealing with process models containing short parallel constructs.

6   Conclusions

The advantage of our GP process mining approach is that it can mine process behaviorally and structurally. This owes to the novel idea of combining the genetic programming with the structuredness metric. Additionally, the extended CM for representing the individuals provides benefits in crossover and mutation. Thus, our GP approach outperforms other process mining approach in some aspects.

References

[1] van der Aalst, W.M.P., van Dongen, B.F., Herbst, J., Maruster, L., Schimm, G., Weijters, A.J.M.M.: Workflow mining: a survey of issues and approaches. Data & Knowledge Engineering 47(2), 237–267 (2003)
[2] Cardoso, J.: Control-flow complexity measurement of processes and Weyuker's properties. Transactions on Enformatika, Systems Sciences and Engineering 8, 213–218 (2005)
[3] Rozinat, A., van der Aalst, W.M.P.: Conformance Testing: Measuring the Fit and Appropriateness of Event Logs and Process Models. In: Bussler, C.J., Haller, A. (eds.) BPM 2005. LNCS, vol. 3812, pp. 163–176. Springer, Heidelberg (2006)
[4] Alves de Medeiros, A.K., Weijters, A.J.M.M.: Genetic Process Mining. Ph.D Thesis, Eindhoven Technical University, Eindhoven, The Netherlands (2006)
[5] Turner, C.J., Tiwari, A., Mehnen, J.: A Genetic Programming Approach to Business Process Mining. In: GECCO 2008, pp. 1307–1314 (2008)
[6] Lassen, K.B., van der Aalst, W.M.P.: Complexity metrics for Workflow nets. Information and Software Technology 51, 610–626 (2009)
[7] van der Aalst, W.M.P., ter Hofstede, A.H.M., Kiepuszewski, B., Barros, A.P.: Workflow patterns. Distributed and Parallel Databases 14(1), 5–51 (2003)
[8] Alves de Medeiros, A.K., Weijters, A.J.M.M., van der Aalst, W.M.P.: Genetic process mining: an experimental Evaluation. Journal of Data Mining and Knowledge Discovery 14(2), 245–304 (2007)

Effects of Diversity on Optimality in GA

Glen MacDonald and Gu Fang

School of Engineering, University of Western Sydney,
Locked Bag 1797, Penrith South DC, NSW 1797, Australia
canti_mac@hotmail.com, g.fang@uws.edu.au

Abstract. Genetic Algorithms (GA) is an evolutionary inspired heuristic search algorithm. Like all heuristic search methods, the probability of locating the optimal solution is not unity. Therefore, this reduces GA's usefulness in areas that require reliable and accurate optimal solutions, such as in system modeling and control gain setting. In this paper an alteration to Genetic Algorithms (GA) is presented. This method is designed to create a specific type of diversity in order to obtain more optimal results. In particular, it is done by mutating bits that are not constant within the population. The resultant diversity and final optimality for this method is compared with standard Mutation at various probabilities. Simulation results show that this method improves search optimality for certain types of problems.

Keywords: Genetic Algorithms, diversity, mutation, optimality.

1 Introduction

Genetic Algorithm (GA) is an Evolutionary Computation (EC) method proposed by Holland [1] which searches the binary space for solutions that satisfy pre-defined criteria. GA is a heuristic search method that contains three operators: *Selection*, *Combination* and *Mutation*. It operates on a collection of individuals (populations). A population contains individuals which have the potential to differ from each other.

The GA works by identifying individuals which most closely satisfy pre-defined criteria, called selection. These selected individuals (the fittest) are combined. Each combination cycle is called a generation. To create new individuals, the combined individuals must differ. The amount of difference in the population is called its *diversity*. Combination occurs by swapping parts between the individuals. The combination methods can vary significantly [2-6]. Goldberg [3] suggested Single Point Crossover (SPC) and Multi-Point Crossover (MPC) as pattern identifying crossover methods. Punctuated Crossover [4] is a deterministic crossover method which alters the probability based on the success of previous crossover points. Random Respectful Crossover [5] produces offspring by copying the parts at positions where the parents are identical and filling the remaining positions with random members from the bit set. Disrespectful Crossover (DC) [6] is a variant on SPC, where the bits in the section below the crossover point are inverted if they are different or have a 50% chance of being in either state if they are the same. DC is essentially a localized high probability mutation algorithm incorporated into a crossover function. Its purpose is in encouraging search divergence, not search convergence. Reduced Surrogate [4] randomly

H. Deng et al. (Eds.): AICI 2009, LNAI 5855, pp. 131–140, 2009.

selects bits which are different as the crossover points for single or double point crossover. This has the advantage of not performing redundant crossovers and is a more efficient crossover as search diversity decreases. Hoshi [2] showed that encouraging similarly ranked individuals to combine resulted in superior solutions.

If the combination of two different individuals can lead to all individuals in the search space then it is possible that the globally optimal solution will be found. However, in GA, like in biology, Combination and Selection are what makes the search converge, as traits in individuals become more common after successive generations. It is believed that these traits are what make the individuals 'fit'. However these 'traits' could be sub-optimal. If this is the case, a population will fail to continually improve and will have reached a locally optimal solution. This means that the GA may not produce an optimal solution for the problem, i.e., the global optima.

To encourage the population to avoid local optima, individuals are altered in a probabilistic manner, called mutation [3]. This is modeled on biological mutation, which is the alteration of individual genes within an organisms DNA. Commonly, mutation is implemented in GA by probabilistically changing bit states, whereby every single bit in an individual has a specific probability of changing state. This type of mutation is also known as De-Jong mutation [7].

The effectiveness of De-Jong mutation has proven to be dependent on the mutation probability and the effectiveness of a mutation probability has been shown to be problem dependent [8-13]. The mutation probability affects the probability of an individual being mutated as well as the number of bits mutated within that individual. A high mutation probability means that an individual has a high chance of having a large number of bits altered; this results in a highly random search which mostly ignores convergence through crossover. A low mutation probability means that an individual has a low chance of having a small number of bits altered; this creates a search which is highly reliant on combination.

Searches which ignore combination do not converge. Searches which are entirely reliant on combination are prone to premature convergence. Mutation alone amounts to a random search and forever diverges. Mutation when combined with selection and combination is a parallel, noise tolerant, hill-climbing algorithm [8]. Hill-climbing algorithms guarantee the location of a local optimum, but will be deceived by a problem with many local optima. Methods of determining the mutation probability based on generation [9], population size [10], length [11], crossover success and the rate of change in fitness have been investigated [12]. A method which prolongs the GA search time by systematically varying the mutation rate based on population diversity has significantly improved the final solution optimality [13]. These mutation strategies have improved the GA in a problem independent fashion. However they do not present any significant improvement to the ideology behind mutation. They do not address the search algorithm, they primarily present methods of altering one of the algorithms parameters in order to prolong the search or more reliably disrupt convergence. This is unlikely to make the search more efficient or more robust but rather just make it more exhaustive. They also do not allow convergence and divergence to be complimentary.

In this paper a new mutation method is introduced to improve the diversity of population during the convergence. This method also intends to ensure that the convergence is not disrupted by the mutation process.

The remainder of the paper is organized as follows: section 2 details the methodology used for maintain the diversity while ensuring convergence. The numerical results are shown in section 3. The conclusions are then given in section 4.

2   The Proposed Mutation Method

2.1   Related Work

The difference between individuals in GA is primarily measured by the Euclidean distance or Hamming Distance [14]. Hamming Distance [15] is the number of bits which differ between the individuals. Diversity for a population is generally the average of the distance measure between all individuals.

Hamming Distance was used as the diversity measure in this paper as it is problem independent. It also takes into account the unique nature of the binary space, which is lost in the Euclidean measurement; the implemented GA is binary coded. In this paper diversity measurement was primarily used as an analysis tool and is the average Hamming distance between all unique individuals in the population.

When a bit has a constant state within the population it is said to have converged, as in all future populations this bit state will never alter without mutation. When a search is converging it has a declining population diversity. When a search has converged the population has no useful diversity as combination will never create differing individuals. Traditional mutation alters every bit with a pre-set probability. Thus it will probabilistically interrupt convergence at some point during the search. Dynamic alteration of this convergence interruption has shown to improve the search result [12], [13].

2.2   Proposed Method

The number of unique individuals which can be created, in a string of length n at a distance $d = [0:n]$, through the combination of two dissimilar individuals can be determined by the combinations equation:

$$N^d = C_n^d = \frac{n!}{D!(n-D)!} .$$ (1)

where N^d is the number of individuals at distance d and D is the Hamming distance between the individuals. D can be calculated as:

$$D = \frac{n}{2} - \left| \frac{n}{2} - d \right| .$$ (2)

In general, mutation will always be a hill-climbing algorithm and a dynamic De-Jong mutation rate can only compromise between hill-climbing and unguided, convergence disturbing divergence.

The proposed mutation method is named *Coherent Diversity Maintenance* (*CDM*). This method maintains population diversity by the following means:

1. Deliberately targets non-unique individuals to be replaced with 'Random Immigrants (RI)'. 'Random Immigrants' are randomly generated individuals.

2. Preserves bit convergence by ensuring the new RI contain the bit states converged within the rest of the population.

Fundamentally, CDM provides the crossover with more samples from the current search space as the search space for each generation is reduced through the convergence of the last generation. However, CDM differs from using RI only, as RI ignores convergence and provides crossover with new samples from the entire search space. The effectiveness of RI style mutation is dependent on the correlation between population size and search space size. If the search space is significantly larger than the population size, a diverse population will give combination a more incoherent and disconnected sample. As the search space reduces, CDM will become more beneficial than RI alone as the CDM samples will be more representative of the converged search space.

To evaluate the effectiveness of the CDM, it was compared with Random Immigration, standard De-Jong mutation of varying probabilities and a GA with no mutation by using many different types of benchmark functions in the following section.

2.3 Mutation Rate

Mutation traditionally requires the specifying and tuning of the rate at which mutation affects the population. The mutation rate for the CDM and the implemented version of RI was determined by the number of repeated individuals in the population. Therefore, this method requires the comparison of all individuals in the population.

The computational cost of this comparison is:

$$(N^2/2 - N) \times L \ . \tag{3}$$

where L is the length of individual and N is the number of individuals in a population. The mutation rate is self regulated by the amount of diversity in the population. The repetitive combination of like individuals has a higher probability of creating identical individuals than the combination of relatively dissimilar individuals. Thus, as the population diversity decreases and combination results in more identical individuals, the mutation rate increases. Replacing only repetitive individuals has the benefit of not interrupting convergence because the product of combination is not ignored or diverted unless combination creates redundant individuals. This ensures that convergence and divergence do not compete within the search.

Even when the mutation rate becomes quite high it will not remove individuals which are useful for selection. The main disadvantage of this is that it does not guarantee the avoidance of a local optima if found.

3 Results

The GA testbed used ranking selection without elitism and was tested on two population sizes, 50 and 100, over 20 generations. The tests were conducted using both single point and dual point crossover. The mutation methods were tested on six (6) benchmark functions with various shapes: Two Multi-Modal functions, two single optima functions and two flat functions. The functions were tested over 3 to 5 dimensions with 6 to 12 Bit resolution per dimension. The mutation methods were compared with respect to diversity, and the final fitness. The results were averaged over 100 tests.

3.1 Benchmark Functions

The following six (6) bench mark functions [16] are used in this paper to evaluate the effectiveness of the proposed CDM. They are:
   Two single-optimum functions:

1. *The Ackley function:*

$$f(x) = -20\exp\left(-0.2\sqrt{\frac{1}{n}\sum_{i=1}^{n}x_i^2}\right) - \exp\left(\frac{1}{n}\sum_{i=1}^{n}\cos(2\pi x_i)\right) + 20 + e \ . \tag{4}$$

$$-32 \le x_i \le 32, \ x_{min} = (0,\cdots,0), \ \text{and} \ f(x_{min}) = 0.$$

2. *The Sphere function:*

$$f(x) = \sum_{i=1}^{n}x_i^2 \ . \tag{5}$$

$$-50 \le x_i \le 50, \ x_{min} = (0,\cdots,0), \ \text{and} \ f(x_{min}) = 0.$$

Two multi-optima functions:

3. *Rastrigin's function:*

$$f(x) = \sum_{i=1}^{n}(x_i^2 - 10\cos(2\pi x_i)) + 10n \ . \tag{6}$$

$$-5.12 \le x_i \le 5.12, \ x_{min} = (0,\cdots,0), \ \text{and} \ f(x_{min}) = 0.$$

4. *Griewank's Function:*

$$f(x) = \frac{1}{4000}\sum_{i=1}^{n}x_i^2 - \prod_{i=1}^{n}\cos\left(\frac{x_i}{\sqrt{i}}\right) + 1 \ . \tag{7}$$

$$-100 \le x_i \le 100, \ x_{min} = (0,\cdots,0), \ \text{and} \ f(x_{min}) = 0.$$

And two flat-optimum functions:

5. *Zakharov's Function:*

$$f(x) = -\sum_{i=1}^{n}x_i^2 + \left(\sum_{i=1}^{n}0.5ix_i\right)^2 + \left(\sum_{i=1}^{n}0.5ix_i\right)^4 \ . \tag{8}$$

$$-10 \le x_i \le 10, \ x_{min} = (0,\cdots,0), \ \text{and} \ f(x_{min}) = 0.$$

6. *Rosenbrock's Function:*

$$f(x) = \sum_{i=1}^{n} (100(x_{i+1} - x_i^2)^2 + (x_i - 1)^2) .$$ (9)

$-10 \le x_i \le 10,\; x_{min} = (1, \cdots, 1),\; \text{and}\; f(x_{min}) = 0.$

3.2 Numerical Results of Fitness

The GA results, in terms of the mean final fitness, are shown in Tables 1 – 6 of the above six benchmark functions. The results were obtained by using different mutation methods – the proposed CDM method, the completely random RI method and De-Jong method with 0.001, 0.002, 0.005 and 0 probability mutation rates.

Table 1. Ackley function final fitness comparison

Cross over points/ no. of bits	CDM	RI	0.001	0.002	0.005	None
Single point/6-Bit	0.38	0.86	1.21	0.62	0.16	1.89
Dual point/ 6-Bit	1.56	3.24	2.86	2.14	0.92	4.16
Single point/ 12Bit	0.3	0.39	0.54	0.36	0.07	0.9
Dual point/ 12Bit	1.58	2.11	1.79	1.26	0.5	2.56

Table 2. Sphere function final fitness comparison

Cross over points/ no. of bits	CDM	RI	0.001	0.002	0.005	None
Single point/ 6-Bit	0.1	0.18	0.24	0.12	0.01	0.62
Dual point/ 6-Bit	0.07	0.08	0.07	0.04	0.01	0.2
Single point/ 12Bit	0.81	1.77	1.31	0.77	0.1	3.22
Dual point/ 12Bit	0.66	0.9	0.65	0.27	0.03	1.43

Table 3. Rastrigin's function final fitness comparison

Cross over points/ no. of bits	CDM	RI	0.001	0.002	0.005	None
Single point/ 6-Bit	1.66	2.34	4.56	4.15	4.04	4.89
Dual point/ 6-Bit	1.59	1.89	2.98	3.04	2.69	3.3
Single point/ 12Bit	6.55	9.76	10.56	9.74	7.38	15.81
Dual point/ 12Bit	5.34	6.06	5.91	5.76	4.72	7.27

Table 4. Griewank function final fitness comparison

Cross over points/ no. of bits	CDM	RI	0.001	0.002	0.005	None
Single point/ 6-Bit	0.04	0.05	0.07	0.05	0.03	0.11
Dual point/ 6-Bit	0.04	0.03	0.05	0.04	0.03	0.07
Single point/ 12Bit	0.08	0.09	0.09	0.07	0.05	0.16
Dual point/ 12Bit	0.06	0.06	0.07	0.05	0.04	0.09

Table 5. Zakharov's function final fitness comparison

Cross over points/ no. of bits	CDM	RI	0.001	0.002	0.005	None
Single point/6-Bit	8.9	29.6	54.3	35.4	19.4	115
Dual point/6-Bit	11.1	17.9	22.7	16.4	10.6	42
Single point/12Bit	294	448	430	136	55.1	1419
Dual point/12Bit	165	257	149	84.2	33.8	305

Table 6. Rastrigin's function final fitness comparison

Cross over points/ no. of bits	CDM	RI	0.001	0.002	0.005	None
Single point/6-Bit	0.01	0.01	0.07	0.06	0.05	0.09
Dual point/6-Bit	0.01	0.01	0.05	0.05	0.04	0.05
Single point/12Bit	0.01	0.03	0.1	0.06	0.03	0.17
Dual point/12Bit	0.03	0.02	0.05	0.03	0.04	0.11

From these results it can be seen that the RI produces worst results than the CDM methods. This is expected as the CDM is designed to encourage diversity whilst still maintaining population convergence.

The more interesting aspects of the results are observed in the fitness results for the two single-optima functions (Tables 1-2) of CDM in comparison with the De-Jong methods. It seems that the CDM is not generating better results than the normal probabilistic mutation methods. It is believed that this is due to the fact that single optima solutions can be found using hill-climbing methods that the De-Jong's method falls into. Therefore, the normal mutation method will generate acceptable results.

The CDM in general will generate better or similar results for the multi-optima functions (Tables 3-4). In these cases, as the CDM is designed to encourage diversity, it is likely to generate better, or at least no worse results than the other mutation methods.

In dealing with the flatter-optima functions (Tables 5-6), it can be seen that the CDM's final results are compatible to those from the normal mutation methods. This is expected as on the flatter surface diversity will not have a significant impact on diversity outcomes.

It is also noted in the computation that the CDM in general performs better in dealing with simpler cases, i.e., with single cross-over point and with shorter bit length in population. CDM performed relatively worse with dual point crossover (DPC) because DPC is better than single point crossover (SPC) in maintaining population diversity. It is believed that CDM performed relatively worse when tested with longer individuals because the search space became too large for the given population size to be effective.

3.3 Numerical Results on Diversity

During the computation, it is noted that the diversity varies significantly with different mutation rates and mutation methods. It was also found that it varies between benchmark functions. However, the most significant effect on diversity came with search space size, which is a combination of parameter number and parameter resolution.

Shown in Fig. 1 is the average population hamming distances for all tests in 6-Bit and 12-Bit parameter resolutions.

It is clear that there are marked differences in search space sizes between 6-Bit and 12-Bit parameter resolutions. The average distance between all solutions at the beginning for a 6-Bit 5-dimensional case is 12 Bits while it is 25 Bits for the 12-Bit case.

In all tests the diversity measurement was made after the crossover. This means that only the useful diversity created through the mutation method is recorded, i.e., when the mutation method creates individuals with above average fitness and a larger bit difference, then it will improve the diversity of the next generation. If mutation does not create fitter individuals with a larger distance from the other relatively fit individuals then it will not improve the next generation's diversity.

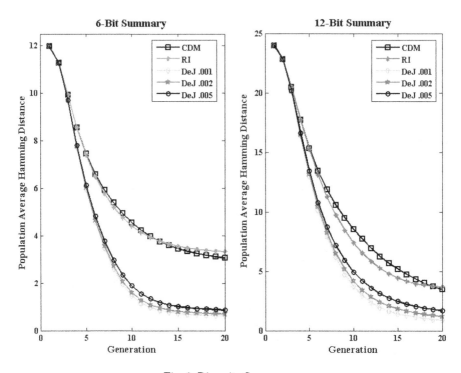

Fig. 1. Diversity Summary

Figure 1 shows the generational diversity summaries for the 6-Bit and 12-Bit tests. In the figure, CDM refers to Coherent Diversity Maintenance mutation. RI is for Random Immigrants mutation and DJ .001 – DJ .005 refer to the De-Jong Mutation with the stated probability.

It can be seen from the figure that the diversity levels of CDM and RI are similar for both parameter resolutions. However CDM consistently creates higher population diversity during the mid generations, whilst tapering off at the final generations. The lower diversity relative to RI in the final generations is to be expected as CDM only creates individuals which are coherent with the converging search space whereas RI disregards the convergence of the search space.

RI and CDM encourage roughly equivalent diversity levels for all tests but produce significantly different final fitness results, thus one can conclude that there is an inconsistent link between diversity and superior convergence. As RI encourages new individuals to come from the entire search space and CDM encourages new individuals to come from the converging search space one can conclude that specific diversity gives better results than unspecific diversity. This occurs because as convergence progresses, CDM is able to provide crossover with a more representative sample of the converging space which enables it to better select the next generation of individuals.

4 Conclusions

In this paper a new mutation scheme, named Coherent Diversity Maintenance (CDM), is introduced. This method is designed to encourage diversity in the GA population by maintaining the convergence trend. It is believed that this method will explore the converging parameter space more fully therefore resulting in better convergence results. Numerical results have shown that the CDM tends to generate better, or at least no worse, results for multi-optima or flatter optima situations than the normal probabilistic mutation method and complete random mutation method.

References

1. Holland, J.H.: Adaption in Natural and Artificial Systems: An Introductory Analysis with Applications to Biology, Control and Artificial Intelligence. MIT Press, Cambridge (1992)
2. Chakraborty, G., Hoshi, K.: Rank Based Crossover – A new technique to improve the speed and quality of convergence in GA. In: Proceedings of the 1999 Congress on Evolutionary Computation, CEC 1999, vol. 2, p. 1602 (1999)
3. GoldBerg, D.E.: Genetic Algorithms in search, optimization and machine learning. Addison-Wesley Publishing Company, Reading (1989)
4. Dumitrescu, D., Lazzerini, B., Jain, L.C., Dumitrescu, A.: Evolutionary Computation. CRC Press, Boca Raton (2000)
5. Radcliff, N.J.: Forma Analysis and Random Respectful Recombination. In: Proceedings of the fourth International Conference on Genetic Algorithms (1991)
6. Watson, R.A., Pollack, J.B.: Recombination Without Respect: Schema Combination and Disruption in Genetic Algorithm Crossover. In: Proceedings of the 2000 Genetic and Evolutionary Computation Conference (2000)
7. De Jong, K.A.: Analysis of the Behaviour of a class of Genetic Adaptive Systems. Technical Report, The University of Michigan (1975)
8. Hoffmann, J.P.: Simultaneous Inductive and Deductive Modeling of Ecological Systems via Evolutionary Computation and Information Theory. Simulation 82(7), 429–450 (2006)
9. Fogarty, T.C.: Varying the Probability of mutation in the Genetic Algorithm. In: The Proceedings of the Third International Conference on Genetic Algorithms 1989, pp. 104–109 (1989)
10. Hesser, J., Manner, R.: Towards an optimal mutation probability for Genetic Algorithms. In: Schwefel, H.-P., Männer, R. (eds.) PPSN 1990. LNCS, vol. 496, pp. 23–32. Springer, Heidelberg (1991)
11. Back, T.: Optimal Mutation Rates in Genetic Search. In: Proceedings of the Fifth International Conference on Genetic Algorithms, June 1993, pp. 2–8 (1993)
12. Davis, L.: Adapting Operator Probabilities in Genetic Algorithms. In: Proceedings of the Third International Conference on Genetic Algorithms, pp. 61–69 (1989)
13. Ursem, R.K.: Diversity-Guided Evolutionary Algorithms. In: Guervós, J.J.M., Adamidis, P.A., Beyer, H.-G., Fernández-Villacañas, J.-L., Schwefel, H.-P. (eds.) PPSN 2002. LNCS, vol. 2439, pp. 462–471. Springer, Heidelberg (2002)
14. Janikow, C.Z., Michalewicz, Z.: An experimental comparison of binary and floating point representations in genetic algorithms. In: Belew, R.K., Booker, J.B. (eds.) Proceedings of the Fourth International Conference Genetic Algorithms, pp. 31–36. Morgan Kaufmann, San Mateo (1991)

15. Gasieniec, L., Jansson, J., Lingas, A.: Efficient Approximation Algorithms for the Hamming Centre Problem. Journal of Discrete Algorithms 2(2), 289–301 (2004)
16. Kwok, N.M., Ha, Q., Liu, D.K., Fang, G., Tan, K.C.: Efficient Particle Swarm Optimization: A Termination Condition Based on the Decision-making Approach. In: 2007 IEEE Congress on Evolutionary Computation, Singapore (2007)

Dynamic Crossover and Mutation Genetic Algorithm Based on Expansion Sampling

Min Dong and Yan Wu

Department of Computer Science and Technology, Tongji University,
1239 Si Ping Road, Shanghai, 200092, China
dongmin_007@163.com

Abstract. The traditional genetic algorithm gets in local optimum easily, and its convergence rate is not satisfactory. So this paper proposed an improvement, using dynamic cross and mutation rate cooperate with expansion sampling to solve these two problems. The expansion sampling means the new individuals must compete with the old generation when create new generation, as a result, the excellent half ones are selected into the next generation. Whereafter several experiments were performed to compare the proposed method with some other improvements. The results are satisfactory. The experiment results show that the proposed method is better than other improvements at both precision and convergence rate.

Keywords: genetic algorithm, expansion sampling, dynamic cross rate, dynamic mutation rate.

1 Introduction

Practical application of genetic algorithm has the problems of premature convergence, less robust in parameter selection, slow convergence speed at anaphase. These deficiencies seriously limit the scope of application of genetic algorithms. How to improve the performance of genetic algorithm to overcome the shortages so that it can be better applied to practical problems, become the main subject of many scientists. In recent years there are many improved genetic algorithms, for example, Self-adaptive simulated binary crossover genetic algorithm [1], fuzzy adaptive genetic algorithm [2], Multi-agent genetic algorithm [3], fast and elitist multi-objective genetic algorithm [4], fully adaptive strategy genetic algorithm [5], quantum clone genetic algorithm [6], improved mutation operator genetic algorithm [7] and so on. These new algorithms make the performance of genetic algorithm improved, but still far from achieving the desired state. In order to maintain the population diversity and avoid the high fitness individual been selected too many times to occupy the entire population, this paper presents a dynamic crossover probability and mutation probability. In order to improve the algorithm's convergence rate, this paper puts the new individuals with the previous generation together and selects the best half individuals into the next-generation. And then this paper proved the effectiveness of the improved algorithm through several experiments.

H. Deng et al. (Eds.): AICI 2009, LNAI 5855, pp. 141–149, 2009.
© Springer-Verlag Berlin Heidelberg 2009

2 Dynamic Crossover Probability

Traditional genetic algorithm uses the fixed cross-rate operator. Because the cross-rate of individuals is the same, it will make all of the contemporary individuals in the cross-operation retained at the same probability, thus the current better individuals will be selected several times at the choice operation of the next round, and the relatively poor individuals in current generation will be eliminated, leading the population quickly evolving to the direction of the current optimal individual. If the current optimal individual is a local optimum, then the entire algorithm can easily fall into local optimum. To avoid this situation and increase the diversity of the population, this paper presents a dynamic cross-rate, namely using the ratio between the Euclidean distance of two chromosomes and the Euclidean distance of the largest and the smallest fitness individual in population as cross-rate:

$$Pc = \frac{|f(a) - f(b)|}{\max(f) - \min(f)} \tag{1}$$

where $f(a)$ is a chromosome's fitness, $f(b)$ for b chromosome's fitness, $max(f)$ and $min(f)$ respectively are the largest and the smallest fitness in the population, P_c is the crossover probability. Such cross-rate will make the individuals in the middle of population retained at a greater chance, and the individuals at both ends of population a greater probably cross, so that the better individuals and the poorer individuals are both changed to avoid too intense competition in the choice operation of next round. The poor individuals' genes can also contribute to the development of the population.

The principle of this improvement is a simple probability problem of average distribution, in which, the crossover probability of a subject to the average distribution of $f(b)$. To illustrate this principle, we assume that $max(f)$ for 2, $min(f)$ for 0, $f(b)$ average distributes in the interval [0,2]. In this case, if an individual a_1 is in the middle of the interval, that is, its $f(a_1)$ is 1, then its cross-rate with b is 1/4, and if an individual $a2$ at both ends of the interval, that is, its $f(a_2)$ is 0 or 2, then its cross-rate with b is 1/2, larger than individuals in the middle. The distribution is shown in Figure 1 below (x-axis is $f(a)$ values, y-axis is crossover rate):

In addition, this algorithm's dynamic crossover probability basically is effective crossover probability (we call crossover between two long distance individuals the effective cross-operation. Because of their relatively big difference between each other, cross-operation changes them comparatively large. On the other side, the cross-operation between very close chromosomes changes individuals very small, the individuals after cross-operation almost unchanged, such cross-operation is ineffective),which can improve the effect of cross-operation and avoid inbreeding and premature convergence effectively. Therefore, although the cross probability of this paper is lower than other genetic algorithms, the overall effect is better. And the lower cross-rate will reduce the computational complexity and accelerate the speed of evolution.

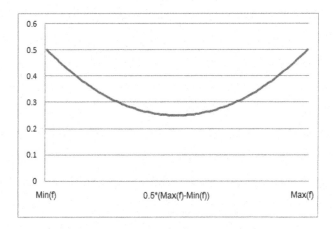

Fig. 1. Crossover rate curve

3   Dynamic Mutation Probability and Expansion Sampling

In order to avoid better individuals rapidly occupied the entire population, this paper proposed a dynamic mutation probability using the ratio between a chromosome's fitness and the largest fitness in population. The mutation probability is defined as:

$$p_m = k * (\frac{f(a)}{\max(f)})^2 \quad 0 < k < 1 \tag{2}$$

Where $f(a)$ is the fitness of an individual a, $max(f)$ is the largest fitness, P_m is the mutation probability, k is a parameter, valid in interval $(0,1)$. This mutation rate allows the better individuals have the bigger probability in order to avoid better individuals rapidly occupying the entire population, lead the population evolving to the direction of diversification. The distribution of mutation rate is shown in Figure 2 below (x-axis is $f(a)$ values, y-axis is mutation rate):

At the same time, this paper uses the expansive optimal sampling, namely puts the new individuals with the previous generation together and chooses the best half individuals into the next-generation. Traditional genetic algorithm puts the new individuals directly into the next generation, which will make the older better individuals which have been crossed or mutated no chance accessing to future generations, thus slows down the convergence rate of the population. And the expansive optimal sampling makes those better individuals enter the next generation, thus rapidly accelerates convergence speed.

The traditional genetic algorithm uses fixed mutation rate in computation. As a result the algorithm is hard to jump out of local optimum because of the relatively small mutation probability when the algorithm gets in local optimum. And sometimes even jumped out, there would be useless because they are not selected at the next choice operation. To avoid this status, this paper combines the dynamic mutation rate with

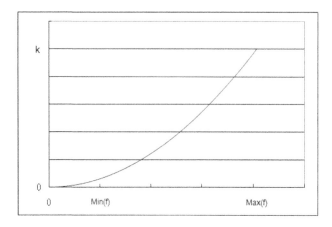

Fig. 2. Mutation rate curve

the expansive optimal sampling. Dynamic mutation rate makes algorithm more easily jumping out of local optimum when the algorithm get in a local optimum, and the expansive optimal sampling ensures this mutated and excellent individuals can be selected into future generations, thereby greatly reducing the probability of getting in local optimum. If the evolution enough, in theory, the genetic algorithm would not get in a local optimum assuredly.

4   Algorithm Processes

The procedures of the proposed method are as follows:

1) Initialize the control parameters: set the population size N, the evolution generation g;

2) g = 0, randomly generate a population of N individuals Po = (x1, x2, ...);

3) Determine the fitness function, calculate the value of individual fitness f(xi), (i = 1,2, ..., N), to evaluate the individuals in population;

4) Determine the choice strategy, the algorithm population in this paper mainly uses the method of roulette, select the advantage individuals from the individuals Pg of generation g to form the mating pool;

5) Determine the dynamic crossover operator. Determine cross-rate of each pair of individuals through the $Pc = | f(a) - f(b) | /(\max(f) - \min(f))$, and then proceed cross-operation to produce the middle results;

6) Determine the dynamic mutation operator. Use the $P_m = k * (f(a)/\max(f))^2$ to determine mutation rate, and then proceed mutation operation;

7) The expansive optimal sampling. Put the individuals after choice, cross and mutation operation with the individuals Pg before operation together and select the best half individuals into the next-generation;

8) Stop the evolution if the results meet the termination criteria, else jump to (3).

5 Simulation Experiments

5.1 Experiment 1

In this paper, the first experiment uses typical complex optimization function as follows:

$$y = x * \sin(10 * pi * x) + 2 \qquad x \in [-1, 2] \qquad (3)$$

Fitness function is as follows:

$$\text{Fitness} = x * \sin(10 * pi * x) + 2 \qquad (4)$$

We use improved mutation operator genetic algorithm (IMGA) in paper [7] and fuzzy adaptive genetic algorithm (FAGA) in paper [2] to compare with this paper's dynamic crossover and mutation genetic algorithm based on expansion sampling (ESDGA), simulating in Matlab7. We change the parameters of genetic algorithm in experiment in order to observe the performance of several algorithms better. In experiment N is the population size, Pc is the crossover probability. In improved mutation operator genetic algorithm the gene discerption proportion L=0.6, the various parameters: a1 = 0.01, b1 = 0.01, a2 = O.01, b2 = 0.05, α = 0.03, g0 = 40. The evolution generation is 500, each kind of parameter repeat 50 times. Min and Max are the minimum and maximum optimal solution in 50 times, Avg is the average, Gen is the convergence generations, local is the number of local optimum solutions. The experiment's results are shown in Table 1, Table 2 and Table 3 (Pc are useless in FAGA and ESDGA, k is used only in ESDGA):

Table 1. 50 independent results for IMGA in Experiment 1

	N=100 P_c=0.75	N=120 P_c=0.9	N=80 P_c=0.8	N=50 P_c=0.5
Min	3.4126	3.7332	3.7871	3.2674
Max	3.8502	3.8500	3.8503	3.8503
Avg	3.8185	3.8425	3.8037	3.8470
gen	145	135	118	142
local	5	7	6	5

Table 2. 50 independent results for FAGA in Experiment 1

	N=100	N=120	N=80	N=50
Min	3.8126	3.8332	3.8071	3.7674
Max	3.8503	3.8503	3.8502	3.8500
Avg	3.8385	3.8425	3.8337	3.8070
gen	22	23	18	14
local	3	6	4	6

Table 3. 50 independent results for ESDGA in Experiment 1

	N=100 k=0.6	N=120 k=0.8	N=80 k=0.4	N=50 k=0.3
Min	3.8503	3.8503	3.8503	3.8503
Max	3.8503	3.8503	3.8503	3.8503
Avg	3.8503	3.8503	3.8503	3.8503
gen	9	10	12	12
local	0	0	0	0

From the tables it is clear that IMGA needs about 130 iterations to converge, with an average of about 12% local optimum. FAGA needs an average of about 20 iterations to converge, 10% local optimum. ESDGA needs only about 12 iterations to converge, and there is no local optimum in total 200 results, better than the two algorithms. Also need to point out that for this paper's algorithm, we have a lot of experiments to select the k value, the table only lists representative values, as proved, the effect of the algorithm are better when k above 0.3.

The speed and accuracy is a contradiction in genetic algorithms. If the algorithm requires high speed, only evolving a few generations, the accuracy of the results will be not satisfactory, and most likely have not convergence or convergence in local optimum. If the algorithm continues the evolution infinitely, it must converge to the global optimum finally, if there is a global optimum. In this paper, we use the latter situation to compare the performance of the proposed algorithm affected by different k values in Experiment 1. This is to say, Algorithm will evolve infinitely until the population converges to the global optimum. We set the population size N=100 and choose different k in the valid interval (0, 1). The result is shown in Figure 3 below (x-axis is k values, y-axis is convergence generations):

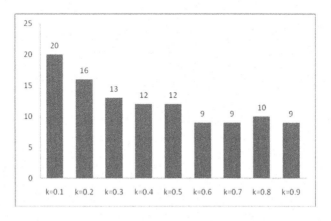

Fig. 3. The influence of different k in experiment 1

From the table we can clearly see that, the bigger k the better performance of this algorithm. 0.3 is a dividing line, and there is not large difference above 0.6. This result subjects to the theory discussed earlier.

5.2 Experiment 2

Then we select a typical needle in haystack problem[5] (Needle-in-haystack, NiH for short) for experiment.

$$\max f(x, y) = (\frac{a}{b+(x^2+y^2)})^2 + (x^2 + y^2)^2 \quad x, y = [-5.12, 5.12] \tag{5}$$

In the function: a = 3.0, b = 0.05; maximum value of the function is 3600 when x = 0, y = 0. And there are four local maximum points (-5.12,5.12), (-5.12, -5.12) , (5.12,5.12), (5.12, -5.12), function value is 2748.78.

This is a typical GA deceptive problem. Here we still use improved mutation operator genetic algorithm (IMGA) in paper [7] and fuzzy adaptive genetic algorithm (FAGA) in the paper [2] to compare with this paper's dynamic crossover and mutation genetic algorithm based on expansion sampling (ESDGA) in Matlab7. The evolution generation is also 500, each parameter repeat 50 times. Min and Max are the minimum and maximum optimal solution in 50 times, Avg is the average, Gen is the convergence generations, local is the number of local optimum solutions. The experiment results are shown in Table 4, Table 5, Table 6 (Pc are useless in FAGA and ESDGA, k is used only in ESDGA):

Table 4. 50 independent results for IMGA in Experiment 2

	N=100 P_c=0.75	N=120 P_c=0.9	N=80 P_c=0.8	N=50 P_c=0.5
Min	3600	3600	2748.8	2748.8
Max	3600	3600	3600	3600
Avg	3583	3600	3583	3511
Gen	278	293	264	260
Local	1	0	1	2

Table 5. 50 independent results for FAGA in Experiment 2

	N=100	N=120	N=80	N=50
Min	3600	3600	3600	2748.8
Max	3600	3600	3600	3600
Avg	3600	3600	3600	3583
Gen	46	57	48	40
Local	0	0	0	1

Table 6. 50 independent results for ESDGA in Experiment 2

	N=100 k=0.6	N=120 k=0.8	N=80 k=0.4	N=50 k=0.3
Min	3600	3600	3600	3600
Max	3600	3600	3600	3600
Avg	3600	3600	3600	3600
Gen	21	21	22	23
Local	0	0	0	0

As can be seen from the tables, IMGA and FAGA appear a few local optimums, in which FAGA need about 50 iterations to converge, IMGA needs about 270 iterations to converge. However ESDGA did not lead to local optimum, and the convergence rate is also very fast, basically as long as 20 on average, better than IMGA and FAGA.

Then we compare the performance of the proposed algorithm affected by different k values in Experiment 2 as same as Experiment 1. We also set the population size N=100 and choose different k in the valid interval (0, 1). The result is shown in Figure 4 below (x-axis is k values, y-axis is convergence generations):

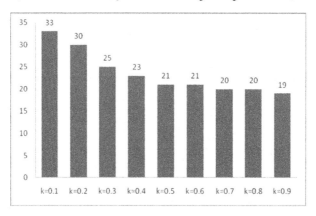

Fig. 4. The influence of different k in experiment 2

The conclusion is similar with Experiment 1 though there are a few differences, the bigger k the better performance of this algorithm.

6 Conclusion

Simple genetic algorithm has defects of slow convergence speed and getting in local optimum easily, making the application of genetic algorithm being limited. To solve the two largest defects, this paper proposed an improved method. As we can see from the experiment results, this paper's dynamic crossover and mutation genetic algorithm based on expansion sampling improve the performance of genetic algorithm, over-come the problems of inbreeding and premature convergence, accelerate the speed of the algorithm convergence, effectively avoid the algorithm falling in local optimum, make the algorithm has a strong optimization capacity.

References

1. Kalyanmoy, D., Karthik, S., Tatsuya, O.: Self-adaptive simulated binary crossover for real-parameter optimization. In: Genetic and Evolutionary Computation Conference, pp. 1187–1194 (2007)
2. Huang, Y.P., Chang, Y.T., Sandnes, F.E.: Using Fuzzy Adaptive Genetic Algorithm for Function Optimization. In: Annual meeting of the North American Fuzzy Information Processing Society, June 3-6, pp. 484–489. IEEE, Los Alamitos (2006)
3. Zhong, W.C., Liu, J., Xue, M.Z., Jiao, L.C.: A Multi-agent Genetic Algorithm for Global Numerical Optimization. IEEE Transactions on Systems, Man, and Cybernetics, Part B 34(2), 1128–1141 (2004)
4. Deb, K., Agrawal, S., Pratap, A., Meyarivan, T.: A fast and elitist multi-objective genetic algorithm: NSGA-II. IEEE Transactions on Evolutionary Computation 6(2), 182–197 (2002)
5. Xiang, Z.Y., Liu, Z.C.: Genetic algorithm based on fully adaptive strategy. Journal of Central South Forestry University 27(5), 136–139 (2007)
6. Li, Y.Y., Jiao, L.C.: Quantum clone genetic algorithm. Computer Science 34(11), 147–149 (2007)
7. Li, L.M., Wen, G.R., Wang, S.C., Liu, H.M.: Independent component analysis algorithm based on improved genetic algorithm. Journal of System Simulation 20(21), 5911–5916 (2008)

Multidisciplinary Optimization of Airborne Radome Using Genetic Algorithm

Xinggang Tang, Weihong Zhang, and Jihong Zhu

Laboratory of Engineering Simulation and Aerospace Computing,
The Key Laboratory of Contemporary Design and Integrated Manufacturing Technology,
Ministry of Education, Northwestern Polytechnical University, Xi'an 710072, China
nwputang@yahoo.com.cn, {zhangwh,JH.Zhu_FEA}@nwpu.edu.cn

Abstract. A multidisciplinary optimization scheme of airborne radome is proposed. The optimization procedure takes into account the structural and the electromagnetic responses simultaneously. The structural analysis is performed with the finite element method using Patran/Nastran, while the electromagnetic analysis is carried out using the Plane Wave Spectrum and Surface Integration technique. The genetic algorithm is employed for the multidisciplinary optimization process. The thicknesses of multilayer radome wall are optimized to maximize the overall transmission coefficient of the antenna-radome system under the constraint of the structural failure criteria. The proposed scheme and the optimization approach are successfully assessed with an illustrative numerical example.

Keywords: multidisciplinary optimization, radome, genetic algorithm, structural failure, transmission coefficient.

1 Introduction

Airborne radomes are often used to protect antennas from a variety of environmental and aerodynamic effects. The design of a high performance airborne radome is a challenging task as the aerodynamic, electromagnetism, structural mechanics, etc requirements are generally involved. The performances of the radome in different disciplines are usually in conflict with each other. A thin and light-weighted radome with excellent electromagnetic transmission ability is apparently structurally unreliable, while a well-designed radome structure with high stiffness and stability will affirmatively have a poor electromagnetic performance. The radome design is hence a multidisciplinary procedure because the analyses of different disciplines are tightly coupled. However, the traditional engineering design approach separates the design procedure into sequential stages. A design failure at certain stage would cause the whole design to start from scratch. This will result in a tremendous cost of design cycle and resources.

The great improvement of the optimization theory and algorithms in the past several decades has provided an efficient solution for radome design. The optimization procedure

H. Deng et al. (Eds.): AICI 2009, LNAI 5855, pp. 150–158, 2009.

can simultaneously take into account the radome characteristics in different disciplines and searches for an optimal design with all design requirements to be well-balanced.

In the earlier researches, the techniques of simulating the electromagnetic characteristics of radomes were well developed, which can be classified into two kinds: (1) high-frequency methods, such as the Ray Tracing (RT) [1] technique based on Geometric Optics (GO), Aperture Integration-Surface Integration (AI-SI) [2] and Plane Wave Spectrum-Surface Integration (PWS-SI) [3] based on Physical Optics (PO); (2) low-frequency methods, such as the Finite Element Method (FEM) [4] and Method of Moment (MoM) [5]. Generally, the high-frequency methods are more suitable for electrically large problems because of high computational efficiency, while the low-frequency methods can provide more accurate analysis with much higher computational complexity. Since the iteration strategy with large computing cost is used in the optimization methods, the high-frequency methods are superior to low-frequency methods for electromagnetic analysis of the radome optimization.

Even though the researches on radome optimization were pioneered early by using a simulated annealing technique [6], it is until recently that more advanced optimization algorithms have been used and promoted the development of the radome design. For example, the particle swarm optimization and the Genetic Algorithm are applied for the radome optimization. The layered thickness of the sandwich radome wall and the shape of the radome are optimized to maximize the overall transmission coefficient for the entire bandwidth [7] or to minimize the boresight error [8]. A multi-objective optimization procedure was further proposed to optimize the boresight error and power transmittance simultaneously using the genetic algorithm and RT method [9]. However, these researches on radome optimization are limited to the electromagnetic characteristics. Recently, the Multidisciplinary Radome Optimization System (MROS) [10] is proposed as a synthesis procedure, in which the material selection, structural analysis, probabilistic fracture analysis and electromagnetic analysis are incorporated to perform the multidisciplinary optimization. This work indicates a new trend of radome optimization which will be more applicable for practical engineering designs.

In this paper, the structural and electromagnetic characteristics are considered simultaneously to perform the radome design. A multidisciplinary optimization procedure is developed based on the finite element model of the radome. The structural analysis and electromagnetic analysis are carried out to obtain the structural failure indexes and the overall transmission coefficient, respectively. The genetic algorithm is employed for the optimization.

2   Multidisciplinary Simulation of Radome

2.1   Simulation Strategy

Normally, the Finite Element Method (FEA) and the Physical Optics (PO) method are the preferred approaches used for the structural analysis and electromagnetic analysis of radome, respectively. However, in the traditional design scheme illustrated in Fig.1, these analyses, as well as the procedure of modeling and postprocessing are implemented separately. There is no collaboration or communion between different analysis procedures, which can no longer match the requirements of rapid design cycle for the modern products.

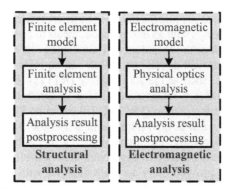

Fig. 1. Traditional radome analysis scheme

A significant improvement for the radome design is to integrate the interested analyses in a multidisciplinary manner. A potential scheme of the multidisciplinary analysis is illustrated in Fig.2. The most important step is to identify what data is to be shared by different sub-analyses. It is known all the geometry and topology information of the radome that has been defined in the finite element model can be shared with the PO analysis. Furthermore, the finite element mesh can also be used as the PO analysis mesh on condition that the element size is adapted to the wavelength according to the Nyquist Sampling Theorem. Therefore, in the proposed analysis scheme shown in Fig.2, the finite element model is defined as the mutual data for both structural and electromagnetic analysis.

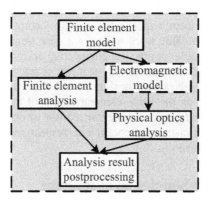

Fig. 2. Multidisciplinary radome analysis scheme

Because of requirements of the low dielectric constant and high mechanical strength, the airborne radome is usually manufactured of sandwich laminate. The choice of materials and wall thickness has a significant influence on the structural strength and electrical performance. The idea in this paper is to find the proper configuration of these parameters to satisfy the design requirements. The validity of the design can be verified with the structural analysis and electromagnetic analysis of the radome respectively.

2.2   Structural Analysis of Radome

The basic issue in radome structural analysis is the material definition. This is very difficult because multilayer thin walled configurations including the A sandwich, B sandwich, and C sandwich are often used as illustrated in Fig.3. Taking A sandwich for example, it consists of three layers: two dense high-strength skins separated by a lower-density, lower-dielectric core material such as foam or honeycomb. This configuration can provide much higher strength for a given weight than a monolithic wall of homogeneous dielectric material. The construction of the rest two kinds of wall configuration can be concluded accordingly. In our demonstrative application, we use A sandwich as the wall configuration which is most frequently used.

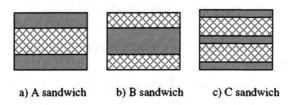

a) A sandwich   b) B sandwich   c) C sandwich

Fig. 3. Multilayer radome wall configuration

Another important aspect of the radome structural analysis is the structural loads and constraints. For common airborne radome (non high-speed radomes), the aerodynamic load due to airflow is the main cause of mechanical stress. The radome is tightly attached to the airframe with special bolts.

The radome structural analysis can performed in Patran/Nastran. The A sandwich configuration is modeled in Patran Laminate Modeler by treating the sandwich layer as an orthotropic layer with the equivalent material properties. The structural validity of the radome can be assessed by the failure analysis of composite materials. Frequently used failure criterions include the Hill, Hoffman, Tsai-Wu, Maximum Stress and Maximum Strain.

2.3   Electromagnetic Analysis of Radome

Based on the equivalence principle, the approach of PWS-SI method can be divided into three steps: 1) a transmitting antenna is considered and the fields transmitted to the radome interior S_1 are computed by the Plane Wave Spectrum method; 2) Transmission through the radome wall is then calculated by the transmission line analogy to obtain the fields over the radome outer surface S_2 ; 3) Radiation to the far field is then determined by the surface integration of the equivalent currents obtained from the tangential field components over the outer surface S_2 .

The antenna-radome system model is illustrated in Fig.4. According to the planar slab approximation technique, the radome wall is treated as being locally planar and modeled as an infinite planar slab with complex transmission coefficients, lying in the tangent plane at the incident points of the radiation rays.

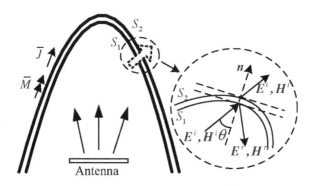

Fig. 4. The antenna-radome system model

According to PWS technique, the transmitted electric field at the near-filed point P can be calculated by

$$E(x,y,z) = \frac{1}{2\pi} \int_{-\infty}^{\infty} \int_{-\infty}^{\infty} A(k_x, k_y) e^{-jk_0 \cdot r} dk_x dk_y \tag{1}$$

where (x, y, z) are the coordinates of P in the antenna coordinate system, $k_0 = ik_x + jk_y + kk_z$ is the radial wave number and $r = ix + jy + kz$ is the vector of point P. $A(k_x, k_y)$ is the so-called plane wave spectrum or the Fourier transform of the antenna aperture, which can be formulated as

$$A(k_x, k_y) = \frac{1}{2\pi} \iint_S E(x, y, 0) e^{j(k_x x + k_y y)} d\xi d\eta \tag{2}$$

If the antenna aperture is circularly symmetric, the surface integral (2) will regress to a linear integral. Thus the computational complexity will be greatly reduced.

The transmission coefficients for perpendicular polarization $T_\perp(\theta_P)$ and parallel polarization $T_{//}(\theta_P)$ of the multilayered radome wall can be determined by the transmission line analogy [11]. Thus the fields over the radome outer surface can be formulated by

$$E_M^t = \left[n_{BM} \cdot E^i \right] \cdot n_{BM} T_\perp(\theta_P) + \left[t_{BM} \cdot E^i \right] \cdot t_{BM} T_{//}(\theta_M)$$
$$H_M^t = \left[n_{BM} \cdot H^i \right] \cdot n_{BM} T_{//}(\theta_P) + \left[t_{BM} \cdot H^i \right] \cdot t_{BM} T_\perp(\theta_M) \tag{3}$$

where

$$n_{BP} = P_M \times n_{RM}, t_{BP} = n_{BP} \times n_{RM}, P_M(x, y, z) = \frac{\text{Re}(E \times H^*)}{\left| \text{Re}(E \times H^*) \right|}$$

Thus, the electric far field can be calculated with a surface integral technique by

$$E_p = R_1 \times \int_{S_2} \left[n \times E_M^t - \sqrt{\frac{\mu}{\varepsilon}} R_1 \overset{\times}{\times} (n \times H_M') \right] \cdot e^{jk_0' \cdot R_1} dS_2 \tag{4}$$

where R_1 is the unit vector of the observation direction, r' is the unit vector of the incidence point on the outer surface S_2 and n is the normal vector of the outer surface S_2 on the incidence point.

With the far field distribution of the antenna-radome system, the electromagnetic performance of the radome, such as the transmission ratio, boresight error, etc.

3 Optimization Using Genetic Algorithm

GA simulates the natural biological evolution strategy based on Darwinian *survival of the best fitness* principal. The theoretical foundations of GA were firstly led by Holland in 1975 [12]. Since then, GA and its many variations have been applied successfully to a wide range of different problems [13]. They provide an alternative way of traditional gradient-based optimization techniques for those mathematically intractable optimization problems such as combinational optimization, discrete-variable optimization, and so on. Generally, GA is accomplished by firstly generating random solutions for a problem. Each randomly generated solution is then evaluated by its fitness and the good solutions are selected and recombined to form new generation solutions. The process of optimization based on fitness and recombination to generate variability is repeated until the best solution or the stop criterion is identified.

The optimization of radome considered in this paper is a multidisciplinary optimization which considers the performances of radomes in both structural mechanics and electromagnetics. The complex coupling between the two disciplines makes the optimization modeling and sensitivity analysis intractable. However, there is no such barrier for applying GA to the radome optimization.

Practically, the shape of the airborne radome often conforms to the unitary aerodynamic performance of the aircraft, which should not be changed in the subsequent design. In our applications, we consider the multilayer sandwich as the radome wall. The radome optimization takes the thicknesses of each layer as design variables. The failure indexes of each layer and the overall transmission coefficient of the antenna-radome system are calculated by the structural analysis and electromagnetic analysis respectively. The optimization objectives are to maximize the overall transmission coefficient and minimize the failure indexes. Since the intractability of handling multi-objective optimization, we consider one objective as the constraint. Thus, the multidisciplinary radome optimization can be formulated as

$$\text{Maximize} \quad Tran(t_1, t_2, \cdots, t_n)$$

$$\text{Subject to} \quad \max(f_{i,j}) \leq f_{\max} \tag{5}$$

$$t_{i,\min} \leq t_i \leq t_{i,\max}$$

where t_i denotes the thickness of the ith layer, *Tran* denotes the overall transmission and $f_{i,j}$ is the failure index of the ith layer of the jth element.

4 Numerical Results

In this section, the proposed optimization approach is demonstrated with a paraboloidal radome. For the structural analysis, the finite element model is illustrated in Fig.5. The A sandwich radome wall is modeled as a 7 layered laminate. Each skin is constructed with 3 layers, while the core is constructed with 1 layer. We suppose the bottom of the radome is fixed to the airframe tightly, and the whole radome is enduring a frontal pressure caused by the relative movement of air or wind. The structural failure of the radome material is verified with Hill Failure Criterion of composite.

For the electromagnetic analysis, we suppose a circular aperture antenna is placed on the bottom of radome which works at 2GHz, and the finite element model contains 2721 quadrangle elements with the maximum element area of 1.04×10^{-3} m^2. Thus, the finite element mesh satisfies the electromagnetic analysis requirements according to the Nyquist Sampling Theorem, which indicates that there are at least 4 elements within an unit area of square of the wavelength.

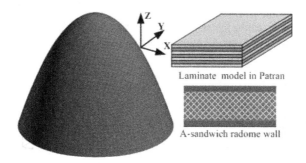

Laminate model in Patran

A-sandwich radome wall

Fig. 5. Finite element model of a paraboloidal radome

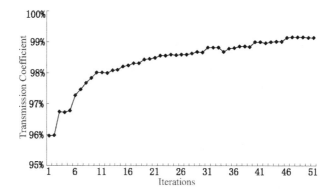

Fig. 6. The iteration history of objective function

Table 1. Optimal layered thickness of the radome wall (mm)

Design Variables	t_1	t_2	t_3	t_4	t_5	t_6	t_7
Lower limit	0.1	0.1	0.1	1	0.1	0.1	0.1
Upper limit	0.5	0.5	0.5	10	0.5	0.5	0.5
Initial value	0.2	0.2	0.2	8.8	0.2	0.2	0.2
Optimal value	0.1	0.1	0.1	9.96	0.1	0.1	0.1

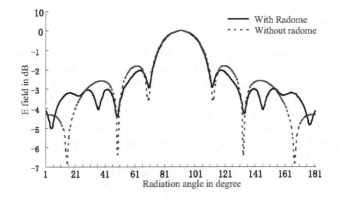

Fig. 7. Radiation patterns of the electrical far field

Table 2. Structural performance of the initial and final design

Structural performance	Maximal Plystress	Maximal Displacement	Maximal Failure index
Initial design	24.2	0.337	0.452
Final design	48.3	0.671	0.403

We consider the radome optimization problem (5) to maximize the transmission coefficient and constrain the failure indexes less than 0.6. The GA is performed with 50 generations and 70 individuals for each generation. The iteration history of the objective function is illustrated by Fig.6.

The optimal design of the radome is achieved at generation 48 with the optimal layer thicknesses listed in Table.1. From the radiation patterns of the electrical far field in Fig.7 and the structural performances comparison in Table.2, we can see that the final design achieve a better electromagnetic performance at the price of the depression of the structural performance.

5 Conclusion

This paper proposes a multidisciplinary optimization scheme for airborne radome design. The design procedure considers the structural and electromagnetic performance of the radome simultaneously. Genetic Algorithm is employed for the optimization to maximize the overall transmission coefficient under constraints of the structural failure of the radome material. The optimization scheme is successfully validated by the design optimization of a paraboloidal radome.

This work provides the new trend of radome optimization which will be more efficient and applicable for the radome design engineering. Even though the results of the demonstration are primitive, the proposed optimization scheme and the solution approach can be easily extended to more complicated applications.

References

1. Tricoles, G.: Radiation Patterns and Boresight Error of a Microwave Antenna Enclosed in an Axially Symmetric Dielectric Shell. J. Opt. Soc. Am. 54, 1094–1101 (1964)
2. Paris, D.T.: Computer-Aided Radome Analysis. IEEE. Trans. on AP 18, 7–15 (1970)
3. Wu, D.C.F., Rudduck, R.C.: Wave Spectrum Surface Integration Technique for Radome Analysis. IEEE Trans. on AP 22, 497–500 (1974)
4. Povinelli, M.J., Angelo, J.D.: Finite Element Analysis of Large Wavelength Antenna Radome Problems for Leading Edge and Radar Phased Arrays. IEEE Trans. on Magnetics 27, 4299–4302 (1991)
5. Chang, D.C.: A Comparison of Computed and Measured Transmission Data for the AGM-88 Harm Radome. Master's thesis, Naval Postgraduate School, Monterey, California, AD-A274868 (1993)
6. Chan, K.K., Chang, P.R., Hsu, F.: Radome design by simulated annealing technique. In: IEEE International Symposium of Antennas and Propagation Society, pp. 1401–1404. IEEE Press, New York (1992)
7. Chiba, H., Inasawa, Y., Miyashita, H., Konishi, Y.: Optimal radome design with particle swarm optimization. In: IEEE International Symposium of Antennas and Propagation Society, pp. 1–4. IEEE Press, New York (2008)
8. Meng, H., Dou, W., Yin, K.: Optimization of Radome Boresight Error Using Genetic Algorithm. In: 2008 China-Japan Joint Microwave Conference, pp. 27–30. IEEE Press, New York (2008)
9. Meng, H., Dou, W.: Multi-objective optimization of radome performance with the structure of local uniform thickness. IEICE Electronics Express 5, 882–887 (2008)
10. Baker, M.L., Roughen, K.M.: Structural Optimization with Probabilistic Fracture Constraints in the Multidisciplinary Radome Optimization System (MROS). In: 48th AIAA/ASME/ASCE/AHS/ASC Structures, Structural Dynamics, and Materials Conference, Honolulu, Hawaii (2007); AIAA-2007-2311
11. Ishimaru, A.: Electromagnetic wave propagation, radiation, and scattering. Prentice Hall, Englewood Cliffs (1991)
12. Holland, J.H.: Adaptation in Natural and Artificial Systems. University of Michigan Press, Michigan (1975)
13. Riolo, R., Soule, T., Worzel, B.: Genetic programming theory and practice IV. Springer, New York (2007)

Global Similarity and Local Variance in Human Gene Coexpression Networks

Ivan Krivosheev[1], Lei Du[2], Hongzhi Wang[1], Shaojun Zhang[1,2], Yadong Wang[1,*], and Xia Li[1,2,*]

[1] School of Computer Science and Technology, Harbin Institute of Technology Harbin, 150001, China
[2] College of Bioinformatics Science and Technology, Harbin Medical University Harbin, 150081, China

Abstract. For the study presented here, we performed a comparative analysis of whole-genome gene expression variation in 210 unrelated HapMap individuals to assess the extent of expression divergence between 4 human populations and to explore the connection between the variation of gene expression and function. We used the GEO series GSE6536 to compare changes in expression of 47,294 human transcripts between four human populations. Gene expression patterns were resolved into gene coexpression networks and the topological properties of these networks were compared. The interrogation of coexpression networks allows for the use of a well-developed set of analytical and conceptual tools and provides an opportunity for the simultaneous comparison of variation at different levels of systemic organization, i.e., global vs. local network properties. The results of this comparison indicate that human co-expression networks are indistinguishable in terms of their global properties but show divergence at the local level.

Keywords: gene coexpression network interrogation.

1 Introduction

In the last few years, gene coexpression networks attracts attention of many researches[1,2]. According to previous studies, these networks are considered as a graph where each node represents a gene, and edges represent statistically high relationship between genes. The interaction between two genes in a gene network does not necessarily imply a physical interaction[4]. For the study presented here, we performed a comparative analysis of whole-genome gene expression variation in 210 unrelated HapMap individuals[7] to assess the extent of expression divergence between 4 human populations and to explore the connection between the variation of gene expression and function. Gene coexpression network could be constructed by using the GeneChip expression profiles data in NCBI GEO (Gene Expression Omnibus repository, database of gene expression data). [3]

---

* Corresponding authors.

H. Deng et al. (Eds.): AICI 2009, LNAI 5855, pp. 159–166, 2009.
© Springer-Verlag Berlin Heidelberg 2009

2 Materials and Methods

2.1 Data Sets

Expression profiles of human gene pairs were compared in order to evaluate the divergence of human gene expression patterns. A total of 47,294 human transcripts for every population were considered. All-against-all gene expression profile comparisons for human populations' matrices (47294*60 CEU, 47294*45 CHB, 47294*45 JPT, and 47294*60 YRI) were used to generate population-specific coexpression networks. For coexpression networks, nodes correspond to genes, and edges link two genes from the same population if their expression profiles are considered sufficiently similar.

2.2 Network Construction

There are a number of existing reverse-engineering methods to construct coexpression network such as Relevance Networks, Bayesian networks etc. Results reported here are for networks constructed using Algorithm for the Reconstruction of Accurate Cellular Networks (ARACNE)[8], a novel information-theoretic algorithm for the reverse-engineering of transcriptional networks from microarray data. ARACNE infers interactions based on mutual information between genes, an information-theoretic measure of pairwise correlation. ARACNE compares favorably with existing methods and scales successfully to large network sizes.

ARACNE relies on a two-step process. First, candidate interactions are identified by estimating pairwise gene-gene mutual information (MI) and by filtering them using an appropriate threshold, I0, computed for a specific p-value, p0, in the null-hypothesis of two independent genes. This step is almost equivalent to the Relevance Networks method. Thus, in its second step, ARACNE removes the vast majority of indirect candidate interactions using a well-known property of mutual information – the data processing inequality (DPI) The DPI states that if genes g1 and g3 interact only through a third gene, g2, (i.e., if the interaction network is $g1<->...<->g2<->...<->g3$ and no alternative path exists between g1 and g3), then the following holds $I(g1,g3) \leq \min[I(g1,g2);I(g2,g3)]$.

For presented study we set mutual information threshold value at 0.001 and DPI value at 0.1. If the P value from this test is smaller than a predefined threshold, the edge will be preserved in the consensus network. When those threshold and DPI values are used, coexpression networks tend to congeal into graphs that are so densely connected as to preclude meaningful analysis of their topological properties.

3 Results and Discussion

3.1 Human Population Networks

Human population coexpression networks were evaluated with respect to a number of parameters describing their global topological properties and found to be highly similar (Table 1). For each network were computed a variety of parameters provided by Cytoscape (open source bioinformatics software platform) plugin, Network Analyzer[9]. The numbers of nodes and edges in each network are comparable, with the CEU population network showing higher values for both. The average degree ($<k>$) is the average number of edges per node and gives rough approximation of how dense the network is. The CEU population network shows a slightly higher $<k>$ which is consistent with the greater numbers of nodes and edges. However, $<k>$ is again similar for all networks and rather high. By way of comparison, typical world-wide web networks have $<k> \approx 7$. The values of $<k>$ might not be particularly relevant because, as will be shown below, the degree distributions are highly skewed.

A more refined notion of network density is given by the average clustering coefficient ($<C>$). The clustering coefficient C of a node i is defined as the fraction of the pairs of neighbors of node i that are linked to each other: $C_i = 2n_i / k_i(k_i-1)$, where n_i is the number of observed links connecting the k_i neighbors of node i and $k_i(k_i-1)/2$ is the total number of possible links. The average clustering coefficient ($<C>$) is the mean of this value for all nodes with at least two neighbors, and for all human population networks $<C> \approx 0.3$ (Table 1). For networks of this size, these $<C>$ values are considered to be quite high. The high density of the coexpression networks is not necessarily surprising because, as one could reasonably expect, co-expression is, largely (but not entirely), transitive. In other words, if gene A is coexpressed with genes B and C, then genes B and C are likely to be coexpressed as well.

Table 1. Network parameters

Parameter	Population			
	CEU	CHB	JPT	YRI
Clustering coefficient	0.350	0.272	0.304	0.320
Network diameter	16	19	24	26
Network centralization	0.047	0.032	0.066	0.042
Average degree	4.576	10.721	16.46	13.781
Number of nodes	5546	3180	3572	3061
Number of edges	72073	17047	29398	21092
Network density	0.005	0.003	0.005	0.005
Network heterogeneity	1.539	1.601	1.778	1.622
Characteristic path length	25.991	6.297	6.733	7.246

3.2 Intersection Network

As described above, the human population gene coexpression networks are closely similar in terms of their global topological characteristics; they share similar node degree (k) distributions and $C(k)$ distributions as well as similar average node degrees ($<k>$), clustering coefficients ($<C>$) and path lengths ($<l>$). Other parameters related to neighborhood, such as network density, network centralization and network heterogeneity are closely similar.

We further sought to evaluate the similarity between the population-specific coexpression networks at a local level. There is as yet no general method for assessing local network similarity (or graph isomorphism). However, in the case of the human population gene coexpression networks generated here, the use of orthologous gene pairs results in a one-to-one mapping between the nodes of the two networks. In this sense, the networks can be considered to be defined over the same set of nodes N, and thus can be directly compared by generating an intersection network. The human population intersection network is defined as the network over the set of nodes N where there is a link between two nodes i and j if i and j denote two pairs of orthologous genes which are connected in every human population network. Thus, the intersection network captures the coexpressed gene pairs conserved between 4 human populations.

The global characteristics of the intersection network are shown in Table 2. The intersection network node degree and $C(k)$ distributions are clearly similar to those of the population-specific networks as are the average clustering coefficient ($<C>$ = 0.213) and average path length ($<l>$ = 3.04). Network diameter equals 10. The network diameter and the average shortest path length, also known as the characteristic path length, indicate small-world properties of the analyzed network. Taken together, these findings indicate that the global structure of the population-specific coexpression networks is preserved in the intersection network. However, the most striking feature of the intersection network is the small fraction of genes (~20%) and edges (~4–16%) that are conserved between populations networks (Table 3). Accordingly, the average node degree is lower ($<k>$ = 7.518) in the intersection network than it is in each of the population-specific networks.

Table 2. Number of nodes and edges in intersection network

	Nodes	Edges
Intersection network	713	2680
CEU	5546 (13%)	72073 (4%)
CHB	3180 (22%)	17047 (16%)
JPT	3572 (20%)	29398 (9%)
YRI	3061 (23%)	21092 (13%)

Table 3. Intersection network parameters

Parameter	Intersection network	Random network
Clustering coefficient	0.213	0.001
Network diameter	10	10
Network centralization	0.061	-
Average degree	7.518	0.0
Number of nodes	713	713
Number of edges	2680	2680
Network density	0.011	-
Network heterogeneity	1.592	-
Characteristic path length	3.040	0.005

Random network generated according Erdos-Renyi model, it has same number of nodes and edges with intersection network.

3.3 Functional Coherence of Gene Coexpression Networks

Genes in the networks were functionally categorized using their Gene Ontology (GO) biological process annotation terms. Overrepresented GO terms were identified with BINGO[14] by comparing the relative frequencies of GO terms in specific clusters with the frequencies of randomly selected GO-terms. The Hypergeometric test was used to

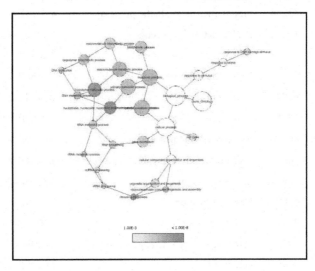

Fig. 1. Overrepresented GO terms

do this with the Benjamini and Hochberg false discovery rate correction for multiple tests and a *P*-value threshold of 0.001. Pairwise similarities between gene GO terms were measured using the semantic similarity method, which computes the relative distance between any two terms along the GO-graph . Result is shown in Table 4.

The graph (Figure 1) visualizes the GO categories that were found significantly over-represented in the context of the GO hierarchy. The size (area) of the nodes is proportional to the number of genes in the test set which are annotated to that node. The color of the node represents the (corrected) *p*-value. White nodes are not significantly over-represented, the other ones are, with a color scale ranging from yellow (*p*-value = significance level, e.g. 0.001) to dark orange (*p*-value = 5 orders of magnitude smaller than significance level, e.g. $10^{-5} * 0.001$). The color saturates at dark orange for *p*-values which are more than 5 orders of magnitude smaller than the chosen significance level.

Table 4. GO process annotation

GO-ID	Description	p-value
43283	biopolymer metabolic process	3.4840E-11
42254	ribosome biogenesis	8.2487E-11
43170	macromolecule metabolic process	7.3685E-9
6259	DNA metabolic process	1.1960E-8
8152	metabolic process	1.3627E-8
44237	cellular metabolic process	2.3559E-8
22613	ribonucleoprotein complex biogenesis and assembly	4.3938E-8
43284	biopolymer biosynthetic process	4.6317E-8
16072	rRNA metabolic process	6.9420E-8
44238	primary metabolic process	7.6631E-8
9058	biosynthetic process	1.9158E-7
6394	RNA processing	1.9444E-7
6365	rRNA processing	4.0186E-7
16070	RNA metabolic process	6.0819E-7
9059	macromolecule biosynthetic process	6.2404E-7
10467	gene expression	9.9307E-7
6260	DNA replication	1.1083E-6
34470	ncRNA processing	3.6057E-6
6974	response to DNA damage stimulus	6.6203E-6
7049	cell cycle	1.0995E-5
6996	organelle organization and biogenesis	1.4177E-5

In fact, from the figure it could be seen that the category 'biopolymer metabolic process' is the important one, and that the over-representation of 'macromolecule metabolic process' and 'metabolic process' categories is merely a result of the presence of those 'protein modification' genes. The fact that both categories are colored equally dark, is due to the saturation of the node color for very low p-values.

4 Conclusion

The global topological properties of the human population gene coexpression networks studied here are very similar but the specific architectures that underlie these properties are drastically different. The actual pairs of orthologous genes that are found to be co-expressed in the different population are highly divergent, although we did detect a substantial conserved component of the co-expression network. One of the most prevalent functional classes that show clear function-expression coherence are genes involved in biopolymer metabolism. Example of these cluster is shown in Figure 2.

The biological relevance of the global network topological properties appears questionable[10]. Of course, this does not prevent network analysis from being a powerful approach, possibly, the most appropriate one for the quantitative study of complex systems made up of numerous interacting parts.

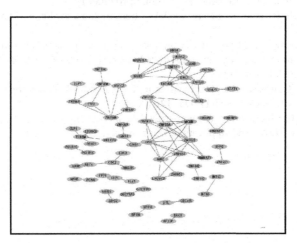

Fig. 2. Cluster of tightly coexpressed and functionally coherent genes

Acknowledgements

This work was supported in part by the National Natural Science Foundation of China (Grant Nos. 30871394, 30370798 and 30571034), the National High Tech Development Project of China, the 863 Program (Grant Nos. 2007AA02Z329), the National Basic Research Program of China, the 973 Program (Grant Nos. 2008CB517302) and the National Science Foundation of Heilongjiang Province (Grant Nos. ZJG0501, 1055HG009, GB03C602-4, BMFH060044, and D200650).

References

1. Horvath, S., Dong, J.: Geometric interpretation of gene coexpression network analysis. PLoS Comput. Biol. 4(8), e1000117 (2008)
2. Carter, S.L., et al.: Gene co-expression network topology provides a framework for molecular characterization of cellular state. Bioinformatics 20(14), 2242–2250 (2004)
3. Bansal, M., et al.: How to infer gene networks from expression profiles. Mol. Syst. Biol. 3, 78 (2007)
4. Potapov, A.P., et al.: Topology of mammalian transcription networks. Genome Inform. 16(2), 270–278 (2005)
5. Yu, H., et al.: The importance of bottlenecks in protein networks: correlation with gene essentiality and expression dynamics. PLoS Comput. Biol. 3(4), e59 (2007)
6. Stranger, B.E., et al.: Relative impact of nucleotide and copy number variation on gene expression phenotypes. Science 315(5813), 848–853 (2007)
7. The International HapMap Project. Nature 426(6968), 789–796 (2003)
8. Margolin, A.A., et al.: ARACNE: an algorithm for the reconstruction of gene regulatory networks in a mammalian cellular context. BMC Bioinformatics 7(suppl. 1), S7 (2006)
9. Vlasblom, J., et al.: GenePro: a Cytoscape plug-in for advanced visualization and analysis of interaction networks. Bioinformatics 22(17), 2178–2179 (2006)
10. Tsaparas, P., et al.: Global similarity and local divergence in human and mouse gene co-expression networks. BMC Evol. Biol. 6, 70 (2006)
11. Khaitovich, P., et al.: A neutral model of transcriptome evolution. PLoS Biol. 2(5), E132 (2004)
12. Yanai, I., Graur, D., Ophir, R.: Incongruent expression profiles between human and mouse orthologous genes suggest widespread neutral evolution of transcription control. OMICS 8(1), 15–24 (2004)
13. Jordan, I.K., Marino-Ramirez, L., Koonin, E.V.: Evolutionary significance of gene expression divergence. Gene 345(1), 119–126 (2005)
14. Babu, M.M.: Introduction to microarray data analysis. In: Grant, R.P. (ed.) Computational Genomics: Theory and Application. Horizon Press, Norwich (2004)

A Grid Based Cooperative Co-evolutionary Multi-Objective Algorithm

Sepehr Meshkinfam Fard[1], Ali Hamzeh[2], and Koorush Ziarati[2]

[1] Department of mathematics, College of Science, Shiraz University, Shiraz, Iran
[2] CSE and IT Department, School of ECE, Shiraz University, Mollasadra Ave, Shiraz, Iran
Sepehr.meshkinfam@gmail.com, ali@cse.shirazu.ac.ir,
ziarati@shirazu.ac.ir

Abstract. In this paper, a well performing approach in the context of Multi-Objective Evolutionary Algorithm (MOEA) is investigated due to its complexity. This approach called NSCCGA is based on previously introduced approach called NSGA-II. NSCCGA performs better than NSGA-II but with a heavy load of computational complexity. Here, a novel approach called GBCCGA is introduced based on MOCCGA with some modifications. The main difference between GBCCGA and MOCCGA is in their niching technique which instead of the traditional sharing mechanism in MOCCGA, a novel grid-based technique is used in GBCCGA. The reported results show that GBCCGA performs roughly the same as NSCCGA but with very low computational complexity with respect to the original MOCCGA.

Keywords: Genetic algorithms, multi-objective, grid based, cooperative, co-evolutionary, GBCCGA.

1 Introduction

The basic idea of Evolutionary Algorithms (EA) [1] is to encode candidate solutions for specific problems into corresponding chromosomes and then evolve this chromosome via some iterative evolutions and selection phases to achieve the best possible chromosome and decode it as the resulted solution.

Genetic algorithms are a well-known family in the area of EA which have been designed and developed by John Holland in 1960s and were developed by Holland and his colleagues and students at the university of Michigan in during the 1960s and the 1970s [2]. In this method, chromosomes are in form of strings form alphabet {0, 1}. There exists three main phases in the evolution process of GA: Crossover or recombination, mutation and the selection phase where crossover and mutation are responsible to produce new chromosomes and selection tries to select best of them.

Genetic algorithms are well-known problem-solver in the area of multi-objective optimization. Fonseca and Fleming [3] represented the idea of the relationship between the fitness function used in GA to discover good chromosomes and the Pareto Optimality concept used in multi-objective optimization. Since then, several methods in usage of genetic algorithm to solve multi-objective problems have been composed such as those mentioned in [1].

H. Deng et al. (Eds.): AICI 2009, LNAI 5855, pp. 167–175, 2009.
© Springer-Verlag Berlin Heidelberg 2009

One of these techniques, called MOCCGA which will be explained further, is based on using the Co-operative Co-evolutionary GA in the search process.

A Co-evolutionary algorithm is a kind of evolutionary algorithm in which fitness of a candidate individual is evaluated based on the relationship between candidate individual and others.

In this paper, MOCCGA approach has been modified by replacing its niching technique and it is proved that its performance is comparable with some recently introduced techniques but in a more efficient manner.

The organization of this paper is as follows: Section 2 introduces the multi-objective optimization problem and the approach of using EA to solve those problems. In section 3, some important related works are mentioned in brief. The original MOCCGA and the proposed modified version are introduced in section 4. The benchmarks and results are given in section 5 following by the conclusion.

2 Multi-Objective Optimization

Mathematical foundations of multi-objective optimization were introduced during 1895-1906 [4]. The concept of vector maximum problem was introduced by Harold W. Kuhn and Albert W. Tucker [5].

Suppose $F = (f_1(x), f_2(x),..., f_m(x)): X \rightarrow Y$ is   a   multi-objective   function where $x = (x_1, x_2,..., x_n) \in X$. Now we can define a multi-objective optimization problem as follow:

$$\min F(x) = (f_1(x), f_2(x),..., f_m(x))$$
$$subject \ \ to: \ \ x \in X \tag{1}$$

Where X is called the decision space and Y is called the objective space.

Another important concept is domination which is defined as follows: A vector x' is said to dominate a vector x'' regarding problem F, if for each $i = 1,2,...,m$, $f_j(x') \leq f_j(x'')$ and there exists at least one $j \in \{1,2,...,m\}$ that $f_j(x') < f_j(x'')$. A non-dominated solution is not dominated by any other solution in the search space. So, in the process of solving multi-objective optimization problems, the goal is to find as much as possible of the existing non-dominate solution. Vilfredo Pareto [6] introduced the concept of optimality in multi-objective optimization. A vector $x_p \in X$ is said to be *Pareto Optimal* with respect to X if there is no vector $x_d \in X$ such that $F(x_d) = (f_1(x_d), f_2(x_d),..., f_m(x_d))$ dominates $F(x_p) = (f_1(x_p), f_2(x_p),..., f_m(x_p))$.

2.1 Multi-Objective Evolutionary Algorithms

Using EA to solve multi-objective optimization problem, called Multi-Objective Evolutionary Algorithms (MOEAs), is based on two main separate phases:

    1. To find non-dominated solutions and assigning rank to individuals.

    2. Maintain diversity of the population.

These phases are explained in the following.

2.1.1 Finding Non-dominated Solutions

This phase can be illustrated easily employing Multi Objective Genetic Algorithm MOGA [3]. One of the most well-known methods in MOEAs to find the non-dominated solutions and assign their rank is designed by Carlos M. Fonseca and Peter J. Fleming [3] which is used in MOGA. To describe this ranking schema, consider the candidate solution s which is going to be assigned a rank. Then, suppose n be the number of all candidate solutions in the current population that dominate s, rank of s will simply be $n+1$. So, it can be seen that rank of any non-dominated solution is one.

2.1.2 Diversity Preserving

Several diversity preserving techniques are examined in MOEAs. Niching methods [7] are some of the most well-known techniques in this domain. Niching techniques have been employed in several algorithms such as MOCCGA [8], NSGA [9], and NSGA-II [10] and so on. In niching techniques, fitness of each candidate solution is changed relative to the number of other solution which resides in its neighborhood which is called the same niche. In the other word, the more candidate solutions crowded in a niche the more their punishment. In the simplest niching techniques, the niche is defined based on a circular, distance-based shape with the radius of σ_{share}. Another novel niching technique is a grid-based one which is used in Pareto Archived Evolutionary Strategy (PAES) which is designed by J. D. Knowles and D. W. Crone [11]. This method is the one used in this paper and will be going in details.

3 Related Works

Co-operative Co-evolutionary effect is originally used in Co-operative Co-evolutionary Genetic Algorithm (CCGA) that was designed by Potter and De Jong [12]. In CCGA, there exist several sub-populations, which each of them contain partial solutions. Then, an individual from each subpopulation is selected and combined with other individuals to form a total solution. Fitness of each individual will be evaluated based on the fitness of the combined solution. After that, each sub-population is evolved using a traditional genetic algorithm.

Keervativuttitumrong et al. [8] employed CCGA [12] in MOEA field and combine it with MOGA [3] to form the Multi-Objective Co-operative Co-evolutionary Genetic Algorithm (MOCCGA) [8]. This approach is described in the next section in more details.

As the next work in the chain, in [9], Srinivas et al. introduced the non-dominated sorting genetic algorithm (NSGA) which is based on sorting a population according to the level of non-domination.

Deb et al. [10] proposed modified version of NSGA and named it NSGA-II which demonstrates better performance using some tricky modification. Finally, Iorio et al. [13] designed a co-operative co-evolutionary multi-objective genetic algorithm using the non-dominated sorting, by combining NSGA-II with CCGA and called NSCCGA. At the best of our knowledge, NSCCGA is the current state of the art based on CCGA in the family of MOEA But same as previous methods it suffers from the high computational complexity due to the non-domination sorting mechanism. In this paper, we are going to introduce a mechanism based on CCGA. It is to be mentioned that it has a comparative performance to NSCCGA but with very lower computational complexity.

4   Grid Based Multi-Objective Co-operative Co-evolutionary Genetic Algorithm

In this paper, MOCCGA has been modified employing a new niching technique adapted from PAES [11] with some modification. This new approach is called Grid Based multi-objective Co-operative Co-evolutionary Genetic Algorithm or GBCCGA. To describe this new method, at first, we must describe PAES and MOCCGA in more details.

4.1   PAES's Diversity Technique

PAES has a novel diversity technique proposed in [11] based on a low computational complexity archiving mechanism. This archive stores and maintains some of non-dominated solutions that have been found by the algorithm from the beginning of the evolution. Also, PAES utilize a novel fitness sharing method in which the objective space is divided into different hyper cubes recursively. Then, each solution, in accordance with its objectives, is inserted into one of those hyper cubes and its fitness is decreased based on the number of previously found solutions reside in the same hypercube. This diversity technique has a lower computational complexity comparing other diversity technique such as fitness sharing in MOGA [3] and clustering in NSGA-II [10] where show promising performance to keep diversity of an evolving population. In this technique, the objective space is divided recursively to form a hyper grid. This hyper grid divides the objective space into hyper cubes where each hyper cube has width of $d_r/2^l$, where d_r is the range of values in objective d of the current solution, and l is an arbitrary value selected by user.

4.2   MOCCGA in More Details

MOGA [3] and CCGA [12] was integrated together to form MOCCGA in [8]. This algorithm decomposes the problem into sub-populations according to the dimension of the search space. MOCCGA assigns rank to each individual of every sub-population with respect to its sub-population. The ranking schema is similar to MOGA's ranking schema [3]. Each candidate individual of a sub-population is combined with the best individual of other sub-populations to form a complete solution. After that, this solution is compared with other solutions and ranked based on Pareto dominance. Finally, this rank is assigned to candidate individual. For the purpose of diversity, fitness sharing mechanism is used and applied in the objective space. Then each sub-population is evolved separately using traditional genetic algorithm.

4.3 The Overall Architecture of GBCCGA

Considering results reported by NSGA-II [10] and NSCCGA [13], they perform better than MOCCGA. All of them use the same niching technique and a very computational complex domination level mechanism. But, based on our analysis, we hypothesize that performance of MOCCGA is heavily depends on its niching technique. So, we decide to adapt PAES's diversity mechanism, as a well-known, well performing niching mechanism, in MOCCGA and investigate its performance. So, GBCCGA is developed based on the original MOCCGA and a fitness sharing mechanism inspired by PAES. Like MOCCGA, GBCCGA generates initial population randomly. It decomposes this population into several sub-populations according to the dimension of the search space. Each candidate individual of every sub-population is combined with the best individuals of other sub-populations to form a complete solution.

Again, it is combined with randomly chosen individuals of other sub-populations to form another complete solution. Then, all of these solutions are ranked based on Pareto non-domination schema [3].

Now, the niching mechanism is activated to calculate the penalty function for each candidate solution. The main difference between GBCCGA and MOCCGA is in this step where the latter uses fitness sharing mechanism from MOGA [3] and former uses modified PAES's niching technique. The first step to adapt this technique for use in GBCCGA is to define a policy to be able to share fitness of all individuals in a different sub-population based on one archive.

1.   **for** g=2 to #of generations do
2.   **for** each sub-population do
3.   Evaluate fitness of each candidate   individual in a sub-population
a.   Combining it with the best individuals of other sub-populations.
b.   Combining it with the random individuals of the other sub-populations.
c.   Rank all generated combined solutions (two for each individual) based on a simple Pareto ranking method.
d.   Assign the better rank (between those two generated combined solutions for each individual) to each individual in the current sub-population.
e.   Apply niching penalty based on the archive grid.
f.   Store rank 1 solution in the grid as described before.
4.   **end for**
5.   **end for**

Fig. 1. The Overall Procedure of GBCCGA

To achieve this, we compare both combined solutions: the one which is based on selection of the best partner and the random-based one, according to Pareto ranking schema. Then, the better one is located in the archive grid, based on the value of its objective function. After that, for each archived solution resides in the same hypercube of that grid, one unit of penalty is applied to the rank of that combined solution. Then, after applying penalty, the rank is assigned to the candidate individual, which participates in combined solution, of the current sub-population as its fitness value. Also, if the rank of the candidate individual is one, the corresponding combined solution is stored in the proper location at the archive grid. So, each candidate individual in any sub-population can participate in forming the same archive.

Now, it can be seen that in GBCCGA, the original fitness sharing of MOCCGA is replaced with PAES based one as described above. In figure 1, the overall procedure of GBCCGA is depicted.

5   Test Problems and Experimental Results

The chosen benchmarks are those which are commonly used in the literature such as [8], [9], [10] and [13]. These problems were originally developed by Zitzler et al. [14]. They are called ZDT1 to ZDT5. ZDT1 has a convex Pareto front, ZDT2 has a non-convex Pareto front, ZDT3's Pareto front is a discrete one (5 discontinues convex parts), ZDT4 with 100 Pareto optimal front where only one of them is the global optimum and finally ZDT5 with a deceptive Pareto front. In this section, GBCCGA's result is compared with the MOCCGA's, because our algorithm is a modification of MOCCGA, and NSCCGA as the current state of the art in CCGA family. Results of NSCCGA are drawn directly from [13]. It must be noted that to justify our implementation[1] the MOCCGA's results are successfully regenerated in our experiments and represented in figures 2 to 6. All results are average over 5 independent runs where all approaches are allowed to run for a maximum number of 40 generations with the population size of 200. Also, the parameter settings of GBCCGA are as follows: $p_c=.7$, $p_m=0.01$ and $l=30$.

As you can see in figures 2 to 6, GBCCGA is tested against MOCCGA and NSCCGA. The outcome indicates that GBCCGA outperforms MOCCGA in all test cases. Also it remains competitive to NSCCGA in ZDT1, ZDT2 and ZDT3 but In the case of ZDT4, NSGA-II finds better convergence and spread of solutions than GBCCGA. It can be seen that, as we hypothesized, the main weakness of MOCCGA is its simple niching mechanism which is replaced by a more sophisticated and effective one adapted from PAES in GBCCGA. It can be seen that using this modification, GBCCGA produced results similar to NSCCGA with less computational complexity. Unreported results depicted that GBCCGA can solve ZDT test suite problem in about 1.8 times faster than NSCCGA (our implementation). Regarding these results and the fact that the leveling mechanism in NSCCGA has a high level polynomial order, it can be concluded that GBCCGA demonstrated promising research direction in the context of MOGA generally and MOCCGA based approaches specially.

---

[1] The implementation is available online at
http://www.cse.shirazu.ac.ir/~ali/GBCCGA/matlab-code.zip

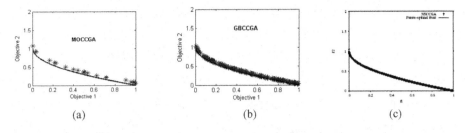

Fig. 2. Non-dominated solutions for ZDT1 by (a) MOCCGA, (b) GBCCGA and (c) NSCCGA

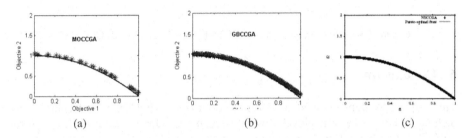

Fig. 3. Non-dominated solutions for ZDT2 by (a) MOCCGA, (b) GBCCGA and (c) NSCCGA

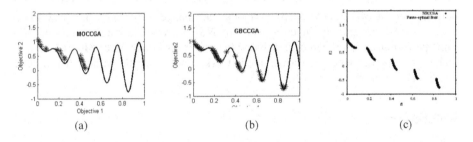

Fig. 4. Non-dominated solutions for ZDT3 by (a) MOCCGA, (b) GBCCGA and (c) NSCCGA

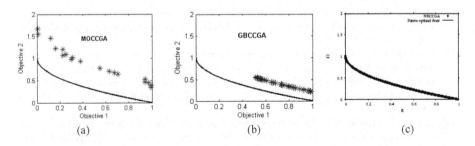

Fig. 5. Non-dominated solutions for ZDT4 by (a) MOCCGA, (b) GBCCGA and (c) NSCCGA

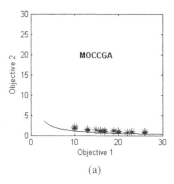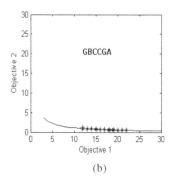

Fig. 6. Non-dominated solutions for ZDT5 by (a) MOCCGA, (b) GBCCGA

6 Conclusion

This paper has demonstrated a modified version of MOCCGA by replacing its niching technique with a Novel diversity technique adopted from PAES to form GBCCGA that has lower computational complexity and remains competitive to current state of the art NSCCGA [13] Since we use a parameter-free niching mechanism, because it is not necessary to select the σ_{share} factor which is highly effective in the performance of MOCCGA. Also, as another advantage, GBCCGA has very lower computational complexity in compare with NSCCGA while roughly produce the same performance. So it can be concluded that the most weakness of MOCCGA is its niching mechanism which is shown in this work to be repairable without heavy computational complexity such as NSCCGA.

References

[1] Coello Coello, C.A., Lamont, G.B., Van Veldhuizen, D.A.: Evolutionary Algorithms for Solving Multi-Objective Problems, 2nd edn. Springer Science+Business Media, LLC, New York (2007)

[2] Holland, J.H.: Adaptation in Natural and Artificial Systems. An Introductory Analysis with Applications to Biology. Control and Artificial Intelligence. University of Michigan Press, Ann Arbor (1975)

[3] Fonseca, C.M., Fleming, P.J.: Genetic Algorithms for Multi-objective Optimization: Formulation, Discussion and Generalization. In: Forrest, S. (ed.) Proceedings of the Fifth International Conference on Genetic Algorithms, San Mateo, California, University of Illinois at Urbana-Champaign, pp. 416–423. Morgan Kaufmann Publishers, San Francisco (1993)

[4] Stadler, W.: Initiators of Multicriteria Optimization. In: Jahn, J., Krabs, W. (eds.) Recent Advances and Historical Development of Vector Optimization, pp. 3–47. Springer, Berlin (1986)

[5] Kuhn, H.W., Tucker, A.W.: Nonlinear Programming. In: Neyman, J. (ed.) Proceedings of the Second Berkeley Symposium on Mathematical Statistics and Probability, pp. 481–492. University of California Press, California (1951)

[6] Pareto, V.: Cours D'Economie Politique, vol. I, II. F. Rouge, Lausanne (1896)

[7] Niching Methods for Genetic Algorithms Samir W. Mahfoud 3665E. Bay Dr. #204-429 Largo, FL 34641 IlliGAL Report No. 95001 (May 1995)

[8] Keerativuttitumrong, N., Chaiyaratana, N., Varavithya, V.: Multi-objective Co-operative Co-evolutionary Genetic Algorithm. In: Guervós, J.J.M., Adamidis, P.A., Beyer, H.-G., Fernández-Villacañas, J.-L., Schwefel, H.-P. (eds.) PPSN 2002. LNCS, vol. 2439, pp. 288–297. Springer, Heidelberg (2002)

[9] Srinivas, N., Deb, K.: Multiobjective Optimization Using Nondominated Sorting in Genetic Algorithms. Evolutionary Computation 2(3), 221–248 (Fall 1994)

[10] Deb, K., Pratap, A., Agarwal, S., Meyarivan, T.: A Fast and Elitist Multiobjective Genetic Algorithm: NSGA–II. IEEE Transactions on Evolutionary Computation 6(2), 182–197 (2002)

[11] Knowles, J.D., Corne, D.W.: Approximating the Nondominated FrontUsing the Pareto Archived Evolution Strategy. Evolutionary Computation 8(2), 149–172 (2000)

[12] Potter, M.A., de Jong, K.: A Cooperative Coevolutionary Approach to Function Optimization. In: Davidor, Y., Männer, R., Schwefel, H.-P. (eds.) PPSN 1994. LNCS, vol. 866, pp. 249–257. Springer, Heidelberg (1994)

[13] Iorio, A.W., Li, X.: A Cooperative Coevolutionary Multiobjective Algorithm Using Nondominated Sorting. In: Deb, K., et al. (eds.) GECCO 2004. LNCS, vol. 3102, pp. 537–548. Springer, Heidelberg (2004)

[14] Zitzler, E., Deb, K., Thiele, L.: Comparison of Multiobjective Evolutionary Algorithms: Empirical Results. Evolutionary Computation 8(2), 173–195 (Summer 2000)

Fuzzy Modeling for Analysis of Load Curve in Power System

Pei-Hwa Huang, Ta-Hsiu Tseng, Chien-Heng Liu, and Guang-Zhong Fan

Department of Electrical Engineering, National Taiwan Ocean University,
2 Peining Road, Keelung 20224, Taiwan
B0104@mail.ntou.edu.tw

Abstract. The main purpose of this paper is to study the use of fuzzy modeling for the analysis of customer load characteristics in power system. A fuzzy model is a collection of fuzzy IF-THEN rules for describing the features or behaviors of the data set or system under study. In view of the nonlinear characteristics of customer load demand with respect to time, the method of fuzzy modeling is adopted for analyzing the studied daily load curves. Based on the Sugeno-type fuzzy model, various models with different numbers of modeling rules have been constructed for describing the investigated power curve. Sample results are demonstrated for illustrating the effectiveness of the fuzzy model in the study of power system load curves.

Keywords: power system, load curve, fuzzy model, Sugeno-type fuzzy model.

1 Introduction

Understanding of system load characteristics is a primary concern in power system planning and operations. The load curve, which is developed from the measured data showing load demand with respect to time, plays an important role for the analysis of an electric power customer. Based on the survey results of system load characteristics, system engineers and operators are able to carry out the studies of load forecasting, generation planning, system expansion, cost analysis, tariff design, and so forth. Due to the inherent nonlinearity, many methods have been proposed for examining customer load curves in power systems [1]-[12].

A fuzzy system model is basically a collection of fuzzy IF-THEN rules that are combined via fuzzy reasoning for describing the features of a system under study [13]-[21]. The method of fuzzy modeling has been proven to be well-suited for modeling nonlinear industrial processes described by input-output data. The fuzzy model describes the essential features of a system by using linguistic expressions. The fuzzy model not only offers the accurate expression of the quantitative information for a studied nonlinear system, but also can provide a qualitative description of the physical feature [22]-[25]. In view of the nonlinear characteristics of a the load curve, the fuzzy model is employed for analyzing the curve in that the method of fuzzy modeling is suitable for modeling nonlinear processes described by input-output data.

The main purpose of this paper is to investigate the application of the Sugeno-type fuzzy modeling method [14] for analyzing a power system daily load curve which

H. Deng et al. (Eds.): AICI 2009, LNAI 5855, pp. 176–184, 2009.

depicts the data of a customer load demand versus the time in a day. Different numbers of modeling rules are taken into account in constructing the fuzzy models for representing a daily load curve. The parameters of the Sugeno-type fuzzy model are calculated by employing the algorithm of ANFIS (Adaptive Neuro-Fuzzy Inference System) [15]-[17]. The IF-THEN structure of a fuzzy model can yield both a numerical (quantitative) approximation and a linguistic (qualitative) description for the studied load curve. It is shown from the computational results that the obtained fuzzy models are capable of providing both quantitative and qualitative descriptions for the power system load curve.

2 Load Curve

A load curve is obtained from the measurements of load demand of a certain customer, region, or utility system with respect to time and thus provides a way of understanding how much electric energy is being consumed at different times of day. Based on the time duration considered, there are various types of load curves: daily load curves, weekly load curves, seasonal load curves, and annual load curves. In the load study of power system, the daily load curve is the first concern since it reveals the most fundamental features of a specific load [1]-[6].

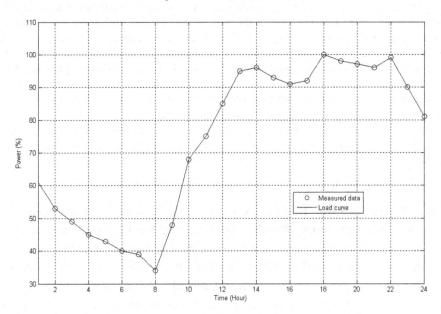

Fig. 1. Residential daily load curve

A typical load curve of a residential customer in a power system is shown in Fig. 1 which is developed from a set of 24 hourly measurements and depicts the load demand variations with the time in a day. It is noted that the curve is normalized with the original power value of each data point converted into its percentage (%) value by choosing the highest (peak) power demand in the curve as the base power value. The highest (peak) demand happened around evening period after all the family returned from work or school. It is evident that the load curve shown in Fig.1 is a nonlinear graph and the method of fuzzy modeling is to be applied for analyzing the load curve.

3 Fuzzy Modeling

The characteristic (membership) function of a crisp (conventional) set assigns each element of the set with a status of membership or non-membership. On the other hand, a set consists of elements with varying degrees of membership associated with the set is referred to as a fuzzy set [13]. Based on fuzzy set theory, the fuzzy model, which consists of a collection of fuzzy IF-THEN rules, is used for describing the behavior or characteristics of the data or system under study [14]-[21]. The fuzzy model expresses an inference mechanism such that if we know a premise, then we can infer or derive a conclusion. In modeling nonlinear systems, various types of fuzzy rule-based system could be described by a collection of fuzzy IF-THEN rules. The objective of this paper is to study an application of the fuzzy model to describe the nonlinear load curve in power system.

Among various types of fuzzy models, the Sugeno-type fuzzy model has recently become one of the major topics in theoretical studies and practical applications of fuzzy modeling and control [14]-[17]. The basic idea of Sugeno-type fuzzy model is to decompose the input space into fuzzy regions and then to approximate the system in every region by a simple linear model. The overall fuzzy model is implemented by combining all the linear relations constructed in each fuzzy region of the input space. The main advantage of the Sugeno-type fuzzy model lies in its form with a linear function representation in each subregion of the system.

In this study, the single-input-single-output (SISO) Sugeno-type fuzzy model consisting of n modeling rules, as described in (1), is adopted for describing a load curve:

$$R_i : \text{IF } T \text{ is } A_i \quad \text{THEN } P = a_i T + b_i, \quad i = 1, 2, \cdots, n. \tag{1}$$

where T (time in a day, Hour) denotes the system input and P (load demand, %) represents the output in the ith subregion. It is noted that the input domain is partitioned into n fuzzy subregions (time periods), each of which is described by a fuzzy set A_i, and the system behavior in each subregion is modeled by a linear equation $P = a_i T + b_i$ in which a_i is the coefficient related with time variation and b_i is a constant. Let m_i denote the membership function of the fuzzy set A_i. If the system input T is at a specific time T^0 then the overall system output P is obtained by computing a weighted average of all the outputs P_i's from each modeling rule R_i as shown in (2), (3) and (4).

$$w_i = m_i(T^0), \quad i = 1, 2, \cdots, n. \tag{2}$$

$$P_i = a_i T^0 + b_i, \quad i = 1, 2, \cdots, n. \tag{3}$$

$$P = (\sum_{i=1}^{n} w_i P_i) / (\sum_{i=1}^{n} w_i). \tag{4}$$

To construct a fuzzy model in (1), the task of model identification is to be performed. Model identification is a process of recognizing the structure in data via comparisons with some known structure. The aim of model identification is to figure out all the parameters of a model. The model identification for the Sugeno-type fuzzy model are usually divided into two tasks: structure identification and parameter identification.

(1) Structure Identification:
    Set the number of modeling rules according to the system characteristics. Different numbers of modeling rules will yield different ways of description for a load curve.

(2) Parameter Identification:
    First select the type of membership functions depending on the shape described in each fuzzy set. Then the parameters of the linear function in each rule are to be identified. In this study, the Gaussian type membership function in (5) is chosen for the fuzzy set in the IF-part of the modeling rule:

$$m_i(x) = \exp\left[-\frac{(x - c_i)^2}{2\sigma_i^2}\right]. \tag{5}$$

    where the variable x stands for the time in a day and the parameters c_i and σ_i define the shape of each membership function $m_i(x)$ for the fuzzy set A_i. Then the data points of load demand measurements are fed into an adaptive network structure, namely the "Adaptive Neuro-Fuzzy Inference System" (ANFIS) [15]-[17], for the purpose of model training to calculate the parameters (c_i, σ_i) of the membership function in the IF-part and the coefficients (a_i, b_i) of the linear function in the THEN-part.

4 Computational Results

The residential daily load curve shown in Fig. 1 is utilized as the modeling study object. The input variable T (Hour) is the time in a day and the output P (%) is the power demand of the customer. Note that the load curve reveals a nonlinear time-dependent characteristic and the approach of fuzzy modeling will be suitable in the analysis of the load curve.

The (SISO) Sugeno-type fuzzy model described in (1) is adopted for describing this load curve. The Gaussian type membership function in (5) is employed for representing the fuzzy set A_i in the IF-part of the modeling rule and the THEN-part is a linear model. Under different preset numbers of modeling rules, the ANFIS algorithm is used for model training to calculate the model parameters. The number of iterations for model training is set to be 1000.

Table 1. Root-mean-squared errors for different numbers of modeling rules

Number of rules	Root-mean-squared error
2	3.7874
3	2.4922
4	1.8353
5	1.4501
6	0.3892

Table 2. Parameters of three-rule model

Rule	(c_i, σ_i)	(a_i, b_i)
1	(1.06, 4.76)	(-4.39, 63.10)
2	(16.97, 3.64)	(-4.32, 139.10)
3	(24.47, 4.74)	(-15.18, 452.90)

Table 3. Parameters of six-rule model

Rule	(c_i, σ_i)	(a_i, b_i)
1	(1.32, 2.53)	(-5.85, 65.52)
2	(5.36, 1.30)	(-2.48, 57.75)
3	(11.35, 0.93)	(5.66, 11.30)
4	(14.13, 1.31)	(-2.47, 130.30)
5	(19.32, 0.88)	(-1.81, 132.80)
6	(24.33, 1.15)	(-9.22, 302.20)

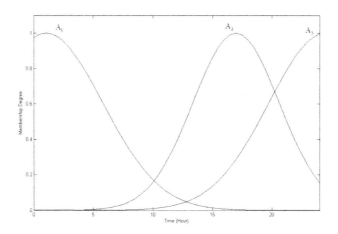

Fig. 2. Membership functions of three-rule model

To better understand the performance of the Sugeno-type fuzzy model, comparative studies are conducted under conditions of different numbers of modeling rules.
The root-mean-squared errors between the fuzzy models and the original measured
load demand data under different numbers (two, three, four, five and six) of modeling
rules are calculated and tabulated as Table 1. The parameters of the fuzzy models with
three modeling rules and six modeling rules are tabulated in Table 2 and Table 3,

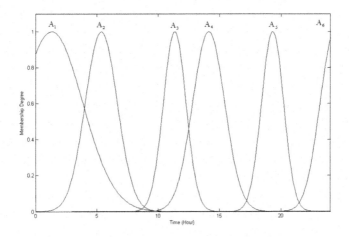

Fig. 3. Membership functions of six-rule model

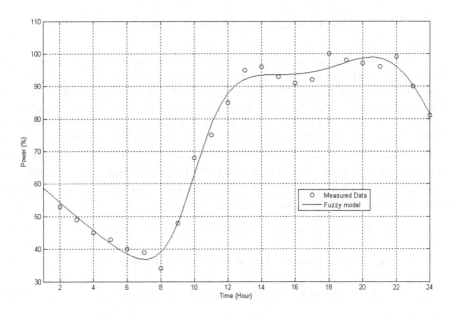

Fig. 4. Three-rule fuzzy model and measured load data

respectively. The fuzzy sets shown in Fig. 2 and Fig. 3 represent the membership functions for the three-rule and six-rule fuzzy models. Moreover, the measured data together with fuzzy modeling output for the three-rule and the six-rule fuzzy models are shown in Fig. 4 and Fig. 5 for the purpose of comparison.

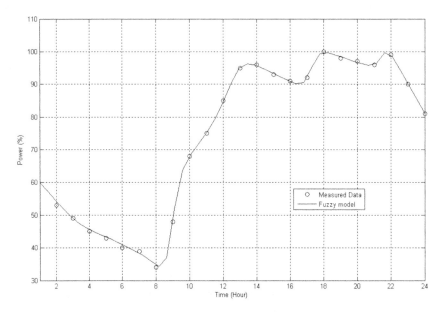

Fig. 5. Six-rule fuzzy model and measured load data

From the root-mean-squared errors in Table 1, it is found that more modeling rules will yield a model with less modeling error while the number of rules vary from two to six. It is also observed from Fig. 4 and Fig. 5 that the fitting accuracy of the six-rule fuzzy model is better than that of the three-rule fuzzy model. Both the three-rule fuzzy model with parameters in Table 2 and the six-rule fuzzy model with parameters in Table 3 can provide IF-THEN rules for representing the studied load curve.

The n-rule fuzzy models in (1) can be converted into models with linguistic rules by giving a proper time label to each fuzzy set A_i in the IF-part of the ith rule. For example, from the parameters in Table 2 and the fuzzy sets in Fig. 2, the three-rule fuzzy model can be converted into the following three linguistic rules in (6)-(8):

$$\text{IF } T \text{ is } \textit{early morning}, \text{ THEN } P = -4.39 \times T + 63.10, \tag{6}$$

$$\text{IF } T \text{ is } \textit{afternoon}, \text{ THEN } P = -4.32 \times T + 139.10, \tag{7}$$

$$\text{IF } T \text{ is } \textit{night}, \text{ THEN } P = -15.18 \times T + 452.90. \tag{8}$$

Then likewise, we can obtain a set of six linguistic rules in (9)-(14) for the six-rule fuzzy model based on the parameters in Table 3 and the fuzzy sets in Fig. 3.

$$\text{IF } T \text{ is } \textit{early morning}, \text{ THEN } P = -5.85 \times T + 65.52, \tag{9}$$

$$\text{IF } T \text{ is } \textit{morning}, \text{ THEN } P = -2.48 \times T + 57.75, \tag{10}$$

$$\text{IF } T \text{ is } \textit{near noon}, \text{ THEN } P = 5.66 \times T + 11.30, \tag{11}$$

$$\text{IF } T \text{ is } \textit{afternoon}, \text{ THEN } P = -2.47 \times T + 130.30, \tag{12}$$

$$\text{IF } T \text{ is } \textit{evening}, \text{ THEN } P = -1.81 \times T + 132.80, \tag{13}$$

$$\text{IF } T \text{ is } \textit{night}, \text{ THEN } P = -9.22 \times T + 302.20. \tag{14}$$

Obviously, both the set of three linguistic rules in (6)-(8) and the set of six linguistic rules in (9)-(14) are capable of providing qualitative descriptions for the load curve.

5 Conclusion

The main purpose of this paper is to present the description and analysis of the power system load curve by fuzzy modeling. In view of the nonlinear characteristic of the power system load curve which depicts the time variations of load demand of a customer, the method of fuzzy modeling is employed for representing the curve. A typical residential daily load curve is employed as the studied load curve. Based on the Sugeno-type fuzzy model, various models with different numbers of modeling rules has been identified to describe the load curve. It is found that such fuzzy model is capable of providing both quantitative and qualitative descriptions for the load curve. The validty of the Sugeno-type fuzzy model in the analysis of power system load curve has been verified in this work.

Acknowledgments. The authors would like to thank Prof. Shu-Chen Wang, Mr. Muh-Guay Yang, and Mr. Min-En Wu for all their help with this work.

References

1. Walker, C.F., Pokoski, J.J.: Residential load shape modeling based on customer behavior. IEEE Trans. on Power Systems and Apparatus 104, 1703–1711 (1985)
2. Talukdar, S., Gellings, C.W.: Load Management. IEEE Press, Los Alamitos (1987)
3. Bjork, C.O.: Industrial Load Management-Theory, Practice and Simulations. Elsevier, Amsterdam (1989)
4. Tweed Jr., N.B., Stites, B.E.: Managing load data at Virginia Power. IEEE Computer Applications in Power 5, 25–29 (1992)
5. Schrock, D.: Load Shape Development. PennWell Publishing Co. (1997)
6. Pansini, A.J., Smalling, K.D.: Guide to Electric Load Management. PennWell Publishing Co. (1998)

7. Chen, C.S., Hwang, J.C., Tzeng, Y.M., Huang, C.W., Cho, M.Y.: Determination of customer load characteristics by load survey system at Taipower. IEEE Trans. on Power Delivery 11, 1430–1435 (1996)
8. Chen, C.S., Hwang, J.C., Huang, C.W.: Application of load survey systems to proper tariff design. IEEE Trans. on Power System 12, 1746–1751 (1997)
9. Senjyu, T., Higa, S., Uezato, K.: Future load curve shaping based on similarity using fuzzy logic approach. IEE Proceedings-Generation, Transmission and Distribution 145, 375–380 (1998)
10. Konjic, T., Miranda, V., Kapetanovic, I.: Prediction of LV substation load curves with fuzzy inference systems. In: 8th International Conference on Probabilistic Methods Applied to Power Systems, pp. 129–134. Iowa State University (2004)
11. Zhang, C.Q., Wang, T.: Clustering analysis of electric power user based on the similarity degree of load curve. In: 4th International Conference on Machine Learning and Cybernetics, vol. 3, pp. 1513–1517. IEEE Press, Los Alamitos (2005)
12. Zhang, J., Yan, A., Chen, Z., Gao, K.: Dynamic synthesis load modeling approach based on load survey and load curves analysis. In: 3rd International Conference on Electric Utility Deregulation and Restructuring and Power Technologies (DRPT 2008), pp. 1067–1071. IEEE Press, Los Alamitos (2008)
13. Zadeh, L.A.: Fuzzy sets. Information and Control 8, 338–353 (1965)
14. Sugeno, M., Kang, G.: Structure identification of fuzzy model. Fuzzy Sets and Systems 28, 15–23 (1988)
15. Jang, J.S.R.: ANFIS: Adaptive-network-based fuzzy inference system. IEEE Trans. on Systems, Man and Cybernetics 23, 665–685 (1993)
16. Jang, J.S.R., Sun, C.T.: Neuro-fuzzy modeling and control. Proceedings of the IEEE 83, 378–406 (1995)
17. Jang, J.S.R., Sun, C.T., Mizutani, E.: Neuro-Fuzzy and Soft Computing. Prentice-Hall, Englewood Cliffs (1997)
18. Bezdek, B.: Fuzzy models-What are they, and why. IEEE Trans on. Fuzzy Systems 1, 1–6 (1993)
19. Sugeno, M., Yasukawa, T.: A fuzzy-logic-based approach to qualitative modeling. IEEE Trans. on Fuzzy Systems 1, 7–31 (1993)
20. Yager, R., Filev, D.: Essentials of Fuzzy Modeling and Control. John Wiley & Sons, Chichester (1994)
21. Ross, T.: Fuzzy Logic with Engineering Applications. Wiley, Chichester (2004)
22. Huang, P.H., Chang, Y.S.: Fuzzy rules based qualitative modeling. In: 5th IEEE International Conference on Fuzzy Systems, pp. 1261–1265. IEEE Press, Los Alamitos (1996)
23. Huang, P.H., Jang, J.S.: Fuzzy modeling for magnetization curve. In: 1996 North American Power Symposium, pp. 121–126. Mass. Inst. Tech. (1996)
24. Huang, P.H., Chang, Y.S.: Qualitative modeling for magnetization curve. Journal of Marine Science and Technology 8, 65–70 (2000)
25. Wang, S.C., Huang, P.H.: Description of wind turbine power curve via fuzzy modeling. WSEAS Trans. on Power Systems 1, 786–792 (2006)

Honey Bee Mating Optimization Vector Quantization Scheme in Image Compression

Ming-Huwi Horng

Department of Information Engineering and Computer Science, National Pingtung Institute
of Commerce, PingTung, Taiwan
horng@npic.edu.tw

Abstract. The vector quantization is a powerful technique in the applications of
digital image compression. The traditionally widely used method such as the
Linde-Buzo-Gray (LBG) algorithm always generated local optimal codebook.
Recently, particle swarm optimization (PSO) is adapted to obtain the near-global
optimal codebook of vector quantization. In this paper, we applied a new swarm
algorithm, honey bee mating optimization, to construct the codebook of vector
quantization. The proposed method is called the honey bee mating optimization
based LBG (HBMO-LBG) algorithm. The results were compared with the other
two methods that are LBG and PSO-LBG algorithms. Experimental results
showed that the proposed HBMO-LBG algorithm is more reliable and the reconstructed images get higher quality than those generated form the other three
methods.

Keywords: Vector quantization, LBG algorithm, Particle swarm optimization,
Quantum particle swarm optimization, Honey bee mating optimization.

1 Introduction

Vector quantization techniques have been used for a number of years for data compression. The operations of VQ include dividing the image to be compressed into
vectors (or blocks) and each vector is compared to the codewords of a codebook to
find its reproduction vector. The codeword, which is most similar to an input vector,
is called the reproduction vector of input vector. In the encoding process, an index,
which points to the closest codeword of an input vector, is determined. Normally, the
size of the codebook is much smaller than the original image data set. Therefore, the
purpose of image compression is achieved. In the decoding process, the associated
sub-image is exactly retrieved by the same codebook which as been used in the encoding phase. When each sub-image is completely reconstructed, the decoding is
completed.

Vector quantization (VQ) algorithms have been performed by many researchers;
new algorithms continue to appear. The generation of codebook is known as the most
important process of VQ. The k-means based algorithms are designed to minimize
distortion error by selecting a suitable codebook. A well-known method is the LBG

H. Deng et al. (Eds.): AICI 2009, LNAI 5855, pp. 185–194, 2009.
© Springer-Verlag Berlin Heidelberg 2009

algorithm [1]. However, the LBG algorithm is a local search procedure. It suffers from the serious drawback that its performance depends heavily on the initial starting conditions. Many studies have been undertaken to solve this problem. Chen, Yang and Gou proposed an improvement based on the particle swarm optimization (PSO) [2]. The result of LBG algorithm is used to initialize global best particle by which it can speed the convergence of PSO. In addition, Wang et al. proposed a quantum particle swarm algorithm (QPSO) to solve the 0-1 knapsack problem [3]. Zhao et al. employed a quantum particle swarm optimization to select the thresholds of the multilevel thresholding [4].

Over the last decade, modeling the behavior of social insects, such as ants and bees, for the purpose of search and problems solving had been the context of the emerging area of swarm intelligence. Therefore, the honey-bee mating may also be considered as a typical swarm-based approach for searching for the optimal solution in many application domains such as clustering [5] and multi-level image thresholding selection [6]. In this paper, the honey bee mating optimization (HBMO) associated with the LBG algorithm is proposed to search for the optimal codebook that minimizes the distortion between the training set and the codebook. In other words, HBMO algorithm is a search technique that finds the optimal codebook for the input vectors. Experimental results have demonstrated that the HBMO-LBG algorithms performed better than the LBG and PSO algorithms consistently.

This work is organized as follows. Section 2 introduces the vector quantization and LBG algorithm. Section 3 presents this proposed method which searches for the optimal codebook using the HBMO algorithm. Performance evaluation is discussed in detail in Section 4. Conclusions are presented in Section 5.

2 Vector Quantization

This section provides some basic concepts of vectors quantization and introduces the traditional LBG algorithm.

2.1 Definition

Pl Vector quantization (VQ) is a lossy data compression technique in block coding. The generation of codebook is known as the most important process of VQ. Let the size of original image $Y = \{y_{ij}\}$ be $M \times M$ pixels that divided into several blocks with size of $n \times n$ pixels. In other words, there are $N_b = \left[\frac{N}{n}\right] \times \left[\frac{N}{n}\right]$ blocks that represented by input vectors $X = (x_i, i = 1,2,..., N_b)$ and n is generally assigned to be 4 in experiments. Let L be $n \times n$. The input vector x_i, $x_i \in \Re^L$ where $\Re^L$ is L-dimensional Euclidean space. A codebook C comprises N_c L-dimensional codewords, i.e. $C = \{c_1, c_2,....,c_{N_c}\}$, $c_j \in \Re^L$, $\forall j = 1,2,..,N_c$. Each input vector is represented by a row vector $x_i = (x_{i1}, x_{i2},....,x_{iL})$ and each codeword in the codebook is denoted as $c_j = (c_{j1}, c_{j2}, ..., c_{jL})$. The VQ techniques assign each input vector to a related codeword, and the codeword will replace the associated input vectors finally to obtain the aim of compression.

The optimization of C in terms of MSE can be formulated by minimizing the distortion function D. In general, the lower the value of D is, the better the quality of C will be.

$$D(C) = \frac{1}{N_b} \sum_{j=1}^{N_c} \sum_{i=1}^{N_b} \mu_{ij} \cdot \left\| x_i - c_j \right\|^2 .$$

(1)

Subject to the following constraints:

$$\sum_{j=1}^{N_c} \mu_{ij} = 1, \quad \forall i \in \{1, 2, ..., N_b\} \quad .$$

(2)

$$\mu_{ij} = \begin{cases} 1 & \text{if } x_i \text{ is in the } j\text{th cluster,} \\ 0 & \text{otherwise} \end{cases} .$$

(3)

and

$$L_k \leq c_{jk} \leq U_k, k = 1, 2, ..., L .$$

(4)

where L_k is the minimum of the kth components in the all training vectors, and U_k is the maximum of the kth components in all training vectors. The $\left\| x - c \right\|$ is the Euclidean distance between the vectors x and c.

Two necessary conditions exist for an optimal vector quantizer.

(1)The codewords c_j must be given by the centroid of R_j:

$$c_j = \frac{1}{N_j} \sum_{i=1}^{N_j} x_i , x_i \in R_j .$$

(5)

where N_j is the total number of vectors belonging to R_j.

(2) The partition R_j, $j = 1, ..., m$ must satisfy

$$R_j \supset \{ x \in X : d(x, c_j) < d(x, c_k) \forall k \neq j \} .$$

(6)

2.2 LBG Algorithm

An algorithm for a scalar quantizer was proposed by Lloyd [7]. Linde et al. generalized it for vector quantization [1]. This algorithm is known as LBG or generalized Lloyd algorithm (GLA). It applies the two following conditions to input vectors for determining the optimal codebooks.

Given input vectors, x_i, $i = 1,2,..,N_b$, distance function d, and an initial codewords $c_j(0)$, $j = 1,...,N_c$. The LBG iteratively applies the two conditions to produce optimal codebook with the following algorithm:

(1). Partition the input vectors into several groups using the minimum distance rule. This resulting partition is stored in a $N_b \times N_c$ binary indicator matrix U whose elements are defined as the following:

$$\mu_{ij} = \begin{cases} 1 & \text{if } d(x_i, c_j(k)) = \min_p d(x_i, c_p(k)) \\ 0 & \text{otherwise} \end{cases} \tag{7}$$

(2). Determine the centroids of each partition. Replace the old codewords with these centroids:

$$c_j(k+1) = \frac{\sum_{i=1}^{N_b} \mu_{ij} x_i}{\sum_{i=1}^{N_b} \mu_{ij}}, \quad j = 1,...,N_c. \tag{8}$$

(3) Repeat steps (1) and (2) until no c_j, $j = 1,...,N_c$ changes anymore.

3   HBMO-LBG Vector Quantization Algorithm

3.1   Honey Bee Mating Optimization

A honeybee colony typically consists of a single egg-laying long-lived queen, anywhere from zero to several thousands drones and usually 10,000-60,000 workers [8]. Queens are specialized in egg laying. A colony may contain one queen or more during its life cycle, which named monogynous and/or polygynous colonies, respectively. A queen bee may live up to 5 or 6 years, whereas worker bee and drones never live more than 6 months. After the mating process, the drones die. The drones are the fathers of colony. They are haploid and act as amplify their mother's genomes without altering their genetic composition, expect through the mutation. The drones practically considered as agents that pass one of their mother's gametes and function to enable females to act genetically as males. Worker bees specialized in brood care and sometimes lay eggs. Broods arise either from fertilized (represents queen or worker) or unfertilized (represents drones) eggs.

In the marriage process, the queen(s) mate during their mating flights far from the nest. A mating flight starts with a dance performed by the queen who then starts a mating flight during which the drones follow the queen and mate with her in the air. In each mating, sperm reaches the spermatheca and accumulates them to form the genetic pool of the colony. Each time a queen lays fertilized eggs, she randomly retrieves a mixture of the sperm accumulated in the spermatheca to fertilize the egg. In practices, the mating flight may considered as a set of transitions in a state-space where the queen moves between the different states in some speed and mates with the drone encountered at each state probability. Furthermore, the queen initializes with some energy content during the flight mating and returns to her nest when the energy is within some threshold from zero to full spermatheca.

In order to develop the algorithm, the capability of workers is restrained in brood care and thus each worker may be regarded as a heuristic that acts to improve and/or take care of a set of broods. An annealing function is used to describe the probability of a drone (D) that successfully mates with the queen (Q) shown in Eq. (9).

$$P(Q,D) = \exp[-\Delta(f)/S(t)] \tag{9}$$

where $\Delta(f)$ is the absolute difference of the fitness of D and the fitness of Q, and the $S(t)$ is the speed of queen at time t. After each transition of mating, the queen's speed and energy are decayed according to the following equation:

$$S(t+1) = \alpha \times S(t) \tag{10}$$

where α is the decreasing factor ($\alpha \in [0,1]$). Workers adopt some heuristic mechanisms such as crossover or mutation to improve the brood's genotype. The fitness of the resulting genotype is determined by evaluating the value of the objective function of the brood genotype. It is important to note that a brood has only one genotype. The popular five construction stages of the HBMO algorithm had been proposed by Fathian et al. [5] that are also used to develop the algorithm of multilevel image thresholding method in this paper. The five stages are described as follow:

(1) The algorithm starts with the mating flight, where a queen (best solution) selects drones probabilistically to form the spermatheca (list of drones). A drone then selected from the list randomly for the creation of broods.
(2) Creation of new broods by crossover the drone's genotypes with the queens.
(3) Use of workers to conduct local search on broods (trial solutions).
(4) Adaptation of worker's fitness, based on the amount of improvement achieved on broods.
(5) Replacement of weaker queen by fitter broods.

3.2 HBMO-Based Vector Quantization

This section introduces a new codebook design of vector quantization algorithm that uses the honey bee mating optimization method. In the HBMO-LBG algorithm, the solutions include the best solution; candidate solution and the trivial solution are represented in the form of codebook. Figure 1 shows the structures of drone set (trivial solution). The same structure is used to represent the queen and breeds. The essential of designed algorithm is based on the Fathian's five stage scheme. Followings are the detail algorithm: The fitness function used also defined in Eq. (11).

$$Fitness(C) = \frac{1}{D(C)} = \frac{N_b}{\sum_{j=1}^{N_c}\sum_{i=1}^{N_b} \mu_{ij} \cdot \left\| x_i - c_j \right\|^2} \tag{11}$$

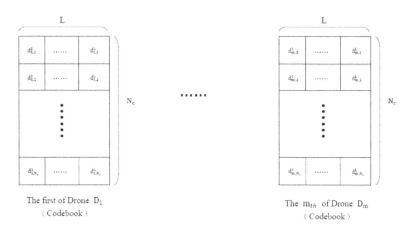

Fig. 1. The structure of solutions (codebooks) of drone set

Stage 1. (Generate the drone set and initial queen)

In this sage, the result of LBG algorithm is first set to the queen Q, and then the set of drones (trivial solutions), $\{D_1, D_2, ..., D_m\}$ are randomly generated. Secondly, the fitness of each drone are computed and the largest one is pick on denoted as D_{best}. If the fitness of D_{best} is larger than the fitness of the Q, the queen Q is replaced by the D_{best}.

Stage 2. (Mating flight)

In this stage, the queen selects the best mating set from the set of drones according to the Eq. (11). The queen's speed are decayed after each mating process until its speed is less than the predefined minimum flight speed or its spermatheca is full. In general, the number of the sperms, n_{sperm}, is smaller than the number of drones m. Let Sperm is a collection of the sperms represented in Eqs.(12) and (13).

$$Sperm = [Sp_1, Sp_2, ..., Sp_{n_{sperm}}] \tag{12}$$

$$Sp_{i,k} = (Sp_{i,k}^1, Sp_{i,k}^2, ..., Sp_{i,k}^L), \ 1 \le k \le N_c \tag{13}$$

where the Sp_i is the ith sperm in the spermatheca and the $Sp_{i,k}$ is the kth codeword of the codebook represented by the Sp_i.

Stage 3 (Breeding Process)

In the breeding process, the queen randomly selects a sperm Sp_j from the spermatheca if the corresponding random parameter R_j is smaller than the predefined parameter P_c. The new brood is generated from the queen Q and the sperm Sp_j by using the Eq. (14).

$$brood_j = Q \pm \beta \times (Sp_j - Q) \tag{14}$$

where β is randomly number ranges from 0 to 1.

Stage 4. Brood mutation with the royal jelly by works
The population of broods is improved by applying the mutation operators as follows:

Step 1. For all broods, the random number R_i of i-th brood is generated.

Step 2. If the R_i is less than the predefined mutation ratio P_m, the one tenth codeword of ith codebook (brood) are randomly select to do mutauion according to the Eq. (15).

$$Brood_i^k = Brood_i^k \pm (\delta + \varepsilon) \times Brood_i^k$$
$$\delta \in [0,1],\ 0 < \varepsilon < 1 \tag{15}$$

The δ is random number ranges from 0 to 1; k is one of the index of selected codewords among the selected and the ε is pre-defined parameter. .

Step 3. The best brood, $brood_{best}$ with maximum fitness v alue is selected as the candidate queen.

Step 4. If the fitness of $brood_{best}$ is superior to the queen, we replace queen by $brood_{best}$.

Stage 5. Check the termination criterion
If the termination criterion is satisfied then finish the algorithm, else discard the all previous trial solutions (brood set). Then go to stage 2 until the assigned iteration completed.

4   Experimental Results and Discussion

We implement the proposed HBMO-LBG algorithm in language of Matlab under a personal computer with 2.4GHz CPU, 1G RAM with window XP system.   In the implementation of HBMO-LBG algorithm, it needs many parameters that are predefined.   Five 512×512 still images named "LENA", "PEPPER", "BABOON", "GOLDHILL", and "Lake" are used for conducting our experiments. These original test images are shown in Fig. 2. The block size is 4×4, resulting in a total of 16384 vectors to be encoded in an image.

The resulting codebooks are generated on the training set by LBG, PSO-LBG, and HBMO-LBG methods respectively. The size of the designed codebook are 1024, 512, 256, 128, 64 and 32. The bit rate is defined in Eq. (16).

$$bit\_rate = \frac{\log_2 N_c}{K} \tag{16}$$

The N_c represents the size of designed codebook and the K is the pixel number of block. Furthermore, the quality of encoded image was evaluated using the peak signal-to-noise ratio (PSNR). The PSNR is defined as

(a). LENA

(b). PEPPER

(c). BABOON

(d). GOLDHILL

(e). LAKE

Fig. 2. The test images: (a) LENA, (b) PEPPER, (c) BIRD, (d) CAMERA and (e)GOLDHILL

$$PSNR = 10 \times \log_{10}(\frac{A^2}{MSE}) \text{ (dB)} \qquad (17)$$

The A is the maximum of gray level and MSE is the mean square between the original image and the decompressed image.

$$MSE = \frac{1}{N_b \times N_c} \sum_{i=1}^{M} \sum_{j=1}^{M} (y_{ij} - \overline{y_{ij}})^2 \qquad (18)$$

where $M \times M$ is the image size, y_{ij} and $\overline{y_{ij}}$ denote the pixel value at the location (i, j) of original and reconstructed images, respectively. The experimental results are shown in Table 1. Tabe 1 show the PSNR values and the execution times of test images by using the three different vector quantization methods. Obviously, the usages of the HBMO-LAB algorithm have highest PSNR value compared with other two methods. Furthermore, the PSNR of LBG algorithm is the worst and the other three algorithms can significantly improve the results of LBG algorithm.

Table 1. The PSNR and execution times for the five test images with different bit rate using the four different vector quantization that are LBG, PSO-LGB, QPSO-LBG and HBMO-LBG algotithms

Image (512×512)	Bit_rate (bit/pixel)	LBG-LBG		PSO-LBG		HBM-LBG	
LENA	0.3125	25.430	3	25.760	478	25.887	298
	0.375	25.657	4	26.567	932	26.830	568
	0.4375	25.740	6	27.508	1828	27.562	1136
	0.5	25.750	9	28.169	3666	28.337	2525
	0.5625	25.758	17	28.994	7406	29.198	4386
	0.625	25.760	36	29.863	14542	29.958	8734
PEPPER	0.3125	25.600	3	25.786	501	26.143	483
	0.375	25.891	4	27.101	928	27.206	951
	0.4375	25.976	6	27.791	1838	28.014	1848
	0.5	25.998	9	28.861	4223	28.951	3680
	0.5625	26.006	17	29.661	7271	29.829	6340
	0.625	26.008	32	30.584	15218	30.869	12475
BABOON	0.3125	19.832	3	20.423	478	20.439	483
	0.375	19.902	4	20.822	934	20.980	939
	0.4375	19.921	6	21.373	1844	21.512	1855
	0.5	19.926	9	22.038	3649	22.052	3590
	0.5625	19.928	17	22.568	7368	22.639	6376
	0.625	19.929	32	23.246	14690	23.276	12660
GOLDHILL	0.3125	26.299	3	26.126	478	26.423	484
	0.375	26.473	4	27.146	926	27.237	934
	0.4375	26.557	6	27.898	1826	28.062	1842
	0.5	26.593	10	28.619	3651	28.745	3686
	0.5625	26.596	19	29.434	7289	29.522	6367
	0.625	26.599	31	30.275	14582	30.378	12656
LAKE	0.3125	23.467	3	23.945	483	24.225	482
	0.375	23.571	4	24.714	926	25.116	935
	0.4375	23.610	6	25.716	1833	25.971	1843
	0.5	23.624	9	26.413	4203	26.509	3678
	0.5625	23.626	17	27.154	7271	27.217	7366
	0.625	23.627	37	28.022	14521	28.113	14900

5 Conclusion

This paper gives a detailed description of how the HBMO (honey bee mating optimization) algorithm can be used to implement the vector quantization and enhance the performance of LBG method. All of our experimental results showed that the proposed algorithm can significantly increase the quality of reconstructive images with respect to other three methods included the traditional LBG, PSO-LBG and QPSO-LBG. The proposed HBMO-LBG algorithm can provide a better codebook with small distortion.

Acknowledgment

The author would like to thank the National Science council, ROC, under Grant No. NSC 97-2221-E-251-001 for supports of this work.

References

1. Linde, Y., Buzo, A., Gray, R.M.: An algorithm for vector quantizer design. IEEE Transaction on Communications 28(1), 84–95 (1980)
2. Chen, Q., Yang, J., Gou, J.: Image Compression Method Using Improved PSO Vector Quantization. In: Wang, L., Chen, K., S. Ong, Y. (eds.) ICNC 2005. LNCS, vol. 3612, pp. 490–495. Springer, Heidelberg (2005)
3. Wang, Y., Feng, X.Y., Huang, Y.X., Pu, D.B., Zhou, W.G., Liang, Y.C., Zhou, C.G.: A novel quantum swarm evolutionary algorithm and its applications. Neurocomputing 70, 633–640 (2007)
4. Zhao, Y., Fang, Z., Wang, K., Pang, H.: Multilevel minimum cross entropy threshold selection based on quantum particle swarm optimization. In: Proceeding of Eight ACIS International Conference on Software Engineering, Artificial Intelligence, Networking and Parallel/Distributed Computing, pp. 65–69 (2007)
5. Fathian, M., Amiri, B., Maroosi, A.: Application of honey-bee mating optimization algorithm on clustering. Applied Mathematics and Computations, 502–1513 (2007)
6. Horng, M.H.: Multilevel Minimum Cross Entropy Threshold selection based on Honey Bee Mating Optimization. In: 3rd WSEAS International conference on Circuits, systems, signal and Telecommunications, CISST 2009, Ningbo, pp. 25–30 (2009)
7. Lloyd, S.P.: Least square quantization in PCM's. Bell Telephone Laboratories Paper, Murray Hill, NJ (1957)
8. Abbasss, H.A.: Marriage in honey-bee optimization (HBO): a haplometrosis polygynous swarming approach. In: The Congress on Evolutionary Computation (CEC 2001), pp. 207–214 (2001)

Towards a Population Dynamics Theory for Evolutionary Computing: Learning from Biological Population Dynamics in Nature

Zhanshan (Sam) Ma

IBEST (Initiative for Bioinformatics and Evolutionary Studies) & Departments of Computer Science and Biological Sciences, University of Idaho, Moscow, ID, USA
ma@vandals.uidaho.com

Abstract. In evolutionary computing (EC), population size is one of the critical parameters that a researcher has to deal with. Hence, it was no surprise that the pioneers of EC, such as De Jong (1975) and Holland (1975), had already studied the population sizing from the very beginning of EC. What is perhaps surprising is that more than three decades later, we still *largely* depend on the experience or ad-hoc trial-and-error approach to set the population size. For example, in a recent monograph, Eiben and Smith (2003) indicated: "*In almost all EC applications, the population size is constant and does not change during the evolutionary search.*" Despite enormous research on this issue in recent years, we still lack a well accepted theory for population sizing. In this paper, I propose to develop a *population dynamics theory* for EC with the inspiration from the population dynamics theory of biological populations in nature. Essentially, the EC population is considered as a *dynamic system* over time (*generations*) and space (*search space* or *fitness landscape*), similar to the spatial and temporal dynamics of biological populations in nature. With this conceptual mapping, I propose to 'transplant' the biological population dynamics theory to EC via three steps: (*i*) experimentally test the feasibility—whether or not emulating natural population dynamics improves the EC performance; (*ii*) comparatively study the underlying mechanisms—why there are improvements, primarily via statistical modeling analysis; (*iii*) conduct theoretical analysis with theoretical models such as *percolation* theory and *extended evolutionary game theory* that are generally applicable to both EC and natural populations. This article is a summary of a series of studies we have performed to achieve the general goal [27][30]–[32]. In the following, I start with an extremely brief introduction on the theory and models of natural population dynamics (Sections 1 & 2). In Sections 4 to 6, I briefly discuss three categories of population dynamics models: deterministic modeling with Logistic chaos map as an example, stochastic modeling with spatial distribution patterns as an example, as well as survival analysis and extended evolutionary game theory (EEGT) modeling. Sample experiment results with Genetic algorithms (GA) are presented to demonstrate the applications of these models. The proposed EC population dynamics approach also makes survival selection largely unnecessary or much simplified since the individuals are naturally selected (controlled) by the mathematical models for EC population dynamics.

H. Deng et al. (Eds.): AICI 2009, LNAI 5855, pp. 195–205, 2009.
© Springer-Verlag Berlin Heidelberg 2009

Keywords: Population Dynamics, Evolutionary Computing, Genetic Algorithms, Logistic Chaos Map, Spatial Distributions, Insect Populations, Survival Analysis, Survival Selection, Fitness Landscape, Power Law.

1 Introduction

In much of today's evolutionary computing, the setting of population is still experience-based, and often the fixed population size is preset manually before the start of computing. This practice is simple, but not robust. The dilemma is that small populations may be ineffective but big populations can be costly in term of computing time and memory space. In particular, in Genetic programming, a big population is often a culprit for the too early emergence of code bloat, which may cause failure or even crash the system. Furthermore, many of the problems we face are NP-hard problems; efficient population sizing may have significant impacts on the success of the heuristic algorithms. Nevertheless, an optimal population size effective in exploring fitness space and efficient in utilizing computing resources is theoretically very intriguing. Achieving the balance in effectiveness and efficiency is such a dilemma that prompted Harik et al's (1999) approach to the problem with the Gambler's ruin random walk model.

If there is an optimal population size, it is very likely to be a *moving target*, which may depend on selection pressure, position in search space, and fitness landscape. Indeed, the population in nature or the biological population is a *moving target*—a self-regulated dynamic system influenced by both intrinsic and environment stochasticity. A natural question can be: what if we emulate the natural population dynamics in EC? A follow-up question is how to emulate? The answer for the latter is actually straightforward because we can simply adopt the mathematical models developed for the natural population dynamics. The discipline that study natural population dynamics is *population ecology*, which is a major branch of *theoretical ecology* (or *mathematical ecology*). Theoretical ecology is often compared with *theoretical physics*, and there is a huge amount of literature on population dynamics, on which computer scientists may draw for developing an EC population dynamics theory. In this paper, I can only present a bird's-eye view of the potential population dynamics models. Furthermore I limit the models to the categories that we have experimentally tested, and the test results demonstrated significant improvement with the dynamic populations that are controlled by the mathematical models.

It should be noted that there are quite some existing studies (e.g.,[8][9][11]) that were conducted to develop alternatives for the fixed-size populations. Due to page space limitation, here I totally skip the review of existing studies on dynamic populations in EC (a review was presented in [23]). To the best of my knowledge, this is the first work that introduces, in a *comprehensive* manner, the natural population dynamics theory to EC. The remainder of this paper is organized as follows: Section 2 is an extremely brief overview of the natural population dynamics theory. Sections 4, 5, and 6 introduce three major categories of the population dynamics models: deterministic, stochastic, and extended evolutionary game theory models. In Sections 4 and 5, I also show some *sample* experiment results with the models introduced. Section 3 briefly introduces the experiment problem used in Sections 4 and 5. Section 7 is a summary.

Overall, this article is a summary of a series of studies with the objective to develop an EC population dynamic theory by emulating natural populations and 'transplanting' mathematical models developed for natural population dynamics in theoretical ecology to EC [27][30]–[32].

2 A Brief Overview on Population Dynamics Theory

2.1 Population Dynamics Modeling

Population dynamics or the spatial-temporal changes of population numbers is the central topic of *population ecology*. The mathematical modeling of population dynamics can be traced back to Thomas Malthus's (1798) "*An Essay on the Principle of Populations*" (cited in [17]). Malthus's model for human population growth in which he argued that human population grow geometrically while resources such as food grows arithmetically, and therefore fast population growth would lead to much human misery. Malthus's conclusion had significant influence on Darwin's formation of struggle for life or natural selection. Pierre Verhulst (1845) derived the famous Logistic population growth model, which was re-discovered by Pearl and Read in 1920 (cited in [17]). Logistic model is the *solution* of a first order nonlinear differential equation [Equ. (1)], whose difference counterpart [Equ. (3), (5)] is found to be able to demonstrate Chaos behavior in the 1970s [35].

Several categories of models exist for population dynamics modeling. Logistic differential equation and its difference equation counterpart form the foundation for studying single species population dynamics deterministically. It has extraordinary rich mathematical properties from stable growth, oscillation, attractors to chaos, and its applications have been expanded well beyond biology. Logistic differential equation is also the basis for forming the famous Lotka-Volterra differential equations system for interspecific interactions such as predation and competition. Differential equation or deterministic modeling is traditionally used to model population dynamics, and is still the dominant approach ([16][17][35]–[37]). An alternative to the traditional differential equation modeling is the stochastic modeling with stochastic process models such as birth-death process, e.g., [18]. It is often more difficult to get analytic solutions in stochastic modeling. The third group of analytic models is the matrix model which can be formed in either deterministic or stochastic form (e.g., [4]), and it turns out that matrix is often equivalent to either the differential equation or stochastic process model (e.g., Markov Chains). The fourth type of analytical model is the optimization and game theory models, including evolutionary game theory that was an integration of three cross-disciplinary fields: Darwin's evolutionary theory, game theory and population dynamics ([12][29][39]). The previous four categories are primarily analytical models, although some of them may need simulation to get the solutions.

Besides analytical modeling approaches, *statistical modeling* (e.g., time-series, spatial statistics) and *simulation modeling* are widely used in modeling population dynamics. Both approaches are extremely important for studying natural population dynamics (e.g., [20][22]); they seem less suitable for controlling population sizes in EC. However, both the approaches can be used to analyze the 'behavior' of EC populations, such as modeling fitness landscape. There is an additional broad category of modeling approaches, mostly from complexity science, such as percolation theory, cellular automata, interacting particle systems (IPS), random graphs theory, agent-based computing, etc (e.g., [2][3]). Among the last category, some models may offer immensely helpful insights if analytical solutions are available (e.g., critical percolation probability and phase transitions with percolation models). However, if analytical solutions are not available, then 'embedding' a simulation model in the EC population is probably too costly to be useful. Of course, these models can still be applied to analyze the behavior of EC populations, and they can be more powerful than statistical modeling because these models usually focus on underlying mechanisms.

2.2 Population Regulation

Population regulation was one of the most contested theories in the history of ecology (e.g., see [16] for a brief history) and the debates started in the 1950s and culminated in the 1960s, and even these days, the antagonistic arguments from both schools occasionally appear in ecological publications (e.g., [1][40]). The debate sounds simple from an engineering perspective. The core of the debate lies in the fundamental question: is population *regulated* by feedback mechanisms such as density-dependent effects of natural enemies or is simply *limited* by its environment constraints. Within the regulation school, there are diverse theories on what factors (intrinsic such as competition, natural enemies, gene, behavior, movement, migration, etc) and how the population is regulated. Certainly, there are mixed hypothesis of two schools. About a dozen hypotheses have been advanced since the 1950s. The debates were "condemned" by some critics as "bankrupt paradigm", "a monumental obstacle to progress" (cited in [1][40]). However, there is no doubt that the debates kept population ecology as the central field for more than three decades, and is critical for shaping population ecology as the most quantitatively studied field in ecology. The *theoretical ecology* is often dominated by the contents of population ecology [17]. In addition, the important advances in ecology such as Chaos theory, spatially explicit modeling, agent or individual based modeling all initiated in population ecology.

The importance of population regulation cannot be overemphasized, since it reveals the mechanisms for population dynamics. Even more important is to treat population dynamics from the time-space paradigm, not just the temporal changes of population numbers. In addition, the concept of metapopulation is also crucial, which implies that local population extinction and recolonization happen in nature [10]. Obviously, population regulation as control mechanisms for population size is very inspiring for the counterpart problem in evolutionary computation. What is exciting is that population regulation and population dynamics can also be unified with evolutionary game theory, and even united with population genetics under same mathematical modeling framework such as Logistic model and Lotka-Volterra systems (e.g., [12][29][39]).

3 The Test Problem

Assume a problem is represented as a string of bits of size L=32. This string is broken up into blocks of length B. The problem is to find blocks in which all bits are set to one. The fitness function is the number of blocks satisfying this requirement. If B does not divide L evenly, the leftmost block will be of different size. The following are four examples of bit streams and their corresponding fitness values (F). The first and third have optimum solutions, and the blocks that counts in the fitness function are underlined.

```
L = 32, B = 4, F=8: 1111 1111 1111 1111 1111 1111 1111 1111
L = 32, B = 4, F=2: 0000 1111 0110 0100 0111 1010 1111 1010
L = 32, B = 7, F=5: 1111 1111111 1111111 1111111 1111111
L = 32, B = 7, F=2: 1111 0000110 0001000 1111111 0011010
```

Three metric are used to evaluate the population-sizing schemes or different population dynamics models. The number of times the optimum solution is found in an experiment run (consisting of a fixed number of trials) is termed *hits* per experiment. The other two metrics are *PopuEvalIndex*, *FirstHitPopuEvlaIndex*, and The small *PopuEvalIndex* and *FirstHitPopuEvlaIndex* indicate that fewer evaluations of fitness function and other associated computations are needed. In addition, it also implies that less memory space is needed for storing the individual information. Of course, the large number of *hits* also indicates the scheme is more robust and efficient.

For more detailed definitions of the metrics and the problem, one may refer to [27]. In the next two sections, all the sample experiment results, which are excerpted from [30][31] to illustrate the corresponding models, are obtained with this test problem and are based on the three metrics discussed above.

4 Deterministic Modeling

4.1 Logistic Differential and Difference Equations

The well-known Logistic model has the differential equation form:

$$\frac{dN(t)}{dt} = rN(t)\left(1 - \frac{N(t)}{K}\right)$$
(1)

Equation (1) has a unique solution:

$$N(t) = \frac{K}{1 + ae^{-rt}}.$$
(2)

This equation (2) is attributed to Verhust (1838) and Pearl (1925) and is referred to as the *Logistic law* in the literature (Cited in [17]).

The difference equation form of Equation (1) is:

$$N(t+1) = N(t) + rN(t)\left(1 + \frac{N(t)}{K}\right)$$
(3)

N is the population number at time *t* or *t*+1, *K* is referred to as *environment capacity* in biological literature. *r* is the per capita growth rate per generation.

4.2 One-Parameter Logistic Difference Equation—Logistic Chaos Map

To facilitate analysis with Equation (2), let

$$x = \frac{Nr}{K(1+r)} \qquad (4)$$

substituting x into equation (3) derives the one-parameter dimensionless Logistic Map, also known as one-hump nonlinear function, or one-dimensional quadratic map.

$$x_{n+1} = ax_n(1-x_n) \qquad (5)$$

where $a = r + 1$. To avoid trivial dynamic behavior, the model requires $1<a<4$ and $0<x<1$. The population size (x) is converted to the (0, 1) interval, and the conversion also eliminates the other parameter K, which makes the analysis more convenient. The extremely rich and complex behavior represented by the deceptively simple equation (5) was discovered by Robert May in 1976 [35]. [38] contains detailed discussion of the Logistic Chaos map model. The one-parameter Logistic map model is particularly convenient for controlling EC population dynamics, and offers extremely rich dynamics. Figures 1 and 2 show the performance of dynamic populations with the Logistic chaos map model vs. that of the fixed-size population.

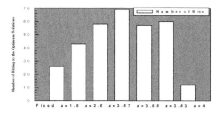
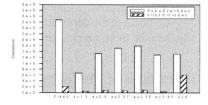

Fig. 1. Total number of *Hits* for each Parameter value of *a* and for fixed population

Fig. 2. *PopuEvalIndex* and *First Hit PopuEvalIndex* for BlockSize=8

Figures 1 and 2 demonstrate that dynamic populations controlled with the Logistic chaos map model under various parameter values all significantly outperform the fixed-size population (the leftmost) except for $a=4$. Parameter $a=4$ is theoretically impossible with the model; it was included to test the boundary case. The detailed experiment results with the Logistic chaos map model are documented in [30].

5 Stochastic Modeling

There are many stochastic models for population dynamics (e.g., [18]). In this section, I introduce a category of probability distribution models that are used to describe the spatial distribution patterns of animal populations. Natural populations are dynamic in both time and space. In nature, the population spatial distribution patterns are considered as the emergent expression (property) of individual behavior at the population level and are *fine-tuned* or optimized by natural selection [33][34]. It is assumed that EC populations emulating the spatial distribution patterns of natural populations should be advantageous.

Generally, there are three types of spatial distribution patterns: *aggregated, random* and *regular*. *Random distribution (sensu* ecology) can be fitted with *Poisson probability distribution (sensu* mathematics), and the *regular distribution* (also termed uniform distribution, sensu *ecology*) is totally regular with even spacing among individuals. *Aggregated* (also termed contagious, congregated, or clustered) *distribution (sensu* ecology) represents nonrandom and uneven density in space. The probability *distributions (sensu* mathematics) for aggregated distributions *(sensu* ecology) are strongly skewed with very long right tails. The most widely used discrete probability distribution *(sensu* mathematics) for the aggregated distribution *(sensu* ecology) pattern is the *Negative Binomial Distribution (NBN)*.

The aggregated distribution or fat-tail distribution is a property of power law. Taylor (1961) discovered that the Power law model fits population spatial distribution data ubiquitously well,

$$V = aM^b,\qquad(6)$$

where M and V are population mean and variance, respectively, and a and b are parameters. According to Taylor (1961), $b>1$ corresponds to aggregated distribution, $b=1$ to random distribution and $b<1$ to regular distribution. Ma (1988, 1991) [33][34] extended Taylor's Power law model with his concept of *population aggregation critical density (PACD)*, which was derived, based on the Power Law:

$$m_0 = \exp[\ln(a)/(1-b)],\qquad(7)$$

where a and b are the parameters of Power Law and m_0 is the *PACD*. According to Ma's (1988, 1991) reinterpreted Power Law, population spatial distributions are population density-dependent and form a continuum on the population density series. The *PACD* is the transition threshold (population density) between the aggregated, random and regular distributions [33][34].

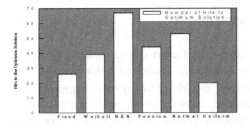

Fig. 3. Number of *Hits* for Each Distribution

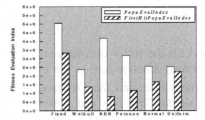

Fig. 4. Fitness Evaluation Indexes

Figures 3 and 4 demonstrate that aggregated distributions [including negative binomial distribution (NBN) and Weibull distributions] perform best, and the fixed-size population performs worst. Even the random population (Poisson distribution) performs better than the fixed-size population. Of course, mathematically, uniform distribution is the same as the fixed-size population, and both perform similarly poor. This result confirms that emulating the spatial distributions of natural populations improves EC performance because aggregated distributions dominate natural populations, while uniform (and even random) distribution is extremely rare in nature.

6 Survival Analysis and the Extended Evolutionary Game Theory

Compared with *evolutionary game theory* (EGT), the *extended evolutionary game theory* (EEGT) modeling by Ma (2009) [29] possesses two new features: (*i*) the extension with *survival analysis* to describe lifetime or fitness of the game players; (*ii*) the extension with *agreement algorithms* to model time and space dependent *frailty* and *deception*. To some extent, EEGT itself is very similar or even equivalent to an EC approach when implemented as a simulation environment. It should be noted that the term *survival* is overloaded: one is *sensu* statistics in *survival analysis*. Survival analysis models (the first extension) can be introduced to control the survival of EC populations. In other words, the first extension alone is sufficient to model EC populations, similar to the modeling of natural insect populations with survival analysis [22][24]. This is because by controlling individual survival probability or lifetime, it is equivalent to controlling population dynamics. Actually, with survival analysis models, *survival selection* in EC is not necessary, because individuals are naturally *selected* by their survival probability (survivor function).

Survivor function is defined as:

$$S(t) = P(T \geq t), \quad 0 < t < \infty \tag{8}$$

which is the probability that an individual will survive beyond time T. Various distribution models (such as Weibull, lognormal, and logistic distributions) can be used as survivor function. These distributions are termed *parametric models*. In addition, *semi-parametric models*, which describe the *conditional* survivor function influenced by environment covariates (z), can be utilized, and the most well-known seems to be Cox's proportional hazards model (PHM):

$$S(t \mid z) = [S_0(t)]^{\exp(z\beta)} \tag{9}$$

where
$$S_0(t) = \exp\left[-\int_0^t \lambda_0(u)du \right]. \tag{10}$$

and z is the vector of *covariates*, which can be any factors that influence the baseline survivor function $S_0(t)$ or the lifetime of *individuals*. Dedicated volumes have been written to extend the Cox model. Therefore, models (8)–(10) offer extremely rich modeling flexibility and power. Furthermore, more complex multivariate survival analysis ([13][28]) and competing risks analysis ([5]) can be introduced to model EC populations.

Besides offering extremely rich and powerful stochastic models, the three 'sister' fields of survival analysis can be utilized to model either individual lifetime (individual level) or population survival distribution (population level). One may conjecture that survival-analysis based EC populations should perform similar to the previous stochastic populations since in both cases, probability distributions models are used to control EC populations. My experiment results indeed confirm this conjecture [32]. However, since survival analysis is advanced to study *time-to-event* random variables (also known as *survival* or *failure* time) and the lifetime of individuals in EC is a typical time-to-event random variable, it does has some unique advantages such as capturing the survival process mechanistically, and uniquely dealing with censoring (incomplete information) and frailty (uncertainty combined with vulnerability). An additional advantage in the case of EC, as argued previously, is that the *survival selection* in EC is not necessary anymore because *survival selection* is essentially equivalent to determine how long an individual can 'live' in the EC population. Therefore, survival analysis modeling offers a unified solution for both *population-sizing* and *survival selection*.

7 Summary—Towards an EC Population Dynamics Theory

The previous sections and the more detailed results ([27][30]–[32]) show that it is advantageous to emulate natural population dynamics. This seems to suggestion that it is feasible to develop a population dynamics theory for EC, by learning from the natural populations dynamics. In my opinion, EC has been primarily influenced by the genetic aspects of the evolutionary theory, and much of the ecological aspects of the evolutionary theory seem still missing. Ecology, which perhaps had equally, if not more, influenced, Darwin's development of his evolutionary theory, should be the next frontier to draw inspiration for EC [23]. It may be unnecessary to coin a new term such as *ecological computing* (which actually already exists in literature, but is largely associated with topics not related to this paper) because ecological aspects and genetic aspects of evolution are hardly separable. Nevertheless, treating the EC population as a dynamic system rather than a fixed-size entity is fully justified; then, the natural population dynamics theory should be a critical inspiration source for building an EC population dynamics theory.

I envision three steps to develop an EC population dynamics theory based on the natural population dynamics theory. The first step is to emulate natural populations and experimentally test the emulation. The second step is to explore the underlying mechanisms why the emulation may improve EC, such as the studies of fitness dynamics (landscape) or scaling law (power law). Figures 5 and 6 are two examples of such studies, which illustrate fitness dynamics with Chaotic populations and Taylor's power law with stochastic populations, respectively. Figure 5 demonstrates that with Chaotic populations, multiple clusters of individuals simultaneously search for the optimum solution and all reach the fitness range of 0.5–0.6. In contrast, with the fixed-size population, there is only one cluster and the fitness only reaches 0.18–0.20 range (not shown here, refer to [30]). Figure 6 demonstrates that Taylor's power law describes the *fitness* distribution in stochastic EC populations very well. Note that both the axes in Fig. 6 are in the log-scale, so the power function is displayed as linear [31]. The third step should be to perform rigorous theoretical analysis, in the context of EC, of the population dynamics models introduced from natural population dynamics. In this step, some of the population dynamics models, such as Percolation theory models and extended evolutionary game theory, can be 'recursively' applied to the theoretical analysis because these models are generally applicable to many phenomena including both natural and EC populations [23]. For example, the application of Percolation theory model for analyzing EC populations should be very similar to Harik et al. (1999) study based on the Gambler's ruin problem [11].

The potential inspiration of ecology to EC is certainly beyond the population dynamics theory. The relationship between ecology and evolution is best described with G. E. Hutchinson's (1966) treatise title–"*The Ecological Theater and the Evolutionary Play*" [14]. Anyway, without the ecological theater, natural selection has no place to act and evolution cannot happen. Perhaps from this broad perspective, a term such as *ecological computing (computation)* may be justified when the ecological principles play more critical roles in EC.

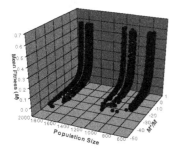

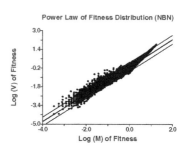

Power Law of Fitness Distribution (NBN)

Fig. 5. The relationship between *mean fitness*, population size and the fitness aggregation index

Fig. 6. Taylor's Power law applied to fitness variance vs. mean fitness

References

1. Berryman, A.A.: Population regulation, emergent properties, and a requiem for density-dependence. Oikos 99(3), 600–606 (2002)
2. Bollobás, B.: Random Graphs, 2nd edn., 500 p. Cambridge University Press, Cambridge (2001)
3. Bollobás, B., Riordan, O.: Percolation, 334 p. Cambridge University Press, Cambridge (2006)
4. Caswell, H.: Matrix Population Models, 2nd edn. Sinauer, Sunderland (2001)
5. Crowder, M.J.: Classical Competing Risks Analysis, 200 p. Chapman & Hall, Boca Raton (2001)
6. DeJong, K.A.: An analysis of the behaviors of genetic adaptive systems. Ph.D. Thesis, University of Michigan, Ann Arbor, MI (1975)
7. Eiben, A.E., Smith, J.E.: Introduction to Evolutionary Computing. Springer, Heidelberg (2003)
8. Goldberg, D.E., Rundnick, M.: Genetic algorithms and variance of fitness. Complex Systems 5(3), 265–278 (1991)
9. Goldberg, D.E., et al.: Genetic algorithms, Noise, and the Sizing of Populations. Complex Systems 6, 333–362 (1992)
10. Hanski, I.: Metapopulation Dynamics. Oxford University Press, Oxford (1999)
11. Harik, G., Cantú-Paz, E., Goldberg, D.E., Miller, B.L.: The Gambler's ruin problem, genetic algorithms, and the sizing of populations. Evol. Comput. 7(3), 231–253 (1999)
12. Hofbauer, J., Sigmund, K.: Evolutionary Games and Population Dynamics, 323 p. Cambridge University Press, Cambridge (1998)
13. Hougaard, P.: Analysis of Multivariate Survival Data, 560 p. Springer, Heidelberg (2000)
14. Hutchinson, G.E.: The Ecological Theater and the Evolutionary Play. Yale University Press (1966)
15. Ibrahim, J.G., Chen, M.H., Sinha, D.: Bayesian Survival Analysis, 481 p. Springer, Heidelberg (2005)
16. Kingsland, S.E.: Modeling Nature, 2nd edn. University of Chicago Press (1995)
17. Kot, M.: Elements of Mathematical Ecology, 453 p. Cambridge University Press, Cambridge (2001)

18. Lande, R., Engen, S.: Stochastic Population Dynamics in Ecology. Oxford University Press, Oxford (2003)
19. Lawless, J.F.: Statistical models and methods for lifetime data, 2nd edn. Wiley, Chichester (2003)
20. Legendre, P., Legendre, L.: Numerical Ecology, 851 p. Elsevier, Amsterdam (1998)
21. Ma, Z.S.: New Approaches to Reliability and Survivability with Survival Analysis, Dynamic Hybrid Fault Models, and Evolutionary Game Theory. PhD Dissertation, University of Idaho, 177 p. (2008)
22. Ma, Z.S., Bechinski, E.J.: Survival-analysis-based Simulation Model for Russian Wheat Aphid Population Dynamics. Ecological Modeling 216(2), 323–332 (2008)
23. Ma, Z.S.: Why Should Populations be Dynamic in Evolutionary Computation? Eco-Inspirations from Natural Population Dynamics and Evolutionary Game Theory. In: The Sixth International Conferences on Ecological Informatics. ICEI-6 (2008)
24. Ma, Z.S., Bechinski, E.J.: Accelerated Failure Time Modeling of the Development and Survival of Russian Wheat Aphid. Population Ecology 51(4), 543–548 (2009)
25. Ma, Z.S., Krings, A.W.: Survival Analysis Approach to Reliability Analysis and Prognostics and Health Management (PHM). In: Proc. 29th IEEE–AIAA AeroSpace Conference, 20 p. (2008a)
26. Ma, Z.S., Krings, A.W.: Dynamic Hybrid Fault Models and their Applications to Wireless Sensor Networks (WSNs). In: The 11-th ACM/IEEE MSWiM 2008, Vancouver, Canada, 9 p. (2008)
27. Ma, Z.S., Krings, A.W.: Dynamic Populations in Genetic Algorithms. In: SIGAPP 23rd Annual ACM Symposium on Applied Computing (ACM SAC 2008), Brazil, March 16-20, 5 p. (2008)
28. Ma, Z.S., Krings, A.W.: Multivariate Survival Analysis (I): Shared Frailty Approaches to Reliability and Dependence Modeling. In: Proc. 29th IEEE–AIAA AeroSpace Conference, 21 p. (2008)
29. Ma, Z.S.: Towards an Extended Evolutionary Game Theory with Survival Analysis and Agreement Algorithms for Modeling Uncertainty, Vulnerability and Deception. In: Deng, H., Wang, L., Wang, F.L., Lei, J. (eds.) AICI 2009. LNCS (LNAI), vol. 5855, pp. 608–618. Springer, Heidelberg (2009)
30. Ma, Z.S.: Chaos Population Chaotic Populations in Genetic Algorithms (submitted)
31. Ma, Z.S.: Stochastic Populations, Power Law and Fitness Aggregation in Genetic Algorithms (submitted)
32. Ma, Z.S.: Survival-Analysis-Based Survival Selections in Genetic Algorithms (in preparation)
33. Ma, Z.S.: Revised Taylor's Power Law and Population Aggregation Critical Density. In: Proceedings of Annual National Conference of the Ecological Society of China, Nanjing, China (1988)
34. Ma, Z.S.: Further interpreted Taylor's Power Law and Population Aggregation Critical Density. Trans. Ecol. Soc. China 1, 284–288 (1991)
35. May, R.M.: Simple mathematical models with very complicated dynamics. Nature 261, 459–467 (1976)
36. May, R.M., McLean, A.R.: Theoretical Ecology. Oxford University Press, Oxford (2007)
37. Pastor, J.: Mathematical Ecology of Populations and Ecosystems. Wiley-Blackwell (2008)
38. Schuster, H.H.: Deterministic Chaos: an introduction, 2nd edn., 269 p. VCH Publisher (1988)
39. Vincent, T.L., Brown, J.L.: Evolutionary Game Theory, Natural Selection and Darwinian Dynamics, 382 p. Cambridge University Press, Cambridge (2005)
40. White, T.C.R.: Opposing paradigms: regulation of limitation of populations. Oikos 93, 148–152 (2001)

Application of Improved Particle Swarm Optimization Algorithm in UCAV Path Planning[*]

Qianzhi Ma and Xiujuan Lei

College of Computer Science of Shaanxi Normal University
Xi'an, Shaanxi Province, China, 710062
zhihui312@163.com
xjlei168@163.com

Abstract. For the calculation complexity and the convergence in Unmanned Combat Aerial Vehicle (UCAV) path planning, the path planning method based on Second-order Oscillating Particle Swarm Optimization (SOPSO) was proposed to improve the properties of solutions, in which the searching ability of particles was enhanced by controlling the process of oscillating convergence and asymptotic convergence. A novel method of perceiving threats was applied for advancing the feasibility of the path. A comparison of the results was made by WPSO, CFPSO and SOPSO, which showed that the method we proposed in this paper was effective. SOPSO was much more suitable for solving this kind of problem.

Keywords: Unmanned Combat Aerial Vehicle (UCAV), Second-order Oscillating Particle Swarm Optimization, path planning.

1 Introduction

The unmanned combat aerial vehicle is an experimental class of unmanned aerial vehicle. It is likely to become the mainstay of the air combat force. In order to improve the overall survival probability and the operational effectiveness of UCAV, an integrated system is needed in the UCAV flight mission and pre-flight mission to make resource coordination and determine the flight path. Flight path planning is one of the most important parts for the UCAV mission planning. That is to generate a path between an initial location and the desired destination that has an optimal or near-optimal performance under specific constraint conditions. The flight path planning in a large mission area is a typical large scale optimization problem. A series of algorithms have been proposed to solve this complicated multi-constrained optimization problem, such as the evolutionary computation[1], genetic algorithm[2],ant colony algorithm[3]. However, those algorithms have their shortcomings, for instance, the results of planning fall into local minimum easily, or they need to be further optimized to meet the needs of aerial vehicle performance constraints. Particle swarm optimization algorithm (PSO), as a new intelligent optimization algorithm has been tried to solve this problem due to the merits

---

[*] This work was partially supported by Innovation Funds of Graduate Programs, Shaanxi Normal University, China. #2009CXS018.

H. Deng et al. (Eds.): AICI 2009, LNAI 5855, pp. 206–214, 2009.

of rapid searching and easier realization[4]. But the result does not approach ideal consequence. In this paper, we propose the method based on second-order oscillating particle swarm optimization algorithm (SOPSO) to improve the optimization results for UCAV path planning. The model for the path planning problem was built, then the design scheme and specific realization were given. Finally the rationality and validity of the algorithm was analyzed based on the simulation experiments and the results.

2 Particle Swarm Optimization

2.1 Standard Particle Swarm Optimization

The development of particle swarm optimization (PSO)[5] was based on observations of the foraging behavior of bird flocking. To an optimization problem, each solution represents the position of a bird, called a 'particle' which flies in the problem search space looking for the optimal position. Each particle in the search space is characterized by two factors: its position and velocity. Mathematically, the particles in PSO are manipulated according to the following equations(1),(2):

$$v_i(t+1) = wv_i(t) + c_1 r_1 (p_i - x_i(t)) + c_2 r_2 (p_g - x_i(t)) \tag{1}$$

$$x_i(t+1) = v_i(t+1) + x_i(t) \tag{2}$$

where t indicates an (unit) pseudo-time increment, r_1 and r_2 represent uniform random numbers between 0 and 1, c_1 and c_2 are the cognitive and social scaling parameters which are usually selected such as $c_1 = c_2 = 2$. $x_i(t) = (x_{i1}, x_{i2}, \cdots, x_{id})$ represents the ith particle. $v_i(t) = (v_{i1}, v_{i2}, \cdots, v_{id})$ represents the rate of the position change (velocity) for particle i . $p_i = (p_{i1}, p_{i2}, \cdots, p_{id})$ represents the best personal position (the position giving the best personal fitness value) of the ith particle, and d is the dimension number of particle. The symbol g represents the index of the global best particle in the population. w is the inertia weight. The PSO algorithm with evolution equation (1) and (2) is called standard particle swarm optimization (SPSO). In standard PSO algorithm, $w = 1$. When w is computed as follows:

$$w = w_{max} - iter \cdot (w_{max} - w_{min}) / Maxiter \tag{3}$$

$iter$ and $Maxiter$ indicate the current iteration and the max iteration respectively. This algorithm is called particle swarm optimization with linear decrease inertia weight (WPSO)[6].

Another classical improved PSO algorithm called particle swarm optimization with constriction factor (CFPSO)[7] updates the evolution equation as equations (4) and (2):

$$v_i(t+1) = \chi(v_i(t) + c_1 r_1 (p_i - x_i(t)) + c_2 r_2 (p_g - x_i(t))) \tag{4}$$

The constraint factor χ helps to ensure the algorithm convergence. Here $\chi = 2/\left|2 - \varphi - \left(\varphi^2 - 4\varphi\right)^{1/2}\right|$, $\varphi = c_1 + c_2$ and $\varphi > 4$.

2.2 Second-Order Oscillating PSO

Second-order oscillating particle swarm optimization[8] integrates the oscillational element into PSO.

Set $\varphi_1 = c_1 r_1$, $\varphi_2 = c_2 r_2$, get the (5),(6) followed based on (1),(2)

$$v_i(t+1) = wv_i(t) + \varphi_1(p_i - x_i(t)) + \varphi_2(p_g - x_i(t)) \tag{5}$$

$$x_i(t+1) = v_i(t+1) + x_i(t) \tag{6}$$

In the view of control theory, equation (1) is a parallel connection of two inertia links the inputs of which are p_i and p_g respectively. In order to increase the diversity of particles, the inertia links are replaced by second-order oscillation links. Thus, the new evolutionary equations of second-order oscillating particle swarm optimization algorithm will be gotten as (7) and (8) based on (1), (2):

$$\begin{aligned} v_i(t+1) = wv_i(t) &+ \varphi_1(p_i - (1+\xi_1)x_i(t) - \xi_1 x_i(t-1)) \\ &+ \varphi_2(p_g - (1+\xi_2)x_i(t) + \xi_2 x_i(t-1)) \end{aligned} \tag{7}$$

$$x_i(t+1) = v_i(t+1) + x_i(t) \tag{8}$$

In the second-order oscillating algorithm, the searching process is divided into two phases, we can set ξ_1, ξ_2 with different values to control that it will begin global search or local search in accordance with the laws of convergence. We can take $\xi_1 < (2\sqrt{\varphi_1} - 1)/\varphi_1$, $\xi_2 < (2\sqrt{\varphi_2} - 1)/\varphi_2$ at prophase, for the oscillation convergence and better global search capability; take $\xi_1 \geq (2\sqrt{\varphi_1} - 1)/\varphi_1$, $\xi_2 \geq (2\sqrt{\varphi_2} - 1)/\varphi_2$ at anaphase for the progressive convergence and better local search capability.

3 The Design and Process of the Algorithm

3.1 Environments and Path Statements

Modeling of the threat sources is the key task in UCAV optimal path planning. The main threats in the flight environment include local fires, radar detection and so on. Assume unmanned aerial vehicles fly in the high-altitude without altitude change. It simplifies path planning to a two-dimensional path planning. Here, we suppose that the threat source is radar detection, and all the radars in task region are same. The aerial vehicles' task region is showed in Figure 1. There are some circular areas around the threat points in target region which represent the threat scope.

The flight task for aerial vehicles is from point A to point B. View the starting point A as the origin point, and the line between the starting point and the target point as X axes, establish coordinate system. Divide AB into m sub-sections equally. There are m-1 vertical lines between point A and point B. A path can be formed by connecting the points on these parallel vertical lines. For example, a collection of points constitutes a route:

$$path = \{A, L_1(x_1, y_1), L_2(x_2, y_2), \cdots, L_{m-1}(x_{m-1}, y_{m-1}), B\} \tag{9}$$

where $L_i(x_i, y_i)$ denotes the point on the i-th vertical line.

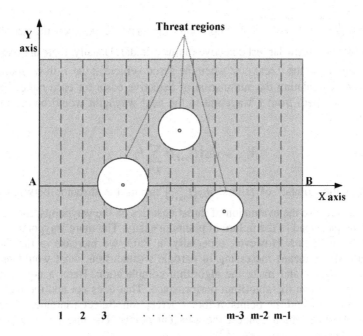

Fig. 1. The task region of UCAV

3.2 Code Design for Particles

Since the distance between adjacent vertical lines are equal, the ordinal path points' longitudinal coordinates just can be used to encode a path. In the SOPSO, we use a particle to indicate a path, which is encoded like this: $y_1 y_2, y_3,..., y_{m-1}$. In this way, the searching space is decreased a half by this encoding method. So it is propitious to speed up the searching process of algorithm.

3.3 Fitness Function

In this paper, only the horizontal path optimization is considered, so the mission can be simplified to find a path which has a high probability of not being detected by enemy radars and shortest flight distance from the start point to the target point. According to the characteristic of radars, if (x_t, y_t) denotes the location of a radar, P as the detective cost for the air vehicle at (x, y) is given as follows:

$$
P = \begin{cases} \dfrac{R_A^4}{R^4 + R_A^4} & R \le R_A \\[2ex] 0 & R > R_A \end{cases} \tag{10}
$$

Where $R = \sqrt{(x - x_i)^2 + (y - y_i)^2}$ represents the distance between the aerial vehicle and the radar. R_A is the largest detective radius of radar. Usually, there are many radars in target region, so the threat cost when the aerial vehicle at the i-th waypoint can be evaluated by calculating the summation of detective costs for each radar. The threat cost on the sub-path from i waypoint to the next waypoint would be calculated approximately as:

$$W_{i,i+1} = dis(L_{i,i+1}) * \sum_{j=1}^{t} P_j \qquad (11)$$

t indicates the number of radars. $dis(L_{i,i+1})$ means the distance between adjacent waypoints. Because the evaluation of threat cost lies on the waypoints, the blind spots on planning path lead to the failure of planning easily. The more waypoints, the better for perceiving threats. However, especially in PSO, we increase in the number of waypoints which means increasing in particle's dimension, what would reduce the accuracy of results and make the algorithm complication. Here, a novel method is proposed to improve the algorithm performance. That does not add the particle's dimension, but only add the points of threat perception. Select a number of perceptual points on the line which connect the adjacent points. The threat cost expressed by equation (11) becomes equation (12):

$$W_{i,i+1} = \sum_{k=1}^{n} dis(L_{k,k+1}) * \sum_{j=1}^{t} P_{j,k} \qquad (12)$$

k indicates the perceiving point on sub-path connected by the i waypoint and the $i+1$ waypoint. n is the number of perceiving points that we will calculate.

Besides threats detection, we also should consider making the flight distance as short as possible. The flight distance is showed to be the sum of line distances between points in the flight line. There are m sections in this path. The distance from point $L_i(x_i, y_i)$ in vertical line i to point $L_{i+1}(x_{i+1}, y_{i+1})$ in vertical line $i+1$ can be described as:

$$dis(L_i, L_{i+1}) = \sqrt{(|AB|/m)^2 + (y_{i+1} - y_i)^2} \qquad (13)$$

so we describe the objective function of aerial vehicle in path planning as follows:

$$J = \sum_{i=1}^{m-1} (\delta W_{i,i+1} + (1 - \delta)dis(L_i, L_{i+1})) \qquad (14)$$

Where δ is coefficient in [0,1].It shows the impact both of avoiding threat and flight distance , and the bigger δ is, the shorter the flight distance would be, but the more dangerous the flight would be.

3.4 Implementation Steps of the Algorithm

After the analysis above, the implementation steps of the SOPSO algorithm are introduced as follows:

STEP 1. Set the relative parameter of the algorithm, such as acceleration coefficients c_1, c_2, the max iteration *Maxiter*, population scale N, dimension of particle d etc.

STEP 2. Initialize the velocity and position of particle: Generate an initial population and a velocity randomly.

STEP 3. Compute the personal fitness value, update the personal best value and global best value.

STEP 4. If current iteration *iter* < *Maxiter*/2, update the current locations of particles according to the equation (7), (8), and set $\xi_1 < (2\sqrt{\varphi_1} - 1)/\varphi_1$, $\xi_2 < (2\sqrt{\varphi_2} - 1)/\varphi_2$; otherwise, Update the current locations of particles according to the equation (7), (8) as well, this time set $\xi_1 \geq (2\sqrt{\varphi_1} - 1)/\varphi_1$, $\xi_2 \geq (2\sqrt{\varphi_2} - 1)/\varphi_2$. Set *iter=iter+1*.

STEP 5. Check the ending conditions. If *iter* > *Maxiter*, exit it and the current global best value is just the global optimal solution; otherwise, turn to step 3.

STEP 6. Stop the computation and output the correlative values of the optimal person.

4 Experiments and Results Analysis

The algorithms were programmed in MATLAB 7.0 and ran on a PC with Intel Core 2 CPU, 1.86 GHz. Experiments based on WPSO, CFPSO and SOPSO were done under the same condition. Set the size of flight region as 160*160, the starting point at (0, 60) and the goal point at (160, 60). The size of population was 50. The dimension number of a particle was 9. The maximum iteration number was 500. The other correlated parameters were set with the best ones obtained by many experiments. w was a random number in [0,1]. In SOPSO, $c_1 = c_2 = 0.4$. Taking into account that the flight distance and avoiding threats have the same degree of influence to the path planning, let $\delta = 0.5$.

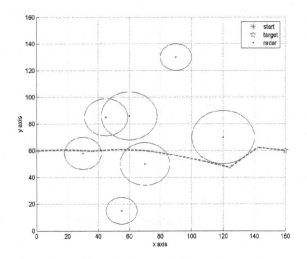

Fig. 2. Path planning by WPSO

Figure 2~4 show the best path found by using WPSO,CFPSO and SOPSO running for 50 times. From the result comparison we can see that the path planed by SOPSO is more accurate and smooth than the other two results and such a path is not only of shorter flight distance but also can avoid the threats effectively, while the path found by WPSO is not feasible.

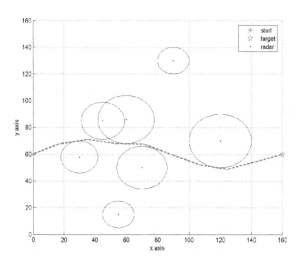

Fig. 3. Path planning by CFPSO

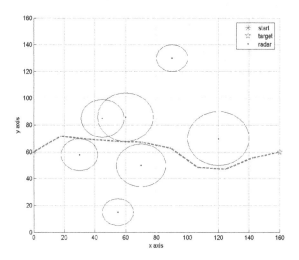

Fig. 4. Path planning by SOPSO

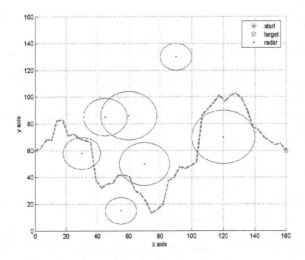

Fig. 5. Path planning by SOPSO with adding particle's dimension

The path showed in Figure 5 is planed by SOPSO in which the particle's dimension is the number of perceiving points in the algorithm before. The iteration number in this algorithm is also 500. Compared the results in Figure 4 and Figure 5, the path showed in Figure 4 is better than the other one in Figure 5 obviously.

Not only the flight distance of the former is shorter than the latter one, but also it has a greater performance in avoiding the threats.

Table 1 gives the data comparison of path planning by different algorithms. Where, fMin, fMax, fMean represent the min value, the max value and the mean value of the results separately which are the min value of the object function obtained by running programs for 50 times. The statistical success rate of UCAV avoiding the radar threats are given through counting the times of avoiding threats successfully during the 50 times running.

Table 1. Results comparison of the route planed by three kinds of algorithm

Algorithm	fMin	fMax	fMean	Success rate
WPSO	103.8151	113.2756	111.7782	0/50
CFPSO	86.0100	95.8890	90.3673	18/50
SOPSO	83.7250	109.9700	91.2832	49/50

Judging from the experimental results, it is obvious that the proposed algorithm SOPSO has a higher success rate and it can find the feasible and optimal path for the aerial vehicle, while the CFPSO cannot effectively escape the threat regions even though the mean objective function value is smaller. From practical considerations, this proposed method provides a new way for path planning of UCAV in exact application in the future.

5   Conclusion

In order to plan the path for UCAV, a model was built to solve this problem. SOPSO algorithm was proposed to plan the path for UCAV in flight missions. Simulation results showed that the proposed method could find a path with shorter flight distance and avoiding the threats effectively, which was much more suitable for practical application. Our future work will focus on the exact application research in path planning with complex conditions in this field.

References

1. Zheng, C.W., Li, L., Xu, F.J.: Evolutionary route planner for unmanned air vehicles. IEEE Transactions on Robotics and Automation, 609–620 (2005)
2. Wang, Y.X., Chen, Z.J.: Genetic algorithms (GA) based flight path planning with constraints. Journal of Beijing University of Aeronautics and Astronautics, 355–358 (1999)
3. Ye, W., Ma, D.W., Fan, H.D.: Algorithm for low altitude penetration aircraft path planning with improved ant colony algorithm. Chinese Journal of Aeronautics, 304–309 (2005)
4. Chen, D., Zhou, D.Y., Feng, Q.: Route Planning for Unmanned Aerial Vehicles Based on Particle Swarm Optimization. Journal of Projectiles, Rockets, Missiles and Guidance, 340–342 (2007)
5. Kennedy, J., Eberhart, R.C.: Particle Swarm Optimization. In: Proceedings of the IEEE International Conference on Neural Network. IEEE Service Center, pp. 1942–1948. IEEE Press, New Jersey (1995)
6. Shi, Y., Eberhart, R.: A Modified Particle Swarm Optimizer. In: Proceedings of the IEEE International Conference on Evolutionary Computation. IEEE Service Center, pp. 69–73. IEEE Press, Piscataway (1998)
7. Clerc, M.: The swarm and the queen: towards a deterministic and adaptive particle swarm optimization. In: Proceedings of the Congress on Evolutionary Computation, pp. 1951–1957. IEEE Service Center, Piscataway (1999)
8. Hu, J.X., Zeng, J.C.: Two-order Oscillating Particle Swarm Optimization. Journal of System Simulation 19, 997–999 (2007)

Qianzhi Ma. She was born in March 1985. Her main research fields involve intelligent optimization and path planning. She is a graduate student in Shaanxi Normal University.

Xiujuan Lei. She was born in May 1975. She is an associate professor of Shaanxi Normal University. She has been published more than 20 pieces of papers. Her main research fields involve intelligent optimization, especially particle swarm optimization, etc.

Modal Analysis for Connecting Rod of Reciprocating Mud Pump

Zhiwei Tong[*], Hao Liu, and Fengxia Zhu

Faculty of Mechanical & Electronic, China University of Geosciences
Wuhan, China, 430070
Tel.:+ 86-27-67883277, Fax: +86-27-67883273
tongzhiwei888@126.com

Abstract. Modal analysis is an effective method to determine vibration mode shapes and weak parts of the complex mechanical system, its main purpose is to use optimal dynamics design method of mechanical structure system instead of the experience analog method. Reciprocating mud pump is the machine that transport mud or water in the process of drilling, which is an important component of the drilling equipment. In order to improve the performance of reciprocating pump and decrease the failure of the connecting rod caused by vibration during running, a modal analysis method is performed. In this paper, a three-dimensional finite-element model of connecting rod was built to provide analytical frequencies and mode shapes, then the modal distribution and vibration mode shapes for connecting rod were obtained by computing. The results showed the weakness of the connecting rod, which would provide the reference to dynamics analysis and structural optimization for connecting rod in the future.

Keywords: Connecting rod, Finite element, Modal analysis.

1 Introduction

Modal analysis is an effective method to determine vibration mode shapes and weak parts of the complex mechanical system. Connecting rod is an important component of the reciprocating mud pump dynamic system, it is not only a transmission component but also a moving parts, at the same time it must withstand variable load such as tensility, compress force and bending in the working process, so its use reliability has a great significance for the normal operation of reciprocating mud pump[1]. Therefore, dynamic characteristics study on the connecting rod has become an important part of design. In this paper, a modal analysis was applied to connecting rod of reciprocating pump by ANSYS software, the main purpose of analysis is to identify the model parameters of connecting rod such as frequency, vibration mode shapes and provide a basis for structural dynamics analysis and the follow-up optimal design of connecting rod.

---

[*] Corresponding author.

H. Deng et al. (Eds.): AICI 2009, LNAI 5855, pp. 215–220, 2009.

2 Modal Analysis Theory

Modal analysis has been used to study the inherent frequencies and vibration mode shapes of mechanical systems, which only relate with the stiffness speciality and mass distribution of structure, and it has nothing to do with external factors [2]. Modal analysis is the starting point of all the dynamic analysis, inherent frequency and vibration mode analysis is also necessary if you want to process the harmonic response or transient analysis. In addition, the structure is supposed to linear in modal analysis which is in line with the actual working conditions of reciprocating pump.

The main purpose of modal analysis is to use optimal dynamics design method of mechanical structure system instead of the experience analog and passive method. By using the modal analysis, every mode shapes of structure can be identified and the weak parts of structural system's dynamic characteristics can be found, then the effective measures were taken to improve the structure and optimize the system.

The movement differential equation of the N freedom degrees linear system can be express as

$$[M]\{\ddot{X}\}+[C]\{\dot{X}\}+[K]\{X\}=\{F(t)\} \tag{1}$$

In (1), [M] is the mass matrix, [C] is the damping matrix, [K] is the stiffness matrix, $\{X\}$ is the displacement response column vector of node, $\{X\}=\{x_1,x_2,\ldots\ldots,x_n\}^T$, $\{F(t)\}$ is the load column vector [3].

The damping and external load were not considered when studying the inherent frequencies and vibration mode shapes in the process of modal analysis, so the vibration differential equations could be further simplified to get the undamped free vibration differential equations of system(as shown in (2)). '

$$[M]\{\ddot{X}\}+[K]\{X\}=0 \tag{2}$$

Any free vibration can be regarded as a simple harmonic vibration, a hypothesis was put forward as shown in the following.

$$\{X(t)\}=\{\phi\}\cos\omega t \tag{3}$$

Put (3) into (2), some equations were shown in the following.

$$([K]-\omega^2[M]\{\phi\})=0$$
$$[K]\{\phi\}=\lambda[M]\{\phi\} \tag{4}$$

In these equations, ω is the Circular frequency; $\{\phi\}$ is the Characteristic column vector; λ ($\lambda=\omega^2$) is the eigenvalue.

In the process of free vibration, not each point amplitude is zero, so (4) can be become into (5).

$$\|[K]-\omega^2[M]\|=0 \tag{5}$$

Equation (5) is the undamped vibration system characteristic equation. Modal analysis can also be considered to solve the eigenvalues ω^2 in (4). In the mechanical engineering, mass matrix and stiffness matrix can be seen as a symmetric positive definite matrix, and the eigenvalue number obtained through equation computing are equal to the order N of matrix, that is, there are N natural frequency ($\omega_i(i = 1, 2, ..., n)$). For each natural frequency, a column vector consisting of N nodes amplitude can be determined by the (4), namely, the vibration mode of structure ($\{\phi_i\}$).

ANSYS software provides some modal extraction methods such as subspace, slock method, curtail method, dynamic power, non-symmetric, damping method, QR damped method and so on. In this paper, the subspace iteration method was chosen which had the advantages of high accuracy and arithmetic stability compared with other methods [5].

3 Constructing the Finite Element Modal for Connection Rod

3.1 Constructing the Three-Dimensional Model for Connection Rod

For the complex mechanical systems, CAD software is commonly used to build model in order to enhance the speed of modeling, and then the model is imported into CAE software to analyze through the appropriate interface file. The CAD model may contain a number of design details which often are not based on the strength of the structure, it would cause model complication if the model retain these details in the process of modeling, at the same time, it is bound to increase the number of units, even cover up the principal contradiction of problem and cause negative impact for the results of the analysis [6]. Therefore, the local structure can be simplified that has less impact on the results.

Reciprocating mud pump commonly uses the crankshaft to facilitate assembly and adjusts the clearance of bushing, connecting rod must be produced into the separate structure. In order to simplify the analysis, the connecting rod and rod cap can be seen as a whole when modeling with Pro/E. A three-dimensional solid model of the reciprocating pump connecting rod was built by Pro / E software platform, then it was saved with IGES format and was imported into finite element analysis software-ANSYS.

3.2 Make the Structure Discrete

The materials of connecting rod are ZG35CrMo, its material properties are shown in table 1. The geometric model was divided into grid with 8-node Solid45 cells, those dividing precision is three levels and it was divided by the ANSYS software automatically, the finite element model of connecting rod is shown in Fig. 1.

Table 1. Material properties of connecting rod

EX (Mpa)	PRXY	DENS (kg/m³)
210000	0.3	7800

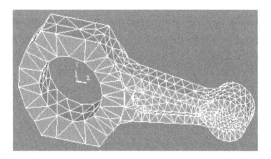

Fig 1. Finite element model of connecting rod

3.3 Boundary Constraints

The assumptions of boundary conditions is very important in modal analysis, it is not only affects the accuracy of the calculation but also decide whether to complete the calculation [7]. The movement of connecting rod is a planar motion which can be regarded as synthetic motion with translational and rotation, both ends of connecting rod are endure the restrictions of crosshead pin and crank pin. In accordance with the actual work of reciprocating pump connecting rod, the cylindrical coordinate system on the holes axis of both ends is defined, the points on the hole ends have freedom of rotating along the axis. Modal are decided by the inherent characteristic of system, which has nothing to do with the external load, therefore, the load boundary condition do not need to be set.

4 Finite Elements Solve of Connecting Rod and the Analysis of Result

Vibration of structure can be expressed as the linear combination of each mode, the low-level inherent modes have a great impact on the vibration of structure and play a decisive role for dynamic characteristics of the structure. A characteristic of modal for the connecting rod by three-dimensional finite element model is established above, and the 20 order inherent frequency of vibration modes (as shown in table 2) are calculated out. The six vibration modes are extracted to analyze. Fig. 2 shows the six order vibration mode shapes.

From the calculation results, connecting rod is prone to cause too large dynamic stress produce the fatigue cracks in its working process as a result of its dense mode. In particular, the first step and second step modal frequencies are relatively low, and it causes the dynamic response easily. It may cause the coupled vibration because of the small frequency difference of adjacent modal [8].

As shown in the Fig. 2, connecting rod has many various vibration forms which focus on the bending vibration. The first and second step vibration mode shapes are bending vibration along with Z and Y-axis respectively, the middle part of connecting rod have a larger vibration displacement and it shows as U form, the third step is the composite of bending and torsional vibration for connecting rod, the fourth and fifth step are the bending vibration around the Z and Y-axis, connecting rod shows S form

Table 2. The Twenty Calculations of Inherent Frequencies for Connecting Rod

N	Frequency (Hz)	N	Frequency (Hz)
1	136.6	11	1175.3
2	186.8	12	1356.3
3	230.5	13	1421.5
4	283.6	14	1518.9
5	435.0	15	1652.4
6	637.2	16	1705.2
7	780.1	17	1870.2
8	865.4	18	1980.8
9	958.2	19	2275.0
10	1073.4	20	2346.2

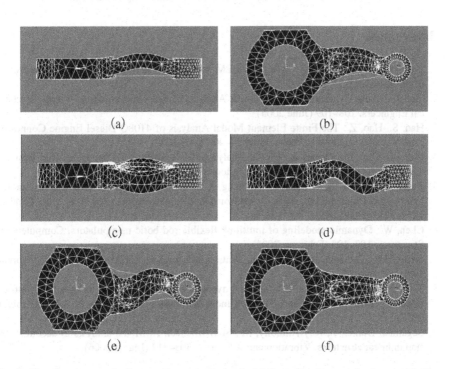

(a) (b) (c) (d) (e) (f)

Fig. 2. Six vibration mode shapes of connecting rod: (a) first vibration mode shape with frequency 136.6 Hz; (b) second vibration mode shape with frequency 186.8 Hz; (c) third vibration mode shape with frequency 230.5 Hz; (d) fourth vibration mode shape with frequency 283.6Hz;(e) fifth vibration mode shape with frequency 435.0Hz; (f) sixth vibration mode shape with frequency 637.2 Hz.

in the process of vibration, the sixth mode shape of vibration is not obvious, it is mainly bending along with the X and Y axis. Bending vibration of the connecting rod will make piston relation to sleeve of cylinder and the neck part of crank relation to bearings deflect, which can bring about additional stress and cause cracks and damage, so this situation should be focused on considering in the design to avoid damage [9].

5 Conclusion

Dynamic characteristic analysis of connecting rod is supplement and development for static design, it is an important means to do rational design and increase using reliability for connecting rod structure. The inherent vibration characteristics of reciprocating pump connected rod were analyzed by ANSYS software, which included the natural frequencies and mode shapes. Finally, more accurate and intuitive results were obtained. These results showed that the stress of connecting rod should be commonly concentrate at connect point of large and small parts, but stress concentration at the central of connecting rod is also more obvious, therefore, the traditional design concepts should be changed and taken fully into account the central of connecting rod design.

References

1. Yang, B., He, Z., Peng, X., Xing, Z.: FEM Analysis on the Valve of Linear Compressor. Fluid Machinery 33, 24–27 (2005)
2. Yu, B., Huang, Z.: Finite Element Modal Analysis of Rotor in Centrifugal Pump. Mechanical Engineers, 108–109 (June 2005)
3. Han, S., Hao, Z.: The Finite Element Modal Analysis of 4108Q Diesel Engine Connecting Rod. Vehicles and Power Technology, 37–40 (April 2002)
4. Dai, W., Fan, W., Cheng, Z.: Modal Analysis of Diesel Engine Connecting Rod Based on ANSYS. Mechanical Engineering and Automation, 39–41 (April 2007)
5. Zhang, G.X., Zhong, Y., Hong, X., Leng, C.L.: Microelectromechanical systems motion measurement and modal analysis based on Doppler interferometry. J. Vac. Sci. Technol., 1251–1255 (May 2009)
6. Chen, W.: Dynamic modeling of multilink flexible rod botic manipulators. Computers and Structures, 183–195 (February 2001)
7. Tlusty, I.F.: Dynamic structural identification task methods. In: Annals of CIRP, February 1980, pp. 260–262 (1980)
8. Kim, D.: Parametric studies on static and dynamic performance of air foil bearings with different top foil geometries and bump stiffness distributions. Tribology, 354–364 (February 2007)
9. Segalman Daniel, J., Roy Anthony, M., Starr Michael, J.: Modal analysis to accommodate slap in linear structures. Vibration and Acoustics, 303–317 (March 2006)

A Time-History Analysis Algorithm of a Non-viscously Damped System Using Gauss Precise Integration

Hongyu Shen and Zhongdong Duan

School of Civil Engineering, Harbin Institute of Technology,
Harbin, Heilongjiang Province, China, 150090
shenhongyu2002@hit.edu.cn
duanzd@hit.edu.cm

Abstract. Time-history analysis of a multiple-degree-of-freedom system with non-viscous damping was considered. It was assumed that the non-viscous damping forces depend on the past history of velocities via convolution integrals over exponentially decaying kernel functions. By transforming the equation of motion into a first-order extended state-space representation and combining the Gauss-Legendre quadrature with the calculation technique of matrix exponential function in the precise time-integration method, a more applicable time-history analysis method of this system was proposed. The main virtues of the proposed method were the avoidance of matrix inversion, the high efficiency and the selectable accuracy. A numerical example was given to demonstrate the validity and efficiency of the method.

Keywords: non-viscous damping, state-space equation, precise integration method, Gauss-Legendre quadrature method, time-history analysis.

1 Introduction

The damping characteristics and dynamic responses of a vibrating system is a key topic in the structural dynamic analysis. The viscous damping model, first advanced by Lord Rayleigh in 1877 [1], is usually used to simulate energy dissipation in a structural dynamic response. It is however, well known that the classical viscous damping model is a mathematical idealization and the true damping model is likely to be different. Moreover, increasing use of modern composite materials and intelligent control mechanisms in aerospace and automotive industries demands sophisticated treatment of dissipative forces for proper analysis and design. As a result, there has been an increase in interest in the recent years on the investigation for other damping models which can better reflect the energy dissipation of damping and many researchers also began to study the dynamic analysis method involved [2-7].

At the same time, the time-history analysis method is more and more widely used due to the rapid construction of high rise buildings and the fast development of the structural dynamic analysis theory and computational technology. At present, the widely used methods include linear accelerating method, Wilson-θ method [8], Newmark-β method [9], etc. Most recently, the precise time-integration method (PIM)

H. Deng et al. (Eds.): AICI 2009, LNAI 5855, pp. 221–230, 2009.

proposed by Zhong et al. attracted lots of attention [10-12]. It can obtain a nearly perfectly accurate solution as a numerical method, and therefore gives a very useful idea in time history analysis. However, an inversion of matrix is needed in solving the inhomogeneous equations during the structural dynamic analysis by PIM [13-14]. It is known that both the accuracy and stability get worse in an inverse calculation, so the application of this method is limited.

The time-history analysis method of a convolution integral non-viscously damped system is studied in this paper. This model is first advanced by Biot about viscous materials [15]. It is assumed that the non-viscous damping forces depend on the past history of velocities via convolution integrals over exponentially decaying kernel functions. The advantage of this model is its generality compared with traditional viscous models. It can describe various damping mechanism by selecting different kernel function types [16] and is more accurate in expressing the time-delay effect. The dynamic analysis of non-viscous damped system of this paper is based on the condition that the kernel function is exponential function [17].

Obviously, the conventional numerical integration methods can not be directly used in this non-viscous damped system due to the change of its mathematic model. Adhikari and Wagner gave a state-space form of this non-viscous damped system [16], and also proposed a mode superposition method of the state-space (MSM) and a direct time-domain integration method (DTIM) to study the system dynamic responses [18]. According to some studied cases, the mode superposition method has a good accuracy, but it needs large computation. The condition of the direct time-domain integration is just inversed. Accordingly, it is significant to propose a method which has both high accuracy and efficiency of computation. In this paper, the dynamic response state equation of the non-viscous damped system in [16] was first transformed. Then a modified precise integration without the inversion of matrix is set up with basic principles of the precise integration and the Gauss-Legendre quadrature. Under the acting of discrete load (such as earthquake wave), the cubic spline interpolation is used to calculate the function values of the Gauss integral point in recursion expressions. Finally, a numerical simulation is given to demonstrate the efficiency of the method.

2   Motion Equations of the System and State-Space Formalism

2.1   Motion Equations of the System

The non-viscous damping model in this paper considers the damping force relates with the velocity time history by a convolution integral between the velocity and a decaying kernel function in its mathematical formulation. The damping force can be expressed as

$$\mathbf{F}_d(t) = \int_0^t \mathbf{G}(t-\tau)\dot{\mathbf{x}}(\tau)d\tau \tag{1}$$

where $\mathbf{G}(t) \in \mathbf{R}^{N \times N}$ is a matrix of kernel functions

$$G(t) = \sum_{k=1}^{n} C_k g_k(t) = \sum_{k=1}^{n} C_k \mu_k e^{-\mu_k t}, \quad t \geq 0 \tag{2}$$

where C_k is a damping coefficient matrix, $g_k(t)$ is a damping function, in this paper, $g_k(t) = \mu_k e^{-\mu_k t}$, $\dot{x}(t)$ is the velocity vector. In the special case when $g_k(t) = \delta_k(t)$, where $\delta(t)$ is the Dirac-delta function, Eq. (1) reduces to the case of viscous damping.

The motion equations of an N-degree-of-freedom (DOF) system with such damping can be expressed as

$$M\ddot{x}(t) + \sum_{k=1}^{n} \int_0^t C_k \mu_k e^{-\mu_k(t-\tau)} \dot{x}(\tau) d\tau + Kx(t) = f(t) \tag{3}$$

where $M, K \in R^{N \times N}$ is mass and stiffness matrices, $f(t) \in R^N$ is the forcing vector, $\mu_k \in R^+$ is a relaxation parameter, n is the number of relaxation parameters of different damping mechanism. The initial conditions of (3) is

$$x(0) = x_0 \in R^N \quad \text{and} \quad \dot{x}(0) = \dot{x}_0 \in R^N \tag{4}$$

2.2 Review of State-Space Formalism

Obviously, Eq. (3) describes a non-viscous damped dynamic system. The state-space method has been used extensively in the modal analysis of the non-proportional viscous damped. We expect to extend the state-space approach to this non-viscous damped system and simplify the expressions of damping and equation, in order to find a step-by-step integral algorithm to analyse the system dynamic responses. Recently Wagner and Adhikari have proposed a state-space method for this damped system [16]. Here we briefly review their main results.

Case A: All C_k Matrices are of Full Rank

We introduce the internal variables

$$y_k(t) = \int_0^t \mu_k e^{-\mu_k(t-\tau)} \dot{x}(\tau) d\tau \in R^N, \quad \forall k = 1, \dots, n \tag{5}$$

Differentiating (5) one obtains the evolution equation

$$\dot{y}_k(t) + \mu_k y_k(t) = \mu_k \dot{x}(t) \tag{6}$$

Using additional state-variables $v(t) = \dot{x}(t)$, Eq. (3) can be represented in the first-order form as

$$B\dot{z}(t) = Az(t) + r(t) \tag{7}$$

where

$$
B = \begin{bmatrix}
\sum_{k=1}^{n} C_k & M & -C_1/\mu_1 & \cdots & -C_n/\mu_n \\
M & O & O & O & O \\
-C_1/\mu_1 & O & C_1/\mu_1^2 & O & O \\
\vdots & O & O & \ddots & O \\
-C_n/\mu_n & O & O & O & C_n/\mu_n^2
\end{bmatrix} \in R^{m \times m}
\tag{8}
$$

$$
A = \begin{bmatrix}
-K & O & O & O & O \\
O & M & O & O & O \\
O & O & -C_1/\mu_1 & O & O \\
O & O & O & \ddots & O \\
O & O & O & O & -C_n/\mu_n
\end{bmatrix} \in R^{m \times m}
\tag{9}
$$

$$
r(t) = \begin{Bmatrix} f(t) \\ 0 \\ 0 \\ \vdots \\ 0 \end{Bmatrix} \in R^m, \quad
z(t) = \begin{Bmatrix} x(t) \\ v(t) \\ y_1(t) \\ \vdots \\ y_n(t) \end{Bmatrix} \in R^m
\tag{10}
$$

where A and B are the system matrices in the extended state-space. They are symmetric matrices, and the order of the system is $m = 2N + nN$; $z(t)$ is the extended state-vector, and $r(t)$ is the force vector in the extended state-space, O is a n-order null matrix, $rank(C_k) = N$, $\forall k = 1, \ldots, n$.

It can be seen that, when $\mu \to \infty$, Eq. (5) change into $y(t) = \dot{x}(t)$, Eq. (3) will change into a equation of viscous damped system.

Case B: C_k Matrices are of Rank Deficient
When C_k matrices are rank deficient, we should transform C_k into lower order full rank matrices; thus, lots of memory storage and time of calculation will be saved.

Let us introduce a matrix $R_k \in R^{N \times r_k}$

$$
R_k^T C_k R_k = d_k \in R^{r_k \times r_k}
\tag{11}
$$

where d_k is a diagonal matrix consisting of only the nonzero eigenvalues of C_k, thus, the columns of matrix R_k are the eigenvectors of C_k. Define a set of variables of reduced dimension $\tilde{y}_k(t) \in R^{N \times r_k}$ using R_k as

$$\mathbf{y}_k(t) = \mathbf{R}_k \tilde{\mathbf{y}}_k(t) \tag{12}$$

Eq. (3) can be represented in a first-order form as

$$\tilde{\mathbf{B}}\dot{\tilde{\mathbf{z}}}(t) = \tilde{\mathbf{A}}\tilde{\mathbf{z}}(t) + \tilde{\mathbf{r}}(t) \tag{13}$$

where

$$\tilde{\mathbf{B}} = \begin{bmatrix} \sum_{k=1}^{n}\mathbf{C}_k & \mathbf{M} & -\mathbf{C}_1\mathbf{R}_1/\mu_1 & \cdots & -\mathbf{C}_n\mathbf{R}_n/\mu_n \\ \mathbf{M} & \mathbf{O}_{N,N} & \mathbf{O}_{N,r_1} & \cdots & \mathbf{O}_{N,r_n} \\ -\mathbf{R}_1^{\mathrm{T}}\mathbf{C}_1/\mu_1 & \mathbf{O}_{N,r_1}^{\mathrm{T}} & \mathbf{R}_1^{\mathrm{T}}\mathbf{C}_1\mathbf{R}_1/\mu_1^2 & \cdots & \mathbf{O}_{r_1,r_n} \\ \vdots & \vdots & \vdots & \ddots & \vdots \\ -\mathbf{R}_n^{\mathrm{T}}\mathbf{C}_n/\mu_n & \mathbf{O}_{N,r_n}^{\mathrm{T}} & \mathbf{O}_{r_1,r_n}^{\mathrm{T}} & \cdots & \mathbf{R}_n^{\mathrm{T}}\mathbf{C}_n\mathbf{R}_n/\mu_n^2 \end{bmatrix} \in \mathbf{R}^{\tilde{m}\times\tilde{m}} \tag{14}$$

$$\tilde{\mathbf{A}} = \begin{bmatrix} -\mathbf{K} & \mathbf{O}_{N,N} & \mathbf{O}_{N,r_1} & \cdots & \mathbf{O}_{N,r_n} \\ \mathbf{O}_{N,N} & \mathbf{M} & \mathbf{O}_{N,r_1} & \cdots & \mathbf{O}_{N,r_n} \\ \mathbf{O}_{N,r_1}^{\mathrm{T}} & \mathbf{O}_{N,r_1}^{\mathrm{T}} & -\mathbf{R}_1^{\mathrm{T}}\mathbf{C}_1\mathbf{R}_1/\mu_1 & \cdots & \mathbf{O}_{r_1,r_n} \\ \vdots & \vdots & \vdots & \ddots & \vdots \\ \mathbf{O}_{N,r_n}^{\mathrm{T}} & \mathbf{O}_{N,r_n}^{\mathrm{T}} & \mathbf{O}_{r_1,r_n}^{\mathrm{T}} & \cdots & -\mathbf{R}_n^{\mathrm{T}}\mathbf{C}_n\mathbf{R}_n/\mu_n \end{bmatrix} \in \mathbf{R}^{\tilde{m}\times\tilde{m}} \tag{15}$$

$$\tilde{\mathbf{r}}(t) = \begin{Bmatrix} \mathbf{f}(t) \\ \mathbf{0}_N \\ \mathbf{0}_{r_1} \\ \vdots \\ \mathbf{0}_{r_n} \end{Bmatrix} \in \mathbf{R}^{\tilde{m}}, \quad \tilde{\mathbf{z}}(t) = \begin{Bmatrix} \mathbf{x}(t) \\ \mathbf{v}(t) \\ \tilde{\mathbf{y}}_1(t) \\ \vdots \\ \tilde{\mathbf{y}}_n(t) \end{Bmatrix} \in \mathbf{R}^{\tilde{m}} \tag{16}$$

It can be seen that the order of the system is $\tilde{m} = 2N + \sum_{k=1}^{n} r_k$, and each symbolic significance is same as case A. when All $\mathbf{C}_k$ Matrices are of Full Rank, each $\mathbf{R}_k$ matrix can be chosen as the identity matrix and (13) is same to (7), so we can consider full rank as a special case.

2.3 Introductions of Analysis Method

After obtaining the system matrices of state-space, the eigenvalue Λ and eigenvector Ψ of system matrices A and B (or $\tilde{\mathbf{A}}$ and $\tilde{\mathbf{B}}$) are solved through the solution method of two order eigenvalue problem and satisfied equation $(\mathbf{A}\Lambda + \mathbf{B})\begin{Bmatrix} \Psi \\ \Psi\Lambda \end{Bmatrix} = \mathbf{0}$

where Λ is a diagonal matrix; then the common mode superposition method is applied to solve the system dynamic time-history analysis considering the initial condition. The time-history analysis result calculated by this method under the ideal state of complete model is the theoretical accurate solution, but the application of the method is limited by inefficiency calculation and demanding computing time.

Adhikari presented the direct time-domain integration method in order to solve the shortcoming of wasting large computing time of the mode superposition method [18]. Through direct integration of (7) or (13) at every time step, the time-domain recursion expression of the state vector $z(t)$ or $\tilde{z}(t)$ is obtained and the explicit recursive equations of the displacement and velocity are derived. The method can significantly increased the efficiency of the system time history analysis due to avoiding the large scale calculation of entire system matrices and state vector, but the direct integration approximate treatment of state equation induces decreasing calculation accuracy which will be discussed in the following numerical example.

Considering computational efficiency and calculation accuracy, the precise time-integration method presented by Mr. Zhong [10] is applied to treat and solve the system state space equation. The basic idea of the precise time-integration method is to change the conventional two order system equation into one order expression, and the general solution of the system equation is subsequently transformed into the exponential matrix and then the 2^N algorithm is adopted to calculate the exponential matrix to complete the system time history analysis by applying the additional theorem of exponential function. The errors of the precise time-integration method are only originated from the exponential matrix expansion besides the arithmetic errors of the matrix multiplication. The more accuracy analysis indicates that the exponential matrix expansion is very small for a damped vibration system [11], so the numerical results of the precise time-integration method can meet the exact solution; moreover, the main body of this method is the add operation applied for solving the exponential matrix and the method can embody the technical characteristic of the computes and increase the computational efficiency of the algorithm. However, there is an inverse matrix in the precise time-integration equation for the nonhomogeneous dynamic equation of the structure under external force [13][14] and there are problems such as loss calculation accuracy, stabilization and non-existence of inverse matrix in the process of matrix inversion. The problems always exist in the multi-DOF system and are particularly serious for matrices with high dimensions. In this paper, a modified precise time-integration method without the inversion of matrix is proposed combining with the basic principles of the precise time-integration and Gauss-Legendre integration.

3   Gauss Precise Integration

Considering the general case when the C_k matrices are of rank deficient, rearrange the state-space Eq. (13) as

$$\dot{V} = HV + F \tag{17}$$

where $V = \tilde{z}(\tau)$, $H = \tilde{B}^{-1}\tilde{A}$, $F = \tilde{B}^{-1}\tilde{r}(\tau)$. In order to describe the algorithm, the exponential matrix is defined as

$$e^{H\tau} \approx I + H\tau + (H\tau)^2/2 + (H\tau)^3/6 + ... + (H\tau)^n/n! + ... \tag{18}$$

Multiplying both sides of (17) by $e^{-H\tau}$ and integrating it from t to t_k after organization yields

$$e^{-H\tau}\left(\dot{V}-HV\right)=e^{-H\tau}F, \quad \int_{t_k}^{t}d\left(e^{-H\tau}V\right)=\int_{t_k}^{t}e^{-H\tau}F(\tau)d\tau \tag{19}$$

then

$$V(t)=e^{H(t-t_k)}V_k+e^{Ht}\int_{t_k}^{t}e^{-H\tau}F(\tau)d\tau \tag{20}$$

where V and V_k represent the solution vector of t and t_k, respectively.

Introducing Gauss-Legendre quadrature

$$\int_{a}^{b}f(x)dx=\frac{b-a}{2}\sum_{j=0}^{m}L_{j}f(\frac{a+b}{2}+\frac{b-a}{2}t_{j})+E(f) \tag{21}$$

where t_j is the coordinate of integral point, L_j is a coefficient of Gauss-Legendre quadrature, $E(f)$ is the Residual error.

Let $\tau=t-t_k$. Substituting (21) into (20) yields

$$V(t)=e^{H\tau}V_k+\frac{\tau}{2}\sum_{j=0}^{m}L_{j}e^{H\frac{\tau}{2}(1-t_j)}F\left(t_k+\frac{\tau}{2}(1+t_j)\right)+O(\tau^{2n+2})$$

$$=T_{\tau}V_k+\frac{\tau}{2}\sum_{j=0}^{m}L_{j}T_{j}F\left(t_k+\frac{\tau}{2}(1+t_j)\right)+O(\tau^{2n+2}) \tag{22}$$

where $T_{\tau}=e^{H\tau}$, $T_{j}=e^{H\frac{\tau}{2}(1-t_j)}$. Eq. (22) is the step-by-step integral recursion expression of solution vector V_t from solution vector V_k of t_k. It suggests that, there is not any inverse matrix in the expression. Under the acting of discrete load (such as earthquake wave), the cubic spline interpolation is used to calculate the $F(t_j)$ of the Gauss integral point.

Now, let's discuss the precise computation of exponential matrix T_{τ}, the additional theorem of the exponential function is given by the identity

$$T_{\tau}=e^{H\tau}=\left(e^{H\tau/M}\right)^{M} \tag{23}$$

where $M=2^{N}$, generally $N=20$, Because τ should be a small time interval, $\Delta t=\tau/M$ is an extremely small time interval. The exponential matrix T_{τ} departs from the unit matrix I into a very small extent. Hence it should be distinguished as

$$e^{H\Delta t}\approx I+H\Delta t+(H\Delta t)^{2}/2=I+T_{a} \tag{24}$$

obviously, for computing the matrix T_{τ}, Eq. (24) should be factored as

$$\mathbf{T}_r = \left[\mathbf{I} + \mathbf{T}_a\right]^{2^N} = \left[\mathbf{I} + \mathbf{T}_a\right]^{2^{N-1}} \times \left[\mathbf{I} + \mathbf{T}_a\right]^{2^{N-1}} \tag{25}$$

Such factorization should be iterated N times, and the high-precision value of $\mathbf{T}_r$ will be obtained, so will the $\mathbf{T}_j$. By substituting them into (22), we can realize an efficient high-precision time-history analysis of this non-viscously damping system.

4 Numerical Example

Here a two-dimension and three-story frame is used. It has 12 DOF, each node has a rotational DOF and each story has a lateral DOF. This example only is applied to test the performance of the proposed method, and the damping model will not be discussed in detail. Therefore, the non-viscously damping model of system can be built by one exponential function, and the motion equation of system is described in (3) (the value of n takes 1) and mechanical model is shown in Fig 1. Otherwise, to test the calculated accuracy and the method of dealing with the discrete load of the proposed method, it is necessary to consider the displacement of the second floor under the acting of sinusoidal wave $\sin(\pi t)$ and El-Centro wave respectively.

For the numerical calculation we have assume $E = 3 \times 10^7 N / m^2$, $m_{\text{column}} = 300 kg$, $m_{\text{beam}} = 100 kg$, $m_{\text{story}} = 2000 kg$, $A_{\text{column}} = 0.6 \times 0.6 m^2$, $A_{\text{beam}} = 0.2 \times 0.4 m^2$ $\mu_1 = 1/\gamma_1 T_{\min}$, $\gamma_1 = 0.05$, $T_{\min} = 2\pi/\omega_{\max}$, $\mathbf{C}_1 = \alpha\mathbf{M} + \beta\mathbf{K}$, $\alpha = 2\omega_1\omega_2 \dfrac{\xi_1\omega_2 - \xi_2\omega_1}{\omega_2^2 - \omega_1^2}$, $\beta = 2\dfrac{\xi_2\omega_2 - \xi_1\omega_1}{\omega_2^2 - \omega_1^2}$, $\xi_1 = 0.02$, $\xi_2 = 0.05$.

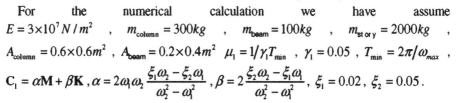

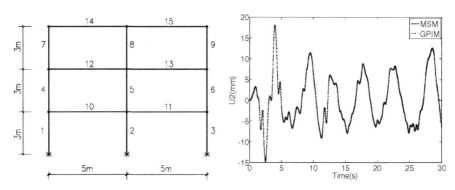

Fig. 1. Twelve-DOF steel frame with non-viscous damping

Fig. 2. Horizontal displacement of $U2$ under El-Centro wave

Firstly, three types of methods are used to calculate the displacement of structure under the acting of sinusoidal wave $\sin(\pi t)$, and the time steps are taken $0.02s$ and $0.5s$, respectively. The number of Gauss integral points is taken 2, that is, $m=2$. Here the results obtained by the modal superposition method (MSM) [16] are deemed as the exact solution. If the step takes small value, the results obtained by all the three methods are similar, as shown in Fig. 3a. If the big step is taken, the results obtained by the direct time-domain integration method (DTIM) and precise time-integration method (PIM) generate big errors, while the proposed method (GPIM) can still obtain the exact results, as shown in Fig. 3b. It is obvious that the order of accuracy of GPIM depends on the number of Gauss integral point and the size of step τ, it is selectable.

However, no matter whether the function fitting or the numerical calculation, bothDTIM and PIM methods may generate big integral error when these two methods are applied to analyse the structural response under the acting of discrete load, since it is inevitable to calculate the integration and difference of loading function in these two methods. The proposed method can overcome the disadvantages of the above two methods because there are no integral and differential items in recursion expression and the cubic spline interpolation is used to calculate the Gauss integral point of load. The displacement of the second floor under the acting of EI-Centro wave is calculated by the proposed method, and the results are shown in Fig. 2.

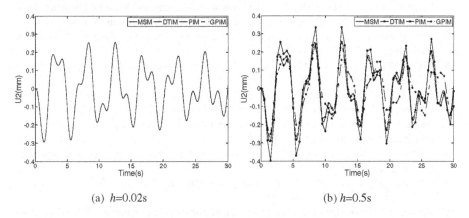

(a) h=0.02s   (b) h=0.5s

Fig. 3. Horizontal displacement of $U2$ under Sine wave

It is obvious that the calculation accuracy and amount of the proposed method depend on the number of Gauss integral point and the size of step τ. Comparing with conventional precise time-integration methods, the proposed method aggravates the calculating burden of exponential matrix, while the calculating accuracy is improved obviously when only $m = 2$, so the calculation amount is not significantly increased. On the other hand, the relative big integral step can be implemented, which can lighten the calculating burden evidently. As a result, the proposed method could improve both the calculating accuracy and the efficiency.

5 Conclusions

The time-history analysis method of a convolution integral non-viscously damped system is studied in this paper, based on the state-space form of this non-viscous damped system given by Adhikari and Wagner. By combining basic principles of the precise integration and the Gauss-Legendre quadrature, a new algorithm to analyze the dynamic response of this damping system in time domain is proposed. Because there are no inverse matrix items in the proposed method, the advantage is more significant when the method is applied to large structures with multi-DOFs in which nonsingular phenomenon of system matrix occurs easily. A numerical example shows that the proposed method can both ensure the high calculating accuracy and lighten the calculating burden by increasing the integral step. For the discrete load, it is more effective and reliable to operate.

References

1. Rayleigh, L.: Theory of Sound, 2nd edn., vol. 2. Dover Publications, New York (1877); (1945 re-issue)
2. Dong, J., Deng, H., Wang, Z.: Studies on the damping models for structural dynamic time history analysis. World Information on Earthquake Engineering 16(4), 63–69 (2000) (in Chinese)
3. Liang, C., Ou, J.: Relationship between Structural Damping and Material Damping. Earthquake Engineering and Engineering Vibration 26(1), 49–55 (2006) (in Chinese)
4. Woodhouse, J.: Linear damping models for structural vibration. Journal of Sound and Vibration 215(3), 547–569 (1998)
5. Maia, N.M.M., Silva, J.M.M., Ribeiro, A.M.R.: On a general model for damping. Journal of Sound and Vibration 218(5), 749–767 (1998)
6. Li, Q.S., Liu, D.K., Fang, J.Q., Jeary, A.P., Wong, C.K.: Damping in buildings: its neural network model and AR model. Engineering Structures 22, 1216–1223 (2000)
7. Adhikari, S.: Damping modelling using generalized proportional damping. Journal of Sound and Vibration 293, 156–170 (2006)
8. Wilson, E.L.: Nonlinear dynamic analysis of complex structures. Earthquake Engineering and Structural Dynamics 1(3), 241–252 (1973)
9. Newmark, N.M.: A method of computation for structural dynamics. Journal of Engineering Mechanics 85(3), 249–260 (1959)
10. Zhong, W.: On precise time-integration method for structural dynamics. Journal of Dalian University of Technology 34(2), 131–136 (1994) (in Chinese)
11. Zhong, W.: On precise integration method. Journal of Computational and Applied Mathematics 163, 59–78 (2004)
12. Zhong, W., Zhu, J., Zhong, X.: A precise time integration algorithm for nonlinear systems. In: Proc. of WCCM-3, vol. 1, pp. 12–17 (1994)
13. Qiu, C., Lu, H., Cai, Z.: Solving the problems of nonlinear dynamics based on Hamiltonian system. Chinese Journal of Computational Mechanics 17(2), 127–132 (2000) (in Chinese)
14. Lu, H., Yu, H., Qiu, C.: An integral equation of non-linear dynamics and its solution method. Acta Mechanica Solida Sinica 22(3), 303–308 (2001) (in Chinese)
15. Biot, M.A.: Linear thermodynamics and the mechanics of solids. In: Proc. of the third US national congress on applied mechanics, pp. 1–18. ASME Press, New York (1958)
16. Wagner, N., Adhikari, S.: Symmetric state-space formulation for a class of non-viscously damped systems. AIAA J. 41(5), 951–956 (2003)
17. Adhikari, S., Woodhouse, J.: Identification of damping: Part 1, viscous damping. Journal of Sound and Vibration 243(1), 43–61 (2001)
18. Adhikari, S., Wagner, N.: Direct time-domain integration method for exponentially damped linear systems. Computers and Structures 82, 2453–2461 (2004)

Application of LSSVM-PSO to Load Identification in Frequency Domain

Dike Hu, Wentao Mao, Jinwei Zhao, and Guirong Yan

MOE key lab for strength and vibration, Xi'an Jiaotong University,
Xi'an, China
{hdkjll,maowt.mail,zjw76888,yanguir}@gmail.com

Abstract. It is important to identify loads on aircraft to facilitate the designing of aircraft and ground environmental experiments. A new approach of load identification in frequency domain based on least squares support vector machine (LSSVM) and particle swarm optimization (PSO) was proposed. The principle of load identification using LSSVM was derived and the corresponding model of relationship between loads and responses was constructed using LSSVM. To get better performance of identification, PSO was adopted to find optimal hyper-parameters of LSSVM with the best generalization ability of the identification model. The results of numerical simulation and random vibration experiments of simplified aircraft show that the proposed approach can identify not only singlesource load, but also multisource loads effectively with high precision, which demonstrate the proposed approach is greatly applicable to engineering project.

Keywords: load identification, LSSVM, PSO, inverse problem.

1 Introduction

There are some loads such as vibration, noise etc. acting on aircraft during the transportation, launch, and flight phase. It is commonly recognized that the acquisition of these loads facilitate not only the designing of strength and rigidity of aircraft and reliability of internal instruments, but also ground environmental experiments. In practice, it is difficult to obtain accurate loads directly by experiments on the ground or theoretical analysis [1] because of the complexity of aircraft structure and boundary conditions of the structure and the corresponding loading conditions. However, the responses of aircraft structure such as acceleration, displacement, strain, etc. can be easily measured by the sensors in applications. Usually, it is using the structural responses that loads are identified. This process is frequently referred to as "load identification".

Because many problems are treated in frequency domain in aerospace industry, the load identification problem in frequency domain is more concerned. The basic idea of the traditional frequency approach of load identification is that [2] the transfer function model of system is established and the loads are identified through the inverse of the transfer function and the responses. However the transfer function matrix is often ill-posed in resonance region, as a result, the precision and stability of load identification are not as good as expected.

H. Deng et al. (Eds.): AICI 2009, LNAI 5855, pp. 231–240, 2009.
© Springer-Verlag Berlin Heidelberg 2009

It is well considered that the relationship between loads and responses only depends on the structure itself and boundary conditions [1], and is inherent once the structure and the boundary conditions are determined. Thus the load identification process can be understood as seeking the relationship between loads and responses when the structure and the boundary conditions are determined.

The support vector machine (SVM) is a new tool to find the complex relationship between variables in recent years. It can get the best compromise between the complexities of the model and the learning ability according to the limited samples. The inner product operation in high-dimensional space is transformed into the kernel function operation in low-dimensional space by the kernel function method, which effectively solves the "dimension disaster" problem. The Least Squares Support Vector Machine (LSSVM) is a new SVM model proposed by Suykens[3]. The LSSVM uses least squares linear constraints as a loss function, which simplifies the convex optimization problem in the traditional SVM to a set of linear equations in LSSVM. It has many advantages [3]: operation simple, fast convergence, high precision, etc. and is widely used in estimating nonlinear function, etc.

Taking advantages of the SVM and LSSVM described above, this paper firstly use LSSVM to search for the relationship between loads and responses of aircraft in frequency domain, and establish a model of load identification. Secondly, because the performance of LSSVM is determined by the hyper-parameters (the regularization parameter and kernel parameters are called hyper-parameters) of LSSVM, the particle swarm optimization (PSO) [5] is used to find optimal parameters based on the best generalization ability of the load identification model. Finally, the identification of load of the simulation model and random vibration experiments are conducted with LSSVM-PSO.

2   Principle of Load Identification Based on LSSVM-PSO in Frequency Domain

2.1   Identification Model of SVM in Frequency Domain

For linear deterministic (or stochastic) system, the relationship between inputs $F(\omega)$ (or $S_{FF}(\omega)$) and outputs $x(\omega)$ (or $S_{xx}(\omega)$) can be described as the following function:

$$x(\omega) = H(\omega)F(\omega)$$
$$S_{xx}(\omega) = H(\omega)S_{FF}(\omega)H^{H}(\omega) \tag{1}$$

where $H(\omega)$ is the transfer function, described as the following function:

$$H(\omega) = (K - \omega^2 M + i\omega C)^{-1} \tag{2}$$

where M, K and C are mass matrix, stiffness matrix and damping matrix.

When the responses $x(\omega)$ (or $S_{xx}(\omega)$) are known, according to Eq.(1), the loads can be obtained as the following form:

$$F(\omega) = H(\omega)^{-1} x(\omega)$$
$$S_{FF}(\omega) = [H^{H}(\omega)H(\omega)]^{-1} H^{H}(\omega)S_{xx}(\omega)H(\omega)[H^{H}(\omega)H(\omega)]^{-1} \tag{3}$$

According to Eq.(3), the identification of loads need the inverse of transfer function, but if the condition number of $H(\omega)$ is large, it would lead to numerical instability of load identification. Because there are noise in the measurement and some nonlinear factors in the structure, the relationship between loads and responses is often nonlinear, written as the following function:

$$F(\omega) = f(\omega, x(\omega))$$
$$S_{FF}(\omega) = f(\omega, S_{xx}(\omega)) \tag{4}$$

The SVM can transform the nonlinear problem in low-dimensional space into a linear problem in high-dimensional space by a nonlinear mapping function. According to the theory of SVM [6], if the $x(\omega)$ (or $S_{xx}(\omega)$) are the inputs of SVM, $F(\omega)$ (or $S_{FF}(\omega)$) are the outputs of SVM, and then the nonlinear function $f(\cdot)$ can be viewed as a regression model of SVM. Application SVM to load identification is seeking the relationship between loads $x(\omega)$ (or $S_{xx}(\omega)$) and responses $F(\omega)$ (or $S_{FF}(\omega)$), and establishing the regression model of $f(\cdot)$.

2.2 Model Selection of LSSVM Based on PSO

LSSVM is a new SVM model, whose advantage has been described above. The regression of LSSVM can be formulated as following: Given a training set $\{x_i, y_i\}_{i=1}^{l}$, where $x_k \in R^n$ is the kth input pattern and $y_k \in R$ is the kth output pattern, the following optimization problem can be written as follows:

$$\min J(w,\xi) = \frac{1}{2}\| w^2 \| + \frac{C}{2}\sum_{i=1}^{l}(\xi_i)^2 \tag{5}$$
$$s.t: y_i = w^T \bullet \phi(x_i) + b + \xi_i, i = 1, 2, ..., l$$

where $\phi(\bullet)$ is a nonlinear function which maps the input space into a higher dimensional space. C is the regularization parameter and ξ_i is the error function.

In order to solve the optimization problem above, constructing Lagrange can be written as follows:

$$L(w,b,\xi,\alpha) = J(w,\xi) - \sum_{i=1}^{l}\alpha_i\{w^T\phi(x_i) + b + \xi_i - y_i\} \tag{6}$$

where $\alpha_i(\alpha_i \in R)$ is the Lagrange multiplier.

Derivation of $L(w,b,\xi,\alpha)$ and elimination of w,ξ_i, the following set of linear equations can be obtained:

$$\begin{bmatrix} K+C^{-1}\mathbf{I} & I \\ I^T & 0 \end{bmatrix}\begin{bmatrix} \alpha \\ b \end{bmatrix} = \begin{bmatrix} y \\ 0 \end{bmatrix} \tag{7}$$

where $I = (1,1,...,1)^T, y = (y_1, y_2,..., y_l)^T, \alpha = (\alpha_1, \alpha_2,..\alpha_l)^T$ $K = \{K_{ij} = K(x_i, x_j)\}_{i,j=1}^l$ is the kernel matrix. where $K_{ij}(x, y)$ is kernel function, described as the following function: .

$$K_{ij}(x, y) = \phi(x_i)^T \bullet \phi(y_i) \tag{8}$$

α and b can be obtained by solving the set of linear Eq.(7). Then the decision function of LSSVM can be obtained as following:

$$y(x) = f(\cdot) = \sum_{i=1}^l \alpha_i K(x, x_i) + b \tag{9}$$

According to Eq.(8) and Eq.(9), the final result is not related to the specific mapping function, but related to the inner product of kernel function. Generally, the kernel function is chosen firstly, then α and b can be calculated, and the regression prediction can be achieved according to Eq.(9).

Commonly practicable kernel functions including:

Gauss kernel: $K(x, x') = exp(-\|x - x'\|^2/\sigma)$

Polynomial kernel: $K(x, x') = ((x \cdot x') + c)^d$

Sigmoid kernel: $K(x, x') = tanh(\kappa(x \cdot x') + v)$

As analyzed above, the hyper-parameters are the key of the achievement of regression prediction. To evaluate the performance of the hyper-parameters, leave-one-out and cross validation are commonly used. However both of these methods require high computational cost when the training set is large. The leave-one-out method has been improved in literature [7], and a new simple generalization bound is obtained, which is the unbiased estimate of the true leave-one-out generalization ability. This bound can be described as following:

$$PRESS(\theta) = \sum_{i=1}^l \left\{ r_i^{(-i)} \right\}^2 \tag{10}$$

where $r_i^{(-i)} = y_i - \overset{\wedge(-i)}{y_i} = \dfrac{\alpha_i}{Q_{ii}^{-1}}$, $Q = \begin{bmatrix} K + C^{-1}I & 1 \\ 1^T & 0 \end{bmatrix}$.

The aim of model selection is to choose the best hyper-parameters by optimizing the generalization ability of LSSVM. In practice, users usually adopt grid search method and gradient-based methods. The former requires re-training the model many times, and the latter commonly falls into the local minimum. As a kind of popular method, swarm intelligence algorithm has been proved to be an efficient and robust method for many non-linear, non-differentiable problems. In this paper, particle swarm optimization (PSO) [5] is applied to select the optimal hyper-parameters based on the best generalization ability of the load identification model. The procedure of model selection using PSO is presented as follows:

Algorithm 1. Model Selection using PSO and Generalization Criteria of LSSVM

Initialize: I : the number of iterations, m : swarm size, swarm: $S = \{x_1, x_2, \cdots, x_m\}$, present location and fitness value of each particle. The equation (10) is taken as the fitness function. $i = 0$

while $i < I$

   for all $x_{1,\cdots,m} \in S$

Calculate velocity and Update its position according to the following equation:

$$v_j(i) = w * v_j(i-1) + c_1 * r_1 * \left(p_{jbest} - x_j(i-1)\right) + c_2 * r_2 * \left(g_{best} - x_j(i-1)\right)$$

$$x_j(i) = x_j(i-1) + v_j(i)$$

where $w \in [0,1]$ is the adaptive inertia weight;

$c_1, c_2 \in R$ are the individual/social behavior weights; $r_1, r_2 \in [0,1]$ are uniformly distributed random numbers; p_{jbest} is the best previous position of the jth particle and g_{best} is the best particle in the swarm.

Evaluate the goodness of x_j

Update p_{jbest} (if necessary)

   end for

Update g_{best} (if necessary)

i ++

end while

In algorithm 1, x_j contains hyper-parameters, and g_{best} is the smallest $PRESS(\theta)$. The optimal hyper-parameters can be obtained through algorithm 1.

2.3 Load Identification Process with LSSVM-PSO

The load identification approach with LSSVM-PSO is shown in Figure 1.

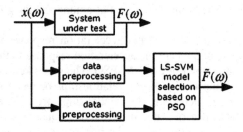

Fig. 1. An approach for load identification using LSSVM-PSO

An approach for load identification using LSSVM-PSO shown in Figure 1 contains the following four sub-processes. Firstly, preparing for loads and responses data required for training, i.e. the structure is added with known loads, and the structural responses and loads are obtained. This step corresponds to the dynamic calibration process of the actual system. White noise is often chosen as loads in the actual dynamic calibration process, because all the characteristics of the system can be excited by white noise. Secondly, training the LSSVM based on PSO. With the responses as inputs of LSSVM, and loads as outputs of LSSVM, the PSO is used to find optimal parameters of LSSVM based on the best generalization ability of the relationship between loads and responses. Then the model of load identification is established. Thirdly, measuring the responses in the operation process of aircraft. Finally, calculating the loads by putting the responses obtained in the third phase into the model established in the second phase.

3 Application of LSSVM-PSO to Load Identification of Flight Simulation Model

Because the identification of flight loads with LSSVM and PSO is a new study, there are many factors that need to be studied before conducting flight load identification. In order to verify the feasibility of the proposed method, a simulation model is chosen, which is simplified as following: without considering the internal structure of aircraft, the aircraft is reduced to a cylindrical structure including a cylindrical shell and a fixture. The cylindrical structure is activated with only one concentrated force. The finite element model is shown in Figure 2. The concentrated force is added near the centre of the fixture. The parameters of the finite element model are described as following: the external diameter, inner diameter, and height of the cylindrical shell is 370mm,360mm and 370mm respectively; the diameter and height of the fixture is 380mm and 17.5mm respectively; the density, modulus of elasticity and Poisson ratio is $7850kg/m^3$, $2.1 \times 10^{11} P_a$ and 0.3 respectively for both of them.

An axial white noise force is added in the concentrated force point, which can excite all the characteristics of the system, meanwhile, the acceleration response of one of the cylindrical shell structure's nodes is obtained. The same process repeated 29 times with the same force point and the same response point but the different amplitude spectrum of white noise force. 29 groups of forces and responses are obtained. It is a white noise force, so the amplitude spectrum in frequency domain is a flat spectrum, but the phase is random. Only the size of amplitude spectrum is changing and the shape of amplitude spectrum is always flat among the 29 groups of white noise force.

The amplitude spectrums of load and acceleration response in a group are constructed as a training sample, where the acceleration response is the input of LSSVM-PSO, and the load is the output of LSSVM-PSO. The gauss kernel is chosen to be the kernel function. According to Eq.(4), the relationships between loads and responses are one to one with frequencies, i.e. the relationships are different in different frequencies. Thus the influence of dates in different frequencies does not need to be considered, and the learning process can be implemented at the corresponding frequency. There are 2000 frequency points in the simulation data, so the total times of learning process is 2000. The relationships are established through the learning of 29 training samples with LSSVM-PSO, and the model of load identification is obtained.

Another force is added in the same force point, whose shape is trapezoidal and parameters are different from the 29 white noise force described above. The amplitude spectrums of trapezoidal spectrum force and acceleration response are obtained. The acceleration response is inputted into the model of load identification established above, and the force can be identified, i.e. the amplitude spectrum of load is calculated according to Eq.(10) with the parameter obtained from the training process. The amplitude spectrum of trapezoidal spectrum force is used to verify the identification results. The load identification results are shown in Figure 4(The blue solid line is actual measured values and the red dot Line is identified values).

<table>
<tr><td>

Fig. 2. The finite element model of cylindrical structure</td><td>**Fig. 3.** The identification results</td></tr>
</table>

The mean square error $RMSE = \sqrt{\frac{1}{n}\sum_{j=1}^{n}\left[P_{real}(\omega_j) - P_{pred}(\omega_j)\right]^2}$ and average relative error

$Er = \frac{1}{n}\sum_{i=1}^{n}\left|P_{i,real} - P_{i,pred}\right| / \left|P_{i,real}\right|$ are chosen to judge the identification error. The identification error of trapezoidal spectrum force are: $RMSE = 0.0027$, $Er = 0.0082\%$.

As shown in Figure 3, the blue solid line almost coincides with the red dot Line, i.e. the amplitude spectrum of identified load corresponds with the measured load spectrum. Because there is no noise in simulation data and the only nonlinear factor of numerical error is very poor, the mean square error and average relative error are both very small in the identification error. The simulation tests demonstrate that the proposed approach based on LSSVM-PSO is an effective way to identify load in frequency domain, and the identification accuracy is high.

4 Load Identification of Random Vibration Experiments of Cylindrical Structure

This section uses the proposed approach to identify multisource loads of random vibration experiments of the cylindrical structure. According to Eq.(4), each load is only related to responses and is nothing to do with other loads. In this way the multisource loads identification problem can be transformed into a number of singlesource load identification problems. Each load identification model can be established by means of that all the responses are the inputs of LSSVM-PSO, and one of loads is the output of LSSVM-PSO. Another load identification model can also be established by the same inputs and the different output of another load. The process repeats times as many as the number of loads. In this way the multisource loads identification problem is solved.

The cylindrical structure system of random vibration experiments is composed of a cylindrical shell, a fixture and a vibration shaker. The fixture is connected with the cylindrical shell by 18 bolts, and the whole cylindrical structure is fixed on the vibration shaker through four force sensors. The forces measured by the four force sensors are the actual loads of the cylindrical structure. The system of random vibration experiments is shown in Figure 5.

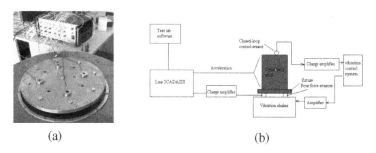

(a)   (b)

Fig. 5. The system of random vibration experiments: (a) The four force sensors; (b) Scheme of the whole system

The driver spectrum shown in Figure 6 is added to drive the vibration shaker. 29 groups of data are obtained by changing the driver spectrum parameter values of a,b,c,d,e,f,g, where a is fixed on 5Hz and g is fixed on 2000Hz. Each group's data contains the power spectral densities (PSD) of four acceleration responses whose positions are located on the cylindrical structure, four forces and drive current.

It brings nonlinear effects during the vibration process because of the connection of bolts. Moreover, the signals are influenced by the noise of the blower of the vibration shaker, while these also bring nonlinear effects. The relationship of PSD between drive current and one of the four acceleration responses in a specific frequency (100Hz) among 29 groups' data is shown in Figure 7. As shown in Figure 7, the relationship between response and drive current is nonlinear, which denotes the relationship between responses and forces is nonlinear. The methods based on linear system used to identify loads will lead to large identification error in this case.

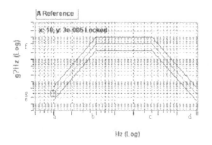

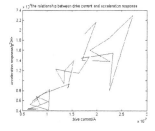

Fig. 6. The driver spectrum

Fig. 7. The relationship between current and response

The gauss kernel is also chosen to be the kernel function. The four PSDs of four acceleration responses are used as inputs of LSSVM-PSO and the PSD of each force is used as output of LSSVM-PSO. Four identification models of load identification are established, where each model is used to identify one force.

Another drive spectrum whose parameters are different from the 29 drive spectrum described above is used to drive the vibration shaker. The PSDs of four acceleration responses at the same points and four forces are obtained. The acceleration responses are inputted into each model of load identification established above to identify loads and the forces are the actual loads of cylindrical structure to compare with the identified forces to verify the validity of the proposed approach and identification accuracy.

The identification process is the same as the process described at section 3. The identification results are shown in Figure 8. The mean square error and average relative error as described above and energy relative error $E_{grms} = |grms_{real} - grms_{pred}| / grms_{real}$ are chosen to judge identification error. The identification errors are shown in Table 1.

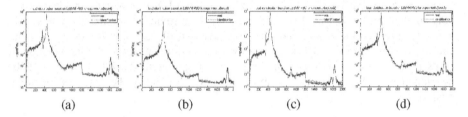

(a)   (b)   (c)   (d)

Fig. 8. The identification results of four forces: (a):Force1; (b):Force2; (c):Force3; (d):Force4

Table 1. The identification error of four forces

	Force 1	Force 2	Force 3	Force 4
RMSE	0.1436	0.1830	0.2595	0.1515
Er	21.35%	22.47%	18.50%	21.75%
Erms	0.11%	0.11%	0.077%	0.086%

As shown in Figure 8, the blue solid line almost coincides with the red dot line, the location and size of peak is also consistent and there are no false peaks in the identification curves, i.e. the PSD of identified load corresponds with the measured PSD of load. These demonstrate that the proposed approach based on LSSVM-PSO can identify the four forces with effective results. The four forces are not exactly same that the individual peaks don't appear in some forces, because the cylindrical structure is deformed during the vibration process. As can be seen from the identification error in Table 1, because the system is nonlinear seriously and the signals are influenced by the noise greatly, the mean square error and average relative error are both larger than the results of the simulation data, but the experimental identification errors are still small in the bound of engineering permissible error .The identified results are almost according to the measured values in terms of energy, which are import to experiment. The experiments demonstrate that the proposed approach based on LSSVM-PSO is an effective way to identify multisource loads in frequency domain, and the identification accuracy is high.

5 Conclusions and Final Remarks

The identification of loads on aircraft is a complex problem. In this paper, a new approach using LSSVM is proposed to seeking the relationship between loads and responses and establishing the load identification model. As the performance of LSSVM is determined by the proper hyper-parameters, how to obtain the proper hyper-parameters is treated as an optimization problem and the PSO is used to solve this problem based on the best generalization ability. To demonstrate the effectiveness of the proposed method, we firstly simplify an aircraft to a cylindrical structure. The results of numerical simulation show that the proposed approach can identify single-source load with high precision. Moreover, the multisource loads identification problem is also transformed into a number of singlesource load identification problems in frequency domain. The results of random vibration experiments of cylindrical structure show that the proposed approach can identify multisource loads and the experimental identification errors are satisfying the demands of engineering permissible error. The approach based on LSSVM-PSO proposed in this paper can identify not only singlesource load, but also multisource loads in frequency domain, which supply a new idea for load identification. The approach was discussed in frequency domain, and can also be used for load identification in time domain. However, the position of loads is determined previously in the proposed approach, the future study will be developed to make this approach more effective to identify the loads of unknown positions.

References

1. Cao, X., Sugiyama, Y., Mitsui, Y.: Application of artificial neural networks to load identification. Computers and Structures 69, 63–78 (1998)
2. Xu, Z.-y., Liao, X.-h.: Dynamic State Loading Identification and Its Development. Journal of Changzhou Institute of Technology 19, 13–18 (2006)
3. Suykens, J.A.K., Vandewalle, J., Moor, D.B.: Optimal Control by Least Square Support Vector Machines. Neural Networks 14, 23–25 (2001)
4. Kennedy, J., Eberhart, R.: Particle swarm optimization. In: Proceedings of the IEEE International Conference on Neural Networks, Perth, Australia, pp. 1942–1984 (1995)
5. Vapnik, V.: The Nature of Statiscal Learning Theory. Springer, New York (1999)
6. Cawley, G.C.: Leave-one-out cross-validation based model selection criteria for weighted LS-SVMs. In: Proceedings of the International Joint Conference on Neural Networks, Vancouver, BC, Canada, pp. 2970–2977 (2006)

Local Model Networks for the Optimization of a Tablet Production Process*

Benjamin Hartmann[1], Oliver Nelles[1], Aleš Belič[2],
and Damjana Zupančič–Božič[3]

[1] University Siegen, Department of Mechanical Engineering,
D–57068 Siegen, Germany
{benjamin.hartmann,oliver.nelles}@uni-siegen.de
[2] University Ljubljana, Department of Electrical Engineering,
SLO–1000 Ljubljana, Slovenia
ales.belic@fe.uni-lj.si
[3] KRKA d.d., SLO–8501 Novo Mesto, Slovenia
damjana.zupancic@krka.biz

Abstract. The calibration of a tablet press machine requires comprehensive experiments and is therefore expensive and time-consuming. In
order to optimize the process parameters of a tablet press machine on the
basis of measured data this paper presents a new approach that works
with the application of local model networks. Goal of the model-based
optimization was the improvement of the quality of produced tablets,
i.e. the reduction of capping occurence and the variation of the tablet
mass as well as the variation of the crushing strength. Modeling and optimization of the tablet process parameters show that it is possible to
find process settings for the tabletting of non-preprocessed powder such
that a sufficient quality of the tablets can be achieved.

1 Introduction

The pharmeceutical industry is increasingly aware of the advantages of implementing a quality-by-design (QbD) principle, including process analytical technology (PAT), in drug development and manufacturing [1], [2], [3]. Although
the implementation of QbD into the product development and manufacturing
inevitably requires large resources, both human and financial, large-scale production can be established in a more cost-effective manner and with improved
efficiency and product quality.

Tablets are the most common pharmaceutical dosage form prepared by compression of a dry mixture of powders consisting of active ingredient and ex-
cipients into solid compacts. The process of tabletting consists of three stages:
a) The powder mixture is filled into the die; b) compaction, where the powder is compressed inside a die by two punches, resulting in plastic and elastic

---

* This work was supported by the German Research Foundation (DFG), grant NE
656/3–1.

H. Deng et al. (Eds.): AICI 2009, LNAI 5855, pp. 241–250, 2009.

deformation and/or particles fragmentation, and c) ejection, where the tablet is ejected from the die and elastic recovery of the tablet may occur. An intensive elastic recovery can lead to separating the upper part of a tablet from the tablet body (capping). The mechanical behavior of powders during tabletting and the quality of tablets depend on the powder characteristics (formulation) and the tabletting parameters on the tablet press machine [4], [5], [6], [7]. In order to assure a high and repeatable quality of products the processes and formulations should be optimized. This is particularly important on high capacity rotary tablet presses which can produce a few hundred thousand tablets per hour and are very sensitive to product and process variables. A special challenge is the large scale production of products in form of tablets where the active drug loading is high (> 50% of the total mass of a tablet). Active drugs as organic molecules very often have inappropriate physical properties which prevent using direct compression of powder mixtures, which is the most economical production method (shortest process time, minimum cleaning and validation need, low energy consumption, environmentally friendly). The problem of capping is often observed in production of tablets with high drug loading. This problem can be solved either by formulation or process optimization. To find the optimal set of the parameters (main compression force, precompression force, compression speed ...) on the tablet press, it is possible to make a few tablets for each set of parameters, and then the operator can decide which settings should work best, and then starts the production. Fine machine setup optimization must be performed, due to some batch to batch variations of raw materials or some other variables. During the setup procedure, many of the tablets do not pass the quality control and must hence be discarded, which can be very expensive, regarding the price of the raw material and the number of trials, needed to find optimal settings. To optimize the quality of tablets and to reduce the number of faulty tablets, modeling and simulation procedures can be used, where a model of quality parameters with respect to process parameters must be developed. The raw materials used in tests should cover all the expected characteristics of raw materials that can be expected to appear in production. With the model, optimal settings of the machine can be found with respect to raw material characteristics, without further loss of raw material. The aim of the present study was the optimization of the tabletting process in order to diminish capping occurrence and the variation of tablet mass and crushing strength. Optimization was performed for the product where 70% of the tablet weight represent active ingredients, on the high capacity rotary-tablet press, in the standard production environment. Local model networks were used to model the relation between quality and process parameters.

2 Production of Model Tablets

Due to the high drug content and chosen manufacturing procedure (direct compression) changes in raw material characteristics can result in variable crushing strength of tablets, large tablet mass variation and most problematic, intense capping problems. If problems in direct tabletting are too severe, dry granulation can be used as an alternative to improve the particle compression characteristics. Dry granulation means aggregation of smaller particles using force. The compacts are milled and sieved afterwards. Next step is tabletting. Three types of mixtures for tabletting were prepared:

A mixture for direct tabletting (Type: "Direkt" - 1 sample),

B granulate prepared by slugging (precompression on a rotary-tablet press), using different parameters of tabletting speed and compression force (Type: "Briket" - 4 samples),

C granulate prepared by precompressing on a roller compactor using different parameters of compacting speed and force (Type: "Kompakt" - 4 samples).

The composition of all powder mixtures was the same, milling and sieving of compacts was performed on the same equipment. After dry granulation, the particle size distribution is normally shifted toward bigger particles, also capping occurence is often decreased due to a change in particle characteristics (flowabilty and compression characteristics). Nine samples were tableted on a rotary-tablet press using different combinations of the following parameters: compresson speed v, precompression force pp, main compression force P. Each of these parameters were set at three levels. Tablets were evaluated according to capping occurence, crushing strength and tablet mass variability [8].

2.1 Capping Coefficient

The capping occurency (capping coefficient, CC) was calculated as the ratio between the number of tablets that have a tendency to cap and the number of tablets that have been observed. It was observed visually on each tablet after a classical tablet hardness test (Erweka). A Tablet was considered to have capping tendency if the upper part of the tablet falled off from the tablet body or if a significant shape (a step-like form) was formed on the breaked surface of the tablet.

2.2 Experiments on the Rotary-Tablet Press

For experiments on the rotary-tablet press three groups of raw materials were used ("Direkt", "Briket", "Kompakt"). The types Kompakt and Briket were further divided into four sub-groups depending on the combination of compression force and speed during slugging or roller compaction. For each of these raw materials several combinations of settings on the rotary-tablet press were set (tabletting speed, main compression force, precompression force). Produced tablets were controlled afterwards: CC, average crushing strength, standard deviation of crushing strength, average tablet mass and standard deviation of tablet mass. For each raw material and parameter setting, 10 tablets were made and analyzed. All together 76 data points were measured.

3 Modeling

To model the relation between process parameters, raw material characteristics, and quality of the tabletting process (CC, standard deviation of mass, standard deviation of crushing strength) it was not reasonable to use physical laws that describe the operation of the tabletting machine, considering the aim. The system characteristics depend on many stochastic processes (particle rearrangement and deformation type, friction, bondforming between particles, ...), therefore, the model would be too complex for process parameter optimization. On the other hand, there is a relatively large database available, hence, it is more appropriate to use data-based computational intelligence methods instead.

3.1 Local Model Networks

To model the system characteristics a local model network approach [9], [10], [11] was used with two to four inputs, depending on the output. For each output (CC, standard deviation of crushing strength, standard deviation of mass) a separate neuro-fuzzy model was constructed.

The output $\hat{y}$ of a local model network with p inputs $\underline{u} = [u_1 \ u_2 \ \cdots \ u_p]^T$ can be calculated as the interpolation of M local model outputs $\hat{y}_i$, $i = 1, \ldots, M$ [12],

$$\hat{y} = \sum_{i=1}^{M} \hat{y}_i(\underline{u})\Phi_i(\underline{u}) \tag{1}$$

where the $\Phi_i(\cdot)$ are called interpolation or validity or weighting functions. These validity functions describe the regions where the local models are valid; they describe the contribution of each local model to the output. From the fuzzy logic point of view (1) realizes a set of M fuzzy rules where the $\Phi_i(\cdot)$ represent the rule premises and the $\hat{y}_i$ are the associated rule consequents. Because a smooth transition (no switching) between the local models is desired here, the validity functions are smooth functions between 0 and 1. For a reasonable interpretation of local model networks it is furthermore necessary that the validity functions form a *partition of unity*:

$$\sum_{i=1}^{M} \Phi_i(\underline{u}) = 1 \,. \tag{2}$$

Thus, everywhere in the input space the contributions of all local models sum up to 100%. In principle, the local models can be chosen of arbitrary type. If their parameters shall be estimated from data, however, it is extremely beneficial to choose a linearly parameterized model class. The most common choice are polynomials. Polynomials of degree 0 (constants) yield a neuro-fuzzy system with singletons or a normalized radial basis function network. Polynomials of degree 1 (linear) yield local linear model structures, which is by far the most popular choice. As the degree of the polynomials increases, the number of local models required for a certain accuracy decreases. Thus, by increasing the local models' complexity, at some point a polynomial of high degree with just one local model ($M = 1$) is obtained, which is in fact equivalent with a global polynomial model ($\Phi_1(\cdot) = 1$).

Besides the possibilities of transferring parts of mature linear theory to the nonlinear world, local *linear* models seem to represent a good trade-off between the required number of local models and the complexity of the local models themselves. This paper deals only with local models of linear type:

$$\hat{y}_i(\underline{u}) = w_{i,0} + w_{i,1}u_1 + w_{i,2}u_2 + \ldots + w_{i,p}u_p. \tag{3}$$

However, an extension to higher degree polynomials or other linearly parameterized model classes is straightforward.

3.2   Data Pre-processing

Each raw material batch has a unique composition of particles, since its production cannot be exactly repeated. The raw material characteristics can be charaterized by the particle size distribution. The model outputs are a function of the tablet composition, the main compression pressure P, the pre-compression pressure pp, and the tabletting speed v. The tablet composition and the raw material characteristics, respictively, are given by the particle size distribution. The particle sizes were separated in the following 8 subgroups: 0-0.045, 0.045-0.071, 0.071-0.125, 0.125-0.25, 0.25-0.5, 0.5-0.71, 0.71-1.0, and 1.0-1.25mm. As training data we had 9 different tablet compositions available. Figure 1 shows the correlation between the standard deviation σ of the particle size dirstribution and the mean values μ. Due to the sparse training data we assumed that the raw material characteristic is uniquely desribed with the mean value μ of the particle size distribution. For simplicity the inputs and outputs had been normalized to the interval $[0, 1]$.

3.3   Model Structure

Due to available prior knowledge it can be assumed that the three different model outputs which represent the tablet quality depend on different sets of

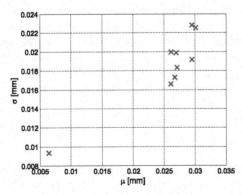

Fig. 1. Correlation between standard deviation σ of the particle size dirstribution and the mean values μ

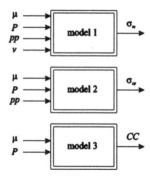

Fig. 2. Model structure. One model for each output dimension.

input variables. Therefore it is advantageous to build up three individual MISO models each with the relevant inputs. Using a single MIMO model would discard this knowledge and thus would yield suboptimal results, compare Fig. 2. The task is to model the following three output dimensions: The capping coefficient CC, the standard deviation of the tablet mass σ_m and the standard deviation of the crushing strength σ_{cs}, respectively.

4   Results

First, the local model networks were designed and validated, next, the models were used to find the optimal process parameters. Furthermore we investigated the robustness of the optimization in terms of varying training data.

4.1   Modeling

The training of the local model networks was performed with the LOLIMOT algorithm proposed in [9]. One advantage of this training algorithm among others is that the training has only to be carried out once to get reliable and reproducible results. For final training the whole data set was used, since, considering the rather small dimensionality of the problem, the nonlinear input-output relationship could be visually monitored in order to check for potential overfitting. The identified relations can be seen in Figs. 4 to 7. For modeling the standard deviation of the mass σ_m 3 local models (LMs) were used. The standard deviation of the crushing strength σ_{cs} was modeled with 4 LMs and the capping coefficient CC with 3 LMs.

4.2   Optimization

Once the models for the tabletting process parameters have been built, the question arises: How they should be utilized for optimization? Figure 3 illustrates the used optimization scheme. Basically, the manipulated variables that are model inputs are varied by an optimization technique until their optimal values are

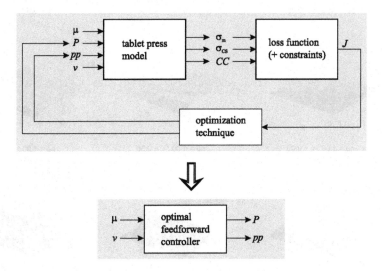

Fig. 3. Optimization of the tablet press parameters

found. Optimality is measured in terms of a loss function that depends on the tabletting process parameters, i.e. the model outputs. When the operating point dependent optimal input values are found, they can be stored in a static map that represents a feedforward controller, which can be realized either as a look-up table (since it has only two inputs) or as a neural network. To convert the multi-objective optimization problem into a single-objective problem the loss function was formulated as follows:

$$J\left(\sigma_m, \sigma_{cs}, \mathrm{CC}\right) = w_1\sigma_m + w_2\sigma_{cs} + w_3\mathrm{CC}$$
$$\longrightarrow \min_{P, pp} \quad \text{with } 0 \leq P \leq 1,\, 0 \leq pp \leq 1\,. \tag{4}$$

This method is called the weighted-sum method, where the coefficients w_i of the linear combination are called weights [13], [14]. All relevant model outputs are directly incorporated and the loss function is chosen as a weighted sum of the tablet process parameters. Due to expert knowledge the parameters were set to $w_1 = 0.25$, $w_2 = 0.25$ and $w_3 = 0.5$. For optimization the BFGS quasi-Newton algorithm [15] was used. The results are summarized in Table 1. Figure 6 shows the shapes of the loss functions for the three tablet types "Direkt", "Kompakt" and "Briket". Figure 8 illustrates the variation of the optimal process parameters, if single data points are omitted from the training dataset. Because of the sparse dataset, omitting critical data points can lead to different shapes of the loss function. Therefore outliers can be observed in the boxplots. However, the comparison of the median values with the results generated with the whole dataset (Table 1) is satisfactory.

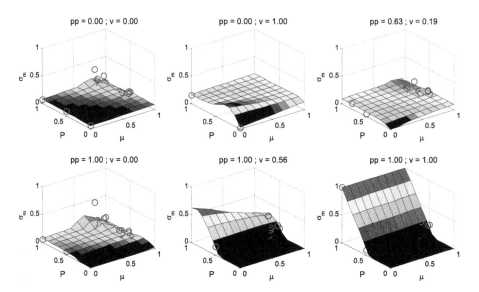

Fig. 4. Model plot that shows the relations between the inputs mean value of the tablet particle sizes μ, main compression force P, pre-compression force pp, the tabletting speed v and the output standard deviation of tablet mass σ_m. The red dots are measured data points.

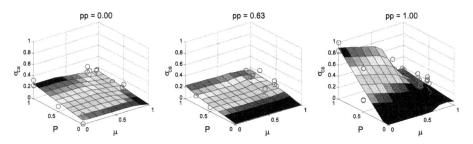

Fig. 5. Model plot that shows the relations between the inputs mean value of the tablet particle sizes μ, main compression force P , pre-compression force pp and the output standard deviation of the crushing strength σ_{cs}

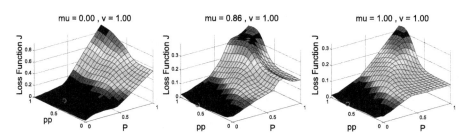

Fig. 6. Loss functions with respect to the raw material parameter $\mu = 0, 0.86, 1$, main compression force P, pre-compression force pp and tabletting speed $v = 1$

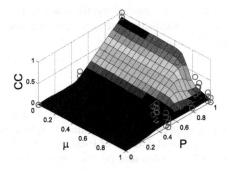

Fig. 7. Model plot that shows the relations between the inputs mean value of the tablet particle sizes μ, main compression force P and the output capping coefficient CC

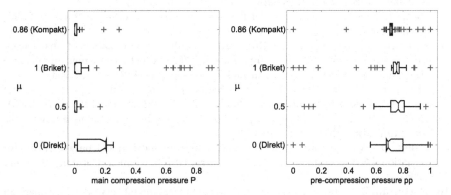

Fig. 8. Boxplots that illustrate the variation of the optimal process parameter values in case of omitting single data points

Table 1. Results of the optimization

μ	P	pp
0	0.21	0.69
0.5	0.04	0.73
0.86	0.01	0.70
1	0.03	0.74

5   Conclusion

This paper presents a local model network approach for the model-based optimization of the parameters of a tablet press machine. Several experiments on the tablet press machine with different tablet and process parameters made it possible to create a well generalizing model. Due to the high flexibility and robustness with respect to the sparse and noisy data, local model networks are well suited for this kind of modeling problem. Model plots show the applicability of the selected approach. Modeling and simulation indicate that it is possible to find process settings for the tabletting of non-preprocessed powder such that a sufficient quality of tablets can be achieved.

References

1. Frake, P., Greenhalgh, D., Gierson, S., Hempenstall, J., Rudd, D.: Process control and end-point determination of a fluid bed granulation by application of near infrared spectroscopy. International Journal of Pharmaceutics 1(151), 75–80 (1997)
2. Informa: PAT - Quality by design and process improvement, Amsterdam (2007)
3. Juran, J.: Juran on Quality by Design. The Free Press, New York (1992)
4. Sebhatu, T., Ahlneck, C., Alderborn, G.: The effect of moisture content on the compression and bond-formation properties of amorphous lactose particles. International Journal of Pharmaceutics 1(146), 101–114 (1997)
5. Rios, M.: Developments in powder flow testing. Pharmaceutical Technology 2(30), 38–49 (2006)
6. Sorensen, A., Sonnergaard, J., Hocgaard, L.: Bulk characterization of pharmaceutical powders by low-pressure compression ii: Effect of method settings and particle size. Pharmaceutical Development and Technology 2(11), 235–241 (2006)
7. Zhang, Y., Law, Y., Chakrabarti, S.: Physical properties and compact analysis of commonly used direct compression binders. AAPS Pharm. Sci. Tech. 4(4) (2003)
8. Belič, A., Zupančič, D., Škrjanc, I., Vrečer, F., Karba, R.: Artificial neural networks for optimisation of tablet production
9. Nelles, O.: Axes-oblique partitioning strategies for local model networks. In: International Symposium on Intelligent Control (ISIC), Munich, Germany (October 2006)
10. Hartmann, B., Nelles, O.: On the smoothness in local model networks. In: American Control Conference (ACC), St. Louis, USA (June 2009)
11. Hartmann, B., Nelles, O.: Automatic adjustment of the transition between local models in a hierarchical structure identification algorithm. In: IFAC Symposium on System Identification (SYSID), Budapest, Hungary (August 2009)
12. Nelles, O.: Nonlinear System Identification. Springer, Berlin (2001)
13. Chong, E., Zak, S.: An Introduction to Optimization. John Wiley & Sons, New Jersey (2008)
14. Boyd, S., Vandenberghe, L.: Convex Optimization. Cambridge University Press, Cambridge (2004)
15. Scales, L.: Introduction to Non-Linear Optimization. Computer and Science Series. Macmillan, London (1985)

Implementation of On/Off Controller for Automation of Greenhouse Using LabVIEW

R. Alimardani[1], P. Javadikia[1], A. Tabatabaeefar[1], M. Omid[1], and M. Fathi[2]

[1] Department of Agricultural Machinery Engineering, Faculty of Biosystems Engineering, University of Tehran, 31587-77871 Karaj, Iran
[2] Department of Electronic and Informatics, University of Siegen, 57074 Siegen, Germany
{Pjavadikia,atabfar,omid,rmardani}@ut.ac.ir
fathi@informatik.uni-siegen.de

Abstract. The present study is concerned with the control and monitoring of greenhouse air temperature, humidity, Light intensity, CO2 concentration and irrigation. A computer-based control and monitoring system was designed and tested and to get this target is used Supervisory Control & Data Acquisition (SCADA) system. The end product is expected to give the farmer or end user a kiosk type approach. Entire greenhouse operation is governed and monitored through this kiosk. This approach is fairly novel considering the unified system design and the SCADA platform, NI LabView 7.1.

Keywords: Control, Lab View, Automation, on/off, Greenhouse.

1 Introduction

Environmental conditions have a significant effect on plant growth. All plants require certain conditions for their proper growth. Therefore, it is necessary to bring the environmental conditions under control in order to have those conditions as close to the idea as possible. To create an optimal environment, the main climatic and environmental parameters need to be controlled. These parameters are nonlinear and extremely interrelated, rendering the problem of management of a greenhouse rather intractable to analyze end control through the classical control methods. An automated management of a greenhouse brings about the precise control needed to provide the most proper conditions of plant growth. Greenhouse Automation has considerably evolved in the past decade with the onus shifting to factors like flexibility, time-to-market and upgradeability. SCADA systems, hence, form an interesting platform for developing systems for greenhouse automation. It allows a 'supervisory level' of control over the greenhouse, facilitates data acquisition and creates a developer-friendly environment for implementing algorithms.

The five most important parameters to consider when creating an ideal greenhouse are temperature, relative humidity, soil moisture, light intensity and Co2 concentration. In order to design a successful control system, it is very important to realize that the five parameters mentioned above are nonlinear and extremely interdependent [2-1]. The computer control system for the greenhouse includes the following components [5]:

H. Deng et al. (Eds.): AICI 2009, LNAI 5855, pp. 251–259, 2009.

1. Acquisition of data through sensors
2. Processing of data, comparing it with desired states and finally deciding what must be done to change the state of the system
3. Actuation component carrying the necessary actions.

This paper describes a solution to the second part of the system. The information is obtained from multi-sensor stations and is transmitted through serial port to the computer. It will then be processed and the orders will be sent to the actuation network.

2 Material and Method

The original system includes four main subsystems, namely, the Temperature and Humidity, Irrigation, light intensity and co2 concentration. The system program contains the running algorithm, which governs the whole subsystems. The real time Data transmit from microcontroller to Lab View through serial port of the computer and the system to the microcontroller. The designed system has follows.

Real time soil moisture
Real time Co2
Real time lighting
Real time temperature
Real time humidity

The seven actuators are as follows.

Water valve
Ventilation Fans
Cooler
Circulate Fans
Sprayer
Heater
Lamps and Shade

The system can be set up for any type of plant and vegetation by simply setting certain parameters on the front panel of Lab View (fig. 1.). The entire system is simulated in Lab View by National Instruments. The main block-diagram of an automated greenhouse is shown in fig. 2.

2.1 Temperature and Humidity System

The temperature and humidity system consists of different subsystems as follows. Special attention must be given to these subsystems for the proper functioning of the overall system.

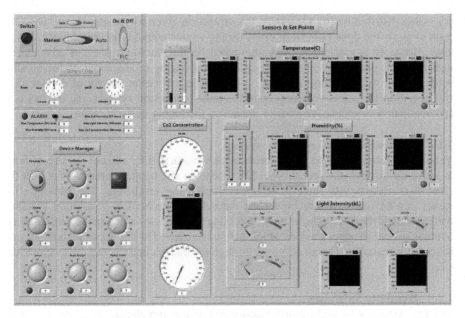

Fig. 1. Front panel of the main system

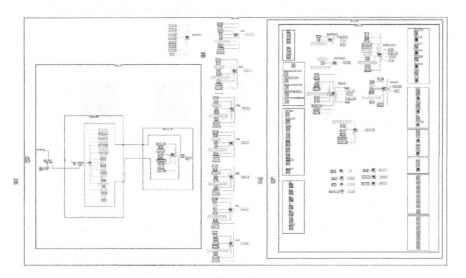

Fig. 2. Block diagram of the main system

2.1.1 Heating Subsystem

Temperature control is necessary for attaining high crop growth, yield and quality. Extreme temperatures may induce stress and associated damage to the plasmatic structures or the photosynthetic apparatus of the plant. Less extreme suboptimal temperatures may delay plant development and affect other plant characteristics such as dry matter distribution [6]. Generally, the protection given by the greenhouse is sufficient to allow the development of crops during winter without the use of heating systems. However, a greenhouse with automated heating facilities presents advantages like increased production speed, possibility of producing products out of season and better control of producing products out of season and better control of diseases[5]. Uniform crop growth is very important for most production systems and the heating and ventilation systems have a major impact on producing uniform crops [7].

2.1.2 Ventilation and Cooling Subsystem

Ventilation systems can be either mechanical or natural (i.e., without the use of fans to move air through the greenhouse) [7]. whenever the natural ventilation is insufficient, it is necessary to force it, which is done using ventilators. The cooling efficiency can be increased combining the natural and mechanical ventilation systems with air humidifiers. A general type of cooling system uses a porous pad installed in one top side of the greenhouse, which maintained wet. On the opposite side, an exhaust fan is installed. The air admitted through the pad becomes cooler by evaporation effect [5].

2.1.3 Humidity Subsystem

The fog system is formed by suspended pipes on the greenhouse structure, spraying tiny drops of water into the greenhouse, contributing to increase air humidity and decrease the air temperature. Dehumidifier system is used for decreasing the air humidity [5].

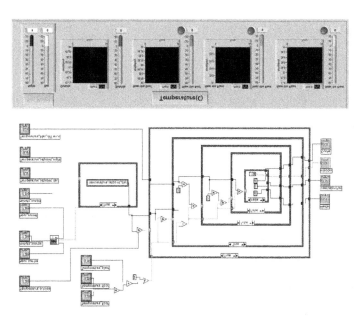

Fig. 3. Front panel and block diagram of the Temperature system

2.1.4 System Design and Simulation

The simulated system in LAbVIEW is described in Fig. 3 and 4 for temperature and humidity systems. The TEMPERATURE PALLETE shows the current temperature and the idea ranges of temperature for desired crop, while the HUMIDITY PALLETE shows the current humidity and the idea ranges of humidity. If the temperature falls below a set value, then heating system is turned on. Similarly, if the humidity rises above a set value, then the dehumidifier is turned on. The method is an on-off control with a certain dead band which can be arbitrarily defined by the user.

2.2 Lighting System

Using the supplemental lighting system is a common way for greenhouse lighting. However it can be done either with photoperiod lighting system or through walkway and security lighting. Greenhouse lighting systems allow us to extend the growing season by providing plants with an indoor equivalent to sunlight [3]. Front panel and block diagram of the light system is shown in figure 5.

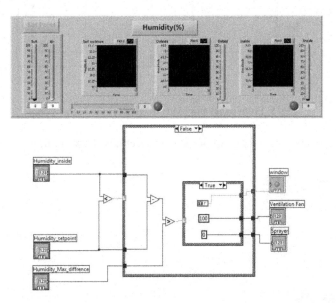

Fig. 4. Front panel and block diagram of the Humidity system

2.3 Co2 System

The fifth major factor, namely the Co2 concentration, plays a very important role in the photosynthesis process. The average Co2 concentration in the atmosphere is approximately 313 ppm, which is enough for effective photosynthesis. A problem arises when a greenhouse is kept closed in autumn or/and winter in order to retain the heat, when not enough air is circulated to have the appropriate Co2 concentration [1, 4]. In order to improve the growing of herbs inside the greenhouse, it is necessary to increase Co2 concentration in company with favorable conditions of temperature and light.

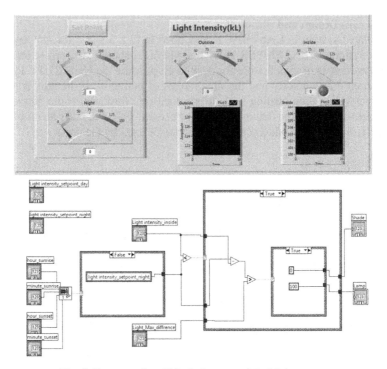

Fig. 5. Front panel and block diagram of the Light system

Front panel and block diagram of the Co2 system is shown in figure 6.

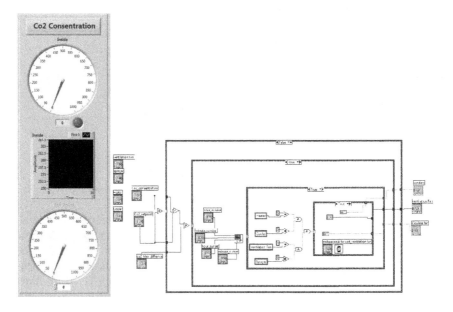

Fig. 6. Front panel and block diagram of the Co2 system

2.4 Irrigation System

The irrigation system consists of a soil moisture control module. For the desired plant, the optimum value of soil moisture is set through the front panel. The values of soil moisture are measured through sensors in real time which is then displayed in the front panel. Depending on this condition, the flow control valve is handled. The RT values show the Real-Time values as obtained from the sensors.

3   Results and Discussion

Performance of the system was evaluated by installing it in the greenhouse. A number of experiments were carried out during the 2008 autumn season. In the next section, one of the experiments which were conducted from 10 to 12 a.m. on December 7, 2008, in the city of Karaj, is presented and discussed. The system performance was tested for two modes of operations: (a) uncontrolled mode of operation, and (b) controlled mode. During the first day the system was tested as a data-acquisition system, i.e. only data from the sensors were recorded and monitored on the screen with no operation of fans, sprayer or other installed environmental systems in the greenhouse. For the second day of the experiment, inside air temperature was controlled by the system.

Fig. 7 shows the results for the uncontrolled experiment. The data were recorded with no operation of fans and sprayer. These show comparison between the inside and the outside condition. The results indicate that the outside temperature is always, about 15°C, less than the temperature inside the greenhouse, because the solar radiation entered the greenhouse through transparent poly carbonate. This is a further confirmation of greenhouse effect. Also, it was found that the rate of change of temperature in the upper part, i.e. near the cover is higher than that of the height of plants. This rise of vertical temperature is due to receiving solar incident radiation. During a

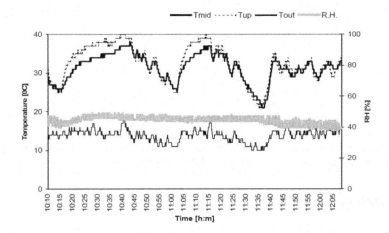

Fig. 7. Measured temperatures and relative humidity in no-control operation

temperature decrease, it was found that the heat exchange near the cover is occurring more rapidly. The fluctuations of temperature and to a lesser extend of RH in the model greenhouse during observation time were affected by natural conditions such as surface evaporation within the greenhouse, solar radiation and ambient temperature. For the controlled mode of operation, two separate experiments are carried out: (i) temperature control only (Fig. 8a), and (ii) both temperature and humidity control (Fig. 8b). When the controller was put into operation, it was found that Tout<Tmid<Tup.

This could be partly due to the use of re-circulating fans.

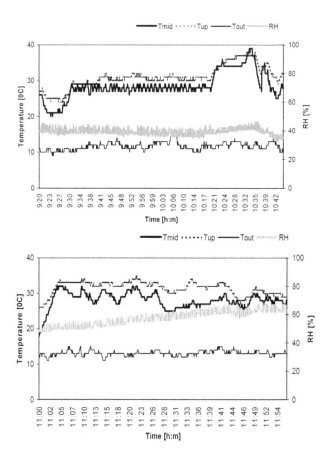

Fig. 8. Measured temperatures and relative humidity under the action of controller: (a) Temperature control (up), and (b) Temperature and RH control (down)

4   Conclusions

This paper described the design and simulation and Implementation of a fully automated greenhouse using LabVIEW. For better simulation results, the entire five major interrelated environment variables in a greenhouse Temperature-Humidity, soil moisture, light intensity and Co2 concentration were considered altogether. The resulting system was designed in such a way that the system could be easily controlled in such a way that the system could be easily controlled and monitored by an amateur user who might have no or little technical background on the subject, the main advantages of this simulation are more facilities in the entire system. By means of this simulation, the optimal level of environment and growth factors inside the greenhouse can be achieved. Once the proposed system is designed, standardized, and implemented, it provides and convenient control over the greenhouse management in order to increase efficiency.

References

1. Putter, E., Gouws, J.: An Automatic Controller for a Greenhouse Using a Supervisory Expert System. In: Electro technical Conference, Melecon 1996, 8th Mediterranean, pp. 1160–1163 (1996)
2. Fourati, F., Chtourou, M.: A Grennhousr Control with Feed-Forward and Recurrent Neural Networks. Simulation Modeling Practice and Theory 15, 1016–1028 (2007)
3. Bowman, G.E., Weaving, G.S.: A Light Modulated Greenhouse Control System. Journal of Agriculture Engineering Research 15(3), 255–258 (1970)
4. Klaring, H.P., Hauschild, C., Heibner, A., Bar-Yosef, B.: Model-Based Control of Co2 Concentration in Greenhouse at Ambient Levels Increases Cucumber Yield. Agriculture for Meteorol. 143, 208–216 (2007)
5. Metrolh, J.C., Serodio, C.M., Couto, C.A.: CAN Based Actuation System for Greenhouse Control. Indusrial Elect., 945–950 (1999)
6. Korner, O., Challa, H.: Design for an Improved Temperature Integration Concept in Greenhouse Cultivation. Computer and Electronic in Agriculture 39, 39–59 (2003)
7. Roberts, W.J.: Creating a Master Plan for Greenhouse Operation, Rutgers, The State University of New Jersey (2005)

Structure Design of the 3-D Braided Composite Based on a Hybrid Optimization Algorithm

Ke Zhang

School of Mechanical and Automation Engineering,
Shanghai Institute of Technology
120 Caobao Road, Shanghai 200235, China
zkwy2004@126.com

Abstract. Three-dimensional braided composite has the better designable characteristic. Whereas wide application of hollow-rectangular-section three-dimensional braided composite in engineering, optimization design of the three-dimensional braided composite made by 4-step method were introduced. Firstly, the stiffness and damping characteristic analysis of the composite is presented. Then, the mathematical models for structure design of the three-dimensional braided composite were established. The objective functions are based on the specific damping capacity and stiffness of the composite. The design variables are the braiding parameters of the composites and sectional geometrical size of the composite. The optimization problem is solved by using ant colony optimization (ACO), contenting the determinate restriction. The results of numeral examples show that the better damping and stiffness characteristic could be obtained. The method proposed here is useful for the structure design of the kind of member and its engineering application.

Keyword: Composite, Structure Design, Ant Colony Algorithm.

1 Introduction

At present, textile composites are being widely used in advanced structures in aviation, aerospace, automobile and marine industries [1]. Textile composite technology by preforming is an application of textile processes to produce structured fabrics, known as performs. The preform is then impregnated with a selected matrix material and consolidated into the permanent shape. Three-dimensional (3-D) braiding method which was invented in 1980s offers a new opportunity in the development of advanced composite technology. The integrated fibre network provides stiffness and strength in the thickness direction, thus reducing the potential of interlaminated failure, which often occurs in conventional laminated composites. Other distinct benefits of 3-D textile composites include the potential of automated processing from preform fabrication to matrix infiltration and their near-net-shape forming capability, resulting in reduced machining, fastening, and scrap rate [2]. The direct formation of the structural shapes eliminates the need for cutting fibres to form joints, splices, or overlaps with the associated local strength loss, and simplifies the laborious hand lay-up composite manufacturing process.

H. Deng et al. (Eds.): AICI 2009, LNAI 5855, pp. 260–269, 2009.

In comparing with two dimensional composite laminates, a key property of the new innovative 3-D braided composite is its ability to reinforced composites in the thickness direction. Braiding with continuous fibres or yarns can place 3-D reinforcements in monocoque structural composites, it includes multi-directional fibre bundle which interconnect layers. Since the braiding procedure dictates the yarn structure in the preform and the yarn structure dictates the properties of the composite, designing the braiding procedure to yield the desired structural shape that is endowed with the desired properties is an important element in textile composite technology [3]. Thus, it is feasible to design the textile structural composites with considerable flexibility in performance based upon a wide variety of preform geometries and structure parameter.

The ant colony optimization (ACO) algorithm is a novel simulated ecosystem evolutionary algorithm. It takes inspiration from the observations of ant colonies foraging behavior with which ants can find the shortest paths from food sources to their nest [4]. Preliminary study has shown that the ant colony algorithm is very robust and has great abilities in searching better solutions. ACO algorithm has been successfully used to solve many problems of practical significance including the Quadratic Assignment Problem [5], Traveling Salesman Problem [6], Single Machine Total Weighted Tardiness Problem [7].

Whereas wide application of hollow-rectangular-section 3-D braided composite in engineering, the 3-D braided composite made by 4-step method is taken as the studied object in this paper. The 3-D braided composite description is given in section 2. The stiffness and damping analysis of the composite were introduced in Section 3. And model of structure optimization designs for the composite were proposed. A hybrid algorithm based on ACO algorithm for solving the optimization problem is presented in section 4. Simulation results and the discussion of the results are presented in section 5. Finally, conclusions are given in Section 6.

2 3-D Braided Composite Description

The 3-D braided fibre construction is produced by a braiding technique which interlaces and orients the yarns by an orthogonal shedding motion, followed by a compacting motion in the braided direction. The basic fibre structure of 4-step 3-D braided composite is four-direction texture. An idealized unit cell structure is constructed based upon the fibre bundles oriented in four body diagonal directions in a rectangular parallelepiped which is shown schematically in Fig.1 [8][9]. The yarn orientation angles α and β are so-called braided angle. From the geometry of a unit cell associated with particular fibre architecture, different systems of yarn can identified whose fibre orientations are defined by their respective interior angle α and cross sectional angle β, as previously show in Fig.1. According to requirement of braid technology, the range of α and β are commonly between 20°and 60°. In Fig.1, geometric parameters of cross section of composite are b_1, b_2, h_1, and h_2 respectively.

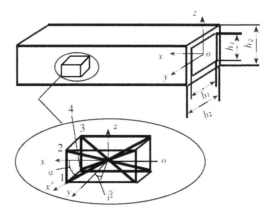

Fig.1. The unit cell structure of a 3-D braided composite

3 Optimization Design Model of 3-D Braided Composite

3.1 The Stiffness Properties of 3-D Braided Composite

A model for grain system interruption has been chosen wherein the stiffness for each system of yarns are superimposed proportionately according to contributing volume of four yarn system to determine the stiffness of the composite. Let volume percent of fibre of the composite be V_f, volume fraction of each system of braiding yarn is equal to $V_f / 4$ [10]. Assuming each system of yarn can be represented by a comparable unidirectional lamina with an elastic matrix defined as C and a coordinate transformation matrix as T, the elastic stiffness matrix Q of this yarn system in the longitudinal direction of the composite can be expressed as $Q = TCT^T$.

Here, we ignore approximately the effect of stress components σ_y, σ_z and τ_{yz}, and only consider the effect of σ_x, τ_{zx} and τ_{xy}.

The constitutive relations of braiding yarn for 3-D braided composite are gotten as follows [11]

$$[N]_i = [K]_i [\varepsilon]_i \quad (i = 1,2,3,4). \tag{1}$$

By stiffness matrix of each system, according to composition principle of composite mechanics, we can obtain the general stiffness matrix of 3-D composite :

$$[K] = \frac{1}{4} \sum_{i=1}^{4} [K]_i , \tag{2}$$

$$\text{where } \lambda_i = V_{f_i} / V_f, \quad [K]_i = \begin{bmatrix} K_{11} & 0 & 0 & K_{14} \\ 0 & K_{22} & K_{23} & 0 \\ 0 & K_{32} & K_{33} & 0 \\ K_{41} & 0 & 0 & K_{44} \end{bmatrix}, \quad K_{11}, K_{22}, K_{33} \text{ and } K_{44} \text{ are}$$

tensile, flexural, torsional and shear stiffness coefficient respectively, the others are the coupling stiffness coefficient..

3.2 The Damping Properties of 3-D Braided Composite

By definition, specific damping capacity can be expressed as

$$\Psi = \Delta U / U_m, \tag{3}$$

where ΔU is dissipation energy in anyone stress cycle, U_m is maximum strain energy in the cycle.

Adams and Bacon had presented cell damping model of laminated composite in reference [12], they think that per unit area dissipation energy of each layer can be broken down into three component relate to direct strain, thus

$$\Delta U = \Delta U_x + \Delta U_y + \Delta U_{xy}. \tag{4}$$

Considering the influence of transverse shear deformation, reference [13] has modified equation (6). In equation (7), they added to two terms dissipation energy resulted by transverse shearing strain, namely

$$\Delta U = \Delta U_x + \Delta U_y + \Delta U_{xy} + \Delta U_{yz} + \Delta U_{xz}. \tag{5}$$

As before, 4-step 3-D braided composite can be regard as superimpose of four unidirectional fibre system, and each yarn system is not in same plane. Thus, unit energy dissipation of each system should break down into six components,

$$\Delta U = \Delta U_{x'} + \Delta U_{y'} + \Delta U_{z'} + \Delta U_{x'y'} + \Delta U_{y'z'} + \Delta U_{x'z'}, \tag{6}$$

where $x'y'z'$ is fibre coordinate system of each system.

Hereinafter, suppose composite only have the action of bending moment M_x. General dissipation energy of composite is the superimpose by dissipation energy of each system according to volume fraction [11] :

$$\Delta U = \frac{1}{4} \sum_{k=1}^{4} \Delta U_k = \frac{1}{4} \sum_{k=1}^{4} \Delta W_k \int_0^{l'/2} M_x^2 dx. \tag{7}$$

where l' is length of the composite. And maximum strain energy of composite is as

$$U_m = \int_0^{l'/2} M_x k_x dx = \frac{1}{k_{22}} \int_0^{l'/2} M_x^2 dx. \tag{8}$$

Therefore, specific damping capacity of the composite can be obtained as follows

$$\Psi = \frac{k_{22}}{4} \sum_{k=1}^{4} \Delta W_k \,. \tag{9}$$

3.3 Optimization Design Model of 3-D Braided Composite

We choose specific damping capacity and some stiffness coefficient simultaneously as maximum optimization objective function, structure parameter of materials (α, β) and geometric parameter of section (b_1, b_2, h_1, h_2) as design variables, constraint condition are numeric area of structure parameters of materials on technological requirements. Here, we design optimization mathematical models of the 3-D braided composite as follows,

$$f = \max(\eta_1 \Psi + \eta_2 K_{11} + \eta_3 K_{22} + \eta_4 K_{33} + \eta_5 K_{44})$$

$$\text{s.t. } \alpha_{\min} \leq \alpha \leq \alpha_{\max}, \beta_{\min} \leq \beta \leq \beta_{\max}$$

$$(h_2 / b_2)_{\min} \leq (h_2 / b_2) \leq (h_2 / b_2)_{\max} \tag{10}$$

$$(b_1 / b_2)_{\min} \leq (b_1 / b_2) \leq (b_1 / b_2)_{\max}$$

$$(h_1 / h_2)_{\min} \leq (h_1 / h_2) \leq (h_1 / h_2)_{\max}$$

Where $\eta_1 + \eta_2 + \eta_3 + \eta_4 + \eta_5 = 1$, and η_1 to η_5 are respectively weight coefficients for specific damping capacity and flexural stiffness coefficient in the unitive objective function. Of course, we also obtain other optimization model according to practical requirement. For the non-linear optimization design problem, optimization solution can be obtained expediently by using the facilities of ant colony optimization.

4   Hybrid Optimization Algorithm

4.1   Optimization Model

Ant colony optimization is a kind of discrete algorithm. There for the optimization problem must be converted in to a discrete one before can be solved by an ant colony algorithm.

First discretize each variable in its feasible region, i. e., divide the variables to be optimized in its value range. The number of the division relates with the range of the variable and the complexity of the problem, etc. Suppose the variable to be optimized is divided into q_i nodes, denote the node value as x_{ij} where i is variable number ($i = 1,2,\ldots n$) and j is node number of x_i ($j = 1,2,\ldots q_{ij}$).

After the division, a network which composed of variables and their division is constructed as depicted in Fig. 2, where "●"is the node of each variable division. The number of the division of each variable can be unequal and the division can be linear or nonlinear.

Suppose using m ants to optimize n variables, for ant k at time $t = 0$, set the ant at origin O. Then the ant will start searching according to the maximum transition probability. The searching starts from x_1 and goes on in sequence. Each ant can only select one path in its division to do transition. The ant k transit from x_i to x_{i+1} are called a step and is denoted as t. The ant k starts searching from the first variable x_1 to the last variable x_n is called an iteration and is denoted as NC. An iteration includes n steps.

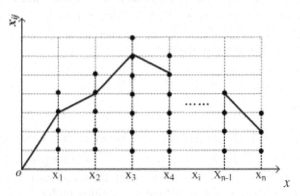

Fig. 2. Variable division and ant search path

At time t, the transition probability of ant k transit from node h of x_i to node l of x_{i+1} is defined as equation (11)

$$p_{hl}^k(t) = \frac{[\tau_{hl}(t)]^\alpha [\eta_{hl}]^\beta}{\sum_{j=1}^{q_{i+1}} [\tau_{hj}(t)]^\alpha [\eta_{hj}]^\beta}, \tag{11}$$

where $\tau_{hl}(t)$ is the pheromone level of node l of x_{i+1} at time t. η_{hl} (NC) is the visibility of node l of x_{i+1} in NC iteration. The denominator represents the sum of pheromone level multiple visibility of all node of x_{i+1} at time t. Given $\alpha, \beta > 0$ are heuristic factors, they will control the relative importance of pheromone level $\tau_{hl}(t)$ and visibility η_{hl} (NC) respectively. While searching, the ant will select the node which has the maximum transition probability node to transit.

The visibility of node j of variable x_i in iteration NC is defined as equation (12)

$$\eta_{ij}(NC) = \frac{(r_{i_{upper}} - r_{i_{lower}}) - |x_{ij} - x_{ij}^*(NC - 1)|}{r_{i_{upper}} - r_{i_{lower}}}, \tag{12}$$

where i is variable number ($i = 1, 2, \ldots n$) and j is node number of x_i ($j = 1, 2, \ldots q_{ij}$). $r_{i_{upper}}, r_{i_{lower}}$ are the upper and lower limit value of range of variable x_i respectively.

$x_{ij}^{*} (NC - 1)$ is the optimal value of the variables (parameter values of datum axes) in iteration $(NC - 1)$. In the first iteration, $x_{ij}^{*} (NC - 1)$ can be obtained by using the least-square method. After that ($NC > 1$), $x_{ij}^{*} (NC - 1)$ will be the node value of the variable relevant the optimal path produced in last iteration.

At the initial time $t = 0$, the pheromone levels of each node are equal, i. e., $\tau_{ij}(0) = c$ and all of the ants are at origin O. After n time units, all of the ants will crawl from start node to terminal node and the pheromone levels of each node can be adjusted according following formation.

$$\tau_{ij}(t+n) = \rho\tau_{ij}(t) + \Delta\tau_{ij}, \tag{13}$$

where i is variable number ($i = 1,2,...n$) and j is node number of variable x_i ($j = 1,2,...q_{ij}$). $(1-\rho)$ represents the decay of pheromone level of node x_{ij} from t to time $t+n$. The evaporation rate ρ $(1 < \rho < 1)$ in equation (13) is an important parameter which is used to ensure that pheromones on unvisited paths evaporate over time and that those paths are affixed low visiting probabilities in the future. A judicious compromise between pheromone evaporation and deposition is necessary.

$$\Delta\tau_{ij} = \sum_{k=1}^{m} \Delta\tau_{ij}^{k}, \tag{14}$$

where $\Delta\tau_{ij}^{k}(t)$ is the pheromone of node x_{ij} left by ant k in this iteration. It can be calculated according to equation (15)

$$\Delta\tau_{ij}^{k} = \begin{cases} Q/f_k, & \text{if kth ant pass through this node in this iteration} \\ 0, & \text{else} \end{cases}, \tag{15}$$

where f_k is the value of objective function of the ant k in this iteration. It is calculated with equation (10). Q is a positive constant.

4.2 Search Algorithm Hybridization

Being a global search method, the ACO algorithm is expected to give its best results if it is augmented with a local search method that is responsible for fine search. The ACO algorithm is hybridized with a gradient based search. The "candidate" solution found by the ACO algorithm at each iteration step is used as an initial solution to commence a gradient-based search using "fmincon" function in MATLAB [14], which based on the interior-reflective Newton method. The "fmincon" function can solve the constraint problem here.

4.3 Hybrid Optimization Algorithm

The procedure of hybrid optimization algorithm is shown as follows.

Step1: Decide the number of variable and discrete the variables by dividing them. Each variable is divided into q_i nodes which constitute the number vector $\mathbf{q}$. Node value is x_{ij}.

Step2: Given the ant number m and define a one dimensional vector **path** $_k$ for ant k which has n elements. The one dimensional vector **path** $_k$ is called path table and used to save the path that the ant k passed through.

Step3: Initialization. Let time counter $t = 0$ and the maximum iteration number is NC_{max}. Let the pheromone level of each node $\tau_{ij}(0) = c$, $\Delta \tau_{ij} = 0$. Set the m ants at origin O. According equation (12), calculate the visibility $\eta_{ij}(1)$ used in the first iteration.

Step4: Let variable counter $i = 1$, variable division number counter $q = 1$.

Step5: Use equation (11) to calculate the transition probability of ant k transit from node h of variable x_i to node l of variable x_{i+1} and select the maximum probability to do transition. Then save node l of x_{i+1} to the ith element of **path** $_k$. Transit all ants from node of variable x_i to node of variable x_{i+1}.

Step6: Let $i = i+1$, if $i \leq n$, then go to Step 5, else, go to Step 7.

Step7: Based on the path that ant k walked, i. e., **path** $_k$, calculate the objective function f_k of ant k according to objective functions. Save the node values relevant optimal path in this iteration.

Step8: Let $t \leftarrow t+1$, $NC \leftarrow NC+1$. Update the pheromone level of each node according to equations (13), (14), (15) and clear all elements in **path** $_k$ ($k = 1 \sim m$). Calculate visibility of each node according to equation (12).

Step9: Let the above solution as an initial solution, and carry out a gradient-based search using "fmincon" function in MATLAB Optimization Toolbox.

Step10: If $NC < NC_{max}$ and the whole ant colony have not converged to the same path yet, then set all of the ants at the origin O again and go to Step 4; If $NC < NC_{max}$ and the whole ant colony have converged to the same path then the iteration complete and the optimal path is the optimal solution.

5 Numerical Examples

In order to test the validity of the proposed optimization design procedure, numerical examples have been preformed. Suppose braiding yarn system (four-direction) of 3-D braided composite be homogeneous carbon fiber, where performance of component material are : E_{f1}=258.2GPa, E_{f2}=18.2 GPa, G_{f21}=36.7 GPa, μ_{f21}=0.25, μ_m=0.35, Ψ_L=0.0045, Ψ_T=0.0422, Ψ_{LT}=0.0705, Ψ_{TT} =0.0421, E_m=3.4GPa, G_m=1.26GPa [11]. The performance of unidirectional composite for each system can be calculated by performance of component material according to mesomechanics. The perimeter of rectangular cross section is equal to 0.08m, and length of the composite is 0.5m.

Here, we choose common constraint condition: $\alpha_{min}=20°$, $\alpha_{max}=60°$, $\beta_{min}=20°$, $\beta_{max}=60°$, $(h_2/b_2)_{min}=0.4$, $(h_2/b_2)_{max}=2.5$, $(b_1/b_2)_{min}=0.2$, $(b_1/b_2)_{max}=0.8$, $(h_1/h_2)_{min}=0.2$, $(h_1/h_2)_{max}=0.8$. Two examples are performed as follows.

(1) Example 1: Let $\eta_1=1$, $\eta_2=\eta_3=\eta_4=\eta_5=0$, maximum of damping Ψ can be calculated by using hybrid optimization algorithm presented, and the optimization results according to optimization model are shown in Table 1.

(2) Example 2: Let $\eta_3=1$, $\eta_1=\eta_2=\eta_4=\eta_5=0$, maximum of stiffness K_{22} can be calculated and the optimization results are shown in Table 1.

(3) Example 3: Let $\eta_1=\eta_3=0.5$, $\eta_2=\eta_4=\eta_5=0$. Having eliminated difference of order of magnitude for Ψ and K_{22}, the optimization results are shown in Table 1.

As shown in Table 1, the optimal objectives of hollow-rectangular-section 3-D braided composite can be improved observably by optimization design for structure parameter of materials and geometric parameter of section.

From the optimization results of Example 1 and 2, if only make a search for maximizing specific damping capacity or stiffness coefficients of the composite, damping or stiffness characteristic of the hollow-rectangular-section 3-D braided composite can be improved obviously . But it may influence on flexural stiffness or damping of composite respectively (may result in a few reducing), thus Example 1 and 2 exist any limitation and shortage. The optimization results of Example 3 show that both of damping and stiffness may attain better maximum at the same time. The optimization effects denote that the optimization models are propitious to engineering application.

Table 1. The results of preliminary design and optimization design

	α	β	h_2/b_2	b_1/b_2	h_1/h_2	Ψ	K_{22} (GPa·m²)
Preliminary design	30°	30°	1.0	0.5	0.5	0.0027	2.51×10^{-5}
	30°	30°	1.0	0.5	0.5	0.0027	2.51×10^{-5}
	30°	30°	1.0	0.5	0.5	0.0027	2.51×10^{-5}
Optimal design	36°	52°	1.04	0.35	0.34	0.0106	1.59×10^{-5}
	23°	28°	1.12	0.29	0.28	0.0022	1.87×10^{-4}
	27°	45°	1.29	0.22	0.22	0.0079	3.92×10^{-5}

6 Conclusions

According to the stiffness and damping properties of hollow-rectangular-section 3-D braided composite, the mathematical model of structure design of the 3-D braided composite was proposed. The results of numeral examples show that the better damping and stiffness characteristic could be obtained by using optimal design, contenting the determinate restriction. The method proposed here is useful for the design and engineering application of the kind of member.

References

1. Huang, G.: Modern Textile Composites. Chinese Textile Press, Beijing (2000)
2. Zheng, X.T., Ye, T.Q.: Microstructure Analysis of 4-Step Three-Dimensional Braided Composite. Chinese Journal of Aeronautics 16(3), 142–149 (2003)
3. Wang, Y.Q., Wang, A.S.D.: On the Topological Yarn Structure of 3-D Rectangular and Tubular Braided Preforma. Composites Science and Technology 51, 575–583 (1994)
4. Colorni, A., Dorigo, M., Maniezzo, V.: Distributed Optimization by Ant Colonies. In: The First European Conference on Artificial Life, pp. 134–142 (1991)
5. Gambardella, L.M., Taillard, E.D., Dorigo, M.: Ant Colonies for the Quadratic Assignment Problem. J. Oper. Res. Soc. 50, 167–176 (1999)
6. Gamardella, L.M., Bianchi, L., Dorigo, M.: An Ant Colony Optimization Approach to the Probabilistic Traveling Salesman Problem. In: Guervós, J.J.M., Adamidis, P.A., Beyer, H.-G., Fernández-Villacañas, J.-L., Schwefel, H.-P. (eds.) PPSN 2002. LNCS, vol. 2439, p. 883. Springer, Heidelberg (2002)
7. Dorigo, M., Stuzle, T.: Ant Colony Optimization. MIT Press, Cambridge (2004)
8. Yang, J.M.: Fiber Inclination Model of Three-dimensional Textile Structural Composites. Journal of Composite Materials 20, 472–476 (1986)
9. Li, W., Hammad, M., EI-Shiekh, A.: Structural Analysis of 3-D Braided Preforms for Composites, Part 1:Two-Step Preforms. J. Text. Inst. 81(4), 515–537 (1990)
10. Cai, G.W., Liao, D.X.: The Stiffness Coefficients of Rectangular Cross Section Beams with 2-step Three-dimensional Braided Composite Beam. Journal of Huazhong University of Science and Technology 24(12), 26–28 (1996)
11. Cai, G.W., Zhou, X.H., Liao, D.X.: Analysis of the Stiffness and Damping of the 4-step Three-dimensional Braided Composite Links. Journal of Mechanical Strength 21(1), 18–23 (1999)
12. Adams, R.D., Bacon, D.G.C.: Effect of Orientation and Laminated Geometry on the Dynamic Properties of CFRP. J. Composite Materials 7(10), 402–406 (1973)
13. Lin, D.X., Ni, R.G., Adams, R.D.: Prediction and Measurement of the Vibrational Damping Parameters of Carbon and Glass Fibre-Reinforced Plastics Plates. Journal of Composite Material 18(3), 132–151 (1984)
14. Grace, A.: Optimization Toolbox for Use with MATLAB, User's Guide. Math Works Inc. (1994)

Robot Virtual Assembly Based on Collision Detection in Java3D

Peihua Chen[1,*], Qixin Cao[2], Charles Lo[1], Zhen Zhang[1], and Yang Yang[1]

[1] Research Institute of Robotics, Shanghai Jiao Tong University
cph@sjtu.edu.cn, charleslo77@gmail.com, zzh2000@sjtu.edu.cn,
iyangyang186@yahoo.com.cn
[2] The State key Laboratory of Mechanical System and Vibration,
Shanghai Jiao Tong University
qxcao@sjtu.edu.cn

Abstract. This paper discusses a virtual assembly system of robots based on collision detection in Java 3D. The development of this system is focused on three major components: Model Transformation, Visual Virtual Assembly, and an XML output format. Every component is presented, as well as a novel and effective algorithm for collision detection is proposed.

Keywords: Virtual Assembly, Collision detection, Service robot, Java 3D, VRML.

1 Introduction

Virtual Assembly (VA) is a key component of virtual manufacturing [1]. Presently, virtual assembly serves mainly as a visualization tool to examine the geometrical representation of the assembly design and provide a 3D view of the assembly process in the field of virtual manufacturing [2-5]. However, in service robots' simulation, we also need to assemble many types of robots in the virtual environment (VE). There are mainly two methods of virtual assembly in these systems, e.g., assembling every single part together via hard-coding using link parameters between each part, and another method is assembling a virtual robot in an XML(eXtensible Modeling Language), both of which are not visual and complicated.

This paper focuses on the visual implementation of making virtual assembly of service robots in Java3D. We choose Java JDK 1.6.0 and Java 3D 1.5.2 as the platform. First we discuss the system architecture and methodology. This is followed by its implementation. After that, an application of this system will be presented and conclusions obtained.

2 System Architecture

2.1 System Architecture

The architecture of this system is presented in Fig.1. It includes three parts: Model Transformation, Visual Virtual Assembly (VVA), and XML Output. Firstly, after we

---

* Corresponding author.

H. Deng et al. (Eds.): AICI 2009, LNAI 5855, pp. 270–277, 2009.

create the CAD models of the robot, we export these models as IGES [7] format to be imported into 3DS Max. In 3DS Max, we transform the Global Coordinate System (GCS) of the IGES model into the coordinate system according to the right-hand rule, and then export the models in VRML format. Then, we load the VRML files into Java 3D. After which, we set the scale factor of the model to 1:1. Then, we set the target model as static, and move the object model towards the target model until the collision detection between them is activated. After which, we attach the object model to the target model, which means appending the object node to the target node. If the virtual assembly is not completed, then import another VRML file into the VE for assembling; if it is finished, then output the assembly result into an XML file.

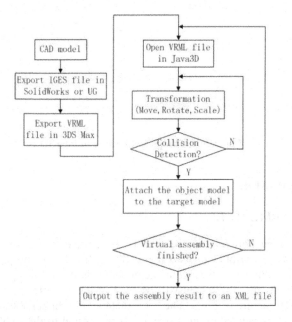

Fig. 1. System Architecture of VA for robots

2.2 Scene Graph Architecture

The use of scene graph architecture in Java 3D can help us organize the graphical objects easily in the virtual 3D world. Scene graphs offer better options for flexible and efficient rendering [8]. Fig.2 illustrates a scene graph architecture of Java 3D for loading the VRML models into the VE for scaling, translation and rotation transformations. In Fig.2, BG stands for BranchGroup Node, and TG, Transform -Group Node.

As shown in Fig.2, the Simple Universe object provides a foundation for the entire scene graph. A BranchGroup that is contained within another sub-graph may be reparented or detached during run-time if the appropriate capabilities are set. The TransformGroup node specifies a single spatial transformation, via a Transform3D object, that can position, orient, and scale all of its children [9], e.g., moveTG, rotTG, scaleTG, etc., as shown in Fig.2. The moveTG handles the translations of the VRML model, rotTG is for rotations, and scaleTG is for scaling.

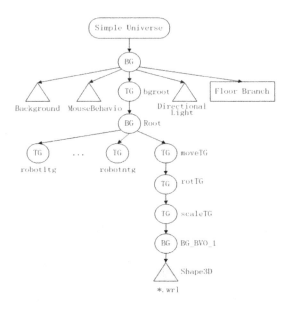

Fig. 2. Scene graph architecture in Java 3D

3 Implementation

3.1 Loading VRML Models

Java 3D provide support for runtime loaders. In this paper, we use the VrmlLoader class in package j3d-vrml97 to load the VRML files. In our project, we rewrite a vrmlload class with the construction method of vrmlload (String filename). After selecting a VRML file, the model will be seen in the 3D Viewer of our project, as shown in the left picture of Fig.3. The coordinate system in 3D Viewer is right-hand, with the orientation semantics such that +Y is horizontal to the right, +X is directly towards the user, and +Z is the local gravitational direction upwards.

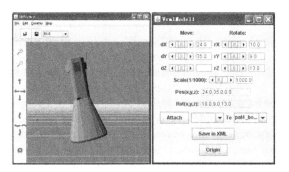

Fig. 3. A repositioned VRML model and its control panel

Fig.3 also shows how a model's position, orientation, and size can be adjusted. Here it shows a robot body that has been moved and rotated. In the control panel of the model, there are several operations. They are: translations of x, y, and z axes; rotations of the model around x, y, and z axes; scaling the model; displaying "Pos(x, y, z)" and "Rot(x, y, z)" at runtime, as shown in Fig.3. In addition, when another model is loaded, a corresponding control panel will be created at the same time.

3.2 Algorithm for Collision Detection

In the VVA system, collision detection plays an important role. Every node in the scene graph contains a Bounding field that stores the geometric extent of the node [10]. Java3D offers two classes "WakeupOn -CollisionEntry" and "WakeupOnCollision -Exit" to carry out collision detection for 3D models. When any objects in Java 3D scene collide with other bounds, "Wakeup -OnCollisionEntry" is triggered and when collision is released, "WakeupOnCollision -Exit" is triggered. So we can use them as triggering conditions and we are able to obtain the control behavior after collision occurs.

This paper proposes a novel and effective algorithm to detect collisions. VRML file format uses IndexedFaceSet node to describe shapes of faces and triangle patches to reveal all types of shapes in 3D models. IndexedFaceSet includes a coord field that contains a series of spatial points, which can be used in the coordIndex field to build the surfaces of 3D models. We conserve the points' position, and establish space for collision detection using the following method.

BoundingPolytope is chosen to use more than three half-space [11] to define a convex and closed polygonal space of a VRML model. We make use it to build the collision detection space. The function that defines every half-space α is: $Ax + By + Cz + D \leq 0$, where A, B, C, and D are the parameters that specify the plane. The parameters are passed into the x, y, z, and w fields, respectively, of a Vector4d object in the constructor of BoundingPolytope (Vector4d[] planes). Thus we must make out the parameters A, B, C, and D of every triangle plane. The following method is to compute the value of A, B, C, and D. We suppose that one triangle is made up of three points: L, M, and N, shown as in Fig.4. And we obtain vector $\overline{LM}$ and $\overline{MN}$ based on a vertex sequence.

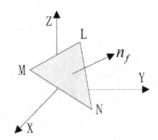

Fig. 4. Triangle plane in the WCS

$$\overline{LM} = (X_M - X_L)i + (Y_M - Y_L)j + (Z_M - Z_L)k \quad . \tag{1}$$

$$\overline{MN} = (X_N - X_M)i + (Y_N - Y_M)j + (Z_N - Z_M)k \quad . \tag{2}$$

The normal vector of plane α is the vector product of $\overline{LM}$ and $\overline{MN}$. That is

$$\overline{n_f} = \overline{LM} \times \overline{MN} = \begin{bmatrix} i & j & k \\ (X_M - X_L) & (Y_M - Y_L) & (Z_M - Z_L) \\ (X_N - X_M) & (Y_N - Y_M) & (Z_N - Z_M) \end{bmatrix} . \tag{3}$$

Once we obtain normal vector $\overline{n_f}$, we get values of A, B, C. Then we can obtain the value of constant D if we substitute any point's coordinate into the equation. After transmitting all plane parameters to one group of Vector4d objects, we can use BoundingPolytope to build a geometrical detection space. Then, we invoke setCollisionBounds(Bounds bounds) to set the collision bounds. After setting a collision bound, Java3D can carry out collision detection in response to the setting bounds.

3.3 Visual Virtual Assembly

While assembling the object model to the target model, we have to rotate the object model around the z-axis, y-axis, and x-axis in this order to correspond with the target Local Coordinate System (LCS). Then we need to move along the x, y, and z axes in both the positive and negative directions. In Fig.5, we present how we assemble the model J4 to model J3. Fig.5 (a) shows the initial positions of the two VRML models. Fig.5 (b) presents the state after the second model (J4) is rotated about the x-axis by -90°. We can also see from this picture that we don't need to make any translations

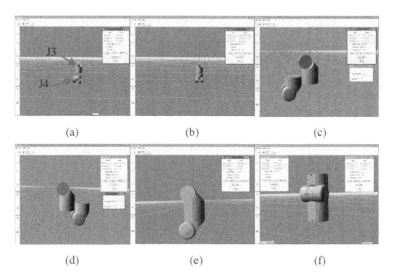

(a)   (b)   (c)

(d)   (e)   (f)

Fig. 5. Example of assembling one VRML model to another

along y-axis anymore, because the second model's y-axis is already corresponding to the first model. From Fig.5(c) and Fig.5 (d), we can obtain the offset values along +x direction and –x direction through the collision detection between the two models. As we have set the threshold value of collision detection at 0.5mm, thus when the distance between the two models is smaller than 0.5, the collision detection will be activated and the actual offset value will be obtained. Here, the offset value in the +x direction is 140.5-0.5=140(mm), and the offset value in the –x direction is -80.5+0.5=-80(mm), as shown in Fig.5 (d). So, in order to make the x-axis of the second model correspond to the first one, the total offset value along the x-axis is: (140 + (- 80)) / 2 = 30(mm), and the result can be seen in Fig.5 (e). Then move along the z-axis, and the operations are similar with the previous method. The final result of this virtual assembly is presented in Fig.5 (f).

After we have moved the second model to the correct position, we attach the second model's node to the first model's node.

3.4 Save the Virtual Assembly Result in XML Files

After we have realized the virtual assembly of the robot, we save the virtual assembly result as XML files, as shown in Fig.1. In this paper, we use JAXB (Java Architecture for XML Binding) [12] to access and process XML data. JAXB makes XML easy to use by compiling an XML schema into one or more Java technology classes. The structure of the XML schema of our project is presented in Fig.6, and the output data could be referred to, in this structure.

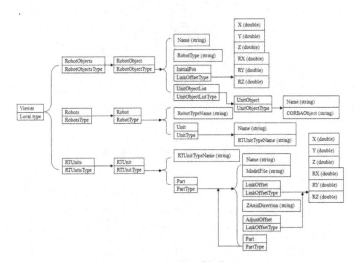

Fig. 6. Structure of the XML schema

4   Application

In this paper, we create the CAD models of the real robot as shown in Fig.7, and then translate them to VRML files according to the right-hand rule.

Fig. 7. A real robot with left arm

Then, we start the visual virtual assembly in Java 3D, as shown in Fig.8. After the virtual assembly is completed, we save the virtual assembly result to an XML file. Then we run the program again, and open the XML file just created, as shown in Fig.9 (a). At the same time, we can control the motion of every joint through the corresponding control panel, as shown in Fig.9 (b).

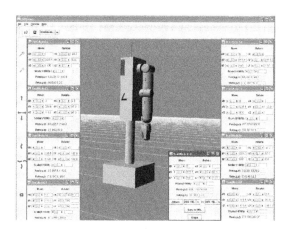

Fig. 8. Virtual assembly of the robot

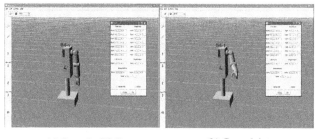

(a) Open the XML file in Java 3D

(b) Control the motions of the robot

Fig. 9. Open the XML file stored the VA data and Control the motions of the robot

5 Conclusions

From the application in section 5, Virtual Assembly of robots is carried out based on Collision Detection, and it proves to work well. Virtual Assembly of robots based on Collision Detection in Java3D allows for visual feedback, and makes the process much easier for the end-users. Through the virtual assembly, users can obtain the assembly result file in XML format that complies with a special XML schema. Then the users can easily load the new assembled robot via the XML file, and operate every joint of the robot, e.g., developing a program for workspace and kinematic analysis.

Acknowledgments. This work was supported in part by the National High Technology Research and Development Program of China under grant 2007AA041703-1, 2006AA040203, 2008AA042602 and 2007AA041602; the authors gratefully acknowledge the support from YASKAWA Electric Cooperation for supporting the collaborative research funds and address our thanks to Mr. Ikuo agamatsu and Mr. Masaru Adachi at YASKAWA for their cooperation.

References

1. Jayaram, S., Connacher, H., Lyons, K.: Virtual Assembly Using Virtual Reality Techniques. CAD 29(8), 575–584 (1997)
2. Choi, A.C.K., Chan, D.S.K., Yuen, A.M.F.: Application of Virtual Assembly Tools for Improving Product Design. Int. J. Adv. Manuf. Technol. 19, 377–383 (2002)
3. Jayaram, S., et al.: VADE: a Virtual Assembly Design Environment. IEEE Computer Graphics and Applications (1999)
4. Jiang-sheng, L., Ying-xue, Y., Pahlovy, S.A., Jian-guang, L.: A novel data decomposition and information translation method from CAD system to virtual assembly application. Int. J. Adv. Manuf. Technol. 28, 395–402 (2006)
5. Choi, A.C.K., Chan, D.S.K., Yuen, A.M.F.: Application of Virtual Assembly Tools for Improving Product Design. Int. J. Adv. Manuf. Technol. 19, 377–383 (2002)
6. Ko, C.C., Cheng, C.D.: Interactive Web-Based Virtual Reality with Java 3D, ch. 1. IGI Global (July 9, 2008)
7. http://ts.nist.gov/standards/iges//
8. Sowizral, H., Rushforth, K., Deering, M.: The Java 3D API specification, 2nd edn. Addison-Wesley, Reading (2001)
9. Sun Microsystems. The Java 3DTM API Specification, Version 1.2 (March 2000)
10. Selman, D.: Java 3D Programming. Manning Publications (2002)
11. http://en.wikipedia.org/wiki/Half-space
12. Ort, E., Mehta, B.: Java Architecture for XML Binding (JAXB). Sun Microsystems (2003)

Robust Mobile Robot Localization by Tracking Natural Landmarks

Xiaowei Feng, Shuai Guo, Xianhua Li, and Yongyi He

College of Mechatronics Engineering and Automation, Shanghai University,
Shanghai 200444, China
xwfeng1982@163.com

Abstract. This article presents a feature-based localization framework to use with conventional 2D laser rangefinder. The system is based on the Unscented Kalman Filter (UKF) approach, which can reduce the errors in the calculation of the robot's position and orientation. The framework consists of two main parts: feature extraction and multi-sensor fusing localization. The novelty of this system is that a new segmentation algorithm based-on the micro-tangent line (MTL) is introduced. Features, such as lines, corners and curves, can be characterized from the segments. For each landmark, the geometrical parameters are provided with statistical information, which are used in the subsequent matching phase, together with a priori map, so as to get an optimal estimate of the robot pose. Experimental results show that the proposed localization method is efficient in office-like environment.

Keywords: Localization, feature extraction, Unscented Kalman Filter, mobile robot.

1 Introduction

Localization is the key for autonomous mobile robot navigation systems. It is well known that using solely the data from odometry is not sufficient since the odometry provides unbounded position error [1]. The problem gives rise to the solution that robot carries exteroceptive sensors (sonar, infrared, laser, vision, etc.) to get landmarks and uses this operation to update its pose estimation.

Instead of working directly with raw scan points, feature based localization first transforms the raw scans into geometric features. Feature based method increase the efficiency and robustness of the navigation and makes the data more compact. Hence, it is necessary to automatically, and robustly, detect landmarks for data association, and ultimately localization and mapping purposes for robots. This approach has been studied and employed intensively in recent research on Robotics [2]. One of the key problems of feature-based navigation is the reliable acquisition of information from the environment. Due to many advantages of laser rangefinder, such as good reliability, high sampling rate and precision, it is wildly used in mobile robot navigation.

UKF[3] is a nonlinear optimal estimator based on given sampling strategy without the linearization steps required by the Extended Kalman Filter(EKF). And it is not

H. Deng et al. (Eds.): AICI 2009, LNAI 5855, pp. 278–287, 2009.

restricted to Gaussian distributions. It has been improved that with appropriate sigma point selection schemes, the UKF performs better estimation by exploiting more information such as the first three moments of an arbitrary distribution or the first four non-zero moments of a Gaussian distribution. In proposed system, UKF fuses the information from odometry with the information from range images to get robot's position estimation.

In this paper, we describe a feature based localization system, which is composed of two main parts: feature extraction and probabilistic localization. For feature extraction, we describe a geometrical feature detection method for use with conventional 2D laser rangefinder. The contribution of this paper is that our method employs the variance function of local tangent associated to laser range scan readings to track multiple types of natural landmarks, such as line, curve and corner, instead of calculating curvature function. The extracted features and their error models are used, together with a priori map, by an UKF so as to get an optimal estimate of robot's current pose vector and the associated uncertainty.

2 Scan Data Clustering and Segmentation

2.1 Adaptive Clustering

Clustering is a key procedure for landmark extraction process. It can roughly segment the range image to some consecutive point clusters, which will also enhance the efficiency of the extraction procedure. Range readings belong to the same segment while the distance between them is less than a given threshold [4].

$$D_{max} = r_{l-1} \left(\sin \Delta\phi / \sin(\lambda - \Delta\phi) \right) + 3\sigma_r \qquad (1)$$

Where $\Delta\phi$ is the laser angular resolution, λ is an auxiliary constant parameter and σ_r the range variance. In our experiments, the parameter values are $\sigma_r = 0.01$ m and $\lambda = 10°$. Clusters having too few numbers of points, such as isolated range readings, are removed as big noise.

2.2 Segmentation Based on MTL

For range data acquired from clustering process, the relationship of adjacent points is, however, totally dependent on the shape of the scanned environment. The difference between neighboring points can be depicted by a function of geometry parameters, such as the direction of the local tangent or curvature of objects in the environment, etc.

The tangent of the points belong to the same features is supposed to be same or change regularly, see Fig.1(a). In this work, we employ the variance function of local tangent associated to laser range scan images to break the segments into several unique features. The tangent of each scan point can be calculated by line-regression of k-neighboring points. In our experiment, k is 7.

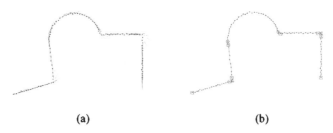

(a)   (b)

Fig. 1. a) MTL of range points; b) Features extracted associated to (a): □ -line, △ - corner and ○ -curve

The variance between two consecutive tangents which belong to point $p(i)$ and $p(i+1)$ is

$$\text{var} = \arctan\left(\left|(k(i)-k(i+1))/(1+k(i)\cdot k(i+1))\right|\right) \qquad (2)$$

Where $k(i)$ and $k(i+1)$ are the slop of two tangents for $p(i)$ and $p(i+1)$. Fig.2 is the tangent variance function with respect to Fig.1(a).

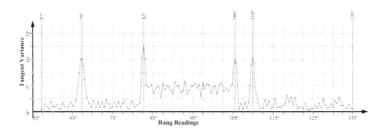

Fig. 2. Tangent variance function with respect to Fig.1(a)

The tangent variance function is searched for geometric features:

1) Line segments. Therefore, they are those sets of consecutive range readings with variance values close to zero. And they have a certain minimum width (l_{min} =5 range points in our experiment).
2) Curve segments. The variance function associated to a curve segment is detected wherever a region of the function has a constant value larger than a preset value. This value can be found over a certain minimum width and the maximum in the middle of the region differs only by a certain amount from the values at the end.
3) Corners. The corner is always defined by a value associated to a local peak of the variance function. It is a region between two segments which have been marked as line or curve and also has a certain minimum width, which can filter out spurious peak value due to random noise which has a relative low value.

Fig.1(b) shows the detected line segments, corners and curve segments associated to Fig.1(a).

3 Landmark Extraction

3.1 Line Segments

The line parameters are (θ, d) from the line equation $x\cos\theta_1 + y\sin\theta_1 = d_1$. Given n points $p(r_i, \phi_i)$, the regression parameters are given by following expressions[5]:

$$\theta = \frac{1}{2}\arctan(\frac{-2\sum_i (\overline{y} - y_i)(\overline{x} - x_i)}{\sum_i \left[(\overline{y} - y_i)^2 - (\overline{x} - x_i)^2\right]}) = \frac{1}{2}\arctan\frac{N}{D} ,$$

$$d = \overline{x}\cos\theta + \overline{y}\sin\theta .$$

(3)

Where $\overline{x} = \sum r_i \cos\phi_i / n$ and $\overline{y} = \sum r_i \sin\phi_i / n$.

The parameters θ and d are a function of all the measured points $p_i, i = 1\cdots n$ belonging to equation (3). Assuming that the individual measurements are independent, the covariance matrix of parameters (θ, d) can be calculated as

$$C_{\theta, d} = \sum_i^n J_i C_{xyi} J_i^T .$$

(4)

Where uncertainty of a measured point p_i is represented by a covariance matrix C_{xyi}, and the terms of Jacobian matrix are obtained as follows

$$\partial\theta/\partial x_i = ((\overline{y} - y_i)D + (\overline{x} - x_i)N)/(N^2 + D^2) ,$$

$$\partial\theta/\partial y_i = ((\overline{x} - x_i)D + (\overline{y} - y_i)N)/(N^2 + D^2) ,$$

$$\partial d/\partial x_i = \cos\theta/n + (\overline{y}\cos\theta - \overline{x}\sin\theta)(\partial\theta/\partial x_i) ,$$

$$\partial d/\partial y_i = \sin\theta/n + (\overline{y}\cos\theta - \overline{x}\sin\theta)(\partial\theta/\partial y_i) .$$

(5)

Where N and D are the numerator and denominator of the expression (3).

3.2 Curve Segments

A circle can be defined by the equation $(x - x_0)^2 + (y - y_0)^2 = r^2$, where (x_0, y_0) and r are the center and the radius of the circle, respectively. For a circle fitting problem, the data set (x, y) is known and the circle parameters need to be estimated.

Assuming that we have obtained M measurements $p(x_i, y_i)$, our objective is to find $p(x_0, y_0, r)$ subjecting to $f_i(x_0, y_0, r) = (x_i - x_0)^2 + (y_i - y_0)^2 - r^2 = 0$ where $i = 1\cdots M$.

We use the Levenberg-Marquardt[6] method to solve this nonlinear least-squares problem.

$$(A_k^T A_k + \lambda_k I)\Delta p_k = -A_k^T \overline{f_k} . \tag{6}$$

Where $A = [\nabla f_1, \nabla f_2, \cdots, \nabla f_M]^T$ is the Jacobian matrix, and $\Delta p_k = p_{k+1} - p_k$ and p_k is the estimate of $p(x_0, y_0, r)$ at the $k-th$ iteration. The initial value of (x_0, y_0, r) is got by using the first, last and middle points. According to[7], an estimation of the curve segment uncertainty can be approximated by three points of the curve segment.

3.3 Corners

Then, real and virtual corners can be obtained from the intersection of the previously detected line segments. Once a corner is detected, its position (x_c, y_c) is estimated as the intersection of the two lines which generate it. Given two lines $x\cos\theta_1 + y\sin\theta_1 = d_1$ and $x\cos\theta_2 + y\sin\theta_2 = d_2$, the equations to calculate corner (x_c, y_c) is:

$$\begin{cases} x_c = (d_1 \sin\theta_2 - d_2 \sin\theta_1)/(\sin(\theta_2 - \theta_1)) \\ y_c = (d_2 \cos\theta_1 - d_1 \cos\theta_2)/(\sin(\theta_2 - \theta_1)) \end{cases} . \tag{7}$$

Where (d_1, θ_1) and (d_2, θ_2) can be get from previous line extraction process. And the variance of (x_c, y_c) is

$$C_{x_c y_c} = JC_{d_1\theta_1 d_2\theta_2} J^T . \tag{8}$$

Where J is the Jacobian matrix of equation (7) to $(d_1, \theta_1, d_2, \theta_2)$. And $C_{d_1\theta_1 d_2\theta_2}$ is the covariance matrix of line parameters $(d_1, \theta_1, d_2, \theta_2)$. For convenience, the line parameters (d_1, θ_1) and (d_2, θ_2) for two different tangent lines are supposed to be independent. Then $C_{d_1\theta_1 d_2\theta_2}$ is simplified to be $diag(C_{d_1\theta_1}, C_{d_2\theta_2})$.

4   Natural Landmark-Based Localization

Wheel slippage causes a drift in the mobile robot, which can be quite disastrous in mobile robot localization and navigation. To provide noise rejection and develop a model dependant estimation of position and orientation, an Unscented Kalman Filter (UKF) is applied in our research.

In this application the initial location of the robot is known, and the robot has a priori map of the locations of geometric landmark $p_i = [p_x, p_y]_i$. Each landmark is assumed to be known. At each time step, observations $z_j(k)$ of these Landmarks are taken. UKF combines observations with a model of the system dynamics which are the kinematics relationships that express the change in position of the robot as a function of the displacement of two wheels to produce an optimal estimate of the robot's position and orientation.

4.1 Kinematic Model

The system model describes how the vehicle's position $x(k)$ changes with time in response to a control input $u(k) = [\Delta s_l, \Delta s_r]_k$ and a noise disturbance $v(k)$. It has the form:

$$x(k+1) = f(x(k), u(k)) + v(k) , \quad v(k) \sim N(0, Q(k)) . \tag{9}$$

Where $f(x(k), u(k))$ is nonlinear state transition function. And noise $v(k)$ is assumed to be zero mean Gaussian with variance $Q(k)$. According to the kinematics of a two wheel differential drive mobile robot using in our experiment, the translation function then has the form:

$$f(\cdot, \cdot) = \begin{bmatrix} x(k) \\ y(k) \\ \theta(k) \end{bmatrix} + \begin{bmatrix} \Delta s \cdot \cos(\theta + \Delta\theta/2) \\ \Delta s \cdot \sin(\theta + \Delta\theta/2) \\ \Delta\theta \end{bmatrix} . \tag{10}$$

Where $\Delta s = (\Delta s_r + \Delta s_l)/2$ and $\Delta\theta = (\Delta s_r - \Delta s_l)/b$ and b is the width between two wheels. The control input $u(k) = [\Delta s_l, \Delta s_r]_k$ are the displacement of two wheels, which can be obtained from odometry.

4.2 Observation Model

The observation model describes how the measurements $z_j(k)$ are related to the vehicle's position and has the form:

$$z_j(k) = h(x(k), p_i) + w_j(k) , \quad w_j(k) \sim N(0, R_j(k)) . \tag{11}$$

The measurement function $h(x(k), p_i)$ expresses an observation $z_j(k)$ from the sensor to the target as a function of the robot's location $x(k)$ and the landmark coordinates p_i. $w_j(k)$ is a zero-mean, Gaussian measurement noise with covariance $R_j(k)$. Both the distance and orientation information measured by the sensors are relative to the robot itself, therefore the observation equation is the function of the robot state and the landmark:

$$h(\cdot,\cdot) = \begin{bmatrix} \sqrt{(p_x - x(k))^2 + (p_y - y(k))^2} \\ \arctan(\dfrac{p_y - y(k)}{p_x - x(k)}) - \theta(k) \end{bmatrix}. \tag{12}$$

4.3 Estimation Cycle

First, using the system model and control input $u(k)$, we predict the robot's new location at time step k+1:

$$\overline{x}(k+1) = \sum_{i=0}^{2n} W_i \cdot \overline{\chi}_i^{k+1}. \tag{13}$$

Where $\overline{\chi}_i^{k+1} = f(\hat{\chi}_i^k, u(k))$ in which W_i is the mean weight with respect to the $i-th$ sigma point $\overline{\chi}_i^{k+1}$ [3], and $\overline{x}(k+1)$ is the expectation of robot predicted pose. And the variance associated with this prediction is

$$P_{\overline{x}_{k+1}} = \sum_{i=0}^{2n} W_i (\overline{\chi}_i^{k+1} - \overline{x}(k+1))(\overline{\chi}_i^{k+1} - \overline{x}(k+1))^T. \tag{14}$$

Next, we use this predicted robot location to generate predicted observations of landmark p_j, $j = 1, \cdots, M$.

$$\overline{z}_j(k+1) = \sum_{i=0}^{2n} W_i \overline{\zeta}_{i,j}^{k+1}. \tag{15}$$

Where $\overline{\zeta}_{i,j}^{k+1} = h(\overline{\chi}_i^{k+1}, p_j)$.

The next step is to actually take a number of observations $z_j(k+1)$ of different landmarks and compare these with our predicted observations. The difference between a prediction $\overline{z}_j(k+1)$ and an observation $z_j(k+1)$ is termed the innovation, and it is written as $v_j(k+1) = z_j(k+1) - \overline{z}_j(k+1)$.

The variances of these innovations $S_j(k+1)$, can be estimated from

$$S_j(k+1) = \sum_{i=0}^{2n} W_i (\overline{\zeta}_{i,j}^{k+1} - \overline{z}(k+1))(\overline{\zeta}_{i,j}^{k+1} - \overline{z}(k+1))^T + R_j(k+1). \tag{16}$$

Where $R_j(k+1)$ is covariance of additive measurement noise for observations $z_j(k+1)$.

The innovations are obtained by matching observations to predicted targets. The matching criterion is given as validation regions in

$$v_j(k+1)S_j^{-1}(k+1)v_j^T(k+1) \le \chi^2 .$$ (17)

This equation is used to test each sensor observation $z_j(k+1)$ for membership in the validation gate for each predicted measurement. When a single observation falls in a validation gate, we get a successful match.

Finally, we use innovation $v_j(k+1)$ to update the vehicle location vector and associated variance. We then utilize the standard result that the Kalman gain can be written as

$$K_j^{k+1} = P_{\bar{x}_{k+1}\bar{z}_j^{k+1}} P_{\bar{z}_j^{k+1}}^{-1} .$$ (18)

Where

$$P_{\bar{z}_j^{k+1}} = \sum_{i=0}^{2n} W_i(\bar{\zeta}_{i,j}^{k+1} - \bar{z}_j(k+1))(\bar{\zeta}_{i,j}^{k+1} - \bar{z}_j(k+1))^T$$ (19)

$$P_{\bar{x}_{k+1}\bar{z}_j^{k+1}} = \sum_{i=0}^{2n} W_i(\bar{\chi}_i^{k+1} - \bar{x}(k+1))(\bar{\zeta}_{i,j}^{k+1} - \bar{z}_j(k+1))^T$$ (20)

Then the updated vehicle position estimate is

$$\hat{x}(k+1) = \bar{x}(k+1) + \sum_{i=0}^{M} K_{k+1} \cdot v_j(k+1)$$ (21)

With associated variance

$$P_{\hat{x}_{k+1}} = P_{\bar{x}_{k+1}} - \sum_{j=0}^{M} K_j^{k+1} \cdot P_{\bar{z}_j^{k+1}}^{-1} \cdot (K_j^{k+1})^T$$ (22)

5 Experimental Results

In this section, we use the mobile robot RoboSHU which is equipped with a laser rangefinder LMS200 as exteroceptive sensor to test the localization capability of our method. The rangefinder is configured to have a view of 180 degrees with a resolution of nearly 0.5°. The robot is running under a RTOS in a 1.6 GHz PC and can be controlled by a remote control module via wireless network. All Algorithms are programmed in C++. The average processing times per scan is 0.3s, being suitable for real time applications. The localization experiment is carried out in a laboratory environment (see Fig.4).

For simplicity, we assumed time-invariant, empirically estimated covariance matrices $Q(k)=Q$ and $R_j(k)=R$, $j=1,\cdots,M$, given by $R=diag(50,50,0.01)$, $Q(1,1)=Q(2,2)=10$, and $Q(1,2)=Q(2,1)=0$. The initial pose (x_0,y_0,θ_0) is assumed Gaussian with initial covariance matrices $P_0 = diag(100,100,0.025)$.

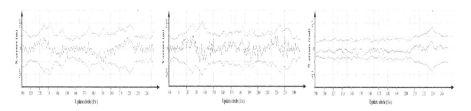

Fig. 3. Pose errors computed with the ground truth and the 1σ confidence bounds computed using the estimate covariance of system in x, y and θ are shown

Fig.3 shows the pose estimate errors of our UKF localization method, respectively in coordinates (x, y) and orientation θ. The 1σ confidence bounds for the estimates are also superimposed. It can be seen that the errors are not divergent.

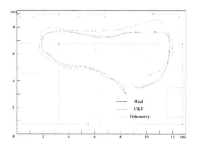

Fig. 4. Localization experiment: odometry(blue line), estimated trajectory(red line), real data(black line)

Fig.4 represents the trajectory estimation obtained using our localization method. The trajectory predicted by the odometry deviates from the true robot's path as time increases. The UKF is capable of tracking the robot's position fairly accurate, even though the trajectory includes abrupt turns. We can see that the robot can localize itself by fusing the sensor information so as to navigate successfully. This experiment proves the validity of the localization algorithm based on UKF.

6   Conclusions

This paper describes the implementation and obtained results of an UKF based Localization framework using a 2D laser rangefinder. A segmentation based on the variance of the local tangent value is introduced. This algorithm can provide line segments, corners and curve segments for mobile robot localization. Landmarks extracted from these segments are not only characterized by their parameters but also uncertainties. The system is based on an UKF that utilizes matches between observed geometric Landmarks and a priori map of Landmarks locations to provide optimal pose estimation for mobile robot. The accuracy and robustness of the proposed method was demonstrated in an indoor environment experiment.

Acknowledgement

We are grateful for the financial support from the State High-Tech Development Plan (863 program) of China under contract No. 2007AA041604.

References

1. Iyengar, S., Elfes, A.: Autonomous Mobile Robots, vol. 1, 2. IEEE Computer Society Press, Los Alamitos (1991)
2. Jensfelt, P., Christensen, H.: Laser Based Position Acquisition and Tracking in an Indoor Environment. In: IEEE Int. Proc. on Robotics and Automation, vol. 1 (1998)
3. Julier, S.J., Uhlmann, J.K., Durrant-Whyte, H.F.: A new approach for filtering nonlinear systems. In: Proc. Am. Contr. Conf., Seattle, WA, pp. 1628–1632 (1995)
4. Arras, K.O., Tomatis, N., Jensen, B.T., Siegwart, R.: Multisensor on-the-fly localization: Precisionand reliability for applications. Robotics and Autonomous Systems 34, 131–143 (2001)
5. Arras, K.O., Siegwart, R.: Feature extraction and scene interpretation for map based navigation and map building. In: Proc. of SPIE, Mobile Robotics XII, vol. 3210 (1997)
6. Nash, J.C.: Compact numerical methods for computers: linear algebra and function minimization. Adam Hilger Ltd. (1979)
7. Nunez, P., Vazquez-Martin, R., del Toro, J.C., Bandera, A., Sandoval, F.: A Curvature based Method to Extract Natural Landmarks for Mobile Robot Navigation. In: IEEE Int. Symposium on Intelligent Signal Processing 2007, October 2007, pp. 1–6 (2007)

Multi-Robot Dynamic Task Allocation Using Modified Ant Colony System

Zhenzhen Xu, Feng Xia, and Xianchao Zhang

School of Software, Dalian University of Technology, 116620 Dalian, China
njxzz@hotmail.com, dr.fxia@hotmail.com, xczhang@dlut.edu.cn

Abstract. This paper presents a dynamic task allocation algorithm for multiple robots to visit multiple targets. This algorithm is specifically designed for the environment where robots have dissimilar starting and ending locations. And the constraint of balancing the number of targets visited by each robot is considered. More importantly, this paper takes into account the dynamicity of multi-robot system and the obstacles in the environment. This problem is modeled as a constrained MTSP which can not be transformed to TSP and can not be solved by classical Ant Colony System (ACS). The Modified Ant Colony System (MACS) is presented to solve this problem and the unvisited targets are allocated to appropriate robots dynamically. The simulation results show that the output of the proposed algorithm can satisfy the constraints and dynamicity for the problem of multi-robot task allocation.

Keywords: multi-robot, task allocation, ant colony, MTSP.

1 Introduction

The multiple robots system has been widely used in the planetary exploration, seafloor survey, mine countermeasures, mapping, rescue and so on. A typical mission is visiting the targets locating on different positions in a certain area. Multi-robot task allocation (MRTA) is a critical problem in multi-robot system. This research introduces a dynamic task allocation algorithm for multiple robots to visit multiple targets.

The existing research on multi-robot task allocation for targets visiting is limited and immature. The waypoint reacquisition algorithm provides a computationally efficient task allocation method [1]. This approach is accomplished by employing a cluster-first, route-second heuristic technique with no feedback or iterations between the clustering and route building steps. However, it will generally not find the high-quality solutions. It also may leave some waypoints (*i.e.* targets) unvisited. The planning of underwater robot group is modeled as a MTSP and implemented by genetic algorithm [2]. The author of this paper has studied the multi-objective task allocation problem for multi-robot system in the literature [3]. However, all the proposed algorithms do not take into account the dynamicity of multi-robot system and the obstacles in the environment. Due to the dynamicity of multi-robot system, when some robot is damaged by a destructive target, the system must reallocate the targets unvisited to the remaining robots on-line. Meanwhile, each robot will change its path during the procedure of avoiding obstacles, thus the initial schedule of targets visiting may not be optimal and rescheduling should be considered.

H. Deng et al. (Eds.): AICI 2009, LNAI 5855, pp. 288–297, 2009.
© Springer-Verlag Berlin Heidelberg 2009

In this paper, the task allocation problem for multiple robots is modeled as a constrained MTSP. The MTSP is an NP-hard problem in combinatorial optimization. It is a generalization of the well-known TSP where more than one salesman is allowed to be used in the solution. Although there is a wide body of the literature for the TSP, the MTSP has not received the same amount of attention. To find a guaranteed optimal solution to the MTSP using exhaustive searches is only feasible for very small number of cities. Many of the proficient MTSP solution techniques are heuristic such as evolutionary algorithm [4], simulated annealing [5], tabu search [6], genetic algorithms [7], neural networks [8] and ant systems [9].

Due to the requirement of targets visiting for multiple robots, robots typically have dissimilar starting and ending locations and the workload of each robot should be balanced. This problem belongs to the fixed destination multi-depot MTSP [10] and the constraints of balancing the targets number visited by each vehicle should also be taken into account. However, the heuristic methods mentioned above can not deal with this constrained MTSP directly. The classical Ant Colony System (ACS) algorithm is used to solve MTSP which can be transformed to TSP widely. In ACS, each ant constructs a solution from a random starting point and visits all targets one by one. However, it can not solve the fixed destination multi-depot MTSP with the constraints mentioned above. In this paper, a new Modified Ant Colony System (MACS) algorithm is presented to solve this constrained MTSP and applied to the multi-robot dynamic task allocation problem.

The reminder of this paper is organized as follows. First, the multi-robot task allocation problem is formulated by integer linear programming formulation. In Section 3, we give a detail description of the Modified Ant Colony System (MACS). Multi-robot dynamic task allocation using MACS is introduced in section 4. Finally, simulation results show that the proposed algorithm can find optimized solutions which can satisfy the constraints and dynamicity for the problem of multi-robot dynamic task allocation.

2 Multi-Robot Task Allocation Problem Statement

2.1 Integer Linear Programming Formulation

The multi-robot task allocation problem can be stated as follows: Given n targets with random locations in a certain area, let there be m robots which typically have dissimilar starting and ending locations outside the area. The m robots must visit the n targets, and each target must be visited exactly once by only one robot. In order to saving energy and time, the total distance of visiting all targets must be minimized. In addition, the targets number visited by each robot should be average due to the requirement of workload balancing.

We propose the following integer linear programming formulation for the constrained MTSP defined above. The distance objective function $f(x)$ can be described as follows:

$$f(x) = \sum_{i=1}^{m} (\sum_{k=1}^{n_i-1} d(T_i^k, T_i^{k+1}) + d(S_i, T_i^1) + d(T_i^{n_i}, E_i)) \tag{1}$$

where T_i^k is the kth target visited by robot i, $d(T_i^k, T_i^{k+1})$ is the distance between T_i^k and T_i^{k+1}, S_i is the starting depot of robot i, E_i is the ending depot of robot i, n_i is the target number assigned to robot i.

The task number constraint $g(x)$ is defined as follows:

$$g(x) = \max(n_i) - \min(n_i) \in \{0,1\} \tag{2}$$

The constrained task allocation problem for multiple robots is formulated as follows:

$$\min f(x) \tag{3}$$

$$\text{subject to } g(x) \tag{4}$$

2.2 Colony Architecture of Multi-Robot System

The multi-robot system discussed in this paper is constructed by the hierarchical colony architecture [11]. Fig. 1 shows the logical hierarchical model of multi-robot system.

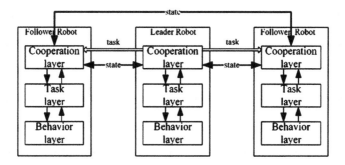

Fig. 1. The hierarchical colony architecture of multi-robot system

In the hierarchical colony architecture, a leader-follower organization structure is adopted. The leader robot supervises the whole system to enhance global coordination and supports the capability of self-organization in case of partial or total failure of one or several robots. Meanwhile, each robot is an autonomous agent and distributed in physics. It can autonomously execute individual tasks and communicate with other robots when necessary. Thus, the hierarchical colony architecture provides multi-robot system with a framework for interaction and coordination among member robots.

The leader robot can be dynamic chosen with any effective strategy. An index-based method is proposed here according to the predefined unique index of each robot. The robot with the index $i+1 (i \in \{1,2,\cdots,m-1\})$ is selected to be the leader when the former leader robot with the index i $(i \in \{1,2,\cdots,m-1\})$ is damaged and out of work.

3 Modified Ant Colony System

In this section, a Modified Ant Colony System (MACS) algorithm which is extended from the classical Ant Colony System (ACS) [12] for solving multi-robot task allocation is proposed. Some modifications are adopted at the steps of initialization and solution construction respectively.

3.1 Initialization

Due to the characteristic of fixed destination multi-depot MTSP, the MACS algorithm is different from the classical ACS in initialization. The dissimilar starting depots and ending depots of robots should be considered. Instead of putting all ants on the targets randomly in ACS, the MACS put all ants on the starting depots or ending depots of robots randomly. That is, we put all ants on the $S_i (i = 1,2,\cdots,m)$ or $E_i (i = 1,2,\cdots,m)$ randomly. All ants will start from one depot and search at the same space.

Similarly, initialization of the pheromone matrixes and cost matrixes of ant colony are modified. The pheromone and the cost from one depot to all targets should be calculated and stored.

3.2 Solution Construction

As far as the multiple depots and workload balancing are concerned, the method of solution construction needs to be modified. According to the theory of classical ACS, each ant runs the steps of solution construction and pheromone updating until all targets have been visited. In MACS, there are three modifications during the procedure of solution construction.

(1) The task number allocation phase is employed to realize the task number constraint. Here, the task number is equal to the number of targets assigned to each robot to visit. This phase focuses on assigning the number of task while the task allocation and optimization will be performed at the following phase. If the number of targets n can be divided exactly by the number of robots m, the task number n_i for robot $i(i = 1,2,\cdots m)$ should be n/m. Otherwise, let s be the quotient and r be the remainder, then n_i can be defined mathematically as:

$$n_i = \begin{cases} s+1, & if\ (u > u_0\ and\ v < r)\ or (u \le u_0\ and\ w \ge m-r) \\ s, & if\ (u \le u_0\ and\ w < m-r)\ or (u > u_0\ and\ v \ge r) \end{cases} \tag{5}$$

where u is a random uniform variable $[0,1]$ and the value u_0 is a parameter to influence the target number of different robots. v denotes the total number of robots whose target number is $s+1$ and w denotes the total number of robots whose target number is s. After this phase, the task number for each robot to visit is clear and we can see $\sum_{i=1}^{m} n_i = n$.

(2) Each ant starts from the initial position, and then selects unvisited targets to construct a route. In MACS, the procedure of solution construction is a little complex compared with the ACS. Suppose that an ant starts from $S_p, p \in \{1,2,\cdots,m\}$, if the number of targets which the ant has visited is equal to n_p, then the ant returns to E_p. After that, the ant chooses an unvisited $S_i (i = 1,2,\cdots p-1, p+1,\cdots,m)$, departs from that starting depot and selected unvisited targets iteratively. The targets which have been visited by an ant are recorded in the ant's tabu table. A valid solute is constructed by an ant until all the targets, starting depots and ending depots are visited. We define the ant number is a, thus there are a solutions constructed by a ants.

Every ant selects the next city independently. The rules of moving from target i to target j for ant k can be formulated as follows:

$$j = \begin{cases} \arg \max_{l \in N_i^k} \left\{ \tau_{il} [\eta_{il}]^\beta \right\} & \text{if } q \le q_0 \\ J, & \text{otherwise} \end{cases} \tag{6}$$

If q is larger than q_0, the probability of moving from target i to target j for ant k can be formulated as follows:

$$p_{ij}^k = \frac{[\tau_{ij}]^\alpha [\eta_{ij}]^\beta}{\sum_{l \in N_i^k} [\tau_{il}]^\alpha [\eta_{il}]^\beta}, \quad \text{if } j \in N_i^k \tag{7}$$

where τ_{il} is equal to the amount of pheromone on the path between the current target i and possible target l. η_{il} is the heuristic information which is defined as the inverse of the cost (e.g. distance) between the two targets. q is a random uniform variable $[0,1]$ and the value q_0 is a parameter to determine the relative influence between exploitation and exploration. α and β are the parameters whose value determine the relation between pheromone and heuristic information. N_i^k denotes the targets which ant k has not visited.

(3) The objective function values of the routes constructed by all ants is computed according to Eq.1 and sorted in increasing order. In order to improve future solutions, the pheromone on the best solution found so far must be updated after all ants have constructed their valid routes. The global pheromone updating is done using the following equation:

$$\tau_{ij} = (1 - \rho)\tau_{ij} + \rho \Delta \tau_{ij}^{best}, \quad \forall (i,j) \in T^{best} \tag{8}$$

where $\rho(0 < \rho \le 1)$ is a parameter that controls the speed of global pheromone evaporation. T^{best} denotes the best solution found so far, and $\Delta \tau_{ij}^{best}$ is defined as the inverse of the objective value of T^{best}. In addition, each ant should execute local pheromone updating as soon as it selects a new target j according to the following equation:

$$\tau_{ij} = (1 - \xi)\tau_{ij} + \xi\tau_0 \tag{9}$$

$$\tau_0 = 1 / nC^{nn} \tag{10}$$

where $\xi(0 < \xi < 1)$ represents the speed of local pheromone evaporation. τ_0 is the initial value of pheromone. C^{nn} is the objective value computed with the Nearest Neighbor Algorithm. The global updating belongs to a positive feedback mechanism, while the local updating can increase the opportunity to choose the targets which have not been explored and avoid falling into the state of stagnation.

After the MACS have met the end condition (*e.g.* the pre-specified number of iterations, denoted as N_c), the optimal solution is obtained. The time complexity of MACS is $O(N_c \cdot a \cdot n^2)$ which is equal to the time complexity of the classical ACS.

4 Multi-Robot Dynamic Task Allocation Using MACS

Multi-robot dynamic task allocation consists of two phases: initial task allocation and online task reallocation. The MACS is a general method and can be applied in both phases. The corresponding optimal results will be obtained with appropriate input data at different conditions.

4.1 Initial Task Allocation

In the first phase, the MACS algorithm is used to solve constrained MTSP and all targets are allocated to appropriate robots initially. The MACS is carried out with the program language of c and embedded in the control system software of each robot. Initial task allocation should be run by a leader robot with centralized mode. The result of allocation will be sent to each robot in the multi-robot system by communication.

4.2 Online Task Reallocation

In the second phase, the robots set off to visit individual targets and run local path planning algorithm to avoid obstacles.

Task reallocation with centralized mode. The leader robot is responsible for colony supervision online to conduct failures. All the follower robots will send state messages to leader robot whenever a timer signal arrived during the procedure of executing tasks. If one robot has been found losing touch with the leader robot. Then it is considered encountering damage and lost. In order to avoid leaving some targets unvisited, the leader robot will reallocate the tasks unvisited to other remaining robots with centralized mode. The MACS algorithm only runs in the control system of the leader robot at this situation. The inputs of algorithm are the current positions of remaining robots and the positions of targets unvisited. The output of task reallocation will be sent to the remaining robots. Each robot which receives this message will update its task list and visit the new target.

Task rescheduling with distributed mode. In order to adapt the unpredictable environment, each robot will reschedule its targets list during the procedure of avoiding obstacles with distributed mode. The MACS algorithm runs in the control system of each robot at this situation. The inputs of algorithm are the current position of the robot avoiding obstacles and the positions of targets unvisited by this robot. The result of task rescheduling is the targets list with new order which is probably different with the former order of visiting.

5 Simulation Experiments

In this section, we discuss the parameter settings for the proposed task allocation algorithm and present simulation results. According to the MACS algorithm, the parameters are set as follows: $a = n$, $\alpha = 1$, $\beta = 2$, $q_0 = 0.9$, $u_0 = 0.5$, $\rho = 0.5$, $\xi = 0.1$, $N_c = 200$. The algorithm was tested in the scenario which contains 20 targets ($n = 20$) and 3 robots ($m = 3$). The targets positions and the obstacles are generated randomly. Assumed that the respective starting and ending depots coordinates of the three robots are $S_1 = (20,0)$, $S_2 = (50,0)$, $S_3 = (80,0)$, $E_1 = (40,100)$, $E_2 = (50,100)$, $E_3 = (60,100)$. Two typical experiments were carried out to verify the validity and rationality of the proposed task allocation algorithm.

Fig. 2 shows the results of experiment A which focuses on verifying task reallocation with centralized mode. The targets layout and the result of initial task allocation are shown in Fig. 2 a). As we have seen, each robot obtains a valid route, and the task number constraint is satisfied. The leader robot is R_1 and it supervises the whole multi-robot system.

In Fig. 2 b), the robot R_2 is damaged and out of work, then the leader robot R_1 detects this failure. The MACS is applied to task reallocation by robot R_1 with the input of coordinates of targets unvisited and current positions of remaining robot (R_1 and R_3). The result of online task reallocation is illustrated in Fig. 2 c). The unvisited targets are allocated to R1 and R3 autonomously.

The coordinates of targets in this experiment are shown in Table 1. However, the process of targets visiting with multi-robot system is unpredictable and the layout of targets is randomly. There are many other simulation experiments with various layouts of targets, and the similar valid solutions are obtained. This experiment results show that the leader robot can perform online task reallocation with centralized mode using MACS algorithm.

Experiment B focuses on the procedure of task rescheduling with distributed mode. Assumed that there are two obstacles with the figure of square in the area, the layouts of targets and obstacles can be seen in fig. 3 a). The two obstacles have the size of 15m×10m and 20m×5m respectively. The coordinates of targets in experiment B are shown in Table 1. According to the initial task allocation, the paths of robot R_2 and R_3 are across the obstacles.

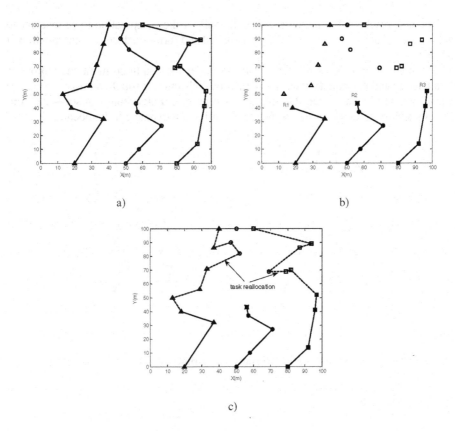

a)   b)

c)

Fig. 2. Results in experiment A

R_2 or R_3 will run local path planning and avoid the obstacles as soon as it detects the obstacle during the procedure of target visiting. Meanwhile, each robot independently runs MACS online to optimize the schedule of targets visiting with the input of its current position and the coordinates of targets unvisited in its task list. The results of task rescheduling of R_2 and R_3 are shown in Fig. 3 b). The travel distance of robot according to the new visiting order of targets is reduced obviously.

Table 1. Coordinates of targets

Experiment	Coordinates
A	{52,82}, {18,40}, {47,90}, {37,32}, {37,86}, {79,69}, {57,37}, {69,69}, {58,10}, {82,70}, {13,50}, {56,43}, {96,41}, {87,86}, {97,52}, {94,89}, {71,27}, {29,56}, {92,14}, {33,71}
B	{28,22}, {59,74}, {46,75}, {47,70}, {77,55}, {42,44}, {67,38}, {66,69}, {35,60}, {21,60}, {20,51}, {48,46}, {60,66}, {15,19}, {37,26}, {63,63}, {22,30}, {50,69}, {17,43}, {61,88}

The task rescheduling method allows the bulk of the computation to be distributed among the robots on their respectively targets list and further reduces the computational burden.

The computing time of MACS algorithm under the condition mentioned above is less than 1s and can satisfy the requirement of real-time for robot control system. The simulation results show that the output of the proposed algorithm can solve the constraints and dynamicity for the problem of multi-robot dynamic task allocation.

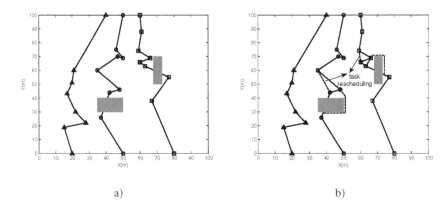

a)   b)

Fig. 3. Results in experiment B

6   Conclusion

In this paper, we considered the situation which is more appropriate in the real multi-robot task allocation problem for task visiting. Besides the objective of minimizing the total distance of multiple robots, we also took into account the constraints of multiple depots and workload balancing.

The task allocation problem for multiple robots was modeled as a constrained MTSP. This is also a fixed destination multi-depot MTSP. As many proficient heuristic methods to solve MTSP can not deal with such problem directly, this research proposed Modified Ant Colony System (MACS) method to solve this constrained MTSP.

The proposed MACS algorithm was used to solve the dynamic task allocation for multiple robots. Combined with the hierarchical colony architecture and the dynamicity of multi-robot system, the MACS algorithm was not only applied to reallocate targets unvisited with centralized mode when some robot encountered damage but also carried out to reschedule individual targets list with distributed mode when some robot avoided obstacles. This research made the first attempt to solve the dynamic task allocation problem.

The simulation results show that the output of the proposed MACS algorithm can not only solve the constraints of multi-robot task allocation problem, but also satisfy the requirement of dynamicity for the problem of online task reallocation.

References

1. Stack, J.R., Smith, C.M., Hyland, J.C.: Efficient reacquisition path planning for multiple autonomous underwater robots. In: Ocean 2004 - MTS/IEEE Techno-Ocean 2004 Bridges across the Oceans, pp. 1564–1569 (2004)
2. Zhong, Y., Gu, G.C., Zhang, R.B.: New way of path planning for underwater robot group. Journal of Harbin Engineering University 24(2), 166–169 (2003)
3. Xu, Z.Z., Li, Y.P., Feng, X.S.: Constrained Multi-objective Task Assignment for UUVs using Multiple Ant Colonies System. In: The 2008 ISECS International Colloquium on Computing, Communication, Control, and Management, pp. 462–466 (2008)
4. Fogel, D.B.: A parallel processing approach to a multiple traveling salesman problem using evolutionary programming. In: Proceedings of the fourth annual symposium on parallel processing, pp. 318–326 (1990)
5. Song, C., Lee, K., Lee, W.D.: Extended simulated annealing for augmented TSP and multi-salesmen TSP. In: Proceedings of the international joint conference on neural networks, vol. 3, pp. 2340–2343 (2003)
6. Ryan, J.L., Bailey, T.G., Moore, J.T., Carlton, W.B.: Reactive Tabu search in unmanned aerial reconnaissance simulations. In: Proceedings of the 1998 winter simulation conference, vol. 1, pp. 873–879 (1998)
7. Tang, L., Liu, J., Rong, A., Yang, Z.: A multiple traveling salesman problem model for hot rolling scheduling in Shangai Baoshan Iron & Steel Complex. European Journal of Operational Research 124, 267–282 (2000)
8. Modares, A., Somhom, S., Enkawa, T.: A self-organizing neural network approach for multiple traveling salesman and robot routing problems. International Transactions in Operational Research 6, 591–606 (1999)
9. Pan, J.J., Wang, D.W.: An ant colony optimization algorithm for multiple traveling salesman problem. In: Proceedings of the first international conference on innovative computing, information and control (2006)
10. Kara, I., Bektas, T.: Integer linear programming formulations of multiple salesman problems and its variations. European Journal of Operational Research, 1449–1458 (2006)
11. Xu, Z.Z., Li, Y.P., Feng, X.S.: A Hierarchical control system for heterogeneous multiple uuv cooperation task. Robot 30(2), 155–159 (2008)
12. Dorigo, M., Gambardella, L.M.: Ant colony system: A cooperative learning approach to the traveling salesman problem. IEEE Transactions on Evolutionary Computation 1(1), 53–66 (1997)

A Novel Character Recognition Algorithm Based on Hidden Markov Models

Yu Wang, Xueye Wei, Lei Han, and Xiaojin Wu

School of Electronics and Information Engineering, Beijing JiaoTong University,
Beijing 100044, China
bjtuwangyu@126.com

Abstract. In this paper, a novel character recognition algorithm based on the Hidden Markov Models is proposed. Several typical character features are extracted from every character being recognized. A novel 1D multiple Hidden Markov models is constructed based on the features to recognize characters. A large number of vehicle license plate characters are used to test the performance of the algorithm. Experimental results prove that the recognition rate of this algorithm is high aiming at different kinds of character.

Keywords: Character recognition, Hidden Markov Models.

1 Introduction

The optical character recognition (OCR) has become an important research orientation in modern computer application domain. The character has many applications, such as the vehicle license plate character recognition, information retrieval and understanding for video images, the character recognition for bills and documents and etc. In the past years, many researches have been devoted to character recognition. Independent component analysis character recognition in [1] recognizes characters using a target function to reconstruct the character being recognized and analyzing the errors; character recognition based on neural network in [2,3] recognize characters with strong anti interference performance and brief program; character recognition based on nonlinear active shape models in [4] has good performance when the character is nonrigid and with normal variation rules. Actually, there are lots of different uncertain information in character recognition such as the variable character shapes, the complex character pattern, so the algorithms in [1-4] have respective limitations in their applications.

It is a good method to adopt the statistical model building in character recognition because of the uncertain information. The Hidden Markov Models (HMMs) has superiority performance to process the serialization dynamic non-stationary signal. It has been widely used in speech recognition [5]. The HMMs has great potential in the character recognition because it has better flexibility in variable pattern processing. One-dimension (1D) HMMs is used to recognize character in [6]. A two-dimension (2D) HMMs is used in [7]. Because a 2DHMMs has a heavier computation than the 1DHMMs, a pseudo 2DHMMs with less computation is applied to recognize characters in [8].

H. Deng et al. (Eds.): AICI 2009, LNAI 5855, pp. 298–305, 2009.

The algorithm proposed in this paper is based on the Freeman chain code. It extracts multiple character features from each character. A 1D-multiple-HMMs is built according these features to recognize characters. A large number of vehicle license plate characters are used to test the performance of the algorithm. The experimental results indicate this algorithm has a good recognition performance aiming at different kinds of character.

2 Hidden Markov Models

A hidden Markov model (HMM) is a statistical model in which the system being modeled is assumed to be a Markov process with unobserved state. The Hidden Markov Model is a finite set of states, each of which is associated with a (generally multidimensional) probability distribution. Transitions among the states are governed by a set of probabilities called transition probabilities. In a particular state an outcome or observation can be generated, according to the associated probability distribution. It is only the outcome, not the state visible to an external observer and therefore states are ``hidden" to the outside; hence the name Hidden Markov Model.

An HMM can be defined as follows:

$$\lambda = (\pi, A, B) \tag{1}$$

where π is the initial state distribution, A is the set of state transition probabilities and B is the probability distribution in each of the states.

Character recognition involves two basic problems of interest in HMM: the learning problem (construct models) and the evaluation problem (recognize character).

The learning problem is how to adjust the HMM parameters, so that the given set of observations (called the training set) is represented by the model in the best way for the intended application. Thus it would be clear that the ``quantity" we wish to optimize during the learning process can be different from application to application. In other words there may be several optimization criteria for learning, out of which a suitable one is selected depending on the application. This problem can be solved by Baum-Welch algorithm in [5].

The evaluation problem is how to recognize characters by HMM. Give the HMM model λ and the observation sequence $X = (X_1, X_2, ..., X_T)$ to calculate the probability of $P = (X \mid \lambda)$. This problem can be solved by forward-backward algorithm in [5].

3 Model Building

This proposed algorithm is based on the Freeman chain code. Several different features of every character are extracted. Slope analysis method is used to extract chain code of each feature than has been projected. These extracted chain code are used as observation sequence to establish model. Freeman 8-direction chain code is shown in figure 1.

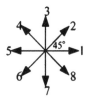

Fig. 1. Freeman 8-direction chain code

3.1 Model Initialization

The number of observation symbols is $M = 8$ because the values of Freeman 8-direction chain code are $(1,2,3,4,5,6,7,8)$ which are used as observation sequence. Considering many redundancy situations that exist in some characters, a typical leap type left-right 8-states Markov chain model is selected as the basic structure of every character. This Markov chain is shown is figure 2.

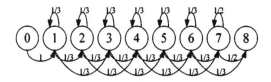

Fig. 2. Left-right 8-states Markov chain model

Each state in the model has a self-circulation process to absorb redundancy strokes. Each state can transfer to the first and the second state behind it to describe the wanting strokes of characters. The initial state distribution π and the set of state transition probabilities A can be defined by this model.

We suppose the probability of different symbols in observation sequence is the same, so observation probability in each of the states can be initialized as $1/M$.

3.2 Character Features Extraction

3.2.1 Horizontal and Vertical Projection Chain Code

Some characters are simple, others are complex. Characters' vertical direction is complex which has lots of horizontal strokes. Characters' horizontal direction is complex which has lots vertical strokes. The sole horizontal projection and the sole vertical projection can not well reflect all features of character individually. Horizontal and vertical projection should be extract separately to express features of characters. The projection modes are shown in figure 3.

After the extraction of horizontal and vertical projection image, Freeman chain codes of projection image contour can be extracted utilizing the slope method [9]

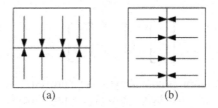

Fig. 3. Horizontal and Vertical Projection mode

(a)   (b)

Fig. 4. Character A and its horizontal projection contour

from the left of the contour. These chain codes can be used qua the horizontal and vertical features observation sequence. A typical 16×16 character ``A`` and its horizontal projection contour image are shown in figure 4. Its chain code is (122...63). In the similar way, vertical projection contour chain codes can be got.

3.2.2 Target Pixels' Centroid Distance Projection Chain Code

Many characters have translations, rotation and scale variation in different extent affine variations, so how to extract appropriate features for computer recognition is hard. In this paper, a novel character feature extraction method, the binary image character centroid distance projection, is proposed. The contour image of character centroid distance projection can be used to extract chain codes to express the observation sequence of character features. This method can express character features although the character is in the affine variations.

Suppose that the origin of image $f(x,y)$ is in the centroid position (a,b) and the image $f(x,y)$ is transformed from image $f(x',y')$ in the transformation $x = x' - a$, $y = y' - b$. The coordinate of centroid (a,b) can be denoted as:

$$a = \frac{\iint x' f(x',y')dx'dy'}{\iint f(x',y')dx'dy'} \qquad (2)$$

$$b = \frac{\iint y' f(x',y')dx'dy'}{\iint f(x',y')dx'dy'} \qquad (3)$$

An image can be got from the image $f(x,y)$ after the affine variations

$$\begin{bmatrix} x_{\bullet} \\ y_{\bullet} \end{bmatrix} = A \begin{bmatrix} x \\ y \end{bmatrix} + \begin{bmatrix} d_1 \\ d_2 \end{bmatrix} \tag{4}$$

The image centroid position is not change can be proved.

PROOF: expression (5) can be got from the expression (4)

$$\begin{bmatrix} x_{\bullet} - d_1 \\ y_{\bullet} - d_2 \end{bmatrix} = A \begin{bmatrix} x \\ y \end{bmatrix} \tag{5}$$

Translations have no influence to the centroid position, so the translations' influence can be neglected. The centroid of image after affine variations can be expressed as:

$$a^{\bullet} = \frac{\iint x f(x_a^{\bullet}, y_b^{\bullet}) dx dy}{\iint f(x_a^{\bullet}, y_b^{\bullet}) dx dy} \tag{6}$$

$$b^{\bullet} = \frac{\iint y f(x_a^{\bullet}, y_b^{\bullet}) dx dy}{\iint f(x_a^{\bullet}, y_b^{\bullet}) dx dy} \tag{7}$$

Let $\begin{bmatrix} x_{\bullet} \\ y_{\bullet} \end{bmatrix} = \begin{bmatrix} x_{\bullet} - d_1 \\ y_{\bullet} - d_2 \end{bmatrix}$, so

$$\begin{bmatrix} x \\ y \end{bmatrix} = A^{-1} \begin{bmatrix} x_{\bullet} \\ y_{\bullet} \end{bmatrix} = \begin{bmatrix} k_{11} & k_{12} \\ k_{21} & k_{22} \end{bmatrix} \begin{bmatrix} x_{\bullet} \\ y_{\bullet} \end{bmatrix} \tag{8}$$

Hence

$$\begin{aligned} a^{\bullet} &= \frac{\iint x f(x_a^{\bullet}, y_b^{\bullet}) dx dy}{\iint f(x_a^{\bullet}, y_b^{\bullet}) dx dy} \\ &= \frac{\iint (k_{11} x_a^{\bullet} + k_{12} y_b^{\bullet}) f(x_a^{\bullet}, y_b^{\bullet}) dx_a^{\bullet} dy_b^{\bullet}}{\iint f(x_a^{\bullet}, y_b^{\bullet}) dx_a^{\bullet} dy_b^{\bullet}} \\ &= k_{11} \frac{\iint x_a^{\bullet} f(x_a^{\bullet}, y_b^{\bullet}) dx_a^{\bullet} dy_b^{\bullet}}{\iint f(x_a^{\bullet}, y_b^{\bullet}) dx_a^{\bullet} dy_b^{\bullet}} \\ &+ k_{12} \frac{\iint y_b^{\bullet} f(x_a^{\bullet}, y_b^{\bullet}) dx_a^{\bullet} dy_b^{\bullet}}{\iint f(x_a^{\bullet}, y_b^{\bullet}) dx_a^{\bullet} dy_b^{\bullet}} \\ &= 0 \end{aligned} \tag{9}$$

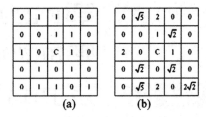

Fig. 5. 5×5 window and targets' centroid distance

In a similar way $b' = 0$.

From the expressions above, we can get that the centroid position of image after affine variations is not changed. So the character target pixels' centroid distance expresses the features of character. Figure 5 shows the window 5×5 with centroid as the origin and the target pixels' centroid distance image.

3.3 Model Building

An observation sequence set for every character $X = \{X_1, X_2, X_3\}$ and sub-models for different character features λ_1, λ_2 and λ_3 can be got from these character features Freeman chain codes. The system model $\lambda = \{\lambda_1, \lambda_2, \lambda_3\}$ can be got. Each system $\lambda = \{\lambda_1, \lambda_2, \lambda_3\}$ corresponding to the character is saved in database for recognition.

4   Character Recognition

The character optimal model $\lambda = \{\lambda_1, \lambda_2, \lambda_3\}$ can recognize characters after parameters estimation. Three observation sequences can be got for each character being recognized. In character recognition, forward-backward algorithm is used to calculate the probability of each sub-model according to the corresponding observation sequence. Three values of probability P_1, P_2 and P_3 where $P_1 = P(X_1 | \lambda_1)$, $P_2 = P(X_2 | \lambda_2)$, $P_3 = P(X_1 | \lambda_1)$. So the probability P of each character being recognized for each character model in database using the weighted method.

$$P = P(X | \lambda) = w_1 P_1 + w_2 P_2 + w_3 P_3 \tag{10}$$

where $w_1 = w_2 = w_3 = 1/3$. The character model in database with the biggest probability P is the character recognized.

5   Experiment Results Analysis

In order to testify the character recognition performance of the proposed algorithm, a lot of vehicle license plate characters are used as the experimental images. All the

Fig. 6. Experimental plates and the characters recognition results

experimental character images have been preprocessed: noise removal, image binarization and character normalization. Some vehicle license plate used for experiment and the characters recognition result are shown in figure 6. Results of recognition for different kinds of characters are shown in table 1.

Tabel 1. Results of recognition for different kinds of characters

	Characters	Recognized	Recognition Rate
Chinese character	197	180	91.3%
Alphabet	366	335	91.5%
Numeral	816	770	94.3%
Total	1379	1285	93.1%

The results in table 1 show that the proposed algorithm has a high recognition rate although there are three different kinds of characters and the results in figure 6 show that characters can be well recognized no matter the characters are inclined or fouled.

6   Conclusion

The character recognition algorithm based on the HMM is a worthy studying subject. In this paper, a recognition algorithm based on the Freeman chain code and HMM is proposed. A 1D-multiple-HMMs is built according multiple features which are extracted from the character to recognize characters. A large number of vehicle license plate characters are used to test the performance of the algorithm. The experimental results show that this proposed algorithm has a well performance in recognition for different kinds of characters.

References

[1] Min, L., Wei, W., Xiaomin, Y., et al.: Independent component analysis based on licence plate recognition. Journal of Sichuan University 43(6), 1259–1264 (2006)

[2] Qingxiong, Y.: Realization of character recognition based on neural network. Information Technology (4), 92–95 (2005)

[3] Tindall, T.W.: Application of neural network technique to automatic license plate recognition. In: Proceedings of European Covention on Security and Detection, Brighton, Enaland, pp. 81–85 (1995)

[4] Shi Da-ming Gunn, S.R., Damper, R.I.: Handwritten Chinese radical recognition using nonlinear active shape models. IEEE Trans. on Pattern Analysis and Machine Intelligence 25(2), 277–280 (2003)

[5] Rabiner, L.R.: A tutorial on hidden Markov models and select applications in speech recognition. Proc. of IEEE 77(2), 257–286 (1989)

[6] Awaidah, S.M., Mahmoud, S.A.: A multiple feature/resolution scheme to Arabic (Indian) numerals recognition using hidden Markov models. Signal Processing 89(6), 1176–1184 (2009)

[7] Yin, Z., Yunhe, P.: 2D-HMM based character recognition of the engineering drawing. Journal of Computer Aided Design and Computer Graphics 11(5), 403–406 (1999)

[8] Li, J., Wang, J., Zhao, Y., Yang, Z.: A new approach for off-line handwritten Chinese character recognition using self-adaptive HMM. In: The Fifth World Congress on Intelligent Control and Automation, 2004 (WCICA 2004), vol. 5, pp. 4165–4168 (2004)

[9] Gang, L., Honggang, Z., Jun, G.: Application of HMM in Handwritten Digit OCR. Journal of Computer Research and Development 40(8), 1252–1256 (2003)

[10] Keou, S., Fenggang, H., Xiaoting, L.: A fast algorithm for searching and tracking object centroids in binary images. Pattern Recognition and Artificial Intelligence 11(2), 161–168 (1988)

New Algorithms for Complex Fiber Image Recognition

Yan Ma and Shun-bao Li

Shanghai Normal University, Dept. of Computer Science, No. 100 GuiLin Road,
200234 Shanghai, China
{ma-yan,lsb}@shnu.edu.cn

Abstract. This paper presents an automatic recognition approach for complex fiber images, including the binarization method based on fiber boundary continuity and variance, the corner detection algorithm based on chain codes, the recognition method based on the curvature similarity in the same fiber. The experimental results show that the most fibers can be recognized by using the proposed automatic recognition methods.

Keywords: Fiber image, Chain code, Binarization, Recognition.

1 Introduction

At present, many approaches have been presented for the fiber image recognition using digital image processing technology. The processes of the fiber image analysis and recognition are: image preprocessing, image analysis and image recognition. It was presented to process data by the principal factor analysis and create experts library[1]. In [2], the method was designed to extract the image fiber edge, and use the curve fitting to make a quantitative analysis of the fiber. The gray value morphological method was given to process the fiber image[3].

The most above methods are designed for simple fiber image, that is, there is only 1-2 fibers and less cross-points in the image. In this paper, the fiber image being recognized is complex, that is, they are different in size, cross-cutting and adhesive in different fibers. This paper presents the binarization method based on fiber boundaries continuity and variance, thinning and pruning for binary fiber image, the corner detection algorithm based on chain codes, the recognition method based on the curvature similarity in the same fiber. The experimental results show that the better results has been obtained by using the proposed automatic recognition methods.

2 Image Preprocessing

2.1 Image Binarization

The input gray image is taken by melectron microscope (As shown in Fig. 1). The fibers are mutual adhesive with different intensity from Fig.1. If we use the traditional differential operators, such as Sobel operator, Canny operator for edge detection, the results are sensitive to the threshold. When selecting a big threshold, high-contrast edge can be detected, and low-contrast edge may be neglected. When selecting a small

H. Deng et al. (Eds.): AICI 2009, LNAI 5855, pp. 306–312, 2009.

threshold, the small stripes may be detected and it is be not beneficial for the following recognition.

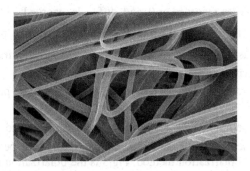

Fig. 1. The original fiber image

The fiber image has the characteristics that the boundary variance is big and non-boundary variance is small. We present the edge detection method based on the variance and reduce the threshold based on the continuity of the fiber boundary until all edge has been detected. Fig.2 is the variance distribution along horizontal direction in the period of Fig.1, x represents the pixel location, y respresents the variance. If we encounter the valley point B while detecting, the variance of B is recorded as Var_B, and the peak point belong to the neighborhood of B is further searched. If we encounter the peak point A in Fig.2, the variance of A is recorded as Var_A. The initial threshold is defined as Th_0. If $Var_A - Var_B > Th_0$, the point A is considered as boundary point and recorded as 1, otherwise recorded as 0. While taking a big Th_0, other peak points with small variance, such as point C, D and E can not be detected.

It is found through out experiments that the distribution of variance of the same fiber boundary has greater difference in the influence of outside light and the distance between the fiber and electron microscopy. But the variance for the local adjacent boundary is related and the image binarization method based on variance and boundary continuity is presented:

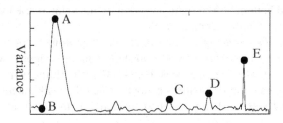

Fig. 2. The variance distribution

Step 1: Define the initial threshold Th_0 ; While the difference of variance $diff_{var}$ between the peak point and the valley point is greater than Th_0, the peak point is recorded as 1, otherwise recorded as 0, and $diff_{var}$ is recorded;

Step 2: Define $step$ as step. Define $Th_0 = Th_0 - step$. The peak point, satisfied that it is recorded as 0, $diff_{var} < Th_0$, and at least one point in its 8-point neighhood has been recorded as 1, is recorded as 1;

Step 3: Return to Step 2, until $Th_0 < Th_{min}$, Th_{min} is the predefined minimum value.

Through our experiments, Th_0 , $step$ and Th_{min} can be set to 20, 1 and 5 respectively. The experimental results show that we can receive the better results by using the proposed image binarization method. Fig.3 is the binarizaiton result for the image shown in Fig.1.In some cases, it is the Contact Volume Editor that checks all the pdfs. In such cases, the authors are not involved in the checking phase.

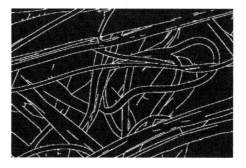

Fig. 3. Binary image

2.2 Thinning and Pruning

In this paper, the binary image is thinned using the look-up table thinning algorithm proposed in the literature [8]. The 8-neighbor look-up table, in which the value of 8-neighbor is corresponding with their value, is predefined. The binary image is scanned from up to bottom, left to right, according to the value in the look-up table to determine whether to delete the point.

The fiber image after thinning will have burr and the wrong recognized fiber may occur due to the burr. The deburr method is to start from the endpoint, walk along the connected pixels and calculate the walk length. The algorithm will continue until encountering the cross point or another endpoint. When the walk length is lower than the threshold (usually set to 10), it can be considered as burr and the values of the walking through pixels are set to 0.

After binarization, thinning and deburr in the image process, it will occur that one continuous fiber is segmented into two parts wrongly. So the filling algorithm is required to reduce measure error. In this paper, the angle between adjacent fiber endpoints is calculated and the filling algorithm proposed in the literature [9] is fulfilled.

2.3 Corner Detection

The thinned, pruned fiber image may be crossed and adhesive in different fibers, and that will occur the wrong connecting point, corner, which should be deleted in order to ensure the accuracy of recognition.

In this paper, a fiber from the starting point to terminal point is assigned 8 neighbor chain codes. The chain code values are 0-7 shown in Fig.4.

Fig. 4. The value of chain codes

Define a chain code as the array $c = \{c_1, c_2, \cdots, c_n\}$ (in this code only consider the chain code values, regardless of the starting point coordinates). Define the difference between the adjacent chain code values in the array as d_i (that is, the relative chain code), d_i is calculated using the following equation:

$$d_i = \begin{cases} c_{i+1} - c_i & |c_{i+1} - c_i| \le 2 \\ c_{i+1} - c_i - 8 & c_{i+1} - c_i \ge 6 \\ c_{i+1} - c_i + 8 & c_{i+1} - c_i \le -6 \end{cases} \tag{1}$$

Since that the largest difference between the directions represented by the adjacent chain codes is $\pm 90°$, the possible values for d_i is 0, ± 1, ± 2. If the value is zero, the direction of fiber is unchanging. If positive, it turns right. If negative, it turns left. The point with $|d_i|$ of 2 may be considered as corner in the theory. But it may not be the real corner due to the noise in the fiber image. To address the problem, the corner detection algorithm based on chain code is given as follows:

(1) Calculate the difference d_i between adjacent chain code values according to formula (2), the difference between equation (2) and equation (1) is that the difference d_i between the values of the adjacent chain codes separated by $length1$ ($length1$ can be set to about 20 based on our experiments), which can avoid the impact of local noise in the fiber image.

$$d_i = \begin{cases} c_{i+length1} - c_i & |c_{i+length1} - c_i| \le 2 \\ c_{i+length1} - c_i - 8 & c_{i+length1} - c_i \ge 6 \\ c_{i+length1} - c_i + 8 & c_{i+length1} - c_i \le -6 \end{cases} \tag{2}$$

(2) Once $|d_i| = 2$, define $n_1 = n\Big|_{|d_{i+j}|=2} j = 1,2,\cdots n$ in the predefined length n, that is, n_1 is the number of $|d_i|$ of 2 among the following n points after i. Assume s be the scale value. While $\dfrac{n_1}{n} > s$, the point corresponding to d_i is considered as corner. The features of the proposed corner detection algorithm is, while $|d_i|$ is 2, the point is not seen as corner immediately, the point that satisfies that, the number of $|d_i|$ of 2 in its following points exceeds the predefined value s (s can be set to 0.4), is seen as corner instead. That can avoid the impact of the local noise.

3 Fiber Recognition

The fiber boundaries are represented by chain codes. The chain codes must be matched each other for the purpose of recognition. Since there exists many different lengths of chain codes in an image, and two boundaries belonging to the same fiber have the similarity in curvature, the fast recognition algorithm is given:

(1) Select a chain code $c_i = \{c_{i1}, c_{i2}, \cdots, c_{in}\}$, assume the cordination of the initial point p_0 is x_i, y_i, calculate the slope k of the point p_0. While $|k| \geq 1$, the direction of point p_0 is up and down, the search zone is $[x_i - h, x_i + h]$, While $|k| < 1$, the direction of point p_0 is left and right, the search zone is $[y_i - h, y_i + h]$, h can be defined as the width of the thickness fiber and generally about 50.

(2) Assume a chain code $c_j = \{c_{j1}, c_{j2}, \cdots, c_{jm}\}$ be found in the search zone, due to unequal length for each chain codes, that is, $n \neq m$. Suppose $n \geq m$, select m chain code values from the start of c_i to compare with c_j one by one. Once the values of chain code are equal, the match value *match* increases 1. While one round comparison stops, the chain codes of c_i are shifted to right by one position and the algorithm continues to compare it with c_j until arriving the position of $n - m + 1$. The $n - m + 1$ number of *match* are obtained and we represent the largest *match* as $match_{max}$. Normalize $match_{max}$, let $\overline{match_{max}} = match_{max} / m$. Define the largest match value of each chain codes as $M = \{\overline{match_{max\,1}}, \overline{match_{max\,2}}, \cdots, \overline{match_{max\,k}}\}$. Find the maximum in M, once the maximum is greater than the predefined value, c_i is considered to be matched with c_j, and they are recognized as the same fiber.

4 The Experimental Results

The above algorithms are achieved by MATLAB. Fig.5 shows the recognized fibers using different gray value.

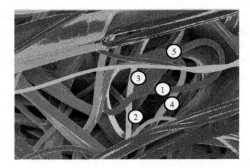

Fig. 5. The recognized fibers with same gray value

The most fibers are recognized correctly using the proposed algorithm as Fig.5. For example, No.1, No.2 and No.3 are cross-cutting in Fig.5, while No.1 is recognized as one fiber, No.2 and No.3 are recognized as the same fiber and the results are consistent with the judge of human eyes. No.1 and No.4 are considered as the same fiber in the position of No.5, since the bifurcation of the two fibers is not obvious in the position of No.5.

5 Summary

This paper presents binarization, prunning, corner detection and recognition methods for the recognition of cross-cutting and winding complex fibers. The experimental results show that the most fibers can be recognized by using the proposed methods. The diameter and length of fibers can be further measured with the help of the recognized fibers.

Acknowledgments. Supported by Leading Academic Discipline Project of Shanghai Normal University (Project Number DZL805, DCL200802).

References

1. Jinhu, S., XiaoHong, W.: The Application of Image Processing in Specialty Animal Hair Fibers Recognition. Journal of Textile Research 25(4), 26–27 (2004)
2. ZaiHua, Y., Yuhe, L.: Animal Fiber Length Measurement Based on Image Processing. Computer Engineering and Applications 41(8), 180–181 (2005)
3. Jian, M., WenHua, Z., WeiGuo, C.: The Application of Gray Value Morphology in Animal Fiber Image Processing. Computer Engineering and Applications 25(5), 42–44 (2004)
4. ShuaiJie, R., WenSheng, Z.: Measuring Diameter and Curvature of Fibers Based on Image Analysis. Journal of Image and Graphics 13(6), 1153–1158 (2008)
5. SuPing, Y., PeiFeng, Z., JianPing, C.: Segmentation on Cotton Fiber Cross Secctions Based on Mask. Computer Engineering 33(13), 188–191 (2007)
6. Bribiesca, E.: A Geometric Structure for Two-Dimensional Shapes and Three-Dimensional Surfaces. Pattern Recognition 25(5), 483–496 (1992)
7. Canny, J.A.: Computational Approach to Edge Detection. IEEE Trans. on PAMI 8(6), 679–698 (1986)

8. Wei, Y., Ke, G., YiKun, W.: An Efficient Index Thinning Algorithm of Fingerprint Image Based on Eight Neighbourhood Points. Journal of SiChuan University of Science & Engineering 21(2), 61–63 (2008)
9. XiaoFeng, H., WenYao, L., ShouDong, X.: The Computer Measurement for Cotton Fiber Length. Cotton Science 15(6), 339–343 (2003)

Laplacian Discriminant Projection Based on Affinity Propagation

Xueping Chang and Zhonglong Zheng*

Department of Computer Science, Zhejiang Normal University, Zhejiang, China
cxp0407@163.com, zhonglong@zjnu.cn

Abstract. The paper proposes a new algorithm for supervised dimensionality reduction, called Laplacian Discriminant Projection based on Affinity Propagation (APLDP). APLDP defines three scatter matrices using similarities based on representative exemplars which are found by *Affinity Propagation Clustering*. After linear transformation, the considered pairwise samples within the same exemplar subset and the same class are as close as possible, while those exemplars between classes are as far as possible. The experiments on several data sets demonstrate the competence of APLDP.

1 Introduction

Linear dimensionality reduction is widely spread for its simplicity and effectiveness in dimensionality reduction. Principal component analysis(PCA), as a classic linear method for unsupervised dimensionality reduction, aims at learning a kind of subspaces where the maximum covariance of all training samples are preserved [2]. Locality Preserving Projections(LPP), as another typical approach for unsupervised dimensionality reduction, seeks projections to preserve the local structure of the sample space [3]. However, unsupervised learning algorithms can not properly model the underlying structures and characteristics of different classes [5]. Discriminant features are often obtained by supervised dimensionality reduction. Linear discriminant analysis(LDA) is one of the most popular supervised techniques for classification [6,7]. LDA aims at learning discriminant subspace where the within-class scatter is minimized and the between-class scatter of samples is maximized at the same time. Many improved LDAs up to date have demonstrated competitive performance in object classification [10,11,12,14,16].

The similarity measure of the scatter matrices of traditional LDA is based on the distances between sample vectors and the corresponding center vectors. Frey proposed a new clustering method, called affinity propagation clustering(APC), to identify a subset of representative examples which is important for detecting patterns and processing sensory signals in data[21]. Compared with center vector, the exemplars are much more representative because the similarities between the samples and exemplars within the same subsection are more closely than those of between the samples and the center vectors. Motivated by APC, traditional LDA, Laplacian Eigenmaps(LE) and the nearest neighborhood selection strategy [4,8,9], we propose a new dimensionality reduction algorithm, Laplacian discriminant projection based on affinity propagation (APLDP),

---

* Corresponding author.

H. Deng et al. (Eds.): AICI 2009, LNAI 5855, pp. 313–321, 2009.
© Springer-Verlag Berlin Heidelberg 2009

for discriminant feature extraction. In our algorithm, we play much emphasis on exemplar based scatter similarity which can be viewed as extensions of within-class scatter and the between-class scatter similarity. We formulate the exemplar based scatter by means of similarity criterions which were commonly used in LE and LPP. The extended exemplar scatter are governed by different Laplacian matrices. Generally, LDA can be regarded as a special case of APLDP. Therefore, APLDP not only conquers the non-Euclidean space problem[5], but also provides an alternative way to find potential better discriminant subspaces.

2 Related Work

2.1 Linear Discriminant Analysis

Let $\mathcal{X} = [x^1, x^2, \ldots, x^n] \in \mathbb{R}^{D \times n}$ denote a data set matrix which consists of n samples $\{x^i\}_{i=1}^n \in \mathbb{R}^D$. Linear dimensionality reduction algorithms focus on constructing a small number, d, of features by applying a linear transformation $W \in \mathbb{R}^{D \times d}$ that maps each sample data $\{x^i\}$ of $\mathcal{X}$ to the corresponding vector $\{y^i\} \in \mathbb{R}^d$ in d-dimensional space as follows:

$$W : \quad x^i \in \mathbb{R}^D \rightarrow y^i = W^T x^i \in \mathbb{R}^d. \tag{1}$$

Assume that the matrix $\mathcal{X}$ contains c classes, and is ordered such that samples appear by class

$$\mathcal{X} = [X_1, X_2, \ldots, X_c] \\ = [x_1^1, \ldots, x_{c_1}^1, \ldots, x_1^c, \ldots, x_{c_c}^c]. \tag{2}$$

In traditional LDA, two scatter matrices, i.e., within-class matrix and between-class matrix are defined as follows [6]:

$$S_w = \frac{1}{n} \sum_{i=1}^c \sum_{x \in X_i} (x - \overline{m}_{(i)})(x - \overline{m}_{(i)})^T \tag{3}$$

$$S_b = \frac{1}{n} \sum_{i=1}^c n_i (\overline{m}_{(i)} - \overline{m})(\overline{m}_{(i)} - \overline{m})^T, \tag{4}$$

where n_i is the number of samples in the $i - th$ class X_i, $\overline{m}_{(i)}$ is the mean vector of the $i - th$ class, and $\overline{m}$ is the mean vector of total samples. It follows from the definition that $trace(S_w)$ measures the within-class compactness, and $trace(S_b)$ measures the between-class separation.

The optimal transformation matrix W obtained by traditional LDA is computed as follows [6]:

$$W_{opt} = \arg \max_w \frac{tr(W^T S_b W)}{tr(W^T S_w W)}. \tag{5}$$

To solve the above optimization problem, the traditional LDA computes the following generalized eigenvalue equation

$$S_b w_i = \lambda S_w w_i, \tag{6}$$

and takes the d eigenvectors that are associated with the d largest eigenvalues $\lambda_i, i = 1, \ldots, d$.

2.2 Affinity Propagation Clustering

Affinity Propagation Clustering(APC), recently proposed by Frey in *Science*, is a kind of clustering algorithm that works by finding a set of exemplars in the data and assigning other data points to the exemplars[21]. The input of APC is pair-wise similarities $s(n, m)$ which are set to be negative squared Euclidean distance:

$$s(n, m) = -\|V_n - C_m\|^2. \tag{7}$$

In APC algorithm, all data points are simultaneously considered as potential exemplars. Therefor, all $s(m, m)$ can be initialized to be the same. In the process of clustering, two kinds of message, the *responsibility* $r(n, m)$ and the *availability* $a(n, m)$, are exchanged among data points by taking competition mechanism into account. To start with, the *availabilities* are set to be zeroes, and the updating rule in the whole process is defined as:

$$r(n, m) = s(n, m) - \max_{m' \, s.t. m' \neq m} \{a(n, m') + s(n, m')\} \tag{8}$$

$$r(m, m) = s(m, m) - \max_{m' \, s.t. m' \neq m} \{a(m, m') + s(m, m')\} \tag{9}$$

$$a(n, m) = \min\{0, r(m, m) + \sum_{n' \, s.t. n' \neq n, m} \max\{0, r(n', m)\}\} \tag{10}$$

$$a(m, m) = \sum_{n' \, s.t. n' \neq m} \max\{0, r(n', m)\} \tag{11}$$

Messages are updated on the above simple formula that search for minima of an appropriately chosen energy function. The message-passing procedure may be terminated under a fixed number of iterations, or after the local decisions keep constant for some number of iterations.

3 Laplacian Discriminant Projection Based on Exemplars

3.1 Discriminant Within-Exemplar Scatter

Let x_i^s denote the $s - th$ sample in $i - th$ class. We formulate Eq(1) as follows:

$$y_i^s = W^T x_i^s \tag{12}$$

Assume that each x_i^s belongs to its exemplar neighborhood within $i - th$ class.

Each $X_j \in \mathscr{X}$ can be divided into k exemplar neighborhoods $\varkappa_i^s, s = 1, ..., k$.

Let $\overline{\varkappa_i^s}$ denote the center vector of exemplar neighborhoods $\varkappa_i^s$. We define the within-exemplar scatter of $\varkappa_i^s$ as $\mathfrak{a}^s$ as follows:

$$\mathfrak{a}^s = \sum_{y_i^s \in \gamma_i^s} \alpha_i^s \|y_i^s - \overline{\gamma_i^s}\|^2, \tag{13}$$

where α_s^i is the weight defined as

$$\alpha_i^s = exp(-\frac{\|x_i^s - \overline{\varkappa_i^s}\|^2}{t}). \tag{14}$$

To obtain the compact expression of Eq(13), let $\mathcal{A}_i^s = diag(\alpha_i^1, ..., \alpha_i^{n_s})$ be a diagonal matrix and $Y_i^s = [y_i^1, ..., y_i^{n_s}]$. In addition, let e_{n_s} denote the all one column vector of length n_s. Then $\overline{\gamma_i^s} = \frac{1}{n_s} Y_i^s e_{n_s}$. Eq(13) can be reformulated as:

$$
\begin{aligned}
\mathfrak{a}_i^s &= \sum \alpha_i^s tr\{(y_i^s - \overline{\gamma_i^s})(y_i^s - \overline{\gamma_i^s})^T\} \\
&= tr\{\sum \alpha_i^s y_i^s (y_i^s)^T\} - 2tr\{\sum \alpha_i^s y_i^s (\overline{\gamma_i^s})^T\} + tr\{\sum \alpha_i^s \overline{\gamma_i^s} (\overline{\gamma_i^s})^T\} \\
&= tr\{Y_i^s \mathcal{A}_i^s (Y_i^s)^T\} - \frac{2}{n_s} tr\{Y_i^s \mathcal{A}_i^s e_{n_s} (e_{n_s})^T (Y_i^s)^T\} \\
&\quad + \frac{(e_{n_s})^T \mathcal{A}_i^s e_{n_s}}{n_s^2} tr\{Y_i^s e_{n_s} (e_{n_s})^T (Y_i^s)^T\} \\
&= tr\{Y_i^s L_i^s (Y_i^s)^T\}
\end{aligned}
\tag{15}
$$

where

$$
L_i^s = \mathcal{A}_i^s - \frac{2}{n_s} \mathcal{A}_i^s e_{n_s} (e_{n_s})^T + \frac{(e_{n_s})^T \mathcal{A}_i^s e_{n_s}}{n_s^2} e_{n_s} (e_{n_s})^T
\tag{16}
$$

The within-exemplar scatter of class i:

$$
\mathfrak{a}_i = \sum_{s=1}^{k_s} \mathfrak{a}_i^s = \sum_{s=1}^{k_s} tr\{Y_i^s L_i^s (Y_i^s)^T\}
\tag{17}
$$

There also exists a 0-1 indicator matrix P_i^s satisfying $Y_i^s = Y_i P_i^s$. Each column of P_i^s records the exemplar information which is derived from the clustering process.

$$
\begin{aligned}
\mathfrak{a}_i &= \sum_{s=1}^{k_s} tr\{Y_i P_i^s L_i^s (P_i^s)^T (Y_i^s)^T\} \\
&= tr\{Y_i L_i Y_i^T\}
\end{aligned}
\tag{18}
$$

where $L_i = \sum_{s=1}^{k_s} P_i^s L_i^s (P_i^s)^T$. The total within-exemplar scatter of all classes:

$$
\mathcal{A} = \sum_{i=1}^c \mathfrak{a}_i = \sum_{i=1}^c tr\{Y_i L_i (Y_i)^T\}
\tag{19}
$$

There exists a 0-1 indicator matrix Q_i satisfying $Y_i = Y Q_i$. Each column of Q_i records the class information which is known for supervised learning. Then Eq(19) can be reformulated as

$$
\begin{aligned}
\mathcal{A} &= \sum_{i=1}^c tr\{Y Q_i L_i (Q_i)^T (Y_i)^T\} \\
&= tr\{Y L_{Exem} Y^T\},
\end{aligned}
\tag{20}
$$

where $L_{Exem} = \sum_{i=1}^c Q_i L_i (Q_i)^T$ can be called the *within-exemplar Laplacian matrix*.

Plugging the expression of $Y = W^T X$ into Eq(20), we obtain the final form of the total within-exemplar scatter:

$$
\mathcal{A} = tr(W^T \mathcal{D}_{Exem} W)
\tag{21}
$$

where $\mathcal{D}_{Exem} = X L_{Exem} X^T$ is the *within-exemplar scatter matrix*.

3.2   Discriminant Within-Class Scatter

Besides the within-exemplar scatter, APLDP also takes the within-class scatter into account. The with-class scatter of $i - th$ class, η_i, is defined as follows:

$$\eta_i = \sum_{s=1}^{k_i} \alpha_i^s \|\overline{y_i^s} - \overline{y_i}\|^2. \tag{22}$$

By the similar deduction, η_i can be formulated as

$$\eta_i = tr\{Y_i L_{\eta_i} Y_i^T\}, \tag{23}$$

where

$$L_{\eta_i} = \mathcal{A}_{\eta_i} - \frac{2}{n_i} \mathcal{A}_{\eta_i} e_{n_i}(e_{n_i})^T + \frac{(e_{n_i})^T \mathcal{A}_{\eta_i}^s e_{n_i}}{n_i^2} e_{n_i}(e_{n_i})^T. \tag{24}$$

The total within-class scatter of all classes:

$$\mathcal{Y} = \sum_{i=1}^{c} \eta_i = tr\{Y_i L_{\eta_i} Y_i^T\} = tr(W^T \mathfrak{D}_{\omega} W), \tag{25}$$

where $\mathfrak{D}_{\omega} = X L_{\eta_i} X^T$ is the *within class scatter matrix*.

3.3   Discriminant Between-Class Scatter

Let $\mathcal{B}$ denote the exemplar based between-class scatter of all classes. $\mathcal{B}$ is defined as

$$\mathcal{B} = \sum_{j=1}^{c} \sum_{i=1}^{c_s} \alpha_j^i \|\overline{y}_j^i - \overline{y}\|^2, \tag{26}$$

where $\overline{y}_j^i$ implies its corresponding original $i - th$ exemplar center of $j - th$ class of X_j. α_j^i is the weight, defined as

$$\alpha_j^i = exp(-\frac{\|\overline{x}_j^i - \overline{x}\|^2}{t}). \tag{27}$$

Let $\overline{Y} = [\overline{y}_1^{k_1}, \ldots, \overline{y}_1^{k_1}, \ldots, \overline{y}_c^{k_c}, \ldots, \overline{y}_c^{k_c}]$ consists of exemplar center vectors of all classes. By the similar deduction, $\mathcal{B}$ can be formulated as follows

$$\mathcal{B} = tr\{\overline{Y} L_b \overline{Y}^T\}, \tag{28}$$

where

$$L_b = \mathcal{A}_b - \frac{2}{n_{exem}} \mathcal{A}_b e_{exem} e_{exem}^T + \frac{e_{exem}^T \mathcal{A}_b e_{exem}}{n_{exem}^2} e_{exem} e_{exem}^T \tag{29}$$

is the exemplar based *between-class Laplacian matrix*.

Taking $\overline{Y} = W^T \overline{X}$ into account, we re-write Eq.(28) as follows

$$\mathcal{B} = tr(W^T \mathfrak{D}_{\mathcal{B}} W). \tag{30}$$

where $\mathfrak{D}_{\mathcal{B}} = \overline{X} L_b \overline{X}^T$ is the *total exemplar based between-class scatter matrix*.

3.4 Discriminant Projection

We construct the following Fisher criterion

$$f(W) = \arg\max_{w} \frac{\mathscr{B}}{\mathscr{A} + \mathscr{D}} = \frac{tr(W^T \mathfrak{D}_\mathscr{B} W)}{tr(W^T (\mathfrak{D}_{\mathfrak{Exem}} + \mathfrak{D}_\mathfrak{B})W)}. \tag{31}$$

Let $\mathfrak{D}_{\mathfrak{EB}} = \mathfrak{D}_{\mathfrak{Exem}} + \mathfrak{D}_\mathfrak{B}$, then

$$f(W) = \arg\max_{w} \frac{\mathscr{B}}{\mathfrak{D}_{\mathfrak{EB}}} \tag{32}$$

To solve the above optimization problem, we take the similar approach used in the traditional LDA. We take the d eigenvectors derived from the following generalized eigenvalue analysis

$$\mathfrak{D}_\mathfrak{B} w_i = \lambda_i \mathfrak{D}_{\mathfrak{EB}} w_i \tag{33}$$

that are associated with the d largest eigenvalues $\lambda_i, i = 1, \ldots, d$.

4 Experiments

In this section, we investigate the use of APLDP on several data sets including UCI[1] and PIE-CMU face data set [13]. The data sets used in the paper belong to different fields in order to test the performance of APLDP algorithm. We compare our proposed algorithm with PCA[2], LDA[7], LPP[3] and Marginal Fisher Analysis (MFA)[1].

4.1 On UCI Data Set

In this experiment, we perform on iris data taken from the UCI Machine Learning Repository. There are 150 samples of 3 classes (50 samples per class) in iris data set. We randomly select 20 samples per class for training and the remaining samples for testing. The average results are obtained over 50 random splits. All algorithms reduce the original samples to 2-dimensional space. The classification is based on k-nearest neighbor classifier. The experimental results are shown in Table1. In terms of APLDP algorithm, there are several parameters which should be set before the experiments. In this experiment, the number of clusters in each class is set $k = 2$, and the time variable $t = 10$.

Table 1. Recognition accuracy of different algorithms

Algorithm	PCA	LDA	LPP	MFA	APLDP
Accuracy	95.112	95.391	95.391	95.383	95.913

To demonstrate the performance of APLDP algorithm, we randomly select one split from the 50 splits. The embedding results of APLDP in $2D$ space, together with other four algorithms, are shown in Fig.1. As illustrated in Table1, APLDP algorithm outperforms other methods with recognition rate of 95.913%. We can find that the within-class embedding result of APLDP is more compact than those of others four methods, as illustrated in Fig.1.

---

[1] Available at http://www.ics.uci.edu/ mlearn/MLRepository.html

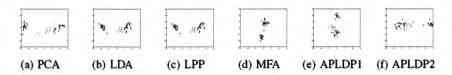

   (a) PCA   (b) LDA   (c) LPP   (d) MFA   (e) APLDP1   (f) APLDP2

Fig. 1. Embedding results in $2 - D$ space of PCA, LDA, LPP, MFA and APLDP1,2(with and without AP)

4.2 On PIE-CMU Face Data Set

The PIE-CMU face data set consists of 68 subjects with 41, 368 face images [13]. In this experiment, we select 40 subjects with 120 face images for each from CMU data set, 60 images for training, and the other 60 images for testing. Before the experiment, faces in the images are detected by the face detection system described in [17][18]. The detected faces are converted to gray scale images and resized to 32×32. Some samples are shown in Fig.2. Totally, there are 2, 400 images in the training set and the testing set, respectively.

It should be mentioned that we take PCA as preprocessing step for LDP. The number of principal components is a free parameter to choose. As pointed out in [19,20], the dimension of principal subspaces significantly affects the performance of recognition tasks. Besides, they confirmed that the optimal number lies in the interval [50, 200].

Fig. 2. Some samples of CMU-PIE face data set

Fig. 3. From the top row to the bottom row, the face-like images are Eigenfaces, Fisherfaces, Laplacianfaces, MFAfaces and APLDPfaces, respectively

Table 2. Best recognition accuracy of different algorithms

Algorithm	PCA	LDA	LPP	MFA	APLDP
Dims	180	39	91	58	112
Accuracy	69.786	79.510	79.533	83.328	84.04

Based on their work, we find the best dimension of PCA is 182. Therefore, we take 182 as the number of principal components in the following experiments.

For the sake of visualization, we illustrate algorithmic-faces derived from different algorithms, such as Eigenfaces from PCA, Fisherfaces from LDA and Laplacianfaces from LPP, in Fig.3. The special face-like images derived from MFA and APLDP can be called MFAfaces and ELDPfaces, respectively.

The average results with corresponding reduced dimensions are obtained over 50 random splits. The classification is also based on k-nearest neighbor classifier. The experimental results are shown in Table2. In the experiment, the parameters of APLDP algorithm are set as $k = 3$, and the time variable $t = 10$.

5   Conclusions

In this paper, based on affinity propagation and LDA algorithms, we propose a new method, exemplar based Laplacian Discriminant Projection(APLDP) for supervised dimensionality reduction. Using similarity weighted discriminant criterions, we define the exemplar based within-class Laplacian matrix and between-class Laplacian matrix. In comparison with the traditional LDA, APLDP focuses more on the enhancement of the discriminability while keeping local structures. Therefore, APLDP has the flexibility of finding optimal discriminant subspaces.

Acknowledgement

The authors confirm that the research was supported by National Science Foundation of China (No.60805001).

References

1. Yan, S., Xu, D., Zhang, B., Zhang, H., Yang, Q., Lin, S.: Graph Embedding and Extension: A General Framework for Dimensionality Reduction. IEEE Transactions on Pattern Analysis and Machine Intelligence 29(1), 40–51 (2007)
2. Turk, M., Pentland, A.: Eigenfaces for recognition. Journal of Cognitive Neuroscience 3(1), 71–86 (1991)
3. He, X., Yan, S., Hu, Y.X., Niyogi, P., Zhang, H.: Face recognition using Laplacianfaces. IEEE Transactions on Pattern Analysis and Machine Intelligence 27(3), 328–340 (2005)
4. Belkin, M., Niyogi, P.: Laplacian Eigenmaps for dimensionality reduction and data representation. Neural Computation 15, 1373–1396 (2003)
5. Zhao, D., Lin, Z., Xiao, R., Tang, X.: Linear Laplacian Discrimination for Feature Extraction. In: CVPR (2007)
6. Fukunaga, K.: Introduction to Statistical Pattern Recognition, 2nd edn. Academic Press, Boston (1990)
7. Belhumeur, P.N., Hespanha, J.P., Kriegman, D.J.: Eigenfaces vs. fisherfaces: Recognition using class specific linear projection. IEEE Transactions on Pattern Analysis and Machine Intelligence 19(7), 711–720 (1997)
8. Weinberger, K., Blitzer, J., Saul, L.: Distance metric learning for large margin nearest neighbor classification. In: NIPS, pp. 1475–1482 (2006)

9. Nie, F., Xiang, S., Zhang, C.: Neighborhood MinMax Projections. In: IJCAI, pp. 993–998 (2007)
10. Howland, P., Park, H.: Generalizing discriminant analysis using the generalized singular value decomposition. IEEE Transactions on Pattern Analysis and Machine Intelligence 26(8), 995–1006 (2004)
11. Liu, C.: Capitalize on dimensionality increasing techniques for improving face recognition grand challenge performance. IEEE Transactions on Pattern Analysis and Machine Intelligence 28(5), 725–737 (2007)
12. Martinez, A., Zhu, M.: Where are linear feature extraction methods applicable. IEEE Transactions on Pattern Analysis and Machine Intelligence 27(12), 1934–1944 (2006)
13. Sim, T., Baker, S., Bsat, M.: The CMU Pose, illumination, and expression (PIE) database. In: IEEE International Conference of Automatic Face and Gesture Recognition (2002)
14. Wang, X., Tang, X.: Dual-space linear discriminant analysis for face recognition. In: CVPR, pp. 564–569 (2004)
15. Yan, S., Xu, D., Zhang, B., Zhang, H.: Graph embedding: A general framework for dimensionality reduction. In: CVPR (2005)
16. Yang, J., Frangi, A., Yang, J., Zhang, D., Jin, Z.: KPCA plus LDA: a complete kernel Fisher discriminant framework for feature extraction and recognition. IEEE Transactions on Pattern Analysis and Machine Intelligence 27(2), 230–244 (2005)
17. Zheng, Z.L., Yang, J., Zhu, Y.: Face detection and recognition using colour sequential images. Journal of Research and Practice in Information Technology 38(2), 135–149 (2006)
18. Zheng, Z.L., Yang, J.: Supervised Locality Pursuit Embedding for Pattern Classification. Image and Vision Computing 24, 819–826 (2006)
19. Wang, X., Tang, X.: A unified framework for subspace face recognition. IEEE Transactions on Pattern Analysis and Machine Intelligence 26(9), 1222–1228 (2004)
20. Wang, X., Tang, X.: Random sampling for subspace face recognition. International Journal of Computer Vision 70(1), 91–104 (2006)
21. Frey, B.J., Dueck, D.: Clustering by Passing Messages Between Data Points. Science 315, 972–994 (2007)

An Improved Fast ICA Algorithm for IR Objects Recognition

Jin Liu and Hong Bing Ji

School of Electronics Engineering, Xidian Univ., Xi'an, China
jinliu@xidian.edu.cn

Abstract. In this paper, an improved algorithm based on fast ICA and optimum selection for IR objects recognition is proposed. Directed against the problem that the Newton iteration is rather sensitive to the selection of initial value, this paper presents a one dimension search to improve its optimum learning algorithm in order to make the convergence of the results independent of the selection of the initial value. Meanwhile, we design a novel rule for the distance function to retain the features of the independent component having major contribution to object recognition. It overcomes the problem of declining of recognition rate and robustness associated with the increasing of training image samples. Compared with traditional methods the proposed algorithm can reach a higher recognition rate with fewer IR objects features and is more robust in different kinds of classes.

Keywords: IR objects recognition, fast ICA, one dimension search, distance function, robustness.

1 Introduction

Objects recognition is one of the key techniques in the field of pattern recognition. The key of objects recognition is feature extraction and dimension reduction. Directed against the lower recognition rate of IR image due to the characteristics of IR image itself, it is necessary to look for a new method for image information processing and feature extraction which can remove redundant information in the object image data and presents object image using feature vector with invariant features. Several characteristic subspace methods have been developed in the field of objects recognition, such as Principal Component Analysis (PCA) characteristic subspace algorithm[1][2], Independent Component Analysis (ICA) characteristic subspace algorithm[3][4] and so on. ICA is different from the PCA in that it does not focus on signals' second-order statistical correlation. Instead it is based on the higher-order statistics of signals, i.e., it studies the independent relationship between signals. Therefore ICA method can reveal the essential structure between image data. Hyvarinen A. [5] [6] etc proposed a fast independent component analysis (Fast ICA) algorithm which is a fast iterative optimization algorithm with fast convergence rate and exempts the need to determine learning step. In recent years, fast ICA has obtained universal attention and has been widely used in feature extraction [7], blind source separation [8], speech signal processing [9], target detection [10] and face recognition [11] [12], etc.

H. Deng et al. (Eds.): AICI 2009, LNAI 5855, pp. 322–329, 2009.

Directed against the problem that the Newton iteration in the fast ICA algorithm is rather sensitive to the selection of initial value and the problem of declining in recognition rate and robustness with the increasing of training image samples, this paper presents an improved fast ICA and features optimization algorithm which can be applied to IR multi-object recognition. The paper is organized as follows. Section 2 introduces ICA theory and fast fixed-point algorithm for ICA. Section 3 IR describes multi-object classification recognition based on improved fast ICA and a feature optimization algorithm and gives the rule for feature optimization. Section 4 gives experimental results produced by using improved fast ICA algorithm and conducts performance evaluation. Section 5 is conclusion.

2 Related Works

2.1 Review of ICA Theory

Suppose $\{x_1, x_2, \cdots, x_n\}$ are n observed random signals formed by linear mixing of m unknown independent sources $\{s_1, s_2, \cdots, s_m\}$. $X = [x_1, x_2, \cdots, x_n]^T$ is the matrix representation of the observed signals. $S = [s_1, s_2, \cdots, s_m]^T$ is the independent source signals. Assuming the observed random signals and independent source signals are zero mean, we have

$$x_i = \sum_{j=1}^{m} a_{ij} s_j \tag{1}$$

where $i = 1, 2, \cdots, n. j = 1, 2, \cdots, m$.

The linear model of ICA can be expressed as

$$X = AS = \sum_{j=1}^{m} a_j s_j \tag{2}$$

where $A = [a_1, a_2, \cdots, a_m]$ is a full rank $n \times m$ mixed matrix, $a_j (j = 1, 2, \cdots, m)$ is the base vector of the mixed matrix. The observed signal X can be viewed as a linear combination of the column vector a_j of the mixing matrix A, and the corresponding weights are just the independent source signals $s_j (j = 1, 2, \cdots, m)$.

2.2 Fast and Fixed-Point Algorithm for ICA

For a random variable x, its entropy is defined as

$$H(x) = -\int f(x) \log f(x) dx \tag{3}$$

where $f(x)$ is the probability density distribution function of the random variable x.

The negative entropy of a random variable x is defined as

$$J(x) = H(x_{gauss}) - H(x) \tag{4}$$

where x_{gauss} is a random variable having Gaussian distribution and the same variance with x.

From (4), we can see that the value of negative entropy is always non-negative, and only when the random variable x is of Gaussian distribution, the negative entropy is zero. The stronger the non-Gaussian random of variable x, the bigger $J(x)$ is. The more effective approximate negative entropy formula is

$$J(w_i) \approx [E\{G(w_i^T X)\} - E\{G(v)\}] \tag{5}$$

where $G(\bullet)$ are some non-quadratic functions, and v is a random variable with standard normal distribution.

By finding the projection direction w_i which makes $J(w_i)$ maximum, an independent component can be extracted. If $J(w_i)$ takes the maximum value, $E\{G(w_i^T X)\}$ will also have the maximum value, and the extreme point w_i of $E\{G(w_i^T X)\}$ is the solution of equation

$$E\{Xg(w_i^T X)\} = 0 \tag{6}$$

where $g(\bullet)$ is the derivative of $G(\bullet)$. According to Newton iteration theorem, we have

$$w_i^+ = w_i - \frac{E\{Xg(w_i^T X)\}}{E\{X^T Xg'(w_i^T X)\}} \approx w_i - \frac{E\{Xg(w_i^T X)\}}{E\{g'(w_i^T X)\}} \tag{7}$$

where $E\{X^T Xg'(w_i^T X)\} \approx E\{X^T X\}E\{g'(w_i^T X)\} = E\{g'(w_i^T X)\}$. Multiply the two sides of (7) with $-E\{g'(w_i^T X)\}$, and let $w_i^* = -E\{g'(w_i^T X)\}w_i^+$, we have

$$\begin{cases} w_i^* = E\{Xg(w_i^T X)\} - E\{g'(w_i^T X)\}w_i \\ \qquad w_i = w_i^* / \|w_i^*\| \end{cases} \tag{8}$$

By conducting iteration according to (8), the w_i^T obtained through convergence corresponds to the row vectors of the separation matrix W, and then an independent component s_i can be extracted.

3   IR Objects Recognition Based on Improved Fast ICA and Features Optimization

3.1   Improved Iteration Learning Algorithm Based on One Dimension Search

Fast ICA algorithm utilized the principle of Newton iteration. However, since the direction of Newton iteration is not necessarily in the descent direction, when the initial value is far from the minimum point, a situation may arise that the object function obtained by iteration cannot reach the best point or even not convergent. Targeted at this problem, this paper proposes a strategy that adds a one-dimensional search along the direction of Newton iteration, and designs the new iteration formula

$$\begin{cases} x^{(k+1)} = x^{(k)} + \lambda_k d^{(k)} \\ d^{(k)} = -\nabla^2 f(x^{(k)})^{-1} \nabla f(x^{(k)}) \\ f(x^{(k)} + \lambda_k d^{(k)}) = \min_\lambda f(x^{(k)} + \lambda d^{(k)}) \end{cases} \tag{9}$$

where λ is the damping factor, ∇ is the first order partial derivative of the function, ∇^2 is the second order partial derivative of the function. The optimum solution of the function can be obtained by iteration.

In the case of $\|w_i\| = 1$, it is transformed into an unconstrained extreme value problem, and the new cost function is

$$F(w_i, c) = E\{G(w_i^T X)\} + c(\|w_i\| - 1) \tag{10}$$

where c is referred to as cost factor.

Therefore, the damping iteration of transformation matrix w is modified to

$$w_i^+ = w_i - \lambda_i \frac{E\{Xg(w_i^T X)\} - cw_i}{E\{g'(w_i^T X)\} - cI} \tag{11}$$

Multiply both sides of (11) by $E\{g'(w_i^T X)\} - cI$, let $w_i^* = w_i^+[E\{g'(w_i^T X)\} - cI]$, and normalize the obtained parameters, we get the new iteration formula

$$\begin{cases} w_i^* = \lambda_i[E\{Xg(w_i^T X)\}] - E\{g'(w_i^T X)\}w_i - c(\lambda_i - I)w_i \\ w_i = w_i^*/\|w_i^*\| \end{cases} \tag{12}$$

where λ_i is the one-dimensional search factor, which makes the cost function enter the convergence region of Newton iteration beginning with a certain w_i in a certain range, and ensures the convergence of the algorithm. When $\lambda_i = 1$, it becomes the original fast ICA algorithm.

3.2 Feature Optimization for IR Object Classification Recognition

The optimal selection of features is a key issue that affects the correctness of object classification and recognition. In feature subspace, the concentrated the feature vectors of the same class and the more far apart the feature vectors of different classes, the easier the classification and recognition. Therefore, when extracting features from IR object images, it is required to extract features that are far different between different classes of objects and are almost the same for one class of objects. With the increasing of the number of observation images, the number of independent components and the corresponding features are also increasing. Among these added features, some will have less contribution to classification recognition due to similarity to existing features; others may even reduce recognition rate due to feature distortion. In order to ensure correctness and efficiency of classification and recognition, we select a minimum number of features from the original features that are most effective in classification. For this purpose, we propose a new feature selection criteria based on intra-class and inter-class distance function.

Suppose observation image matrix F is composed of M classes and there are $N_i(i = 1, 2, \cdots, M)$ observations for each class. Column vector A_j of mixed matrix A corresponds to independent component s_j, and a_{ij} is the jth feature of the ith observation. U_j is defined as the mean value of intra-class distance of the jth feature in M classes. This can be expressed as

$$\begin{cases} U_j = \dfrac{1}{MN(N-1)} \sum_{c=1}^{M} \sum_{s=1}^{N_i} \sum_{\substack{p=1 \\ p \neq s}}^{N_i} \left| a_{q_c+s,j} - a_{q_c+p,j} \right| \\ q_c = \sum_{i=0}^{c-1} N_i, \text{ where } N_0 = 0; i = 1, 2, \cdots, M \end{cases} \quad (13)$$

Similarly, V_j is called the mean value of inter-class distance of the jth feature in M classes, and can be expressed as

$$\begin{cases} V_j = \dfrac{1}{M(M-1)} \sum_{l=1}^{M} \sum_{\substack{k=1 \\ k \neq l}}^{M} \left| \dfrac{1}{N_l} \sum_{s=1}^{N_l} a_{q_l+s,j} - \dfrac{1}{N_k} \sum_{p=1}^{N_k} a_{q_k+p,j} \right| \\ q_l = \sum_{i=0}^{l-1} N_i, q_k = \sum_{i=0}^{k-1} N_i, \text{ where } N_0 = 0; i = 1, 2, \cdots, M \end{cases} \quad (14)$$

Effective classification factor β_j is defined as

$$\beta_j = \frac{U_j}{V_j} \quad (15)$$

which reflects the difficulty degree of classification of M classes by the jth feature. The smaller β_j is, the more sensitive the jth feature is, and the more easy and effective the classification of the M classes.

We also have the relationship in practical situations

$$0 < U_j < V_j \tag{16}$$

In accordance with the definition of effective classification factor, we can further obtain

$$0 < \beta_j < 1 \tag{17}$$

We choose the independent components corresponding to β_j satisfying (17) to form the optimal feature subspace, and carry out object recognition and classification in this optimal characteristic subspace by minimum distance classifier.

4 Experiments and Analysis

Multi-object classification/recognition experiments have been made by using real infrared object image database. This database contains nine classes of infrared objects, and the images for each class were taken each time when the targeted object has rotated 10 degrees in front of the IR camera. So there are 36 photos for each class (324 IR object images for 9 classes). The training set consists of 162 images (18 photos for each class), and image size is 200×100. Particularly, the 18 photos for each class were taken each time when the object has rotated 20 degrees, such as at 5 degrees, 25 degrees, 45 degrees etc., and 9 photos are chosen randomly to add Gaussian noise to. So the testing set is made up of 162 frames of images from 9 classes. The 18 images for each class were taken each time when the object has rotated 20 degrees, such as at 15 degrees, 35 degrees, 55 degrees etc., and 9 photos are chosen randomly to add Gaussian noise to. Part of the infrared object images in the training set are shown in Fig. 1.

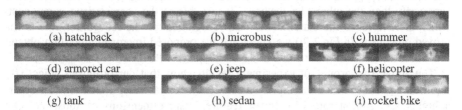

(a) hatchback	(b) microbus	(c) hummer
(d) armored car	(e) jeep	(f) helicopter
(g) tank	(h) sedan	(i) rocket bike

Fig. 1. Part of the IR object images in the training set

The experiment verifies the feasibility and effectiveness of the algorithm through IR database built by ourselves. We use PCA feature extraction algorithm (Traditional PCA) in [2], Fast ICA feature extraction algorithm (Traditional FastICA) in [6] and our proposed object recognition algorithm based on improved fast ICA and feature

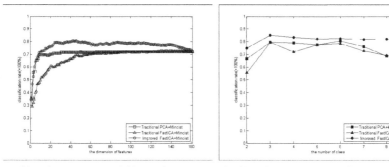

(a) Classification rate in different dimension   (b) Classification rate in different classes

Fig. 2. Performance comparison with different algorithms

optimization (Improved FastICA) to conduct feature extraction of independent components, and use minimum distance classifier (Mindist) in optimal feature subspace for classification. The respective ratio of correct classification of the above three methods is tested by using different classes. In order to test the performance of the algorithms more accurately, the experimental results are obtained by running the algorithms 10 times and using independent data sets. Each algorithm is measured by correctness rate of classification (correct classification sample numbers divided by all tested sample numbers). Experimental results are shown in Fig. 2.

In Fig.2.(a) we can see that the classification rate of the proposed Improved FastICA + Mindist algorithm can reach 75% under the condition when the dimension of characteristics is 20, meanwhile the Traditional FastICA + Mindist and Traditional PCA + Mindist algorithm is 60% and 70% respectively. When the dimension is 40, the recognition rate of our proposed Improved algorithm achieves maximum 81%, and with the increasing of dimensions the classification rate has a little reduction, which shows that some characteristics are not helpful to the classification; the recognition rate of Traditional FastICA + Mindist algorithm is only 66%; the recognition rate of Traditional PCA + Mindist algorithm remains unchanged in 71%. Because of the incompleteness and uncertainty of infrared images information, the improved method has more clear superiority than the other two methods. It just retains the characteristic components which are helpful to the classification and recognition, so the recognition rate is improved obviously. In Fig.2.(b) we know that the proposed improved algorithm obtains more robustness in terms of recognition rate compared with the other two methods.

The above two groups of experiment results show that our proposed object recognition algorithm based on improved fast ICA and feature optimization can obtain a higher recognition rate with fewer features   and dimensions of subspace. Our proposed algorithm also has the advantage of great robustness in different classes, so it lays a foundation for the engineering application of the recognition algorithm.

5   Conclusions

In this paper, we develop an IR object recognition technique employing an improved fast ICA and feature optimization algorithm. Targeted at the problem that the Newton iteration in the fast ICA algorithm is sensitive to the selection of initial value, one dimension search is imposed on the direction of Newton iterative in order to ensure the convergence of the results and the robustness to initialization. Meanwhile, a novel rule based on distance function is designed to select the optimal features favorable for recognition according to the characteristics of infrared image. It overcomes the problem of declining in recognition rate and robustness along with the incensement of the number of training image samples. Experimental comparison on IR object database under similar conditions shows that our proposed method can obtain highest recognition rate. Theoretical derivation and experimental results show that the improved fast ICA features optimization algorithm can improve recognition accuracy and efficiency. In the future, we aim to further improve the recognition rate in combination with other feature information.

Acknowledgement. This paper is supported by the National Natural Science Foundation of China (NSFC) under Grant Number 60677040.

References

1. Oja, E.: Neural Networks, Principle Components and Subspaces. International Journal of Neural Systems 1(1), 61–68 (1989)
2. Jolliffe, I.T.: Principal Component Analysis, 2nd edn. American Statistical Association, American (2003)
3. Hyvarinen, A., Karhunen, J., Oja, E.: Independent Component Analysis. Wiley, New York (2001)
4. Karhunen, J., Hyvarinen, A., Vigario, R., et al.: Applications of Neural Blind Separation to Signal and Image Processing. In: IEEE International Conference on ICASSP, vol. 1(1), pp. 131–134 (1997)
5. Hyvarinen, A.: Blind Source Separation by Nonstationarity of Variance: a Cumulant-based Approach. IEEE Transactions on Neural Networks 12(6), 1471–1474 (2001)
6. Hyvarinen, A.: Fast and Robust Fixed-point Algorithms for Independent Component Analysis. IEEE Transactions on Neural Networks 10(3), 626–634 (1999)
7. Jouan, A.: FastICA (MNF) for Feature Generation in Hyperspectral Imagery. In: 10th International Conference on Information Fusion, pp. 1–8 (2007)
8. Shyu, K.-K., Lee, M.-H., Wu, Y.-T., et al.: Implementation of Pipelined FastICA on FPGA for Real-Time Blind Source Separation. IEEE Transactions on Neural Networks 19(6), 958–970 (2008)
9. Shen, H., Huper, K.: Generalised Fastica for Independent Subspace Analysis. In: IEEE International Conference on Acoustics, Speech and Signal Processing, vol. 4, pp. 1409–1412 (2007)
10. Ming-xiang, W., Yu-long, M.: A New Method for The Detection of Moving Target Based on Fast Independent Component Analysis. Computer Engineering 30(3), 58–60 (2004)
11. Mu-chun, Z.: Face Recognition Based on FastICA and RBF Neural Networks. In: 2008 International Symposium on Information Science and Engineering, vol. 1, pp. 588–592 (2008)
12. Li, M., Wu, F., Liu, X.: Face Recognition Based on WT, FastICA and RBF Neural Network. In: Third International Conference on Natural Computation, vol. 2, pp. 3–7 (2007)

Facial Feature Extraction Based on Wavelet Transform

Nguyen Viet Hung

Department of Mathematics and Computer Science,
Ho Chi Minh City, University of Pedagogy, Viet Nam
nvhung@math.hcmup.edu.vn

Abstract. Facial feature extraction is one of the most important processes in face recognition, expression recognition and face detection. The aims of facial feature extraction are eye location, shape of eyes, eye brow, mouth, head boundary, face boundary, chin and so on. The purpose of this paper is to develop an automatic facial feature extraction system, which is able to identify the eye location, the detailed shape of eyes and mouth, chin and inner boundary from facial images. This system not only extracts the location information of the eyes, but also estimates four important points in each eye, which helps us to rebuild the eye shape. To model mouth shape, mouth extraction gives us both mouth location and two corners of mouth, top and bottom lips. From inner boundary we obtain and chin, we have face boundary. Based on wavelet features, we can reduce the noise from the input image and detect edge information. In order to extract eyes, mouth, inner boundary, we combine wavelet features and facial character to design these algorithms for finding midpoint, eye's coordinates, four important eye's points, mouth's coordinates, four important mouth's points, chin coordinate and then inner boundary. The developed system is tested on Yale Faces and Pedagogy student's faces.

Keywords: Computer vision, face recognition, wavelet transform, facial feature extraction.

1 Introduction

In recent years, computer vision has rain-storm development. Computer vision concerns with developing systems that can interpret the content of natural scenes [6, 7]. Like the aims of all applications, to make computers more user-friendly and to increase computer ability to identify learn and gain knowledge, computer vision systems begin with the simple process of detecting objects and locating objects, continue with the higher process of detecting and locating features to gain more information, and then proceed to the most advanced process of extracting meaningful information, to apply to intelligent systems.

Face recognition and face expression are one part of computer vision. To recognize one person, we must have three main processes. One of the main processes is facial feature extraction. In a computer vision society, a feature is defined as a function of one or more measurements, the values of some quantifiable property of an object, computed so that it quantifies some significant characteristics of the object [6,7]. One

H. Deng et al. (Eds.): AICI 2009, LNAI 5855, pp. 330–339, 2009.

of the biggest advantages of feature extraction lies in that it significantly reduces the information (compared to the original image) to represent an image for understanding the content of the image [12]. Basically, the extraction of facial feature points (eyes, mouth, nose, chin, inner boundary) plays an important role in various applications such as face detection, face recognition, model based image coding, and expression recognition, and has become a popular area of research due to emerging applications in human-computer interface, surveillance systems, secure access control, video conferencing, financial transaction, forensic applications, pedestrian detection, driver alertness monitoring systems, image database management system and so on.

In space problem of facial feature extraction, we have some obstacles given as follow: input images are so blur and have a lot of noise, face sizes and orientations. Sometimes facial features may be covered by other things, such as a hat, a pair of glasses, a hand, etc.

Many approaches to facial feature extraction have been reported in literature over the past few decades, ranging from the geometrical description of salient facial features to the expansion of digitized images of the face on appropriate basis of images [8]. Different techniques have been introduced recently, for example, principal component analysis [1], geometric modeling [2], auto-correlation [13], deformable template [14], neural networks [15], elastic bunch graph matching [16], color analysis [17] and so on. Lam and Yan [3] have used snake model for detecting face boundary. Although the snake provides good results in boundary detection, the main problem is to find the initial position [4].

We can divide these approaches to four main parts: geometry-based, template-based, color segmentation, appearance-based approach. Generally geometry-based approaches extract features using geometric information such as relative positions and sizes of the face components. Template-based approaches match facial components to previously designed templates using appropriate energy functional. The best match of a template in the facial image will yield the minimum energy. Color segmentation techniques make use of skin color to isolate the face. Any non-skin color region within the face is viewed as a candidate for eyes and mouth. The performance of such techniques on facial image databases is rather limited, due to the diversity of ethnical backgrounds [18]. In appearance-based approaches, the concept of "feature" differs from simple facial features such as eyes and mouth. Any extracted characteristic from the image is referred to as a feature [5]. Methods such as principal component analysis (PCA), independent component analysis, and Gabor wavelets [21] are used to extract the feature vector.

This paper explores facial feature extraction such as eyes, mouth, chin, and inner boundary. Facial feature extraction is established by two approaches: (i) edge detection based on wavelet transform and (ii) geometry property. Based on wavelet property, we can make the edge of an image bolder and reduce the noise from that image. Appling geometry property of human face, eyes, and mouth, we give some thresholds. Experimental results indicate that the algorithm can be applied to image having noises, human having beard and human expression.

2   Background Wavelets and Edge Detection Based on Wavelets

The concept of wavelet analysis has been developed since the 1980s. It is a relatively recent development of applied mathematics. Independent from its developments in harmonic analysis, the wavelet transform has been studied in its continuous form and initially applied to analyze geological data by A. Grossmann, J.Morlet and their co-workers. However, at that time, the roughness of wavelets made mathematics suspect the existence of a "good" wavelet basic until two great events took place in 1988, namely:

Daubechies, a female mathematician, constructed a class of wavelet bases, which are smooth, compactly supported and orthonormal. They are referred to as Daubechies bases, and are successfully applied to many fields today.

French signal analysis expert, S.Mallat, with mathematician, Y.Meyer, proposed a general method to construct wavelet bases. It's termed multiresolution analysis (MRA) and is intrinsically consistent with sub-band coding in signal analysis [9].

In 1985, Y.Meyer [22,23], discovered that one could obtain orthonormal bases for $L^2(R)$ of the type:

$$\psi_{j,k}(x) = 2^{j/2}\psi(2^j x - k), \quad j,k \in Z. \tag{1}$$

and that the expression

$$f = \sum_{j,k \in Z}\langle f,\psi_{j,k}\rangle\psi_{j,k}. \tag{2}$$

for decomposing a function into these orthonormal wavelet converged in many function spaces.

In 1987, Stéphane Mallat [22,23] made a giant step in the theory of wavelets when he proposed a fast algorithm for the computation of wavelet coefficients. He proposed the pyramidal schemes that decomposed signals into subbands. These algorithms and theory of wavelets is the concept of multi-resolution analysis (MRA) [10].

2.1   Wavelets

Wave is an oscillation function in the time domain or frequency domain. What is a wavelet? The simplest answer is a "short" wave (wave + let =>wavelet). The suffix "let", in the mathematical term "wavelet", comes from "short", which indicates that the duration of function is very limited. In other words, the wavelet is a special signal, in which, two conditions have to be satisfied:

1. First, the wavelet must be oscillatory (wave).
2. Second, its amplitudes are nonzero only during a short interval (short).

A signal or function f(t) can be analyzed, described or processed if we give linear analysis as follows

$$f(t) = \sum_l a_l\psi_l(t). \tag{3}$$

where l is infinite or finite summary, a_l is real coefficient of expansion and $\psi_l(t)$ is real function, an expansion. If (1) is unique, the set of $\psi_l(t)$ is called base of function class.

If base is orthogonal, i.e.

$$\langle \psi_k(t), \psi_l(t) \rangle = \int \psi_k(t)\psi_l(t)\,dt = 0, (k \neq l). \tag{4}$$

then coefficient can be calculated by product

$$a_k = \langle f(t), \psi_k(t) \rangle = \int f(t)\psi_k(t)\,dt. \tag{5}$$

In Fourier series, the orthogonal functions, $\psi_l(t)$, are $\sin(kw_0t)$ and $\cos(kw_0t)$

In Wavelets, we have

$$f(t) = \sum_k \sum_l a_{j,k}\psi_{j,k}(t). \tag{6}$$

where

$$\psi_{j,k}(t) = 2^{\frac{j}{2}}\psi(2^j t - k), \ k, j \in Z. \tag{7}$$

In general, $g(t) \in L^2(R)$ can be written as follows:

$$g(t) = \sum_{k=-\infty}^{\infty} c(k)\varphi_k(t) + \sum_{j=0}^{\infty}\sum_{k=-\infty}^{\infty} d(j,k)\psi_{j,k}(t). \tag{8}$$

where

$$c(k) = c_0(k) = \langle g(t), \varphi_k(t) \rangle = \int g(t)\varphi_k(t)\,dt. \tag{9}$$

$$d(j,k) = d_j(k) = \langle g(t), \psi_{j,k}(t) \rangle = \int g(t)\psi_{j,k}(t)\,dt. \tag{10}$$

$\varphi_k(t), \psi_{j,k}(t)$ are scaling function, and wavelet function, respectively.

$$\psi_{j,k}(t) = 2^{\frac{j}{2}}\psi(2^j t - k), \ \varphi_{j,k}(t) = 2^{\frac{j}{2}}\varphi(2^j t - k). \tag{11}$$

2.2 Edge Detection Based on Wavelets

The edge of an image is a boundary which the brightness of image changes abruptly. In image processing, an edge is often interpreted as a class of singularities. In a function, singularities can be presented easily as discontinuities where the gradient approaches infinity. However, image data is discrete, so edges in an image are often defined as the local maxima of the gradient [10].

Edge detection is a technique related to gradient operators. Because an edge is characterized by having a gradient of large magnitude, edge detectors are approximations of gradient operators. Noise influences the accuracy of the computation of gradients. Usually an edge detector is a combination of a smooth filter and a gradient operator. First, an image is smoothed by the smooth filter and then its gradient is computed by the gradient operator.

The 2-norm of the gradient is

$$\|\nabla f\|_2 = \sqrt{f^2_x + f^2_y}. \tag{12}$$

In oder to simplify computation in image processing, we often use the 1-norm instead.

$$\|\nabla f\|_2 = |f_x| + |f_y|. \tag{13}$$

Their discrete forms are

$$\|\nabla f\|_2 = \sqrt{\left(f(i-1,j) - f(i+1,j)\right)^2 + \left(f(i,j-1) - f(i,j+1)\right)^2}. \tag{14}$$

and

$$\|\nabla f\|_1 = \frac{1}{2}\left(|f(i-1,j) - f(i+1,j)| + |f(i,j-1) - f(i,j+1)|\right). \tag{15}$$

1. Local maxima definition: Let $f(x,y) \in H^f(\Omega)$. A point $(\bar{x},\bar{y}) \in \Omega^0$ is called an edge point of the image f(x,y), if $\|\nabla f\|$ has a local maximum at $(\bar{x},\bar{y})$ i.e. $\|\nabla f(\bar{x},\bar{y})\| > \|\nabla f(x,y)\|$ in a neighborhood of $(\bar{x},\bar{y})$. An edge curve in an image is a continuous curve on which all points are edge point. The set of all edge points of f(x,y) is called an edge image of f(x,y).

2. Threshold definition: Let $f(x,y) \in H^2(\Omega)$. Assume that

$$\max_{(x,y)\in\Omega}\|\nabla f(x,y)\| = M. \tag{16}$$

Choose K, $0 < K < M$, as the edge threshold. If $\|\nabla f(\bar{x},\bar{y})\| \geq K$ the $(\bar{x},\bar{y})$ is called an edge point of f(x,y) [19].

Fig. 1. Lena image and Lena image with noise

Edge detection is a very important process in image processing. Basing on this process, we can do recognition pattern, image segmentation, scene analysis, etc. The traditional methods to detect edge fail to deal with images having noise. Noise is something we don't expect in the original image.

Maxima of wavelet transform modulus proved by Mallat, Hwang, and Zhong [19,20] can detect the location of the irregular structures. Edge of the image data f(x,y) is the singularities of f(x,y) so they relate to local maximal modulus of wavelets.

Basing on wavelet theory, we use a filter J-steps including two-channels, lowpass and highpass channel to filter low frequency and high frequency of the signal. After we use analysis bank which has J steps, we separate image frequency to each parts of frequency. Depending on goal requirement, we keep or remove parts of frequency, then we inverse these by synthesis bank. In edge detection, we are interested in high frequency (details of image) and omit low or medium frequency (the coarse of image). Depending on frequency and J-step, we can get a different output image. For example, if we keep many parts of high frequency, the output image will blur with boundary, but if we keep few parts, the output image will lack information. Therefore, the number of parts we keep and omit depends on each aims.

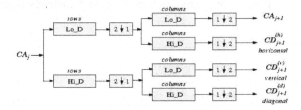

Fig. 2. Decomposition Steps

After analyzing an image by wavelet transform, we will omit some unnecessary coarses. We have some solutions and predictable results of these solutions.

- If we do not omit any parts, the reconstruction image will be an original image.
- If we keep HH part and omit the others, the reconstruction image will be an image lacking in information.
- If we keep HL part or LH part and omit the others, the reconstruction image will be an image having a respectively thicker vertical or horizontal boundary than if we keep LH part or HL part. And if we keep both HL and LH part, the output image will be blurred.
- If we need more image information, we apply analysis bank to LL part to get more information in J+1 step.

3 Algorithms for Extracting Facial Feature

The proposed facial feature extraction method consists of two main steps: preprocess, and main process. In preprocess, we use wavelet properties to reduce noise and finding space. Noise can make it difficult to find facial feature. The feature we need to extract is always on the small part of the human face, so unnecessary information must be removed in advance so as to decrease finding space, which increases accuracy and lessens running time.

In the main process, we have four mains steps:

Step 1 (Find eyes) In this step, we build some algorithms to get coordinates of two eyes, coordinates of eye's top and bottom, and coordinates of eye corners.

Step 2 (Find mouth) We have 3 algorithms to gain coordinate of mouth, and coordinates of mouth corners.

Step 3 (Find chin) To obtain coordinate of bottom chin, we build one algorithm.

Step 4 (Find inner boundary) After using several algorithms, we get a set point of inner boundary.

Step one is the most important in this paper. First, we apply MES algorithm. The aim of this algorithm is to reduce eye space by finding 2 lines which are parallel with vertical axis. The left line is a line closest to the left of left eye. The right line is similar. We use one algorithm to find the eye's region. The next step, we find approximate middle point of two eyes, left and right eyes, find top and bottom point of eyes and then find coordinate of two eye corner. Step two includes: find coordinate of mouth, two corners of mouth, top point of upper lip and bottom point of lower lip. Step three has one algorithm, approximately chin coordinate. Step 4 includes algorithm of the inner boundary.

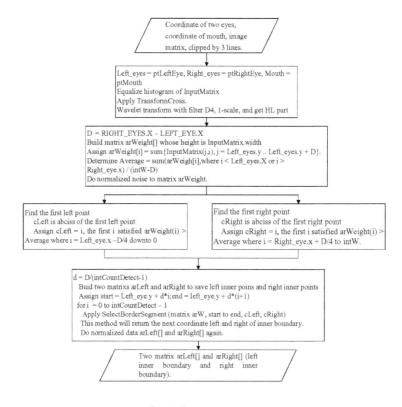

Fig. 3. Inner boundary

Fig. 4. All features

4 Experimental Results

The facial feature extraction method was implemented in C#, Visual Studio 2007, Windows XP and Dot.net framework 2.0.

To evaluate the accuracy in finding these facial features, we compare facial features obtained by hand and facial feature gained by these algorithms. The mean of error of point to point is calculate by this formula $m_e = \dfrac{1}{ns}\sum\limits_{i=1}^{i=n} d_i$

where d_i is an error of point to point, s is the distance between two eyes and n is the number of features.

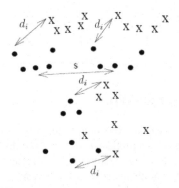

Fig. 5. Mean of error of point to point

The accurate scale of eyes is 93.54% with $m_{e2} = 0.0534$ in 124 Yale images (front human face) and 91.17% with $m_{e2} = 0.043$ in 34 student's images.

The accurate scale of mouth is 91.93% with $m_{e1} = 0.068$ in 124 Yale images and 88.23% with $m_{e1} = 0.054$ in 34 student's images.

The accurate scale of chin is 91.93% with $m_{e1} = 0.0017$ in 124 Yale images and 88.23% with $m_{e1} = 0.0034$ in 34 student's images.

The rate between the inner boundary we obtain by these algorithms and by hand is 90.32 in 124 Yale images and 85.29% in 34 student's images.

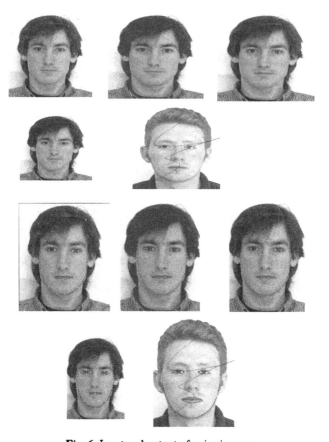

Fig. 6. Input and output of noise images

5 Conclusion and Future Work

Facial feature extraction is one of the most interesting research studies and there are many approaches to it. In our paper, we proposed some algorithms based on wavelet transform and geometry features to extract eye features, mouth features, chin and inner boundary. However, this program may not always give accurate results due to several issues, including some lossy information, concealing, and different orientations. In the future, we can improve these algorithms so that they can work well with different orientations and build face recognition system or facial feature analysis relying on these features.

References

1. Turk, M., Pentland, A.: Eigenfaces for Recognition. Journal of Cognitive Neuroscience 3(1), 71–86 (1991)
2. Ho, Y.Y., Ling, H.: Facial Modeling from an Uncalibrated Face Image Using a Coarse-to-Fine Genetic Algorithm. Pattern Recognition 34(8), 1015–1031 (2001)

3. Lam, M.K., Hong, Y.: Locating and Extracting the Eye in Human Face Images. Pattern Recognition 29(5), 771–779 (1996)
4. Bhuiyan, A., Ampornaramveth, V., Muto, S., Ueno, H.: Face Detection and Facial Feature Localization for Human-Machine Interface. NII Journal 5 (2003)
5. Bagherian, E., Rahmat, R., Udzir, N.: Extract of Facial Feature Point. JCSNS 9(1) (January 2009)
6. Castleman, R.: Digital Image Processing. Prentice-Hall, Englewood Cliffs (1996)
7. Van Vliet, J.: Grey-Scale Measurements in Multi-Dimensional Digitized Images. Doctor Thesis. Delft University Press (1993)
8. Hjelmas, E., Low, K.B.: Face detection: A Survey. Computer Vision and Image Understanding 83(3), 236–274 (2001)
9. Tang, Y., Yang, L., Liu, J., Ma, H.: Wavelet Theory and Its Application to Pattern Recognition. World Science Publishing Co. (2000) ISBN 981-02-3819-3
10. Li, J.: A Wavelet Approach to Edge Detection. Master Thesis of Science in the Subject of Mathematics Sam Houston State University, Huntsville, Texas (August 2003)
11. Nixon, S., Aguado, S.: Feature Extraction and Image Processing (2002) ISBN 0750650788
12. Lei, B., Hendriks, A., Reinders, M.: On Feature Extraction from Images. Technical Report on Inventory Properties for MCCWS
13. Goudail, F., Lange, E., Iwamoto, T., Kyuma, K., Otsu, N.: Face Recognition System Using Local Autocorrelations and Multiscale Integration. IEEE Trans. Pattern Anal. Machine Intell. 18(10), 1024–1028 (1996)
14. Yuille, A., Cohen, D., Hallinan, P.: Feature Extraction from Faces Using Deformable Templates. In: Proc. IEEE Computer Soc. Conf. on computer Vision and Pattern Recognition, pp. 104–109 (1989)
15. Rowley, H., Beluga, S., Kanade, T.: Neural Network-Based Face Detection. IEEE Transactions on Pattern Analysis and Machine Intelligence 20(1), 23–37 (1998)
16. Wiskott, L., Fellous, J.-M., Kruger, N., Von der Malsburg, C.: Face Recognition by Elastic Bunch Graph Matching. IEEE Trans. Pattern Anal. & Machine Intelligence 19(7), 775–779 (1997)
17. Bhuiyan, M., Ampornaramveth, V., Muto, S., Ueno, H.: Face Detection and Facial Feature Extraction. In: Int. Conf. on Computer and Information Technology, pp. 270–274 (2001)
18. Chang, T., Huang, T., Novak, C.: Facial Feature Extraction from Colour Images. In: Proceedings of the 12th IAPR International Conference on Pattern Recognition, vol. 2, pp. 39–43 (1994)
19. Mallat, M., Zhong, S.: Characterization of Signals from Multiscale Edges. IEEE Trans. Pattern Anal. Machine Intell. 14(7), 710–732 (1992)
20. Mallat, S., Hwang, W.L.: Singularity Detection and Processing with Wavelets. IEEE Trans. Inform. Theory 38(2), 617–643 (1992)
21. Tian, Y., Kanade, T., Cohn, J.: Evaluation of Gabor Wavelet-Based Facial Action Unit Recognition in Image Sequences of Increasing Complexity. In: Proceedings of the Fifth IEEE International Conference on Automatic Face and Gesture Recognition, pp. 218–223 (2002)
22. Meyer, Y.: Ondelettes, fonctions splines et analyses graduées. Rapport Ceremade 8703 (1987)
23. Meyer, Y.: Ondelettes et Opérateurs. Hermann, Paris (1987)
24. The Yale database, http://cvc.yale.edu/

Fast Iris Segmentation by Rotation Average Analysis of Intensity-Inversed Image

Wei Li[1,**] and Lin-Hua Jiang[2,*,**]

[1] Institute of Digital Lib., Zhejiang Sci-Tech University, Zhejiang, China
[2] Department of BFSC, Leiden University, Leiden, The Netherlands
L.Jiang@science.leidenuniv.nl

Abstract. Iris recognition is a reliable and accurate biometric technique used in modern personnel identification system. Segmentation of the effective iris region is the base of iris feature encoding and recognition. In this paper, a novel method is presented for fast iris segmentation. There are two steps to finish the iris segmentation. The first step is iris location, which is based on rotation average analysis of intensity-inversed image and non-linear circular regression. The second step is eyelid detection. A new method to detect the eyelids utilizing a simplified mathematical model of arc with three free parameters is implemented for quick fitting. Comparatively, the conventional model with four parameters is less optimal. Experiments were carried out on both self-collected images and CASIA database. The results show that our method is fast and robust in segmenting the effective iris region with high tolerance of noise and scaling.

Keywords: Biometrics, Iris segmentation, Iris location, Boundary recognition, Edge detection.

1 Introduction

Iris recognition is known as one of the most reliable and accurate biometric techniques in recent years [1-5]. Iris, suspending between the cornea and lens and perforated by the pupil, is a round contractile membrane of the eye (A typical iris image and related medical terminology are shown in Figure 1a). As a valuable candidate which could be used in highly reliable identification systems, the iris owns many advantages [4]: The patterns of iris are rather stable and not easy to change throughout one's whole life; Iris patterns are quite complex and can contain many distinctive features; An image of iris is much easier to be captured, it can be collected within a distance as far as one meter, and the process would be much less invasive. With all these advantages, Iris recognition is more and more widely used in modern personnel identification system.

The process of iris recognition mainly consists of six steps: image acquisition, segmentation, normalization, encoding, matching, and evaluation.

---

[*] Correeponding Author.
[**] Both authors equally contributed to this work.

H. Deng et al. (Eds.): AICI 2009, LNAI 5855, pp. 340–349, 2009.

Iris segmentation is the first substantial step of image processing to subtract the effective texture region of iris from the acquired images. The latter steps, such as iris feature encoding and evaluation, are directly based on the result of iris segmentation.

The main task of iris segmentation is to locate the inner and outer boundaries of iris, which is so called iris location. Dozens of papers (e.g. [6-11]) can be found in this active research field. Most of the cases, eyelids/eyelashes may strongly affect the final accuracy of iris recognition. For a complete segmentation, eyelids and eyelashes also need to be detected and removed. Besides the problem of iris location, eyelids and eyelashes detection are also major challenges for effective iris segmentation.

Eyelid detection is a challenging task due to the irregular shape of eyelids and eyelashes. Normally, much computational cost is needed and higher chances to be failed in eyelid detection. The corresponding research was addressed in many works (e.g. [12-15]), however it is less discussed and resolved compared to the research of iris location.

In the remainder of this paper, we will introduce the new method to do iris segmentation by rotation average analysis of intensity-inversed image in a detail. For iris location, in fact, circular edge detection, we first locate the inner boundary (as a circle) and then fitting the outer boundary (as a circle) by least-square non-linear circular regression. A simplified mathematical model of arc for detecting the eyelids is created to speed up the conventional process of eyelid detection. Other conventional methods are compared and discussed.

2 Segmentation Methods

There are two steps to complete the iris segmentation. First, locating the circular boundaries of the iris, which includes fitting the inner boundary of the iris (the pupillary boundary) and fitting the outer boundary of the iris (the limbus boundary). The fitted two circles are not necessarily to be concentric. According to the fact that the distance between the centers of them is rather small, it will be much easier to locate one after the other.

Normally the iris images in use are near infrared (NIR) images which contain abundant texture features in dark and brown color iris. In the NIR iris images, the sclera is not white, high contrast level can be found at the boundary between pupil and iris (inner boundary), while on the other hand the contrast at the boundary between iris and sclera (limbus, outer boundary) is much less (figure 1a). So, we started the whole process from locating the pupil.

2.1 Locate the Circular Boundaries of the Iris

Find the Pseudo Center and Radius of the Inner Boundary
An iris image consists of an array of pixels with positions (x,y). For an image, we assume the top left pixel with the coordinate (1,1) and the bottom right pixel (n,n).

To find the center of the pupil (figure 1b), an image is scanned twice and the most consecutive high-intensity pixels are individually counted both vertically and horizontally. A pixel is taken as 'high-intensity pixel' only if its intensity is higher than a user

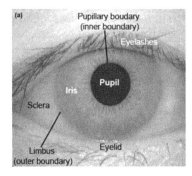 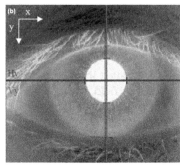

Fig. 1. (a) A NIR Iris image and related medical terminology. (b) Locate the pupil in the intensity-inversed iris image. Calculate the center coordinates by averaging Hx and Vx, as well as Hy and Vy. Scan the image vertically to get the most consecutive high intensity pixels on thick red line (Vx). Do the same horizontally showing as the thick blue line (Hy).

defined threshold, which is normally close to the global maximum of intensity, e.g. 85% of the maximum.

First, scan the image vertically, for each fixed x, scan along y axis. The x-coordinate - for which the most consecutive high intensity pixels in the y-direction are found - is stored (thick red line, Vx). This step is repeated for fixed y, and scan along x axis, and a y-coordinate is stored (thick blue line, Hy).

It is not necessarily true that (Vx,Hy) is the correct center. Therefore the second step is to calculate the middle point of the thick red and blue line. For example, if the first high-intensity pixel on thick red line is found at (Vx,100) and the last high-intensity pixel at (Vx,180), then the middle point (Vx, Vy) is (Vx, 140).

In a final step, Vx is averaged with Hx (thin blue line), and Hy is averaged with Vy (thin red line),which result in the mediated center coordinate. We call it pseudo center, because it may be affected by reflections and noise or irregular edge of the pupil. It is necessary to finely locate the center in the following process. The average length of the consecutive high intensity pixels is the diameter of the pupil.

Refine the Pupillary Center and Rotation Average Analysis

For an accurate refinement of the pseudo center, the center of mass of the pupil can be quickly calculated within a narrow range to avoid the effect of eyelid/eyelash noise. Formula to calculate the center of mass in a digital image is:

$$\overline{M} = \frac{1}{N} \sum_{1,2}^{m,n} I(i,j) * \overline{X} \tag{1}$$

Where, $\overline{M}$ is the coordinate of centroid, N is the number of total pixels, I is the intensity of each pixel, $\overline{X}$ is the coordinate of each pixel. The area of summation can be a limited region near the estimation.

Normally the pseudo center of the pupil slightly runs off the real center, and this refinement may finely calibrate it in several pixels. Figure 2a shows the inversed image, centered with the detected coordinate of the center of the pupil.

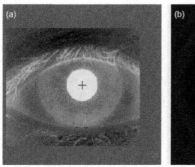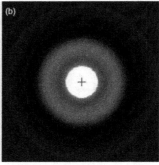

Fig. 2. (a) The intensity-inversed iris image is centered in a square panel. (b) The rotation average image of (a).

The pupil is thus quickly located. The time-consuming procedure of 'integro- differential' of intensities [2] and Canny edge detection [16] are not needed at all.

After an accurate center of the pupil and its radius are determined, we can next center the iris image in a larger-size square panel so that a rotation average image and curve can be calculated. Figure 2b shows the rotation average image according to the refined center, and figure 3a shows the rotation average curve.

Different averaging method can be used to calculate the rotation average curve:

A) Mean average

It is the most common way to calculate the mean average of all the pixels in a circular shell. It runs rather quickly, and most of the time, the result is stable.

B) Median average

Median is the number in the middle, when all numbers are listed in order. The statistic calculation of median average makes it more stable and less sensitive to singular points. Disturbance from the reflection spots and the spike noise can be reduced.

Mean and median are both types of averages, although mean is the most common type of average and usually refers to the arithmetic mean. Here, we choose the median average in calculating the rotation average curve.

From the rotation average curve (figure 3a), we clearly see an oscillation section near the middle-peak, which corresponds to the iris region containing abundant texture information. The curve is smooth and continuous, so that it is easy to find the middle-valley and middle-peak on the curve. The mean-value of the middle-peak and maximum intensity on the curve just corresponds with the papillary boundary, while the mean-value of the middle-peak and minimum intensity can be used to roughly segment the limbus boundary.

This limbus boundary calculated by the rotation average analysis is a rough value, assuming the pupil and limbus share the same center. Since the limbus boundary and pupillary boundary are not necessarily concentric, the pupillary boundary center we have found can not be used as the center of limbus boundary directly (though it's very close to the truth). A refinement step of fitting the outer boundary is necessary.

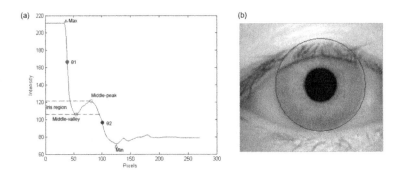

Fig. 3. (a) Rotation average curve and analysis of the intensity-inversed image. (b) Intermediate result of iris segmentation.

Fitting the Outer Boundary by Circular Regression

Based on this rotation average analysis of intensity-inversed image, rough center and radius of the outer boundary of iris are known from the former step. Now we are trying to refine the circle to fit the outer boundary.

In mathematics, "circle fitting" is a question of non-linear circular regression: Given a set of measured (x, y) pairs that are supposed to reside on a circle, but with some added noise. A circle to these points, i.e. find the center (x_c, y_c) and radius r, such that:

$$(x - x_c)^2 + (y - y_c)^2 = r^2 \tag{2}$$

The threshold used for roughly segmenting the limbus boundary can be utilized to binarize the iris image and generate a simple edge map for this circle fitting (figure 4a). This process is much faster than classical Canny edge detection.

Direct mathematical calculation of circular regression requires a good initial guess and less massive noise (e.g. eyelashes). Since we already had a very close estimation of the outer boundary, it will be much easier to accurately locate the limbus boundary by limiting the circle fitting in a certain range of estimation (for instance, +-10%).

The circle fitting is based on orthogonal distance regression. Most of the known fitting routines rely on the Levenberg-Marquardt optimization routine. The Levenberg-Marquardt algorithm requires an initial guess as well as the first derivatives of the distance function. In practice it converges quickly and accurately even with a wide range of initial guesses. For two-dimensional circle fitting, we give the distance, objective and derivatives functions:

Distance equation:

$$d(x, y) = \sqrt{(x - x_c)^2 + (y - y_c)^2} - r \tag{3}$$

Objective function:

$$E(x_c, y_c, r) = \sum \left(\sqrt{(x - x_c)^2 + (y - y_c)^2} - r \right)^2 \tag{4}$$

Derivatives:

$$\frac{\partial d}{\partial x_c} = -(x - x_c)/(d + r) \; ; \frac{\partial d}{\partial y_c} = -(y - y_c)/(d + r) \; ; \frac{\partial d}{\partial r} = -1 \qquad (5)$$

Good circular regression algorithms can readily be found to minimize the objective function E. Usually such an algorithm requires an initial guess, along with partial derivatives, either of E itself or of the distance function d. Of course, one can implement a least-squares algorithm which uses a different optimization algorithm.

We recommend and use Taubin's method [17] for circle fitting (see figure 4 for the result). It is a robust and accurate circle fitting method, which works well even if data points are observed only within a small arc. It is more stable than the other simple circle fitting methods, e.g. by Kasa [18].

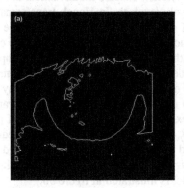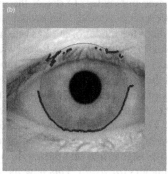

Fig. 4. (a)Simple binarized edge map for circle fitting (b) Circular regression and result of fitting the outer boundary

To be compared with, the known Hough transform algorithm [19] needs to calculate a Canny edge map first, and then use edge points to vote for particular model. The idea of both the Hough transform and non-linear circular regression is to find the contours which can fit in a circle with its center as (x_c, y_c) and the radius as r. Hough transform uses "voting" mechanism, while circular regression uses directly mathematical calculation.

Another method is to use integro-differential method to search for the circle of outer boundary, due to the limited searching range, this method is also fast. Here we do not intend to go into the details.

2.2 A New Mathematical Model of Arc for Detecting the Eyelid

The integro-differential approach can be adapted to detect the eyelid, when the contour model of circle is replaced by an arc model. This arc edge detector can be described as:

$$\max_{d,a,\theta} \left| G_\sigma(r) * \oint_{arc(d,a,\theta)} \frac{I(x,y)}{L(arc)} \right| \qquad (6)$$

G is a Gaussian low-pass filter, I is the intensity of pixels on arc, L is the length of arc. d, a, and θ are the three parameters of our new arc model, which will be introduced latter.

The contours of upper/lower eyelid can be regarded as a piece of parabolic arc approximately. Both Hough transformation and integro-differential method could be applied in this task. The methods used in finding the upper and lower eyelid arcs are in fact pattern fitting algorithms. In our implementation, we chose integro-differential method to avoid the calculation of edge maps. A standard mathematical model of arc (equation 7 & figure 5a) has four parameters for the j'th image: the apex coordinate (h_j, k_j), the curvature a_j, and the rotation angle θ_j. To fit the data with the mathematical model of arc is a time-consuming searching problem in a 4D space.

$$(-(x-h_j)\sin\theta_j + (y-k_j)\cos\theta_j)^2 = a_j((x-h_j)\cos\theta_j + (y-k_j)\sin\theta_j) \qquad (7)$$

To speed up this process, we proposed an efficient adapted mathematical model of arc with three free parameters to detect the eyelid/eyelash. In Hough transformation and integro-differential methods, the model of circular iris boundaries has three parameters, while here the model of arc has one more parameter. Consequently, the computation complexity would be much higher with the arc contours. In order to improve the performance, we simplified the eyelid model in our implementation by assuming that the axis of the arc goes through the center of pupillary boundary. In this way, the two parameters of the apex coordinate (h_j, k_j) can be replaced by a single parameter d_j, which refers to the distance from the apex to the original point (center of pupillary boundary). That is, $h_j = d_j\cos\theta_j$; $k_j = d_j\sin\theta_j$, and equation 7 is simplified to equation 8. Practically, the curvature a_j can be replaced by the distance from a predefined point on the arc to the tangent line that passes through the apex. Now we have only three easy-to-get parameters (d_j, a_j, θ_j). This new model is illustrated in figure 5b.

$$(-x\sin\theta_j + y\cos\theta_j)^2 = a_j(x\cos\theta_j + y\sin\theta_j - d_j) \qquad (8)$$

By simplifying the fitting problem from a 4D space to a 3D space, the processing speed is greatly improved. In order to gain more improvement of speed, one can simply constrain the searching range of these parameters. (e.g. to constrain the 'd' as the value between the inner and outer radius of the iris.)

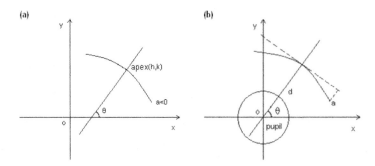

Fig. 5. (a) Conventional mathematical model of parabolic arc with 4 parameters. (b) Novel model of arc with 3 parameters and central axis passing through the center of pupil.

3   Results and Discussion

The new method for iris segmentation has been implemented in Matlab. To evaluate their performances, both self-obtained iris images and the CASIA iris image database are used. Figure 6 gives some examples of the segmentation results.

The time performance is listed in Table 1. Table 2 presents the comparison of accuracy and speed performance of iris location between our new method and other conventional methods. These results showed significant improvements when applying our new method.

Compared with other methods implemented based on Hough transformation or the integro-differential method [6-11], the new method of locating the pupil is different from conventional 'model fitting' methods in the sense that the time to locate the pupil is a constant for images with the same size, no matter how big the actual areas of pupils occupy in the images. If the pupils are quite different in size, conventional 'model fitting' methods usually consume much more time in searching a wider range, thus slow down the entire process.

Why we inverse the intensity of the image in the analysis? By inversing the intensity of the image, the circular pupil becomes the highest intensity area. It allows locating the pupil by checking the consecutive high intensities in the inversed image, and

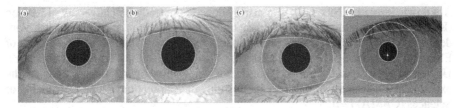

Fig. 6. Examples of iris segmentation results of the proposed method. (a)-(c) Iris segmentation of images from CASIA iris image database. (d) Segmentation of self-obtained iris image. A dummy arc of lower eyelid is drawn on the edge, if the eyelid and eyelash do not cover the iris.

Table 1. Typical time cost in each steps of segmentation for the new method presented

Step	Task	Time cost (s)
1.	Locate pupil and coarse limbus boundary.	0.16
2.	Fitting limbus boundary by circular regression.	0.05
3.	Detect upper or lower eyelid.	1.5

Table 2. Comparison of performance of iris boundaries location (not include eyelid detection)

Method	Success rate	Time cost (s)
Doughman's	96.7%	2.83
Wildes'	94.2%	5.74
Proposed	98.1%	0.21

refining the pupillary center with the centroid (the center of mass) of the pupil. If not inverse the intensity, the calculated center of mass will not be stable, since the surrounding area has higher intensity than the center area, the center of mass will easily move away from the real center.

4   Conclusions

In this paper, we present a novel iris segmentation method. The effective iris is subtracted in two steps: Iris boundaries location and eyelid detection. Intensity analysis of rotation average of inversed image is applied to quickly locate the pupil and give a close initial guess in circular regression of the outer boundary. The outer boundary is refined by least-square non-linear circular regression. A new simplified mathematical model of arc is created and implemented, which also works fast and efficiently compared to the conventional methods. Our method provides a new idea to do iris segmentation based on statistic analysis of rotation average.

In the experiment based on both self-acquired images and the public database (CASIA iris image database), the new method used in the iris segmentation shows fast and robust performance. Prospectively, there are still spaces to be improved in our method: automatic eyelashes detection and removal.

Acknowledgments. This work was mainly done in the Catholic University of Leuven (K.U.Leuven), Belgium. Thanks very much for the instruction of Prof. Dr. Dirk Vandermeulen in the Center for Processing Speech and Images (PSI) at the Department of Electrical Engineering (ESAT). Thanks for Ying Zhang in the Terry Fox Laboratory, University of British Columbia, Canada for acquisition of iris images.Parts of the experimental data were supplied by the Institute of Automation, Chinese Academy of Sciences (CASIA) [20].

References

1. Daugman, J.: New Methods in Iris Recognition. IEEE Trans. System, Man, and Cybernetics–Part B: Cybernetics 37(5), 1167–1175 (2007)
2. Daugman, J.: How Iris Recognition Works. IEEE Trans. Circuits and Systems for Video Technology 14(1), 21–30 (2004)
3. Daugman, J.: The Importance of being Random: Statistical Principles of Iris Recognition. Pattern Recognition 36(2), 279–291 (2003)
4. Daugman, J.: Statistical Richness of Visual Phase Information: Update on Recognizing Persons by Iris Patterns. Int. J. Computer Vision 45(1), 25–38 (2001)
5. Wildes, R.: Iris Recognition: An Emerging Biometric Technology. Proc. IEEE 85(9), 1348–1365 (1997)
6. Trucco, E., Razeto, M.: Robust iris location in close-up images of the eye. Patter Anal. Applic. 8, 247–255 (2005)
7. Tang, R., Han, J., Zhang, X.: An Effective Iris Location Method with High Robustness. Optica Applicata. 37(3), 295–303 (2007)
8. He, Z., Tan, T., Sun, Z.: Iris Localization via Pulling and Pushing. In: ICPR, vol. 4, pp. 366–369 (2006)

9. Yuan, W., Xu, L., Lin, Z.: An Accurate and Fast Iris Location Method Based on the Features of Human Eyes. In: Wang, L., Jin, Y. (eds.) FSKD 2005. LNCS (LNAI), vol. 3614, pp. 306–315. Springer, Heidelberg (2005)
10. Sun, C., Zhou, C., Liang, Y., Liu, X.: Study and Improvement of Iris Location Algorithm. In: Zhang, D., Jain, A.K. (eds.) ICB 2005. LNCS, vol. 3832, pp. 436–442. Springer, Heidelberg (2005)
11. Lee, J.C., Huang, P.S., Chang, C.P., Tu, T.M.: Novel and Fast Approach for Iris Location. IIHMSP 1, 139–142 (2007)
12. He, X.F., Shi, P.F.: An efficient iris segmentation method for recognition. In: Singh, S., Singh, M., Apte, C., Perner, P. (eds.) ICAPR 2005. LNCS, vol. 3687, pp. 120–126. Springer, Heidelberg (2005)
13. Arvacheh, E.M., Tizhoosh, H.R.: Iris Segmentation: Detecting Pupil, Limbus and Eyelids. In: ICIP, pp. 2453–2456 (2006)
14. He, Z., Tan, T., Sun, Z., Qiu, X.: Towards Accurate and Fast Iris Segmentation for Iris Biometrics. IEEE Trans. Pattern Analysis and Machine Intelligence 99(2008), doi:10.1109/TPAMI.2008.183
15. Jang, Y.K., Kang, B.J., Park, K.R.: Study on eyelid localization considering image focus for iris recognition. Pattern Recognition Letters 29(11), 1698–1704 (2008)
16. Canny, J.: A Computational Approach to Edge Detection. IEEE Trans. Pattern Analysis and Machine Intelligence 8, 679–714 (1986)
17. Taubin, G.: Estimation Of Planar Curves, Surfaces And Nonplanar Space Curves Defined By Implicit Equations, With Applications To Edge And Range Image Segmentation. IEEE Trans. PAMI 13, 1115–1138 (1991)
18. Kasa, I.: A curve fitting procedure and its error analysis. IEEE Trans. Inst. Meas. 25, 8–14 (1976)
19. Duda, R.O., Hart, P.E.: Use of the Hough Transformation to Detect Lines and Curves in Pictures. Comm. ACM 15, 11–15 (1972)
20. CASIA iris image database, http://www.sinobiometrics.com

A New Criterion for Global Asymptotic Stability of Multi-delayed Neural Networks[*]

Kaiyu Liu[1] and Hongqiang Zhang[2]

[1] College of Mathematics and Econometrics Hunan University,
Changsha, Hunan 410082, China
[2] College of Mathematics and Compution Science,
Changsha University of Science and Technology,
Changsha, Hunan 410076, China
liukyhnu@yahoo.com.cn, hqzhangcs@gmail.com

Abstract. This Letter presents some new sufficient conditions for the uniqueness and global asymptotic stability (GAS) of the equilibrium point for a class of neural networks with multiple constant time delays . It is shown that the use of a more general type of Lyapunov-Krasovskii functional enables us to establish global asymptotic stability of a class of delayed neural networks than those considered in some previous papers. Our results generalize or improve the previous results given in the literature.

Keywords: Delay neural Networks, Global asymptotic stability, Equilibrium analysis, Lyapunov-Krasovskii functional.

1 Introduction

In the last few years, stability of different classes of neural networks with time delays, such as Hopfield neural networks, cellular neural networks, bidirectional associative neural networks, has been extensively studied[1-7,10-12], particularly regarding their stability analysis. Recently, LMI-based techniques have been successfully used to tackle various stability problems for neural networks,mainly with a single delay(see, for example[8,9,15]) and a few with multi-delays[13,14].In the present paper, motivated by [14],by employing a more general Lyapunov functional and the LMI approach , we establish some sufficient conditions for the uniqueness and global asymptotic stability for neural networks with multiple delays . The conditions given in the literature are generalized or improved .

Consider the following neural networks with multiple delays

$$\dot{u}(t) = -Au(t) + Wg(u(t)) + \sum_{j=1}^{m} W_j g(u(t - \tau_j)) + I, \qquad (1)$$

where $u(t) = [u_1(t), \cdots, u_n(t)]^T$ is the neuron state vector, $A = diag(a_1, \cdots, a_n)$ is a positive diagonal matrix, $W, W_j = (\omega_{ik}^{(j)})_{n \times n}, j = 1, 2, \cdots, m$ are the interconnection

[*] Project supported Partially by NNSF of China (No:10571032).

H. Deng et al. (Eds.): AICI 2009, LNAI 5855, pp. 350–360, 2009.
© Springer-Verlag Berlin Heidelberg 2009

matrices, $g(u) = [g_1(u_1), \cdots, g_n(u_n)]^T$ denotes the neuron activation, and $I = [I_1, \cdots, I_n]^T$ is a constant external input vector, while $\tau_j > 0$ $(j = 1, 2, \cdots, m)$ being the delay parameters.

The usual assumptions on the activation functions are as follows

$$(H) \qquad 0 \le \frac{g_j(\xi_1) - g_j(\xi_2)}{\xi_1 - \xi_2} \le \sigma_j, \qquad j = 1, 2, \cdots, n,$$

for each $\xi_1, \xi_2 \in R, \xi_1 \ne \xi_2$, where σ_j are positive constants. In the following, we will shift the equilibrium point $u^* = [u_1^*, \cdots, u_n^*]^T$ of system (2) to the origin. The transformation $x(\cdot) = u(\cdot) - u^*$ puts system (3) into the following form

$$\dot{x}(t) = -Ax(t) + Wf(x(t)) + \sum_{j=1}^{m} W_j f(x(t - \tau_j)), \tag{2}$$

where $x = [x_1, \cdots, x_n]^T$ is the state vector of the transformed system, and $f_j(x_j) = g_j(x_j + u_j^*) - g_j(u_j^*)$ with $f_j(0) = 0, \forall j$. Note that the functions $f_j(\cdot)$ satisfy the condition (H), that is

$$0 \le \frac{f_j(\xi_1) - f_j(\xi_2)}{\xi_1 - \xi_2} \le \sigma_j, \qquad j = 1, 2, \cdots, n \tag{3}$$

for each $\xi_1, \xi_2 \in R, \xi_1 \ne \xi_2$, where σ_j are positive constants.

2 Main Stability Results

We first introduce the following **Lemma** 1, which will be used in the proof of our main results.

Lemma 1. (Schur complement)The linear matrix inequality (LMI)

$$S = \begin{pmatrix} S_{11} & S_{12} \\ S_{21} & S_{22} \end{pmatrix} > 0$$

is equivalent to the following conditions $S_{11} > 0$ (respectively, $S_{22} > 0$) and $S_{22} - S_{21}S_{11}^{-1}S_{12} > 0$, (respectively, $S_{11} - S_{12}S_{22}^{-1}S_{21} > 0$) where $S_{11}^T = S_{11}$, $S_{22}^T = S_{22}, S_{12}^T = S_{21}$.

Theorem 1. The The origin of (2) is the unique equilibrium point and it is globally asymptotically stable if there exists a positive diagonal matrix $P = diag(p_i > 0)$, and positive definite matrix $Q_j (j = 1,2,\cdots,m)$ such that

$$
\begin{bmatrix}
\Omega & -W_1 & \cdots & -W_m \\
-W_1^T & Q_1 & \cdots & 0 \\
\vdots & \vdots & \cdots & \vdots \\
-W_m^T & 0 & \cdots & Q_m
\end{bmatrix} > 0,
\tag{4}
$$

where $\Omega = 2A\sum^{-1} P - WP - PW^T - P\sum_{j=1}^{m} Q_j P$ ($\sum = diag(\sigma_i > 0)$).

Proof. We will first prove the uniqueness of the equilibrium point. To this end, let us consider the equilibrium equation of (2) as follows

$$
Ax^* - Wf(x^*) - \sum_{j=1}^{m} W_j f(x^*) = 0,
\tag{5}
$$

Where x^* is the equilibrium point. We have to prove $x^* = 0$, for this purpose, multiplying both sides of (5) by $2f^T(x^*)P^{-1}$, we obtain

$$
2f^T(x^*)P^{-1}Ax^* - 2f^T(x^*)P^{-1}Wf(x^*) - 2\sum_{j=1}^{m} f^T(x^*)P^{-1}W_j f(x^*) = 0, \tag{6}
$$

Using (3) in (6) results in

$$
f^T(x^*)(2P^{-1}A\sum^{-1} - P^{-1}W - W^T P^{-1})f(x^*)
$$

$$
-2\sum_{j=1}^{m} f^T(x^*)P^{-1}W_j f(x^*) \le 0.
$$

Adding and subtracting the term $f^T(x^*)\sum_{j=1}^{m} Q_j f(x^*)$ in the left side of the above inequality yields

$$
UBU^T \le 0,
\tag{7}
$$

where $U = [f^T(x^*)P^{-1}, f^T(x^*),\cdots, f^T(x^*)]$,

$$
B = \begin{bmatrix}
\Omega & -W_1 & \cdots & -W_m \\
-W_1^T & Q_1 & \cdots & 0 \\
\vdots & \vdots & \cdots & \vdots \\
-W_m^T & 0 & \cdots & Q_m
\end{bmatrix}.
\tag{8}
$$

Condition (4) in **Theorem** 1 is $B > 0$. Thus it follows from (7) that $f(x^*) = 0$, substitutes this into (5), we get $Ax^* = 0$ which implies $x^* = 0$. This has shown that the origin is the unique equilibrium point for every I.

We will now prove the GAS of the origin of (2). To this end, let us consider the following Lyapunov-Kraasovskii functional

$$V(x(t)) = x^T(t)Cx(t) + 2\sum_{i=1}^{n} p_i^{-1} \int_0^{x_i(t)} f_i(s)ds$$

$$+ \sum_{j=1}^{m} \int_{-\tau_j} f^T(x(s))Q_j f(x(s))ds,$$

where $C, P = diag(p_i > 0)$ and $Q_j (j = 1,2,\cdots,m)$ are positive definite matrices with C being being chosen appropriately later on. Clearly, $V(x(t))$ is positive except at the origin and it is radially unbounded. Evaluating the time derivative of V along the trajectory of (2), we have

$$\dot{V}(x(t)) = -2x^T(t)CAx(t) + 2x^T(t)CWf(x(t))$$

$$+ 2\sum_{j=1}^{m} x^T(t)CW_j f(x(t-\tau_j)) - 2f^T(x(t))P^{-1}Ax(t)$$

$$+ 2f^T(x(t))P^{-1}Wf(x(t)) + \sum_{j=1}^{m} f^T(x(t))Q_j f(x(t))$$

$$+ 2\sum_{j=1}^{m} f^T(x(t))P^{-1}W_j f(x(t-\tau_j))$$

$$- \sum_{j=1}^{m} f^T(x(t-\tau_j))Q_j f(x(t-\tau_j)).$$

Since $f^T(x(t))P^{-1}A\Sigma^{-1}f(x(t)) \le f^T(x(t))P^{-1}Ax(t)$, we can write

$$\dot{V}(x(t)) \le -x^T(t)(CA + AC)x(t) + 2x^T(t)CWf(x(t))$$

$$+ 2\sum_{j=1}^{m} x^T(t)CW_j f(x(t-\tau_j)) + f^T(x(t))\sum_{j=1}^{m} Q_j f(x(t))$$

$$- f^T(x(t))(2P^{-1}A\Sigma^{-1} - P^{-1}W - W^T P^{-1})f(x(t))$$

$$+ 2\sum_{j=1}^{m} f^T(x(t))P^{-1}W_j f(x(t-\tau_j))$$

$$- \sum_{j=1}^{m} f^T(x(t-\tau_j))Q_j f(x(t-\tau_j)),$$

which can be written in the form

$$\dot{V}(x(t)) \le U_1 S U_1^T,$$ (9)

where $U_1 = [x^T(t)C, f^T(x(t)P^{-1}, f^T(x(t-\tau_1)), \cdots, f^T(x(t-\tau_m))]$,

$$S = \begin{bmatrix} AC^{-1} + C^{-1}A & -WP & -W_1 & \cdots & -W_m \\ -PW^T & \Omega & -W_1 & \cdots & -W_m \\ -W_1^T & -W_1^T & Q_1 & \cdots & 0 \\ \vdots & \vdots & \vdots & & \vdots \\ -W_m^T & -W_m^T & 0 & \cdots & Q_m \end{bmatrix}.$$

In the following , we will choose a positive-definite matrix C such that $S > 0$. For this purpose, let

$$S = \begin{pmatrix} AC^{-1} + C^{-1}A & D \\ D^T & B \end{pmatrix},$$ (10)

where $D = [-W_1, \cdots, W_m]$ and B is given in (8). In view of (4) and (8), we have $B > 0$. Now, choose a positive definite matrix G such that

$$G > DB^{-1}D^T.$$ (11)

Since $-A = diag(-a_i < 0)$ is negative definite, there must be a positive-definite matrix H such that

$$H(-A) + (-A)H = -G.$$ (12)

Let $C = H^{-1}$, then C is positive definite and we have by (11) and (12)

$$AC^{-1} + C^{-1}A - DB^{-1}D^T > 0,$$

from which and **Lemma** 1 we conclude $S > 0$, and it follows by (9) that $\dot{V}(x(t)) \le 0$. Moreover $\dot{V}(x(t)) = 0$ if and only if

$$[x^T(t)C, f^T(x(t)P^{-1}, f^T(x(t-\tau_1)), \cdots, f^T(x(t-\tau_m))] = 0,$$

i.e., $\dot{V}(x(t)) = 0$ if and only if

$$x(t) = f(x(t)) = f(x(t-\tau_1)) = \cdots = f(x(t-\tau_m)) = 0.$$

Hence, $\dot{V}(x(t))$ is negative definite and therefore, the origin of (2) or equivalently the equilibrium x^* of (1) is globally asymptotically stable. This completes the proofs.

In view of **Lemma** 1, the following **Theorem** 1' and **Theorem** 1" are both equivalent descriptions of **Theorem** 1.

Theorem 1'. The origin of (2) is the unique equilibrium point and it is globally asymptotically stable if there exists a positive diagonal matrix P and positive definite matrices $Q_j (j = 1,2,\cdots,m)$ such that

$$2A\Sigma^{-1}P - WP - PW^T - P\sum_{j=1}^{m}Q_jP - \sum_{j=1}^{m}WQ_j^{-1}W_j^T > 0.$$

Theorem 1". The origin of (2) is the unique equilibrium point which is globally asymptotically stable if there exists a positive diagonal matrix P and positive definite matrices $Q_j (j = 1,2,\cdots,m)$ such that

$$B_1 = 2A\Sigma^{-1}P - WP - PW^T - P\sum_{j=1}^{m}Q_jP > 0$$

and

$$\begin{bmatrix} Q_1 - W_1^T B_1^{-1} W_1 & \cdots & -W_1^T B_1^T W_m \\ \vdots & \vdots & \vdots \\ -W_m^T B_1^T W_1 & \cdots & Q_m - W_m^T B_1^T W_m \end{bmatrix} > 0.$$

In the following, we will give further result for GAS of the origin of (2)(or equivalently the equilibrium point of (1)). We need the following

Lemma 2. Let $Q_j, S, S_j \in R^{n\times n} (j = 1,2,\cdots,m)$ be positive-definite matrices and let $W_j \in R^{n\times n}$. Then the following LMI

$$\begin{bmatrix} Q_1 - W_1^T S W_1 & \cdots & -W_1^T S W_m \\ \vdots & \vdots & \vdots \\ -W_m^T S W_1 & \cdots & Q_m - W_m^T S W_m \end{bmatrix} > 0 \qquad (13)$$

hold if the following LMIs hold

$$Q_j - W_j^T S_j W_j > 0, \quad j = 1,2,\cdots,m \qquad (14)$$

and

$$S^{-1} > \sum_{j=1}^{m} S_j^{-1}. \qquad (15)$$

Proof. Let A denote the matrix given in the left side of (14), i.e.,

$$A = \begin{pmatrix} Q_1 & & \\ & \ddots & \\ & & Q_m \end{pmatrix} - \begin{pmatrix} W_1^T SW_1 & \cdots & W_1^T SW_m \\ \vdots & \cdots & \vdots \\ W_m^T SW_1 & \cdots & W_m^T SW_m \end{pmatrix},$$

consider the quantic form associate with A as follows

$$f(x) = x^T A x = \sum_{j=1}^m x_j^T Q_j x_j - \sum_{j=1}^m \sum_{i=1}^m x_j^T W_j^T SW_i x_i,$$

where $x = (x_1^T, x_2^T, \cdots, x_m^T) \in R^{nm}$ with $x_j \in R^n, j = 1,2,\cdots,m$ .Adding and

subtracting the term $\sum_{j=1}^m x_j^T W_j^T S_j W_j x_j + \sum_{j=1}^m \sum_{i=1}^m x_j^T W_j^T SW_i x_i$, we have

$$f(x) = x^T A x = \sum_{j=1}^m x_j^T (Q_j - W_j^T S_j W_j) x_j$$

$$+ \sum_{j=1}^m [x_j^T W_j^T S_j W_j x_j - 2x_j^T W_j^T S \sum_{i=1}^m W_i x_i] + \sum_{j=1}^m x_j^T W_j^T S \sum_{i=1}^m W_i x_i . \quad (16)$$

Note that

$$x_j^T W_j^T S_j W_j x_j - 2x_j^T W_j^T S \sum_{i=1}^m W_i x_i$$

$$= [S_j W_j x_j - \sum_{i=1}^m SW_i x_i]^T \cdot S_j^{-1} \cdot [S_j W_j x_j - \sum_{i=1}^m SW_i x_i]$$

$$- \sum_{i=1}^m x_i^T W_i^T SS_j^{-1} SW_i x_i$$

for $i = 1,2,\cdots,m$. This implies that

$$\sum_{j=1}^m [x_j^T W_j^T S_j W_j x_j - 2x_j^T W_j^T S \sum_{i=1}^m W_i x_i]$$

$$\geq -\sum_{i=1}^m x_i^T W_i^T S \sum_{j=1}^m S_j^{-1} SW_i x_i .$$

Substituting this into (16) result in

$$f(x) = x^T A x \geq \sum_{j=1}^m x_j^T (Q_j - W_j^T S_j W_j) x_j$$

$$+ \sum_{i=1}^m [x_i^T W_i^T [S(-\sum_{i=1}^m S_j^{-1} + S^{-1})S] \sum_{i=1}^m W_i x_i ,$$

Which, by using the condition (14) and (15), implies that $f(x) = x^T A x > 0$ for all $x \neq 0$. Therefore we have $A > 0$ and the proofs is completed.

Theorem 2. The origin of (2) is the unique equilibrium point which is globally asymptotically stable if there exists a positive diagonal matrix P and positive definite matrices $S_j (j = 1,2,\cdots,m)$ such that

$$2A\Sigma^{-1}P - WP - PW^T - P\sum_{j=1}^m W_j^T S_j W_j P - \sum_{j=1}^m S_j^{-1} > 0. \tag{17}$$

Proof. In view of (17), there exists a sufficiently small positive constant $\delta > 0$ such that

$$2A\Sigma^{-1}P - WP - PW^T - P\sum_{j=1}^m (W_j^T S_j W_j + \delta E)P - \sum_{j=1}^m S_j^{-1} > 0, \tag{18}$$

where E is the appropriate unit matrix. Let

$$Q_j = W_j^T S_j W_j + \delta E, \quad j = 1,2,\cdots,m \tag{19}$$

and

$$B_2 = 2A\Sigma^{-1}P - WP - PW^T - P\sum_{j=1}^m Q_j P. \tag{20}$$

Then $Q_j > W_j^T S_j W_j \geq 0$ ($j = 1,2,\cdots,m$) and by (18)-(20), we have

$$B_2 > \sum_{j=1}^m S_j^{-1} > 0. \tag{21}$$

It follows from **Lemma 2** that

$$\begin{bmatrix} Q_1 - W_1^T B_2^{-1} W_1 & \cdots & -W_1^T B_2^T W_m \\ \vdots & \vdots & \vdots \\ -W_m^T B_2^T W_1 & \cdots & Q_m - W_m^T B_2^T W_m \end{bmatrix} > 0 \tag{22}$$

In view of (20)-(22), the conclusion of **Theorem 2** follows from **Theorem 1''**. This completes the proofs.

Remark 1. Theorem 1 and **Theorem** 2 generalize **Theorem** 1 in [14] .

The following **Theorem** 3 give explicit condition which ensure the GAS of the origin of (2), or equivalently the equilibrium point of (1).

Theorem 3. The origin of (2) is the unique equilibrium and it is globally asymptotically stable if there exists a positive constant $\alpha > 0$ such that the following conditions hold

(a) $-(2A\Sigma^{-1} + W + W^T + \alpha E)$ is positive definite;

(b) $\sum_{j=1}^{m} \|W_j\|_2 \leq \dfrac{\alpha}{2}$.

Proof. In view of condition (a), there must be a positive constant $\delta > 0$ such that

$$-(2A\Sigma^{-1} + W + W^T + \alpha E) > 2\delta E .\tag{23}$$

Take $P = \lambda E$ and $S_j = \lambda_j E (j = 1,2,\cdots,m)$,where λ and λ_j are positive constant and observe that

$$2A\Sigma^{-1}P - WP - PW^T - \sum_{j=1}^{m} S_j^{-1} - P\sum_{j=1}^{m}(W_j^T S_j W_j + \delta^2 E)P$$

$$= 2\lambda A\Sigma^{-1} - \lambda W - \lambda W^T - \sum_{j=1}^{m}\lambda_j^{-1}E - \lambda^2\sum_{j=1}^{m}(\lambda_j W_j^T W_j + \delta^2 E)$$

$$\geq 2\lambda A\Sigma^{-1} - \lambda W - \lambda W^T - \lambda^2\sum_{j=1}^{m}(\lambda_j\|W_j\|_2^2 + \delta^2 + \lambda^{-2}\lambda_j^{-1})E$$

$$= -\lambda(-2A\Sigma^{-1} + W + W^T)$$

$$- \lambda^2\sum_{j=1}^{m}[\lambda_j(\|W_j\|_2 + \delta - \lambda^{-1}\lambda_j^{-1})^2 + 2\lambda(\|W_j\|_2 + \delta)]E .$$

Let $\lambda^{-1}\lambda_j^{-1} = \|W_j\|_2 + \delta$. Then

$$2A\Sigma^{-1}P - WP - PW^T - \sum_{j=1}^{m} S_j^{-1} - P\sum_{j=1}^{m}(W_j^T S_j W_j + \delta^2 E)P$$

$$\geq -\lambda(-2A\Sigma^{-1} + W + W^T) - 2\lambda\sum_{j=1}^{m}(\|W_j\|_2 + \delta)E$$

$$= -\lambda[(-2A\Sigma^{-1} + W + W^T + \alpha E) - 2\delta E] + 2\lambda(\dfrac{\alpha}{2} - \sum_{j=1}^{m}\|W_j\|_2)E$$

Using (23) and conditions (b) in **Theorem** 3 , we get

$$2A\Sigma^{-1}P - WP - PW^T - \sum_{j=1}^{m} S_j^{-1} - P\sum_{j=1}^{m}(W_j^T S_j W_j + \delta^2 E)P > 0 .$$

Moreover we have

$$2A\Sigma^{-1}P - WP - PW^T - \sum_{j=1}^{m} S_j^{-1} - P\sum_{j=1}^{m} W_j^T S_j W_j P > 0.$$

Thus the proof is completed by **Theorem** 2.

Let $A = E$ and $\alpha = 2 + \beta$ in **Theorem** 3, we have the following **Theorem** 4 for DCNN

$$\dot{x}(t) = -x(t) + Wg(x(t)) + \sum_{j=1}^{m} W_j g(x(t - \tau_j)), \qquad (24)$$

Theorem 4. The origin of **(24)** is the unique equilibrium point which is globally and asymptotically stable if there exists a real constant $\beta > -2$ such that following conditions hold

(I) $-(W + W^T + \beta E)$ is positive definite;

(II) $\sum_{j=1}^{m} \|W_j\|_2 \le 1 + \dfrac{\beta}{2}$.

Remark 2. Theorem 4 improve the results in [15] which $\beta > 0$.

3 Conclusions

The equilibrium and global asymptotic stability properties of neural networks with multiple constant delays have been studied. Some new stability criteria have been derived by employing a more general type of Lyapunov-Krasovskii functional. The conditions are expressed in terms of the linear matrix inequality, which can be verified efficiently with the Matlab LMI Control Toolbox.

References

1. Hopfield, J.: Neurons with Graded Response Have Collective Computational Properties Like Those of Two-State Neurons. Proc. Natl. Acad. Sci. USA 81, 3088–3092 (1984)
2. Marcus, C.M., Westervelt, R.M.: Stability of Analog Neural Network with Delay. Phys. Rev. A 39, 347–359 (1989)
3. Belair, J.: Stability in A Model of A Delayed Neural Network. J. Dynam. Differential Equations 5, 607–623 (1993)
4. Gopalsamy, K., Leung, I.: Delay Iinduced Periodicity in A Neural Network of Excitation and Inhibition. Physica D 89, 395–426 (1996)
5. Ye, H., Michel, N., Wang, A.: Global Stability of Local Stability of Hopfield Neural Networks with Delays. Phys. Rev. E 50, 4206–4213 (1994)
6. Liao, X.F., Yu, J., Chen, G.: Novel Stability Criteria for Bi-directional Associative Memory Neural Networks with Time Delays. Int. J. Circuit Theory Appl. 30, 519–546 (2002)
7. Gopalsamy, K., He, X.Z.: Stability in A Asymmetric Hopfield Networks with Transmission Delays. Physica D 76, 344–358 (1994)

8. Liao, X.F., Chen, G., Sanchez, E.N.: LMI-Based Approach for Asymptotically Stability Analysis of Delayed Neural Networks. IEEE Trans. Circuits Syst. 49(7), 1033–1039 (2002)
9. Liao, X., Chen, G., Sanchez, N.: Delayed-Dependent Exponential Stability Analysis of Delayed Neural Networks: an LMI Approach. Neural Networks 15, 855–866 (2002)
10. Liao, X.F., Wu, Z., Yu, J.: Stability Analysis of Cellular Neural Networks with Continuous Time Dlays. J. Comput. Appl. Math. 143, 29–47 (2002)
11. Arik, S.: Stability Analysis of Delayed Neural Networks. IEEE Trans. Circuits Syst. I. 47, 1089–1092 (2000)
12. Liao, X.X., Wang, J.: Algebraic Criteria for Global Exponential Stability of Cellular Neural Networks with Multiple Time Delays. IEEE Trans. Circuits Syst. 50, 268–275 (2003)
13. Wu, W., Cui, B.T.: Global Robust Exponential Stability of Delayed Neural Networks. Chaos, Solitons and Fractals 35, 747–754 (2008)
14. Liao, X.F., Li, C.: An LMI Approach to Asymptotical Stability of Multi-Delayed Neural Networks. Physica D 200, 139–155 (2005)
15. Arik, S.: Global Asymptotic Stability of A Larger Class of Neural Networks with Constant Time Delay. Phys. Lett. A 311, 504–511 (2003)

An Improved Approach Combining Random PSO with BP for Feedforward Neural Networks

Yu Cui, Shi-Guang Ju, Fei Han*, and Tong-Yue Gu

School of Computer Science and Telecommunication Engineering, Jiangsu University,
Zhenjiang, Jiangsu, China
allen790336@hotmail.com, jushig@ujs.edu.cn, hanfei@ujs.edu.cn,
gtyglx@tom.com

Abstract. In this paper, an improved approach which combines random particle swarm optimization with BP is proposed to obtain better generalization performance and faster convergence rate. It is well known that the backpropagation (BP) algorithm has good local search ability but it is easily trapped to local minima. On the contrary, the particle swarm optimization algorithm (PSO), with good global search ability, converges rapidly during the initial stages of a global research. Since the PSO suffers from the disadvantage of losing diversity, it converges more slow around the global minima. Hence, the global search is combined with local search reasonably in the improved approach which is called as RPSO-BP. Moreover, in order to improve the diversity of the swarm in the PSO, a random PSO (RPSO) is proposed in the paper. Compared with the traditional learning algorithms, the improved learning algorithm has much better convergence accuracy and rate. Finally, the experimental results are given to verify the efficiency and effectiveness of the proposed algorithm.

Keywords: Particle swarm optimization, backpropagation, feedforward neural networks.

1 Introduction

Feedforward neural networks(FNNs) have been widely used to approximate arbitrary continuous functions [1,2], since a neural network with a single nonliner hidden layer is capable of forming an arbitrarily close approximation of any continous nonliner mapping [3]. There have been many algorithms used to train the FNN, such as backpropagation algorithm (BP), particle swarm optimization algorithm (PSO) [4], simulating annealing algorithm (SSA) [5], gentic algorithm(GA) [6,7] and so on. Regarding the FNN training algorithms, the BP algorithm is easily trapped into a local minima and converge slowly [8]. To solve this problem, many improved BP algorithm has been proposed [9-11]. However these algorithms have not removed the disadvantages of the BP algorithm in essence.

In 1995, inspired from complex social behavior shown by the natural species like flock of birds, PSO was proposed by James Kennedy and Russell Eberhart. Different

---

* Corresponding author.

H. Deng et al. (Eds.): AICI 2009, LNAI 5855, pp. 361–368, 2009.

from the BP algorithm, the PSO algorithm has good ability of global search. Although the PSO algorithm has shown a very good performance in solving many problems, it suffers from the problem of premature convergence like most of the stochastic search techniques.

In order to make use of the advantages of PSO and BP, many researchers have concentrated on hybrid algorithm such as PSO-BP, CPSO-BP and so on [12-14]. In these algorithms, the PSO is used to do global search first and then switch to gradient descending searching to do local search around global optimum. It is proved to be better in convergence rate and convergence accuracy. However, premature convergence still exists in the PSO-BP and CPSO-BP. During the global search, every particle has to follow both it's best historic position (pbest) and the best position of all the swarm (gbest). Therefore, pbest will get close to gbest, and then the current particle swarm may lose its diversity.

In this paper an improved PSO-BP approach named RPSO-BP is proposed to solve the problems mentioned above. In this algorithm, there are three methods to update each particle and each particle selects one randomly. In another word, each particle does not have only one update method that following its best historic position and the best position of the swarm, any more. In this way the gbest can avoid getting closed to pbest and the likelihood of premature convergence can be reduced. Moreover, the experiment results show that the proposed algorithm has better generalization performance and convergence performance than other traditional algorithms.

2 Particle Swarm Optimization

PSO can be stated as initializing a team of random particles and finding the optimal solutions by iterations. Each particle will update itself by two optimal values pbest and gbest which are mentioned above.

The original PSO algorithm is described as follows:

$$V_i(t+1) = V_i(t) + c1 * r1 * (P_i(t) - X_i(t)) + c2 * r2 * (P_g(t) - X_i(t)) \tag{1}$$

$$X_i(t+1) = X_i(t) + V_i(t+1) \tag{2}$$

where V_i is the velocity of the ith particle; X_i is the position of the ith particle; P_i is the best position achieved by the particle so far ; P_g is the best position among all particles in the population; $r1$ and $r2$ are two independently and uniformly distributed random variables with the range of [0,1]; $c1$ and $c2$ are positive constant parameters called accelerated coefficients, which control the maximum step size.

The adaptive particle swarm optimization (APSO) [15,16] algorithm is proposed by Shi & Eberhart in 1998. This algorithm can stated as follows:

$$V_i(t+1) = w * V_i(t) + c1 * r1 * (P_i(t) - X_i(t)) + c2 * r2 * (P_g(t) - X_i(t)) \tag{3}$$

$$X_i(t+1) = X_i(t) + V_i(t+1) \tag{4}$$

where w is called the inertia weight that controls the impact of the previous velocity of the particle on its current. Several selection strategies of inertial weight w have been given. Generally, in the beginning stages of algorithm, the inertial weight w should be reduced rapidly, and the inertial weight w should be reduced slowly when around global optimum.

Another important variant of standard PSO is the CPSO, which was proposed by Clerc and Kennedy [17]. The CPSO ensures the convergence of the search producers and can generate higher-quality solutions than standard PSO with inertia weight on some studied problems.

3 The Proposed Algorithm (RPSO-BP)

In this section, RPSO-BP is proposed to solve the problems mentioned above. In order to reduce the likelihood of the premature convergence, a random PSO (RPSO) is proposed. In the RPSO-BP, the FNN is trained by RPSO first and then trained by BP algorithm.

The searching process of the RPSO is started from initializing a group of random particles, and a random parameter f1. On the meantime, Each particle has a random parameter f2. The update of each particle is determined by the following rules:

(1) If f2 is greater than f1, all particles will be updated according to Eqs.(3)- (4).

(2) If f2 is less than f1 and the pbest is nearest to the gbest, a particle j will be selected randomly and the particle will be updated according to Eqs.(4)-(5).

$$V_i(t+1) = w*V_i(t) + c1*r1*(P_j(t) - X_i(t)) + c2*r2*(P_g(t) - X_i(t)) \tag{5}$$

(3) If f2 is less than f1 and the pbest of the particle h is nearest to the gbest, the particle will be updated according to Eqs.(6) and (4).

$$V_i(t+1) = w*V_i(t) + c1*r1*(P_h(t) - X_i(t)) + c2*r2*(P_g(t) - X_i(t)) \tag{6}$$

where P_j and P_h are the personal (local) best position achieved by the particle j and particle h so far respectively.

Similar to the PSO, the parameter w in the above RPSO-BP algorithm reduces gradually as the iterative generation increases as follows[16]:

$$W(iter) = W_{max} - (W_{max} - W_{min})/_{iterma * iter} \tag{7}$$

where W_{max} is the initial inertial weight, W_{min} is the inertial weight of liner section ending, $iterma$ is the used generations that inertial weight, is reduced linearly, $iter$ is a variable whose range is [1,$iterma$].

4 Experimental Results

In this paper, some experiments are conducted to verify the effectiveness of the proposed learning approach. In the following experiments, the performances of the BP, PSO, CPSO, PSO-BP and CPSO-BP are compared with that of the RPSO-BP. All the experiments are carried out in MATLAB 6.5 environment running in a Pentium 4, 2.40 GHZ CPU.

4.1 Function Approximation Problem

The function f(x), y=(sin(x)*cos(x.^3)+(x.^2/9+x/3+cos(x.^3))*exp((-0.5)*(x.^2)))/2, is approximated by FNN in this experiment. For all the six algorithms, the activation functions of the hidden neurons are tansig function, while the one in the output layers is purline function.

First, assume that the number of the hidden neurons is 8, and the number of the total training data is 126 which are selected at an identical spaced interval from [0,3.138], while the 125 testing data are selected at an identical spaced interval from [0.0125,3.1249].

Table 1. The approximation accuracies and iteration number with six learning algorithms for approximating the function f(x)

Learning algorithms	Training MSE	Testing MSE	Iteration number
BP	0.0660675	0.0661	30000
PSO	0.0565	0.0573	20000
CPSO	0.0978	0.2104	20000
PSO-BP	9.99206e-005	6.9930e-004	15000
CPSO-BP	9.9592e-005	6.9584e-004	15000
RPSO-BP	6.59849e-005	4.8331e-004	15000

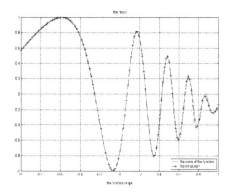

Fig. 1. The approximating result of the RPSO-BP algorithm for approximating the function f(x)

Second, we set the maximal iteration number as 30000 for the BP algorithm, 20000 for the PSO and CPSO algorithms. In the later three algorithms, the maximal iteration number as 300 for PSO, and the one as 15000 for BP algorithm. The corresponding results are summarized in Table 1 and Figs. 1-2.

It can be found from Table 1 and Figs.1-2 that the values of MSE of the new algorithm are always less than those of the other five algorithms for approximating the function. The results also support the conclusion that the generalization performance of the

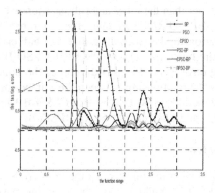

Fig. 2. The testing error curves of six learning algorithms for approximating the function f(x)

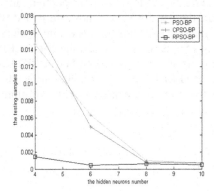

Fig. 3. The relationship between the number of hidden neurons and testing error for approximating the function f(x) with three algorithms

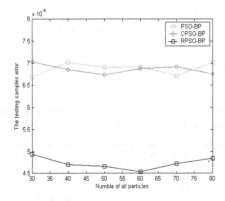

Fig. 4. The relationship between the number of PSO and testing error for approximating the function f(x) with three learning algorithms

proposed algorithm is better than the other algorithms. Moreover, the hybrid algorithms such as PSO-BP, CPSO-BP and RPSO-BP converge faster than other algorithms.

Fig.3. shows the relationship between the number of hidden neurons and testing error for approximating the function f(x) with PSO-BP, CPSO-BP and RPSO-BP. It can be seen that the testing error has a downward trend as the number of hidden neurons increases. Fig. 4 shows the relationship between the number of PSO and testing error for approximating the function f(x) with PSO-BP, CPSO-BP and RPSO-BP. The most suitable number of PSO can be selected as 60 for RPSO-BP.

4.2 Oil Refining Problem

In petrochemical industry, the true boiling point curve of crude oil reflects the composition of the distilled crude oil. To build a model that takes the mass percentage of distilled component as the independent variable and the distilled temperature as the dependent variable is an important problem in petrochemical industry. In this subsection, we use FNN to build nonparametric models to resolve this problem. In this subsection, the number of hidden neurons and the size of PSO we chose are 10 and 50 respectively.

Regarding the three hybrid algorithms in this example, the maximal iteration number of the corresponding PSO is set as 500, and the one for BP is set as 5000. For the BP, the maximal iteration is assumed as 10000. For the PSO and CPSO, the maximal iteration number is assumed as 10000. The simulation results are summarized in the table 2 and the comparison result of the three hybrid algorithms are showed in Fig. 5.

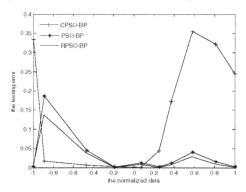

Fig. 5. The testing error curves of six learning algorithms for modeling the true boiling point curve of crude oil

Table 2. The approximation accuracies and iteration number with six learning algorithms for modeling the true boiling point curve of crude oil with six algorithms

Learning algorithms	Training MSE	Testing MSE	Iteration number
BP	0.0865	0.2668	10000
PSO	0.0737	0.2644	10000
CPSO	0.0908	0.2685	10000
PSO-BP	9.8348e-005	0.0604	5000
CPSO-BP	9.85675e-005	0.0610	5000
RPSO-BP	6.02109e-005	0.0296	5000

From the Table 2 and Fig.5 the same conclusions as the experiment of the function approximation problem can be drawn.

5 Conclusions

In this paper, an improved PSO-BP approach which combines RPSO with BP is proposed to train FNN. PSO has good global search ability, but the swarm loses its diversity easily. However, the gradient descent algorithm has good local search ability. In this paper, global search algorithm—RPSO is combined with BP reasonably. Moreover, in order to improve the diversity of the swarm in the PSO, RPSO are proposed in the paper. The RPSO-BP could not only improve the diversity of the swarm but also reduce the likelihood of the particles being trapped into local minima on the error surface. Finally, the experimental results were given to verify that the proposed algorithm has better generalization performance and faster convergence rate than other traditional algorithms. In future research works, we shall focus on how to apply this improved algorithm to solve more practical problems.

Acknowledgements. This work was supported by the National Natural Science Foundation of China (No.60702056), Natural Science Foundation of Jiangsu Province (No.BK2009197) and the Initial Funding of Science Research of Jiangsu University (No.07JDG033).

References

1. Homik, K.: Mulitilayer feedforward networks are universal approximators. Neural Networks 2, 359–366 (1989)
2. Chen, D.S., Jain, R.C.: A robust Back-propagation Algorithm for Function Approximation. IEEE Trans. on Neural Network 5, 467–479 (1994)
3. Meng, J.L., Sun, Z.Y.: Application of combined neural networks in nonlinear function approximation. In: Proceedings of the Third World Congress on Intelligent Control and Automation, Hefei, pp. 839–841 (2000)
4. Kennedy, J., Eberhart, R.: Particle Swarm Optimization. In: IEEE International Conference on Neural Networks, Perth, IEEE Service Cente, Piscataway, NJ, pp. 1942–1948 (1995)
5. Li, X.P., Zhang, H.Y.: Improvement of Simulated Annealing Algorithm. Software Guide 7(4), 47–48 (2000)
6. Angeline, P.J., Sauders, G.M., Pollack, J.B.: An evolutionary algorithm that constructs recurrent neural networks. IEEE Trans. Neural Networks 5(1), 54–65 (1994)
7. Yao, X.: A review of evolutionary artifical neural networks. Int. J. Intell. Syst. 8(4), 539–567 (1993)
8. Han, F., Ling, Q.H., Huang, D.S.: Modified Constrained Learning Algorithms Incorporating Additional Functional Constraints Into Neural Networks. Information Sciences 178(3), 907–919 (2008)
9. Wang, X.G., Tang, Z., Tamura, H., Ishii, M.: A modified error function for the backpropagation algorithm. Neurocomputing 57, 477–484 (2004)
10. Liu, Y.H., Chen, R., Peng, W., Zhou, L.: Optimal Design for Learning Rate of BP Neutral Network. Journal of Hubei University of Technology 22(2), 1–3 (2007)

11. Ma, Y.Q., Huo, Z.Y., Yang, Z.: The Implementation of the Improved BP Algorithm by Adding the Item of the Momentum. Sci-Tech Information Development & Economy 16(8), 157–158 (2006)
12. Zhang, J.-R., Zhang, J., Lok, T.-M., Lyu, M.R.: A hybrid particle swarm optimization-back-propagation algorithm for feedforward neural network training. Applied Mathematics and Computation 185(2), 1026–1037 (2007)
13. Guo, W., Qiao, Y.Z., Hou, H.Y.: BP neural network optimized with PSO algorithm and its application in forecasting. In: Proc. 2006 IEEE International Conference on Information Acquisition, Weihai, Shandong,China, August 2006, pp. 617–621 (2006)
14. Han, F., Ling, Q.H.: A New Approach for Function Approximation Incorporating Adaptive Particle Swarm Optimization And A Priori Information. Applied Mathematics and Computation 205(2), 792–798 (2008)
15. Shi, Y.H., Eberhat, R.C.: Parameter selection in particle swarm optimization. In: Proc. of 1998 Annual conference on Evolutionary Programming, San Diego, pp. 591–600 (1998)
16. Shi, Y.H., Eberhat, R.C.: A modified particle swarm optimizer. In: Proc.of IEEE World Conf. on Computation Intelligence, pp. 69–73 (1998)
17. Clerc, M., Kennedy, J.: The particle swarm:explosion,stability,and convergence in a multi-dimensional complex space. IEEE Trans. Evolut. Comput. 6(1), 58–73 (2002)

Fuzzy Multiresolution Neural Networks

Li Ying[1,4], Shang Qigang[2], and Lei Na[3,4]

[1] Symbol Computation and Knowledge Engineer Lab of Ministry of Education,
College of Computer Science and Technology
liying@jlu.edu.cn
[2] Department of Mechanical and Engineering, Academy of Armored Force
Engineering, Beijing, Academy of Armored Force Technology, Changchun
[3] College of Mathematics
[4] Jilin University, Changchun 130021, P.R. China

Abstract. A fuzzy multi-resolution neural network (FMRANN) based on particle swarm algorithm is proposed to approximate arbitrary nonlinear function. The active function of the FMRANN consists of not only the wavelet functions, but also the scaling functions, whose translation parameters and dilation parameters are adjustable. A set of fuzzy rules are involved in the FMRANN. Each rule either corresponding to a subset consists of scaling functions, or corresponding to a sub-wavelet neural network consists of wavelets with same dilation parameters. Incorporating the time-frequency localization and multi-resolution properties of wavelets with the ability of self-learning of fuzzy neural network, the approximation ability of FMRANN can be remarkable improved. A particle swarm algorithm is adopted to learn the translation and dilation parameters of the wavelets and adjusting the shape of membership functions. Simulation examples are presented to validate the effectiveness of FMRANN.

Keywords: fuzzy multiresolution neural network, wavelet neural network, particle swarm optimization.

1 Introduction

Artificial neural networks (ANN) and wavelet theory become popular tools in various applications such as engineering problems, pattern recognition and non-linear system control. Incorporating ANN with wavelet theory, wavelet neural networks (WNN) was first proposed by Zhang and Benveniste [1] to approximate nonlinear functions. WNNs are feedforward neural networks with one hidden layer, where wavelets were introduced as activation functions of the hidden neurons instead of the usual sigmoid functions. As a result of the excellent properties of wavelet theory and the adaptive learning ability of ANN, WNN can make an remarkable improvement for some complex nonlinear system control and identification. Consequently, WNN was received considerable attention [2,3,4,5,6,7].

Especially, inspired by the fuzzy model , Daniel [8] put forward a fuzzy wavelet model. The FWN consists of a set of fuzzy rules. Each rule corresponding to a

H. Deng et al. (Eds.): AICI 2009, LNAI 5855, pp. 369–378, 2009.

sub-wavelet neural network consists of single fixed scaling wavelets, where the translation parameters are variable. Such sub-WNNs at different resolution level can capture the different behaviors of the approximated function. The role of the fuzzy set is to determine the contribution of different resolution of function to the whole approximation. The fuzzy model can improve function approximation accuracy given the number of wavelet bases. But such FWNN require a complex initialization and training algorithm. During the initialization step, the dilation values of such FWNN is fixed beforehand through experience and a wavelet candidate library should be provided by analysis the training samples.

As we know, the scaling function corresponds to the global behavior (low frequency information) of the original function, while the wavelets corresponds to the local behavior (high frequency information) of the original function. The FWNN in [8] loses sight of the low frequency, though which may be approximated by wavelet functions. But due to the theory of MRA, it would need much more wavelet nodes.

Incorporating the idea of FWNN with pre-wavelet, we propose a fuzzy pre-wavelet neural network. The hidden of layer of FMRANN consist of not only the scaling function nodes, but also wavelet function nodes. The structure of FM-RANN includes two sub-neural networks: sub-scaling function neural networks and sub wavelet neural networks. The contribution of such sub-neural networks is determined fuzzy rules.

In the learning of artificial neural network (ANN), the back propagation algorithm by Rumelhart, Hinton, and Williams has long been viewed as a landmark event. With the increasing of complexity of the practical problems, the standard back propagation algorithm based the single gradient has ability not equal to its ambition, which motivated many researchers to develop enhanced training procedures with exhibit superior capabilities in terms of training speed, mapping accuracy, generalization, and overall performance than the standard back propagation algorithm. The training method based upon second-order derivative information exhibit better efficient and promising. Primary second-order methods are the back propagation algorithm based on quasi-Newton, Levenburg-Marquardt, and conjugate gradient techniques. Although these methods have shown promise, they often lead to poor local optima partially attributed to the lack of a stochastic component in training procedure. Particle swarm optimization (PSO) in [9,13] is a stochastic population-based optimization technique. The simplicity of implementation and weak dependence on the optimized model of PSO make it a popular tool for a wide range of optimization problems.

A fuzzy multi-resolution neural network (FMRANN) based on modified particle swarm algorithm is proposed to approximate arbitrary nonlinear function in this paper. The basic concepts of Multi-resolution are introduced in Section 1. In section 2, the two different structure of fuzzy multi-resolution neural network (FMRANN) is presented and Aiming at the features of FMRANN, a modified particle swarm optimization is introduced to update the parameters. Two simulation examples are utilized to illustrate the good performance of the FMRANN compared with other methods in Section 4 .

2 Multi-Resolution Analysis (MRA)

The main characteristics of wavelets are their excellent properties of time-frequency localization and multi-resolution. The wavelet transform can capture high frequency (local) behavior and can focus on any detail of the observed object through modulating the scale parameters, while the scale function captures the low frequency (global behavior). In this sense wavelets can be referred to as a mathematical microscope. Firstly, we give a simplified review of wavelets and multiresolution analysis of $L^2(\mathbf{R})$.

A series sequence $V_j, j \in \mathbf{Z}$ of closed subspaces in $L^2(\mathbf{R})$ is called a multiresolution analysis (MRA) if the following holds:

(1) $\cdots \subset V_j \subset V_{j+1} \cdots$.

(2) $f(\cdot) \in V_j$ if and only if $f(2\cdot) \in V_{j+1}$.

(3) $\bigcap_{j \in \mathbf{Z}} V_j = \{0\}, \overline{\bigcap_{j \in \mathbf{Z}} V_j} = L^2(\mathbf{R})$.

(4) There exists a function $\phi \in V_0$, such that its integer shift $\{\phi(\cdot - k), k \in \mathbf{Z}\}$ form a Riesz basis of V_0.

It is noted that the function ϕ is a scaling function, which satisfies the following refinement equation:

$$\phi(x) = \sqrt{2} \sum_n h_n \phi(2x - n),$$

where $\{h_n\}$ is low frequency filter (scaling fiter). For $j \in \mathbf{Z}, k \in \mathbf{Z}$, denote

$$\phi_{j,k}(\cdot) = 2^{j/2} \phi(2^j \cdot -k).$$

For $j \in \mathbf{Z}, \{\phi_{j,k}, k \in \mathbf{Z}\}$ forms a Riesz basis of V_j. The associated wavelet function $\psi(x)$ satisfies the following equation:

$$\psi(x) = \sqrt{2} \sum_n g_n \phi(2x - n),$$

where $\{g_n\}$ is a high frequency filter (wavelet filter). For $j, k \in \mathbf{Z}$, denote

$$\psi_{j,k}(\cdot) = 2^{j/2} \psi(2^j \cdot -k),$$

then $\{\psi_{j,k}, k \in \mathbf{Z}\}$ forms a Riesz basis of W_j, where W_j is the orthogonal complement of V_j in V_{j+1}, i.e.

$$V_{j+1} = V_j \oplus W_j,$$

where V_j is scaling function space and W_j represents difference between spaces spanned by the various scales of scaling function, referred to as detail space or wavelet function space. Obviously,

$$L^2(\mathbf{R}) = \mathbf{V_J} \bigcup_{j=J}^{\infty} \oplus \mathbf{W_j}, \quad \mathbf{L^2(R)} = \bigcup_{j=-\infty}^{\infty} \oplus \mathbf{W_j}.$$

For any function $f(x) \in L^2(\mathbf{R})$ could be written

$$f(x) = \sum_{k \in \mathbf{Z}} c_{J,k} \phi_{J,k}(x) + \sum_{j \geq J}^{\infty} \sum_{k \in \mathbf{Z}} d_{j,k} \psi_{j,k}(x), \tag{1}$$

as a series expansion in terms of the scaling function and wavelets, i.e. as a combination of global information and local information at different resolution levels. On the other hand, $f(x)$ could also be represented as

$$f(x) = \sum_{j \in \mathbf{Z}} \sum_{k \in \mathbf{Z}} d_{j,k} \psi_{j,k}(x), \qquad (2)$$

as a linear combination of wavelets at different resolutions levels, i.e. the combination of all local information at different resolution levels.

The above equations indeed embody the concept of multiresolution analysis and imply the ideal of hierarchical and successive approximation, which is the main motivation of our work.

The common used wavelets are continuous wavelets, Daubecies wavelets and spline wavelets and so on. In WNN, the frequently adopted active functions are continuous wavelets such as Morlet, Mexican Hat, Gaussian derivatives wavelets, which do not exist scaling function.

Daubechies compact support orthonormal wavelets constructed by Irigrid Daubechies become a milestone of wavelet fields, which make wavelet rapidly overspread almost all engineer fields from theory to application.

Daubecies wavelets is provided by filters. The birth of Daubecies wavelets make a more extensive application. But unfortunately, they have not analytic formula, which limited some application of demanding analytic formula such as WNN.

Here the cascade algorithm i.e. the single-level inverse wavelet transform repeatedly [11], is employed to approximate the corresponding scale function and wavelet function, which are adopted as active functions in our proposed fuzzy multi-resolution neural networks.

In common WNN, taking Daubechies orthonormal wavelets as active functions, due to the un-redundancy of orthonormal wavelet, such kind of WNN can provide a unique and efficient representation for the given function, but which highly relies on the selected wavelets and lead to poor robustness. In our work we use not only Daubechies orthonormal wavelets but also the according scaling function to construct our proposed Fuzzy multi-resolution neural network. Obviously, the adoption of the fuzzy mechanism and the introduction of scaling function can provide better robustness.

3   Fuzzy Multiresolution Neural Network

In this subsection, a fuzzy multiresolution neural network (FMRANN) is proposed to approximate arbitrary nonlinear function. The active function of the FMRANN consists of not only the wavelet functions, but also the scaling functions, whose translation parameters and dilation parameters are adjustable. A set of fuzzy rules are involved in the FMRANN. Each rule either corresponding to a subset consists of scaling functions, or corresponding to a sub-wavelet neural network consists of wavelets with same dilation parameters. Incorporating the time-frequency localization and multi-resolution properties of wavelets with the ability of self-learning of fuzzy neural network, the approximation ability of FMRANN can be remarkable improved.

Artificial neural networks mimic the structure and function of neural models of human brain. The significant characteristics of ANN are known as the parallel processing ability and learning ability through examples. ANN have shown its outstanding performance on many complex nonlinear problems even though ANN is far away from the elaborating structure and function mechanism of real human brain.

Fuzzy logic is based on the attempt to emulate the human reasoning mechanism, which is not numbers but rather indicators fuzzy sets.

The property of Multi-resolution analysis of wavelets conform with the nature of audition and vision of human.

In this section, we mainly propose the models of Fuzzy Multi-resolution Neural Network which contains two fuzzy boxes referred as scale fuzzy box and wavelet fuzzy box.

Let $\mathbf{x} = (x_1, x_2, \cdots, x_N) \in R^N, y \in R$. $\Phi(\mathbf{x})$ and $\Psi(\mathbf{x})$ are respectively scaling function and wavelet function of $L^2(R^N)$. For simplification, denote

$$\Phi_k = \det(D_k^0)\Phi(D_k^0\mathbf{x} - \mathbf{t_k^0}),$$

and

$$\Psi_k = \det(D_k^1)\Psi(D_k^1\mathbf{x} - \mathbf{t_k^1}).$$

Note that for i=0,1, $D_k^i = diag(a_{k,1}^i, \cdots, a_{k,N}^i)$, $a_{k,1}^i > 0$, is dilation matrix. $\mathbf{t_k^i} = (b_{k,1}^i, \cdots, b_{k,N}^i)$, $b_{k,1}^i \in R$, is translation vector.

Incorporate with the idea of multi-resolution analysis and fuzzy logic, we give the description of FMRANN as follows:

$$y = \sum_{i=1}^{c} \hat{\mu}_i(\mathbf{x})\hat{y}_i(\mathbf{x}), \tag{3}$$

where

$$\hat{\mu}^i(\mathbf{x}) = \frac{\mu^i(\mathbf{x})}{\sum\limits_{i=0}^{c} \mu^i(\mathbf{x})},$$

$$\hat{y}^0(\mathbf{x}) = \sum_{k=1}^{M^0} w_k^0 \Phi_k(\mathbf{x}),$$

$$\hat{y}^1(\mathbf{x}) = \sum_{k=1}^{M^1} w_k^0 \Psi_k(\mathbf{x}).$$

From figure 1 to figure 2, we respectively show the graphs of scaling functions and wavelet functions with different support length and vanishing moment which are used for our FMRANN.

3.1   Training Algorithm of Fuzzy Multi-Resolution Neural Network Based on Modified Particle Swarm Optimization

Particle swarm optimization (PSO) is a stochastic and population-based optimization technique, which first introduced by Kennedy and Eberhart [9,13] in

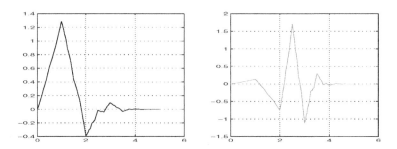

Fig. 1. The graph of scaling function ϕ and wavelet function ψ according to Daubecies wavelet with support width 5 and vanishing moment 3

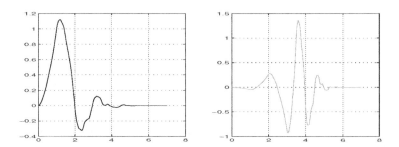

Fig. 2. The graph of scaling function ϕ and wavelet function ψ according to Daubecies wavelet with support width 7 and vanishing moment 4

1995. Particle swarm optimization method mimics the development of this technique was inspired by the animal social behaviors such as school of fish, flock of birds etc and mimics the way they find food sources. PSO has gained much attention and wide applications in different fields. Compared with genetic algorithm (GA), PSO employs a cooperative strategy while GA utilizes a competitive strategy. In nature, cooperation is sometimes more benefit for survival than competition. For example,the PSO has shown its superior ability than CA in some problems such as optimizing the parameters of artificial neural network (ANN) [14]. Gradient decent is another popular algorithm for the learning of ANN, which is a good choice when the cost function is unimodal. But the majority of the cost functions are multimodal, gradient decent might lead to convergence to the nearest local minimum. But PSO has the ability to escape from local minima traps due to its stochastic nature.

In PSO, each particle represents an alternative solution in the multi-dimensional search space. PSO can find the global best solution by simply adjusting the trajectory of each particle towards its own best location and towards the best location of the swarm at each time step (generation).

The implement of PSO can be categorized into the following main steps:
(1). Define the Solution Space and a Fitness Function:
(2). Initialize Random Swarm Position and Velocities:
(3). Updating particle Velocities and Position

$$v(k+1) = w * v(k) + c_1 * rand()(pbest - x)$$
$$+c_2 * rand() * (gbest - x);$$
$$x(k+1) = x(k) + v(k+1)$$

where w is inertial weight, c_1 and c_2 are constriction factors. Use of inertia weight and constriction factors has made the original implementation of the technique very efficient [9,13].

The inertial weight factor provides the necessary diversity to the swarm by changing the momentum of particles and hence avoids the stagnation of particles at local optima. The empirical investigations shows the importance of decreasing the value of from a higher valve (usually 0.9 to 0.4) during the search.

The majority problem of PSO is premature convergence of particle swarm. How to efficiently avoid such case is considerably important. Various mechanisms have been designed to increase the diversity among the particles of a swarm.

Here we propose an mechanism called forgetfulness to improve the the diversity of particle the premature convergence. The basic ideal is that we impose each particle and the swarm slightly amnesia when the particle swarm trend to premature convergence. This mechanism give the particles restarting chance based on the existing better condition. Firstly, define a merit to reflect the convergent status of the particle swarm as follows:

$$\tau = \frac{\sigma_f^2}{max_{1 \leq j \leq N}\{(f_i - \bar{f})^2\}}$$

Where

$$\bar{f} = \frac{1}{N} \sum_{i=1}^{N} f_i, \sigma_f^2 = \frac{1}{N} \sum_{i=1}^{N} (f_i - \bar{f})^2$$

Where f_i is the fitness of the i-th particle, N is the size of the particle swarm, $\bar{f}$ is the average fitness of all particles, and σ_f^2 is the covariance of fitness. For a given small threshold, if τ is less than this threshold and the expected solution haven't reached, then we think that this particle swarm tends to premature convergence. Under this condition, we impose each particle and the swarm slightly amnesia. Each particle forget its historical best position and consider the current position as its best position. Similarly, the swarm don't remember its historical global best position and choose the best position from the current positions of all particles.

4 Simulation

In the following, the concrete numerical examples are given to demonstrate the validity of the presented FMRANN. In this section, for evaluating the performance of the network, we take the following merits defined in [1] as a criterion to compare the performance of various methods.

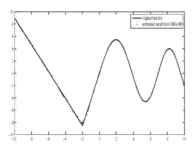 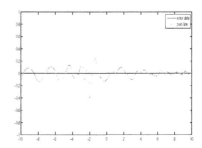

Fig. 3. The left figure is the comparison between original function and estimated result from FMRANN with performance index 0.019912. The right figure is the error between original function and estimated result from FMRANN with performance index 0.019912.

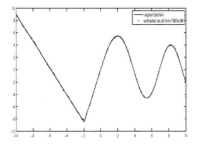 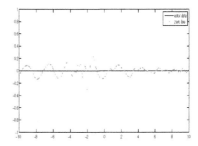

Fig. 4. The left figure is the comparison between original function and estimated result from FMRANN with performance index 0.015591. The right figure is the error between original function and estimated result from FMRANN with performance index 0.015591.

$$J = \sqrt{(\sum_{l=1}^{N} y_l - \hat{y}_l)/(\sum_{l=1}^{N} y_l - \bar{y})}, \bar{y} = \frac{1}{N}\sum_{l=1}^{N} y_l$$

Where y_l is the desired output and $\hat{y}_l$ is the estimated output from the constructed neural networks.

We respectively selected spline wavelets with order 3 and Daubechies wavelet with support width 2N-1 and vanishing moment N. We compare our FMRANN with other works in [1], [7] and [8], which take the Mexican Hat function as wavelet function.

Example I - Approximation of univariate piecewise function:

$$f(x) = \begin{cases} -2.186x - 12.864, -10 \le x < -2 \\ 4.246x, -2 \le x < 0 \\ 10e^{-0.05x-0.5}sin[(0.03x+0.7)x], 0 \le x \le 10 \end{cases} \tag{4}$$

Table 1. Comparison of our FMRANN with others work for Example I

Method	Number of unknown parameters used for training in different networks	Performance index J
FMRANN-prewavlets	23	0.016547
FWN[8]	28	0.021
WNN [1]	22	0.05057
WNN [7]	23	0.0480

Table 2. Comparison of our FMRANN with others work for Example II

Method	Number of unknown parameters used for training in different networks	Performance index J
FMRANN-prewavlets	45	0.018023
FWN [8]	107	0.0147
WNN [1]	442	0.03395

Example II - Approximation of bivariate function:

$$y = (x_1^2 - x_2^2) * sin(0.5 * x_1), x_1, x_2 \in [-10, 10]. \qquad (5)$$

In the simulation, we sampled 200 points distributed uniformly over $[0, 10]$ and $[-10,10]$ respectively as training data. In Table I, we compare the performance of our FMRANN with other WNNs, here the number of rules used in FWN is 6. In Table 1. and Table 2., the compared results with others work are given. From the Fig.3 to Fig.6., the approximation results and error for example I are provided. It should be noticed that the performance of FMRANN shown in the Table 1 and Table 2 is the average of the nine simulations. Each simulation is computed on uniformly sampled test data of 200 points. Among the results of FMRANN for example I, the maximal Error value is 0.019912 and the minimal Error value 0.015591. Among the results of FMRANN for example II, the maximal Error value is 0.0326 and the minimal Error value 0.01320.

Figure. 3 shows the excellent performance of our FMRANN, which corresponds to the worst performance value among the nine simulations for example I. Figure. 4 shows the excellent performance of our FMRANN, which corresponds to the best performance value among the nine simulations for example I. Obviously, the performance of our FMRANN with considerable smaller parameters is superior to that of other WNNs. Our constructed FMRANN is significant and effective.

References

1. Zhang, Q.H., Benveniste, A.: Wavelet networks. IEEE Trans. Neural Netw. 3(6), 889–898 (1992)
2. Pati, Y.C., Krishnaprasad, P.S.: Analysis and synthesis of feedforward neural networks using discrete affine wavelet transformation. IEEE Trans. Neural Networks 4, 73–85 (1993)

3. Zhang, J., Walter, G.G., Lee, W.N.W.: Wavelet neural networks for function learning. IEEE Trans. Signal Processing 43, 1485–1497 (1995)
4. Zhang, Q.: Using wavelet networks in nonparametric estimation. IEEE Trans. Neural Networks 8, 227–236 (1997)
5. Alonge, F., Dippolito, F., Mantione, S., Raimondi, F.M.: A new method for optimal synthesis of wavelet-based neural networks suitable for identification purposes. In: Proc. 14th IFAC, Beijing, P.R. China, June 1999, pp. 445–450 (1999)
6. Li, X., Wang, Z., Xu, L., Liu, J.: Combined construction of wavelet neural networks for nonlinear system modeling. In: Proc. 14th IFAC, Beijing, P.R. China, June 1999, pp. 451–456 (1999)
7. Chen, J., Bruns, D.D.: WaveARX neural network development for system identification using a systematic design synthesis. Ind. Eng. Chem. Res. 34, 4420–4435 (1995)
8. Daniel, J., Ho, W.C., et al.: Fuzzy wavelet networks for function learning. IEEE Trans. on Fuzzy Systems 9(1), 200–211 (2001)
9. Eberhart, R.C., Kennedy, J.: A new optimizer using particle swarm theory. In: Proc. 6th Int. Symp. Micro Machine and Human Science, Nagoya, Japan, pp. 39–43 (1995)
10. Daubechies, I.: Ten lectures on wavelets CBMS. SIAM 61, 194–202 (1994)
11. Strang, G., Nguyen, T.: Wavelets and Filter Banks. Wellesley-Cambridge Press (1996)
12. Chui, C., Wang, J.: A general framework for compactly supported splines and wavelets. Proceedings of Americal Mathematical Soceity 113, 785–793
13. Eberhart, R.C., Simpson, P., Dobbins, R.: Computational Intelligence PC Tools: Academic, ch. 6, pp. 212–226 (1996)
14. Eberhart, R.C., Shi, Y.: Comparison between genetic algorithms and particle swarm optimization. In: Porto, V.W., Saravanan, N., Waagen, D., Eiben, A.E. (eds.) EP 1998. LNCS, vol. 1447, pp. 611–618. Springer, Heidelberg (1998)

Research on Nonlinear Time Series Forecasting of Time-Delay NN Embedded with Bayesian Regularization

Weijin Jiang[1], Yusheng Xu[2], Yuhui Xu[1], and Jianmin Wang[1]

[1] School of computer, Hunan University of Commerce, Changsha 410205, China
jwjnudt@163.com
[2] China north optical-electrical technology, Beijing 100000, China
yshxu520@163.com

Abstract. Based on the idea of nonlinear prediction of phase space reconstruction, this paper presented a time delay BP neural network model, whose generalization capability was improved by Bayesian regularization. Furthermore, the model is applied to forecast the imp&exp trades in one industry. The results showed that the improved model has excellent generalization capabilities, which not only learned the historical curve, but efficiently predicted the trend of business. Comparing with common evaluation of forecasts, we put on a conclusion that nonlinear forecast can not only focus on data combination and precision improvement, it also can vividly reflect the nonlinear characteristic of the forecasting system. While analyzing the forecasting precision of the model, we give a model judgment by calculating the nonlinear characteristic value of the combined serial and original serial, proved that the forecasting model can reasonably 'catch' the dynamic characteristic of the nonlinear system which produced the origin serial.

1 Introduction

Basically, the traditional time serial forecasting model was built on the foundation of statistic technology, they are mainly subjected to linear model, which has advantage of simple nature and strong explanation ability, but as for the forecasting of tremendous system's evolution serial, it fails, especially when the change of real economic system is strong, high level nonlinear characteristic and classical chaos appears in the system, the common forecasting method can hardly handle it. Chaos theory is an important discovery in nonlinear dynamic, which come into being from 1960's, until 1980's, it have developed to be a new study with special concept system and method frame. Because of this, people have some new acknowledgement on the complexity of time serial. According to utilize the phase space reconstruction technology of chaotic theory, economic time serial is embedded in the reconstructed phase space, and with the help of fractal theory and symbol dynamics, the complex moving characteristic of economic dynamic system can be described, furthermore, the inner rule can be found and the forecasting report is produced. Therefore, it has a realistic meaning and scientific value[1-6] to find the evolution rule of economic time serial on the stand of nonlinear dynamics and study the chaotic characteristic and nonlinear forecasting of complex economic dynamic system.

H. Deng et al. (Eds.): AICI 2009, LNAI 5855, pp. 379–388, 2009.

Artificial neural network is a kind of nonlinear dynamics system, which possesses strong robustness and error-tolerance. The powerful ability to draw on nonlinear reflection guarantee it has special advance in solving nonlinear problems, and this make it suitable for the modeling and forecasting of changeable complex economic system and economic serial. According to some certain structure setting, the study has proved that neural network forecasting had obvious advantage when the system possessed nonlinear characteristic. Reference [1] firstly gives a try on using neural network to study and forecast the tome serial data which the computer produced, then neural network forecasting application like Matsuba is used in stock market, after that, there were a lot of product of neural network application appeared in neural network international meeting. Up to this day, the scholars has applied neural network in every occupation of economic, such as securities, exchange ratio, GDP forecasting, financial alarm and market demand forecasting[2]. Paper [3,4] gives a demonstration about the relation between phase space reconstruction and forecasting, theoretically, it is reasonable, but the calculation of embedded dimension is usually be fake nearest spot method, and the way of embedded time-delay calculation is average information method. Paper [5,6] discusses respectively about both their advantages and disadvantages, and comes a conclusion that there maybe some certain error by using these methods, which has been mentioned before, to calculate embedded dimension and time-delay because of short actual data and noise, as a results, reconstruction state vector based on these method may receive a bad forecasting report. In paper [7]' selection about embedded dimension follows the rule that the dimension possess minimum error, whose appearance come along with tremendous times of forecasting, can be promoted to embedded dimension, however, the real dimension may not show up because the times of common is limited, and this method can only cut down the error to a local minimum, furthermore, many aimless experiment will waste time, finally[8], the data is always divided into two parts (training set and forecasting set), determining embedded dimension directly by the forecasting error in forecasting set is not useful, because in actual forecasting, forecasting error can not be gained.

These studies usually use BP neural network model—error reverse transform neural network, which contains a great many parameters, besides, these parameters always judged by experience, as a result, the model is hard to establish. The large repetitious training and experiment is a kind of 'Over Fitting' which may give the training sample a fine combination effect, but as for new input sample, the output may have big difference with the objective value, which has no popularization capability. Conversely Bayesian Regularization can overcome this problem, and it can improve the generalization capability of network. That why this paper founded a new model for nonlinear time serial forecasting, which is called Time Delay Neural Network (acronym: TDBP) forecasting model by some kinds of tricks, such as following phase space reconstruction theory, utilizing G-P algorithm to calculate the saturation embedded dimension—the number of input-level's spot, combining Bayesian Regularization to certain the number of the hidden-lever's spot and improving the generalization capability of network. In this model, when the forecasting error of certain set is relatively small, the forecasting error of forecasting error will be accordingly tiny. The results of experiment proved that chaos time serial forecasting and control system based on this method possessed fine precision, furthermore, it can track the original time serial quickly, besides, it has faster response capability and smaller number of iterate than other similar methods.

As for the nonlinear characteristic of trade serial, this paper put the established TDBP model into some certain occupation's trade forecasting, according to adopt the actual amount of imp&exp of 1/1989-5/2003 as the ample of training network, a new model was founded[9,10], which successfully forecast and combine this occupation's international trade affairs[8]. Meanwhile, on the foundation of analyzing forecasting precision, this paper gave a capability evaluation for the nonlinear forecasting model according the maximum Lyapunov index and the relation dimension between combined serial and original data.

2 Phase Space Reconstruction and BP Model Based on Bayesian Regularization

The intention of forecasting is to deduce object's future by its own develop rule. According to Kolmogorov theorem, all time serial can be regarded as a substitute of input-output system which is certain by nonlinear mechanism. Accompany with the development and maturity of neural network theory and phase space reconstruction technology, a new road is provided for nonlinear forecasting of economic time serial.

Suppose disperse time serial is $\{X(t)\}(t=1, 2, ..., T)$, select proper embedded dimension m, time-delay τ and reconstruction phase space $\{Y(j)\}(j=1, 2, ..., N, N=T-(m-1))$, and:

$$Y(j)=[X(j),X(j-\tau),X(j-2\tau),...,X(j-(m-1)\tau)] \tag{1}$$

According to Takens embedding theorem, there is reflection $F : R^m \rightarrow R^m$ makes Y $(j+\tau) = F (Y (j))$, it is:

$$\begin{bmatrix} X(j+\tau) \\ X(j) \\ \vdots \\ X(j-(m-2)\tau) \end{bmatrix} = F\left(\begin{bmatrix} X(j) \\ X(j-\tau) \\ \vdots \\ X(j-(m-1)\tau) \end{bmatrix}\right) \tag{2}$$

Then forecasting reflection $\overline{F} : R^m \rightarrow R^m$ can be denoted as:

$$X(j+\tau)= \overline{F} (X(j),X(j-\tau)...,X(j-(m-1)\tau)) \tag{3}$$

Here, it is a nonlinear needed to evaluate, definitely, to find a concrete function expression is difficult, But the function can be obtained by utilizing neural network's capability of approaching nonlinear reflection.

Usually, when the size of training set is certain, the generalization capability of network directly concern with its scale, if the scale of neural network is far smaller than the size of training sample set, then the change of 'over-training' is small. Based on this idea, reference [8] change neural network's object function to:

$$F(w) = \beta sse + \alpha ssw \tag{4}$$

And sse denotes the error's square sum, ssw denotes the square sum of all power coefficient's valve value in the network, α, β are the coefficient of objective function. Suppose that network's power value is random variable, after the output data is figured out, we can use Bayesian formula to revamp the ratio density function of power value:

$$P(w \mid D, \alpha, \beta, M) = \frac{P(D \mid w, \beta, M) P(D \mid w \mid \alpha, M)}{P(D \mid \alpha, \beta, M)} \tag{5}$$

And w denotes network power value's vector, D denotes data set, M denotes network model. $P(D \mid \alpha, \beta, M)$ is full probability, $P(w \mid \alpha, M)$ denotes the first check density function of power vector, $P(D \mid w, \beta, M)$ is similar function of timing output, commonly supposed that noise and power vector exist in data follow the Gauss distribution, it means:

$$P(D \mid w, \beta, M) = e^{-\beta sse} / Z_D(\beta) \tag{6}$$

$$P(w \mid \alpha, M) = e^{-\alpha ssw} / Z_w(\alpha) \tag{7}$$

And $Z_D(\beta) = (\pi / \beta)^{n/2}$, $Z_w(\alpha) = (\pi / \alpha)^{N/2}$, n denotes the number of sample, N denotes the number of power value, put (6), (7) in formula (5), noticed that the standard factor $P(D \mid \alpha, \beta, M)$ had no relation with power vector w, so:

$$P(w \mid D, \alpha, \beta, M) = \frac{e^{-(\alpha ssw + \beta sse)} / [Z_w(\alpha) Z_D(\beta)]}{P(D \mid \alpha, \beta, M)} \tag{8}$$

$$= e^{-F(w)} / Z_F(\alpha, \beta)$$

From formula (8), we can see that the optimal power value posses the biggest last check probability, and maximum last validation probability equals to the minimum regularization objective function $F(w) = \beta sse + \alpha ssw$. If $\alpha \ll \beta$, network training can reduce error to minimum; and if $\alpha \gg \beta$, network training will automatically reduce the efficient network parameters, which is to make up to big network error. On the condition that the error of network training is as small as possible, it is evident that new capability objective function is efficient, which can minimize the efficient parameter and lessen the scale of network.

Here, we adopt Bayesian method to certain the optimal regularization parameter α, β, suppose α, β 's first validation function is $P(\alpha, \beta, M)$, then the last is:

$$P(\alpha, \beta \mid D, M) = \frac{P(D \mid \alpha, \beta, M) P(\alpha, \beta, M)}{P(D \mid M)} \tag{9}$$

Formula (9) illustrate that the last validation of maximum regularization parameter α, β is equal to maximum similar function $P(D \mid \alpha, \beta, M)$. Combining (5) and (8), formula (10) can be concluded:

$$P(D \mid \alpha, \beta, M) = \frac{Z_F(\alpha, \beta)}{Z_w(\alpha) Z_D(\beta)} \tag{10}$$

In order to figure $Z_F(\alpha, \beta)$ out, unfolding F (w) on the minimum point w^*, because gradient is 0, then:

$$F(w) = F(w^*) + (w-w^*)^T H(w^*)(w-w^*)/2 \tag{11}$$

At least, Hison matrix $H(w^*) = \nabla^2 F(w^*)$ is semi-symmetry, so it can be written as $H = (H^{\frac{1}{2}})^T H^{\frac{1}{2}}$, suppose that $u = H^{\frac{1}{2}}(w - w^*)$. Putting formula (11) into formula (8), and integrating the two sides:

$$1 = \frac{1}{Z_F(\alpha, \beta)} e - F(w^*) \int_{R^{Ne}} -u^T u/2 [\det(H(w^*)^{-1})]^{1/2} du \tag{12}$$

$$= \frac{(2\pi)^{N/2}}{Z_F(\alpha, \beta)} e^{-F(w^*)} [\det(H(w^*)^{-1})]^{1/2}$$

Then:

$$ZF(\alpha, \beta) = (2\pi)^{N/2} e^{-F(w^*)} [\det(H(w^*)^{-1})]^{1/2} \tag{13}$$

Putting formula (13) into formula(10), obtaining logarithm from both sides, besides, utilizing the one-power condition of optimal value to obtain the optimal regularization parameter:

$$\alpha^* = \frac{\gamma}{2E_w(w^*)}, \quad \beta^* = \frac{n-\gamma}{2E_D(w^*)} \tag{14}$$

And $\gamma = N - 2atr(H)^{-1}$ denotes the number of efficient network parameter, which reflect the actual scale of network. Paper [8] pointed out that: as for Hison matrix H, when using quick algorithm Levenberg-Marquardt to train the network, it was easy to approach by Gauss-Network method. In the process of training, sse and ssw can be the factor to figure out the number of nerve cell of hidden level (marked as N_{hl}) by following the efficient parameter γ, after certain times of iterate, when these three parameters is in a forever status or smaller, it gives a sign that this network is in convergence status, and it is the time to stop training; selecting a smaller N_{hl}, continuously increasing the number of hidden nerve cell until current N_{hl} makes network convergence, then the state of sse and ssw will be unchangeable, as a result, this N_{hl} can be regarded as the spot number of hidden-level.

This paper adopts single-output three-level BP neural network, separately, the stimulating function of these three levels is logsig, logsig, pureline. The nerve cell number of input level equals to the embedded dimension number of nonlinear time serial, besides, many study has proved that the number of maturity embedded dimension of time serial can be the nerve cell number of network input level. Here, we obtained the maturity embedded dimension number by utilizing G-P algorithm. we use MatLab language to build the concrete model and on the aid of tool box function trainbr() Bayesian Regularization algorithm is realized. Fig.1 gives out the basic frame of the model.

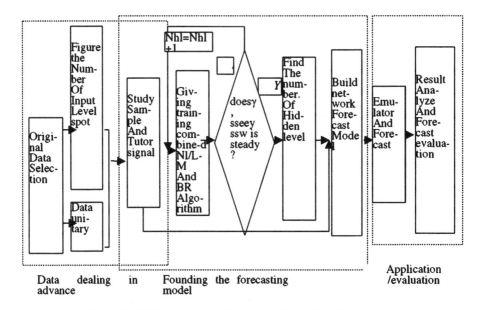

Fig. 1. TDBP forecasting model based on Bayes Regularization

3 The Selection of Trade Data and the Build of Forecasting Model

Considering that the scale of international trade in our country is small before 1990, because the opening strategy for world was still in the tentative stage. But there were lots changes in the international trade environment and trade system after 1990, every kinds of nonlinear factor came out off water, therefore, we selected the month data of one corporation's imp&exp business from January 1st 1989 to June 2003 as the training sample. Before the network study, we did unitary operation for the original data.

$$X(t) = \frac{X'(t) - Min\{X'(t)\}}{Max\{X'(t)\} - Min(X'(t)\}}$$
(15)

And $X'(t)$ denotes original serial, $X(t)$ denotes unitary serial.

We utilized G-P algorithm to confirm the relation dimension of imp&exp serial, and then obtained the maturity embedded dimension: when the embedded dimension of imp&exp trade is 13, relation dimension is mature; when it embedded dimension

Table 1. Import trade

Nhl	sse	ssw	N	γ
2	0.2773	38.9909	213	27.3
4	0.2780	35.8799	243	27.0
5	0.2780	36.8654	258	27.0
6	0.2780	35.7243	273	26.9
8	0.2780	35.6423	303	26.9
10	0.2780	35.7176	333	26.9

Table 2. Export trade

Nhl	sse	ssw	N	γ
4	0.2164	22.1673	345	17.7
6	0.2160	22.1450	381	17.6
7	0.2160	22.1971	399	17.6
8	0.2159	22.2076	417	17.6
10	0.2159	22.2452	453	17.6
12	0.2159	22.2283	489	17.6

comes to 16, relation dimension is still mature. The time-delay τ is 1. When training the network, we adopted unitary data of imp&exp trade's month data from 1/1989 to 5/2003 to do the phase space reconstruction, and then respectively obtained 13-dimension matrix or 16-dimension matrix which is be the study sample; the tutor signal of network model is the unitary data of imp&exp trade's month data from 2/1990 to 6/2003 and from 5/1990 to 6/2003 separately. As for different N_{hl}, utilizing these study samples and tutor signals to do network training until the network is in the status of convergence, the obtained parameter is illustrated by table.1 and table 2.

The network training parameter in table.1 demonstrates that: when network model of imp&exp trade is under the condition that $N_{hl} \geq 6$ or $N_{hl} \geq 8$, γ, *sse* and *ssw* is steady, as a results, the hidden level nerve cell number of nonlinear forecasting model of imp&exp trade are 6 and 8. On the foundation of these parameters, TDBP network model of imp&exp trade can be established, which can give nonlinear forecast for future development.

4 Example Verification

This paper used multi-steps forecasting method, which feedback the forecasting value to the input end, reconstruct input matrix and do next step forecasting. Here the 12-step of 7/2003-6/2004 is given out: as for the data of 6/2003-12/2003, we can compare it directly from the original data, from which we can judge the generalization capability of neural network and the forecasting precision of non-linear model; the forecasting results of 1/2004-6/2004 shows the develop trend of our country's

Table 3. The actual month data of imp&exp trade and forecasting results from 7/2003 to 12/2003 (10 thousand dollar)

	Export trade data			import trade data		
	Actual value	Forecasting value	Relative (error%)	Actual value	Forecasting value	Relative error%
July	380.9	377.02	-1.0	365.1	367.92	0.8
August	374	382.69	2.3	346.2	379.89	9.7
September	419.3	405.26	-3.3	416.5	389.23	-6.5
October	409.1	387.8	-5.2	351.9	377.42	7.3
November	417.6	397.62	-4.8	368.9	404.14	9.6
December	480.6	411.95	-14.3	423.4	429.32	1.4

Table 4. Imp&exp trade forecasting results from 1/2004 to 6/2004 (10 thousand dollar)

	January	February	March	April	May	June	sum
export	395.3	347.64	417.67	458.92	441.58	444.94	2506.05
import	448.37	405.84	469.43	489.43	486.12	486.04	2785.23

Table 5. Nonlinear characteristic calculation result of original month serial and forecast serial

characteristic sample	import		export	
	Original serial	Forecast serial	Original serial	Forecast serial
Relation dimension	1.4576	1.4206	1.6819	1.9697
Maximum Lyapunov index	0.0061	0.0060	0.0115	0.0167

imp&exp trade. The actual month data of imp&exp trade and forecasting results is generalized in table.3 and table.4.

The results of table.3 shows that forecasting value is very close to the actual value except the data of 1/2003's export trade, and the relative error is under 10 percent, single step forecasting relative error is below 1 percent. Meanwhile combining the figure and forecasting data, apparently, the seasonal change trend of imp&exp trade can be predicted, and the forecasting point it out that the amount of imp&exp trade of the second half year of 2003 is larger than the first half one, when it come to December, the volume steps to the maximum number, then imp&exp volume of January and February suffer a slow down, but January is little bigger than February, all the change is in accordance with the actual development. Usually, the corporation like to finish the good exchange at the end of one year, and the custom's statistic is balanced before the new year, because China's spring festival is always in January or February, as a results, the amount of December is usually the largest one, January and February the least, owing to the lag factor of last year, amount of January will bigger than February. The forecasting demonstrated that: there will be a trade deficit partly because of model factor; actually, it is also the results of environment change. On the stand of international, the value of dollar may rise, which can hoist the value of RMB, unfortunately, this can sharpen down the capability of China's export trade and prompts import deals. After the Cancun meeting, EU and USA will give more attention on business coordination of bilateral and distribute economic, which can prompt the coordination of distribute economic, strengthen the international trade protectionism and descent a blow for our country's strategy of broadening export volume and absorbing foreign investment. As for the domestic aspect, the drop of custom will promote China's import volume; every blade has two curves, proper unfavorable balance of trade will buffer the pressure of RMB appreciation. Base on these factors, the forecasting that the amount of export trade in 2004 will change more, and the import trade may increase faster than export, In a word, emulation and forecasting basically combined the training data and develop trend.

Table.5 shows that the nonlinear characteristic value of the two import serials almost identical; as for export trade, the forecasting serial and original serial have tiny difference, but it is not a big deal, besides, when the embedded dimension is 16, these two serial's relation dimension reach the point of maturity. Actually, many scientific studies had proved that there are more factors are influencing the export trade than import, which produces more noise in the export trade serial and fades the nonlinear characteristic of actual system. Fortunately, the model of this paper is insensitive of noise, and it has a good capability of generalization, which can enable the model obtain a direct nonlinear expression form from the dynamic system, produce the combination serial and efficiently 'catch' the dynamics characteristic of nonlinear system that produced the original serial.

5 Conclusions

As a kind of nonlinear dynamics method, neural network has a strong capability to close nonlinear reflection, and that is why it is so suitable in modeling and forecasting of nonlinear economic system. Up to this day, the scholar has reaped enormous actual fruit from applying neural network in the problems of economic management. Based on nonlinear forecasting theory of phase reconstruction, this paper has built a time-delay BP neural network model. Considering the weak generalization capability of common BP neural network, this paper combined the Bayesian Regularization to improve the objective function and the generalization capability of neural network. In order to explain more vividly, this paper then adopted some certain occupation's actual monthly imp&exp data of 1/1989-5/2003 as training sample, finally, it built a multi-steps nonlinear forecasting model. The results of imp&exp trade's multi-step forecast showed that the model could not only reasonably forecast the develop trend of actual serial, but also combined the actual data, besides, the relative error of single forecast is in the scope of 1 percent. In addition, nowadays, the forecast evaluation only focus on the improvement of the combination of comparing data point and the forecasting precision, which lead to the overlook of the difference between combined serial and original serial. When the system is nonlinear, it got a problem, because in nonlinear system, even two identical serial may come from two different sub-systems.

For this problem, this paper made an improvement: utilizing the relation dimension of combined serial and original data and the maximum Lyapunov index, comparing the memory structure of combined serial and original serial, and then finding that the model can efficiently catch up the inner structure and rule of the nonlinear system that produce the original serial. The model of this paper has something really special from others, its free parameter is fewer, it has a strong capability of robustness and generalization, which make it very suitable for the forecasting of nonlinear chaos time serial.

Acknowledgement

This paper is supported by the Society Science Foundation of Hunan Province of China No. 07YBB239.

References

1. Ott, E., Grebogi, C., Yorke, J.A.: Controlling chaos. Physical Review Letters 64(11), 1196–1199 (1990)
2. Islam, M.N., Sivakumar, B.: Characterization and prediction of runoff dynamics: a nonlinear dynamical view. Advances in Water Resources (25), 179–190 (2002)
3. Ramirez, T.A., Puebla, H.: Control of a nonlinear system with time-delayed dynamics. Phys. Lett. A 262, 166–173 (1999)
4. Dongmei, L., Zhengou, W.: Small parameter perturbations control algorithm of chaotic systems based on predictive control. Information and Control 32(5), 426–430 (2003)
5. Weijin, J.: Non-Linear chaos time series prediction & control algorithm based on NN. Journal of Nanjing University (NaturalSciences) 38(Computer issue), 33–36 (2002)
6. Weijin, J.: Research of ware house management chaotic control based on alterable. Micro Electronics & Computer 19(11), 55–57 (2002)
7. Sivakumar, B., Jayawardena, A.W., Fernando, T.M.K.G.: River flow forecasting: use of phase-space reconstruction and artificial neural networks approaches. Journal of Hydrology (265), 225–245 (2002)
8. Islam, M.N., Sivakumar, B.: Characterization and prediction of runoff dynamics: a nonlinear dynamical view. Advances in Water Resources (25), 179–190 (2002)
9. Jayawardena, A.W., Li, W.K., Xu, P.: Neighborhood selection for local modeling and prediction of hydrological time series. Journal of Hydrology (258), 40–57 (2002)
10. Kim, H.S., Eykholt, R., Salas, J.D.: Nonlinear dynamics, delay times, and embedding windows. Phy. D (127), 48–60 (1999)

An Adaptive Learning Algorithm for Supervised Neural Network with Contour Preserving Classification

Piyabute Fuangkhon and Thitipong Tanprasert

Distributed and Parallel Computing Research Laboratory
Faculty of Science and Technology, Assumption University
592 Soi Ramkamhang 24, Ramkamhang Road, Huamak, Bangkapi, Bangkok, Thailand
piyabute@yahoo.com, t_tanprasert@yahoo.com

Abstract. A study of noise tolerance characteristics of an adaptive learning algorithm for supervised neural network is presented in this paper. The algorithm allows the existing knowledge to age out in slow rate as a supervised neural network is gradually retrained with consecutive sets of new samples, resembling the change of application locality under a consistent environment. The algorithm utilizes the contour preserving classification algorithm to pre-process the training data to improve the classification and the noise tolerance. The experimental results convincingly confirm the effectiveness of the algorithm and the improvement of noise tolerance.

Keywords: supervised neural network, contour preserving classification, noise tolerance, outpost vector.

1 Introduction

It is known that repetitive feeding of training samples is required for allowing a supervised learning algorithm to converge. If the training samples effectively represent the population of the targeted data, the classifier can be approximated as being generalized. However, there are many times when it is impractical to obtain such a truly representative training set. Many classifying applications are acceptable with convergence to a local optimum. For example, a voice recognition system may be customized for effectively recognizing voices of a group of limited number of users, so the system may not be practical for recognizing any speaker. As a consequence, this kind of application needs occasional retraining when there is sliding of actual context locality.

Our focus is when only part of the context is changed; thereby establishing some new cases, while inhibiting some old cases, assuming a constant system complexity. The classifier will be required to effectively handle some old cases as well as new cases. Assuming that this kind of situation will occur occasionally, it is expected that the old cases will age out, the medium-old cases are accurately handled to a certain degree, and new cases are most accurately handled. Since the existing knowledge is lost while retraining new samples, an approach to maintain old knowledge is required. While the typical solution uses both prior samples and new samples on retraining, the major drawback of this approach is that all the prior samples for training must be maintained.

H. Deng et al. (Eds.): AICI 2009, LNAI 5855, pp. 389–398, 2009.

Research works related to the proposed algorithm are in the field of adaptive learning [1], [2], incremental learning [16], and contour preserving classification [3]. The prior one can be categorized into three strategies [3]. The first strategy (increasing neurons) [4], [5], [6], [7] will increase number of hidden nodes when the error is excessively high. These algorithms will adapt weights of neuron that is closest with input sample and neighbors only. However, the increasing size of neural networks is a cause of accuracy tradeoff. The second strategy (rule extraction) [8], [9], [10], [11] will interchange between rules and weight of neuron. The accuracy of networks depends on discovered rules. The translation of weight vector into rules also partly suppresses certain inherited statistical information. The last strategy (aggregation) [12], [13], [14], [15] will allow existing weights to change in bounded range and will always add new neurons for learning samples of new context. This method requires two contexts be similar and network's size is incrementally larger.

In this paper, an alternative algorithm is proposed for solving adaptive learning problem for supervised neural network. The algorithm improved from [17], [18] is able to learn new knowledge while maintaining an old knowledge through the decay rate while allowing the adjustment of the number of new samples. In addition, the improvement of the classification and the noise tolerance is archived by utilizing the contour preserving classification algorithm which helps expanding the territory of both classes while maintaining the shape of both classes.

Following this section, section 2 summarizes the outpost vector model. Section 3 describes the methodology. Section 4 demonstrates the experimental results of a 2-dimension partition problem. Section 5 discusses the conclusion of the paper.

2   The Outpost Vector Model

This section summarizes the outpost vector method as originally published in [3]. Fig. 1 illustrates the concept of outpost vector. Each input vector is modeled to span its territory as a circle (sphere in case of 3-dimensional space or hyper sphere in case of more-dimensional space) until the territories collide against one another.

In Fig. 1, the territory of input vector k of class A (denoted by A_k) is founded by locating the input vector in class B which is nearest to A_k (denoted by $B^*(A_k)$) and declaring the territory at half way between A_k and $B^*(A_k)$. Accordingly, the radius of A_k's territory is set at half of the distance between A_k and $B^*(A_k)$. This is to guarantee that if $B^*(A_k)$ sets its territory using the same radius, then the distance from the hyper plane to either A_k or $B^*(A_k)$ will be at maximum.

For $B^*(A_k)$, the nearest input vector of class A to $B^*(A_k)$ is not necessarily A_k. $B^*(A_k)$ may not place its outpost vector against A_k although A_k has placed an outpost vector against it. However, $B^*(A_k)$ places its outpost vector against A_j because it is nearest to $B^*(A_k)$. Since it is attempted to carve the channel between the two classes in such a way that its concavity or convexity can influent the learning process, it is desirable to generate an additional set of outpost vectors. As illustrated in Fig. 1, an additional outpost vector of $B^*(A_k)$ (designated as "class B") is placed on $B^*(A_k)$'s territory in the direction of A_k in response to the existence of an outpost vector of A_k in that direction. In [3], a theorem was provided for proving that the territory created will not overlap across different classes. All the outpost vectors are combined with the original input vectors for training the network.

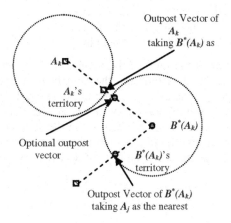

Fig. 1. Territory and Outpost Vectors

3 Methodology

The algorithm [17], [18] utilizes the concept of adaptive learning and outpost vectors in modeling the new training samples. The adaptive learning algorithm will maintain limited number of training samples from the previous training session (decayed prior samples) to be used in the next training session while the outpost vectors will help expanding the territory of both classes and maintaining the shape of the boundary between both classes.

There are three parameters in the algorithm: new sample rate, outpost vector rate, and decay rate. Firstly, the new sample rate is the ratio of the number of selected new samples over the number of new samples. It determines the number of selected new samples to be included in the final training set. The larger new sample rate will cause the network to learn new knowledge more accurately. The number of selected new samples is calculated by formula:

$$nss = nw \times ns \qquad (1)$$

where   nss   is the number of selected new samples,
            nw   is the new sample rate $[0, \infty)$,
            ns   is the number of new samples

Secondly, the outpost vector rate is the ratio of the number of generated outpost vectors over the number of new samples. It determines the number of outpost vectors to be included in the final training set. Using larger outpost vector rate will generate more outpost vectors at the boundary between both classes. When outpost vector rate is equal to 1.0, the number of outpost vectors will be at maximum. The number of outpost vectors is calculated by formula:

$$nov = ov \times ns \qquad (2)$$

where   nov is the number of outpost vectors,
         ov   is the outpost vector rate [0,1],
         ns   is the number of new samples

Lastly, the decay rate is the ratio of the number of decayed prior samples over the number of new samples. It determines the number of decayed prior samples to be included in the final training set. The larger decay rate will cause the network to forget old knowledge slower. When decay rate is greater than 1.0, more than one instance of some prior samples will be included in the decayed prior sample set. There is an exception for the first training session where the new samples will also be used as the decayed prior samples. The number of decayed prior samples is calculated by formula:

$$ndc = dc \times ps \qquad (3)$$

where   ndc is the number of selected prior samples,
         dc   is the decay rate $[0, \infty)$,
         ps   is the number of prior samples

After the final training set is available, it will be trained with a supervised neural network. The whole process will be repeated when new samples have arrived.

Algorithm. Sub_Trainning _with_Outpost_Vectors
1 Set *new sample set* as *prior sample set* for the first training session
2 *for each* training session
3   Construct *selected new sample set* by
3   If new sample rate is equal to 1.0
4   Select all new samples from *new sample set*
5   Else
6   Calculate the number of select new samples (**1**)
7   Randomly select samples from *new sample set*
8   End if
9   Construct *outpost vector set* by
10   Calculate the number of outpost vectors (**2**)
11   Generate outpost vectors from *new sample set*
12   Construct *decayed prior sample set* by
13   Calculate the number of decayed prior samples (**3**)
14   Randomly select samples from *prior sample set*
15   Construct *final sample set* by
16   UNION (*selected new sample set, outpost vector set ,decayed prior sample set*)
17   Train the network with the *final sample set*
18   Set *final sample set* as *prior sample set* for the
       next training session
19 **end for**

In order to study the noise tolerance of the algorithm, a random noise radius is used to move the samples to their nearby location. Samples at the boundary can even be moved across their boundary. However, the new location must also be outside the territory of the other class.

4 Experiment

The experiment was conducted on a machine using Intel® Pentium D™ 820 2.4 GHz with 2.0 GB main memory running Microsoft® Windows™ XP SP3. Feed-forward backpropagation neural network running under MATLab 2007a is used as the classifier.

The proposed algorithm was tested on the 2-dimension partition problem. The distribution of samples was created in limited location of 2-dimension donut ring as shown in Fig. 2. This partition had three parameters: Inner Radius (R1), Middle Radius (R2) and Outer Radius (R3). The class of samples depended on geometric position. There were two classes which were designed as one and zero.

Inner Radius	$0 < Radius \leq R1$	*Class 0*
Middle Radius	$R1 < Radius \leq R2$	*Class 1*
Outer Radius	$R2 < Radius \leq R3$	*Class 0*

The context of the problem was assumed to shift from an angular location to another while maintaining some overlapped area between consecutive contexts as shown in Fig. 3. The set numbers shown in Fig. 3 identify the sequence of training and testing sessions.

In the experiment, each training and testing set consisted of eight sub sets of samples generated from eight problem contexts. Each sub set of samples consisted of 400 new samples (200 samples from class 0 and 200 samples from class 1). The samples from class 0 were placed in the outer radius and inner radius. The samples from class 1 were placed in the middle radius.

In the sample generation process, the radius of the donut ring was set to 100. Placement of the samples from both classes was restricted by the gap or empty space introduced between both classes to test the noise tolerance of the network.

There were two training sets having the gap of size 5 (Set A) and 10 (Set B) as shown in Fig. 4. The gap of size 0 was not used to generate the training set because there was no available space at the boundary between both classes to generate outpost vectors.

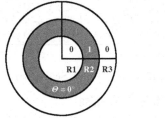

Fig. 2. Shape of Partition Area

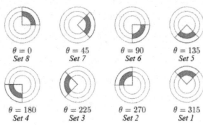

Fig. 3. Shifting of Problem Context

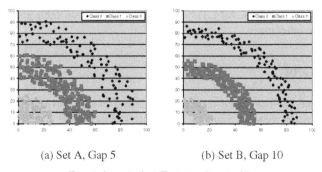

(a) Set A, Gap 5   (b) Set B, Gap 10

Fig. 4. Sample Sub Training Sets in S8

There were six main testing sets having the gap and noise radius 5:0 (Set M), 5:5 (Set N), 5:10 (Set O), 10:0 (Set P), 10:10 (Set Q), and 10:20 (Set R) as shown in Fig. 5. To test the noise tolerance of the training sets, four testing sets having the gap and noise radius 5:5 (Set S), 5:10 (Set T), 10:10 (Set U), and 10:20 (Set V) were generated by adding noise radius to all samples in the two training sets. The noise samples introduced into the testing set (Set N, O, Q, R, S, T, U, V) were intended to test the noise tolerance of the network when outpost vector was applied.

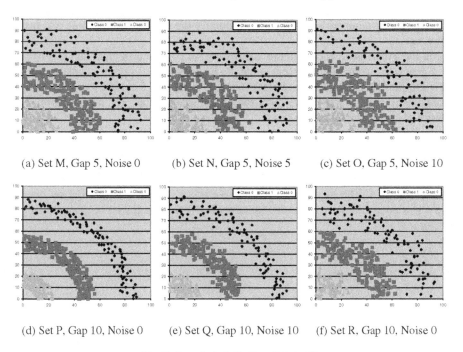

(a) Set M, Gap 5, Noise 0   (b) Set N, Gap 5, Noise 5   (c) Set O, Gap 5, Noise 10

(d) Set P, Gap 10, Noise 0   (e) Set Q, Gap 10, Noise 10   (f) Set R, Gap 10, Noise 0

Fig. 5. Sample Sub Testing Sets in S8

In the training process, the sub training (training on sub set of training set) was conducted eight times with eight sub training sets to cover eight problem contexts in a training set. Each final sub training set was composed of three components:

1. Selected new samples taken from sub training set
2. Outpost vectors generated from sub training set
3. Decay prior samples randomly selected from final sub training set from the previous training session

The numbers of vectors in each part of the final sub training set were determined by the new sample rate, the outpost vector rate, and the decay rate. Table 1 shows the number of vectors in a sample final sub training set when new sample set consisted of 200 samples, new sample rate was equal to 1.0, outpost vector rate was equal to 0.5 and decay rate was equal to 1.0. Fig. 6 shows samples of final training set for the last training session when decay rate is equal to 1.0 and 2.0 respectively.

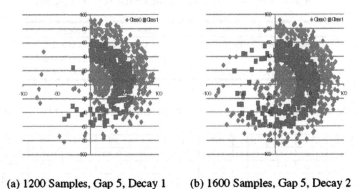

   (a) 1200 Samples, Gap 5, Decay 1   (b) 1600 Samples, Gap 5, Decay 2

Fig. 6. Sample Last Sub Training Set

Because the prior sample set was constructed at the end of the algorithm, there was no prior sample set to make decayed prior sample set for the first sub training session. An additional step for solving this problem is to also use the new sample set as the prior sample set.

The procedure for the experiment started from the feed forward back-propagation neural network being trained with the following parameters:

1. network size = [10 1]
2. transfer function for hidden layer = "logsig"
3. transfer function for output layer = "logsig"
4. max epochs = 500
5. goal = 0.001

After the first training session (S1), seven sub training sessions followed. At the end of the eighth sub training session (S8), the supervised neural network was tested with the sub testing samples from every context to evaluate the performance of testing data from each context.

The testing results are shown in Table 2, 3, 4, 5, 6, 7, and 8. For the testing sets without noise samples (Set M, P), applying outpost vector can lower the mean square error (MSE) effectively. For the testing sets with noise samples and small gap (Set N, O), medium outpost vector rate (OV 0.5) give better results. For the testing sets with noise samples and large gap (Set Q, R), large outpost vector rate (OV 1.0) give better results. For the testing sets with noise samples generated from the training sets (Set S, T, U, V), large outpost vector rate (OV 1.0) generally give better results.

The testing results show that the proposed algorithm can classify samples in the newer contexts (S1, S7, S8) accurately while the accuracy of classifying samples from the older contexts (S2, S3, S4, S5, S6) is lower because its old knowledge is decaying. The proposed algorithm presents some level of noise tolerance because the difference between the mean square errors (MSEs) of the classification of testing sets with and without noise samples is insignificant.

Table 1. The Number of Vectors in a Final Sub Training Set

Type	Rate	Vectors
Selected New Samples	1.0	200 (200 × 1.0)
Outpost Vectors	0.5	100 (200 × 0.5)
Decayed Prior Samples	1.0	200 (200 × 1.0)

Table 2. MSEs from Training Set A with Testing Set M, N, O and Decay Rate 1

SET	OV	NS	S1	S2	S3	S4	S5	S6	S7	S8
M	0.0	00	0.13	0.21	0.17	0.16	0.25	0.14	0.01	0.03
N	0.0	05	0.14	0.19	0.12	0.17	0.26	0.14	0.01	0.04
O	0.0	10	0.14	0.21	0.16	0.19	0.25	0.14	0.03	0.04
M	0.5	00	0.02	0.14	0.15	0.02	0.00	0.00	0.00	0.00
N	0.5	05	0.02	0.14	0.15	0.03	0.00	0.00	0.00	0.00
O	0.5	10	0.02	0.15	0.14	0.03	0.01	0.00	0.00	0.00
M	1.0	00	0.20	0.38	0.27	0.14	0.06	0.02	0.01	0.01
N	1.0	05	0.21	0.38	0.27	0.13	0.06	0.03	0.01	0.01
O	1.0	10	0.21	0.37	0.27	0.14	0.07	0.03	0.02	0.01

Table 3. MSEs from Training Set A with Testing Set M, N, O and Decay Rate 2

SET	OV	NS	S1	S2	S3	S4	S5	S6	S7	S8
M	0.0	00	0.27	0.30	0.28	0.28	0.28	0.23	0.21	0.21
N	0.0	05	0.28	0.30	0.28	0.27	0.28	0.25	0.20	0.22
O	0.0	10	0.27	0.30	0.27	0.28	0.27	0.23	0.21	0.21
M	0.5	00	0.12	0.23	0.13	0.02	0.01	0.00	0.00	0.00
N	0.5	05	0.12	0.23	0.13	0.03	0.01	0.00	0.00	0.00
O	0.5	10	0.12	0.24	0.13	0.03	0.02	0.01	0.01	0.01
M	1.0	00	0.18	0.32	0.31	0.20	0.11	0.08	0.07	0.05
N	1.0	05	0.18	0.32	0.31	0.20	0.10	0.09	0.07	0.06
O	1.0	10	0.18	0.32	0.31	0.20	0.11	0.09	0.08	0.06

Table 4. MSEs from Training Set B with Testing Set P, Q, R and Decay Rate 1

SET	OV	NS	S1	S2	S3	S4	S5	S6	S7	S8
P	0.0	00	0.17	0.30	0.26	0.19	0.21	0.16	0.08	0.06
Q	0.0	10	0.18	0.31	0.27	0.20	0.21	0.16	0.08	0.07
R	0.0	20	0.17	0.30	0.27	0.20	0.21	0.17	0.10	0.09
P	0.5	00	0.27	0.45	0.38	0.39	0.35	0.19	0.08	0.04
Q	0.5	10	0.27	0.45	0.38	0.39	0.34	0.20	0.08	0.04
R	0.5	20	0.27	0.45	0.37	0.38	0.33	0.21	0.09	0.07
P	1.0	00	0.22	0.31	0.22	0.26	0.25	0.23	0.12	0.01
Q	1.0	10	0.25	0.34	0.22	0.27	0.25	0.23	0.13	0.01
R	1.0	20	0.24	0.34	0.25	0.29	0.24	0.25	0.14	0.05

Table 5. MSEs from Training Set B with Testing Set P, Q, R and Decay Rate 2

SET	OV	NS	S1	S2	S3	S4	S5	S6	S7	S8
P	0.0	00	0.22	0.32	0.35	0.33	0.28	0.25	0.20	0.17
Q	0.0	10	0.24	0.32	0.35	0.33	0.29	0.25	0.21	0.18
R	0.0	20	0.25	0.32	0.35	0.34	0.29	0.25	0.21	0.20
P	0.5	00	0.16	0.31	0.34	0.29	0.19	0.08	0.03	0.03
Q	0.5	10	0.16	0.31	0.34	0.29	0.19	0.08	0.03	0.04
R	0.5	20	0.16	0.31	0.33	0.29	0.20	0.10	0.05	0.06
P	1.0	00	0.17	0.28	0.15	0.03	0.09	0.10	0.01	0.00
Q	1.0	10	0.17	0.30	0.16	0.04	0.09	0.09	0.01	0.01
R	1.0	20	0.18	0.30	0.17	0.08	0.11	0.11	0.02	0.02

Table 6. MSEs from Training Set A with Testing Set S, T and Decay Rate 1

SET	OV	NS	S1	S2	S3	S4	S5	S6	S7	S8
S	0.0	05	0.13	0.19	0.14	0.15	0.24	0.14	0.01	0.02
T	0.0	10	0.13	0.25	0.18	0.14	0.24	0.14	0.08	0.02
S	0.5	05	0.01	0.14	0.15	0.02	0.00	0.00	0.00	0.00
T	0.5	10	0.02	0.15	0.14	0.02	0.00	0.00	0.02	0.00
S	1.0	05	0.19	0.37	0.27	0.14	0.06	0.03	0.01	0.01
T	1.0	10	0.18	0.39	0.26	0.14	0.07	0.02	0.02	0.01

Table 7. MSEs from Training Set B with Testing Set U, V and Decay Rate 1

SET	OV	NS	S1	S2	S3	S4	S5	S6	S7	S8
U	0.0	10	0.18	0.30	0.27	0.20	0.21	0.16	0.07	0.08
V	0.0	20	0.20	0.30	0.26	0.21	0.21	0.16	0.09	0.07
U	0.5	10	0.29	0.43	0.41	0.39	0.35	0.17	0.07	0.05
V	0.5	20	0.31	0.44	0.37	0.42	0.35	0.17	0.08	0.06
U	1.0	10	0.24	0.32	0.22	0.27	0.24	0.21	0.11	0.01
V	1.0	20	0.28	0.29	0.23	0.27	0.25	0.21	0.14	0.03

Table 8. MSEs from Training Set B with Testing Set U, V and Decay Rate 2

SET	OV	NS	S1	S2	S3	S4	S5	S6	S7	S8
U	0.0	10	0.23	0.37	0.30	0.33	0.29	0.24	0.20	0.19
V	0.0	20	0.28	0.26	0.37	0.29	0.28	0.24	0.21	0.18
U	0.5	10	0.18	0.33	0.34	0.29	0.19	0.07	0.03	0.04
V	0.5	20	0.20	0.30	0.33	0.29	0.19	0.07	0.03	0.05
U	1.0	10	0.18	0.31	0.17	0.03	0.08	0.10	0.01	0.01
V	1.0	20	0.21	0.26	0.15	0.08	0.09	0.10	0.02	0.01

5   Conclusion

A study of noise tolerance characteristics of an adaptive learning algorithm for supervised neural network is presented. The noise samples are used to test the noise tolerance of the algorithm. Overall result shows that combining adaptive learning algorithm with contour preserving classification yields effective noise tolerance, better learning capability, and higher accuracy.

References

[1] Tanprasert, T., Kripruksawan, T.: An approach to control aging rate of neural networks under adaptation to gradually changing context. In: ICONIP 2002 (2002)

[2] Tanprasert, T., Kaitikunkajorn, S.: Improving synthesis process of decayed prior sampling technique. In: Tech 2005 (2005)

[3] Tanprasert, T., Tanprasert, C., Lursinsap, C.: Contour preserving classification for maximal reliability. In: IJCNN 1998 (1998)

[4] Burzevski, V., Mohan, C.K.: Hierarchical growing cell structures. In: ICNN 1996 (1996)

[5] Fritzke, B.: Vector quantization with a growing and splitting elastic net. In: ICANN 1993 (1993)

[6] Fritzke, B.: Incremental learning of local linear mappings. In: ICANN 1995 (1995)

[7] Martinez, T.M., Berkovich, S.G., Schulten, K.J.: Neural-gas network for vector quantization and it application to time-series prediction. IEEE Transactions on Neural Networks (1993)

[8] Chalup, S., Hayward, R., Joachi, D.: Rule extraction from artificial neural networks trained on elementary number classification tasks. In: Proceedings of the 9th Australian Conference on Neural Networks (1998)

[9] Craven, M.W., Shavlik, J.W.: Using sampling and queries to extract rules from trained neural networks. In: ICML 1994 (1994)

[10] Setiono, R.: Extracting rules from neural networks by pruning and hidden-unit splitting. Neural Computation (1997)

[11] Sun, R.: Beyond simple rule extraction: Acquiring planning knowledge from neural networks. In: ICONIP 2001 (2001)

[12] Thrun, S., Mitchell, T.M.: Integrating inductive neural network learning and explanation based learning. In: IJCAI 1993 (1993)

[13] Towell, G.G., Shavlik, J.W.: Knowledge based artificial neural networks. Artificial Intelligence (1994)

[14] Mitchell, T., Thrun, S.B.: Learning analytically and inductively. Mind Matters: A Tribute to Allen Newell (1996)

[15] Fasconi, P., Gori, M., Maggini, M., Soda, G.: Unified integration of explicit knowledge and learning by example in recurrent networks. IEEE Transactions on Knowledge and Data Engineering (1995)

[16] Polikar, R., Udpa, L., Udpa, S.S., Honavar, V.: Learn++: An incremental learning algorithm for supervised neural networks. IEEE Transactions on Systems, Man, and Cybernetics (2001)

[17] Tanprasert, T., Fuangkhon, P., Tanprasert, C.: An Improved Technique for Retraining Neural Networks In Adaptive Environment. In: INTECH 2008 (2008)

[18] Fuangkhon, P., Tanprasert, T.: An Incremental Learning Algorithm for Supervised Neural Network with Contour Preserving Classification. In: ECTI-CON 2009 (2009)

Application Study of Hidden Markov Model and Maximum Entropy in Text Information Extraction

Rong Li[1,*], Li-ying Liu[2], He-fang Fu[1], and Jia-heng Zheng[3]

[1] Department of computer, Xinzhou Teachers' College,
Xinzhou, Shanxi 034000, China
[2] Weather bureau of Xinzhou City, Xinzhou, Shanxi 034000, China
[3] School of Computer and Information Technology, Shanxi University,
Taiyuan, Shanxi 030006, China
{Rong Li,Li-ying Liu,He-fang Fu,
Jia-heng Zheng,lirong_1217}@126.com

Abstract. Text information extraction is an important approach to process large quantity of text. Since the traditional training method of hidden Markov model for text information extraction is sensitive to initial model parameters and easy to converge to a local optimal model in practice, a novel algorithm using hidden Markov model based on maximal entropy for text information extraction is presented. The new algorithm combines the advantage of maximum entropy model, which can integrate and process rules and knowledge efficiently, with that of hidden Markov model, which has powerful technique foundations to solve sequence representation and statistical problem. And the algorithm uses the sum of all features with weights to adjust the transition parameters in hidden Markov model for text information extraction. Experimental results show that compared with the simple hidden Markov model, the new algorithm improves the performance in precision and recall.

Keywords: text information extraction, maximum entropy, hidden markov model, generalized iterative scaling algorithm.

1 Introduction

The universal application of WWW causes on-line text quantity to increase exponentially, therefore how to deal with these huge amounts of on-line text information becomes the present important research subject. The automatic text information extraction is an important link of text information processing [1].Text information extraction refers to extract automatically the related or specific type information from text. At present, the text information extraction model mainly has three kinds: the dictionary-based extraction model[2], the rule-based extraction model[3] and the extraction model based on Hidden Markov Model (HMM)[4-8].

The text information extraction using HMM is one kind of information extraction method based on statistics machine learning. HMM is easy to establish, does not need

* This work is supported by the Natural Science Foundation of China (Grant #60775041).

H. Deng et al. (Eds.): AICI 2009, LNAI 5855, pp. 399–407, 2009.

the large-scale dictionary collection and rule set, its compatibility is good and the extraction precision is high, thus HMM obtains researchers' attention. For example, In References [4], Kristie Seymore et al. used HMM to extract forehead information of computer science research paper such as title, author and abstract. In References [5], Dayne Frietag and Andrew McCallum adopted one kind of "shrinkage" technology to improve the probability estimate of the HMM information extraction model. In References[6], Freitag D and McCallum A used the stochastic optimization technique is to select automatically the most suitable HMM model structure for information extraction. Souyma Ray and Mark Craven[7] applied phrase structure analysis technology in the natural language processing to the HMM text information extraction. T Scheffer et al.[8] used the active learning technology to reduce the flag data which is needed when training HMM information extraction model. The HMM method doesn't consider the characteristic information of text context and the characteristic information contained in text word itself, but these information are very useful to realize the correct text information extraction. Freitag D et al.[9]proposed a Maximum Entropy Markov Model(MEMM) for the segmentation of questions and answers in the FAQs(Frequently Asked Questions) text. MEMM is one kind of exponential model. It mainly takes the text abstract characteristic as the input and selects the next state in the base of the Markov state transition, and on this point, it is near to finite state automaton. Since MEMM incorporates the text context characteristic information and the characteristic information contained in text word itself into Markov model, it can improve the performance of information extraction. But it hasn't concrete text word count and only considers the abstract characteristic, which causes its performance to be inferior to that of HMM in some certain circumstances.

In the paper, we combine the advantage of maximum entropy model, which can integrate and process rules and knowledge efficiently, with that of hidden Markov model, which has powerful technique foundations to solve sequence representation and statistical problem, and present a Maximal Entropy-based Hidden Markov Model(ME-HMM) for text information extraction. The algorithm uses the sum of all features with weights to adjust the transition parameters in hidden Markov model. Experimental results show that compared with the simple hidden Markov model, the new algorithm improves the performance in precision and recall.

2   Text Information Extraction Based on HMM

2.1   Hidden Markov Model

An HMM includes two layers: one observation layer and one hidden layer. The observation layer is the observation sequence for recognition and the hidden layer is a Markov process, (i.e. a limited state machine), in which each state transition all has transition probability.

An HMM is specified by a five-tuple $(S, V, A, B, \prod)$:

$$S = \{S_1, S_2, \ldots, S_N\}$$

$$V = \{V_1, V_2, \ldots V_M\}$$

$$A = \{a_{ij} = p(q_{t+1} = S_j | q_t = S_i), 1 \leq i, j \leq N\}$$

$B=\{b_j(V_k)=P(V_k \text{ at } t|q_t=S_j), 1\le i,j\le N, 1\le k\le M\}$

$\prod=\{\pi_i=P(q_1=S_i), 1\le i\le N\}$

Among the notation, S denotes the state set of N states, V represents the word set of M possible output words, and A, B, $\prod$ are the probabilities for the state transition, the observation emissions, and the initial state, respectively.

2.2 Text Information Extraction Based on HMM

Hidden Markov model is mainly used to solve the following three fundamental problem: evaluation problem, study problem and decoding problem. For the commonly used algorithm, see also Reference [10-12]. Text information extraction needs to solve the study problem and the decoding problem of HMM. The purpose of text information extraction is to extract those specific and compelling information from large quantity of information, namely to extract some information such as author, publishing press, publication time, affiliation and so on from the "word string "array composed of different text information, which is similar to part-of-speech tagging in the field of Chinese information processing. When extracting text information using HMM, Maximum Likelihood(ML) algorithm for marked training sample set or Baum-Welch algorithm for unmarked training sample set is adopted generally to obtain HMM parameters. Then Viterbi algorithm is used to find out the state label sequence of the maximum probability from input text information and the state label is the content label to be extracted that is we define beforehand. In the paper, the manually marked training sample is adopted to train model parameters in the text information extraction model based on HMM. Information extraction is a two-stage process:

1) Obtain HMM parameters by using the statistical method from the training sample. ML algorithm is adopted to construct HMM model and obtain model parameters a_{ij}, $b_j(V_k)$, π_i by the methods of statistic. ML algorithm demands to have sufficient marked training array, and each word in each array uses corresponding class label. Therefore, the transition number of times from state S_i to S_j may be calculated and noted as C_{ij}. The number of times of outputting word V_k can be calculated in the specific state, noted as $E_j(K)$.Similarly, the number of times of which a sequence starts from a certain specific state may be calculated, noted as $Init(i)$.Therefore, the probability can be described as follows:

$$\pi_i = \frac{Init(i)}{\sum_{j=1}^{N} Init(j)}, 1\le i\le N \tag{1}$$

$$a_{ij} = \frac{C_{ij}}{\sum_{k=1}^{N} C_{ik}}, 1\le i, j\le N \tag{2}$$

$$b_j(k) = \frac{E_j(k)}{\sum_{i=1}^{M} E_j(i)}, 1\le i\le N, 1\le j\le M \tag{3}$$

2) Apply the established HMM for text information extraction. The text information extraction process based on HMM is that state sequence Q^* generating the maximum probability symbol sequence should be sought if given HMM and a symbol sequence and then the observation text for marked target state label is the content of information extraction. Viterbi algorithm is the classical approach to solve the HMM decoding problem. To avoid data underflow problem, an improved Viterbi algorithm is brought forwarded in the paper. The place of improvement is that all the probabilities in Viterbi formulas are multiplied by proportion factor 10^2, then the logarithm is taken to both sides of formulas so as to obtain improved Viterbi formulas.

3 Text Information Extraction Using ME-HMM

3.1 Maximum Entropy Principle

Maximum entropy principle, as a very important principle in thermodynamics, has widespread application in many other domains, and it is also one main processing method in the natural language processing aspect [13-15].

If the natural language is regarded as a stochastic process, we will construct a stochastic process model p, $p \in P$. The set of output value is Y, $y \in Y$. The set of N samples is $S=\{(x_1, y_1),(x_2, y_2)...,(x_n, y_n)\}$, where (x_i, y_i) is an observed event, the event space is X^*Y; The language knowledge is represented with the characteristic which is a two-valued function f: $X^*Y \rightarrow \{0,1\}$. Entropy describes the uncertainty of the random variable. The greater the random variable's uncertainty, the bigger is its entropy.

The entropy of model p may be described as :

$$H(p) \equiv -\sum_{x,y} p(x,y) log\ p(x,y) \tag{4}$$

The maximum entropy model is represented as:

$$p^* = arg\ \max_{p \in C} H(p) \tag{5}$$

where C is the model set which satisfies the restraint. The remaining problem is to seek p^* in C. p^* can be described as the following form:

$$p^*(y|x) = \frac{1}{Z(X)} exp[\sum_i \lambda_i f_i(x,y)] \tag{6}$$

where $Z(x)$ is a normalized constant.

$$Z(X) = \sum_y exp[\sum_i \lambda_i f_i(x,y)] \tag{7}$$

Where λ_i is the model parameter and also can be seen as the characteristic weight.

In 1972, Darroch and Ratcliff proposed a called GIS algorithm(generalized iterative scaling algorithm), the algorithm is a general iterative algorithm [16]. An advantage of the ME method is that the experimenter only needs to concentrate on what characteristic should be selected, but does not need to consider how to use these characteristics.

Each characteristic's contribution is decided by the corresponding weight, and these weights may be obtained automatically by the GIS learning algorithm.

3.2 Text Information Extraction Based on ME-HMM

In the maximum entropy method, we call a rule as a characteristic. The idea of the maximum entropy method is to find a characteristic set, and determine the important degree of each characteristic. The maximum entropy model can integrate each kind of characteristic and rule into a unified frame, and hidden Markov model has the very strong superiority in sequence representation problem and the statistical learning aspect. So, if the maximum entropy method is incorporated into the hidden Markov model for text information extraction, both can we solve the knowledge expression problem, and may momentarily add the newly gained language knowledge to the model, and the combination of the two methods is that of the regular method and the statistical method union. In accordance with the idea, the paper adopts hidden Markov model based on maximum entropy principle for text information extraction. For ease of comparison, three models including HMM, MEMM and ME-HMM proposed in the paper are shown in Figure 1, respectively.

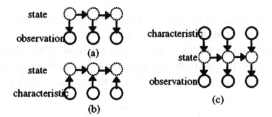

Fig. 1. (a)HMM; (b)MEMM; (c)ME-HMM

Given the characteristic set, we define a two-valued function:

$$f_{i,j}(o_t,s_t)=\begin{cases} 1, \text{ if the observation } o_t \text{ has charastic } i, \\ \quad \text{ and the marked state } s_t = s_j \\ 0, \qquad\qquad\qquad\qquad otherwise \end{cases} \tag{8}$$

The ME-HMM model constructs a characteristic - state transition probability matrix $M= \{M_{i,j}\}$, where the matrix unit $M_{i,j}$ is the transition probability from the characteristic i to the state j, and $M_{i,j}$ satisfies the following condition:

$$\sum_j M_{i,j} =1, \ 1\le i\le N_F, 1\le j\le N_S \tag{9}$$

In the above formula, N_F is the chosen characteristic number, N_S is the state number of the model. In the training stage, each observation is performed for feature extraction. And each observation in the training data set implicitly corresponds to a flag state, therefore we count the relation between the state and the characteristic. By using the GIS algorithm [16], the probability matrix of the characteristic–state transition can be obtained. The GIS algorithm is as follows:

1) Calculate the average value of each characteristic-state in the training data. When the total length of the observation sequence is m_s, for the ith characteristic, the jth state, its average value is

$$F_{i,j} = \frac{1}{m_s} t \sum_{k=1}^{m_s} f_{i,j}(o_{t_k}, s_{t_k})$$ (10)

2) Take as the 0th iteration of the GIS algorithm by the beginning of the random parameter and make $M_{i,j}^{(0)} = 1$.

3) In the nth iteration, use the current $M_{i,j}^{(n)}$ value and calculate the the expected value of each characteristic-state:

$$E_{i,j}^{(n)} = \frac{1}{m_s} \sum_{k=1}^{m_s} \sum_{k \in s} P_{s'}^{(n)}(s \mid o_{t_k}) f_{i,j}(o_{t_k}, s)$$ (11)

where s' is the previous state of s, and

$$P_{s'}^{(n)}(s \mid o_{t_k}) = \frac{1}{Z(o,s')} exp[\sum_i M_{i,j}^{(n)} f_{i,j}(o_{t_k}, s)]$$ (12)

$Z(o,s')$ is the normalization constant which can guarantee that the value is one by method of summation for all next state s of all s' state.

4) In the situation of satisfying the constraints limit, combine the expected value with the average value of the training data to decide how to adjust the parameter. Designate a constant C, the adjustment formula is as follows.

$$M_{i,j}^{(n+1)} = M_{i,j}^{(n)} + \frac{1}{C} log[\frac{F_{i,j}}{E_{i,j}^{(n)}}]$$ (13)

5) If parameter values converge, then terminate the algorithm, else go to Step (3).

In order to reduce the iterative number of times, we use the statistical method to assign the initial value for $M_{i,j}^{(0)}$. When utilizing the Viterbi algorithm, the state at time t is determined together by the probability of the state at time t -1 and the observation characteristic at time t.

$$p(s_t = s_j \mid s_{t-1}, o_t) = \frac{1}{\gamma}(\lambda \cdot \alpha_{t-1,j} + (1-\lambda) * \sum_i (M_{i,j} * f_{i,j}(o_t, s_t)))$$ (14)

Where $\alpha_{t-1,j}$ is the transition probability from the known state at time t-1 to state j. γ is a normalized parameter.

$$\gamma = \sum_j (\lambda \alpha_{t-1,j} + (1-\lambda) * \sum_i (M_{i,j} * f_{i,j}(o_t, s_t)))$$ (15)

here λ is the weight of adjusting the relatively important degree of the characteristic-state transition probability and state transition probability.

4 Experiments

For ease of comparison, we conduct the experiment by using the standard data set which is provided by Carnegie Mellon University for computer scientific research paper forehead information extraction. In order to reduce the time complexity of the characteristic selection process in the maximum entropy method, we artificially select some useful characteristics. The concrete state can be divided into the positive characteristic and negative characteristic according to the contribution of the characteristic attribute relative to the state.

The positive characteristic of the state indicates that when observation presents this characteristic, the characteristic tends to shift to this state. The negative characteristic of the state indicates that when observation presents this characteristic, the probability of shifting to the state can decrease. In the analysis whether to contain the personal name characteristic, we use the American personal name dictionary downloaded from on-line to match.

When extracting information from the computer scientific research paper forehead, owing to considering the useful information such as the typeset format, newline character, separating character, the text preprocessing first needs to be conducted by adopting the HMM text information method based on text blocks. In the training stage, the initial probability and the transition probability take the block as the Fundamental unit, and formula (1) and (2) are used. The output probability takes the word as the Fundamental unit, and formula (3) is used. In the extraction stage, after the text to be extracted is partitioned into small blocks, we calculate the output probability of each block. The text sequence to be extracted is converted to the block sequence and the block emission probability is the sum of the emission probability of each word in the block. Suppose the block observation sequence is $O = O_1 O_2...O_T$, if the length of tth block is K, namely including K words, noted as $O_{t1} O_{t2}...O_{tK}$, the probability that state j outputting the tth block is:

$$b_j(o_t) = \sum_{k=1}^{K} b_j(o_{tk}) \tag{16}$$

We only use a few characteristics such as personal name, all digits, month or its abbreviation, email symbol "@"(the block without containing the symbol can't be email.). In the experiment, by changing the value of parameter λ, the relatively important degree of the characteristic-state transition probability and state transition probability. When $\lambda = 0.6$ and $train\text{-}set = 400$, experimental results are shown in Table 1.

When using MEMM for text information extraction without considering the statistical probability of words, the performance in the experimental data set in the paper is poor and experimental results are not convenient for combination, therefore we don't list the related experimental results. As shown in Table 1, the text information extraction algorithm based on ME-HMM improves in some degree in the precision and recall rate. When using the person name dictionary to match, Table 1 shows that the recall rate of state "author" increases approximately 8%, which illustrates that the joined related knowledge plays a large role.

Table 1. Precision and recall comparison of information extraction concrete domaion using HMM and ME-HMM

domain	HMM		ME-HMM	
	Precision(%)	Recall(%)	Precision(%)	Recall(%)
title	0.820347	0.830480	0.838308	0.864868
author	0.823954	0.916533	0.898551	0.955185
affiliation	0.872078	0.914193	0.887361	0.916536
address	0.902686	0.840000	0.925010	0.853007
email	0.888176	1.000000	0.988436	1.000000
note	0.880784	0.715228	0.921323	0.692785
web	1.000000	0.546508	1.000000	0.546508
phone	0.986667	0.888878	0.986301	0.912470
date	0.660505	0.990808	0.697607	0.990808
abstract	0.953434	1.000000	0.910071	1.000000
Intro	0.872307	1.000000	0.872307	1.000000
keyword	0.824168	0.932762	0.824168	0.931662
degree	0.512413	0.838224	0.616536	0.812032
pubnum	0.884603	0.642828	0.954045	0.703434
page	1.000000	1.000000	1.000000	1.000000

5 Conclusions and Further Work

The maximum entropy model provides one natural language processing method, and can integrate each kind of characteristic and rule into a unified frame. Incorporating the maximum entropy method into hidden Markov model for text information extraction can solve the problem of knowledge representation problem and may momentarily add the newly gained language knowledge to the model. Experimental results show that this method is effective.

In the follow-up work, we plan that the system can select automatically characteristic from a lot of candidate sets. Moreover, how to reduce the algorithm complexity is also worth the further research.

Acknowledgments. We would like to thank to the reviewers for their helpful comments. This work has been supported by the Natural Science Foundation of China (Grant #60775041).

References

1. Lawrence, S., Giles, L., Bollacker, K.: Digital libraries andautonomous citation indexing. Computer 32(6), 67–71 (1999)
2. Riloff, E., Jones, R.: Learning dictionaries for information extraction by multi-level bootstrapping. In: Proceedings of the Sixteenth National Conference on Artificial Intelligence, pp. 811–816. AAAI Press, Orlando (1999)
3. Kushmerick, N.: Wrapper induction: Efficiency and expressiveness. Artificial Intelligence 118(12), 15–68 (2000)

4. Seymore, K., McCallum, A., Rosenfel, R.: Learning hidden Markov model structure for information extraction. In: Proceedings of the AAAI 1999 Workshop on Machine Learning for Information Extraction, pp. 37–42. AAAI Press, Orlando (1999)
5. Frietag, D., McCallum, A.: Information extraction with HMMs and shrinkage. In: Proceedings of the AAAI 1999 Workshop on Machine Learning for Information Extraction, pp. 31–36. AAAI Press, Orlando (1999)
6. Freitag, D., McCallum, A.: Information extraction with HMM structures learned by stochastic optimization. In: Proceedings of the Eighteenth Conference on Artificial Intelligence, pp. 584–589. AAAI Press, Edmonton (2002)
7. Ray, S., Craven, M.: Representing sentence structure in hidden Markov models for information extraction. In: Proceedings of the Eighteenth Conference on Artificial Intelligence, pp. 584–589. AAAI Press, Edmonton (2002)
8. Scheffer, T., Decomain, C., Wrobel, S.: Active Hidden Markov Models for Information Extraction. In: Hoffmann, F., Adams, N., Fisher, D., Guimarães, G., Hand, D.J. (eds.) IDA 2001. LNCS, vol. 2189, p. 309. Springer, Heidelberg (2001)
9. Freitag, D., McCallum, A., Pereira, F.: Maximum entropy Markov models for information extraction and segmentation. In: Proceedings of The Seventeenth International Conference on Machine Learning, pp. 591–598. Morgan Kaufmann, San Francisco (2000)
10. Rabiner, L.E.: A tutorial on hidden Markov models and selected application in speech recognition. Proceedings of the IEEE 77(2), 257–286 (1989)
11. Kwong, S., Chan, C.W., Man, K.F.: Optimization of HMM topology and its model parameters by genetic algorithms. Pattern Recognition, 509–522 (2001)
12. Hong, Q.Y., Kwong, S.: A Training Method for Hidden Markov Model with Maximum Model Distance and Genetic Algorithm. In: IEEE International Conference on Neural Network & Signal Processing (ICNNSP 2003), Nanjing,P.R.China, pp. 465–468 (2003)
13. Berger, A.L., Stephen, A., Della Pietra, V.J.: A maximum entropy approach to natural language processing. Computational Linguistics 22(1), 39–71 (1996)
14. Chieu, H.L., Ng, W.T.: Named entity recognition with a maximum entropy approach. In: Daelemans, W., Osborne, M. (eds.) Proc. of the CoNLL 2003, pp. 160–163. ACL, Edmonton (2003)
15. Xiao, J.-y., Zhu, D.-h., Zou, L.-m.: Web information extraction based on hybrid conditional model. Journal of Zhenzhou University 40(3), 52–55 (2008)
16. Darroch, J., Ratcli, D.: Generalized iterative scaling for log-linear models. Annals of Mathematical Statistics 43(5), 1470–1480 (1972)

Automatic Expansion of Chinese Abbreviations by Web Mining*

Hui Liu[1], Yuquan Chen[2], and Lei Liu[2]

[1] Shanghai Institute of Foreign Trade,
1900 Wenxiang Rd., Shanghai, China
liuh@shift.edu.cn
[2] Shanghai Jiao Tong University,
800 Dongchuan Rd., Shanghai, China
{yqchen,liu-lei}@sjtu.edu.cn

Abstract. Abbreviations are common in everyday Chinese. For applications like information retrieval, we want not only to recognize the abbreviations, but also to know what they stand for. To tackle the emergence of all kinds of new abbreviations, this paper proposes a novel method that expands an abbreviation to its full name employing the Web as the main information source. Snippets containing full names of an abbreviation are obtained through a search engine by learned "help words". Then the snippets are examined using linguistic heuristics to generate a list of candidates. We select the optimal candidate according to a kNN-based ranking mechanism. Experiment shows that this method achieves satisfactory results.

Keywords: Abbreviation Expansion, Web Mining.

1 Introduction

Abbreviations take an important part in modern Chinese language. It is so common that native Chinese speakers will not even notice them in a newspaper or a book. But for natural language processing, the wide spread of abbreviations is not ideal. For computers, abbreviations are unrecognizable strings until they are further processed. Even worse, the formation of Chinese abbreviations is not quite like English ones[1]. Therefore, we cannot simply port an English analyzer into Chinese.

A lot of works[2][3][4] treat abbreviations like out-of-vocabulary words, which means they put an emphasis on how to tag a string as an abbreviation, just like the recognition of a location name, a person name, etc. But for practical applications like information retrieval and machine translation[5], we should learn what an abbreviation stands for: what is their original word form (called *full name*). The relationship is crucial for better relevance evaluation of two documents.

---

* This paper is supported in part by Chinese 863 project No. 2009AA01Z334 and the Shanghai Municipal Education Commission Foundation for Excellent Young University Teachers.

H. Deng et al. (Eds.): AICI 2009, LNAI 5855, pp. 408–416, 2009.

Chang and Lai[6] regard abbreviation expansion as an "error recovery" problem in which the suspect full names are the errors to be recovered from a set of candidates. They apply an HMM-based generation model for identification and recovery of abbreviations. In subsequent works, Chang and Teng[1] use a similar model to mine atomic Chinese abbreviation pairs. Fu et al.[7] suggest a hybrid approach to the abbreviation expansion of Chinese. They generate candidates using a mapping table and an abbreviation dictionary. Then the candidates are selected also using an HMM model. Li and Yarowsky[5] combine abbreviation expansion into a machine translation system to achieve better translation results. Their unsupervised expansion strategy is based on data occurrence.

One problem of these current approaches is data sparseness. Chang and Lai note that in the expansion experiment 42% of test cases cannot be recovered due to the sparseness of the training corpus. For statistical methods, sparseness is always *the* problem. With the emergence of new abbreviations, using a static corpus of texts from the last century is not ideal at all.

Therefore, in this paper we focus on expanding abbreviations to their full names by a novel approach using World Wide Web as the main source of information. We resort to the Web to find co-occurrence of abbreviations and their full names in returned snippets. The candidates are extracted and selected based on a combination of linguistic heuristics and Web-based features. The advantage of using the Web as a corpus is that it is huge and full of redundancy, which is nice for mining tasks and thus it can alleviate sparseness. With Web mining method, we can keep up with new abbreviations by relatively simple methods.

This paper is structured as following. In the next section we will talk about some main ideas of our approach. We will introduce the construction of Chinese abbreviations briefly and show the motivation and outline of our approach. In section 3, we will probe into our method in detail. We will first discuss the generation of candidates and then the selection of them. Section 4 will be devoted to the experiments. The last section is the conclusion.

2 The Main Idea

2.1 The Construction of Chinese Abbreviations

There are three means of constructing abbreviations in Chinese, namely

1. Truncation;
2. Condensation;
3. Summarization.

In truncation, a integral part of a full name is taken as the abbreviation, like "清华" is the abbreviation of "清华大学" (Tsinghua University). It is the simplest approach but also the rarest. By contrast, condensation is the most common. Condensation takes out one morpheme for every selected word in the phrase to form the abbreviation. For example, "北京大学"(Beijing University) is abbreviated as "北大". Summarization is also very common in Chinese, especially in slogans. In summarization, the full name is a phrase or a sentence which contains several words or clauses sharing a same constituent. The full name is abbreviated by a number denoting the number of words/clauses plus the constituent. For instance, "两广"(literally: Two Guang) is the abbreviation of "广东"(Guangdong) and "广西"(Guangxi).

According to linguists[8], the formation of abbreviation conforms to some weak rules. Abbreviations are as short as possible, in case that there will be no ambiguities. The morphemes are taken sequentially to form the abbreviation. The first morpheme is favored in selection for abbreviation over other morphemes in a word. These rules show light on further selection of candidates of full names.

2.2   Outline of Our Approach

The crux of the expansion problem is to find possible candidates of full names. In scientific writing, a full name, together with the proposed abbreviation, must be presented before the use of the abbreviation. In this case, if having caught the related context of both full name and abbreviation, we can decipher the full name. Unfortunately, it is not the case in daily usage like newspaper articles, since some abbreviations are far more common than full names. For example, we will hardly find "美利坚合众国"(The United States of America) in news except for government documents, but the use of "美国"(America) is very extensive. Full names and abbreviations co-occur in some contexts after all, though the co-occurrence is hard to find in articles or static corpus, for it is scarce. Web, on the contrary, contains thousands of terabytes of information, chances are that we can find the co-occurrence in some web documents. Moreover, with the wider and wider usage of the Web, new abbreviations will definitely be recorded in the Web.

Our approach then consists of two parts naturally (Fig. 1). First, we extract the possible contexts from the Web. Second, we recognize the full name from the context. Since we have only a limited access to the Web, that is to say, through a search engine like Google or Baidu[1], which will give us less than 1000 results per query, in most times even less, we cannot happily expect a full name to emerge in a single query like "美国"(America). In contrast, we have to tweak the queries to have the results containing possible full names. After the snippets are extracted, we move on to an extraction module in which we use linguistic heuristics to generate a list of possible candidates. The list is further pruned using co-occurrence data (*local* Web data) in the snippets for efficiency's sake. Finally, we rank the candidates in the pruned list to find the optimal result by features acquired from the Web (*global Web data*) using the k nearest neighbor method[9].

---

[1] http://www.google.cn, http://www.baidu.com

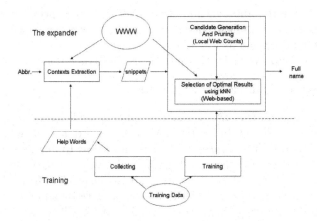

Fig. 1. The overview of this paper's system

3 The Web-Based Full Name Expansion

3.1 Query Learning and Candidate Extraction

First, we have to find text segments containing abbreviations together with their full names through a search engine. If we construct the query using the target abbreviation only, like "美国", the returned snippets will not possibly contain any full names in a single search, because abbreviations appear far more frequently than full names, as mentioned before. Therefore we shall tweak the queries to find needed snippets. In other words, we will learn some types of queries containing specific "help words" that will lead to snippets containing full names.

The learning method is as following. Having a list of training examples of pairs of <abbreviation, full name>, which will also be used in later stages of processing, we first search both the abbreviation and the full name in a search engine using the following query.

"abbreviation" "full name"

The resulting snippets are segmented and some words are taken as potential help words. The helpfulness of a word w for an abbreviation a are calculated as:

$$h(w) = \begin{cases} 0 & w \in STOP \\ |Q| & else. \end{cases} \qquad (1)$$

In Equation 1, Q is a set of queries whose results s_q contain both w, a and its full name f. Moreover, we hope that the occurrence of (w, a, f) is above some threshold γ.

$$Q = \{q | s_q \supset w \wedge s_q \supset a \wedge s_q \supset f \wedge Occ(w, a, f) > \gamma\} \qquad (2)$$

Note that if w occurs in some query's results for over γ times, we add $h(w)$ for 1, rather than count the occurrences of w directly from all returned snippets, because for a specific abbreviation there may be some special word that will appear again and again in the results. For generality's sake, we will consider the count of related queries only.

With all the help words H, the queries are constructed as

"abbreviation" "full name" "help word"

Putting these queries through a search engine, we will have a list of returned snippets that may contain full names. We will then extract all the possible candidates of full names. The strategy is simple: according to linguistic observations in section 2.1, we assume that a full name will always include all characters in the abbreviations.[2] Therefore, if locating all the characters from the abbreviations in the snippets, we can find the first candidate by simply extracting the shortest string containing them. For example, for "美国" we will first find "美利坚合众国". For "上汽" we will first extract "上海汽车". Starting from the first candidate, we scan in both directions to construct more candidates until we meet a "boundary word". A boundary word is either a word in a given stop word list, or a word having appeared next to the abbreviation in question. Namely, the set of candidate C is a set of word sequences defined as following:

$$C = \{s | \forall s' \subset s \; s' \notin STOP' \land \forall ch \in a, ch \in s\} \qquad (3)$$

In the above equation, s and s' are strings, $STOP'$ is a list of boundary words, ch a character.

3.2 Candidate Selection

After the candidates are extracted, we have to decide which one is the full name. Before selection, we will first prune the list to exclude less likely candidates, since selection approach, which is based on Web features, is quite time intensive.

In a straight-forward manner we can just count the occurrences of the candidate strings in the snippets and delete ones with the lowest count. This approach is somewhat clumsy in the sense that c_1 which is a substring of c_2 will always has higher counts than c_2, because each occurrence of c_2 contains a c_1.

Though the occurrence count is not applicable directly as the sole evidence, we can use it as a basis for further judgment. We can model the scenario as a comparison problem. If two strings do not have a subsumption relation between them, the above count is the measurement. Otherwise, we will subtract the co-occurrence count of c_2, the super-string from that of the substring, c_1.

Definition 1. *Let $\preceq$ be a relation on the Cartesian product of two string sets, $S \times S$, in which $\forall s_1, s_2 \in S$, $s_1 \preceq s_2$ iff s_1 is more likely to be a full name.*

For any s_1 and s_2 we can draw the relation $\preceq$ between them according to the following:

---

[2] The assumption is not always true as we show in section 4.

Proposition 1. $\forall s_1, s_2 \in S, s_1 \preceq s_2$ iff $\begin{cases} O(s_1) > 2O(s_2) & s_1 \subseteq s_2 \\ O(s_1) > \frac{O(s_2)}{2} & s_2 \subseteq s_1 \\ O(s_1) > O(s_2) & \textit{Otherwise.} \end{cases}$

With $\preceq$ defined on any string pairs of S, we know that $\preceq$ is a total order. Therefore, we can have some top candidates for further selection.

The selection of an optimal full name from the pruned list is seen as a ranking problem. The ranking values are obtained using kNN. The training examples are transformed into a real value vector $\vec{x}$ of features. The ranking value of a candidate c for an abbreviation a is the average Euclid distance from k nearest neighbors in the examples. Candidates with the lowest value (the nearest) is taken as the final result.

Two kinds of real value features are used here, the structural features and the Web-based features. The structural features represent how the abbreviation is constructed, i.e. the distribution of abbreviation characters (characters in the abbreviation word) in full names. The Web-based features show the occurrences of the candidates in Web context. Table 1 shows the features and some formulae for computing them. In the table Len is a function for a string's length, Ch is the set of abbreviation characters, $I(t)$ is a function that maps an abbreviation character t to its index in the full name, and E is the set of characters that are expanded into 2-character words in the full name.

4 Experiments and Discussions

The test set of our experiment consists of 480 abbreviations. The data are from two sources. One half of the abbreviations (Data Set 2) are extracted randomly

Table 1. Features used

Type	Name	Formula and Description
Structural	density	$\frac{Len(a)}{Len(c)}$ Length of abbreviation characters versus the candidate
	center	$\sum_{t \in Ch} I(t)/Len(c)$ Position of characters in the candidate.
	diversity	$\sqrt{\sum_{0 < i < Len(a)} (I(t_i) - I(t_{(i-1)}))^2}$ How abbreviations are scattered in the candidate;
	span	$\frac{\max_{t,k \in Ch} I(t) - I(k)}{Len(c)}$ The length of span occupied by abbreviation characters.
	expand	$\frac{|E|}{|Ch|}$ Number of characters that are expanded into words in the candidate.
Web	wordcount	Occurrence of candidate in Web; Query: "c"
	titlecount	Occurrences of candidate in titles of Web pages; Query: intitle:"c"
	titledensity	$\frac{1}{N_{snippet}} \sum \frac{Len(c)}{Len(Title)}$ Density of candidate in titles; Query: intitle:"c"
	indcount	Occurrence of the candidate in whole if search has no quotation; Query: c

from the PFR People's Daily Corpus[3], which contains articles from People's Daily in 1998. We take 2-character words tagged as "j" for abbreviations. We do not include single character words because they are often covered in an abbreviation dictionary. Another half of the data (Data Set 1) are hand selected from the Web by an editor who is specialized in neither computer science nor linguistics. The part of information represents the usage of abbreviations in our daily life.

The training examples contain 50 abbreviations from the PFR corpus, together with their full names. In each query throughout the expansion we retrieve 100 snippets from Baidu. Since Web-based expansion is time intensive, we only use 5 help words from learning. We evaluate the accuracy of expansion of Data set 1, Data set 2 and the two sets in general. The overall results are shown in Table 2. Apart from our final system, we present several other baseline systems, B_DICT, B_NOKNN and B_NOHELPWORD. B_DICT relies on an abbreviation dictionary for expanding. The dictionary is generated from the Modern Chinese Standardized Dictionary[10]. The other baseline systems B_NOKNN and B_NOHELPWORD only utilize parts of the final system. B_NOKNN uses only local Web count (occurrence counts from snippets of candidate generation) for selecting the optimal result without the kNN ranker, while B_NOHELPWORD do not use help words but only the abbreviation itself as a query to generate snippets.

Our approach does not perform so well on data set 2 as on data set 1. It is partly due to the fact that the PRF corpus contains texts ten years ago, so the abbreviations are somewhat too "old-fashioned" to appear on the Web. Moreover, some abbreviations tagged in the corpus are used as common words now. From the Table 2 we can also see that B_DICT has the lowest performance, since the dictionary does not cover many abbreviations in use. Both B_NOKNN and B_NOHELPWORD have similar performance, about 0.3 lower than our final system. This inferiority shows the indispensability of the two main components: help words and the kNN based selector.

Table 2. Results of Evaluation: Comparison of baseline systems

Systems	Data set 1	Data set 2	All
B_DICT	0.1625	0.1	0.1313
B_NOKNN	0.5125	0.3458	0.4292
B_NOHELPWORD	0.5833	0.4042	0.4938
Final	0.8792	0.8033	0.8412

We also look into the contribution of each feature used in the kNN-based selector. We evaluate the contribution in a "negative" way, i.e. excluding one feature each time to see the impact on the evaluation result. The results are shown in Table 3. We can see that all the features contribute to the final result,

---

[3] Available at http://icl.pku.edu.cn/icl_groups/corpus/dwldform1.asp

Table 3. Contribution of Features

Feature Excluded	Data set 1	Data set 2	All
None	0.8792	0.8033	0.8412
-titlecount	-0.0833	-0.0875	-0.0854
-titledensity	-0.075	-0.06667	-0.0708
-wordcount	-0.2917	-0.3167	-0.3042
-idcount	-0.125	-0.1667	-0.1458
-density	-0.2583	-0.3167	-0.2875
-span	-0.0958	-0.125	-0.1104
-diversity	-0.15	-0.175	-0.1625
-center	-0.2333	-0.3208	-0.2771
-expand	-0.1583	-0.2125	-0.18542

with the contribution ranging from 8% to more than 30%. For Web generated features, one interesting thing is that features about the title of snippets (title-count and titledensity) have less contribution than others. This may be due to the fact that for some long full names, people tend to use their abbreviations in titles. Another interesting finding is that all the features except for one contribute more to data set 2 than data set 1. This is because it is more difficult to generate full names for abbreviation from Data set 2 from the Web than those in data set 1. Therefore, the candidate list of full names is more error prone, which leads to the larger contribution of the kNN-based selector together with involving features.

One surprise for us is that our approach can handle abbreviations for very complex and long slogans, for example, "四个确保"'s full name is "确保经济平稳较快发展，确保民生持续得到改善，确保社会和谐稳定，确保世博会筹办工作有序进行". This shows the merit of help words, for if we search "四个确保" only, we will not find the complete slogan in the snippet.

The errors of expansion can be sorted in several categories. First, some abbreviations coincide with some common expressions. For example, "上图" is the abbreviation of "上海图书馆" (Shanghai Library), while it also means "the above graph". Therefore, we are not able to get the desired full name through Web search due to the overflow of the more common usage. Second, for some long full names the later parts of them do not appear in the abbreviations, thus often neglected in both snippets and our processing. For example, "中甲" is expanded into "中国足球甲级" instead of "中国足球甲级联赛" (Chinese Football League A). Third, some abbreviations, especially those are not named entities, are always used as a single word. Consequently, it is hard to find its original form, even for a man. For example, "中青年" which is labeled as an abbreviation in PRF corpus, is the abbreviation of "中年和青年" (the young and the middle aged). However, "中青年" is so widely used that we can hardly find the expression of "中年和青年".

5   Conclusion

In this paper we suggest a novel method to expand Chinese abbreviations into their full names. Unlike previous works, we use World Wide Web as the main source of information. The approach consists of two parts: first, we conjure queries using learned help words to extract related Web contents containing both the abbreviation and full name through a search engine. Second, we extract candidates of full names from the contents and select the best result. The extraction is based on linguistic heuristics. The final selection consists of two parts: a pruner based on occurrence information within obtained snippets and a kNN-based selector utilizing a set of Web-based and structural features. Experiment shows that our approach enjoys satisfactory results. We also show the indispensability of help words and kNN-based ranker through experimental results. Moreover, the contributions of different features are discussed and compared.

For further work, we are to compare the role of search engines in our task. We are curious that whether different search engine will have different impact on our task, or other Web mining tasks. Another interesting topic is to apply content information for refined expansion. In this paper we have not considered polysemy, which also exists in abbreviations. Disambiguation is a tough topic to deal with in the future.

References

1. Chang, J., Teng, W.: Mining atomic Chinese abbreviation pairs: A probabilistic model for single character word recovery. Language Resources and Evaluation 40(3/4), 367–374 (2007)
2. Chen, K., Bai, M.: Unknown word detection for Chinese by a corpus-based learning method. Computational Linguistics 3(1), 27–44 (1998)
3. Sun, J., Gao, J., Zhang, L., Zhou, M., Huang, C.: Chinese named entity identification using class-based language model. In: COLING 2002, pp. 24–25 (2002)
4. Sun, X., Wang, H.: Chinese abbreviation identification using abbreviation-template features and context information. In: Matsumoto, Y., Sproat, R.W., Wong, K.-F., Zhang, M. (eds.) ICCPOL 2006. LNCS (LNAI), vol. 4285, pp. 245–255. Springer, Heidelberg (2006)
5. Li, Z., Yarowsky, D.: Unsupervised Translation Induction for Chinese Abbreviations using Monolingual Corpora. In: Proceedings of ACL, pp. 425–433 (2008)
6. Chang, J., Lai, Y.: A preliminary study on probabilistic models for Chinese abbreviations. In: Proceedings of the Third SIGHAN Workshop on Chinese Language Learning, pp. 9–16 (2004)
7. Fu, G., Luke, K., Zhang, M., Zhou, G.: A hybrid approach to Chinese abbreviation expansion. In: Matsumoto, Y., Sproat, R.W., Wong, K.-F., Zhang, M. (eds.) ICCPOL 2006. LNCS (LNAI), vol. 4285, pp. 277–287. Springer, Heidelberg (2006)
8. Huang, L.: More on the construction of modern Chinese abbreviations. Journal of Suihua University (004) (2008)
9. Mitchel, T.: Machine Learning 48(1) (1997)
10. Li, X.: Modern Chinese Standardized Dictionary. Foreign Language Teaching and Researching Press, Language and Literature Press, Beijing (2004)

Adaptive Maximum Marginal Relevance Based Multi-email Summarization

Baoxun Wang, Bingquan Liu, Chengjie Sun, Xiaolong Wang, and Bo Li

School of Computer Science and Technology, Harbin Institute of Technology, Harbin 150001,
China
{bxwang,liubq,cjsun,wangxl,lib}@insun.hit.edu.cn

Abstract. By analyzing the inherent relationship between the maximum marginal relevance (MMR) model and the content cohesion of emails with the same subject, this paper presents an adaptive maximum marginal relevance based multi-email summarization method. Due to the adoption of approximate computing of email content cohesion, the adaptive MMR is able to automatically adjust the parameters according to the changing of the email sets. The experimental results have shown that the email summarizing system based on this technique can increase the precision while reducing the redundancy of the automatic summary results, consequently improve the average quality of email summaries.

Keywords: multi-email summarization, maximum marginal relevance, content cohesion, adaptive model.

1 Introduction

With the popularization of email service, multi-email summarization is highly demanded. As an information fusion technique based on email contents, multi-email summarization can supply users with direct and compact summaries. Since users are used to talking about a specific subject via emails [1], a set of emails of the same subject can be clustered together in the email client, and users from client side usually would like to obtain a panoramic view of these emails for further analysis and decision.

Multi-email summarization is a kind of operation on email contents, which have some characteristics that web page contents do not possess. We take the following two facts as examples: firstly, email contents are usually very short, so there is much less information in email contents for language understanding. Secondly, when writing emails, people tend to use a casual tone, which is more closed to their oral habits, so parsing the contents with a set of rules becomes very tough. The two aspects above introduce the main difficulty to multi-email summarization. Besides, some other techniques should be considered in order to build a multi-email summary system, such as Base64 and Quoted Printable decoding, time and title extracting, and so on.

In this paper, we present an adaptive maximum marginal relevance (AMMR) based approach to generate multi-email summaries, which takes the extracting summary way and is able to automatically adjust the parameters according to the content cohesion of

H. Deng et al. (Eds.): AICI 2009, LNAI 5855, pp. 417–424, 2009.
© Springer-Verlag Berlin Heidelberg 2009

the email collections. Experimental results show that our method can improve the average quality of multi-email summaries and meet the need of system application.

The rest of this paper is organized as follows: Section 2 discusses the related work. Section 3 presents the proposed adaptive MMR model and its application in multi-email summarization. We evaluate our work in section 4. Section 5 concludes this paper and discusses the future work.

2 Related Work

The researches on multi-email summarization have just started now, so most people take the extracting approach. Wan et al. [2] extract question-answer pairs to form email summaries, and a similar method is also discussed in [3], in which Shrestha et al. generate email summaries by joining extracted question-answer sentences. Since QA pairs can reflect the semantic clue of a given email thread, such simple strategies improve the performance of summarization systems to a certain extent, however the disadvantage of this kind of approach is obvious, that is, email summaries based on this method are lack of consistency. Moreover, not all the QA sentences are related to the topic of the email thread.

Rambow et al. [4] introduce machine learning techniques to multi-email summarization. They first extract features for every sentence, and then decide whether a given sentence should be chosen as a part of the summary. In fact, this strategy has been applied in traditional multi document summarization. Ntt [5] employs SVM to extract summary sentences, and their basic idea is to find the hyperplane to separate summary sentences from the sentence set. A summary achieved by classify methods has a higher precision, however its redundancy is also higher, because the sentences extracted by classifier are similar to each other.

Comparing with multi-email summarization, summarization on multi documents is developing better, and many approaches are used to solve the problem. On summarization based on the extraction strategy, besides the method in [4], O'Leary et al. [6] compute the probability of the sentence becoming a member of the summary by using HMM. Kraaij et al. [7] extract summary sentences based on the term frequency. By introducing the deep mining of document relation, people have explored information fusion based summarization. Radev [8] presents a cross-document structure, so as to fuse information of multi documents on every semantic level. Xu [9] proposes a multi-document rhetorical structure (MRS), and has done many researches on summary sentence extracting, ranking, redundancy reducing, and summary generating. Although multi document summarization is different from multi-email summarization on the object to summarize, there is still some experience for the latter to learn.

Our approach employs the improved extracting strategy to generate multi-email summaries. By adopting the maximum marginal relevance (MMR) model, the sentences extracted to form a summary are more relative with the subject, and the redundancy can be reduced in our summary system. However MMR has an inherent disadvantage that it is not able to adjust its parameters automatically for different email sets, which makes the average quality of email summaries decline. In this paper, we analyze the inner relationship between MMR and the content cohesion of emails of the same subject, and propose an AMMR model, which improves the average quality of email summaries.

3 AMMR Based Multi-email Summarization

3.1 MMR Model

Maximum marginal relevance (MMR) model is proposed by Golestein et al. [10], which was employed to improve the performance of search engine by reducing the redundancy of documents retrieved. This thought of lowering redundancy while maintaining relative degree is also the objective of summary sentence selection. Obviously, the sentences related to the subject of emails are repeated in semantic, so they are not suitable to be selected as summary sentences. The sentences that are not related to the subject have lower redundancy, but they can not be chosen to form a summary.

To some extent, subject relevance and redundancy form a pair of contradictory concepts, which can be integrated by MMR model. The model is defined as:

$$MMR(R,S) \overset{def}{=} Arg \max_{s_i \in R \setminus S} \left[\lambda Sim_1(s_i, q) - (1 - \lambda) \max_{s_j \in S} Sim_2(s_i, s_j) \right]. \tag{1}$$

Where R stands for the set of all the sentences belonging to a given subject, S denotes the set of sentences in the summary, and s_i indicates a sentence. Sim_1 and Sim_2 are two functions for describing, which can be either the same or different. Sim_1 is used to depict the similarity of a given sentence and the subject, while Sim_2 is used to describe the similarity of two sentences. λ is a factor of linear interpolation.

From equation 1 we can see that the key of MMR based summarization is to balance between subject relevance and redundancy by tuning factor λ. Obviously, the sentence which makes the result of Sim_1 larger and that of Sim_2 smaller is likely to be selected as the summary sentence.

The subject of emails is another factor for consideration, which is normally presented by a set of terms and plays an important role in sentence selection. We discover that users' queries can reflect the subjects of the emails retrieved. Consequently, we take users' queries as subjects in the use of our AMMR model.

3.2 AMMR Based Multi-email Summarization

As mentioned above, λ is an important changeable parameter in MMR model, which is the key factor to balance between subject relevance and redundancy. In traditional MMR, this parameter is usually a human set value ranging from 0 to 1. Our experimental results show that, however, with λ changing in $[0,1]$, the quality of the summary fluctuates significantly, and for different email sets, the values assigned to λ corresponding the maximum values of summary quality are different. Adjusting λ for different email sets automatically to make the summary quality to approach to the maximal is an effective method to improve the performance of multi-email summarization.

To explain AMMR, we first propose the idea of content cohesion, which is able to reflect the cohesion degree of sentences belonging to a subject. Given a collection of emails and their subject, we assume that all the sentences revolve around this subject, with different contributions. Assuming that there is a way to model the subject correctly and the reasonable features of sentences can be extracted to convert them to vectors, we can regard content cohesion as the variance of the sentence vectors.

However, the content cohesion is difficult to compute since the variance is hard to compute. Thus, we present an approximate method and the idea is as follows. The content cohesion is closed to sentence selection during summary generation. Supposing that the content cohesion of a given email selection is large, we can draw the conclusion that the percentage of sentences related to their subject is large, so we only have to select one or two most relative sentences as representative, and extract proper sentences which are not very relative as the supplement of the summary content. Otherwise we should choose more sentences that are related to the subject to generate a summary, and reduce the number of irrelative sentences. In this paper, we introduce content cohesion into MMR to build an adaptive model, which is defined as follows:

$$AMMR(R,S) \overset{def}{=} Arg \max_{s_i \in R \backslash S} \left[\left(1 - N_q / N \right) Sim_1 \left(s_i, q \right) - N_q / N \max_{s_j \in S} Sim_2 \left(s_i, s_j \right) \right]. \quad (2)$$

Where N_q stands for the number of sentences containing the words in the query, and N denotes the number of all the sentences in the email collection. We estimate the content cohesion by N_q / N, for the more query words a sentence containing, the more relative it is to the subject. And the estimated content cohesion is adopted to replace λ, in order to adjust the parameter of the model according to different email collections automatically.

The proposed AMMR based multi-email summarization algorithm is depicted in Figure 1.

Input: query q and a set of email with the same subject
Output: the email summary
Algorithm:
1. divide the contents of the emails into sentences to form a candidate sentence set R;
2. count the number of sentences containing the words in query q, and compute N_q / N;
3. compute the similarity of sentences in R and q, take the most similar sentence as the first member of the summary, remove it from R and put it to the summary sentence set S;
4. use equation 2 to compute the AMMR value of the remaining sentences, choose the sentence with the highest value and put it into S;
5. loop step 4 until the length of the summary is enough;
6. return the sentences in S ranked by time feature;

Fig. 1. The algorithm based on AMMR

3.3 Sentence Similarity Computing Based on HowNet

Sentence similarity computing plays an important role in both AMMR and the evaluation of summary results. VSM is widely employed to compute similarity in NLP, however it does not perform very well in our experiments. According to our investigation, this problem is caused by the individuation and informality of people while writing emails, which directly lead to the existence of word variants. As we know, the fundamental idea of VSM is the words' cooccurrence, but in email contents there are many words with the same meaning but totally different morphology.

In this paper, we adopt the method based on HowNet to compute semantic similarity of sentences, which is built on the computation of word semantic similarity [11]. Our strategy is to convert the sentence similarity to the weighted mean of word similarities. A greedy algorithm is employed to match the most similar words in two sentences to collect word pairs, and take inverse document frequency as the weight for computing the weighted mean of the words. Experimental results have shown that the average quality of summary has increased by 4% to 6% after introducing this strategy, which will be given in detail in section 4.

4 Experiments

4.1 Evaluation

In this paper we adopt direct evaluation [12] to evaluate the performance of an email summarization system. We first choose several typical user queries as the subjects, and then acquire the top k emails respectively from the email retrieval system to form email collections. For every group of emails, 5 summaries are generated manually by different people as benchmark, with which we can evaluate the machine summaries.

We compute the precision and the redundancy for each email summary and get the quality of the summary according to them. Precision can be computed as follows:

$$precision = Sim(SP, SA) . \tag{3}$$

Where SP stands for the summary generated by human, and SA stands for the summary generated automatically. Sim is a similarity computing function.

We consider redundancy as the semantic similarity of the sentences within the summary, so it can be obtained by computing the average similarity of the sentence pairs within the machine summary, which is defined by:

$$redundancy = \frac{1}{n} \sum_{i,j=1,\dots n, i \neq j} max\_sim(s_i, s_j) . \tag{4}$$

And the quality of the summary is defined as follows:

$$quality = precision - redundancy . \tag{5}$$

4.2 Experimental Results

We choose 6 groups of real emails as test corpus, and each group contains 15 to 20 members. Since the emails are obtained from our email retrieval system, the subject of every group can be expressed by the user's query. The experimental results are shown in figure 2, where the horizontal axis represents the value of linear interpolation factor λ, and vertical axis stands for the quality of summaries defined in (5).

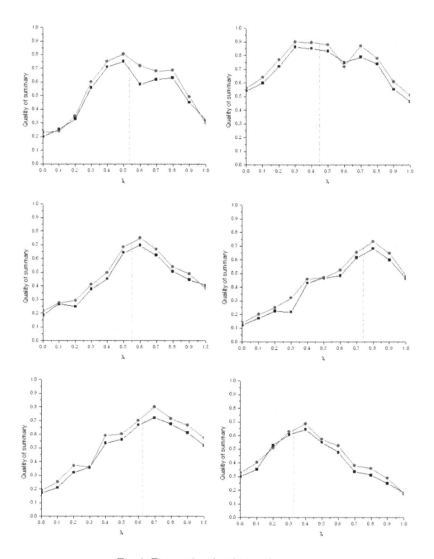

Fig. 2. The results of multi-email summary

From the results we can see the general trend is that the quality can reach its maximum value while λ ranging from 0 to 1. But for different subjects, λ is not at a fixed value or interval when the maximum value appears, which is reasonable according to our discussion on the content cohesion of emails, so the summarization system based on traditional MMR model is unable to achieve a high average quality.

In figure 2, the position of λ marked by the dashed line represents the value of linear interpolation introduced by AMMR, and the corresponding value on vertical axis is the quality AMMR can reach. We can see that our method is able to approach the maximum value of the summary quality on all the 6 groups, especially for the last group, the maximum quality appears when λ is equal to about 0.4, which significantly deviates from the rest 5 groups, and the AMMR model performs well despite of this situation.

In Figure 2, the red dotted line represents the results of the model taking HowNet based sentence similarity computing method, and the black squared line stands for that using VSM, we can see the quality has increased by 4% to 6%, which means that the summary sentences selected tend to be more reasonable after introducing this semantic based strategy.

5 Conclusion

Aiming at increasing the average quality of multi-email summaries, we propose a novel multi-email summarization technique based on adaptive MMR model. We analyze the relationship between content cohesion and the parameters of MMR and find out that the content cohesion is one of the main factors that lead to the decline of summary quality. This motivates us to present an adaptive summarization model to improve the quality of email summaries. The experimental results show that our model can effectively improve the average quality of email summaries. Our work can be applied to build an automatic summary module on the email client.

In the future, we will focus on the following problems: 1) to model email content cohesion in an exact way; 2) to adopt more linguistic features to make the computation of sentence relevance reasonable; 3) to develop algorithms for improving the readability of summary generated.

Acknowledgments. This investigation was supported by the project of the High Technology Research and Development Program of China (grants No. 2006AA01Z197 and 2007AA01Z172), the project of the National Natural Science Foundation of China (grant No. 60673037) and the project of the Natural Science Foundation of Heilongjiang Province (grant No. E200635).

References

1. Fisher, D., Moody, P.: Studies of automated collection of email records. University of Irvine ISR Technical Report UCI-ISR-02-4 (2002)
2. Wan, S., McKeown, K.: Generating overview summaries of ongoing email thread discussions. In: Proceedings of COLING 2004, the 20th International Conference on Computational Linguistics, pp. 745–751 (2004)

3. Shrestha, L., McKeown, K.: Detection of question-answer pairs in email conversations. In: Proceedings of COLING 2004, pp. 889–895 (2004)
4. Rambow, O., Shrestha, L., Chen, J., et al.: Summarizing email threads. In: Proceedings of HLT/NAACL, Boston, USA (2004)
5. Hirao, T., Sasaki, Y., Isozaki, H., et al.: NTT's Text Summarization System for DUC 2002. In: Proceedings of the workshop on automatic summarization, Philadelphia, Pennsylvania, USA, pp. 104–107 (2002)
6. Zajic, D.M., O'Leary, D.P.: Sentence Trimming and Selection: Mixing and Matching. In: Proceedings of the 2006 Document Understanding, New York (2006)
7. Kraaij, W., Spitters, M., van der Heijden, M.: Combining a Mixture Language Model and Naive Bayes for Multi-document Summarization. In: Proceedings of the DUC 2001 workshop (SIGIR 2001), New Orleans (2001)
8. Radev, D.R.: A Common Theory of Information Fusion from Multiple Text Sources Step One: Cross-Document Structure. In: Proceedings of the 1st ACL SIGDIAL Workshop on Discourse and Dialogue, Hong Kong, pp. 74–83 (2000)
9. Xu, Y.: Research of Multi Document Automatic Summarization. Dissertation of the Doctoral Degree in Engineering, Harbin Institute of Technology (2007)
10. Carbonell, J., Goldstein, J.: The use of MMR, diversity-based reranking for reordering documents and producing summaries. In: Proceedings of the 21st annual international ACM SIGIR conference on research and development in information retrieval, Melbourne, Australia, pp. 335–336 (1998)
11. Li, S.J., et al.: Semantic computation in a Chinese question-answering system. Journal of Computer Science & Technology 17(6), 933–939 (2002)
12. Jones, K.S., et al.: Automatic Summarizing Factors and Directions. In: Advance in Automatic Text Summarization. MIT Press, Cambridge (1998)

Semantic Relation Extraction by Automatically Constructed Rules

Rongfeng Huang

Shanghai Jiaotong University, Shanghai, China
silverfox@sjtu.edu.cn

Abstract. This paper presents an approach to construct rules automatically to identify semantic relations, such as Hypernym, Part_Of and Material from Modern Chinese Standard Dictionary (MCSD)[1]. This approach combines all the useful information such as part of speeches, syntactic information, and extracted semantic relations together in order to identify as many semantic relations as possible. However, not all the information provided, like syntactic information, is reliable. A method is required to ensure that the "right" information up to a point can be picked. And then all possible semantic relations are constructed as a net after being identified. Finally, we evaluate the net by determining word similarity.

Keywords: semantic relation,machine-readable dictionary, word similarity.

1   Introduction

Semantic resource is very useful in many Natural Language Processing (NLP) tasks like Information Retrieval, Machine Translation and so on. Some semantic resources have been constructed manually like WordNet [1] and HowNet [2]. And they have showed their immense values. However, manual construction of such resources is time-consuming, high cost and susceptible to arbitrary decision of human. Therefore, lots of automatic approaches to extract semantic relations have been developed. One method is to identify a semantic relation between two words in a sentence. Usually, words between two destination words are key information. For example, Dmitry Zelenko et al [3] use a kernel method to identify relations. Chinatsu Aone et al [4] use a pattern-based method.

Another method, usually taking dictionary as source, is to identify semantic relations between a head word and words in its definition. The main difference between these two methods is head word does not exist in sentences of definition in the latter one. Therefore, we have to consider whole context of a word in definition. In English, the most impressive achievement is MindNet, which are generated by applying rules on results proviede by a broad-coverage parser [5].

---

[1] Xingjian Li, Modern Chinese Standard Dictionary, Foreign Language Teaching and Research Press, Beijing, 2004.

H. Deng et al. (Eds.): AICI 2009, LNAI 5855, pp. 425–434, 2009.

However, an excellent syntactic parser is usually inaccessible. As a way, a special one may be designed. Barriere [6] implements a chart parser to aid his extraction task from a children's dictionary. Others use patterns, most of which are at string level. Markowitz [7] gives several general patterns.

From recent works, string pattern matching seems an alternative way to extract semantic relations from Chinese dictionary, as a syntactic parser for Chinese usually has a much lower precision than one for English. However, complex structural information cannot be expressed adequately in string patterns [8]. In other words, we should make use of more information besides string patterns.

This paper describes an approach which exploits MCSD to identify semantic relations by automatically constructed rules. We use various types of information to capture as much information as possible. And we try to uniform representations of different types' information to make sure they work well with each other. We also try to use a way to avoid abusing unreliable information. One hypothesis used here is "wrong" information occurs occasionally while "right" information doesn't.

In the rest of this paper, we describe types of information used and a way of using them to construct rules (Sect. 2). And then we introduce types of identified semantic relations and a way of constructing all identified semantic relations as a net (Sect. 3). We evaluate the result by hand-checking a scale of random samples (Sect. 4). However, in order to get a more objective evaluation, we evaluate the net by determining word similarity (Sect. 5). Also, we discuss which aspects of the approach should be improved (Sect. 6). Finally, we give a conclusion (Sect. 7).

2   Identification Method

We use various types of information called features and we uniform their representation to make them work well with each others. And then we construct rules with a list of features.

2.1   Features

There are five types of features, including words, part of speeches, syntactic information, positions, and extracted semantic relations. And we use a "Feature-Value" structure to uniform their representations.

Types. In order to be referred easily, a word in definition is called a "candidate word" if its part of speech matches the requirement of a target relation. For example, if and only if a word is a noun, it can be a candidate word that holds a Hypernym relation with the headword. Except for a candidate word, rest words of a sentence are called "definition words" of this candidate word.

Five feature types are described as below.

Firstly, definition words and part of speeches of them are considered just like most other ways do, no matter they are rule-based or pattern-based. However, we do not plan to encode any definition word and its part of speech as only one unit feature. In fact, there is a transition sacrificing precision for recall from words with part of speeches to only part of speeches. Therefore, it's necessary to decide which words should be kept in their forms while others are better to be kept as part of speeches only. In this paper, all decisions about these are made depending on the distribution of words in corpus.

Secondly, result of syntactic analysis is another kind of important useful information. But unfortunately, a reliable syntactic parser like the broad-coverage parser used in MindNet is inaccessible at most of the time, especially when we deal with Chinese. So, we should decide which syntactic information provided by a not very reliable parser is useful for us. All relevant decisions about these are also made depending on the distribution of result of syntactic analysis in corpus. In this paper, a dependency parser is used. Based on the syntactic results, a further analysis is applied. For example, "Is X a subject of a sentence? (X is a definition word)".

Thirdly, the positions of definition words are also taken into account. This is an assistant strategy in fact. One reason is the unreliability of the syntactic parser. And our suspicions will be raised when the number of syntactic relationships of the path linking from a definition word to a candidate word is getting bigger. Another reason is that a certain word may play different syntactic roles in different sentences and its roles are related to its positions more or less. For example, a subject tends to be in a position of the first several ones of a sentence while an object prefers to appear in the tail of a sentence.

Finally, as a special resource, we use extracted semantic relations to aid in the identification of other semantic relations. In this paper, about half of extracted relations are type of Hypernym and the precision of it is 95% with a margin error of +/-5% with 99.5% confidence. So a double-step pass strategy is applied. Hypernym relations are identified first and then they are encoded as a type of feature to be used in a second pass. Making use of extracted semantic relations to improve precisions of other semantic relations' identification is also mentioned in [9].

Uniform Representation. In this paper, any feature is attached with a value derived from a definition to form a "Feature-Value" structure. And then, any candidate word can be attached with a list of such structures. In fact, if we take "Feature-Value" structure as a type of "Feature", value range of which is true, false, then regardless of the types and value ranges of features in a "Feature-Value" structures, we can treat them as the same. We refer "Feature-Value" structures as features in the rest of this paper.

2.2 Construction of Rules

A rule consists of a list of features. If a candidate word owns all the features of a rule, then we say that this candidate word satisfies this rule.

Feature Selection. To balance the precision and the recall, we tend to make sure that each rule provides high precision and acceptable recall and all rules working as a whole provide a good recall. The procedure similar to the one of constructing a decision tree is showed below:

Firstly, select the best feature from features of candidate words by means of a kind of evaluation way which is described later. And then candidate words will be split into two sets, words in one of which own this feature and words in another don't.

Secondly, if the precision of the former is higher than a predefined value, then a new rule is constructed. Otherwise, select more features and split candidate words into more sets like the first step until the precision is acceptable, or give up if the number of selected features exceeds a predefined value.

Thirdly, after a new rule is constructed, any candidate words satisfying this new rule will be removed. And then repeat the procedure until no more new rules can have a precision higher than the predefined value.

Feature Evaluation. There has been a lot of effective ways to choose features. However, different evaluation ways have different trends on precision and recall. And we notice that very few candidate words have relevant semantic relation to a head word. For example, a candidate word has a probability of only 1.5% , estimated from a random small set, to hold Part_Of relation with the head word. And this fact causes severer different trends of different evaluation ways. In experiments of this paper, Information Gain and $\mathcal{X}^2$ Statistic usually give good recalls while perform poorly in precisions. In contrast, Odds Ratio shows an excellent precision but gives an unsatisfactory recall. Therefore, it's necessary to combine them to get a good precision as well as an acceptable recall. In order to make it, we choose Information Gain (see (1),(2)) as the evaluation way for selection of the first feature of a rule.

$$Entropy(C) = -p_+ log_2 p_+ - p_- log_2 p_- \tag{1}$$

Where C is a set of candidate words, p_+ is the proportion of words holding target relation with head words and p_- is the proportion of the rest words.

$$InfoGain(C, F) = Entropy(C) - \sum_{v \in \{true, false\}} \frac{|C_v|}{|C|} Entropy(C_v) \tag{2}$$

Where F is a feature, C_{true} is the set of words with feature F in C and C_{false} is the set of the rest words in C .

And we choose odds ratio (see (3)) to evaluate the rest features of a rule.

$$OddsRatio(C, F) = log \frac{P(F|C_+)(1 - P(F|C_-))}{P(F|C_-)(1 - P(F|C_+))} \tag{3}$$

Where C_+ is the set of words holding target relation with head words, C_- is the set of the rest words in C, $P(F|C_+)$ is the proportion of words holding Feature F in C_+ and $P(F|C_-)$ is similar to $P(F|C_+)$.

3   Semantic Relation

We apply our approach to identify ten types of semantic relations (see Table 1. in Sect. 4.2). And we construct the identified relations as a net to make them more useful.

Three examples, covering from abstract concept to concrete concept and from substance to creature, are showed in Fig. 1, Fig. 2 and Fig. 3.

Head word: 本质(essence)
Definition: 指事物本身所固有的根本属性,它对事物的性质,状况和发展起决定作用(跟"现象"相区别) (The inherent nature of a thing or a class of things, determines the character, state and development of that thing or that class of things. (opposite to phenomena))

本质(essence) –<Hypernym>→ 属性(nature)
| –<Means>→ 决定(determine)
| –<Antonym>→ 现象(phenomena)

Fig. 1. Essence's Definition and Semantic Relations

Head word: 斑鸠(culver)
Definition: 鸟，形状像信鸽，羽毛多为灰褐色，颈后多有白色或黄褐色斑点。 (A bird, looking like homing pigeon, usually has taupe feathers and a neck with white or filemot spots.)

斑鸠(culver) –<Hypernym>→ 鸟(bird)
| –<Part>→ 羽毛(feather)
| –<Part>→ 颈(neck)

Fig. 2. Culver's Definition and Semantic Relations

Head word: 钛(titanium)
Definition: 金属元素，符号Ti。银灰色，质硬而轻，延展性强，熔点较高，耐腐蚀。可用来制造特种合金钢。 (A kind of metal element, signed with "Ti", is silver gray, light, hard, ductile and corrosion-resistant. It can be use to make special alloy steel.)

钛(titanium) –<Hypernym>→ 元素(element)
| –<Color>→ 银灰色(silver gray)
| –<Material>→ 合金钢(alloy steel)

Fig. 3. Titanium's Definition and Semantic Relations

3.1 Constructing Semantic Relations as a Net

It won't be very useful if we isolate every small set of semantic relations extracted from a definition of a word. Therefore, we need to construct all semantic relations as a net. Every word with its part of speech has and only has one node in the net. In other words, every word has not only semantic relations to some words in its definition, but can have additional semantic relations to other words once it is used to explain them. That is because when we look up a word, we may not only consult the definition of this word, but also the definitions of any word which mentions it [9]. Chodorow et al. [10] exploit such insight in developing a tool for helping human users disambiguate hyper/hyponym links among pairs of lexical items. MindNet implements it with full inversion of structures [5].

4 Evaluation

4.1 Experimental Settings

Preprocessing. Before extracting semantic relations, it's necessary to process raw text first, including word segment, part of speech tagging and dependency parsing. All these are done by LTP v2.0[2]. Precision of dependency parsing is 78.23% without labeling type of dependency and 73.91% with labeling.

Dictionary. Rounding number of entries to the nearest thousand, 44000 nouns with their definitions in MCSD are parsed with a set of rules automatically constructed from a small corpus which was also made from MCSD.

4.2 Results

After a double-step pass, 43000 relations are identified. For each type of semantic relations, a random sample of 400 relations is hand-checked and the precisions of all types are listed in Table 1. Using common statistical techniques, we estimate that these rates are representatives of each type with a margin error of +/-5% with 99.5% confidence. From this result, not only a large number of relations have been extracted but also high precisions of each type have been achieved. However, such a "hand-check" evaluation way may not be objective enough as some of the identified relations are somewhat fuzzy to discriminate. Also, the recall is hard to estimate. Further, it's necessary to see the relations and the net they form are really useful. Therefore, we suggest a more objective way of evaluation.

Table 1. Set of semantic relation types

Type	Precision	Type	Precision	Type	Precision	Type	Precision
Antonym	0.96	Color	0.96	Domain	0.88	Hypernym	0.95
Hyponym	0.81	Material	0.84	Means	0.92	Member	0.93
Part	0.88	Synonym	0.99				

---

[2] LTP is a language platform based on XML presentation, http://ir.hit.edu.cn/demo/ltp/Sharing_Plan.htm

5   Evaluating by Determining Word Similarity

To get a more objective evaluation, we evaluate the net by determining word similarity. We generate automatically word pairs labeled with similarity levels as corpus. Then, we use one half to identify path patterns and use another half to test them.

5.1   Generation of Word Pairs

A thesaurus[3] is used to generate similar word pairs and dissimilar word pairs. Similar strategy is taken by Stephen [11]. And here, we generate six similarity levels of word pairs.

Words in the thesaurus are classified in five levels, including Big Class, Middle Class, Tiny Class, Word Group and Atom Group. Every item in this thesaurus consists of a class code and one or more words. For example:

Aa01A08= 每人,各人, 每位(All mean everyone)
Ba01B10# 导体,半导体,超导体(conductor,semiconductor,superconductor)
Ad02F01@ 外星人(alien)

Fig. 4. Items in Thesaurus

The class code in the head of the first item in Fig. 4 means that these three words are classified in A [Big Class] a [Middle Class] 01[Tiny Class] A [Word Group] 08 [Atom Group]. And "=" means that these three words are the same. If they are not completely the same, then they are denoted by "#". If only one word contained in an item, then a tag "@" is used. The procedure of generating word pairs is showed below:

Firstly, randomly select word pairs from the thesaurus.

Secondly, look up the already selected word pairs to ensure the newly selected pair doesn't exist in them.

Thirdly, look up the net to find out if they exist and connect to each other within 18 arcs.

Fourthly, look up the thesaurus to see if they belong to the same atom group, then they are labeled with "Atom Group", if they belong to the same word group but not the same atom group, then they are labeled with "Word Group", or they are labeled with Tiny Class, Middle Class or Big Class by analogy. And finally if they are from different Big Classes then they are labeled with "General".

Fifthly, repeat the above procedure until 20000 word pairs for each level are available. And then they are split into two sets equally. One set is used for training and another is used for testing.

---

[3] HIT-IR Tongyici Cilin (Extended), http://ir.hit.edu.cn/demo/ltp/Sharing_Plan.htm

5.2 Using Path Patterns to Determine Similarity

Paths that one word travels through to another word are identified. For example, given a word pair showed in Fig. 5, there are several paths exist, and two of them are showed in Fig. 5. From these two paths, we can get two path patterns <Synonym, Hypernym, Hypernym Of> and <Synonym, Hypernym Of, Hypernym>.

Word Pairs: <学子(student),徒弟(apprentice)>
Path pattern1: 学子(student) – <Synonym> → 学生(student) – <Hypernym> → 人(person) ← <Hypernym> – 徒弟(apprentice)
Path Pattern2: 学子(student) – <Synonym> → 学生(student) ← <Hypernym> – 高徒(excellent student) – <Hypernym> → 徒弟(apprentice)

Fig. 5. Two Paths of Word pair <Student,Apprentice>

Stephen [11] also uses path patterns to determine similarity. However, the method adopted here is different. In this paper, if the precision and frequency of a pattern are both high, then it will be used to identify similar word pairs later.

5.3 Result

In experiments of this paper, there are three different definitions of similarity. In the first experiment, only word pairs labeled with "Atom Group" are considered as similar pairs. In the second experiment, word pairs labeled with "Word Group" are also considered. In the third experiment, word pairs labeled with "Tiny Class" are also considered. And the result is showed in Table 2.

The result shows high precision and low recall. The possible reasons may be:

Firstly, the high precision is ensured by the high precision of extraction.

Secondly, the low recall of determining word similarity reflects the low recall of the semantic relations extraction. And there is no doubt that the insufficient semantic relations cause two similar words can't find a stable and short path to reach each other although they may be still connected by a long path. And the longer the path is, the more instable and unreliable the path will be. Also it will become sparser when the length of path grows up.

Thirdly, the way of determining similarity may still need to be improved.

Finally, the knowledge contained in the MCSD may not consist very well with the one contained in the thesaurus.

Table 2. Result of Similarity Experiment

No	Similar Word Pairs	Dissimilar Word Pairs	Precision	Recall
1	10000	50000	95.02%	14.14%
2	20000	40000	91.76%	16.6%
3	30000	30000	87.26%	25.66%

6   Discussion

Although a large number of semantic relations have been identified, the recall still may be low, reflected by the result of determining word similarity. And lots of other knowledge in definitions is still unexploited. Take the definition of culver (see Fig. 2) for example. In Fig. 6, only top level semantic relations linking from head words to words in definitions are identified, and lower level semantic relations are still unexploited.

Top Level: 斑鸠(culver) – <Hypernym> → 鸟(bird)
Lower Level: 羽毛(feather) – <Color> → 灰褐色(taupe)

Fig. 6. Two different level Semantic Relations

7   Conclusion

This paper describes a strategy to construct rules automatically to identify semantic relations from machine readable dictionary, and a conceptual net is constructed from the extracted relations. In order to estimate the precision, a random sample of 400 relations for each type has been hand-checked. However, it is still possibly subjective. Therefore, a more objective way, that is determining word similarity, is applied. And the result shows again that the precision of extracted semantic relations is high. But what frustrates us is the result also reflects indirectly the possible low recall of semantic relation extraction. Our future work is to improve the ability of the approach to identify more semantic relations including top level ones and lower level ones. Also, we will try our approach to extract semantic relations from cyclopedias such as Chinese Wikipedia.

Acknowledgement

Supported by National Natural Science Foundation of China (NSFC) (No. 60873135).

References

1. Miller, G.A., Beckwith, R., Fellbaum, C., Gross, D., Miller, K.J.: Introduction to wordnet: An on-line lexical database. Int. J. Lexicography 3(4), 235–244 (1990)
2. Zhendong, D., Qiang, D.: Hownet, http://www.keenage.com
3. Zelenko, D., Aone, C., Richardella, A.: Kernel methods for relation extraction. J. Mach. Learn. Res. 3, 1083–1106 (2003)

4. Aone, C., Ramos-Santacruz, M.: Rees: a large-scale relation and event extraction system. In: Proceedings of the sixth conference on Applied natural language processing, Morristown, NJ, USA, pp. 76–83. Association for Computational Linguistics (2000)
5. Richardson, S.D., Dolan, W.B., Vanderwende, L.: Mindnet: acquiring and structuring semantic information from text. In: Proceedings of the 17th international conference on Computational linguistics, Morristown, NJ, USA, pp. 1098–1102. Association for Computational Linguistics (1998)
6. Barriere, C.: From a children's first dictionary to a lexical knowledge base of conceptual graphs. PhD thesis, Burnaby, BC, Canada, Adviser-Popowich, Fred. (1997)
7. Markowitz, J., Ahlswede, T., Evens, M.: Semantically significant patterns in dictionary definitions. In: Proceedings of the 24th annual meeting on Association for Computational Linguistics, Morristown, NJ, USA, pp. 112–119. Association for Computational Linguistics (1986)
8. Vanderwende, L.H.: The analysis of noun sequences using semantic information extracted from on-line dictionaries. PhD thesis, Washington, DC, USA, Mentor-Loritz, Donald (1996)
9. Dolan, W., Vanderwende, L., Richardson, S.D.: Automatically deriving structured knowledge bases from on-line dictionaries. In: Proceedings of the First Conference of the Pacific Association for Computational Linguistics, pp. 5–14 (1993)
10. Chodorow, M.S., Byrd, R.J., Heidorn, G.E.: Extracting semantic hierarchies from a large on-line dictionary. In: Proceedings of the 23rd annual meeting on Association for Computational Linguistics, Morristown, NJ, USA, pp. 299–304. Association for Computational Linguistics (1985)
11. Richardson, S.D.: Determining similarity and inferring relations in a lexical knowledge base. PhD thesis, New York, USA (1997)

Object Recognition Based on Efficient Sub-window Search

Qing Nie[1], Shouyi Zhan[2], and Weiming Li[1]

[1] School of Information and Electronic, Beijing Institute of Technology, Beijing 100081, China
[2] School of Computer Science, Beijing Institute of Technology, Beijing 100081, China
qingnie@bit.edu.cn

Abstract. We propose a new method for object recognition in natural images. This method integrates bag of features model with efficient sub-window search technology. sPACT is introduces as local feature descriptor for recognition task. It can capture both local structures and global structures of an image patch efficiently by histogram of Census Transform. An efficient sub-window search method is adapted to perform localization. This method relies on a branch-and-bound scheme to find the global optimum of the quality function over all possible sub-windows. It requires much fewer classifier evaluations than the usually way does. The evaluation on PASCAL 2007 VOC dataset shows that this object recognition method has many advantages. It uses weakly supervised training method, yet has comparable localization performance to state-of-the-art algorithms. The feature descriptor can efficiently encode image patches, and localization method is fast without losing precision.

Keywords: object recognition, sPACT, object detection, bag of features.

1 Introduction

Object class recognition is a key issue in computer vision. In the last few years, Object class recognition has received a lot of attentions. It has been approached in many ways in literatures. These approaches can typically be divided into two types. One type is training a part-based shape model, the model describes objects by simpler, robust and stable features connected in a deformable configuration. These features are usually generic, which can be edge points, lines or even visual words, and features must be grouped to describe a given object. Some typical part-based shape models include constellation model [1], implicit shape model (ISM) [2], k-fans model [3], pair-wise relationships model [4] and star model [5,6]. When detect an object, it first detect individual object parts, which be group together to reason about the position of the entire objects. This is accomplished via generalized Hough transform, where the detections of individual object parts cast probabilistic votes for possible locations of the whole object. These votes are summed up into a Hough image, the peaks of it being considered as detection hypotheses.

A shortcoming of part-based shape model is the need for clean training shapes. Most of this type approaches need pre-segmentation training images or label the ob-

H. Deng et al. (Eds.): AICI 2009, LNAI 5855, pp. 435–443, 2009.

ject boundary box. This quickly becomes intractable when hundreds of categories are considered. Some weakly supervised training methods exist, but with high computation cost.

Another type for object localization is commonly performed using sliding window classifiers [7]. This approach trains a quality function (classifier) and then scans over the image and predicts that the object is present in sub-windows with high score. Siding window object localization has been shown to be very effective in many situations, but suffers a main disadvantage: it is computationally inefficient to scan over the entire image and test every possible object location. To overcome this shortcoming, researchers typically use heuristics to speed up the search, which introduces the risk of missing target object. Lampert C.H. et al.[8] proposes an Efficient Subwindow Search (ESS) to solve this problem. It relies on a branch-and-bound scheme to find the global optimum of the quality function over all possible sub-images, thus returning the same object locations that an exhaustive sliding window approach would. At the same time it requires less classifier evaluations and typically runs in linear time or faster.

Motivated by Lampert C.H.' work, we proposed a multi-object recognition method. This method integrates bag of features model with efficient sub-window search technology. Bag of features model is a very powerful and effective classification model. It allows weakly supervised training method. And ESS is applied in our recognition frame for fast localization. In order to improve the recognition performance, we also designed a new local feature descriptor suitable for our recognition task.

The rest of the paper is organized as follows. We first introduce a modified spatial PACT for our local feature descriptor in Section 2. Then in Section 3 we explain how to use an efficient sub-window search method to perform localization. In Section 4 we present our experiments results on PASCAL 2007 dataset. And come to the conclusions in section 5.

2 Local Feature Descriptor

Bag of features model contains five major steps: feature extraction and description, codebook creation, image feature description, model training and classification. In order to learn and detect object classes in cluttered images of real-world scenes, some kind of efficient local feature descriptors are necessary. Many local feature descriptors have been proposed in the last decade. Some descriptors are appearance-based, such as SIFT, GLOH, Spin image, PCA-SIFT. Some descriptors are structure-based, such as PCBR [9], KAS[10] and so on. Structure-based descriptors are preferred in object detection task. But most of structure-based descriptors need pre-segment images and segmentation itself is an open question in computer vision field. To address this problem, Wu J. [11] proposed to use spatial histogram of Census Transform values (sPACT) to represent a whole image. Census Transform (CT) compares the intensity values of a pixel with its eight neighboring pixels. The eight bits generated from intensity comparisons can be put together and converted to a base-10 number. Census Transform can capture local structures of an image, while large-scale structures are captured by the strong correlation between neighboring CT values and the histogram. The PCA operation results in a compact representation, and Spatial PACT further incorporates global structures in the image. SPACT demonstrates high performance on commonly used datasets. Besides, it need not pre-segment image and evaluates extremely fast.

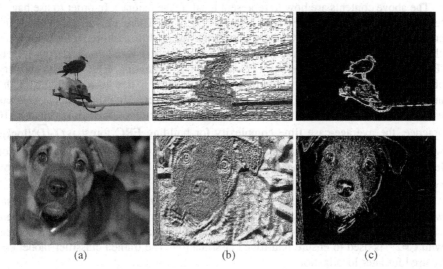

(a)   (b)   (c)

Fig. 1. Examples of CT images. (a) Original images (b) CT images(c) Modified CT images.

Our local feature descriptor is motivated by Wu J.'s work. But we have modified it to fit our situation. And trim it to fit well in our bag of feature model. According to Wu J.'s analysis, the powerful representation capacity of CT histogram is mostly depend on its successful encoding shapes information. But from some examples of CT images shown in fig.1(b), we find that Census Transform is sensitive to neighbor pixels variation. The CT image is unnecessary noisy even in consistent region. In order to remove these noises and keep shape information more clearly, we modified the CT by introducing a threshold CTVALUE. Only when the difference between two neighbor pixels great than CTVALUE, the CT value is 1, otherwise the CT value is 0. The modified CT images are shown in fig.1(c). They are obviously clearer than standard CT images and keep the shape structures better.

In order to remove pixels correlation effects and get a more compact representation, Principal Component Analysis (PCA) operation is used to CT histograms data. The key steps for PCA operation are: ①Random select some training images, convert them to gray images, and calculate the modified CT values. ②Use regular grids sample patches. The sampling interval set to 32, the size of image patches set to 32X32. ③Calculate CT histograms of these patches and normalize the CT histograms. When calculate CT histogram, remove two bins with CT = 0, 255. ④Perform PCA operation on these CT histograms. The PCA results show that the first 40 eigenvectors are enough to cover 92% information. So we get the first 40 eigenvectors as main components.

To add spatial information, an image patch is split into 2x2 blocks, add the center block, the normalized PACT histograms of these blocks are concatenated to form a feature vector which has 40x5= 200 dimensions. Together with the gray histogram of image patch, which result in a feature vector with an overall 200+256= 456 dimension.

The above contents are how to represent local image patches. In order to use bag of feature model, the second step need create a codebook. There are various methods for creating visual codebooks. K-means clustering, mean-shift and hierarchical k-means clustering are the most common methods. Yet Extremely Random clustering forests (ERC forests)[12,13] have recently attracted a lot of attention. ERC Forests are very efficient at runtime, since matching a sample against a tree is logarithmic in the number of leaves. And they can be trained on large, very high-dimensional datasets without significant over fitting and within a reasonable amount of time (hours). In practice, the average-case time complexity for building ERC forest is $O(\sqrt{D}N\log k)$, where D is the feature dimension, N is the number of patches and k is the number of leaf nodes. In contrast, k-means has a complexity of $O(DNk)$. In our recognition system, ERC forest is applied to create visual codebook.

After creating a codebook, we can represent an image as histogram of visual words. And the training images' histograms of visual words are fed to SVM to train a classifier model. LIBSVM[14] is selected for training classifier model. Linear kernels with $C=2^{-5}$ is used to ensure real-time classification. The trained classifier model can be used for later localization.

3   Using ESS to Localize Objects

To perform localization, one can take a sliding window approach, but this strongly increases the computational cost, because the classifier function f has to be evaluated over a large set of candidate sub-windows. In order to overcome this shortcoming, several approaches have been proposed to speed up the search. One way is reducing the number of evaluations by searching only over a coarse grid and by allowing only rectangles of certain fixed sizes. Another way is applying local optimization methods instead of global ones. These techniques sacrifice localization robustness to achieve acceptable speed. It can lead to missing of the objects.

In order to overcome this shorting, we adopt an Efficient Sub-window Search (ESS) technical proposed by Lampert C.H. et al.[8]. ESS is guaranteed to find the locally maximal region. At the same time ESS is very fast, because it relies on a branch-and-bound search instead of an exhaustive search.

The principle of ESS is the following: when searching for object regions, only very few of candidate regions actually contain object instances. It is wasteful to evaluate the quality function for all candidate regions. One should target the search directly to identify the regions of highest score and ignore the rest of the search space where possible. The branch-and-bound framework plits the parameter space into disjoint subsets, while keeping bounds of the maximal quality on all of the subsets. By this way, large parts of the parameter space can be discarded early if their upper bounds are lower than a guaranteed score.

In our localization case, the parameter space is the set of all possible rectangles in an image. These rectangles are parameterize by their top, bottom, left and right coor- dinates (t, b, l, r). We represent set of rectangles as tuples [T,B,L,R], where T = [tlow, thigh] etc.. For each rectangle set, the highest score that the quality function could take on any of the rectangles in the set are calculated. And this highest score is the upper bound of the rectangle set.

Before giving the algorithm of Efficient Sub-window Search, we first discuss how to construct a bound function $\hat{f}$. Denoting rectangles by R and sets of rectangles by $\Re$, the bound function $\hat{f}$ has to fulfill the following two conditions:

$$1)\ \hat{f}(\Re) \geq \max_{R \in \Re} f(R)$$

$$2)\ \hat{f}(\Re) = f(R),\ \text{if}\ R\ \text{is the only element in}\ \Re$$

Condition 1) ensures that $\hat{f}$ is an upper bound to f, and condition 2) guarantees the optimality of the solution to which the algorithm converges. The possible bounds $\hat{f}$ are not unique. A good bound $\hat{f}$ should be fast to evaluate but also tight enough to ensure fast convergence. In our experiments, we adopt bag of features model for object classification, and we use SVM with a linear kernel as classifier. The corresponding SVM decision function is as formula (1)

$$f(I) = \beta + \sum_i \alpha_i \langle h, h^i \rangle \tag{1}$$

where $\langle .,. \rangle$ denotes the scalar product. h^i is the histogram of the training examples i, and α_i and β are the weight vectors and bias that were learned during SVM training. Because of the linearity of the scalar product, we can rewrite this expression as a sum over per-point contribution with weights $w_j = \sum_i \alpha_i h_j^i$

$$f(I) = \beta + \sum_{j=1}^n w_{cj} \tag{2}$$

Here c_j is the cluster index belonging to the feature point x_j and n is the total number of feature points in I. This form allows us to evaluate f over sub-images $R \subset I$ by summing over the feature points that lie within R. Since we are only interested in the argmax of f over all $R \subset I$, we can ignore the bias term β. Set $f = f^+ + f^-$, where f^+ contains only the positive summands of Equation 2 and f^- only the negative ones. If we denote by R_{max} the largest rectangle and by R_{min} the smallest rectangle contained in a parameter region $\Re$, then formula (3)

$$\hat{f}(\Re) = f^+(R_{max}) + f^-(R_{min}) \tag{3}$$

has the desired properties 1) and 2). Using integral images we can evaluate f^+ and f^- at constant time complexity, thus making the evaluation of $\hat{f}$ a constant time operation.

Suppose we use a priority queue P to hold the search states, the pseudo-code which use ESS to locate an object class instance is like follows.

①   Initialize P as empty priority queue
        set [T,B,L,R] = [0, n] × [0, n] × [0,m] × [0,m] (m, n is the size of image I)

②   repeat

        split [T,B,L,R]→[T1,B1,L1,R1] ∪ [T2,B2,L2,R2]

        push([T1,B1,L1,R1], $\hat{f}$ ([T1,B1,L1,R1]) into P

        push([T2,B2,L2,R2], $\hat{f}$ ([T2,B2,L2,R2]) into P

        retrieve top state [T,B,L,R] from P

        until [T,B,L,R] consists of only one rectangle

③   set $(t_{max}, b_{max}, l_{max}, r_{max})$ = [T,B,L,R]

This algorithm always examines the most promising(with highest upper bound) rectangle set. The candidate set is split along its largest coordinate interval into halves, thus forming two smaller disjoint candidate sets. The search is stopped if the most promising set contains only a single rectangle with the guarantee that this is the rectangle of globally maximal score.

4 Experiments

We experiments our methods on PASCAL VOC 2007 dataset [15]. The images in this dataset contain objects at a variety of scales and in varying context. Multiple objects from multiple classes may be present in the same image. The complexity of these images is very high. The data has been split into 50% for training/validation and 50% for testing. We choose ten object classes for our recognition task. We use approximate 1000 training images train classification models, and use approximate 2500 test images test detection performance.

The experiment has three major steps. The first step is codebook creation. ERC forest is applied to create visual codebook. Regular grids sampling method is used to extract features from training images. The sampling interval is set to 32, the size of image patches is set to 32. After extract local features in all training images, describe them using modified sPCAT descriptors as discussed in section 2. These feature vectors are used to train an ERC Forest. We use 4 assembling trees and they contain 4000 leaf nodes in total. The leave nodes are numbered and each leaf node is assigned a distinct visual word index.

The second step is training a classifier. Regular grids sampling method also be selected to extract local features. But in this phase, more patches need be sampled than cookbook creation phase. The sampling interval is set to 8, the size of image patches set to different scales. After getting the feature descriptors, query the corresponding visual words. During a query, for each descriptor tested, each tree is traversed from the root down to a leaf and the returned leaf index is the visual word index. After getting all visual words in an image, calculate the 4000-D histogram of visual words. These histograms are fed to SVM to train classifier model. After get SVM decision function, calculate the weight of each word $w_j = \sum_i \alpha_i h_j^i$, rewrite decision function as a sum over per-point contribution according to formula (2).

The third step is object localization. Only test images that contain object instances are chosen in this step. For each image, image features are extracted as the way in training phase. After getting the corresponding visual words, visual words integral image are calculated. The integral image can be used to speedup the calculation of

f^+ and f^-. Then we can localize object instance using the algorithm discussed in section 3. To find multiple object locations in an image, the best-first search can be performed repeatedly. Whenever an object is found, the corresponding region is removed from the image and the search is restarted until a threshold is not satisfied.

In our experiments, we adapt average precision (AP) scores to compare the performance of detection methods. Detections are considered true or false positives based on the area of overlap with ground truth bounding boxes. To be considered a correct detection, the area of overlap a_0 between the predicted bounding box BB_p and ground truth bounding box BB_g must exceed 50% by the formula:

$$a_0 = \frac{area\ (BB_p \cap BB_g)}{area\ (BB_p \cup BB_g)} \tag{4}$$

Fig. 2. shows the AP scores on the test data for 10 of the categories..

For illustration, we also compare our AP scores with the maximum results and median results in the PASCAL VOC 2007 challenge. From the table we can see that although our results are not prior than best results, our results are better than median results of PASCAL VOC 2007 in most of 10 categories. We believe that our method is general for object categories with different features.

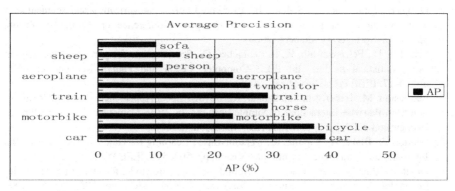

Fig. 2. AP scores on the 10 categories of PASCAL VOC 2007

Table 1. Average Precision(AP) of our method and PASCAL2007 results

Object classes	Our	Max	Median
car	0.39	0.43	0.29
bicycle	0.37	0.41	0.27
motorbike	0.23	0.38	0.22
horse	0.29	0.34	0.20
train	0.29	0.34	0.16
Tv/monitor	0.26	0.29	0.24
aeroplane	0.23	0.27	0.15
person	0.11	0.23	0.12
sheep	0.14	0.18	0.05
sofa	0.10	0.15	0.10

5 Conclusions

We have proposed a new method for object localization in natural images. It uses a weakly supervised training method, yet can get state-of-the-art localization result. It achieves this in several ways. First, we designed a modified sPACT patch descriptor which is suitable for representing local features. Then, ERC forest is applied to speedup both training and testing process. And we further improve the localization performance by employing an Efficient Sub-window Search technical. In future work, we will explore more powerful local feature descriptors, and adapt our method to different SVM kernels.

Acknowledgments. The authors also would like to thank the researchers who kindly published their datasets and software packages online.

References

1. Fergus, R., Perona, P., Zisserman, A.: A visual category filter for google images. In: Pajdla, T., Matas, J.G. (eds.) ECCV 2004. LNCS, vol. 3021, pp. 242–256. Springer, Heidelberg (2004)
2. Leibe, B., Leonardis, A., Schiele, B.: Combined object categorization and segmentation with an implicit shape model. In: 8th European Conference on Computer Vision, pp. 17–32. Springer, Heidelberg (2004)
3. Crandall, D., Felzenszwalb, P., Huttenlocher, D.: Spatial priors for part-based recognition using statistical models. In: IEEE Computer Vision and Pattern Recognition 2005, pp. 10–17. IEEE Press, San Diego (2005)
4. Leordeanu, M., Heber, M., Sukthankar, R.: Beyond Local Appearance: Category Recognition from Pairwise Interactions of Simple Features. In: IEEE Computer Vision and Pattern Recognition 2007, pp. 1–8. IEEE Press, Minnesota (2007)
5. Shotton, J., Blake, A., Cipolla, R.: Contour-Based Learning for Object Detection. In: 10th International Conference on Computer Vision, pp. 503–510. IEEE Press, Beijing (2005)
6. Opelt, A., Pinz, A., Zisserman, A.: A Boundary-Fragment-Model for Object Detection. In: Leonardis, A., Bischof, H., Pinz, A. (eds.) ECCV 2006. LNCS, vol. 3952, pp. 575–588. Springer, Heidelberg (2006)
7. Dalal, N., Triggs, B.: Histograms of Oriented Gradients for Human Detection. In: IEEE Computer Vision and Pattern Recognition 2005, pp. 886–893. IEEE Press, San Diego (2005)
8. Lampert, C.H., Blaschko, M.B., Hofmann, T.: Beyond Sliding Windows: Object Localization by Efficient Subwindow Search. In: IEEE Computer Vision and Pattern Recognition 2008, pp. 1–8. IEEE Press, Anchorage (2008)
9. Deng, H.L., Zhang, W., Mortensen, E.: Principal Curvature-Based Region Detector for Object Recognition. In: IEEE Computer Vision and Pattern Recognition 2007, pp. 1–8. IEEE Press, Minnesota (2007)
10. Ferrari, V., Fevrier, L., Jurie, F., Schmid, C.: Groups of Adjacent Contour Segments for Object Detection. IEEE Trans. Pattern Anal. Machine Intell. 30, 36–51 (2008)
11. Wu, J., James, M.R.: Where am I: Place instance and category recognition using spatial PACT. In: IEEE Computer Vision and Pattern Recognition 2008, pp. 1–8. IEEE Press, Anchorage (2008)

12. Geurts, P., Ernst, D., Wehenkel, L.: Extremely randomized trees. Machine Learning Journal 63, 3–42 (2006)
13. Moosmann, F., Triggs, B., Jurie, F.: Fast Discriminative Visual Codebooks using Randomized Clustering Forests. In: Advances in Neural Information Processing Systems, vol. 19, pp. 985–992 (2006)
14. LIBSVM: a library for support vector machines,
    http://www.csie.ntu.edu.tw/cjlin/libsvm
15. PASCAL 2007 VOC dataset, The PASCAL Visual Object Classes Challenge (2007),
    http://www.pascal-network.org/challenges/VOC/voc2007/

A Multi-Scale Algorithm for Graffito Advertisement Detection from Images of Real Estate

Jun Yang and Shi-jiao Zhu

School of Computer and Information Engineering, Shanghai University of Electric Power,
Shanghai 200090, China
{mail_yangjun, zhusj707}@hotmail.com

Abstract. There is a significant need to detect and extract the graffito advertisement embedded in the housing images automatically. However, it is a hard job to separate the advertisement region well since housing images generally have complex background. In this paper, a detecting algorithm which uses multi-scale Gabor filters to identify graffito regions is proposed. Firstly, multi-scale Gabor filters with different directions are applied to housing images, then the approach uses these frequency data to find likely graffito regions using the relationship of different channels, it exploits the ability of different filters technique to solve the detection problem with low computational efforts. Lastly, the method is tested on several real estate images which are embedded graffito advertisement to verify its robustness and efficiency. The experiments demonstrate graffito regions can be detected quite well.

Keywords: Multi-Scale, Graffito Advertisement Detection, Gabor Filter.

1 Introduction

With the development of network and multimedia technology, people are used to retrieval housing images from Internet to get real estate information instead of text description. However, a large number of housing images are embedded graffito advertisement by real estate agents, for instance, an agent often write a contact telephone number or embed a logo in the images, it is a tedious job for people to select the polluted images one by one from a large scale web image database, therefore, there is great demand for detecting and extracting the graffito advertisement from the images automatically by computer.

Since the background of housing images are very complex, it is difficult to detect and extract these polluted information well. Fortunately, from consistency and robustness of observations from these images, we find that images include graffito advertisement are more relevant than other images and the graffito advertisement is composed of some connected region that we can make good use of these traits to detect and extract the polluted information.

Multi-scale frequency filter is a popular method in image analysis. This technique was inspired by a biological system [1] which has been studied in the past decades. The results of physiological studies have supported that visual processing in the biological visual system at the cortical level

H. Deng et al. (Eds.): AICI 2009, LNAI 5855, pp. 444–452, 2009.

involves a set of different scale filtering mechanisms [2][3]. Some frequency mathematic models have been studied, such as Differences of Gaussian[4] , Gabor functions[5] and Wavelet functions[6].In the frequency domain, it is generally observed that a clutter field in an image has higher frequency content compared to other scatter regions. Therefore, filtering with frequency analysis is an attractive approach for the task of image region analysis.

In this paper, firstly terms used in the proposed multi-scale frequency analysis are defined and invariant operators are introduced to represent a housing image. Since multi-scale method has been studied for its localized frequency filters alike human's visual cortical neurons [3] and used for representation of an image's opponent features [7], a multi-scale algorithm using frequency domain filters to detect polluted regions in a housing image is then proposed. In this approach, because frequency domain features of an image are divided into different subgroups by analyzing regions' distribution, it is more suitable and robust than other feature detection models and it can be used for detecting words, numbers and company's logo which embedded by a real estate agent without any explicit knowledge. The experiments results indicate that images have graffito advertisement with complex background can be detected quite well.

The remainder of this paper is organized as follows: necessary definitions and operators for images which have graffito advertisement are presented in section 2, then a multi-scale algorithm is proposed and some important issues are discussed in section 3. Section 4 provides the experimental comparison results between the proposed algorithm with others algorithm and the concluding remarks are given in Section 5.

2   Graffito Region Detection Using Multi-Scale Filtering

2.1   Definitions for Advertisement Region

Since the advertisement embedded in an image usually is not an isolated line or an edge but a specific text region or a logo, as shown in Figure 1, in order to describe the issues briefly, the necessary symbols and definitions are given as follows.

(a) broker's telephone number   (b) company's logo.

Fig. 1. Examples of images embedded advertisement

Definition 1. For an image distributed in a discrete domain $I \subset Z^2$, a binary image is defined as

$$Bin(x, y) = \begin{cases} 1 & I(x, y) \geq Thr \\ 0 & otherwise \end{cases} \tag{1}$$

Here $I(x, y)$ denotes a pixel point position, L is gray level and Thr presents a specific threshold between 0 and L.

Definition 2. The region which excludes advertisement is presented as B, so the possible advertisement region in an image is defined as

$$\sum_{i=1..n} S_i = I - B \tag{2}$$

Definition 3. For a subset $\Lambda \subseteq S_i$, a candidate advertisement region is a region has two properties as follows:

(1) For a point in the region's neighborhood area $I_\Lambda(x, y) \in S_i$ of Λ, it can be combined into Λ.

(2) For a subarea $Area_\Lambda$ of Λ, it is define as

$$Area_\Lambda = \cup \{I(x, y) | \quad \| I(x, y) - \bar{I}_\Lambda \| < \delta \} \tag{3}$$

The ratio of sub region can be defined as

$$Ratio \, ^A\!/_S = Area_\Lambda / S_{i\Lambda} \tag{4}$$

where T_0 is a threshold and $Ratio \, ^A\!/_S > T_0$, we can regard $\Lambda \subseteq S_i$ as a candidate region.

Definition 4. For multi-scale frequency, a multi-scale operator for the region Λ is defined as

$$F_m = I_\Lambda \otimes G_{m,\theta}, m = m_1...m_n, \theta = \theta_1...\theta_N \tag{5}$$

Where $G_{m,\theta}$ is the m scale with θ direction operator which is described as Gabor filter as below.

Definition 5. In frequency domain, the different covariance of same m is defined as

$$\sigma_{F_m} = \frac{1}{M} \sum_{\theta=\theta_1}^{\theta_M} | (F_{m,\theta} - \overline{F_m})^2 | \tag{6}$$

When $\sigma_{F_m} < \sigma_o$, F_m has isotropy we can get σ_{F_θ} at the same θ value with different frequency.

Definition 6. For the different scale m or σ, a region has advertisement is defined as

$$P = \begin{cases} 1 & 1/M \, | \sigma_{F_{m\theta}} - \overline{\sigma_{F_{m\theta}}} \, \mathsf{k} \, \overset{\Delta}{\sigma}_{m\theta} \\ 0 & otherwise \end{cases} \qquad (7)$$

Where $\overset{\Delta}{\sigma}_{m\theta}$ is an assigned threshold value.

The other property of the polluted region is that the region is continuous not only in the image but also in its corresponding frequency domain. We can verify the region by comparing it with the x-y axes after getting the candidate regions by multi-scale transformation in frequency domain.

2.2 Multi-filters Bank

For implement, Gabor filter is chosen for transforming an image to a frequency domain. The transformation of a 2-D image is modulated by a 2-D Gabor function. A Gabor filter is consisted of two functions having phase by 90 degree, conveniently located in the real and imaginary parts in complex field. Gabor is a function modulated by a complex sinusoid which can be rotated in the x-y plan by θ, it can be symbolized as x' and y'.

$$x' = x*\cos(\theta) + y*\sin(\theta) \qquad (8)$$

$$y' = x*\cos(\theta) - y*\sin(\theta) \qquad (9)$$

The even symmetric Gabor is describe as

$$G(x,y,\theta,f) = \exp(-\frac{1}{2}[(\frac{x'}{s_x})^2 + (\frac{y'}{s_y})^2]) \cos(2*p_i*f*x') \qquad (10)$$

s_x and s_y are variances along x and y-axes respectively, f is the frequency of the sinusoidal function, θ gives the orientation of Gabor filter. By changing the radial frequency and the orientation, the filter makes different target features. The result of spatial domain by Gabor functions is shown in Fig. 2.

After filtering the input image with filters under various frequency and orientation of a channel, these paralleling channels can be analyzed. Since the polluted regions have some common properties under different channels, we can extract these special regions by analyzing the relationship of the channels. The proposed paradigm is shown in Fig.3.

The analysis can be performed in each filter channel using different scale and theta, so it is possible to mark polluted regions in an output image.

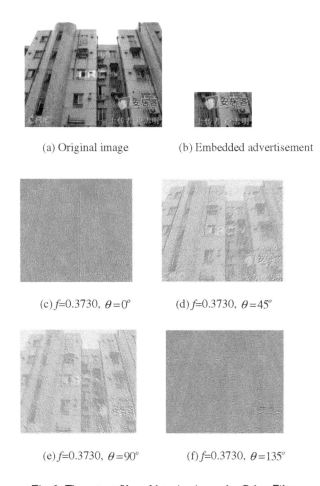

(a) Original image   (b) Embedded advertisement

(c) f=0.3730, $\theta=0^{o}$ (d) f=0.3730, $\theta=45^{o}$

(e) f=0.3730, $\theta=90^{o}$ (f) f=0.3730, $\theta=135^{o}$

Fig. 2. The output filtered housing image by Gabor Filters

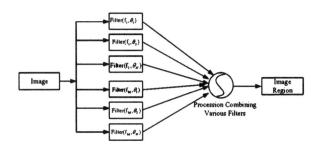

Fig. 3. Filtering bank process paradigm

3 The Proposed Method

From large amounts of observation, we find a housing image has some direction traits even if it has a complex background, for instance it has many straight lines and right angles. These properties can help us separate embedded advertisement from the image using   Gabor function. Here we combine the filtering results from original image with each region's location to find the candidate polluted regions. The detection process is described as bellow.

Step 1. Operation of Multi-Scale Gabor Filters.

For the properties of housing image, it adopts different orientations with variant frequencies. For theta, there are six angles which is $0°,30°,45°,90°,120°,135°$ respectively. In order to find regions contain straight lines in resulting image, we set $\theta = 0°,90°$. For the adaptive of scale, we use the relative size defined as below

$$f = (2^w - 0.5) / W, w=[0...\log_2^{(W/8)}] \tag{11}$$

Where W is the width of an image. Therefore the total number of channel is $6*sizeof(f)$.

Step 2. Channels Combination computation.

After getting the filter results, we can combine the different channels to mark the polluted regions as follows:

(1) Computing the set of Λ in different regions of an original image.

(2) Using the frequency results to find the σ_{F_m} to mark and erase some dynamic value regions.

(3) Using the frequency results to find the different frequency at the same theta σ_{F_θ} .

(4) Getting the transformed gray image under σ_{F_m} and σ_{F_θ} , and then using K-Mean method to cluster the gray image. After that, histogram equalization is performed to scale values of processed gray images.

(5) Calculating by formula (7), a region is marked by $P_{m\cap\theta} = P_m \cap P_\theta$.

(6) Calculating $P_{m\cap\theta}$ and Λ to mark the region.

$$Region = \begin{cases} 1 & (P_{m\cap\theta} = 1) \wedge (\Lambda = 1) \\ 0 & otherwise \end{cases} \tag{12}$$

By formula (12), we can get a 0-1 array which can describe the region is polluted region or not well.

In addition, many other features can be identified using transformed. In our implementation, we used the orientation and frequency dimensions. In order to prove the effectiveness of the method, in the next section, we detect graffito regions from housing images downloaded from Internet.

4 Experimental Results

In order to verify the correctness and effectiveness of our proposed solution, a prototype system is developed by MATLAB. The scheme is tested on the real housing images included advertisement from Web page. The theta equals $0°,30°,45°,90°,120°,135°$ respectively, the frequency is selected by $f = (2^w - 0.5) / W$, $w=[0...\log_2^{(W/8)}]$, where the test W of image is scaled the range of 0 to 255. The comparison results between the proposed method and Summing Squared Response of filters (SSR) (where threshold is 0.5 of range of total summing squared value) are shown in Figure.4. The method of summing squared response is to find a threshold of filters to separate the original image [8]. From Fig.4, we can find that the different level of multi-scale has its own characteristic. When performed by the algorithm, image (d) is obtained which contains graffito features. From the comparison shown in Fig.5, we can find that multi-scale method can gives more graffito regions than the summing squared responses, which can gives more supports for the next step of classification. The result indicates that the detecting of regions from an image having embedded advertisement based on multi-scale algorithm is more suitable for housing images and it is robust for these types of images with more complex background. In the point of image procession, the method also has the ability of eliminating light shadows. Therefore, the result of filtered image cannot give more details of buildings which often have more lines in an image. Additionally, the computation complexity of this method can performed using fast FFT technique which is a critical factor for some real-time applications.

(a) Original image   (b) $\theta = 45°$, $f=0.2537$

(c) $\theta = 135°$, $f=0.3278$ (d) The result

Fig. 4. The filtering results

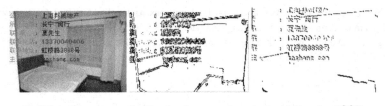

(a) Original image   (b) Result of proposed method   (c) Result of SSR

Fig. 5. The comparison result between proposed method and SSR

5 Conclusions

In this paper, we present a method for detecting graffito polluted regions from an image based on multi-scale algorithm. The main contribution of this paper is that the proposed algorithm can find the relationship of channels in different frequency, therefore, more effective method can be proposed based on its characteristics. On the other hand, the method using fast FFT transform can make algorithm more efficient which can solve problems in the real application of housing images embedded advertisement. Experimental results indicate our method is superior in terms of accuracy, robustness and stability in detection of advertisement in housing images. More studies in the aspect of basic theory and its application should be studied in the near future.

Acknowledgement

This work is supported by Shanghai Education Commission Research Project under Grant 09YZ344, and by Shanghai Special Research Fund of Young College Teacher under Grant sdl08026. The authors are grateful for the anonymous reviewers who made constructive comments.

References

1. De Valois, R.J., Albracht, D.G., Thorell, L.G.: Spatial-frequency selectivity of cells in macaque visual cortex. Vision Research 22, 545–559 (1982)
2. Wilson, H.R.: Psychological evidence for spatial channels. In: Braddick, O.J., Sleigh, A.C. (eds.) Physical and Biological Processing of Images. Springer, Berlin (1983)
3. Watt, R.J., Morgan, M.J.: Spatial lters and the localisation of luminance changes in human vision. Vision Res. 24(24), 1387–1397 (1984)
4. Petrosino, A., Ceccarelli, M.: A scale-space approach to preattentive texture discrimination. In: International Conference on Image Analysis and Processing, 1999. Proceedings, pp. 162–167 (1999)
5. Kamarainen, J.-K., Kyrki, V., Kalviainen, H.: Invariance properties of Gabor filter-based features-overview and applications. IEEE Trans. on Image Processing 15(5), 1088–1099 (2006)
6. Jain, A., Healey, G.: A multiscale rep resentation including opponent color features for texture recognition. IEEE Transactions on Image Processing 7(1), 124–128 (1998)

7. Pollen, D.A., Ronner, S.E.: Visual cortical neurons as localized spatial frequency filters. IEEE Transactions on System, Man and Cybernetics 13(15), 907–916 (1983)
8. Jung, K.: Neural network-based text location in color images. Pattern Recognition Letter 22(14), 1503–1515 (2001)

An Improved Path-Based Transductive Support Vector Machines Algorithm for Blind Steganalysis Classification

Xue Zhang[1] and ShangPing Zhong[1,2]

[1] College of Mathematics and Computer Science, Fuzhou University,
Fuzhou, 350108, China
[2] Fujian Supercomputing Center, Fuzhou, 350108, China
fjzx33@163.com, spzhong@fzu.edu.cn

Abstract. With the technologies of blind steganalysis becoming increasingly popular, a growing number of researchers concern in this domain. Supervised learning for classification is widely used, but this method is often time consuming and effort costing to obtain the labeled data. In this paper, an improved semi-supervised learning method: path-based transductive support vector machines (TSVM) algorithm with Mahalanobis distance is proposed for blind steganalysis classification, by using modified connectivity kernel matrix to improve the classification accuracy. Experimental results show that our proposed algorithm achieves the highest accuracy among all examined semi-supervised TSVM methods, especially for a small labeled data set.

Keywords: Path-based TSVM, Semi-supervised learning, Blind steganalysis.

1 Introduction

Techniques for semi-supervised learning (SSL) are becoming increasingly sophisticated and widespread. Semi-supervised learning is halfway between supervised and unsupervised learning, using a relatively small labeled data set and a large unlabeled data set to obtain the classification. With the rising popular application of support vector machines (SVM), one of the semi-supervised method, transductive SVM [1](TSVM), becomes more popular nowadays. TSVM, maximizes the margin using both labeled data set and unlabeled data set, which is not used in standard SVM. However, the drawback of TSVM is that the objective function is non-convex and thus difficult to minimize. Joachims designs a heuristically optimization algorithm for TSVM called SVMlight [2], which has a wide range of applications in diverse domains. Chapelle and Zien in [3] represent a path-based gradient descent method on the primal formulation of TSVM objective function, called Low Density Separation algorithm (LDS algorithm for short). A novel large margin methodology SSVM and SPSI is proposed by Wang and Shen [4].

Information hiding techniques have recently received a lot of attention. With digital images as carriers, the goal of blind steganalysis is to classify testing images by any steganographic algorithm into two categories: cover images and stego images. Nowadays, most blind steganalysis methods use supervised learning technologies,

H. Deng et al. (Eds.): AICI 2009, LNAI 5855, pp. 453–462, 2009.
© Springer-Verlag Berlin Heidelberg 2009

which require a large amount of labeled data for training. In fact, labeled samples are often difficult, expensive, or time consuming to obtain, while unlabeled data may be relatively easy to collect: we can download a large amount of unlabeled images from the internet. It is of great interest both in theory and in practice because semi-supervised learning requires less human effort and gives higher accuracy [5]. Researches for semi-supervised blind steganalysis classification are still on the initial stage. Our research group once proposed semi-supervised blind steganalysis classification [6] based on SVMlight, but it has high complexity. Thus we consider another semi-supervised algorithm, path-based TSVM algorithm (LDS algorithm [3]) for blind steganalysis. As far as we know, this algorithm is not yet applied to blind steganalysis classification.

In this paper, we build up on the work on LDS algorithm for blind steganalysis classification. But its classification accuracy is not high enough, partly because of the special distribution of steganalysis feature data set, and the lack of utilizing labeled data set information to build the graph. Thus, an improved LDS algorithm for blind steganalysis classification is proposed. It uses Mahalanobis distance instead of Euclidean distance to measure the distance between samples, and modifies the edges between labeled samples when modeling initial weight-graph.

The rest of this paper is organized as follows. Section 2 reviews LDS algorithm and shows its drawbacks. In Section 3, we propose an improved LDS algorithm to overcome those drawbacks. Experimental results on blind steganalysis classification are presented in Section 4. Final conclusions are in Section 5.

2   Review of Path-Based TSVM Algorithm

Path-Based TSVM algorithm is a way to enforce the decision boundary lying in low density regions, and thus not crossing high density clusters, that is why it is equally called Low Density Separation (LDS) algorithm. LDS algorithm first builds a weight-graph derived from the data, using path-based dissimilarity measure to compose connectivity kernel matrix; and then takes manifold mapping method for dimensionality reduction; finally, minimizes TSVM objective function by gradient descent. The first part is the most significant process, and we will show some details.

2.1   Connectivity Kernel on Path-Based Dissimilarity Measure

Connectivity kernel is firstly proposed by Fischer, whose main idea is to transform elongated structures into compact ones [7]. Let a labeled data set $T_l = \{(x_1, y_1), \cdots, (x_l, y_l)\} \in (X \times Y)^l$, where $x_k \in X = R^m$, $y_k \in Y = \{1, -1\}$, $k = 1, \ldots, l$, an unlabeled data set $T_u = \{x_{l+1}, \cdots, x_{l+u}\} \in X^u$. Given the undirected graph $G = (V, E, D)$, where $V = \{x_i\}$ is the set of nodes (samples), E denotes the edges between nodes. Each edge $e(i, j) \in E$ is weighted by Euclidean distance $\|x_i - x_j\|_2$. G can be a full-connected graph or k-nearest neighbors (k-NN) graph.

Path-based dissimilarity measure matrix D is a matrix to measure the dissimilarity of two nodes based on graph-path. Let us denote P_{ij} by all paths from node x_i to node x_j, the dissimilarity measure $d_{ij} \in D$ from node x_i to node x_j is defined as [3],

$$d^\rho(i,j) = \frac{1}{\rho^2} \ln\left(1 + \min_{p \in P_{ij}} \sum_{k=1}^{|p|-1} \left(\exp\left(\rho e\left(p_k, p_{k+1}\right)\right) - 1\right)\right)^2 . \tag{1}$$

d_{ij} denotes the minimum of all path-specific maximum weight. If two nodes are in the same cluster, their pairwise dissimilarity will be small; by contrast, if two nodes are in different clusters, at least one large edge of all paths from node x_i to node x_j crosses low density regions, which will cause large dissimilarity value. Then we can get connectivity kernel K as [3],

$$K(i,j) = \exp\left(-\frac{d^\rho(i,j)}{2\sigma^2}\right) . \tag{2}$$

The linear case corresponds to $\sigma = \infty$, that is $K_{ij} = d_{ij}^\rho$.

2.2 Gradient Descent on TSVM

Chapelle and Zien [3] represent a path-based gradient descent on the primal formulation of TSVM objective function, which can be rewritten as follows,

$$\min \ \frac{1}{2}w^2 + C\sum_{i=1}^{l} L\left(y_i\left(w \cdot x_i + b\right)\right) + C^* \sum_{i=l+1}^{l+u} L^*\left(\left|w \cdot x_i + b\right|\right)$$

$$s.t. \begin{cases} L(t) = \max(0, 1-t) \\ L^*(t) = \exp\left(-3t^2\right) \\ \frac{1}{u}\sum_{i=l+1}^{l+u} w \cdot x_i + b = \frac{1}{l}\sum_{i=1}^{l} y_i \end{cases} \tag{3}$$

The last constraint is to enforce that all unlabeled data are not put in the same class. Unlike the traditional SVM learning algorithms, Chapelle and Zien solve the problem by gradient descent on equation (4), and this primal method has obvious lower time complexity than SVMlight algorithm. We can see the contrast experimental results in Section 4. The implemental details of original LDS algorithm can be found in [3].

2.3 Drawbacks of LDS Algorithm

Although LDS algorithm has the advantages of low time complexity, easy implementation, its classification accuracy is not high enough to some extent. As we mentioned

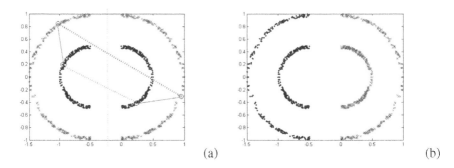

Fig. 1. An example of showing the drawback of LDS algorithm: (a) a synthetic data set is constructed by Delta data set and its mirror points; (b) the classification result done by LDS algorithm

in Section 2.1, the graph is built as all nodes are equal, no matter labeled or unlabeled. Here is a synthetic data set constructed by Delta data set and its mirror points, with four labeled samples, see fig.1(a). We can see that there is a wide gap between Delta data set and its mirror points, thus, if we don't use labeled sample information to build the graph, the classification will be done like fig.1(b).

Fig.1 shows that although two labeled samples are in the same cluster, they sometimes produce a large dissimilarity calculated by Euclidean distance. Chang and Yeung in [8] shows a way of robust path-based semi-supervised clustering, thus we can take it in LDS algorithm. In the synthetic Delta data set, the same cluster samples can be linked together by shrinking the distance of purple red dotted lines in fig.1(a), and the different cluster samples distance (green dotted line in fig.1(a)).can be enlarged to distinguish obviously.

Furthermore, LDS algorithm uses the most common Euclidean distance to build original graph, but the results obtained from Euclidean distance often comes with a deviation from the actual distance, because of the various weights of different features and the mutual influences among the features [9]. On the contrary, Mahalanobis distance can overcome these shortcomings. Mahalanobis distance utilizes the covariance matrix to eliminate the mutual influences among the features, thus can capture the distance more correctly than Euclicean distance. We can use the Within-and-Between-Class Distribution Graph (B-W Graph) [10] to compare the difference between Euclidean distance and Mahalanobis distance for steganalysis features. In B-W Graph, the horizontal axis denotes the between-class distribution B, which provides the main information for classification; and the vertical axis denotes the with-in distribution W. B$\geq$0 means correct classified samples, B<0 means error classified samples.

Fig.2(a) shows the B-W Graph of steganalysis feature set (extracted from 300 original samples and 300 stego samples with Shi's method [11]) by Euclidean distance. In this figure, a lot of stego and original samples are set on the region of B<0, which affects the accuracy of classifier. But if Mahalanobis distance is used instead of Euclidean distance, the result will be better, see Fig.2(b).

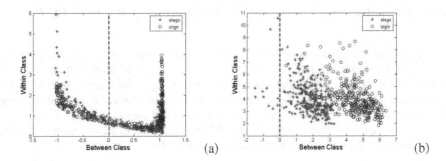

Fig. 2. The steganalysis feature set B-W Graph: (a) B-W Graph is drawn using Euclidean distance; (b) B-W Graph is drawn using Mahalanobis distance

3   A Novel Path-Based TSVM Algorithm for Blind Steganalysis Classification

3.1   Improved Connectivity Kernel on Path-Based Dissimilarity Measure

Now we define an improved path-based dissimilarity measure to construct connectivity kernel. All symbols are the same with that mentioned in Section 2.1. We replace Euclidean distance e in Equation (1) with Mahalanobis distance e_m,

$$e_m(i,j) = \sqrt{(x_i - x_j)^T S^{-1}(x_i - x_j)} \ . \tag{4}$$

Where S denotes the sample covariance matrix for all samples $x \in X^l \bigcup X^u$. Then we add the information of labeled samples by shrinking the edges of two samples in the same clusters to the minimum e_m, and expanding the edges of two samples in different clusters to the maximum e_m,

$$e_m'(i,j) = \begin{cases} \min(e_m) & \text{if } x_i, x_j \in X^l \text{ AND } y_i \times y_j > 0 \\ \max(e_m) & \text{if } x_i, x_j \in X^l \text{ AND } y_i \times y_j < 0 \\ e_m(i,j) & \text{otherwise} \end{cases} \tag{5}$$

Then we use novel e_m' to calculate the linear case connectivity kernel ($\sigma = \infty$),

$$d_m^\rho(i,j) = \frac{1}{\rho^2} \ln\left(1 + \min_{p \in P_{i,j}} \sum_{k=1}^{|p|-1}\left(\exp\left(\rho e_m'(p_k, p_{k+1})\right) - 1\right)\right)^2 \ . \tag{6}$$

$$K_m(i,j) = d_m^\rho(i,j)$$

3.2 Blind Steganalysis Classification Framework and Path-Based TSVM Algorithm

The process of blind steganalysis is actually the process of pattern recognition. It includes three steps: First, do image preprocessing, such as transforming the images into gray scale. Second, extract features and select important features to construct the classifier. Finally, classify the testing data set. For blind steganalysis, testing images can be classified into two categories: cover images and stego images. See Fig.3.

In this paper, we focus on path-based TSVM algorithm in step 3 and want to find a suitable semi-supervised algorithm for blind steganalysis classification.

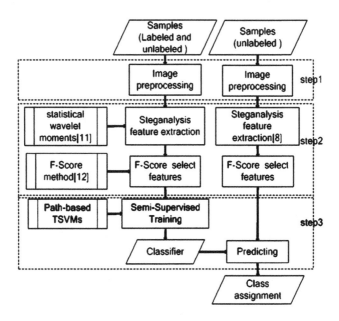

Fig. 3. Blind steganalysis framework based on Path-based TSVM

So combining with the characteristics of blind steganalysis data set, we proposed an improved Path-Based TSVM algorithm.

Algorithm 1. A novel Path-based TSVM algorithm for blind steganalysis

Step 1: Form $(l+u)\times(l+u)$ connectivity kernel matrix by equation (4)-(6).

  a) Build a fully connected graph with standard Mahalanobis distance (equation (4));
  b) Modify the edge lengths between labeled samples (equation (5));
  c) Calculate kernel matrix K_m (equation (6)).

Step 2: Perform multidimensional scaling (MDS) of $(l+u)\times(l+u)$ matrix K_m, and obtain $(l+u)\times p$ matrix K'_m.

Step 3: Train TSVM using gradient descent.

a) Set original $C^* = 2^{index - nofiter} C$, *index=0*.
b) Minimize by gradient descent on the objective function (3).
c) Increase index by one, if *index* equals *nofiter*, goto (d), else goto (b).
d) Print the classification result.

Actually, algorithm 1 can be used in other domains for its flexibility and good performance. We redo the synthetic data set experiment mentioned in Section 2.3 with the same four labeled samples, and our proposed path-based TSVM algorithm gets satisfied result, see fig.4.

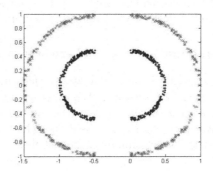

Fig. 4. The classification result done by proposed path-based TSVM algorithm

3.3 Analysis of Algorithmic Time Complexity

The analysis of time complexity for our proposed algorithm is shown as follows: It starts with the $l \times m$ matrix X^l and $u \times m$ matrix X^u, and then builds a fully connected graph with standard Mahalanobis distance. The computation of calculating covariance matrix is $O\left((l+u)^2 m\right)$. For each two samples, the Mahalanobis distance costs $O\left(m^2 + (l+u)^2 m\right)$. Next, we modify the edge lengths that costs $O\left(l^2\right)$. And Dijkstra's algorithm with a priority queue based on a binary heap allows equation (6) to have $O\left((l+u)^2 \left[(l+u) + \log(l+u)\right]\right)$ time complexity [3]. That is, step 1 costs $O\left(m^2 + (l+u)^2 m + (l+u)^3\right)$.

The time complexity of MDS is approximately equal to $O\left((l+u)^3\right)$ in step 2. The kernel matrix K'_m calculated by MDS sharply reduces the time complexity of gradient descent algorithm, which is approximately equal to $O\left((l+u)^3\right)$. Thus, the time complexity of whole algorithm is $O\left(m^2 + (l+u)^2 m + (l+u)^3\right)$. It nearly equals to the time complexity of primal LDS algorithm.

4 Experimental Result

The original image database comes from all the 1096 sample images contained in the CorelDraw Version 10.0 software CD#3. The stego image database comes from those

1096 images embedded by five typical data embedding methods [11]: the non-blind spread spectrum (SS) method (α=0.1) by Cox et al., the blind SS method by Piva et al., the 8×8 block based SS method by Huang and Shi, a generic quantization index modulation (QIM, 0.1 bpp (bit per pixel)), and a generic LSB (0.3 bpp). The hidden data are a random number sequence obeying Gaussian distribution with zero mean and unit variance.

We randomly choose 300 original images as positive samples and 300 stego images as negative samples (totally 600) for testing in our experiment. They also compose the unlabeled data set X^u. And we randomly choose some other samples from original image database and stego image database respectively for the labeled data set X^l. In experiments, we set the default values of parameters as recommended in [3], while setting soft margin parameter C=50 by cross-validation.

4.1 Comparative Experiment for Supervised and Semi-supervised Learning

In this experiment, we randomly choose 10 to 100 labeled samples and combine with 600 samples unlabeled data set for training. The experiment is done for 20 times and then calculates the mean accuracy and variance. The results of semi-supervised learning algorithm (LDS algorithm and our proposed algorithm) and the result of supervised learning algorithm (SVM algorithm) are shown below.

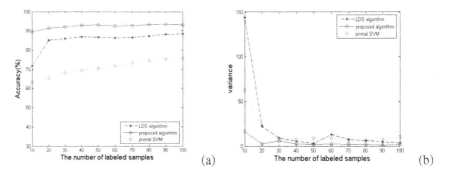

Fig. 5. Comparative Experiment for supervised and semi-supervised learning: (a) the accuracy of three algorithms; (b) the variance of three algorithms

Fig.5(a) illuminates that our proposed algorithm achieves the highest accuracy, even with 10 labeled samples; our proposed algorithm can reach almost 90% accuracy. Fig.5(b) shows the variance of three algorithms. Algorithm with larger variance is usually more unstable. Variance becomes much larger when the number of labeled samples is small, especially for LDS algorithm. This phenomenon might be caused by the biased selection of labeled samples, but we can see that our proposed algorithm still has the best stability.

4.2 Comparative Experiment for Path-Based TSVM Algorithm and SVMlight Algorithm

SVMlight algorithm has a wide range of applications despite its high time complexity, while our proposed algorithm has lower time complexity. SVMlight software tool can be obtained at http://www.cs.cornell.edu/People/tj/. In this experiment, we also randomly choose 10 to 100 labeled samples and combine with 600 samples unlabeled data set for training. The experiment is done for 10 times and then calculates the mean accuracy and cost time. It is worth mentioning that the cost time of building graph in our proposed method is included. The testing results are shown in Fig. 6 and table 1. From the results, we can make the conclusion that our proposed algorithm has much lower time complexity and higher accuracy than SVMlight algorithm.

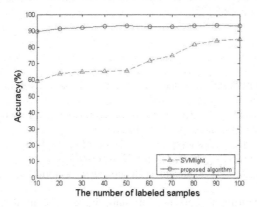

Fig. 6. The accuracy of SVMlight and proposed algorithm with 10 to 100 labeled samples

Table 1. Average time of proposed algorithm and SVMlight

No. of labeled	Proposed (s)	SVMlight(s)	No. of labeled	Proposed (s)	SVMlight(s)
10	4.50	277	60	7.12	601
20	5.25	357	70	7.39	532
30	5.51	520	80	7.54	553
40	6.06	541	90	7.74	408
50	6.52	593	100	8.12	436
average	5.568	457.6	average	7.582	506

5   Conclusions

In this paper, an improved path-based TSVM algorithm is proposed for blind steganalysis classification. It uses Mahalanobis distance instead of Euclidean distance to measure the distance between samples, and adds the information of labeled samples by shrinking the edges of two samples in the same clusters, and expanding the edges of two samples in different clusters. It improves the accuracy and time complexity. Experiments show that our proposed path-based TSVM algorithm performs well even in a small labeled data set. But it still has some drawbacks. With the increasing size of labeled data set, the result is not significantly increasing accordingly. It might be caused by the gradient descent method which will converge to a local minimal value sometimes. Another drawback is, when the size of unlabeled samples becomes very large, the time complexity will be extremely high.

In the future, we will try to do more comparative experiment for other semi-supervised algorithms and find a low time complexity algorithm to deal with the large size of unlabeled samples.

Acknowledgments. This paper is supported by the National Natural Science Foundation of China under Grant NO.10871221, Fujian Provincial Young Scientists Foundation under Grant No.2006F3076.

References

1. Chapelle, O., Scholkopf, B., Zien, A. (eds.): Semi-Supervised Learning. MIT Press, Cambridge (2006)
2. Joachims, T.: Transductive inference for text classification using support vector machines. In: Proceeding of ICML, Bled, pp. 200–209 (1999)
3. Chapelle, O., Zien, A.: Semi-Supervised Classification by Low Density Separation. In: Proceeding of the AISTAT, Barbados, pp. 57–64 (2005)
4. Wang, J., Shen, X.: Large margin semi-supervised learning. J. Mach. Learn. Res. 8, 1867–1891 (2007)
5. Zhu, X.: Semi-Supervised Learning Literature Survey. Technical report, University of Wisconsin-Madison,
   http://pages.cs.wisc.edu/~jerryzhu/pub/ssl_survey.pdf
6. Zhong, S.P., Lin, J.: An Universal Steganalysis Method for GIF Image Based On TSVM. J. Beijing Jiaotong University 33, 122–126 (2009)
7. Fischer, B., Roth, V., Buhmann, J.M.: Clustering with the connectivity kernel. In: Thrun, S., Saul, L., Scholkopf, B. (eds.) Advances in Neural Information Processing Systems, vol. 16. MIT Press, Cambridge (2004)
8. Chang, H., Yeung, D.Y.: Robust Path-Based Spectral Clustering. J. Pattern Recognition 41, 191–203 (2008)
9. Chen, Y.L., Wang, S.T.: Application of WMD Gaussian kernel in spectral partitioning. J. Computer Applications 28, 1738–1741 (2008)
10. Tong, X.F., Teng, J.Z., Xuan, G.R., Cui, X.: JPEG Image Steganalysis Based on Markov Model. J. Computer Engineering 34, 217–219 (2008)
11. Shi, Y.Q., Xuan, G.R., Zou, D., Gao, J.J.: Steganalysis Based on Moments of Characteristic Functions Using Wavelet Decomposition, Prediction-Error Image, and Neural Network. In: Proceeding of ICME 2005, Netherlands (2005)
12. Chen, Y.W., Lin, C.J.: Combining SVMs with various feature selection strategies, http://www.csie.ntu.edu.tw/~cjlin/

Resolution with Limited Factoring*

Dafa Li

Dept of mathematical sciences
Tsinghua University, Beijing 100084, China

Abstract. The resolution principle was originally proposed by J.A. Robinson. Resolution with factoring rule is complete for the first-order logic. However, unlimited applications of factoring rule may generate many irrelevant and redundant clauses. Noll presented resolution rule with half-factoring. In this paper, we demonstrate how to eliminate the half-factoring.

Keywords: factoring rule, linear resolution, resolution.

1 Introduction

Resolution with factoring rule is complete[1]-[7]. There are four kinds of resolvents[1]:

(1). a binary resolvent of clause C_1 and clause C_2,
(2). a binary resolvent of clause C_1 and a factor of C_2,
(3). a binary resolvent of a factor of C_1 and clause C_2,
(4). a binary resolvent of a factor of C_1 and a factor of C_2.

For the definition of the binary resolvent, see [1]. However, unlimited applications of factoring rule may generate many irrelevant and redundant clauses. For example, if a clause contains m literals with the same sign, then we need to test $2^m - (m+1)$ cases to decide which literals of the m literals have a mgu (most general unifier) to find all the factors of the clause. This will require exponential time.

Noll [5] showed that it is sufficient to factorize only one of two parent clauses of a resolvent. See the Basic Lemma in [5] and also see [4]. He called the refinement of the resolution as the resolution rule with half-factoring. Note that lemma 4 in this paper also reports the similar result. Also, in this paper, furthermore, we investigate how to eliminate the half-factoring. In which cases can factoring be eliminated? In lemmas 1, 2, 3 and 4, we discuss all the cases, where factoring is applied. We show that factoring in lemmas 1, 2 and 3 can be eliminated and factoring in lemma 4 is inevitable. Conclusively, we show that if one can binary resolve two clauses on literals that might also both factor in their clauses, then also allow factoring of only one of the clauses on those literals. Otherwise, factoring can be ignored.

---

* The paper was supported by NSFC(Grants: No. 60673034 and 10875061).

H. Deng et al. (Eds.): AICI 2009, LNAI 5855, pp. 463–468, 2009.

We need the following notation and definitions.

Notation: Let θ be a substitution and by the notation in [1], we can write $\theta = \{t_1/x_1, t_2/x_2, ..., t_n/x_n\}$. $VAR\{\theta\} = \{x_1, .., x_n\}$.

Definition
If literals $L^{(1)}, L^{(2)}, ...,$ and $L^{(m_1)}$ of clause C have a mgu σ, then $C\sigma$ is called a factor of $C[1]$ and $L^{(1)}\sigma$ the factored literal.

Definition
Linear binary resolution is a linear deduction, of which each resolvent is a binary resolvent.

2   Factoring Can Be Eliminated in the Following Cases

2.1   Case 1. A Resolvent Has Only One Factored Parent

Lemma 1.
Let $C_1 = L_1^{(1)} \vee L_1^{(2)} \vee ... \vee L_1^{(m_1)} \vee C_1'$ and C_2 be parent clauses. Assume that $L_1^{(1)}, L_1^{(2)}, ...,$ and $L_1^{(m_1)}$ have a mgu α, and C_1 and C_2 have no common variables. Then the binary resolvent of the factor $C_1\alpha$ of clause C_1 and clause C_2, where the factored literal $L_1^{(1)}\alpha$ is the literal resolved upon, is an instance of the resolvent obtained by linear binary resolution with C_1 as a top clause and C_2 as side clauses. See Fig. 1.

Proof.
Let α be a mgu of $L_1^{(1)}, L_1^{(2)}, ...,$ and $L_1^{(m_1)}$. Then,

$$L_1^{(1)}\alpha = L_1^{(2)}\alpha = ... = L_1^{(m_1)}\alpha. \qquad (1)$$

Then $L_1^{(1)}\alpha \vee C_1'\alpha$ is a factor of C_1. Assume that $C_2 = L_2 \vee C_2'$, where L_2 is the literal resolved upon. Let σ be a mgu of $L_1^{(1)}\alpha$ and $\neg L_2$. Then

$$L_1^{(1)}\alpha\sigma = \neg L_2\sigma. \qquad (2)$$

Let R be the resolvent of C_2 and factor $C_1\alpha$ of C_1. Then

$$R = C_1'\alpha\sigma \vee C_2'\sigma. \qquad (3)$$

Since C_1 and C_2 have no common variables, we can choose α such that no variables in $VAR(\alpha)$ occur in C_2. It means $C_2\alpha = C_2$. Especially $L_2\alpha = L_2$. Thus, by Eq. (2)

$$L_1^{(1)}\alpha\sigma = \neg L_2\alpha\sigma. \qquad (4)$$

This says that $\alpha\sigma$ is a unifier of $L_1^{(1)}$ and $\neg L_2$. Let β be a mgu of $L_1^{(1)}$ and $\neg L_2$. Then there is a substitution γ such that

$$\alpha\sigma = \beta\gamma. \qquad (5)$$

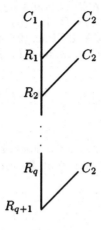

Fig. 1.

We take C_1 as a top clause and C_2 as a side clause and let R_1 be the binary resolvent of C_1 and C_2, where $L_1^{(1)}$ and L_2 are the literals resolved upon. Then

$$R_1 = L_1^{(2)}\beta \vee ... \vee L_1^{(m_1)}\beta \vee C_1'\beta \vee C_2'\beta. \tag{6}$$

Now R_1 is considered as a center clause and C_2 is taken a side clause again, let us compute a binary resolvent of R_1 and C_2. Notice that from Eq. (1) and Eq. (4), $L_1^{(1)}\alpha\sigma = L_1^{(2)}\alpha\sigma = ... = L_1^{(m_1)}\alpha\sigma = \neg L_2\alpha\sigma$. From Eq. (5),

$$L_1^{(2)}\beta\gamma = \neg L_2\beta\gamma. \tag{7}$$

By notation in [1], we write C_2 as $L_2[x_1, ..., x_l] \vee C_2'[x_1, ..., x_l]$, where $x_1, ..., x_l$ are all the variables which occur in C_2. After renaming all the variables in C_2, we obtain $C_2^* = L_2^* \vee C_2'^* = L_2[y_1, ..., y_l] \vee C_2'[y_1, ..., y_l]$, where $y_1, ...$ and y_l are new variables which do not occur in C_1, C_2, β, γ, α or σ. Clearly, $L_2\beta\gamma = L_2[x_1, ..., x_l]\beta\gamma = L_2[x_1\beta\gamma, ..., x_l\beta\gamma]$. Let $\lambda = \{x_1\beta\gamma/y_1, ..., x_l\beta\gamma/y_l\}$. Then,

$$L_2^*\lambda = L_2\beta\gamma = L_2\beta\gamma\lambda, \tag{8}$$

$$C_2'^*\lambda = C_2'\beta\gamma = C_2'\alpha\sigma. \tag{9}$$

So λ is a unifier of L_2^* and $L_2\beta\gamma$, and λ is also a mgu of L_2^* and $L_2\beta\gamma$. From Eqs. (7) and (8), $L_1^{(2)}\beta\gamma\lambda = \neg L_2\beta\gamma\lambda = \neg L_2^*\lambda = \neg L_2^*\gamma\lambda$ since $y_1, ...,$ and y_l do not occur in $VAR(\gamma)$. Therefore $\gamma\lambda$ is a unifier of $L_1^{(2)}\beta$ and $\neg L_2^*$. Let η be a mgu of $L_1^{(2)}\beta$ and $\neg L_2^*$. Then for some substitution ρ,

$$\gamma\lambda = \eta\rho. \tag{10}$$

Let R_2 be the binary resolvent of R_1 and C_2^*, where $L_1^{(2)}\beta$ and L_2^* are the literals resolved upon. Then $R_2 = L_1^{(3)}\beta\eta \vee ... \vee L_1^{(m_1)}\beta\eta \vee C_1'\beta\eta \vee C_2'\beta\eta \vee C_2'^*\eta$. For simplicity, here let $m_1 = 2$. Then R_2 becomes $R_2 = C_1'\beta\eta \vee C_2'\beta\eta \vee C_2'^*\eta$. By Eq. (10), $R_2\rho = C_1'\beta\gamma\lambda\vee C_2'\beta\gamma\lambda\vee C_2'^*\gamma\lambda$. By Eq. (5), $R_2\rho = C_1'\alpha\sigma\lambda\vee C_2'\alpha\sigma\lambda\vee C_2'^*\gamma\lambda$. Thus, $R_2\rho = C_1'\alpha\sigma \vee C_2'\alpha\sigma \vee C_2'^*\lambda$. By Eqs. (9) and (3), $R_2\rho = C_1'\alpha\sigma \vee C_2'\alpha\sigma = C_1'\alpha\sigma \vee C_2'\sigma = R$ since C_1 and C_2 have no common variables. This implies that R is an instance of R_2.

Example 1. The deduction of $\square$ can be obtained from $P(a)$ and $\neg P(x) \vee \neg P(y)$ by binary resolution.

2.2   Case 2. A Resolvent Has Two Factored Parents, Where the Two Factored Literals Are Not Resolved Upon

Lemma 2. Let $C_1 = L_1 \vee L_1^{(1)} \vee L_1^{(2)} \vee ... \vee L_1^{(m_1)} \vee C_1'$ and $C_2 = L_2 \vee L_2^{(1)} \vee L_2^{(2)} \vee ... \vee L_2^{(m_2)} \vee C_2'$. Assume that $L_1^{(1)}, L_1^{(2)}, ...,$ and $L_1^{(m_1)}$ have a mgu α, $L_2^{(1)}, L_2^{(2)}, ...,$ and $L_2^{(m_2)}$ have a mgu β, and C_1 and C_2 have no common variables. Then the binary resolvent of the factor $C_1\alpha$ of C_1 and the factor $C_2\beta$ of C_2, where $L_1\alpha$ and $L_2\beta$ are the literals resolved upon, is an instance of the binary resolvent of C_1 and C_2, where L_1 and L_2 are the literals resolved upon.

Proof.
Let $\mu = \alpha \cup \beta$. Then factor $C_1\alpha = C_1\mu = L_1\mu \vee L_1^{(1)}\mu \vee C_1'\mu$ and factor $C_2\beta = C_2\mu = L_2\mu \vee L_2^{(1)}\mu \vee C_2'\mu$. Let σ be a mgu of $L_1\alpha$ ($= L_1\mu$) and $\neg L_2\beta$ ($= \neg L_2\mu$) and R_f be the binary resolvent of factor $C_1\alpha$ of C_1 and factor $C_2\beta$ of C_2. Then $R_f = L_1^{(1)}\mu\sigma \vee C_1'\mu\sigma \vee L_2^{(1)}\mu\sigma \vee C_2'\mu\sigma$.
   Since $L_1\mu\sigma = \neg L_2\mu\sigma$, $\mu\sigma$ is a unifier of L_1 and $\neg L_2$. Let v be a mgu of L_1 and $\neg L_2$. Then $\mu\sigma = v\gamma$ for some substitution γ. Let R_b be the binary resolvent of C_1 and C_2, where L_1 and L_2 are the literals resolved upon. Then $R_b = L_1^{(1)}v \vee L_1^{(2)}v \vee ... \vee L_1^{(m_1)}v \vee C_1'v \vee L_2^{(1)}v \vee L_2^{(2)}v \vee ... \vee L_2^{(m_2)}v \vee C_2'v$. Since $\mu\sigma = v\gamma$, $R_b\gamma = L_1^{(1)}\mu\sigma \vee C_1'\mu\sigma \vee L_2^{(1)}\mu\sigma \vee C_2'\mu\sigma = R_f$. It means that R_f is an instance of R_b.

Example 2. Let $C_1 = P(x) \vee \neg Q(x) \vee \neg Q(y)$ and $C_2 = \neg P(a) \vee R(u) \vee R(v)$. Then instead of computing the resolvent of factors of the two clauses, we just compute the binary resolvent of the two clauses.

2.3   Case 3. A Resolvent Has Two Factored Parents, Where Only One Factored Literal Is the Literal Resolved Upon

Lemma 3. Let $C_1 = L_1^{(1)} \vee L_1^{(2)} \vee ... \vee L_1^{(m_1)} \vee C_1'$ and $C_2 = L_2 \vee L_2^{(1)} \vee L_2^{(2)} \vee ... \vee L_2^{(m_2)} \vee C_2'$. Assume that $L_1^{(1)}, L_1^{(2)}, ...,$ and $L_1^{(m_1)}$ have a mgu α_1, $L_2^{(1)}, L_2^{(2)}, ...,$ and $L_2^{(m_2)}$ have a mgu α_2, and C_1 and C_2 have no common variables. Then the binary resolvent of the factor $C_1\alpha_1$ of C_1 and the factor $C_2\alpha_2$ of C_2, where the literal $L_2\alpha_2$ and the factored literal $L_1^{(1)}\alpha_1$ are the literals resolved upon, is an instance of the resolvent obtained by linear binary resolution with C_1 as a top clause and C_2 as side clauses. See Fig. 1.

Proof

Let $\alpha = \alpha_1 \cup \alpha_2$. Then $L_1^{(1)}\alpha \vee C_1'\alpha$ is a factor of C_1 and $L_2\alpha \vee L_2^{(1)}\alpha \vee C_2'\alpha$ is a factor of C_2. Let σ be a mgu of $L_1^{(1)}\alpha$ and $\neg L_2\alpha$, and R be the resolvent of the factors of C_1 and C_2. Then $R = C_1'\alpha\sigma \vee L_2^{(1)}\alpha\sigma \vee C_2'\alpha\sigma$. For simplicity, here let $m_1 = 2$. Then, by using the method used in Lemma 1, we can obtain a deduction of R_2 by linear binary resolution with C_1 as a top clause and C_2 as side clauses, and we can show that for some substitution ρ, $R_2\rho = C_1'\alpha\sigma \vee (L_2^{(1)} \vee L_2^{(2)} \vee \ldots \vee L_2^{(m_2)} \vee C_2')\alpha\sigma = C_1'\alpha\sigma \vee L_2^{(1)}\alpha\sigma \vee C_2'\alpha\sigma = R$. It means that R is an instance of R_2.

Example 3. Let $C_1 = \neg P(a) \vee \neg P(u) \vee R(v)$ and $C_2 = P(x) \vee \neg Q(x) \vee \neg Q(y)$. Then instead of computing the resolvent of factors of the two clauses, we derive the deduction by linear binary resolution with C_1 as a top clause and C_2 as side clauses.

3   When a Resolvent Has Two Factored Parents, Where Two Factored Literals Are the Literals Resolved Upon, Also Allow Factoring Only One Parent

Lemma 4. Let $C_1 = L_1^{(1)} \vee L_1^{(2)} \vee \ldots \vee L_1^{(m_1)} \vee C_1'$ and $C_2 = L_2^{(1)} \vee L_2^{(2)} \vee \ldots \vee L_2^{(m_2)} \vee C_2'$. Assume that $L_1^{(1)}, L_1^{(2)}, \ldots$, and $L_1^{(m_1)}$ have a mgu α_1, $L_2^{(1)}, L_2^{(2)}, \ldots$, and $L_2^{(m_2)}$ have a mgu α_2, and C_1 and C_2 have no common variables. Then the binary resolvent of the two factors, where the factored literals $L_1^{(1)}\alpha_1$ and $L_2^{(1)}\alpha_2$ are the literals resolved upon, is an instance of the resolvent obtained by linear binary resolution with C_1 as a top clause and the factor $C_2\alpha_2$ of C_2 as side clauses. See Fig. 1.

Proof

let $\alpha = \alpha_1 \cup \alpha_2$. Then $C_1\alpha = L_1^{(1)}\alpha \vee C_1'\alpha$ and $C_2\alpha = L_2^{(1)}\alpha \vee C_2'\alpha$. Let σ be a mgu of $L_1^{(1)}\alpha$ and $\neg L_2^{(1)}\alpha$ and R be the binary resolvent of the factors $C_1\alpha$ and $C_2\alpha$, where the factored literals $L_1^{(1)}\alpha_1$ and $L_2^{(1)}\alpha_2$ are the literals resolved upon. Then $R = C_1'\alpha\sigma \vee C_2'\alpha\sigma$. Now let us consider C_1 as a top clause and the factor $C_2\alpha$ of C_2 as side clauses. For simplicity, here let $m_1 = 2$. By lemma 1, we can derive $R_2\rho = C_1'\alpha\sigma \vee C_2'\alpha\sigma = R$ for some substitution ρ, where the deduction of R_2 is obtained by linear binary resolution with C_1 as a top clause and the factor $C_2\alpha_2$ of C_2 as side clauses. See Fig. 1. It implies that R is an instance of R_2.

We can find proofs of the following examples by using the limited factoring.

Example 4. $S = \{(1). \ P(x) \vee P(y), (2). \ \neg P(x) \vee \neg P(y)\}$.

Proof

(3). $\neg P(z)$ a factor of (2)
(4). $P(y)$ a binary resolvent of (1) and (3)
(5). $\square$ a binary resolvent of (3) and (4)

Example 5. $S = \{(1).\ P(z,y) \lor P(x,g(x)),(2).\ Q(b) \lor \neg P(u,v),\ (3).\ \neg P(w,g(b)) \lor \neg Q(w)\}[7]$.

Proof

(4). $\neg P(u,v) \lor \neg P(b,g(b))$ a binary resolvent of (2) and (3)

(5). $\neg P(b,g(b))$ a factor of (4)

(6). $P(z,y)$ a binary resolvent of (1) and (5)

(7). $\square$ a binary resolvent of (5) and (6)

4   Summary

Noll [5] showed that it is sufficient to factorize only one of two parent clauses of a resolvent. He called the refinement of the resolution as the resolution rule with half-factoring. In this paper, furthermore, we demonstrate how to eliminate the half-factoring. We show that if one can binary resolve two clauses on literals that might also both factor in their clauses, then also allow factoring of only one of the clauses on those literals. Otherwise, factoring can be ignored. For example, lemmas 2 and 3 declare that it is not necessary to factorize the literals which are not resolved upon.

References

[1] Chang, C.C., Lee, R.C.T.: Symbolic logic and mechanical theorem proving. Academic Press, San Diego (1973)

[2] Duffy, D.: Principles of automated theorem proving. John Wiley & Sons, Chichester (1991)

[3] Fitting, M.: First-order logic and automated theorem proving. Springer, New York (1990)

[4] Kowalski, R.: Studies in completeness and efficiency of theorem proving by resolution. Ph.D. thesis, University of Edingburgh (1970)

[5] Noll, H.: A Note on Resolution: How to get rid of factoring without loosing completeness. In: Bibel, W. (ed.) CADE 1980. LNCS, vol. 87, pp. 250–263. Springer, Heidelberg (1980)

[6] Robinson, J.A.: A machine-oriented logic based on the rosolution principle. J. ACM 12(1), 23–41 (1965)

[7] Socher-Ambrosius, R., Johann, P.: Deduction systems. Springer, New York (1997)

Formal Analysis of an Airplane Accident in $N\Sigma$-Labeled Calculus

Tetsuya Mizutani[1], Shigeru Igarashi[1], Yasuwo Ikeda[2], and Masayuki Shio[3]

[1] Department of Computer Science, University of Tsukuba, Tsukuba, Japan
{mizutani,igarashi}@cs.tsukuba.ac.jp
[2] Mejiro University, Tokyo, Japan
yikeda@krd.biglobe.ne.jp
[3] College of Community Development, Tokiwa University, Mito, Japan
shio@tokiwa.ac.jp

Abstract. $N\Sigma$-labeled calculus is a formal system for representation, verification and analysis of time-concerned recognition, knowledge, belief and decision of humans or computer programs together with related external physical or logical phenomena. In this paper, a formal verification and analysis of the JAL near miss accident is presented as an example of cooperating systems controlling continuously changing objects including human factor with misunderstanding or incorrect recognition.

Keywords: Formal representation and analysis, Time-concerned program systems, Belief and decision, Near miss accident.

1 Introduction

$N\Sigma$-labeled calculus [9], [12] is a formal system, which is a kind of "logic of knowledge and thought [11]", in order to describe time-concerned recognition, knowledge, belief and decision of humans or computer programs together with related external physical or logical phenomena. Formal verification and analysis of the JAL near miss accident [1], [10] in this formalism is presented as an example of cooperating systems controlling continuously changing objects including human factor with misunderstanding or incorrect recognition. In this analysis, some sufficient conditions that the accident would not occur are also shown.

Through this example, relationship among artificial intelligence, external environment and human factors are investigated.

There are many related works in this area. There are many reasoning systems for temporal or time-related logic of knowledge and belief of human, e.g. [3], [5], [7], etc. And also, airplane controlling systems as well as vehicle controlling systems are formally treated in [4]. All of these systems use *model checking* method, while $N\Sigma$-labeled calculus does the *proof-theoretic* method on the *natural number theory*, which is one of the remarkable feature of this calculus.

H. Deng et al. (Eds.): AICI 2009, LNAI 5855, pp. 469–478, 2009.
© Springer-Verlag Berlin Heidelberg 2009

2 $N\Sigma$-Labeled Calculus

A brief explanation of $N\Sigma$-labeled calculus is introduced along with [9] and [12], whose semantics is described in [9] with its soundness. We will partially follow [14] in which *PA* stands for Peano arithmetic, while an occurrence of the variable x in a formula **A** will be indicated explicitly as **A**$[x]$, and $\mu x\mathbf{A}[x]$ will designate the least number satisfying **A**, in this paper.

Peano arithmetic (*PA*) is extended to $PA(\infty)$ called *pseudo-arithmetic* [8], including the *infinity* (∞) and the *minimalization* (μ).

Infinite number of *special constants*, J, J_1, J_2, ... and *labels*, ℓ, ℓ_1, ℓ_2, ... $\in \Sigma$ are added, where Σ is *the set of labels*. The special constants correspond to program variables taking natural number values and possibly ∞, and the change of their values along the natural number time is expressed by the change of local models, or worlds, and vice versa. It must be noticed that a program variable is guarded against quantification in the calculus, since each J is not a variable but a constant. A label indicates a *personality* that is a generalization of an *agent* of multi-agent systems, or an *observer* in physics, including notion of subjectivity.

A *tense* means the time relative to a reference 'observation time' called *now*, which is taken to be 0 throughout the calculus, where ∞ indicates the tense when **false** holds. The value of a special constant may change along with tense.

@-mark as a logical symbol is called the *coincidental* operator. A formula of the form $\mathbf{A}@\langle a, \ell\rangle$ intuitively means that the personality designated by the label ℓ believes at the tense designated by **a** the fact that "**A** holds *now*", while **A**@a and **A**@ℓ mean that **A** holds at tense **a**, and that ℓ believes that **A** holds *now*, respectively. The logical symbol " ; " called the *futurity* operator moves the observation time toward a future time-point. a; b is the tense of b observed at the tense designated by **a**. a; A means the least, i.e. the earliest, time when **A** comes to hold, or *rises*, after or exactly at the tense **a**, which will be called the *ascent* of **A** at **a**.

3   Representation of Cooperative Continuous Programs

Actions or changes of states of programs for verification are represented by *program axioms*, used as the nonlogical axioms in the proofs of theorems for the particular programs. Some primitive notions for program axioms and for representing cooperative continuous programs are introduced.

3.1   Spur and Program Labels

Definition 1. A *spur* α is defined as a term by

$$\alpha = \mu y(0 < y \& 0 < J@y),$$

for an arbitrarily chosen and fixed special constant J.   □

Spurs serve as the generalization of schedulers and the next operator in temporal logic. α, β, γ, ... , κ are used as the metasymbols for spurs. Each component process, or even certain external object, of a multi-CPU parallel program system is assigned a distinct spur usually, and the whole program is described in accordance with the causality, i.e. the relationships between the spurs as motives and the changes of values.

Besides the spur, to each process is assigned a distinct special constant L_i, the formula $L_i = j$ expressing that the control of the i-th process in the parallel program is at its j-th block, statement, etc.

Definition 2. $a_{i,j}$ $(i, j = 0, 1, \ldots)$ is a *program label*, expressed by mutually exclusive special boolean constants defined as:

$$a_{i,j} \equiv L_i = j \quad (i, j = 0, 1, \ldots).$$

where L_i is an arbitrary special constant.   □

3.2 Conservation Axioms

We adopt the following two axiom schemata [8].

$$(CA1) J = z \supset x < \alpha \ \& \ x < \beta \ \& \ \ldots \ \& \ x < \kappa \supset J = z@x,$$

for each special constant J and all spurs $\alpha, \beta, \ldots, \kappa$ in the program system. This axiom means that the value of J does not change until any spur rises, i.e. the next step of any process rises, where each spur $\alpha, \beta, \ldots, \kappa$ indicates the tense when the next step of the process corresponding to the spur rises.

$$(CA2) J = z@a \supset \alpha < \beta \ \& \ \ldots \ \& \ \alpha < \kappa \supset a \le x \le \alpha \supset J = z@x,$$

for each α, a and J, such that J does not occur in the 'action' part of a corresponding program axiom in the axiom tableau (section 4), where the action part means the part of the program axiom denoting the change of program variables. This axiom says that the value of the special constant J does not change until the next step α rises, when the control of the corresponding process is on the program block labeled by a. It means that the value of J does not change within the block corresponding to a, i.e. it does not change if the program axiom does not denote the change of its value.

3.3 Approximation of Continuous System

For representation, analysis and verification of cooperating systems with continuity, the notion of differentiation must be dealt with. In this paper, the first-order derivatives, i.e. speed, are treated as special constants. The distance are defined by the integral using Euler's approximation.

Definition 3. For a term $\mathbf{a}[x]$, $\sum_{0 \le x < y} \mathbf{a}[x]$ is inductively defined as follows.

$$\sum_{0 \le x < 0} \mathbf{a}[x] = 0, \qquad \sum_{0 \le x < y+1} \mathbf{a}[x] = \sum_{0 \le x < y} \mathbf{a}[x] + \mathbf{a}[y],$$

$$\sum_{0 \le x < \infty} \mathbf{a}[x] = \begin{cases} \sum_{0 \le x < \mu y (\forall z (y \le z \supset \mathbf{a}[z]=0))} \mathbf{a}[x], \\ \qquad \text{if } \exists y (\forall z (y \le z \supset \mathbf{a}[z] = 0)), \\ \infty, \qquad \text{otherwise.} \end{cases}$$

□

Definition 4. For a term $\mathbf{a}[x]$, $\int_{x=b}^{b+t} \mathbf{a}[x]$, that is, an approximation of $\int_{x=b}^{b+t} \mathbf{a}[x]dx$, is defined as follows.

$$\int_{x=b}^{b+t} \mathbf{a}[x] \overset{\text{def}}{=} \sum_{0 \leq y < n+1} h \cdot \mathbf{a}[b + yh],$$

where $t = nh$.   □

It must be noted that the error of this approximation can be made arbitrary small by choosing a sufficiently small h.

Definition 5. Let $\dot{A}$ be a special constant. A is defined as follows.

$$A[t] \overset{\text{def}}{=} \int_{x=0}^{t} \mu y(y = \dot{A}[x]@x).$$ □

4   Representation by Axiom Tableaux

By *axiom tableaux* [12], the program axioms representing actions or changes of states of program systems to be verified are easily and readably expressed. The following abbreviations are used in the program tableaux.

Abbreviation 1. 1. Each axiom is numbered and divided into '*condition/ prefix*', '*action*', '*tense*' and '*label*' parts.
2. "&" is replaced by comma " , " whenever there is no confusion.
3. Double line, in contrast to single ones, delimit the scope of free variables, which are quantifiable by "∀". If **"forsome"** occurs in the condition/prefix cell, the variables in the row is bound by "∃".
4. '*' mark in the label cell means the corresponding axiom is not only a belief by the personalities indicating the label cell but also a fact, i.e. it holds without the belief by the personalities.

5   Analysis of JAL Near Miss Accident

5.1   Outline of the Accident

The outline of the JAL near miss accident is as follows [1], [10].

On Jan. 31, 2001, JAL flight 907, called A in this paper, as well as in [10], had departed Tokyo-Haneda for a flight with destination Naha. JAL Flight 958, called B, was en route from Pusan to Tokyo-Narita. A trainee controller at Tokyo Area Control Center (*ACC*) cleared A to climb to FL390[1] at 15:46, Two minutes later B reported at FL370. Both flights were on an intersecting course near the Yaizu NDB. At 15:54 the controller noticed this, but instead of ordering B to descend, he ordered A to descend. Immediately after this instruction, the

---

[1] FL stands for *Flight Level*, and FL1, written as 1[FL] in formulas in this paper, is 100 [ft] in a standard nominal altitude.

Table 1. Program Axioms

index	condition/prefix	action	tense	label
1		$\alpha = CNF = D_0$	$\uparrow anticipateNM(A, B)$ @$Monitor$	$*$, $Monitor$
2		$anticipateNM(A, B) \equiv CNF$		ACC
3		$\beta = (CNF$@$ACC) = D_1$	$\uparrow CNF$	$*$, ACC
4	**forsome** X, Y	$\beta = (cntl(A, X) \lor cntl(B, Y),$ $cntl(A, X) \equiv Ecntl$@$A,$ $cntl(B, X) \equiv Ecntl$@$B) = D_2$	$\uparrow anticipateNM(A, B)$ @ACC	$*$, ACC
5	**forsome** X, Y	$cntl(A, X) \lor cntl(B, Y) \supset$ $\Box_{180} Separation_{7.5}$		$*$, ACC
6		$\gamma_P = \neg cntl(P, X)$@$P = 15sec$	$\uparrow cntl(P, X)$@P	$*$, P, ACC
7	**forsome** X	$\rho_A = RA(A, X) = D_3$	$\uparrow anticipateNM(A, B)$ @$TCAS_A$	$*$, A, $TCAS_A$
8	**forsome** X	$\rho_B = RA(B, X) = D_4$	$\uparrow anticipateNM(A, B)$ @$TCAS_B$	$*$, B, $TCAS_B$
9		$\gamma_P = ope(P, X) = D_5$	$\uparrow cntl(P, X)$@P	$*$, P, ACC
10	$\neg Ecntl$	$\gamma_P = ope(P, X) = D_6$	$\uparrow RA(P, X)$@P	$*$, P, ACC
11		$\gamma_A =$ $(|f_v(A, v, h, s)| \le |\dot{a}_v|,$ $0 \le \dot{a}_v \cdot f(A, v, h, s))$ $= D_7$	$\uparrow ope(A, \langle v, h \rangle)$	$*$, A, B, ACC

crew of A were given an aural TCAS Resolution Advisory (RA) to climb in order to avoid a collision. At the same time the crew of B were given an aural RA to descend. The captain of A followed the instructions of the controller by descending. A now approaching close to B, because captain of B descended as well, following the advisory of his TCAS. A collision was averted when the pilot of A then put his airplane into a nosedive. A missed B by 105 to 165 meters in lateral distance and 20 to 60 meters in altitude difference. About 100 crew and passengers on A sustained injuries due to emergency moreover, while no one was injured on B.

5.2 Formalization

In this paper, the decision and order of the controller at Tokyo ACC, the decision of the crews of the 2 airplanes, and the order from TCASs installed in the airplanes caused the near miss are formulated, verified and analyzed in the calculus.

Only the changes of the vertical positions of the airplanes are dealt with in the analysis, for the simplicity. To do it, the projection of the positions of the airplanes to a vertical plain is considered.

Definition 6. The following schemata are defined, where $\mathcal{P}_i$ ($i = 1, 2$) is a meta-expression representing the airplane A and B, respectively, and $p_i = \langle p_{iv}, p_{ih} \rangle$ is a pair of the special constants p_{iv} representing the vertical position and p_{ih} representing the horizontal position of $\mathcal{P}_i$.

$$\diamond(t, A) \equiv \exists u \le t(A@u), \tag{1}$$

$$hold(\mathcal{P}) \equiv \forall x_v x_h (\dot{p}_v = x_v \& \dot{p}_h = x_h \supset \forall t((\dot{p}_v = x_v \& \dot{p}_h = x_h)@t)), \tag{2}$$

$$Separation(\mathcal{P}_1,\ \mathcal{P}_2,\ v,\ h) \stackrel{\text{def}}{=} v[\text{FL}] < |\boldsymbol{p}_{1v} - \boldsymbol{p}_{2v}| \vee h\ [\text{nm}] < |\boldsymbol{p}_{1h} - \boldsymbol{p}_{2h}|, (3)$$

$$\diamond_{180}\mathbf{A} \equiv \diamond(180[s],\ \mathbf{A}), \tag{4}$$

$$Separation_{7,5} \stackrel{\text{def}}{=} Separation(\mathcal{P}_1,\ \mathcal{P}_2,\ 7,\ 5), \tag{5}$$

$$anticipateNM(\mathcal{P}_1,\ \mathcal{P}_2) \stackrel{\text{def}}{=} hold(\mathcal{P}_1)\&hold(\mathcal{P}_2) \supset \diamond_{180}\neg Separation_{7,5}, \tag{6}$$

where the unit nm is the initials of *nautical mile*. $\square$

Separation is a minimal distance between two airplanes served by the controller to keep safe and ordered traffic, defined as the vertical distance is greater than $v[\text{FL}]$ or the horizontal distance is greater than $h[\text{nm}]$. $Separation_{7,5}$ expresses that the vertical distance is greater than $7[\text{FL}]$ or the horizontal distance is greater than $5[\text{nm}]$. The near miss condition is defined when it does not hold. $anticipateNM(A, B)$ expresses that a near miss will occur within 180 seconds if the speed or the direction of the airplane A or B does not change.

The program axioms representing the actions of ACC, A, B, $TCAS_A$ and $TCAS_B$ by the axiom tableau are listed in Table 1, together with the facts what and when they actually acted, or ordered, when the accident occurred, in Table 2. In these tables, D_i ($0 \le i \le 5$) indicate *delays*. α, β, γ_A, γ_B are spurs for the monitor, the controller, airplanes A and B, respectively. $TCAS_A$ and $TCAS_B$ are TCASs installed in A and B, respectively. ρ_A and ρ_B are spurs of $TCAS_A$ and $TCAS_B$, respectively.

The intended meaning of these program axioms and facts are as follows.

- Axiom 1 (the row indicated by number 1 in Table 1) represents the recognition of the monitor, where *CNF* is a special constant over Boolean values indicating a conflict alert occurs or not. If the monitor anticipates a near miss, a conflict alert will occur with D_0 second delay.
- Axiom 2-5 indicate the recognition of the controller.

Table 2. Facts

index	condition/prefix	action	tense	label
12		↑ $anticipateNM(\boldsymbol{A},\ \boldsymbol{B})$ $= 15{:}54'15''$	S	∗, Monitor
13		↑ $cntl(\boldsymbol{A}, \langle \perp,\ 350[\text{FL}] \rangle)$ $= 15{:}54'27''$	S	∗, A
14		↑ $(cntl(\boldsymbol{A}, \langle \perp,\ 350[\text{FL}] \rangle) \equiv Ecntl)$ $= 15{:}54'27''$	S	∗, A
15		↑ $RA(\boldsymbol{A}, \langle 1500\text{ft/min},\ \perp \rangle)$ $= 15{:}54'35''$	S	∗, A
16		↑ $RA(\boldsymbol{B}, \langle -1500\text{ft/min},\ \perp \rangle)$ $= 15{:}54'34''$	S	∗, B
17		$\neg Ecntl, \neg CNF$	S	∗, A, B, ACC
18		↑ $cntl(\boldsymbol{B}, \langle \perp,\ 350[\text{FL}] \rangle)$ $= 15{:}54'27''$	S	ACC
19	$ope(\boldsymbol{B},\ \langle \perp,\ 350[\text{FL}] \rangle)\ @D_6,$ $hold(\boldsymbol{A})$	$\square_{180}Separation_{7,5}$	$S;\ 15{:}54'27''$	∗, ACC
20		$hold(\boldsymbol{A})$		ACC

- • Axiom 2: The controller recognizes that anticipating the near miss and the corresponding alert will occur simultaneously.
- • Axiom 3: The controller recognizes the alert D_1 seconds after it actually occurs.
- • Axiom 4: The controller operates some emergency controlling order indicated by *Ecntl* to each airplane A or B with D_2 second delay after he notices the near miss danger.
- • Axiom 5: There exist some suitable orders that the two airplanes will not in the near miss state, namely, they will keep a safety separation for 180 seconds.
- – Axiom 6: The order will be voided after 15 seconds.
- – Axiom 7: TCAS of the airplane A outputs resolution advisory, and orders to climb or descend some value X, after D_3 seconds when it recognizes the near miss danger.
- – Axiom 8 is similar. It is for the airplane B.
- – Axiom 9: Each airplane operates to climb or descend some value X with D_5 seconds delay after the controller at the ACC sends the control order.
- – Axiom 10: The airplane must follow the order from its own TCAS if there is no emergency control from ACC.
- – Axiom 11 is the differential equation of the actual movement when the airplane operates the value X, i.e. the pair of values, v and h, where v is the vertical speed to climb or descend and h is vertical position to go.
- – Fact 12: The monitor on the panel of the controller anticipated the danger of near miss at 15:54′15″, where S indicates the actual starting time of the whole system.
- – Fact 13: The pilot or the crew of the airplane A recognized at 15:54′27″ that the controller ordered to descend to FL350.
- – Fact 14: The crew of the airplane recognized that the order was emergent.
- – Fact 15: The crew, on the other hand, recognized at 15:54′35″ that TCAS of A output resolution advisory to climb at 1500ft per minute. Of course these two operations contradicts, and the crew followed that of ACC for the axioms 9 and 10.
- – Fact 16: At the same time, the crew of B recognized the order from its own TCAS to descend at 1500 ft per minute.
- – Fact 17 is the initial condition. At the start time of the system, there are no emergency control and no conflict alert.
- – Facts 18-20 are recognition by the controller.
  - • Fact 18: Though the Fact 13 (and 14), the controller believed, or misrecognized, that he ordered B, instead of A, to descend to FL350.
  - • Fact 19: The controller believed that if he has ordered as Fact 18, the near miss accident would occur.
  - • Fact 20: These two airplane held their vertical and horizontal speed.

In these axioms and facts, the program labels exist implicitly and do not represent explicitly.

5.3 Verification and Analysis

Lemma 1.

$$\exists XY(\uparrow (cntl(A, X) \vee cntl(B, Y)) = 15{:}54'15'' + D_0 + D_1 + D_2@S). \quad (7)$$

Intuitively, it means that there existed some operation from the controller to A or B at 15:54'15" with $D_0 + D_1 + D_2[s]$ delay.

[Proof.] From Axiom 1 and Fact 12,

$$\uparrow CNF = 15{:}54'15'' + D_0@S. \quad (8)$$

From Axiom 3,

$$\uparrow CNF@ACC = 15{:}54'15'' + D_0 + D_1@S. \quad (9)$$

From Axiom 2,

$$\uparrow anticipateNM(A, \ B)@ACC = 15{:}54'15'' + D_0 + D_1@S.$$

Therefore (7) is obtained from Axiom 4.   □

Actually, putting $D_0 + D_1 + D_2 = 8[s]$, Fact 13 satisfies Lemma 1.

Lemma 2.

$$ope(A, \langle \bot, 350[FL] \rangle)@15{:}54'27'' + D_5@S \quad (10)$$
$$\& \, ope(B, \langle -1500ft/min., \bot \rangle)@15{:}54'34'' + D_6@S.$$

It means that the airplane A operated to go (descend) to 350[FL] at 15:54'27" with $D_5[s]$ delay, and B did to descend at 1500ft per minute at 15:54'34" with $D_6[s]$ delay.

[Proof.] From Axiom 9 and Fact 13,

$$\uparrow ope(A, \langle \bot, \ 350[FL] \rangle) = D_5@15{:}54'27''@S.$$

On the other hand, from the conservation axioms,

$$\neg Ecntl@B@15{:}54'34''@S. \quad (11)$$

Thus from Axiom 10 and Fact 16,

$$\uparrow ope(B, \langle -1500ft/min., \bot \rangle) = D_6@ \, 15{:}54'34''@S. \quad (12)$$

□

Let us analyze the accident for applying the actual values to Lemma 2 and Axiom 11. The actual values of the vertical positions a_v and b_v of these planes at 15:54'34" are both FL370. From Axiom 2, both airplanes descended between 15:54'27" and 15:54'34". And also, when they started to descend, the horizontal positions and speed of these planes were $a_h = 5[nm]$, $b_h = -5[\text{nm}]$, $\dot{a}_h = 500[kt]$ and $\dot{b}_h = -500[kt]$, where the origin is the position that the near miss occurred. Applying these values to Axiom 11 and integrating it, the following theorem representing that the near miss really occurred is obtained.

Theorem 1.

$$\diamond_{180} \neg Separation_{7,5}. \quad (13)$$

Table 3. Axiom when TCAS are given priority over ACC

index	condition/prefix	action	tense	label
9'		$\gamma_P = ope(P, X) = D_5$	$\uparrow RA(P, X)@P$	$*,\ P,\ ACC$
10'	$\neg RA(P, Y)$	$\gamma_P = ope(P, X) = D_6$	$\uparrow cntl(P, X)@P$	$*,\ P,\ ACC$

5.4 Some Other Cases

1. *Defending the controller:*
   As Facts 13 and 18, the trainee controller wanted to order B to descend, but he actually did to A. If he ordered correctly, the near miss did not occur, i.e.

$$\Box_{180} Separation_{7,5}, \tag{14}$$

   which is the negation of (13), is derived.
   On the other hand, the defense of the controller argues in court that the order by the controller is still safe, since no one know the order by TCASs, and if they ordered nothing, then the near miss would not occur. Namely, Facts 15-16 do not hold. From the conservation axiom [12], $\dot{b}_v = 0$ holds, therefore (14) is also obtained.

2. *A carried out RA instead of ACC:*
   If the crew of A carried out RA by $TCAS_A$ instead of ACC, i.e. axioms 9' and 10' (Table 3) held instead of 9 and 10, if then A climbed while B carried out the order of $TCAS_B$ to descend, thus (14) also holds.

6 Conclusion

This paper has demonstrated an analysis of a concrete serious near miss accident of airplanes by $N\Sigma$-labeled calculus that is a formal system for complicated control systems involving human factor, especially misunderstanding and inappropriate decision.

If the order from the controller at ACC contradicts that from TCAS, the report from the Investigating Committee for Avion and Traffic Accident [10] recommends that the crew must follow the latter, as mentioned in the Section 5.4. And also, on Jul. 1, 2002, a similar airplane accident occurred at Überlingen [2]. A crew of one airplane followed the descending order from the controller, though there was a climbing order from its own TCAS. On the other hand, the crew of another airplane descended to follow the order from its own TCAS. Consequently, these two airplanes collided.

From the investigations of these accidents, every crew now must follow orders from TCAS if orders from it and from controllers contradicts. Namely, axioms 9' and 10' in Table 3 are substituted for axioms 9 and 10. Following to these new rules, the fact that there is no near miss accident by such a conflict is proved.

One of our future studies is to construct some safety designs to control airplanes, trains, vehicles, etc., against misunderstanding and incorrect decision.

References

1. Avion Safety Network (ed.): ASN Aircraft accident McDonnell Douglas DC-10-40 JA8546 off Shizuoka Prefecture (2005), http://aviation-safety.net/database/record.php?id=20010131-2
2. Avion Safety Network (ed.): ASN Aircraft accident Tupolev 154M RA-85816 Überlingen (2007), http://aviation-safety.net/database/record.php?id=20020701-0
3. Curzon, P., Ruksenas, R., Blandford, A.: An Approach to Formal Verification of Human-Computer Interaction. Formal Aspect of Computing 19, 513–550 (2007)
4. Damm, W., Hungar, H., Olderog, E.-R.: Verification of Cooperating Traffic Agents. International Journal of Control 79, 395–421 (2006)
5. Fagin, R., Halpern, J.Y., Moses, Y., Vardi, M.Y.: Reasoning About Knowledge. The MIT Press, Cambridge (1995)
6. Gentzen, G.: Untersuchungen über das logische Schließen. Mathematische Zeitschrift 39, 176–210, 405–431 (1935); Investigations into Logical Deduction. In: Szabo, M.E. (ed.) The Collected Papers of Gerhard Gentzen, Series of Studies in Logic and the Foundations of Mathematics, pp. 68–131. North-Holland Publ. Co., Amsterdam (1969)
7. Halpern, J.Y., Vardi, M.Y.: The Complexity of Reasoning about Knowledge and Time. I. Lower Bounds. Journal of Computer and System Sciences 38, 195–237 (1989)
8. Igarashi, S., Mizutani, T., Ikeda, Y., Shio, M.: Tense Arithmetic II: @-Calculus as an Adaptation for Formal Number Theory. Tensor, N. S. 64, 12–33 (2003)
9. Mizutani, T., Igarashi, S., Ikeda, Y., Shio, M.: Formal Analysis of an Airplane Accident in $N\Sigma$-labeled Calculus. In: Deng, H., Wang, L., Wang, F.L., Lei, J. (eds.) AICI 2009. LNCS (LNAI), vol. 5855, pp. 469–478. Springer, Heidelberg (2009)
10. Investigating Committee for Avion and Traffic Accident (ed.): Report of the Near Miss Accident by JA8904 belonging to Japan Airlines with JA8546 belonging to the Same Company (2002) (in Japanese)
11. McCarthy, J., Sato, M., Hayashi, T., Igarashi, S.: On the Model Theory of Knowledge. Stanford University Technical Report, STN-CS-78-657 (1979)
12. Mizutani, T., Igarashi, S., Shio, M., Ikeda, Y.: Human Factors in Continuous Time-Concerned Cooperative Systems Represented by $N\Sigma$-labeled Calculus. Frontiers of Computer Science in China 2, 22–28 (2008)
13. Presburger, M.: Über die Vollständigkit eines gewissen Systems der Arithmetik ganzer Zahlen in welchem die Additon als einzige Operation hervortritt. In: Comptes-Rendus du Congres des Mathematicians des Pays Slaves, pp. 92–101 (1930)
14. Shoenfield, J.R.: Mathematical Logic. Addison-Wesley Publishing Company, Reading (1967)

Using Concept Space to Verify Hyponymy in Building a Hyponymy Lexicon

Lei Liu[1], Sen Zhang[1], Lu Hong Diao[1], Shu Ying Yan[2], and Cun Gen Cao[2]

[1] College of Applied Sciences, Beijing University of Technology
[2] Institute of Computing Technology, Chinese Academy of Sciences
{liuliu_leilei,zhansen,diaoluhong}@bjut.edu.cn,
{yanshuying,caocungen}@ict.edu.cn

Abstract. Verification of hyponymy relations is a basic problem in knowledge acquisition. We present a method of hyponymy verification based on concept space. Firstly, we give the definition of concept space about a group of candidate hyponymy relations. Secondly we analyze the concept space and define a set of hyponymy features based on the space structure. Then we use them to verify candidate hyponymy relations. Experimental results show that the method can provide adequate verification of hyponymy.

Keywords: hyponymy, relation acquisition, knowledge acquisition, hyponymy verification.

1 Introduction

Automatic acquisition of semantic relations from text has received much attention in the last ten years. Especially, hyponymy relations are important in accuracy verification of ontologies, knowledge bases and lexicons [1][2].

Hyponymy is a semantic relation between concepts. Given two concepts X and Y, there is the hyponymy between X and Y if the sentence "X is a (kind of) Y" is acceptable. X is a hyponym of Y, and Y is a hypernym of X. We denote a hyponymy relation as hr(X, Y), as in the following example:

中国是一个发展中国家 ---hr(中国,发展中国家)

(China is a developing country ---hr(China, developing country))

Human knowledge is mainly presented in the format of free text at present, so processing free text has become a crucial yet challenging research problem. The error hyponymy relations in the phase of acquiring hyponymy from free text will affect the building of hyponymy lexicon.

In our research, the problem of hyponymy verification is described as follows:

Given a set of candidate hyponymy relations acquired based on pattern or statistics, we denoted these relations as CHR= {(c_1, c_2), (c_3, c_4), (c_5, c_6), ...}, where c_i is the concept of constituting candidate hyponymy relation. The problem of hyponymy verification is how to identify correct hyponymy relations from CHR using some specific verify methods.

H. Deng et al. (Eds.): AICI 2009, LNAI 5855, pp. 479–486, 2009.

In this paper, we present a method of hyponymy verification based on concept space. The rest of the paper is organized as follows. Section 2 describes related work in the area of hyponymy acquisition, section 3 gives the definition of concept space and analyzes the structure of concept space, section 4 gives a group of hyponymy features and presents how to verify candidate hyponymy relations, section 5 conducts a performance evaluation, and finally section 6 concludes the paper.

2   Related Work

There are two main approaches for automatic/ semi-automatic hyponymy acquisition. One is pattern-based (also called rule-based), and the other is statistics-based. The former uses the linguistics and natural language processing techniques (such as lexical and parsing analysis) to obtain hyponymy patterns, and then makes use of pattern matching to acquire hyponymy, and the latter is based on corpus and statistical language model, and uses clustering algorithm to acquire hyponymy.

At present the pattern-based approach is dominant, and its main idea is the hyponymy can be extracted from text as they occur in detectable syntactic patterns. The so-called patterns include special idiomatic expressions, lexical features, phrasing features, and semantic features of sentences. Patterns are acquired by using the linguistics and natural language processing techniques.

There have been many attempts to develop automatic methods to acquire hyponymy from text corpora. One of the first studies was done by Hearst [3]. Hearst proposed a method for retrieving concept relations from unannotated text (Grolier's Encyclopedia) by using predefined lexico-syntactic patterns, such as

...NP_1 is a NP_2...   ---hr(NP_1, NP_2)
...NP_1 such as NP_2...   ---hr(NP_2, NP_1)
...NP_1 {, NP_2}*{,} or other NP_3 ...   ---hr (NP_1, NP_3), hr (NP_2, NP_3)

Other researchers also developed other ways to obtain hyponymy. Most of these techniques are based on particular linguistic patterns.

Morin and Jacquemin produced partial hyponymy hierarchies guided by transitivity in the relation, but the method works on a domain-specific corpus [4].

Llorens and Astudillo presented a technique based on linguistic algorithms, to construct hierarchical taxonomies from free text. These hierarchies, as well as other relationships, are extracted from free text by identifying verbal structures with semantic meaning [5].

Sánchez presented a novel approach that adapted to the Web environment, for composing taxonomies in an automatic and unsupervised way. It uses a combination of different types of linguistic patterns for hyponymy extraction and carefully designed statistical measures to infer information relevance [6].

Elghamry showed how a corpus-based hyponymy lexicon with partial hierarchical structure for Arabic can be created directly from the Web with minimal human supervision. His method bootstraps the acquisition process by searching the Web for the lexico-syntactic patterns [7].

3 Concept Space

In Chinese, one may find several hundreds of different hyponymy relations patterns based on different quantifiers and synonymous words, which is equivalent to the single hyponymy pattern (i.e. (<?C1> is a <?C2>), (<?C3> such as <?C1>,<?C2>)) in English. Fig. 1 depicts a few typical Chinese hyponymy relation patterns.

Pattern 1: <?C1><是一I为一><种I个I名I篇I片I块I部I颗I本I...><?C2>
(Pattern: 〈?C1〉 is a 〈?C2〉)

Pattern 2: <?C3><如I象I包括I包含I囊括I涵盖><?C1>{<或I及I以及I和I与I、 ><?C2>}*<等>
(Pattern: 〈?C3〉 such as 〈?C1〉,〈?C2〉...)

Fig. 1. Defining Chinese hyponymy patterns

In Fig. 2, Pattern 1 means "Pattern: <?C1> is a <?C2>". Pattern 2 means "Pattern: <?C3> such as <?C1>,<?C2>...". These items of the pattern are divided into constant item and variable item. Constant item is composed of one or more Chinese words or punctuations. Variable item is a non-null string variable. "<?C1>" is a variable item in the pattern. "I" expresses logical "or". In pattern1,"是一I为一"means "is a"; "种I个I名 I篇I片I块I部I颗I本..." is a group of quantifiers. In pattern 2, "如I象" means "such as"; "包括I包含I囊括I涵盖" means "include"; "和I与" means "and"; "或" means "or"; "等 " means "etc."; "及I以及" denotes "as well as"; Chinese dunhao "、" is a special kind of Chinese comma used to set off items in a series.

Chinese hyponymy relation patterns will be used to capture concrete sentences from Chinese free corpus. In this process, variables <?C> will be instantiated with words or phrases in a sentence, in which real concepts may be located. Let c and c' be the real concept in <?C>. If hr(c, c') is true, then we tag c by c_L, and c' by c_H, as shown below.

{众所周知，{中国}$_{c_L}$<?C1>/是一个/{社会主义{国家}$_{c_H}$}<?C2>

({It is well-known that {China}$_{c_L}$ <?C1>/is a/{ socialist nation}$_{c_H}$ }<?C2>)

{{农作物}$_{c_H}$主要}<?C5>/有/{{水稻}$_{c_L}$}<?C1>、{{玉米}$_{c_L}$}<?C2>、{{红薯}$_{c_L}$}<?C3>、{{烟叶}$_{c_L}$}<?C4>/等/

(The farm crop mainly includes paddy rice, corn, sweet potato, tobacco leaves etc.)

We can acquire hr(中国,国家), hr(水稻,农作物), hr(玉米,农作物), hr(红薯,农作物) and hr(烟叶,农作物) from the above example.

As we know, Chinese is a language different from any western language. A Chinese sentence consists of a string of characters which do not have any space or delimiter in between [8]. So there are still many error relations in the acquired hyponymy relations from free text. They must be verified further for the building of hyponymy lexicon.

Firstly we initially acquire a set of candidate hyponymy relation from large Chinese free text based on Chinese lexico-syntactic patterns. Then we build concept space using those candidate hyponymy relations.

Definition 1: The concept space is a directed graph $G = (V, E, W)$ where nodes in V represent concepts of the hyponymy and edges in E represent relationships between concepts. A directed edge (c_1, c_2) from c_1 to c_2 corresponds to a hyponymy from concept c_1 to concept c_2. Edge weights in E are used to represent varying degrees of certainty.

For a node c in a graph, we denote by $I(c)$ and $O(c)$ the set of in-neighbors and out-neighbors of v, respectively. Individual in-neighbors are denoted as $I_i(c)$, for $<=i<=$ $|I(c)|$, and individual out-neighbors are denoted as $O_i(c)$, for $1 <= i <= |O(c)|$.

The basic control process about building concept space is shown in Algorithm 1.

Algorithm 1. The basic process of building concept space

Input: the set of candidate hyponymy relations CHR from large Chinese free text based on Chinese lexico-syntactic patterns;

Output: the concept space G.

Step1: Initialize $G = (V, E, W)$, let $V=\varnothing$, $E=\varnothing$, $W=\varnothing$;

Step2: For each $(c_1,c_2)\in$ CHR, continue Step3-Step4 ;

Step3: If $c_1\notin V, c_2\notin V$, then $V=V\cup\{c_1,c_2\}$; $E=E\cup\{(c_1,c_2)\}$; If $c_1\notin V$, $c_2\in V$, then $V=V\cup\{c_1\}$; $E=E\cup\{(c_1,c_2)\}$; If $c_1\in V$, $c_2\notin V$, then $V=V\cup\{c_2\}$; $E=E\cup\{(c_1,c_2)\}$;

Step4: CHR= CHR-$\{(c_1,c_2)\}$;

Step5: For each $r\in E$, set its $w(r) \in W$ to be 0.

Step6: return G ;

----

4 Space Structure Features

When a group of candidate hyponymy relations are correct or error, they often satisfy some space structure feature. If a candidate hyponymy satisfies a certain threshold with matching those features, we think that it is a real hyponymy.

We analysed the influence of the space structure on verification of hyponymy relations. In space structure analysis, we use the coordinate relation between concepts. The coordinate relations are acquired using a set of coordinate relation patterns including "、". Chinese dunhao "、" is a special kind of Chinese comma used to set off items in a series. For example:

In a sentence of matching a coordinate pattern, if there exists concept c_1 and concept c_2 divided by "、", then c_1 and c_2 are coordinate, denoted as $cr(c_1, c_2)$. An example is as shown below.

农作物主要有{水稻}c_1、{玉米}c_2、{红薯}c_3、{烟叶} c_4等

(The farm crop mainly includes paddy rice, corn, sweet potato, tobacco leaves etc..)

cr(水稻, 玉米, 红薯, 烟叶) (cr(paddy rice, corn, sweet potato, tobacco leaves)) is acquired from the above example.

Fig. 2 depicts a few typical structure features of hyponymy.

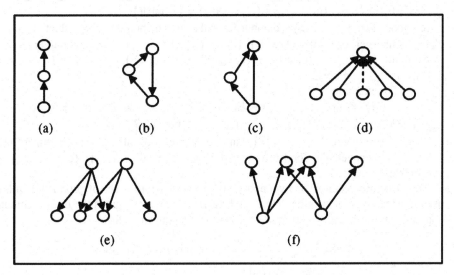

Fig. 2. Typical structure features of hyponymy

Structure (a): $(c_1, c_2), (c_2, c_3)$.
  For example:
    (苹果, 水果), (水果, 食品) ((apple, fruit), (fruit, foodstuff))

Structure (b): $(c_1, c_2), (c_2, c_3), (c_3, c_1)$.
  For example:
(游戏, 生活), (生活, 童话), (童话, 游戏)
    ((game, life), (life, fairy tale), (fairy tale, game))

Structure (c): $(c_1, c_2), (c_2, c_3), (c_1, c_3)$.
  For example:
    (番茄, 蔬菜), (番茄, 食品), (蔬菜, 食品)
    ((tomato, vegetable), (tomato, foodstuff), (vegetable, foodstuff))

Structure (d): $(c_1, c), (c_2, c), ..., (c_m, c), cr(c_1, c_2, ..., c_m)$,
    $(c'_1, c), (c'_2, c), ..., (c'_n, c), cr(c'_1, c'_2, ..., c'_n)$,
    $\{c_1, c_2, ..., c_m\} \cap \{c'_1, c'_2, ..., c'_n\} \neq \varnothing$.
  For example:
c=职业, cr(时装, 休闲装, 礼服), cr(裤装, 时装, 休闲服) $\{c_1,..., c_m\} \cap \{c'_1,..., c'_n\}$
= {时装, 休闲装}
    (c= clothing, cr(fashionable dress, sportswear, full dress), cr(trousers, fashionable dress, sportswear), $\{c_1,..., c_m\} \cap \{c'_1, ..., c'_n\}$ = {fashionable dress, sportswear})

  Structure (e): $(c_1, c), (c_2, c), ..., (c_m, c), (c'_1, c'), (c'_2, c'), ..., (c'_n, c'), \{c_1, c_2, ..., c_m\} \cap \{c'_1, c'_2, ..., c'_n\} \neq \varnothing$.
  For example:

c=食品, {c_1,..., c_m}={牛肉饼, 蛋糕, 面包, 奶油}, c'=产品, {c'_1,..., c'_n}={牛肉饼, 牛肉干, 牛肉汤}, {c_1,..., c_m}∩{c'_1,..., c'_n} = {牛肉饼}

(c= foodstuff, {c_1,..., c_m}={hamburger, cake, bread, butter}, c'=product, {c'_1,..., c'_n}={hamburger, beef jerky, brewis}, {c_1,..., c_m}∩{c'_1, ..., c'_n} = {hamburger})

Structure (f): (c, c_1,), (c, c_2), ..., (c, c_m), (c, c'_1), (c, c'_2), ..., (c, c'_n) {c_1, c_2, ..., c_m}∩{c'_1, c'_2, ..., c'_n}≠∅.

For example:

c=西红柿, {c_1,..., c_m}={植物, 蔬菜, 食品, 果实}, c'=茄子, {c'_1,..., c'_n}={蔬菜, 食品, 食材}, {c_1,..., c_m}∩{c'_1,..., c'_n} = {蔬菜, 食品}

(c=tomato, {c_1,..., c_m}={plant, vegetable, foodstuff, fruit}, c'= aubergine, {c'_1,..., c'_n}={vegetable, foodstuff, food for cooking}, {c_1,..., c_m}∩{c'_1, ..., c'_n} = {vegetable, foodstuff})

The structure features of hyponymy are converted into a set of production rules used in uncertainty reasoning. We use CF (certainty factors) that is the most common approach in rule-based expert system. The CF formula is as follows:

$$CF(CHR, f) = \begin{cases} \dfrac{P(CHR|f) - P(CHR)}{1 - P(CR)}, & P(CHR|f) \geq P(CHR) \\ \dfrac{P(CHR|f) - P(CHR)}{P(CHR)}, & P(CHR|f) < P(CHR) \end{cases} \qquad (1)$$

Where CHR is a set of candidate hyponymy, which has a precision P(CHR). P(CHR|f) is the precision of a subset of CHR satisfying feature f. CF is a number in the range from -1 to 1. If there exists CF(CHR, f)≥0, then we denote f as positive feature and CF(CHR, f) denotes the support degree of feature f; if there exists CF(CHR, f)<0, then we denote f as negative feature and CF(CHR, f) denotes the no support degree of feature f.

For example, P(CHR)=0.71, the precision of candidate hyponymy relations satisfying the structure feature (b) is 23%, namely P(CHR|f)=0.23, the result of CF is (0.23-0.74)/ 0.74= -0.69. The f is a negative feature.

The precision of candidate hyponymy relations satisfying the structure feature (c) is 96%, namely P(CHR|f_b)=0.94, the result of CF is (0.96-0.71)/(1- 0.71)=0.86. The f is a positive feature.

After those structrue features are converted into a set of production rules, we can carry uncertainty reasoning in concept space.

5   Evaluation

5.1   Evaluation Method

We adopt three kinds of measures: R (Recall), P (Precision), and F (F-measure). They are typically used in information retrieval.

Let h be the total number of correct hyponymy relations in the CHR.
Let h_1 be the total number of hyponymy relations in the classify set.
Let h_2 be the total number of correct hyponymy relations in the classified set.

(1) Recall is the ratio of h_2 to h, i.e. $R = h_2/h$
(2) Precision is the ratio of h_2 to h_1, i.e. $P = h_2/h_1$
(3) F-measure is the harmonic mean of precision and recall. It is high when both precision and recall are high. $F = 2RP/(R+P)$

5.2 Experimental Results

We used about 8GB of raw corpus from the Chinese Web pages. Raw corpus is preprocessed in a few steps, including word segmentation, part of speech tagging, and splitting sentences according to periods. Then we acquired candidate hyponymy relations CHR from processed corpus by matching Chinese hyponymy patterns. For analyzing the influence of threshold a, we choose several different values. We manually evaluated 10% initial set CHR and 10% final result. The detailed result is shown in Table 1.

Table 1. The result of Verification

Concept space: 62,265 hyponymy relations P(CHR)=71%				
	The result of verified hyponymy			
α	The number	P	R	F
α=0	56,155(90.2%)	76%	97%	0.85
α=0.2	46,574(74.8%)	82%	86%	0.84
α=0.4	34,365(55.2%)	87%	68%	0.76
α=0.6	20,102(32.3%)	92%	42%	0.58
α=0.8	12,510(20.1%)	95%	27%	0.42

As we can see from table 1, there are 62,265 hyponymy relations in concept space initially. With the increase of threshold α, the precision is also increase. If we want to increase the precision, we can augment γ value. For example, when α=0.8, the precision is up to 95%, and but its recall decreases to 27%. That is to say, when threshold α is a small value, our methods can throw away many error hyponymy relations under the condition of skipping a few correct relations. But when threshold α is a large value, our methods can throw away many error hyponymy relations and also skip many correct relations at same time.

6 Conclusion

For the verification of candidate hyponymy relations, we present a method of hyponymy verification based on concept space. Experimental results demonstrate good performance of the method. It will raise the precision of hyponymy relations and benefit to the building of ontologies, knowledge bases and lexicons.

There are still some inaccurate relations in the result. In future, we will combine some methods (such as web page tag etc.) to the further verification of hyponymy.

Acknowledgments. This work is supported by the National Natural Science Foundation of China under Grant No.60573064, and 60773059; the National 863 Program under Grant No. 2007AA01Z325, and the Beijing University of Technology Science Foundation (grant nos. X0006014200803, 97006017200701).

References

1. Beeferman, D.: Lexical discovery with an enriched semantic network. In: Proceedings of the Workshop on Applications of WordNet in Natural Language Processing Systems, ACL/COLING, pp. 358–364 (1998)
2. Cao, C., Shi, Q.: Acquiring Chinese Historical Knowledge from Encyclopedic Texts. In: Proceedings of the International Conference for Young Computer Scientists, pp. 1194–1198 (2001)
3. Hearst, M.A.: Automated Discovery of WordNet Relations. In: Fellbaum, C. (ed.) To Appear in WordNet: An Electronic Lexical Database and Some of its Applications, pp. 131–153. MIT Press, Cambridge (1998)
4. Morin, E., Jacquemin, C.: Projecting corpus-based semantic links on a thesaurus. In: Proceedings of the 37th Annual Meeting of the Association for Computational Linguistics, pp. 389–396 (1999)
5. Lloréns, J., Astudillo, H.: Automatic generation of hierarchical taxonomies from free text using linguistic algorithms. In: Advances in Object-Oriented Information Systems, OOIS 2002 Workshops, Montpellier, France, pp. 74–83 (2002)
6. Sánchez, D., Moreno, A.: Pattern-ed automatic taxonomy learning from the Web. AI Communications 21(3), 27–48 (2008)
7. Elghamry, K.: Using the Web in Building a Corpus-Based Hypernymy-Hyponymy Lexicon with Hierarchical Structure for Arabic. Faculty of Computers and Information, 157–165 (2008)
8. Zhang, C.-x., Hao, T.-y.: The State of the Art and Difficulties in Automatic Chinese Word Segmentation. Journal of System simulation 17(1), 138–143 (2005)

Advanced Self-adaptation Learning and Inference Techniques for Fuzzy Petri Net Expert System Units*

Zipeng Zhang[1], Shuqing Wang[1], and Xiaohui Yuan[2]

[1] Hubei University of Technology, Wuhan, 430068, China
[2] Huazhong University of Science and Technology, Wuhan, 430074, China
zhzip1@163.com

Abstract. In a complicated expert reasoning system, it is inefficient for commonly fuzzy production rules to depict the vague and modified knowledge. Fuzzy Petri nets are more accurate for dynamic knowledge proposition in describing expert knowledge. However, the bad learning ability of fuzzy Petri net constrains its application in dynamic knowledge expert system. In this paper, an advanced self-adaptation learning way based on error back-propagation is proposed to train parameters of fuzzy production rules in fuzzy Petri net. In order to enhance reasoning and learning efficiency, fuzzy Petri net is transformed into hierarchy model and continuous functions are built to approximate transition firing and fuzzy reasoning. Simulation results show that the designed advanced learning way can make rule parameters arrive at optimization rapidly. These techniques used in this paper are quite effective and can be applied to most practical Petri net models and fuzzy expert systems.

Keywords: Fuzzy Petri net, dynamic fuzzy reasoning, neural network, self-adaptation learning.

1 Introduction

In intelligent diagnoses system, since most knowledge is fuzzy and general logic rules cannot describe it effectively. Fuzzy Petri net (FPN), which is based on fuzzy production rules (FPRs), can provide us a useful way to properly represent fuzzy knowledge [1], [2]. Many results prove that FPN is suitable to represent and reason misty logic implication relations [3],[4].

However, Knowledge in expert systems is updated or modified frequently and expert systems may be regarded as dynamic systems. The models must have the ability to adjust themselves according to the systems' changes. But general FPN which lacks learning ability cannot cope with potential changes of actual systems[5]. Artifical neural network(ANN) has strong adaptation and learning ability[6]. It is effective to combine the superiority of ANN into FPN to make FPN has self-adaptation learning ability to adapt dynamic expert systems.

* This work is supported by NNSFC Grant 50539140 & 50779020 and NSF of Hubei Grant 2008CDB395.

H. Deng et al. (Eds.): AICI 2009, LNAI 5855, pp. 487–496, 2009.

In order to overcome the disadvantage of learning speed slow and local optimization in general back-propagation learning way, adaptive learning techniques are used to learn and train parameters of FPRs in FPN. After a training process, an excellent input-output map of the knowledge system can be gained.

2   Fuzzy Petri Nets

It is difficult to directly monitor the contamination degree of insulator. Here, an indirect measurement method is used to diagnose insulator running state. In the diagnosing, the easy measured leakage current, pulse frequency, the equivalent environment humidity are used to analyse and diagnose insulator running state[8]. Those signals are measured with on-line mode and then are used to analysis with off-line mode. The correlative data may be analysed or processed and display through graphics in the back computer.

FPRs may depict the fuzzy relationships between many propositions. Let R be a set of FPRs, $R = \{R_1, R_2, \cdots R_n\}$. The general formulation of the jth layer is given as follows: R_j : IF d^j THEN $d^{j+1}(CF = \mu^j)$

Where, d^j and d^{j+1} are propositions, the truth of each proposition is a real value between zero and one. μ^j is the value of the certainty factor(CF), $\mu^j \in [0,1]$ [7].

If the antecedent portion or consequence portion of a FPR contains "and" or "or" connectors, then it is called a composite FPRs[8],[9]. The structure of composite FPRs is shown in Fig. 1.

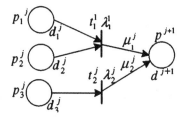

Fig. 1. Structure a composite fuzzy production rule

A FPN is a bipartite directed graph containing two types of nodes: places and transitions, where circles represent places and bars represent transitions. Every FPR may be expressed as a transition of FPN. The proposition of production rule is expressed as place relatively. In a FPN, the relationships from places to transitions and from transitions to places are represented by directed arcs.

Definition 1: A FPN has n layers noses and every layer has different places and transitions. The FPN may be described using a 10-tuple[10],[11].

$$FPN = (P, T, D, I, O, f, M, W, Th, \beta) \tag{1}$$

Where, $P = \{p_1^1, \cdots, p_{a_1}^1, p_1^2, \cdots, p_{a_2}^2, \cdots, p_1^n, \cdots, p_{a_n}^n\}$ is a finite set of places.

$T = \{t_1^1, \cdots, t_{b1}^1, t_1^2, \cdots, t_{b2}^2, \cdots, t_1^n, \cdots, t_{bn}^n\}$ is a finite set of transitions.

$D = \{d_1^1, \cdots, d_{a_1}^1, d_1^2, \cdots, d_{a_2}^2, \cdots, d_1^n, \cdots, d_{a_n}^n\}$ is a finite set of propositions, and $P \cap T \cap D = \Phi$, $|P| = |D|$.

$I : T \rightarrow P^\infty$ is the input function, a mapping from transitions to bags of places, which determines the input places to a transition, and P^∞ denotes bags of places.

$O : T \rightarrow P^\infty$ is the output function, a mapping from transitions to bags of places, which determines the output places form a transition.

$f : T \rightarrow [0,1]$ is an association function, a mapping from transitions to real values between 0 and 1, $f(t_b^j) = \mu_b^j$.

$M : P \rightarrow [0,1]$ is an association function, every place $p_i \in P$ has a sign $m(p_i)$, which replaces the true degree.

$W = \{\omega_1^1, \cdots, \omega_{b_1}^1, \omega_1^2, \cdots, \omega_{b_2}^2, \cdots, \omega_1^n, \cdots, \omega_{b_n}^n\}$ is a set of input weights and output weights which assign weights to all the arcs of a net.

$Th(t_b^j) = \{\lambda_1^1, \cdots, \lambda_{b1}^1, \lambda_1^2, \cdots, \lambda_{b2}^2, \cdots, \lambda_1^n, \cdots, \lambda_{bn}^n\}$ is a set of threshold value from zero to one to transition.

$\beta : P \rightarrow D$ is an association function, a relationship mapping from places to propositions.

3 Dynamic Reasoning Process of FPN

3.1 Fuzzy Reasoning Algorithm

Firstly, some basic definitions are given to explain the transition firing rule of FPN.

Definition2: if $\forall p_i^j \in I(t^j)$, $m(p_i^j) > 0 (i=1,2,\cdots;a \quad j=1,2,\cdots n)$ then $\forall t \in T, t$ is enabled.

Definition3: $\forall t^j \in T$, $\forall p_i^j \in I(t^j)$, if $\sum_i m(p_i^j) \cdot \omega_i^j > Th(t^j)$ then transition fires, at the same time, token transmission takes place.

Definition4: When transition is enabled, it produces a new certainty factor $CF(t)$.

$$CF(t^j) = \begin{cases} \sum_i m(p_i^j) \cdot \omega_i^j, & \sum_i m(p_i^j) \cdot \omega_i^j > Th(t^j) \\ 0, & \sum_i m(p_i^j) \cdot \omega_i^j < Th(t^j) \end{cases} \tag{2}$$

Sigmoid function $F(x)$ is used to approximates the threshed of t^j.

$$F(x) = 1/(1 + e^{-b(x-Th(t^j))})$$ (3)

Where, $x = \sum_i m(p_i^j) \cdot \omega_i^j$, b is instant. When b is chosen properly, if $x > Th(t^j)$, then

$e^{-b(x-Th(t^j))} \approx 0$ and $F(x) = 1$, which denotes transition t is enable. In the other hand,

if $x < Th(t^j)$, then $e^{-b(x-Th(t^j))} \approx 1$ and $F(x) = 0$, which denotes transition t is not

enable and the sign of output place is 0.

3.2 Making Continuous Function of Fuzzy Reasoning

In order to implement learning and fuzzy reasoning effectively, continuous function is used in FPN. Where, $CF(t)(x)$ is utilized to approximate the new certainty factor of output place.

$$CF(t)(x) = x \cdot F(x)$$ (4)

Where, $x = \sum_i m(p_i^j) \cdot \omega_i^j$, $F(x) = 1/(1 + e^{-b(x-Th(t^j))})$

Definition 5: If transition fires, the token of input places does not vary and output place produces new token.

  1) If a place only has one input transition, a new token with certainty factor CF is put into each output place, new token sign is given as:

$$m(p^{j+1}) = f(t^j) \times CF(t^j)(x^j)$$ (5)

Where $p_i^j \in I(t^j) i = 1, 2 \cdots m, p^{j+1} \in O(t^j)$

  2) If a place has more than one input transitions $t_z^j (z = 1, 2, \cdots, c)$ and more than one route are active at the same time, then the new certainty factor is decided by the maximal passed sign of the fired transitions:

$$m(p^{j+1}) = \max(f(t_1^j) \times CF(t_1^j)(x_1^j), f(t_2^j) \times CF(t_2^j)(x_2^j), \\ \cdots, f(t_c^j) \times CF(t_c^j)(x_c^j))$$ (6)

3.3 Process of Fuzzy Reasoning

Definition 6 (Source Places, Sink Places): A place p is called a source place if it has no input transitions. It is called a sink place if it has no output transitions. A source place corresponds to a precondition proposition in FPN, and a sink place corresponds to a consequent.

Definition 7(input place, output place): The set of places P is divided into three parts $P = P_{UI} \cup P_{int} \cup P_O$, where P is the set of places of FPN, $P_{UI} = \{p \in P | p = \Phi\}$, $p \in P_{UI}$ is called a user input place; $P_{int} = \{p \in P | p \neq \Phi \text{ and } p \neq \Phi\}$, $p \in P_{int}$ is called an interior place; $P_O = \{p' = \Phi\}$, $p \in P_o$ is called an output place. Where, Φ is an empty set.

Definition 8(initially enabled transition, current enabled transition.): Let $T_{initital} = \{t \in T | t \cap P_{UI} \neq \Phi \ \ and \ \ t \cap P_{int} = \Phi\}$, $t \in T_{int\ itial}$ is called an initially enabled transition. Let $T_{currem\overline{n}\overline{t}} \{t \in T_{initial} | \forall p_i \in t, m(p_i) > 0, \ \ CF(t) > Th(t)\}$, $t \in T_{current}$ is called a current enabled transition [10],[11].

In order to describe the reasoning process of FPN, an FPN example is given here and its structure is shown in Fig.2. The reasoning system contains three layers. $d_1^1, d_2^1, d_1^2, d_2^2, d_3^2, d_1^3, d_2^3, d_3^3, d_4^3, d^4$ are related propositions of the expert system. Between them there exist the following weighted FPRs.

R^1: IF d_1^1 or d_2^1 THEN d_1^2 ($\lambda_1^1, \mu_1^1, \lambda_2^1, \mu_2^1$)

R_1^2: IF d_1^2 and d_2^2 THEN d_2^3 ($\lambda_1^2, \mu_1^2, \omega_{11}^2, \omega_{12}^2$)

R_2^2: IF d_1^2 and d_3^2 THEN d_4^3 ($\lambda_2^2, \mu_2^2, \omega_{21}^2, \omega_{22}^2$)

R^3: IF d_1^3 and d_2^3 and d_3^3 THEN

d^4 ($\lambda^3, \mu^3, \omega_1^3, \omega_2^3, \omega_3^3$)

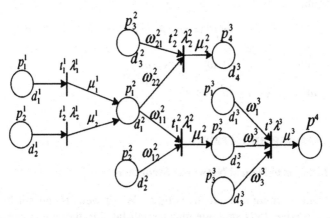

Fig. 2. Structure of the conceived FPN

The conceived FPN according above FPRs is described as follow.
$FPN = (P, T, D, I, O, f, M, W, Th, \beta)$

In this FPN, $p_1^1, p_2^1, p_2^2, p_3^2, p_1^3, p_3^3$ are initial places, p_4^3 and p^4 are output places. The first layer transition set contains t_1^1, t_2^1, the second layer transition set contains t_1^2, t_2^2 and the third layer transition set contains t^3. The process of fuzzy reasoning is given as follow.

1) If $m(p_1^1) > \lambda_1^1$ then t_1^1 fire, or if $m(p_2^1) > \lambda_2^1$ then t_2^1 fire and then the true degree of p_1^2 may be calculated using follow formula.

$$m(p_1^2) = \max(f(t_1^1) \times x_1^1 \times F(x_1^1),$$
$$f(t_2^1) \times x_2^1 \times F(x_2^1)) \tag{7}$$

Where, $x_1^1 = m(p_1^1)$, $x_2^1 = m(p_2^1)$, $F(x) = 1/(1 + e^{-b(x - T(t))})$.

2) The second layer transitions fire, if t_1^2, t_2^2 fires singly, the true degree of p_2^3, p_4^3 may be calculated singly through follow formula.

$$m(p_2{}^3) = x_2^2 \times F(x_2^2) \tag{8}$$

$$m(p_4{}^3) = x_4^2 \times F(x_4^2) \tag{9}$$

Where $x_2^2 = m(p_1^2) \cdot \omega_{11}^2 + (p_2^2) \cdot \omega_{12}^2$, $\omega_{11}^2 + \omega_{12}^2 = 1$,

$F(x_2^2) = \mu_1^2 / (1 + e^{-b(x_2^2 - \lambda_1^2)})$, $x_4^2 = m(p_3^2) \cdot \omega_{21}^2 + (p_1^2) \cdot \omega_{22}^2$,

$\omega_{21}^2 + \omega_{22}^2 = 1$, $F(x_4^2) = \mu_2^2 / (1 + e^{-b(x_4^2 - \lambda_2^2)})$.

3) The last layer transitions fire, the true degree of p^4 may be calculated through follow formula.

$$m(p^4) = x^3 \times F(x^3) \tag{10}$$

Where $x^3 = m(p_1^3) \cdot \omega_1^3 + m(p_2^3) \cdot \omega_2^3 + m(p_3^3) \cdot \omega_3^3$,

$\omega_1^3 + \omega_2^3 + \omega_3^3 = 1$, $F(x^3) = \mu^3 / (1 + e^{-b(x^3 - \lambda^3)})$.

4 FPN Learning and Training

4.1 Translating into Neural Networks Structure

ANN has strong learning and adaptive ability. When weights can not be ascertained via expert knowledge, FPN structure may be translated further into a neural networks-like structure. The back-propagation way with multi-layered feed-forward networks of ANN is used to training FPN model. The trained weight threshold value, and true degree of FPRs may be used to analyze fuzzy reasoning system.

Supposing there are n layers networks, if t^j is the jth layer transition of FPN, the weights of input arc are supposed as ω_1^j, $\omega_2^j \cdots$, ω_m^j, the threshold value of t^j is λ^j and the true degree is μ^j. the network output of nth layer node is expressed as:

$$G(x^j) := F(x^j) \cdot x^j \tag{11}$$

Where, $x^j = W^T \times M = \sum_{i=1}^{m} m(p_i^j)\omega_i^j$, $j = 1,2 \cdots, n$,

$F(x^j) = \mu^j /(1 + e^{-b(x^j - \lambda^j)})$, $W = [\omega_1^j, \omega_2^j, \cdots \omega_m^j]$,

$M = [m(p_1^j), m(p_2^j), \cdots m(p_m^j)]$.

The last layer output may be computed via the last node of network.

$$O(R) = G(x^n) = M(p^n) \tag{12}$$

4.2 Improved Learning Algorithm

The error function E is defined as:

$$E = \frac{1}{2} \sum_{l=1}^{r} \sum_{o=1}^{b} (O(R) - O^*)^2 \tag{13}$$

Where, r is number of samples and b is output places. The learn ways are given as follow.

$$\omega_i^j(k+1) = \omega_i^j(k) - \eta \frac{dE}{d\omega_i^j}, i = 1,2 \cdots m-1, j = 1,2 \cdots n \tag{14}$$

$$\omega_m^j(k+1) = 1 - \sum_{i=1}^{m-1} \omega_i^j(k+1) \tag{15}$$

Where, $\dfrac{dE}{d\omega_i^j}$ may be computed through the follow way.

If the FPN model has n layers, the last layer weight varying $\dfrac{dE}{d\omega_i^n}$ may be computed as follow.

$$\frac{dE}{d\omega_i^n} = \frac{dE}{dO(R)} \times \frac{dO(R)}{d\omega_i^n} = \frac{dE}{dG(x^n)} \times \frac{dG(x^n)}{d\omega_i^n}$$

$$= \sum_{l=1}^{r} \sum_{o=1}^{b} (O(R) - O^*) \times \frac{dG(x^n)}{dx^n} \frac{dx^n}{d\omega_i^n} \tag{16}$$

Where, $\dfrac{dG(x^n)}{dx^n}$ and $\dfrac{dx^n}{d\omega_i^n}$ may be calculated using follow formula:

$$\frac{dG(x^n)}{dx^n} = \frac{\mu^n}{1 + e^{-b(x^n - \lambda^n)}} + \frac{\mu^n x^n b e^{-b(x^n - \lambda^n)}}{(1 + e^{-b(x^n - \lambda^n)})^2} \tag{17}$$

$$\frac{dx^n}{d\omega_i^n} = m(p_i^n) \tag{18}$$

The error term of the other layers may be computed through the same back-propagation method.

In training, the choice of the learning rate η has an important effect on weights convergence. If it is set too small, too many steps are needed to reach an acceptable solution. On the contrary, a large learning rate will possibly lead to oscillation, preventing the error to fall below a certain value. To overcome the inherent disadvantages of pure gradient-descent, an adaptation way of the weight-updates according to the behavior of the error function is applied to training weights.

Here, Individual update-value $\Delta_i(k)$ is introduced for the weight-updating. The adaptive update-value $\Delta_i(k)$ evolves during the learning process based on its local sight on the error function $\frac{dE}{d\omega_i^j}(k)$ and $\frac{dE}{d\omega_i^j}(k-1)$. $\Delta\omega_i^j(k)$ and $\omega_i^j(k+1)$ are obtained according to the following formula.

$$\Delta\omega_i^j(k) = -\frac{dE}{d\omega_i^j}(k) * \Delta_i^j(k) \tag{19}$$

$$\omega_i^j(k+1) = \omega_i^j(k) + \Delta\omega_i^j(k) \tag{20}$$

At the beginning, initial value $\Delta_i^j(0)$ should be set to all update-values $\Delta_i^j(k)$. For $\Delta_i^j(0)$ directly determines the size of the first weight-step, it is preferably chosen in a reasonably proportion to the size of the initial weights. Where, $\Delta_i^j(0) = 0.19$ is very good.

5 Training Example

In this section, the above given fuzzy expert reasoning system (shown as Fig.2) is used as example to illustrate the learning effect of above introduced way.

In the example, assume the ideal parameters are given as:

$\lambda_1^1 = \lambda_2^1 = \lambda_1^2 = \lambda_2^2 = \lambda^3 = 0.5$, $\mu_1^1 = 0.8$, $\mu_2^1 = 0.9$, $\mu_1^2 = 0.9$, $\mu_2^2 = 0.8$, $\mu^3 = 0.9$, $\omega_{11}^2 = 0.35$, $\omega_{12}^2 = 0.65$, $\omega_{21}^2 = 0.25$, $\omega_{22}^2 = 0.75$, $\omega_1^3 = 0.15$, $\omega_2^3 = 0.55$, $\omega_3^3 = 0.3$.

If weights are unknown, neural networks technique are used to estimate the weights. The learning part of the FPN may be formed as a standard sub single layer ANN and a two layers sub neural network. Fig.3 and Fig.4 show the perfection of adaptation learning techniques.

From the simulation example, we can see that the fuzzy reasoning algorithm and the ANN training algorithm are very effectively when weights of FPN are not known. Fig.3 and Fig.4 show that self-adaptation learning way is very quick to find optimization weights to make error function E to least. Self-adaptation learning way overcomes the disadvantage of learning speed slow and local optimization in common learning way.

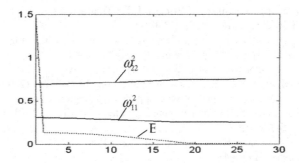

Fig. 3. Training perfection using adaptation way

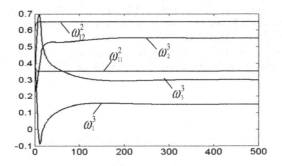

Fig. 4. Training perfection using adaptation way

6 Conclusion

In expert system, the fuzzy reasoning process based on FPRs is difficult to obtain effective knowledge. In order to produce effective reasoning and learning ability, FPN without loop is transformed into hierarchy model in this paper. Through building continuous functions, the approximate transition firing and fuzzy reasoning may be acquire effectively, which provides powerful facility for FPN to reason and learn both forward and backward. In training, self-adaptation learning way has been used in learning and training the parameters of FPN. Simulation results show that designed self-adaptation learning way is very quick to find optimal weights. Therefore the designed reasoning and learning way of FPN is effective for dynamic knowledge inference and learning in expert system.

References

[1] Chen, S.M., Ke, J.S., Chang, J.F.: Knowledge representation using fuzzy Petri nets. IEEE Trans. Knowledge Data Engineering 2, 311–319 (1990)

[2] Cao, T., Sanderson, A.C.: Representation and Analysis of Uncertainty Using Fuzzy Petri Nets. Fuzzy System 3, 3–19 (1995)

[3] Lee, J., Liu, K.F.R., Chiang, W.: A Fuzzy Petri Net-Based Expert System and Its Application to Damage Assessment of Bridges. IEEE Transactions on Systems, Man, and Cybernetics -Part B: Cybernetics 29(3), 350–369 (1999)

[4] Scarpelli, H., Gomide, F., Yager, R.R.: A Reasoning Algorithm for High-level Fuzzy Petri Nets. IEEE Trans. Fuzzy System 4(3), 282–293 (1996)

[5] Li, X., Lara-Rosano, F.: Adaptive Fuzzy Petri Nets for Dynamic Knowledge Representation and Inference. Expert Systems with Applications 19, 235–241 (2000)

[6] Zengren, Y.: Artificial Neural Network Application. Qinghua University Press (1999)

[7] Bugarn, A.J., Barro, S.: Fuzzy reasoning supported by Petri nets. IEEE Trans. Fuzzy System 2(2), 135–150 (1994)

[8] Pedrycz, W., Gomide, F.: A Generalized Fuzzy Petri Nets Model. IEEE Trans. Fuzzy System 2, 295–301 (1994)

[9] Li, X., Yu, W.: Object Oriented Fuzzy Petri Net for Complex Knowledge System Modeling. In: IEEE international conference on control applications, September 2001, pp. 476–481 (2001)

[10] Yeung, D.S., Tsang, E.C.C.: A Multilevel Weighted Fuzzy Reasoning Algorithm for Expert Systems. IEEE Transactions on Systems, Man, and Cybernetics-Part A: Systems and Humans 28(2), 149–158 (1998)

[11] Scarepelli, H., Gomide, F.: A High Level Net Approach for Discovering Potential Inconsistencies in Fuzzy Knowledge Bases. Fuzzy Sets System 62(2), 175–193 (1994)

MIMO Instantaneous Blind Identification Based on Second-Order Temporal Structure and Newton's Method

Xizhong Shen[1,2,*] and Guang Meng[2]

[1] Mechanical & Electrical department, Shanghai Institute of Technology, Shanghai, China 200235
[2] State Key Laboratory for Mechanical System and Vibration, Shanghai Jiao Tong University, Shanghai, China 200030
shenxizhong@online.sh.cn, gmeng@sjtu.edu.cn

Abstract. This paper presents a new MIMO instantaneous blind identification algorithm based on second order temporal property and Newton's Method. Second order temporal structure is reformulated in a particular way such that each column of the unknown mixing matrix satisfies a system of nonlinear multivariate homogeneous polynomial equations. The nonlinear system is solved by Newton's method for the equations. Our algorithm allows estimating the mixing matrix for scenarios with 4 sources and 3 sensors, etc. Simulations and comparisons show its effective with more accurate solutions than the algorithm with homotopy method.

Keywords: instantaneous blind identification, homogeneous system, second order temporal statistics, Newton's method, nonlinear.

1 Introduction

Blind signal processing (BSP) is an important task for numerous applications such as speech separation, dereverberation, communications, signal processing and control, etc. Its research has obtained extensive interests all over the world and many exciting results have been reported, [1] and references therein. Its task is to recover the source signals only given the observations in the sense of some uncertain problems such as scaling, permutation and/or delay.

Multiple-input multiple-output (MINO) instantaneous blind identification (MIBI) is one of the attractive BSP problems, where a number of source signals are mixed by an unknown MIMO instantaneous mixing system and only the mixed signals are available, i.e., both the mixing system and the original source signals are unknown. The goal of MIBI is to recover the instantaneous MIMO mixing system from the observed mixtures of the source signals [1][2]. In this paper, we focus on developing a new algorithm to solve the MIBI problem by using second-order statistics and Newton's method.

---

* Shen Xizhong, Ph.D, born in 1968, as a professor in Shanghai Institute of Technology, and also as a visitor/post-doc in Shanghai Jiao Tong University. He is going on studying blind signal processing, neural network, etc.

H. Deng et al. (Eds.): AICI 2009, LNAI 5855, pp. 497–506, 2009.

Many researchers have investigated the use of second-order temporal structure (SOTS) for MIBI, or BSP [1][3][4][5][6]. The greater majority of the available algorithms are based on the generalized eigenvalue decomposition or joint approximate diagonalization of two or more sensor correlation matrices for different lags and/or times arranged in the conventional manner. An MIBI based on second order temporal structure (SOTS) [2] has been proposed, which arrange the available sensor correlation values in a particular fashion that allows a different and nature formulation of the problem, as well as the estimation of the more columns than sensors. Our work is a continuation and improvement of their work presented in [2].

In this paper, we further develop the algorithm proposed in [2] to obtain more accurate and robust solution. At first, we exploit the SOTS by considering the sensor correlation functions on a noise-free region of support (ROS). Then we project the MIBI problem on the system of homogeneous polynomial equations of degree two. At last Newton's method is applied to estimate the columns of the mixing matrix, which is different from the algorithm in [2] which applied homotopy method. The MIBI method presented in this paper allows estimating the mixing matrix for several underdetermined mixing scenarios with 4 sources and 3 sensors. Simulations and comparisons show its effectivity with more accurate and robust solutions.

2   MIBI Model and Its Assumptions

Let us use the usual model [1][2] in MIBI problem as follows

$$\mathbf{x}(t) = \mathbf{As}(t) + \mathbf{v}(t). \tag{1}$$

where $\mathbf{A} = [\mathbf{a}_1, \cdots, \mathbf{a}_m] \in \mathbb{R}^{n \times m}$ is an unknown mixing matrix with its n-dimention array response vectors $\mathbf{a}_j = \begin{pmatrix} a_{1j} & \cdots & a_{nj} \end{pmatrix}^{\mathrm{T}}, j = 1, 2, \cdots, m$, $\mathbf{s}(t) = \begin{bmatrix} s_1(t), s_2(t), \cdots, s_m(t) \end{bmatrix}^{\mathrm{T}}$ is the vector of source signals, $\mathbf{v}(t) = \begin{bmatrix} v_1(t), \cdots, v_n(t) \end{bmatrix}^{\mathrm{T}}$ is the vector of noises, and $\mathbf{x}(t) = \begin{bmatrix} x_1(t), x_2(t), \cdots, x_n(t) \end{bmatrix}^{\mathrm{T}}$ is the vector of observations.

Without knowing the source signals and the mixing matrix, the MIBI problem is to identify the mixing matrix from the observations by estimating $\mathbf{A}$ as $\hat{\mathbf{A}}$, and if we set the following linear transformation

$$\mathbf{y}(t) = \hat{\mathbf{A}}^+ \mathbf{x}(t). \tag{2}$$

where $\mathbf{y}(t) = \begin{bmatrix} y_1(t), y_2(t), \cdots, y_m(t) \end{bmatrix}^{\mathrm{T}}$ is m-dimensional output vector at time t, whose elements are the estimations of sources, and $\hat{\mathbf{A}}^+$ is the Moore-Penrose inverse of $\hat{\mathbf{A}}$, which is a demixing matrix named in BSP.

The mixing matrix is identifiable in the sense of two indeterminacies, which are unknown permutation of indices of each column of the matrix and its unknown magnitude. When signal s_i is multiplied by a scalar, this is equivalent to rescaling the corresponding column of $\mathbf{A}$ by the scalar. Therefore, the scalar of each column remains undetermined. The usual convention is to assume that each column satisfy the normalization conditions, i.e., on the unit sphere,

$$\sum_{i=1}^{n} a_{ij}^{2} = 1, j = 1, 2, \cdots, m. \tag{3}$$

We set it as the following form for the usage in section IV,

$$S_j(\mathbf{a}_j) = \sum_{i=1}^{n} a_{ij}^{2} - 1 = 0; j = 1, 2, \cdots, m. \tag{4}$$

It should be noted that m sources cannot be determined in their exact order. It is also unknown which the first column of $\mathbf{A}$ is and which the second is, and thus its permutation of indices of each column of the matrix is indeterminate.

To solve the MIBI problem, we first define the following concepts Def 1~2 for the derivation of the algorithm, and then make the following assumptions AS 1~4 [2].

Definition 1. Autocorrelation function $r_u(t, \tau)$ of $u(t)$ at time instant t and lag τ is defined as

$$r_u(t, \tau) \triangleq E\left[u(t)u(t-\tau)\right], \forall t, \tau \in \mathbb{Z}. \tag{5}$$

Definition 2. Cross-correlation function $r_{u,v}(t, \tau)$ of $u(t), v(t)$ at time instant t and lag τ is defined as

$$r_{u,v}(t, \tau) \triangleq E\left[u(t)v(t-\tau)\right], \forall t, \tau \in \mathbb{Z}. \tag{6}$$

AS 1. the source signals have zero cross-correlation on the noise-free ROS Ω:

$$r_{s, j_1 j_2}(t, \tau) = 0, \forall 1 \le j_1 \ne j_2 \le m. \tag{7}$$

AS 2. the source autocorrelation functions are linearly independent on the noise-free ROS Ω

$$\sum_{j=1}^{m} \xi_j r_{s, jj}(t, \tau) = 0 \Rightarrow \xi_j = 0, \forall j = 1, 2, \cdots, m \tag{8}$$

AS 3. the noise signals have zero auto- and cross- correlation functions on the noise-free ROS Ω:

$$r_{n, j_1 j_2}(t, \tau) = 0, \forall 1 \le j_1, j_2 \le m. \tag{9}$$

AS 4. the cross-correlation functions between the source and noise signals are zero on the noise-free ROS Ω:

$$r_{vs,ij}\left(t,\tau\right)=r_{sv,ji}\left(t,\tau\right)=0,\forall 1\leq i\leq n,1\leq j\leq m. \tag{10}$$

Here, no assumptions are made on the mixing matrix, and even rank-deficient one can be identified. This is a significant advantage with respect to other methods because of the two reasons as follows: (1) most IBSS may not only require that the number of sensors is larger than the number of sources, but also that the mixing matrix is full rank; (2) no assumptions are made on the probability density functions of the noise and source signals.

The procedure of our proposed algorithm includes two steps, that is, step 1 is that the problem of MIBI is formulated as the problem of solving a system of homogeneous polynomial equations; and step 2 is that Newton's method is applied to solve the system of polynomial equations. We detail these steps respectively in section III and IV.

3 Projection on Homogeneous Polynomial Equations

In this section, we will review the algebraic structure of MIBI problem derived under the above assumptions, and some details can be referred to [2]. The correlation values of the observations are stacked as

$$\mathbf{R}_{x,\Diamond}\triangleq\left[\mathbf{r}_x\left(t_1,\tau_1\right) \quad \cdots \quad \mathbf{r}_x\left(t_N,\tau_N\right)\right], \tag{11}$$

where $\mathbf{r}_x\left(t_N,\tau_N\right)=\mathrm{E}\left[\mathbf{x}(t_N)\otimes\mathbf{x}(t_N-\tau_N)\right]$, and $\otimes$ denotes the Kronecker product. Consider the eq. (1), we get

$$\mathbf{R}_{x,\Diamond}=\mathbf{A}_{\Diamond}\mathbf{R}_{s,\odot}. \tag{12}$$

Here, $\mathbf{A}_{\Diamond}$ is the second-order Khatri-Rao product of $\mathbf{A}$, which is defined as $\mathbf{A}_{\Diamond}\triangleq\left[\mathbf{a}_1\otimes\mathbf{a}_1 \quad \cdots \quad \mathbf{a}_m\otimes\mathbf{a}_m\right]$, and $\mathbf{R}_{s,\odot}\triangleq\left[\mathbf{r}_s\left(n_1,k_1\right) \quad \cdots \quad \mathbf{r}_s\left(n_N,k_N\right)\right]$ with $\mathbf{r}_s\left(n_q,k_q\right)\triangleq\mathrm{E}\left[\mathbf{s}(n_q)\odot\mathbf{s}(n_q-k_q)\right],q=1,\cdots,N$ where $\odot$ denotes the elementwise product. The homogeneous polynomial equations of degree two are expressed as

$$\mathbf{\Phi}\mathbf{A}_{\Diamond}=\mathbf{0}. \tag{13}$$

Here, $\mathbf{\Phi}=\left(\varphi_{q,i_1i_2}\right)_{q=1,\cdots,Q;i_1,i_2=1,\cdots,n;i_1\leq i_2}$ is a matrix with $\left(\dfrac{1}{2}n\left(n+1\right)-rank\left[\mathbf{R}_{x,\Diamond}\right]\right)\times\left(n^2\right)$ dimensions, of which its rows form a basis for the nonzero left null space $N\left(\mathbf{R}_{x,0}\right)$. Therefore, there are Q equations about each column of $\mathbf{A}$ in (13).

$\mathbf{R}_{x,0}$ is splitted into signal and noise subspace parts as $\mathbf{R}_{x,0} = \mathbf{U}_s\boldsymbol{\Sigma}_s\mathbf{V}_s^{\mathsf{T}} + \mathbf{U}_v\boldsymbol{\Sigma}_v\mathbf{V}_v^{\mathsf{T}}$. $\boldsymbol{\Phi}$ can be obtained by SVD of $\mathbf{R}_{x,0}$, that is, the left null space of $\mathbf{R}_{x,0}$ is $\boldsymbol{\Phi} = \mathbf{U}_v^{\mathsf{T}}$.

The zero contour level of a n-variate homogeneous function is a $(n-1)$-dimension surface embedded in an n-dimension Euclidian space. In an n-dimension Euclidian space, at least $(n-1)$ surfaces of degree $(n-1)$ are required to uniquely define one-dimensional space intersections. The dimension of the solution set of a system of $(n-1)$ homogeneous polynomials in n variables has one-dimension. By eq.(13), the maximum number $M_{\max}$ of columns that can be identified with n sensors equals

$$M_{\max} = \frac{1}{2}n(n+1)-(n-1). \tag{14}$$

$M_{\max}$ is the maximum number of the sources to be identified by SOTS.

4 Newton's Method

In this section, we summarize the main ideas behind the so-called Newton's method that transforms a system of nonlinear equations into a convergent fixed-point problem in a general way.

Newton's method for nonlinear system is able to give quadratic convergence, provided that a sufficiently accurate starting value is known and the inversion of the Jacobian matrix of the nonlinear equations exists [7]. Certainly, we could use the quasi-Newton's method to avoid the calculation of inversion; however we use Newton's method in our algorithm for the number of equations and variables involved are small and its inversion of the corresponding Jacobian matrix is simple.

We expand the expression in (13) as

$$f_q\left(\mathbf{a}_j\right) = \sum_{i_1 \le i_2 ; i_1 i_2 = 1,\cdots,n} \varphi_{q;i_1 i_2} a_{i_1 j} a_{i_2 j} = 0;$$
$$q = 1,\cdots,Q; \forall j = 1,\cdots,m \tag{15}$$

and set $\mathbf{F}\left(\mathbf{a}_j^{(k)}\right) = \begin{bmatrix} f_1\left(\mathbf{a}_j\right) & \cdots & f_Q\left(\mathbf{a}_j\right) \end{bmatrix}^{\mathsf{T}}$. By Newton's method, for each column of the mixing matrix we have

$$\mathbf{a}_j^{(k+1)} = \mathbf{a}_j^{(k)} - \mathbf{J}^{-1}\left(\mathbf{a}_j^{(k)}\right)\mathbf{F}\left(\mathbf{a}_j^{(k)}\right). \tag{16}$$

where $\mathbf{J}\left(\mathbf{a}_j\right)$ is the Jacobian matrix of $\mathbf{F}\left(\mathbf{a}_j^{(k)}\right)$.

We employ the initial solutions as equal distributed vectors in the super sphere defined on the super space of $\mathbf{a}_j$, for example, in our simulation of mixing matrix with 3×4 sizes,

$$\mathbf{A} = \begin{bmatrix} 1 & 0 & 0 & 1 \\ 0 & 1 & 0 & 1 \\ 0 & 0 & 1 & 1 \end{bmatrix}. \tag{17}$$

Thus, the different solutions are easily obtained. Alternatively, to get all the solutions of (16), we set the initial values given by 2^{N+1} vectors with its entries random normally distributed values in [-1,1]. And then, after running the algorithm, we get 2^{N+1} solutions and select the centers of all the solutions.

The procedure of Newton's method for nonlinear equations can be described as follows:

Step 1. Set initial approximation in (17) of an initial solutions $\mathbf{a}_j^{(0)}, j = 1, 2, \cdots, m$;

Step 2. Set tolerance TOL and maximum number of iterations maxIteration;

Step 3. For each solution $\mathbf{a}_j^{(k)}$, calculate $\mathbf{a}_j^{(k+1)}$ by (16) until $\left\|\mathbf{a}_j^{(k+1)} - \mathbf{a}_j^{(k)}\right\|_2$ is less than a certain tolerance or the number of iterations reaches to the maximum value maxIteration.

5   Examples with Speech and Three Sensors

To compare our proposed algorithm with the algorithm proposed in [2], we adopt three mixtures of four speech signals the same example as in [2]. For convenience of comparison, we name our improved algorithm in (16) as MIBI_NWT; and the algorithm in [2] as MIBI_Homotopy. We also develop the algorithm by applying steepest descent method [7] to solve the equations in (15), in which we construct objective function as the sum of square of each function in (15). We name the algorithm with steepest descent method as MIBI_SD, and it is not detailed here for we think it as an intermediate result.

The speech source signals are sampled as 8kHz, consist of 10,000 samples with 1,250ms length, and are normalized to unit variance $\sigma_s = 1$. The signal sequences are partitioned into five disjoint blocks consisting of 2000 samples, and for each block, the one-dimensional sensor correlation functions are computed for lags 1, 2, 3, 4 and 5. Lag zeros are omitted because the corresponding correlation values are noise contaminated. Hence, in total for each sensor correlation functions 25 values are estimated and employed, i.e., the employed noise-free ROS in the domain of block-lag pairs is given by

$$\Omega = \{(1,1), \cdots, (1,5), (2,1), \cdots, (5,5)\},$$

where the first index in each pair represents the block index and the second the lag index. The sensor signals are obtained from (1) with 3×4 mixing matrix,

$$\mathbf{A} = \begin{bmatrix} 0.6749 & 0.4082 & 0.8083 & -0.1690 \\ 0.5808 & -0.8165 & 0.1155 & -0.5071 \\ 0.4552 & -0.4082 & -0.5774 & 0.8452 \end{bmatrix}. \tag{18}$$

The noise signals are mutually statistically independent white Gaussian noise sequences with variances $\sigma_v^2 = 4$. We set the maximum iterative number is 20, and stop the iteration step if the correction of the estimated is smaller than a certain tolerance 10^{-4}.

The signal-to-noise ratio (SNR) is defined as the ratio of the variance of the noise-free part of the sensor observation vector to the variance of the noise vector, i.e.,

$$SNR \triangleq 10 \log \frac{\sum_i \sigma_{s_i}^2}{\sum_i \sigma_{v_i}^2}. \tag{19}$$

Hence, the SNR is -4.7881dB, which is quite bad.

Fig. 1 depicts the intersections of each zero contour level on the unit sphere. There are two blue solid curves as the spherical ellipses (1 and 2 in the figure) corresponding to the case $q = 1$ in eq.(15), and two magenta solid curves (3 and 4 in the figure) to the case $q = 2$. There are also the column vectors of estimated mixing matrix $\hat{\mathbf{A}}$ indicated by solid lines and dots, and the column vectors of $\mathbf{A}$ by dashed lines and stars.

Let θ_j, $j = 1, 2, 3, 4$ be the included angle between the j-th column of $\mathbf{A}$ and its estimate. The estimated mixing matrix is

$$\hat{\mathbf{A}} = \begin{bmatrix} 0.6347 & 0.3784 & 0.8379 & -0.1198 \\ 0.5652 & -0.8785 & 0.2034 & -0.5111 \\ 0.5270 & -0.2915 & -0.5066 & 0.8511 \end{bmatrix},$$

and the included angles are 4.8002, 7.7664, 6.6909 and 2.8494. We see that the estimated columns approximately equal the ideal ones by comparison with the matrix in (18).

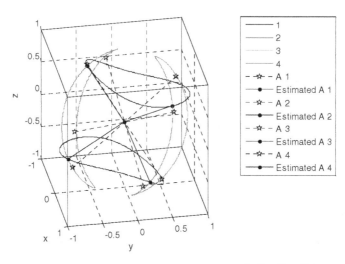

Fig. 1. Zero contour spherical ellipses of $q = 1, 2$, $Q = 2$

Solving a system consisting of trivariate homogeneous polynomial equations in (13) of degree two is equivalent to finding the intersections between the corresponding two-dimensional quadric cones embedded in a three-dimensional Euclidian space. Because such cones can intersect in at most four directions, the maximum number of columns that can be identified with three sensors equals four, and is equal to the maximum number of different intersections between two spherical ellipses, depicted in Fig. 1.

Fig.2 shows the Comparisons of MIBI_NWT with MIBI_Homotopy and MIBI_SD. TL, TR, BL and BR in Fig.2 are respectively the estimated included angles along different running times between the first, second, third and fourth columns and their estimates. Red dot indicates the result of MIBI_NWT; magenta plus the results of MIBI_SD and blue circle indicates the results of MIBI_homotopy. We run them by 10 times with random generated noises each time, and use the same initial values given by 2^{N+1} vectors with its entries random normally distributed values in [-1,1]. The parameters in each algorithm are adjusted for comparison to get the most accurate solution. In MIBI_NWT, we set the maximum iterative number is 20, and stop the iteration step if the correction of the estimated is smaller than a certain tolerance 10^{-4}. In MIBI_homotopy, we set the start system as in [2], and $\lambda_s = 0$,

$\lambda_e = 1 + j$ and $\Delta\lambda = \dfrac{1}{10}(1 + j)$. From Fig.2, we see that the estimated included

angles by MIBI algorithm with Newton's method are smaller than MIBI _homotopy and MIBI_SD. Therefore, MIBI_NWT has better performance than MIBI_homotopy and MIBI_SD.

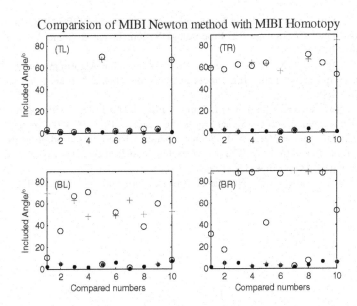

Fig. 2. Comparisons of MIBI_NWT with MIBI_SD and MIBI_Homotopy. TL, TR, BL and BR are respectively the estimated included angles along different running times between the first, second, third and fourth columns and their estimates. Red dot indicates the result of MIBI_NWT; magenta plus the results of MIBI_SD and blue circle indicates the results of MIBI_homotopy.

6 Conclusion

In this paper, we further develop the algorithm proposed in [2] to obtain more accurate solution. The SOTS is considered only on a noise-free region of support (ROS). We project the MIBI problem in (1) onto the nonlinear system of homogeneous polynomial equations in (13) of degree two. The nonlinear system is solved by Newton's method for the equations, which is quite different from the algorithm in [2] which applied homotopy method. Our algorithm allows estimating the mixing matrix for scenarios with 4 sources and 3 sensors, etc. Simulations and comparisons show its effective with more accurate solutions than the algorithm with homotopy method.

Acknowledgments. This paper is supported by the National Science Foundation of China with the project number 10732060, and Shanghai Leading Academic Discipline Project directed by Hu Dachao with the project number J51501, and also Shanghai Education with No. ZX2006-01.

References

1. Cichocki, A., Amari, S.I.: Adaptive Blind Signal and Image Processing: Learning Algorithms and Applications. Wiley, New York (2002)
2. van de Laar, J., Moonen, M., Sommen, P.C.W.: MIMO Instantaneous Blind Identification Based on Second-Order Temporal Structure. IEEE Transactions on Signal Processing 56(9), 4354–4364 (2008)

3. Shen, X., Shi, X.: On-line Blind Equalization Algorithm of an FIR MIMO channel system for Non-Stationary Signals. IEE Proceedings Vision, Image & Signal Processing 152(5), 575–581 (2005)

4. Shen, X., Haixiang, X., Cong, F., et al.: Blind Equalization Algorithm of FIR MIMO System in Frequency Domain. IEE Proceedings Vision, Image & Signal Processing 153(5), 703–710 (2006)

5. Hua, Y., Tugnait, J.K.: Blind identifiability of FIR-MIMO systems with colored input using second order statistics. IEEE Signal Processing Letters 7(12), 348–350 (2000)

6. Lindgren, U., van der Veen, A.-J.: Source separation based on second order statistics—An algebraic approach. In: Proc. IEEE SP Workshop Statistical Signal Array Processing, Corfu, Greece, June 1996, pp. 324–327 (1996)

7. Burden, R.L., Faires, J.D.: Numerical Analysis, pp. 600–635. Thomson Learning, Inc. (2001)

Seed Point Detection of Multiple Cancers Based on Empirical Domain Knowledge and K-means in Ultrasound Breast Image

Lock-Jo Koo*, Min-Suk Ko, Hee-Won Jo, Sang-Chul Park, and Gi-Nam Wang

Department of Industrial Engineering, Ajou University,
442-749, Suwon, Kyunggido, Korea
{lockjo9,sebaminsuk11,happy6654,scpark,gnwang}@ajou.ac.kr

Abstract. The objective of this paper is to remove noises of image based on the heuristic noises filter and to automatically detect seed points of tumor region by using K-MEANS in breast ultrasound. The proposed method is to use 4 different kinds of process. First process is the pixel value which indicates the light and shade of image is acquired as matrix type. Second process is an image preprocessing phase that is aimed to maximize a contrast of image and to prevent a leak of personal information. The next process is the heuristic noise filter which is based on the opinion of medical specialist and it is applied to remove noises. The last process is to detect a seed point automatically by applying K-MEANS algorithm. As a result, the noise is effectively eliminated in all images and an automated detection is possible by determining seed points on each tumor.

Keywords: Decision Support System, Breast ultrasound image, Seed points, Computer Aided Diagnosis.

1 Introduction

Breast cancer has been becoming a growth rate of 2% in the world since 1990. According to the Ministry of Health & Welfare, South Korea and American Cancer Society, breast cancer ranks second in the list of women's cancers. Moreover, in most countries of the world, it is on a higher position of women's cancers [1]-[3].

Early detection of breast cancer is important for the identifiable treatment. However, the segmentation and interpretation of ultrasound image is based on the experience and knowledge of a medical specialist and speckles make a more difficult to distinguish a ROI (Region Of Interests). As a result, more than 30% of masses referred for surgical breast biopsy were actually malignant [4, 5].

CAD (Computer Aided Diagnosis) is defined as a diagnosis that is made by a radiologist, who uses the output from a computerized analysis of medical images, as a 'second opinion' in detecting lesions and in making diagnostic decisions. Most of the previous research related with CAD of breast cancer has been focused on segmentation of lesion and eliminate noises on image. To remove noises, anisotropic diffusion filter,

---

* Corresponding author.

H. Deng et al. (Eds.): AICI 2009, LNAI 5855, pp. 507–516, 2009.
© Springer-Verlag Berlin Heidelberg 2009

second order Butterworth filter, watershed transform etc. are used [4-7]. Each method for removal is used by considering five features of ultrasound images which are consisted area, circularity, protuberance, homogeneity, and acoustic shadow. For segmentation of lesion, algorithms based on region growing method are used because most tumors are included in breast ultrasound images have a multiplicity of shapes & colors, difference of pixel value between tumor and background image. According to the region growing methods, in a case where an objective pixel and the adjacent pixels have an identical feature using histogram, gradient based pixel value of image sequentially executed processing integrates the pixels to form an area. Areas having the same feature are gradually grown to achieve segmentation of the entire image. However, Most of the previous researches have some limitation as follows.

1. The pixel value and tumor shape on ultrasound image are only considered for noise removal, except for location of tumor on the image.
2. Only one ROI (Region of Interests) is segmented, or manual tasks are needed for detection of plural tumors because one seed point is considered for region grow method.

In this paper, the noise removal method based on the heuristic noise filtering related to the lesion position on image and the detection method of one or more seed points using K-MEANS algorithm is proposed. For accomplishment of this goal, the organization of paper is as follows. Section II is the introduction of the each step that is used for noises removal. The detection of seed points in the breast ultrasound image is present in Section III. Then the experimental results and conclusion are discussed in Section IV.

2 Noise Removal

Our database consists of 40 consecutive ultrasound cases, being represented by 64 images. These images were obtained during diagnostic breast exams at the Park Breast Clinic in Korea. Most of the cases were included either benignancy or malignancy. In addition, the included information of these images is outpatient's name, distinction of sex, size of lesion and so on. Out of the images of 40 cases, 13 were benign lesions; the other 27 were malignant lesions which have been chosen for the purpose of testing. The size of the tumors were between 0.5~3cm.

In this section, methods for noises removal on this paper are briefly described by step-by-step process. The proposed method is consisted of four steps. Each step is explained as follows.

Before proceeding further, first, the mean (μ) variance (σ), minimum (Min) and maximum (Max) pixel value are defined as follows. In Eq.(1), (2), (3), (4), I (i,j) defines the i,jth pixel value the n_i by n_j matrix which is transformed from the ultrasound image and this definition is applied to all equations of this paper.

$$\mu = \frac{1}{n_i n_j} \sum_{i=1}^{n_i} \sum_{j=1}^{n_j} I(i,j) \tag{1}$$

$$\sigma^2 = \frac{1}{n_i n_j} \sum_{i=1}^{n_i} \sum_{j=1}^{n_j} |I(i,j) - \mu|^2 \qquad (2)$$

$$Min(I) = \min_{i \in n_i, j \in n_j} (I(i,j)) \qquad (3)$$

$$Max(I) = \max_{i \in n_i, j \in n_j} (I(i,j)) \qquad (4)$$

2.1 Data Acquisition

The original breast ultrasound image that was used in this work is like a Fig. 1(a). The level of pixel value information from each original image is transformed into matrix for applying a CAD. The transformed matrix value that was shown in Fig. 1(b) represents the light and shade value of each pixel on original image and extra information such as outpatient's name, distinction of sex, size of lesion. However, cutting of extra information is needed for protection of personal information. So the extra information in images that were used in this work was cut as a result we could only use breast ultrasound images without extra information.

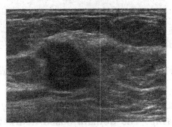

(a) Original image

	1	2	3	4	5	6	7	8	9
1	81	83	84	84	83	79	76	73	67
2	71	68	65	65	65	67	68	72	73
3	73	75	76	73	71	71	72	73	76
4	68	73	78	82	82	82	81	79	78
5	67	69	73	77	79	79	76	73	71
6	66	65	64	63	63	64	66	67	69
7	72	75	81	89	95	101	106	109	111
8	113	115	123	128	135	141	149	151	153
9	148	153	153	149	139	135	135	136	139

(b) The transformed matrix

Fig. 1. Original image and pixel information

2.2 Image Enhancement

Most pixel value of ultrasound image has all range of light and shade that distribute from 0 to 255 values. But the distribution is a little too much to specific pixel value. So, in this paper, ends-in search that makes more vivid the difference of the background and the Region Of Interest (ROI) is used for a image enhancement. A formula of this method is as follow and the result is like Fig. 2.

$$
Low = Min(I_{org}) + \sigma_{org}
$$
$$
high = Max(I_{org}) - \sigma_{org}
$$
$$
I_{Endin}(i,j) = \begin{cases} 0 & I_{org}(i,j) \leq low \\ 255 \times (\dfrac{I_{org}(i,j) - Low}{High - Low}) & Low < I_{org}(i,j) > High \\ 255 & I_{org}(i,j) \geq High \end{cases} \tag{5}
$$

Where, I_{org} is the n_i by n_j matrix which is transformed from the original image and σ_{org} is the standard deviation of I_{org} acquired by Eq. (2).

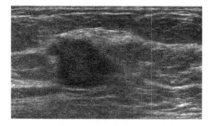

Fig. 2. Applied Image of End-In search

The result of ends-in search was used for unnecessary image removal in domain based heuristic noise filtering. The domain based heuristic noise filtering was basically designed by two domain known of ultrasound image as follows.

1. The shown lesion image on breast ultrasound is represented with a whole feature.
2. The pixel value of noises related to unnecessary image and lesion has similar range.

The proposed noise filter is based on distinct blocks operation of 13 by 13 matrixes as shown in Fig. 2 and ostensive processes is operated by Eq. (6) as next step.

Step 1: The starting points of distinct blocks are four corners in Fig. 3.
Step 2: The mean of distinct block and the standard deviation of whole image applied End-In search are computed and compared.
Step 3: If former is smaller than latter, all pixel values of distinct blocks are transformed to 255. In against case, they are not changed by Eq. (6).
Step 4: If pixel values mentioned above are changed, the distinct blocks started by 1, 4 of Fig. 3 shift in the row direction. In against case, the progress direction of blocks is opposite. Blocks have other starting points are operated in reverse.

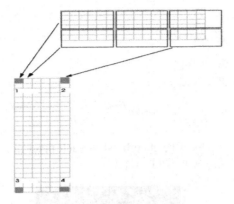

Fig. 3. The Principle of distinct blocks operation

$$\mu_{Endin} = \frac{1}{NM} \sum_{i, j \in \eta} I_{Endin}(i, j)$$

$$I_{Noi}(i, j) = \begin{cases} 255 \\ I_{Endin}(i, j) \end{cases}, \text{ where } \begin{matrix} \mu_{Endin} < \sigma_{Endin} \\ \mu_{Endin} \geq \sigma_{Endin} \end{matrix} \qquad (6)$$

Where, η is the N by M district block of each pixel in the I_{Endin}.
As a result, most shadowing artifacts are removed as like Fig. 4.

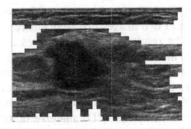

Fig. 4. Applied Image of noise filtering

2.3 Matrix Search Filtering

Generally, pixel value of lesion and surgical breast biopsy has low range but other images that are not interfered have high range value. So we carry out Image negative transformation for computing easily and clearly. A basic formula of Image negative transformation is as follows.

$$I_{Neg}(i, j) = 255 - I_{Noi}(i, j) \qquad (7)$$

By executing the matrix search filtering, small noise in result image conducted image negative transformation is removed. This filter used Sliding Neighborhood Operations as 13 by 13 matrix of neighborhood blocks. Algorithms of this method are as follow and the result is shown in Fig. 5.

$$\mu_{Nei} = \frac{1}{RC} \sum_{i,j \in \rho} I_{Noi}(i,j)$$

$$C_{low} = \mu_{Neg} - \sigma_{Neg}$$

$$C_{high} = \mu_{Neg} + \sigma_{Neg} \tag{8}$$

$$I_{Mat}(i,j) = \begin{cases} 0 \\ I_{Neg}(i,j) \end{cases}, where \begin{matrix} \mu_{Nei} < C_{low} \ or \ \mu_{Nei} < C_{high} \\ C_{low} \le \mu_{Nei} \le C_{high} \end{matrix}$$

Where, ρ is the R by C local neighborhood of each pixel in the I_{Noi} and σ_{Neg} is the standard deviation of I_{Neg} acquired by Eq. (2).

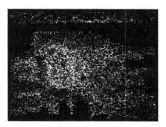

Fig. 5. Applied Image of matrix search filtering

2.4 Differential Image and 2D Adaptive Noise Removal Filtering

The differential image and final filtering are also used for noise removal. The differential image is the difference of two images that were conducted by matrix search filtering and ends-in search. The result is eliminated noises that existed near the lesion because the pixel values of the lesion image of matrix search filtering were high near to 255 but those of the lesion image of ends-in search were low near to 0. In this paper, image of ends-in search is multiplied by 1.5 for elimination of noise around lesion.

The pixel value of ROI and background are clear certainly through the differential image. Consequently, two-dimensional (2D) adaptive noise removal filter and local mean filter is conducted for noise removal in this study. 2D adaptive noise removal filter is methodology to remove noises using local mean and standard deviation [8]. In this paper, pixel-wise Wiener filter is conducted including 100 by 100 neighborhood blocks. The reason applying 100 by 100 matrix is that noises scatter but ROI concentrate. Pixel-wise Wiener filter is applied by Eq. (9) and the result is as Fig. 6.

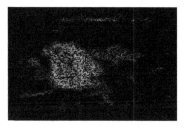

Fig. 6. Applied image of pixel-wise Wiener filter

$$\mu_{Diff} = \frac{1}{RC} \sum_{i,j \in \rho} I_{Diff}(i,j)$$

$$\alpha^2_{Diff} = \frac{1}{RC} \sum_{i,j \in \rho} I_{Diff}(i,j)^2 - \mu_{Diff}^2 \qquad (9)$$

$$I_{2D}(i,j) = \mu_{Diff} + \frac{\alpha^2_{Diff} - \nu^2}{\alpha^2_{Diff}} \|I_{Diff}(i,j) - \mu_{Diff}\|$$

Where, ν^2 is the average of all the local estimated variances and ρ is the R by C local neighborhood of each pixel in the I_{Diff}.

All noises are not entirely removed by pixel-wise Wiener filter. So local-mean filter is conducted once again. The local-mean filter is operated by computing the local mean and standard deviation based pixels that value was higher than 1 in 2D adaptive noise removal filtered image. The process of this method is that the sum (C_S) and difference (C_D) of the mean and standard deviation is compared with each pixel value of image. Then, the pixel value determined based on compared result. The algorithms of determining pixel value were as Eq. (10) and the result is like a Fig. 7.

$$I_{Lcl}(i,j) = \begin{cases} 0 \\ 255 \end{cases}, where \quad \begin{matrix} I_{2D}(i,j) < C_D \, or \, I_{2D} > C_S \\ C_S \geq I_{2D} \geq C_D \end{matrix} \qquad (10)$$

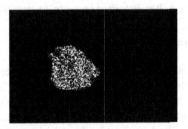

Fig. 7. Applied image of pixel-wise Wiener filter

3   Detection of Seed Points

3.1   Recovery of Tumor Area with Morphology

Original image of breast ultrasound is degraded through Mentioned steps above. So we approached morphological combination of dilation and erosion technique for the recovery of image, because dilation and erosion are often used in combination to implement image processing operations with the same structuring element for removal of some small size of noise. Dilation can be said to add pixels to an object, or to make it bigger, and erosion will make an image smaller. In the simplest case, binary erosion will remove the outer layer of pixels from an object. An image Dilation is using a structuring element. In this paper, we apply morphology combination as the order of erosion and dilation after dilation and erosion.

In Dilation case, the origin of the structuring element is placed over the first white pixel in the image, and the pixels in the structuring element are copied into their corresponding positions in the result image. Then the structuring element is placed over the next white pixel in the image and the process is repeated. This is done for every white pixel in the image.

In Erosion case, the structuring element is translated to the position of a white pixel in the image. In this case, all members of the structuring element correspond to white image pixels so the result is a white pixel. Now the structuring element is translated to the next white pixel in the image, and there is one pixel that does not match. The result is a black pixel. The remaining image pixels are black and could not match the origin of the structuring element.

The result of morphology is shown Fig. 8.

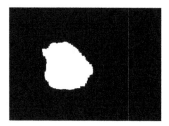

Fig. 8. Reconstructed image using morphology

3.2 The Auto Detection of Seed Point Using K-MANS

In this section, seed points are detected for segmentation of lesion automatically. Previous studies related to the detection of seed points had limitations that the number of legions was one and the place of legion was center in breast ultrasound image. So, K-MEANS algorithm was applied for solving these limitations above mentioned, because this clustering method can best be described as a partitioning algorithm and is suitable for auto detection of seed points if the number of tumors is set. The application of this method is conducted under assumptions are as follows [9].

1. The number of tumors in breast ultrasound image was less than 10.
2. The pixel value of tumors after noise removal is close to 0.

K-means is conducted depending on the coordinates of image pixel, steps are as follows.

Step 1: Coordinates of pixels more than 1 value are detected.
Step 2: The number of clustering group is defined as 1 and K-means conducted based coordinates of step 1.
Step 3: If pixel value of the centroid coordinate is more than 1, K-means is stopped. If pixel value indicated to result of step 2 is 0, K-means is iterated from step 1 by adding the number of clustering group automatically.

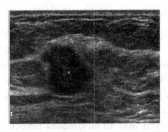

Fig. 9. Seed point detection using K-Means

The result is like as Fig. 9. In this paper, seed points mentioned results above are used to improve boundary of ROI and background image because the segmentation of ROI from seed points without additional processes take a long computing time.

4 Result and Conclusion

In this paper, the number of images applied to the proposed methods is 40, obtained from two hospitals as shown Fig.10. As a result, the most of seed points are detected by proposed methods.

Ultrasound creates a lot of noises and it was quite difficult to get an understandable ROI. However, the image is extremely important and segmentation of ROI is necessary for a surgical operation of breast cancer. So, many previous works have been studied for solving problems mentioned above but it is really difficult to find out ROI due to much noise generated.

CAD system has been developed for early diagnosis of cancers and improvement of diagnostic exactitude. CAD based system is also used as an assistant of diagnosis which is co-developed with medical specialists.

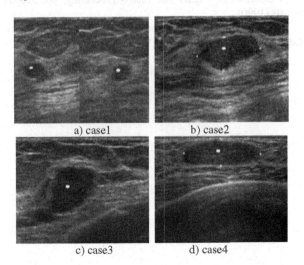

a) case1   b) case2

c) case3   d) case4

Fig. 10. Examples applying to other original images

The objective of this paper is to design noises removal methods and detect automatically seed points for applying the breast ultrasound image to CAD. So, the whole processes are conducted under the discussion with medical specialist and we presented auto detection of seed points preceded for segmentation with K-means algorithm to find out the estimated central area of the tumor. Sometimes it is difficult to differentiate between background color and tumor color, so we applied domain based heuristic noise filter. As a result of applying proposed algorithm, the noise is effectively eliminated in all images and an automated detection is possible by determining seed points on each tumor.

References

1. Minister for Health: Welfare and Family Affairs (MIHWAF). Statics of Cancer, Republic of Korea (2007)
2. Paulo, S.R., Gilson, A.G., Marcia, P., Marcelo, D.F., Chang, R.F., Suri, J.S.: A New Methodology Based on q-Entropy for Breast Lesion Classification in 3-D Ultrasound Images. In: Conf. Proc. IEEE Eng. Med. Biol. Soc., New York, vol. 1, pp. 1048–1051 (2006)
3. Moore, K.S.: Better Breast Cancer Detection. IEEE Spectrum 38, 51–54 (2001)
4. Madabhushi, A., Metaxas, D.N.: Automatic boundary extraction of ultrasonic breast lesions. In: 2002 IEEE Int. Symp. Biomedical Imaging, pp. 601–604 (2002)
5. Huang, Y.L., Chen, D.R.: Automatic Contouring for Breast Tumors in 2-D Sonography. In: Conf. Proc. IEEE Eng. Med. Biol. Soc., vol. 3, pp. 3225–3228 (2005)
6. Madabhushi, A., Metaxas, D.N.: Combining Low-, High-Level and Empirical Domain knowledge for Automated Segmentation of Ultrasonic Breast Lesions. IEEE Trans. Medical Imaging 22(2), 155–169 (2003)
7. Pietro, P., Jitendra, M.: Scale-Space and Edge Detection Using Anisotropic Diffusion. IEEE Trans. Pattern Anal. Machine Intell. 12(7), 629–639 (1990)
8. Lim, J.S.: Two-Dimensional Signal and Image Processing, pp. 536–540. Prentice-Hall, Englewood Cliffs (1990)
9. Guralnik, V., Karypis, G.: A scalable algorithm for clustering protein sequences. In: Proc. Workshop Data Mining in Bioinformatics (BIOKDD), pp. 73–80 (2001)

A Controlled Scheduling Algorithm Decreasing the Incidence of Starvation in Grid Environments

Minna Liu, Kairi Ou, Yuelong Zhao, and Tian Sun

School of Computer Science and Engineering, South China University of Technology,
Guangzhou, 510006, China
liuminna@sina.com

Abstract. A fair scheduling algorithm accounting for the weight and execution time of tasks is critical in the Grid environment. The MTWCT (Minimize Total Weighted Completion Time) has been proved to minimize the total weighted completion time of a set of independent tasks in a processor, but it results in another problem that the response time of some tasks is far longer. To decrease the incidence of the starvation phenomena, an improved algorithm named CSA (Controlled Scheduling algorithm) based on MTWCT is proposed, which computes the ρ factors of tasks by the execution time and weight of step chains, and selects the unexecuted step chain in terms of ρ factor and the executed time of task. Experimental results exhibit that CSA compared with the MTWCT, decreases the completion time of short tasks and the average turnaround time by sacrificing a little total weighted completion time.

Keywords: grid environment, scheduling algorithm, starvation phenomena, step chains.

1 Introduction

Efficient task scheduling can contribute significantly to the overall performance of the Grid and Cluster computing systems [1] [2] since an inappropriate scheduling of tasks can fail to exploit the true potential of the system [3]. Scheduling a set of independent tasks in grid environment is NP-hard [4]. Among the algorithms that have been proposed in the past, the Sufferage algorithm [5] has better performance. However, it deals with one objective which minimizes the makespan in the set of independent tasks [6]. In complex grid environment, the weight and execution time of takes are different. It is unreasonable that assuming the importance and urgencies of all tasks are the same. A scheduling algorithm called MTWCT (Minimize Total Weighted Completion Time) [7], in the accordance to the fact that a task is made up of step chains of precedence constraints, divides steps of each task into step chains by the ratios of weight and execution time of step chains and the step chain is to be executed as a whole. The ratios of weight and execution time of step chains accounts for the importance of the task and the length of execution time, which overcomes the limitation of Sufferage one-sided pursuit of minimal makespan regardless of the importance of the task, but it results in another problem that short tasks with a bit smaller ratios have to wait for long term. That is because MTWCT deals with only one objective which is minimizing the total

H. Deng et al. (Eds.): AICI 2009, LNAI 5855, pp. 517–525, 2009.

weighted completion time, neglecting the average turnaround time of tasks. When the amount of long tasks is very great, it will exacerbate the extent of starvation [8]. Obviously it is unfair to the short tasks with a bit smaller ratios of weight and execution time. In addition, it isn't efficient for the system that all short tasks with a bit smaller ratios haven't chances to be executed until all long tasks have been completed; especially the execution time of long tasks is much longer than that of short tasks.

According to the above analysis, we propose a meaningful modified algorithm based on MTWCT named CSA (Controlled Scheduling Algorithm) which is able to leverage the strengths of MTWCT and avoid its weaknesses. CSA supports the following features.

(1) Avoiding short tasks with a bit smaller ratios waiting for long term.

(2) Decreasing the average turnaround time of all tasks. Namely, it is beneficial to the majority of users and the system.

(3) Being able to make the total weighted completion time approximately optimal.

The rest of the paper is organized as follows: In section 2, we describe the strategies of setting thresholds. The Controlled Scheduling Algorithm is introduced and we present the analysis of the Controlled Scheduling Algorithm. Section 3 provides numerical results to compare the task completion time, the average turnaround time and the total weighted completion time under CSA and under MTWCT. Conclusions are drawn in Section 4.

2 Controlled Scheduling Algorithm

2.1 Model Description

In this paper, we employ the triple used firstly by Graham [9] to depict the scheduling problem. $1|\text{chains}|(\sum w_j p_j)_\pi \cup \frac{1}{n} t_i$: where 1 denotes all tasks are supposed to be executed in a single processor and chains denotes a set of independent tasks. $(\sum w_j p_j)_\pi \cup \frac{1}{n} t_i$ is the objective of CSA that the total weighted completion time and the average turnaround time are taking into account. Let $(\sum w_j p_j)_\pi$ denotes the total weighted completion time and $\frac{1}{n} t_i$ denotes the average turnaround time.

We summarize the notation used throughout the paper in Table 1.

Table 1. Notation

parameters	descriptions
n	Number of tasks in the task queue
k	Number of steps every task
T_{ij}	the jth step of the ith task
p_{ij}	execution time required by step T_{ij}
w_{ij}	weight of step T_{ij}
ρ	the ratio of step chain's weight and execution time
t	the limit executed time of a task
$\Delta\rho$	the range of ρ factor

2.2 Definitions and Theorems Related with CSA

In this section, two definitions and theorems related with CSA are given.

Definition 1 : For a task is composed by k steps at most, the rule of its ρ factors is defined as follows:

$$\frac{\sum_{j=1}^{l^*} w_{ij}}{\sum_{j=1}^{l^*} p_{ij}} = \max_{1 \leq l \leq k} \left\{ \frac{\sum_{j=1}^{l} w_{ij}}{\sum_{j=1}^{l} p_{ij}} \right\} \tag{1}$$

namely, $\rho_{i1} (1,2,...,l^*)$, where l^* is the deadline step of ρ_{i1}. In a similar manner, $\rho_{i2}...\rho_{ik}$ can be obtained.

Definition 2 : Assuming that there are two tasks i and j, task i contains k steps denoted by $T_{i1} \rightarrow T_{i2} \rightarrow ... \rightarrow T_{ik}$, and task j is made up of k steps denoted by $T_{i1} \rightarrow T_{i2} \rightarrow ... \rightarrow T_{ik}$. Let their steps' sequence denoted by $\pi(T_{i1},T_{i2},...,T_{ik},T_{j1},...T_{jk})$, the total weighted completion time is defined to be:.

$$(\sum p_{ij} w_{ij})_\pi = w_{i1} p_{i1} + ... + w_{ik} \sum_{m=1}^{k} p_{im} + w_{j1}(p_{j1} + \sum_{m=1}^{k} p_{im}) + ... + w_{jk}(\sum_{m=1}^{k} p_{im} + \sum_{f=1}^{k} p_{jf}) \tag{2}$$

Theorem 1: If $\dfrac{\sum_{j=1}^{m} w_{ij}}{\sum_{j=1}^{m} p_{ij}} > \dfrac{\sum_{a=1}^{r} w_{fa}}{\sum_{a=1}^{r} p_{fa}}$, and the step chain made up of $T_{i1},T_{i2},...,T_{im}z$ is executed prior to the step chain made up of $T_{f1},T_{f2}...T_{fr}$, which makes the total weighted completion time of steps denoted by $T_{i1},T_{i2},...,T_{im},T_{f1},T_{f2}...T_{fr}$ minimal.

Theorem 2: If T_{i^r} is the deadline step of ρ factor about the step chain denoted by $T_{i1},T_{i2},...,T_{ik}$, a step chain denoted by $T_{i1},T_{i2},...,T_{i^r}$ must be uninterrupted.

The proof of theorem 1 and theorem 2 is given in [7].

2.3 Strategy of Setting the $\Delta\rho$ and t Thresholds

Appropriate t and $\Delta\rho$ are critical to CSA. The objective of setting t is to make a step chain represented by a smaller $\Delta\rho$ factor have a chance to be executed earlier than that of a bigger ρ factor. Setting $\Delta\rho$ is aim to choose other ρ factors in $[\max\_\rho - \Delta\rho, \max\_\rho]$, where max_ρ denotes the maximum ρ factor, to ensure the weighted total completion time as optimal as possible.

2.3.1 The Strategy of Setting t Thresholds
It should be noted that t thresholds come from the long tasks with bigger weights due to the reason that setting t is aim to delay being executed of long tasks.

We define the following arrays that are used in the description of setting t thresholds.

$t[][]$: array of step chains execution time. where $t[i][j]$ denotes the execution time required by the jth step chain of the ith task.

$sum\_t[][]$: array of total execution time of step chains, where $sum\_t[i][j] = \sum_{f=1}^{j} t[i][f]$.

$tt[]$: array of total thresholds.

The strategy of setting t thresholds is described as follows:

```
initialize t[][], sum_t[][], tt[];
for i=1 to n do
    for j=2 to k do
        sum_t[i][j] = sum_t[i][j-1]+t[i][j];
for i=1 to k do
    for j=1 to n do
        If(sum_t[i][j]>max) then max = sum_t[i][j];
        If(sum_t[i][j]<min) then min = sum_t[i][j];
tt[++count]=max;
tt[++count]=min;
```

2.3.2 The Strategy of Setting $\Delta\rho$ Thresholds

Meaningful $\Delta\rho$ thresholds should make the short tasks with smaller weights execute earlier which is helpful to decrease the average turnaround time while long tasks are more than short tasks.

We define the following array and variants that are used in the description of setting $\Delta\rho$ thresholds.

$\rho[][]$: array of ρ factors. Where $\rho[i][j]$ denotes the ρ factor of the task.

$shorttasks\_min\_\rho$: the minimum ρ factor in all the ρ factors from short tasks.

$shorttasks\_max\_\rho$: the maximum ρ factor in all the ρ factors from short tasks.

The description of strategy of setting $\Delta\rho$ thresholds as follows:

```
initialize ρ[][], shorttasks _ min_ ρ, shorttasks _ max_ ρ;
max = 0; min = ∞;
for i=1 to m do
  for j=1 to k do
     if(ρ[i][j]<0) break;
     if(ρ[i][j]>max) max =ρ[i][j] ;
     if(ρ[i][j]<min) min =ρ[i][j];
shorttasks _ min_ ρ = min;
shorttasks _ max_ ρ = max;
count = 0;
for i=1 to h do
  for j=1 to k do
```

$$\Delta \rho[+ + count] = \rho[i][j] - short tasks \_ min\_ \rho;$$
$$\Delta \rho[+ + count] = \rho[i][j] - short tasks \_ max\_ \rho;$$

2.4 Description of CSA

Step 1: choose two critical values $\Delta \rho[i]$ and $\Delta \rho[j]$ from the one-dimensional vector $\Delta \rho[]$ arbitrarily, and select a constant from the interval $(\Delta \rho[i], \Delta \rho[j])$.

Step 2: Similar to step 1, set the initial limit executed time allocate to a task which is called t.

Step 3: Sort all the ρ factors by non-ascend order, after that save them in the one-dimensional vector $\rho \rho[]$.

Step 4: Set a flag to every ρ factor and initialize flag=0. Where flag=0 denotes the step chain represented by ρ factor hasn't been executed; otherwise, means it has been executed.

Step 5: Set a temporary limit executed time $temp\_t$, and initialize $temp\_t=t$.

Step 6: Choose the maximum ρ factor whose value of flag is zero from one-dimensional vector $\rho \rho[]$, which is called $max\_ \rho$.

Step 7: Save all the ρ factors between $max\_ \rho - \Delta \rho$ and $max\_ \rho$ whose values of flags are zero in a one-dimensional vector $candi \_ \rho[]$.

Step 8: If executed time of the task represented by $max\_ \rho$ less than $temp\_t$, execute the step chain represented by $max\_ \rho$, after that the value of $max\_ \rho$'s flag is set to 1 and then go to step 9. Otherwise, if $candi \_ \rho[]$ is an empty vector, execute the step chain represented by $max\_ \rho$, else select a $candi \_ \rho[i]$ with the minimum task executed time from $candi \_ \rho[]$ and process the step chain represented by $candi \_ \rho[i]$.

Step 9: Set $temp \_ t = t$, go to step 11.

Step 10: Update $temp \_ t = temp \_ t + min\_ t$, go to step 11.

Step 11: Repeat the whole procedures from step 5 through step 10 until all step chains have been executed.

2.5 Analysis of CSA

The time complexity of CSA is mainly composed of three parts: setting t thresholds, setting the $\Delta \rho$ thresholds and the choice of step chains. we can obtain that it takes $O(2nk)$ to set t thresholds and $O(mk+hk)$ to set $\Delta \rho$ thresholds according to (3.3). the choice of step chains costs $O((mk+hk)*(mk+hk))$. CSA costs $O(2nk)+O(mk+hk)$ more than MTWCT, which is ignored when m and h is very big.

According to the analysis of the CSA, we can obtain the following deductions:

Deduction 1: if $t \rightarrow +\infty$, then CSA is equivalent to MTWCT.

Proof: Accord to step 8 of the description of CSA, if $t \rightarrow +\infty$, the task executed time of max_ρ always less than $temp\_t$, so the step chain represented by max_ρ will be executed at all times. $\Delta\rho$ and t has no impact on selecting unexecuted step chains, then CSA is equivalent to MTWCT.

Deduction 2: if $\Delta\rho \rightarrow -\infty$, then CSA is equivalent to MTWCT.

Proof: Accord to step 7 of the description of CSA, if $\Delta\rho \rightarrow -\infty$, $candi\_\rho[]$ is an empty vector. So the execution right is always given to the step chain represented by max_ρ according to step 8 of the description of CSA, then CSA is equivalent to MTWCT.

Thanks to the task executed time limit t, CSA overcomes the starvation phenomena under MTWCT, which chooses the unexecuted step chain represented by maximum ρ factor every time resulting in other step chains denoted by a bit smaller ρ factors non-execution long term.

3 Simulations and Analysis

The execution time and weights of steps can't be identified in advance, which need predicting. In this paper, experimental parameters are assumed to be obtained. We focus on comparing MTWCT and CSA in terms of the task completion time, the average turnaround time and the total weighted completion time.

The execution time, weights and ρ factors of step chains of every task are given in Table 2.

From Table 3, we can see that if $t = 0.5$, $\Delta\rho = 0.32$, $t = 0.5$, $\Delta\rho = 0.005$ and $t = 0.06$, $\Delta\rho = 0.02$, CSA is equivalent to MTWCT. From deduction 1, it is obvious that t is equivalent to $+\infty$ while $t > 0.489$ because of the maximum threshold t for 0.489 in this experiment. According to deduction 2, when $\Delta\rho < 0.05$, $\Delta\rho$ is equivalent to $-\infty$ due to the fact that the minimum threshold $\Delta\rho$ is 0.05 in this experiment.

Table 2. Experimental Parameters

category	task	$p_{ij}(us)$	w_{ij}	ρ_{ij}
		0.163	0.25	1.53
	task1	0.116	0.16	1.38
		0.112	0.14	1.25
category1		0.098	0.12	1.22
		0.078	0.11	1.41
	task2	0.045	0.06	1.32
		0.056	0.07	1.24
category2	task3	0.042	0.05	1.19
		0.035	0.04	1.15

Table 3. The ρ factors sequence of MTWCT and CSA

	$t, \Delta\rho$	Sequences of ρ factors
	MTWCT	$\rho_{11}, \rho_{21}, \rho_{12}, \rho_{22}, \rho_{13}, \rho_{23}, \rho_{14}, \rho_{31}, \rho_{32}$
	$t = 0.06, \Delta\rho = 0.02$	$\rho_{11}, \rho_{21}, \rho_{12}, \rho_{22}, \rho_{13}, \rho_{23}, \rho_{14}, \rho_{31}, \rho_{32}$
	$t = 0.06, \Delta\rho = 0.3$	$\rho_{11}, \rho_{21}, \rho_{31}, \rho_{32}, \rho_{22}, \rho_{12}, \rho_{13}, \rho_{23}, \rho_{14}$
	$t = 0.11, \Delta\rho = 0.15$	$\rho_{11}, \rho_{21}, \rho_{22}, \rho_{23}, \rho_{12}, \rho_{31}, \rho_{32}, \rho_{13}, \rho_{14}$
CSA	$t = 0.11, \Delta\rho = 0.1$	$\rho_{11}, \rho_{21}, \rho_{22}, \rho_{12}, \rho_{31}, \rho_{23}, \rho_{13}, \rho_{32}, \rho_{14}$
	$t = 0.162, \Delta\rho = 0.25$	$\rho_{11}, \rho_{21}, \rho_{31}, \rho_{12}, \rho_{22}, \rho_{32}, \rho_{23}, \rho_{13}, \rho_{14}$
	$t = 0.5, \Delta\rho = 0.32$	$\rho_{11}, \rho_{21}, \rho_{12}, \rho_{22}, \rho_{13}, \rho_{23}, \rho_{14}, \rho_{31}, \rho_{32}$
	$t = 0.5, \Delta\rho = 0.005$	$\rho_{11}, \rho_{21}, \rho_{12}, \rho_{22}, \rho_{13}, \rho_{23}, \rho_{14}, \rho_{31}, \rho_{32}$

The number of tasks is fixed to 100, the category1 tasks are assumed to long tasks and the category2 task are supposed to short tasks. The percentage of the category1 tasks are in [10%,90%]. Since $t = 0.5, \Delta\rho = 0.32$, $t = 0.5, \Delta\rho = 0.005$ and $t = 0.06, \Delta\rho = 0.02$, CSA is equivalent to MTWCT, only $t = 0.06, \Delta\rho = 0.03$, $t = 0.11, \Delta\rho = 0.15$, $t = 0.11, \Delta\rho = 0.1$,and $t = 0.152, \Delta\rho = 0.25$ are adopted to test the performance of CSA respectively.

3.1 Comparison of CSA and MTWCT in Terms of Task Completion Time

With the increasing of long tasks, the results of the short tasks completion time is depicted in Fig.1. According to the simulation results, the performance of CSA varies with different t and $\Delta\rho$.

From Fig.1, Note that the short tasks completion time under CSA is obviously smaller than that using MTWCT.

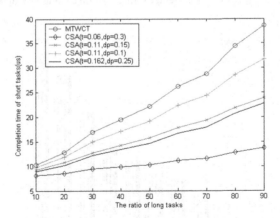

Fig. 1. Comparison of short tasks completion time using two algorithms

3.2 Comparison of CSA and MTWCT in Terms of the Average Turnaround Time

From Fig.1 it can be obtained easily that when short tasks are executed earlier, long tasks are inevitably delayed. As we know, the average turnaround time can reflect the overall performance of a scheduling algorithm. The average turnaround time is short in favor of the majority of users, also reflects high system throughput, and high system utilization rates. The simulation results is showed in Fig.2.

From Fig.2, we can see that either CSA or MTWCT increases the average turnaround time with the increasing of long tasks. The values of $t, \Delta\rho$ have great impact on CSA in the performance of average turnaround time. For $t = 0.11, \Delta\rho = 0.15$, CSA is extremely efficient and when the percentage of long tasks is over 13%, CSA is superior to MTWCT.

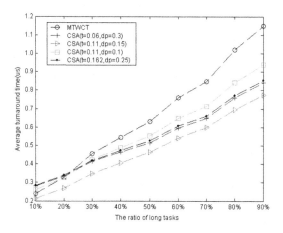

Fig. 2. Comparison of the average turnaround time using two algorithms

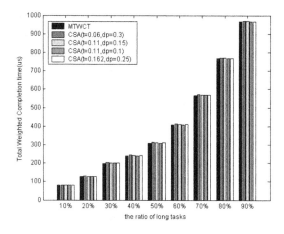

Fig. 3. Comparison of the weighted total completion time using two algorithms

3.3 Comparison of CSA and MTWCT in Terms of the Total Weighted Completion Time

We can see that the weighted total completion time using CSA is definitely not better than that using MTWCT by the previous theoretical analysis. The performance of the total weighted completion time comparison of the CSA and MTWCT is depicted in Figs.3.

From Fig.3, we can see that either CSA or MTWCT increases the weighted total completion time in parallel to the increase in the number of long tasks. The weighted total completion time using CSA is a bit greater than that using MTWCT.

4 Conclusion

In this paper, we presented a meaningful modified scheduling algorithm named Controlled Scheduling Algorithm for a set of independent tasks in a processor of heterogeneous systems, which supports three features proposed in the section I. CSA takes into account of fairness and effectiveness, decreasing the incidence of the starvation phenomena produced by MTWCT. In addition, CSA has the flexibility to set t and $\Delta\rho$ to meet various applications. A large number of experimental results indicate that in general compared with MTWCT, the average turnaround time under CSA decreases around 20% at the price of increasing 3% in the total weighted completion time.

References

1. Foster, I., Kesselman, C.: The Grid2, Blueprint for a New Computing Infrastructure. Morgan Kaufmann, San Francisco (2004)
2. Abraham, A., Buyya, R., Nath, B.: Nature's Heuristics for scheduling Jobs on Computational Grids. In: ADCOM 2000, Cochin, India, pp. 45–52 (2000)
3. Kwok, Y.-K., Ahmad, I.: Static Scheduling Algorithms for Allocating Directed Task Graphs to Multiprocessors. In: ACM Computing Surveys, New York, pp. 406–471 (1999)
4. Maheswaran, M., Ali, S., Siegel, H.: Dynamic Matching and Scheduling of a Class of Independent Tasks onto Heterogeneous Computing Systems. In: Eigth Heterogeneous Computing Workshop, San Juan, Pueto Rico (April 1999)
5. Maheswaran, M., Ali, S., Siegel, H.J., Hensgen, D., Freund, R.F.: Dynamic Mapping of a Class of Independent Tasks onto Heterogeneous Computing Systems. Journal of Parallel and Distributed Computing 59(2), 107–131 (1999)
6. SaiRanga, P.C., Baskiyar, S.: A low complexity algorithm for dynamic scheduling of independent tasks onto heterogeneous computing systems. In: Proceedings of the 43rd annual Southeast regional conference, New York, pp. 63–68 (2005)
7. Pinedo, M.: Scheduling-theory, algorithms, and systems, ch. 4. Prentice Hall, Englewood Cliffs (1995)
8. Cuesta, B., Robles, A., Duato, J.: An Effective Starvation Avoidance Mechanism to Enhance the Token Coherence Protocol. In: IEICE Transactions on Fundamentals of Electronics, Communications and Computer Sciences, Washington, pp. 47–54 (2007)
9. Graham, R.L., Lawler, E.L., Lenstra, J.K., et al.: Optimization and Approximation in Determi-nistic Sequencing and Scheduling. Annals of Discrete Mathematics 5, 287–326 (1979)

A Space Allocation Algorithm for Minimal Makespan in Space Scheduling Problems

Chyuan Perng[1], Yi-Chiuan Lai[2], Zih-Ping Ho[1], and Chin-Lun Ouyang[1]

[1] Department of Industrial Engineering and Enterprise Information
Tunghai University, Taichung City 407, Taiwan
[2] Department of Business Administration
Providence University, Shalu, Taichung County 433, Taiwan
ylai@pu.edu.tw

Abstract. The factory space is one of the critical resources for the machine assembly industry. In machinery industry, space utilizations are critical to the efficiency of a schedule. The higher the utilization of a schedule is, the quicker the jobs can be done. Therefore, the main purpose of this research is to derive a method to allocate jobs into the shop floor to minimize the makespan for the machinery industry. In this research, we develop an algorithm, Longest Contact Edge Algorithm, to schedule jobs into the shop floor. We employed the algorithm to allocate space for jobs and found that the Longest Contact Edge Algorithm outperforms the Northwest Algorithm for obtaining better allocations. However, the Longest Contact Edge Algorithm results in more time complexity than that of the Northwest Algorithm.

Keywords: Space Allocation Algorithm, Resource Constrained Scheduling Problem, Longest Contact Edge Algorithm, Dispatching Rules, Space Scheduling Problem.

1 Introduction

Scheduling is an important tool for manufacturing and engineering, as it can have a major impact on the productivity of a process. In manufacturing, the purpose of scheduling is to minimize production time and costs by arranging a production facility regarding what and when to make, with which staff, and on which equipment. Production scheduling aims to maximize the efficiency of the operation and reduce costs [13]. In general, the scheduling problem could be very complicated. Perng *et al.* [2][3][4][5][6] proposed a space resource constrained job scheduling problem. In their researches, utilization of a shop floor space was an important issue in a machinery assembly factory. The assembly process of a machine required a certain amount of space on a shop floor in the factory for a period of time. The sizes of the shop floor and machines would determine the number of machines which can be assembled at the same time on the shop floor. The sequences of jobs would affect the utilization of the shop floor. The space on the shop floor could be divided into several chunks and some of these jobs could be allocated for simultaneous assembling. When a new job

H. Deng et al. (Eds.): AICI 2009, LNAI 5855, pp. 526–534, 2009.

arrived, the factory has to offer a complete space to allocate the new job based on its space requirement. If the factory was at its capacity, the new order must wait for the space taken by jobs previously assigned to complete production. Due to space capacity constraints, we have to schedule the sequence of jobs carefully in order to maximize space utilization.

The makespan is important when the number of jobs is finite. The makespan is defined as the time the last job leaves the system, that is, the completion time of the last job [8]. The makespan shows within how many days the factory can complete these jobs. Nichols *et al.* [14] was the first paper to discuss scheduling with minimum makespan as an objective. In their research, *n* independent, single operation jobs, all available at time zero, on *m* identical processors; each job must be processed by exactly one of these processors. Computational experience with the procedure indicated that good solutions to large scale problems were easily obtained. Gallo and Scutella [9] tried to minimize the makespan of scheduling problem with tree-type precedence constraints. The assembly factories wanted to select a sequence from which all jobs can be completed in time. Liao and Lin [1] considered the two-uniform-parallel-machine problem with the objective of minimizing makespan. The problem can be transformed into a special problem of two identical parallel machines from the viewpoint of workload instead of completion time. An optimal algorithm was developed for the transformed special problem. Although the proposed algorithm had an exponential time complexity, the results showed that it could find the optimal solution for large scale problems in a short time. Lian *al.* [15] presented a swarm optimization algorithm to solve a job shop scheduling problem. The objective was to find a schedule that minimizes the makespan. Perng *et al.* [5] proposed an algorithm based on a container loading heuristic (CLH) approach [7] to a space scheduling problem with minimal makespan. However, the original CLH could not find all the possible free space, while this research could. Zobolas *et al.* [10] proposed a hybrid metaheuristic for the minimization of makespan in permutation flow shop scheduling problems. A genetic algorithm (GA) for solutions was selected. Computational experiments on benchmark data sets demonstrated that the proposed hybrid metaheuristic reached high-quality solutions in short computational times. Although numerous scheduling problems discussed makespan, there is only a limited amount of literature available on the space scheduling field.

One of the real world applications is the efficient space allocation algorithm. This research proposes a new space allocation algorithm to minimize the makespan for a space scheduling problem. In the other words, the main purpose of this research is to derive a method to allocate the jobs into a shop floor more efficiently.

2 Problem Formulation

Let N denote a set of n jobs. Let s_1, s_2,..., s_n denote the start time for each job. Let p_1, p_2,..., p_n denote the processing time for each job. Let dd_1, dd_2,...,dd_n denote the due date for each job. Let σ denote an arbitrary sequence. The rest of notations is shown as follows:

C_j: completion time of job j

$$C_j = s_j + p_j$$

C_{max}: makespan

$$C_{max} = Max\{s_j + p_j\}$$

σ : an arbitrary sequence

$f(\sigma)$: makespan of σ

The objective is to find a sequence to minimize the makespan.

$$\textbf{min}\ \ f(\sigma) = C_{max} \tag{1}$$

The function $f(\sigma)$ denotes the makespan of σ processing sequence. We assume that all jobs are available at day one, all of the orders are rectangles, a job will not be moved until it is done, there is no constraint on job's height, and the buffer or storage is available to fit in any number or any shape of jobs.

3 Longest Contact Edge Algorithm

The northwest algorithm (NWA) is a basic space allocation approach used in Perng *et al.* [2][3][6]. It was used to allocate jobs into a factory. In this study, we intend to propose a better space allocation algorithm, namely, longest contact edge algorithm (LCEA), which differs from the northwest approach. The flow chat of LCEA is shown in Figure 1.

The foundation of the Longest Contact Edge Algorithm is based on Perng *et al.* [2][3][6]. The difference between LCEA and NWA is that LCEA evaluates every free grid instead of finding only one northwest point and allocating the job at the northwest points. The procedure of LCEA is as follows:

k: job number ($k = 1, 2,\ldots, n$)
a_k: the length of job k
b_k: the width of job k
c: a candidate solution
(rpx_c, rpy_c): the coordinate of c
S: a set of c

$$grid(i, j) = \begin{cases} 1, \text{if the coordinate (i, j) is occupied} \\ 0, \text{otherwise} \end{cases}$$

Step1: Search starting grids
At the beginning, the algorithm searches all free grids in the factory. The algorithm in this step employs NWA in literature [2][3][6]. It scans $(1,1)$, $(1,2)$,... ,(m, n) for starting points. The factory size is m x n grids.

Step 2: Search a job's working space
The algorithm then searches a_k x b_k complete free grids for job k based on the points obtained in step 1. If the a_k x b_k complete grids could be found, the starting point then is defined as a reference point. It is also called a candidate solution c.

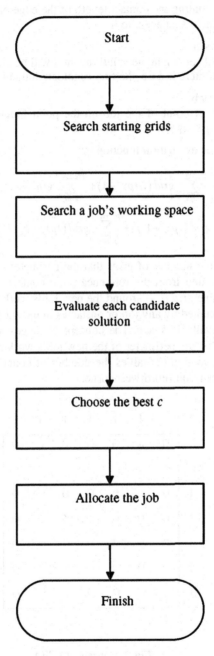

Fig. 1. Flowchart of LCEA

Step 3: Evaluate each candidate solution
Based on Longest Contact Edge evaluation function (2), we tally the value of every c found in step 2 by counting the contact length of the edge of the new job with the job on the shop floor which is adjacent to it.

Step 4: Choose the best c
The largest value z (2) of candidate solution c in S will be chosen as a northwest point to allocate the job. If there is a tie, the first candidate found in step 1 will be chosen.

Step 5: Allocate the job
Allocate the northwest point of the job on the point found in step 4. The reference point of this job is (rpx_c, rpy_c).

The value z of the evaluation function is:

$$z = \underset{c \in S}{\text{Max}} \left\{ \sum_{i=rpx_c}^{rpx_c+a_k-1} grid\left(i, rpy_c - 1\right) + \sum_{i=rpx_c}^{rpx_c+a_k-1} grid\left(i, rpy_c + b_k\right) \right.$$
$$\left. + \sum_{j=rpy_c}^{rpy_c+b_k-1} grid\left(rpx_c - 1, j\right) + \sum_{j=rpy_c}^{rpy_c+b_k-1} grid\left(rpx_c + a_k, j\right) \right\} \qquad (2)$$

In LCEA, we tally the number of grids that the perimeter of the new job we tend to allocate if it is allocated from the reference point contacts with the jobs previously allocated. We examine $grid$ (i, j) around the new job's working space. Let $grid$ $(i, j) = 1$, if $grid$ (i, j) is occupied by any job or obstacles in the shop floor. Let $grid$ $(i, j) = 0$, if $grid$ (i, j) is an available space. The concept of longest contact edge is shown in Figure 2. It shows that the perimeter of the new job's working space is surrounded by unavailable grids. Equation (2) tallies the number of contact grids if the job is allocated at the reference point (northwest corner).

1	1	1	1	1	1	1	1	1
1	0	0	0	0	0	0	0	1
⋮	⋮	⋮	⋮	⋮	⋮	.·˙	0	1
1	0	0	0	0	0	⋯	0	1
1	0	0	0	0	0	⋯	0	1
1	0	0	0	0	0	⋯	0	1
1	0	0	0	0	0	⋯	0	1
1	0	0	0	0	0	⋯	0	1
1	1	1	1	1	1	⋯	1	1

Fig. 2. Concept of LCEA

4   Computation Results and Discussion

In order to evaluate the performance of the proposed algorithm under traditional dispatching rules, namely SPT, LPT, EDD, FCFS, SSR, and LSR, the algorithm was developed by using PHP and Microsoft Visual Basic languages. Integrated software was also developed for solving the space allocation problem. The current version of the software includes both Northwest and Longest Contact Edge Algorithms. The interface of the software is shown in Figure 3. We used a Pentium IV (Celeron CPU 2.40GHz) computer for the computations. We computed the makespans for different number of jobs (25, 50 and 75). Thirty different data sets were obtained from the OR-Library [1] [2] due to real data being too few for statistical analysis. The space requirements of a job and order information were obtained from a company located in central Taiwan. The results are listed in Table 1 to 3.

We employ *t*-test for matched samples of dispatching rules to compare the makespans between the NWA and LCEA. The results indicate that LCEA outperforms NWA for every dispatching rule in makespan. However, LCEA consumes more time complexity T [5]. All results can be obtained within 5 seconds. The LCEA could find any possible free grids in the factory. Hence, it shortens the makespan in a space scheduling problem.

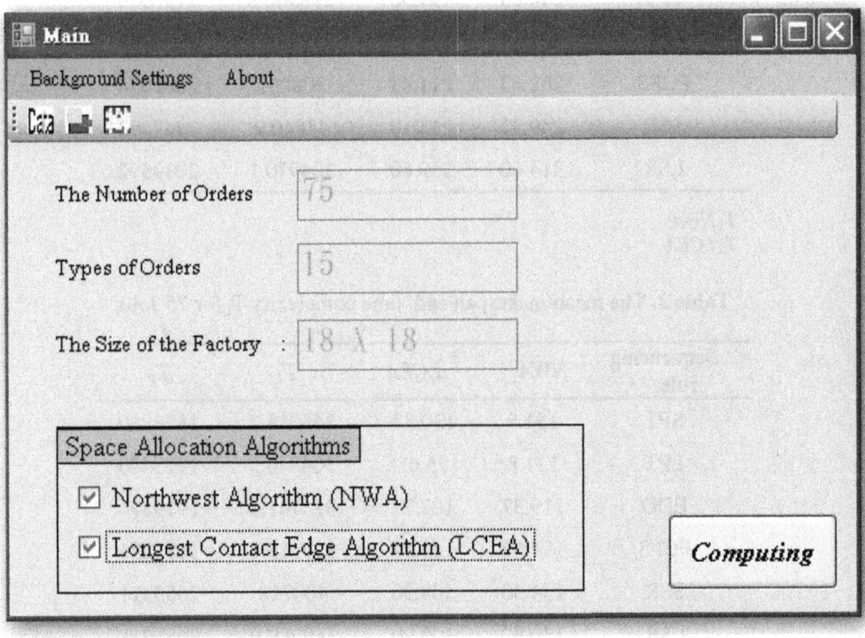

Fig. 3. Interface of Program

Table 1. The mean makespan and time complexity T_i for 25 Jobs

Sequencing rule	NWA	LCEA	T_1	T_2
SPT	85.20	75.67	142455	1221082
LPT	67.43	63.90	140084.4	1232975
EDD	72.53	67.90	124962.9	1293922
FCFS	79.93	72.40	143633.4	1238529
SSR	80.37	75.47	141535.8	1679210
LSR	77.30	72.60	171410.2	921304.1

T_1 NWA
T_2 LCEA

Table 2. The mean makespan and time complexity T_i for 50 Jobs

Sequencing rule	NWA	LCEA	T_1	T_2
SPT	306.33	232.73	324420.2	2012006
LPT	244.63	199.53	298566.8	2150939
EDD	262.53	198.60	296033.1	2056968
FCFS	282.63	211.43	308262	2081974
SSR	249.77	218.10	43540.9	2347393
LSR	314.60	259.60	304970.1	2039592

T_1 NWA
T_2 LCEA

Table 3. The mean makespan and time complexity T_i for 75 Jobs

Sequencing rule	NWA	LCEA	T_1	T_2
SPT	151.5	130.83	345015.2	1873230
LPT	111.8	95.63	304846.5	1923400
EDD	119.37	103.77	315741.6	1919576
FCFS	134.57	115.57	327115.7	1888276
SSR	134.30	108.90	400728	2655211
LSR	130.80	108.90	369289.9	2655211

T_1 NWA
T_2 LCEA

5 Concluding Remark

A space scheduling problem is an important issue in a machinery assembly factory. In this study, we compared NWA with proposed LCEA. We employed the makespan as the performance measure in the space scheduling problem. We employed the LCEA to allocate space for jobs. We found that the LCEA can do the better assignments than the NWA for all schedules obtained from traditional dispatching rules. However, the computing complexity of the LCEA is a little worse than that of the NWA.

There are some assumptions in this study, for example, all jobs are available at day one, shapes of the orders are all rectangles, a job will not be moved until it is done, there is no constraint on job's height, and the buffer or storage is available to fit in any number or any shape of jobs. It may result in different conclusions if some of assumptions are relaxed.

References

1. Liao, C.J., Lin, C.H.: Makespan minimization for two uniform parallel machines. International Journal of Production Economics 84(2), 205–213 (2003)
2. Perng, C., Lai, Y.C., Zhuang, Z.Y., Ho, Z.P.: Application of scheduling technique to job space allocation problem in an assembly factory. In: The Third Conference on Operations Research of Taiwan, vol. 59, pp. 1–7. Yuan-Zhe University, Tao-Yuan County (2006)
3. Perng, C., Lai, Y.C., Zhuang, Z.Y., Ho, Z.P.: Job scheduling in machinery industry with space constrain. System Analysis Section. In: The Fourth Conference on Operations Research of Taiwan, vol. 5, pp. 1–11. National Dong-Hwa University, Hwa-Lian City (2007)
4. Perng, C., Lai, Y.C., Ho, Z.P.: Jobs scheduling in an assembly factory with space obstacles. In: The 18th International Conference on Flexible Automation and Intelligent Manufacturing, Skovde, Sweden, vol. 4B, pp. 1–9 (2008)
5. Perng, C., Lin, S.S., Ho, Z.P.: On space resource constrained job scheduling problems- A container loading heuristic approach. In: The 4th International Conference on Natural Computation, vol. 7, pp. 202–206. Shandong University, Jinan (2008), doi:10.1109/ICNC.2008.419
6. Perng, C., Lai, Y.C., Ho, Z.P.: A space allocation algorithm for minimal early and tardy costs in space scheduling. In: 3rd International Conference on New Trends in Information and Service Science papers (NISS), Sec.1(6), Beijing Friendship Hotel, Beijing, China, pp. 1–4 (2009)
7. Pisinger, D.: Heuristics for the container loading problem. European Journal of Operational Research 141(2), 382–392 (2002)
8. Sule, D.R., Vijayasundaram, K.: A heuristic procedure for makespan minimization in job shops with multiple identical processors. Computers and Industrial Engineering 35(3-4), 399–402 (1998)
9. Gallo, G., Scutella, M.G.: A note on minimum makespan assembly plans. European Journal of Operational Research 142(2), 309–320 (2002)
10. Zobolas, G.I., Tarantilis, C.D., Ioannou, G.: Minimizing makespan in permutation flow shop scheduling problems using a hybrid metaheuristic algorithm. Computers & Operations Research 36(4), 1249–1267 (2009)
11. Beasley, J.E.: OR-Library: distributing test problems by electronic mail. Journal of Operational Research Society 41(11), 1069–1072 (1990)

12. Beasley, J.E.: OR-Library (5/1/2009),
    http://people.brunel.ac.uk/~mastjjb/jeb/info.htm
13. Pinedo, M.: Scheduling theory, algorithms and systems. Prentice Hall Press, New Jersey (2002)
14. Nichols, R.A., Bulfin, R.L., Parker, R.G.: An interactive procedure for minimizing makespan on parallel processors. International Journal of Production Research 16(1), 77–81 (1978)
15. Lian, Z., Jiao, B., Gu, X.: A similar particle swarm optimization algorithm for job-shop scheduling to minimize makespan. Applied Mathematics and Computation 183(2), 1008–1017 (2006)

Biography of Authors

Chyuan Perng is an associate professor in the Department of Industrial Engineering and Enterprise Information at Tunghai University, Taiwan. He received his Ph.D. degree in Industrial Engineering from Texas Tech University, USA. He has also participated in numerous industrial and governmental projects in Taiwan.

Yi-Chiuan Lai is an assistant professor in the Department of Business Administration at Providence University, Taiwan. He received his Ph.D. degree in Industrial Engineering from Iowa State University, USA. His primary research interests lie in simulation, scheduling, analysis of stochastic system, and product realization processes.

Zih-Ping Ho is currently a Ph.D. candidate in the Department of Industrial Engineering and Enterprise Information at Tunghai University, Taiwan. He also owns a system development Studio in the Innovation Center of the same school. He has also participated in several industrial and governmental projects in Taiwan.

Chin-Lun Ouyang graduated from the Department of Industrial Engineering and Enterprise Information at Tunghai University, Taiwan. He received a Master and Bachelor degree. He currently serves at the Conscription Agency Ministry of the Interior in Taiwan.

A New Approach for Chest CT Image Retrieval[*]

Li-dong Wang[1,2] and Zhou-xiang Shou[1]

[1] College of Computer Science and Technology, Zhejiang University, Hangzhou, Zhejiang, 310027, China
[2] Hangzhou Normal University, Hangzhou, ZheJiang, 310012, China
Violet_wld@163.com, shoupan@hz.cn

Abstract. A new approach for chest CT image retrieval is presented. The proposed algorithm is based on a combination of low-level visual features and high-level semantic information. According to the new algorithm, wavelet coefficients of the image are computed first using a wavelet transform as texture feature vectors. The zernike moment is then used as an effective descriptor of global shape of chest CT images in database, and the semantic information is extracted to improve the accuracy of retrieval. Finally, index vectors are constructed by the combination of texture, shape and semantic information, and the technique of relevance feedback is used in the algorithm to enhance the effectiveness of retrieval. The retrieval results obtained by application of our new method demonstrate an improvement in effectiveness compared to other kinds of retrieval techniques.

Keywords: Medical image retrieval, Wavelet transform, Zernike moment, Semantic information, Chest CT image.

1 Introduction

With DICOM (Digital Image Communication and Medician), information of patient can be stored with the actual images. The digitally produced medical images are produced in everincreasing quantities and used for therapy and diagnostics. The medical imaging field has generated additional interest in methods and tools for the management, analysis, and communication of these medical images. Many diagnostic imaging modalities are routinely used to support clinical decision making. It is important to extend such applications by supporting the retrieval of medical images by content.

In recent years, much research has been done into specific medical image retrieval system[1-4], and the research for medical image retrieval is based on the same anatomic region. Chest CT is gray-scale image, so the retrieval method based on color feature can't attain effective performance. Meanwhile, extensive experiments in content-based image retrieval(CBIR) systems have demonstrate that low-level features can not always describe high-level semantic concepts in the user's mind[5]. Therefore, this paper presents a new approach based on combination of low-level visual features and high-level semantic information. Firstly, we extract stexture feature of images by wavelet transform. The

---

[*] Sponsored by the Science and Technology Foundation of Hangzhou Normal University (2009XJ065).

wavelet transform based texture analysis method has proven to be an efficient method for texture analysis due to its property of both space and frequency localization in this specified manner. Then a region-based shape[6] descriptor is presented, which utilizes a set of the magnitudes of Zernike moments. We show that the Zernike moment descriptor can be used effectively as a global shape descriptor of an image, especially for a large medical image database, and then describe the process of extracting Zernike moment descriptor. Finally, in order to improve the retrieval performance, texture and shape features are combined with the high-level semantic feature to reduce the "semantic gap" between visual features and the richness of human semantics, and the technique of relevance feedback is used in the algorithm to enhance the effectiveness of retrieval.

Thus, based on the above technology, a simple prototype system is developed to compare the retrieval accuracies. Experimental results show that the method discussed in this paper is much more effective.

The paper is organized as follows. Section 2 gives a review of feature extraction, including wavelet transform, Zernike moments and semantic feature. In section 3, we proposed a method to combine texture, shape features with semantic information. In section 4, experiments have been conducted and the performance of the proposed algorithm is analyzed. The conclusions are presented in section 5.

2   Features

2.1   Wavelet Transform

It is found that various approaches to texture analysis are very diverse and in this respect, four categories can be defined, which are namely statistical, geometrical, model based and signal processing. The first three texture analysis methods are found suitable for the regular and near regular methods, and the texture information of chest CT images is focused on the middle region. Therefore, we utilize the signal processing algorithm. The wavelet transform based technique performs the space-frequency decomposition with low computational complexity, it has been proven to be an efficient method for texture analysis[7]. In this paper, Daubechies' wavelet is selected as bases due to its orthonormal characteristics.

The experimental procedure for chest CT image texture feature extraction method is as follows:

(1) Chest CT images are preprocessed and the sizes of the images are made to be same to make the system computationally less complex.
(2) A two dimensional Daubechies' wavelet transform is applied to decompose the image into its sub-band images. Three-level pyramidal decomposition is used in this work.
(3) All the sub-band images are stored for calculating the statistical features from the sub-bands.

Fig.1 shows Daubechies' wavelet based decomposed three sub-bands that are created from 512×512 chest CT image. In order to extract texture feature, all sub-band images from three levels at different resolutions will be calculated for feature vector. During statistical feature calculation, if the decomposed sub-band image$f(x,y)$ is with dimension(M,N), then the various statistical features like mean, variance, energy and entropy feature vectors of sub-band images are calculated using the notations[8] which are as follows:

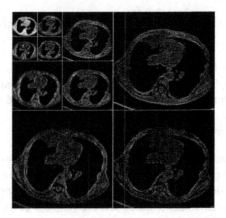

Fig. 1. 3-scale 2D wavelet transform from chest CT image

$$Mean = \frac{1}{MN}\sum_{x=1}^{M}\sum_{y=1}^{N} f(x,y) \tag{1}$$

$$Variance = \frac{1}{(MN)^2}\sum_{x=1}^{M}\sum_{y=1}^{N} |f(x,y) - Mean|^2 \tag{2}$$

$$Energy = \frac{1}{(MN)^2}\sum_{x=1}^{M}\sum_{y=1}^{N} |f(x,y)|^2 \tag{3}$$

$$Entropy = -\sum_{x=1}^{M}\sum_{y=1}^{N} f(x,y)\log f(x,y) \tag{4}$$

These four values are considered to be beneficial to be used as texture features and then calculated from all the sub-bands of chest CT images. Then the similarity between two images can be measured by Euclidean distance.

2.2 Zernike Moment Descriptor

The Zernike moment descriptor has such desirable properties: rotation invariance, robustness to noise, expression efficiency, fast computation and multi-level representation for describing the various shapes of patterns[9-10]. Therefore, we apply the Zernike moments to chest CT image retrieval for shape feature extraction. We can calculate the Zernike moment feature by (7) and (8).

$$C_{nl} = \frac{2n+2}{\pi} \int_0^1 \int_{-\pi}^{\pi} R_{nl}(r)\cos(l\theta) f(r,\theta) r dr d\theta \qquad (7)$$

$$S_{nl} = \frac{2n+2}{\pi} \int_0^1 \int_{-\pi}^{\pi} R_{nl}(r)\sin(l\theta) f(r,\theta) r dr d\theta \qquad (8)$$

where $l \geq 0$ and $n > 0$.

In our experiment, 9 Zernike moments of order 0 to 4 in n and l are extracted from the images, and the similarity between two images is measured by Euclidean distance.

2.3 Semantic Feature Extraction

Images in medical image database are in high similarity, it is difficult to retrieve the images with the same pathologic feature. The low-level feature extraction can not attain effective result. In order to retrieval accuracy of retrieval system, we try to reduce the "semantic gap" between visual features and the richness of human semantics. In this paper, we propose a method to extract semantic information.

Because the descriptions are subjective, different doctors have different diagnosis words for each image, but they have the similar meaning. Considering the doctor's diagnostic description (Table 1) for the images, we can extract keywords from these descriptions with doctor's instruction. According to medical knowledge, we derive the keyword sets(Table 2) as semantic information. Let L_i be the semantic feature vectors, $L_i =(W_{i1}, W_{i2}, \ldots W_{im})$, where m represents the number of keyword derived from description, W_{ij} is the weight of keyword j from the semantic information of image i. In order to reduce the computation complexity, the weight is equal to the number of times that keyword appears in the semantic information, if it does not appear, $W_{ij}=0$.

Table 1. Doctor's diagnostic description of image

Image	Doctor's diagnostic description of Chest CT image
6003	superior lobe of right lung Ca with review of middle lobe of right lung, metastatic carcinoma of mediastinal lymph node: chest symmetry, more textures in pulmones, disorder, apicale segmentum of right lung's superior lobe and right lung's middle lobe both have a nodular contour, no block in all levels of bronchus, mediastina have multi-auxe lymphoid nodes, no adhaesio, effusion in bilateral pleura.

Table 2. Keywords set

Keywords	The number of images			
	0030	0031...	1001	1002
chest symmetry	1	1	1	1
more textures	0	0	1	0
shadow	1	0	0	0
nodular contour	0	0	1	1
thoracic deformity	0	0	0	0

Let p, q represent the texture features of the query image and the database image respectively. The distance between them can be computed as:

$$D(p,q) = 1 - \frac{\sum_{k=1}^{m} W_{qk} \times W_{pk}}{\sqrt{(\sum_{k=1}^{m} W_{qk}^{2})(\sum_{k=1}^{m} W_{pk}^{2})}} \tag{9}$$

where m represents the number of keywords.

3 Feature Fusion

The retrieval method based on low-level features can not attain efficient performance. To improve the effectiveness and accuracy, we propose a method to combine low-level feature with high-level feature. The low-level features contain texture and shape. In order to enhance the effectiveness of retrieval, the technique of relevance feedback is used in the algorithm. The Rocchio algorithm[12], which uses user feedback on the relevancy of retrieved images, has been shown to improve query results. Fig.2 illustrates the major steps of the integrated retrieval.

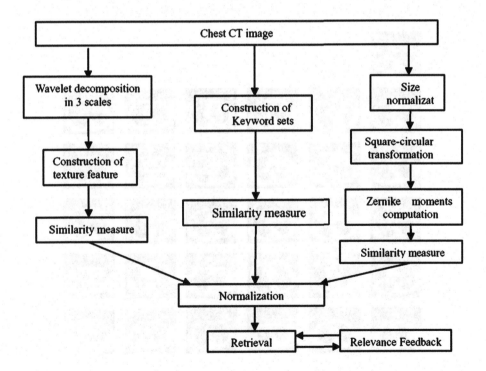

Fig. 2. Integrated retrieval algorithm

Since different feature's distance has different standard, before integrated retrieval we should have a normalization.

4 Results and Discussion

The method described above has been extensively applied on 217 chest CT images, which are from different patients in BMP format, represented by 512×512 size. In particular, a retrieved image is considered a match if it is in the same pathological modality as the query. This assumption is reasonable, since the images in same anatomic region of the same disease are chosen so that it can support clinical decision making.

Fig.3 illustrates a result based on wavelet transform. The retrieval result returns 30 similar images. The pathological modality of the query image is lung cancer. According to the medical knowledge, there are 24 returned images are in the same disease. Meanwhile, Fig.4 demonstrates an example query based on feature fusion. It shows that there are 27 similar images returned.

In system evaluation, the retrieval precision and average-r are calculated as statistical parameters.

$$precision = \frac{a}{a+b} \qquad (12)$$

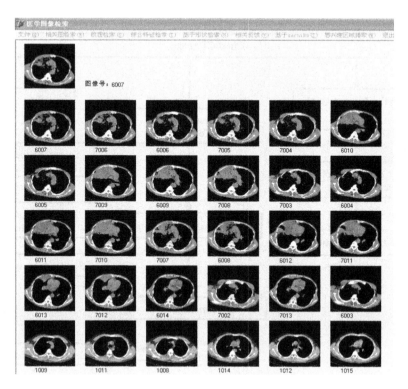

Fig. 3. Example of a query based on wavelet transform

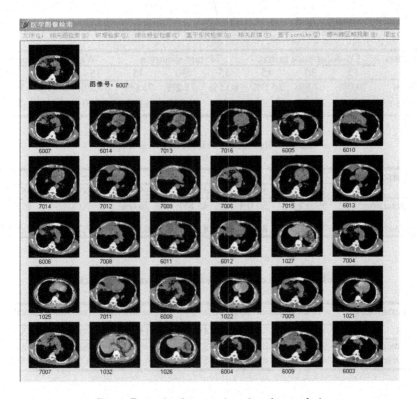

Fig. 4. Example of a query based on feature fusion

where a represents the number of similar images and b is the number of un-similar images in the results. That is, the percentage of similar images retrieved with respect to the number of retrieved images.

$$Average - r = \frac{1}{m}\sum_{r=1}^{m} \rho_r \qquad (13)$$

where m is a given constant, and in this paper $m = 8$, ρ_r is the rank of rth image in the first m similar images. Therefore the ideal value of $\rho_r = 4.5$, and the lower numbers indicate better results. In our case, the retrieval results return 15, 30, 45 images for each query. Tab.3-5 show the average precision, rank of different methods.

Tab.3 compares the results obtained by three texture extraction algorithms, which are co-occurrence matrix, texture spectrum and wavelet transform. A total average 84.5% matched retrieved images is achievable using the wavelet transform algorithm. The same experiment, performed using co-occurrence or texture spectrum, demonstrated lower performance on average precision and average-r.

Table 3. Results of texture indexing retrieval algorithms

parameters	Co-occurrence matrix			Texture spectrum			Wavelet transform		
	15	30	45	15	30	45	15	30	45
precision	70.0%	67.8%	65.0%	80.5%	78.2%	75.0%	87.0%	84.5%	79.2%
Average-r	6.900	7.250	7.200	6.750	6.875	6.900	5.000	5.000	5.750

Table 4. Results of retrieval based on shape

parameters	Hu moments			Zernike moments		
	15	30	45	15	30	45
precision	59.0%	50.0%	45.8%	76.0%	70.8%	69.0%
Average-r	7.170	8.520	8.640	6.900	7.250	7.200

Table 5. Results of retrieval based on relevance feedback

parameters	Feature fusion(first retrieval)			Feature fusion(second retrieval)		
	15	30	45	15	30	45
precision	92.5%	88.2%	83.0%	93.3%	90.2%	85.0%
Average-r	4.500	4.500	5.500	4.500	4.500	5.000

Tab.4 shows the quantitative results obtained by the application of integrated retrieval and shape algorithms. As illustrated, the feature fusion algorithm has a fairly high precision and small average-r, while demonstrating higher retrieval accuracy. Meanwhile, Zernike moment performed better than Hu moment. This is expected and mostly due to its property of multi-level representation for describing the various shapes of patterns and its rotation invariance.

Tab.5 shows the precision of second retrieval is higher than original retrieval. And the average rank reaches ideal value. Therefore the method of relevance feedback is successful with high retrieval efficiency.

5   Conclusion

In this paper, a new approach for chest CT image is presented. Texture and shape features are combined with semantic information for image retrieval. The experimental results show that the proposed algorithm is effective. Applying this method to medical image retrieval will support the clinical decision making and have an enormous potential.

However, further developments should be made in order to improve the retrieval accuracy. The semantic feature vector should be modified to reduce its size and improve its discriminative property.

Acknowledgement

The authors would like to thank Radiotherapic Department, Intervention Radiotherapic Department and Endoscopic Department of Inner Mongolia Hospital.

References

[1] Wan, h.-l., et al.: Texture feature and its application in CBIR. Journal of Computer-aided Design and Computer Graphics 15(2), 195–199 (2003)

[2] Shyu, C., Brodley, C., Kak, A., et al.: ASSERT, A physician-in-the-loop content-based image retrieval system for HRCT image databases. Computer Vision and Image Understanding 75(1/2), 111–132 (1999)

[3] Aisen, A.M., Broderick, L.S., Winer-Muram, H., et al.: Automated storage and retrieval of thin section CT images to assist diagnosis: System description and preliminary assessment. Radiology 228(1), 265–270 (2003)

[4] Sun, J., Zhang, X., Cui, J., Zhou, L.: Image retrieval based on color distribution entropy. Pattern Recognition Letters 27(10), 1122–1126 (2006)

[5] Liu, Y., Zhang, D., Lu, G., Ma, W.-Y.: A survey of content-based image retrieval with high-level semantics. Pattern Recognition (2006)

[6] Wen, C.-Y., Yao, J.-Y.: Pistol image retrieval by shape representation. Forensic Science International 155(1), 35–50 (2005)

[7] Paschos, G.: Fast Color: Texture Recognition Using Chromaticity Moments. Pattern Recognition Letters 21, 837–841 (2000)

[8] Borah, S., Hines, E.L., Bhuyan, M.: Wavelet transform based image texture analysis for size estimation applied to the sorting of tea granules. Journal of Food Engineering 79(2), 629–639 (2006)

[9] Mehrotra, R., Gary, J.E.: Similar-shape retrieval in shape data management. IEEE Comput. 28(9), 57–62 (1995)

[10] Kim, W.-Y., Kim, Y.-S.: A region-based shape descriptor using Zernike moments. Signal Processing: Image Communication 16, 95–102 (2000)

[11] Ortega, M., et al.: Supporting similarity queries in MARS. In: ACM Conf. on Multimedia, pp. 403–413 (1997)

[12] Baeza-Yates, R., Ribeiro-Neto, B.: ModernInformation Retrieval. Addison Wesley, Reading (1999)

Depth Extension for Multiple Focused Images by Adaptive Point Spread Function[*]

Yuanqing Wang and Yijun Mei

Box 1209, Department of Electronic Science & Engineering, Nanjing University No. 22,
Hankou Road, Nanjing 210093, P.R. China
yqwang@nju.edu.cn

Abstract. A novel depth fusion algorithm is proposed for multi-focused images based on point spread functions (PSF). In this paper, firstly, a discrete PSF model is present and functions with different parameters are prepared for the proposed algorithm. Based on the analysis of the effect of PSF, the detail process to fuse the depth of multi-focused images is described. By employing PSF to convolve each original images and comparing with its adjacent one in the image sequence, the focused and defocused region in the original images may be located. Combining the the focused region in the images according to Choose Max rules, an all-in-focus image may be got. The complexity of the algorithm for an image series with more than two original images is discussed at the end of this paper. Experimental results show that the image is distinctly segmented into multi-regions and the image edge is legible as well. The proposed algorithm based on PSF convolvetion as a focus measure has been shown to be experimentally valid. The fusion results are satisfactory with smooth transitions across region boundaries.

Keywords: Point spread functions (PSF), depth fusion, multi-focused images, image fusion.

1 Introduction

Depth of field is the area in front of and behind the specimen that will be in acceptable focus. In any imaging system(e.g. video camera, light optical microscope, etc.), as an inherent feature, image blurring is inevitable for all of the lens. The system works with finite depth of field which is relative to the numerical aperture (NA) and can not image all objects at various distances with the same clarity. Everything immediately in front of or in back of the focusing distance begins to lose sharpness.

Depth fusion is performed at three different processing levels according to the stage at which the fusion takes place: pixel, feature and decision[1]. In general, the techniques for multi-focused image fusion can be grouped into two classes: (1) Color related techniques, and (2) Statistical/numerical methods. Selecting the appropriate approach depends strongly on the actual applications. Some commonly

---

[*] This work is Funded by the key program from the National Natural Science Foundation of China(60832003).

H. Deng et al. (Eds.): AICI 2009, LNAI 5855, pp. 544–554, 2009.

used techniques in pixel-level fusion are: weighted combination; ptimization approach and biologically-based approaches such as neural networks and multi-resolution decompositions.[2]

Physically relevant features of an image often exist at its different scales or resolutions and may be exploit by multi-scale and multi-resolution approaches. The basic idea in multi-scale and multi-resolution techniques is to decompose the source images at first by applying the Laplacian pyramid[3], wavelet[4] or complex wavelet[5] transform, then the fusion operation on the transformed images is performed and finally the fused image is reconstructed by inverse transform. These techniques may obtain spatial and frequency domain localization of an image and can provide information on the sharp contrast changes. Multi-scale methods involve processing and storing of scaled data at various levels which are of the same size as that of the original images. This results in a huge amount of memory and time requirement [2]. The support vector machines (SVM) is a classification technique that has outperformed many conventional approaches in various applications. Shutao Li use the discrete wavelet frame transform(DWFT) for the multi-resolution decomposition, and then replace the choose-max rule by support vector machines (SVM) for fusing the wavelet coefficients.[6] Vivek Maik present a method to extend depth of field.[7] A single fused image is obtained by decomposing the source images using filter subtract decimate in Laplacian domain, and then extract the depth of focus in the source images using sum-modified-Laplacian. K. Aizawa[8] present an algorithm by convolving the different focused image with blurring functions (Gaussian functions). After the nth iteration, the sharp image may be obtained. This method is formulated based on a linear imaging model and does not need any object segmentation or complex scene content-dependent processing. But the iterative generation algorithm will be complex and time-consuming will be great for multi-focused images more than three images.

We have research a novel algorithm based on Point Spread Function (PSF) which has strong immunity from noise. In our work, PSF is applied as a tool to purposely blur some specified in-focus region in an image, just like a typical lens system to image the objects in the scene out of focus. With proposed method, the in-focus region in an image may be accurately located. In this paper, in Section 2, we gave a model of a PSF along with analysis of its effect to a focused image, and the criterion is described to detect the blurring region and estimate the blurring status. In Section 3, the proposed algorithm to obtain a depth fusion image from a series of original images is described in detail. Experimental results and conclusion are presented in Section 4.

2 Detecting the In-focused Image by PSF

2.1 Basic Idea

Most lenses are not perfect optical systems. The light from a single object point passing through a lens, which with aperture of finite size in diameter, is converged into a cone with its tip at the CCD (Charge Couple Device) plane, if the point is perfectly in focus; or slightly in front of or behind the CCD plane, if the object point is somewhat out of focus. In the out-of-focus case the point is rendered as a circle where the CCD plane cuts the converging cone or the diverging cone on the other side of the image point. This circle of confusion will introduce a certain amount of blurring.

Blurring can be numerically characterized by a Point Spread Function (PSF). PSF is the output of the imaging system for an input point source. Assuming that intensity distribution of scene, I(xs, ys, zs), is recorded by a typical lens system. For the object out of focus with the point spread function, Pz, the image function, f(xi, yi), may be calculated as below:

$$f(x_i, y_i) = I(x_s, y_s, z_s) * P_z \qquad (1)$$

where, * is the convolving operator.

In the similar way, a blurry image may be reconstructed from a well focused image. The discrete blurry image f(k, l) for digital image may be obtained from the original sharp image g(n, m) by convolving it with discrete PSF:

$$f(k,l) = \sum_{i=-\infty}^{+\infty} \sum_{j=-\infty}^{+\infty} g(k+i, l+j) h(-i, -j) \qquad (2)$$

where the function, h, is the discrete PSF.

Comparing equation (2) with (1), they are different in approach but equivalent in result. If the discrete PSF, h, is suitably choiced, blurry image, f(k, l), will be the same blurry as image function, f(xi, yi).

2.2 Series of PSF

According to the isoplanatism condition, PSF has the same shape over the whole field of view at the same focal plane. The complicated PSF may be simplified to a concise expression as:

$$P(r) = \frac{1}{4r^2} \qquad (3)$$

where, r is the radius to the center of the blur circle, and for discrete case, it may be denoted as a symmetric matrix, h, which's diagonal elements are confirmed as:

$$h_{ii} = \begin{cases} i^2 & ,i \le (n+1)/2 \\ (n+1-i)^2 & ,i > (n+1)/2 \end{cases} \qquad (4)$$

where, i=1,2,3,......,n, and n is odd number.

Define a PSF vector, a, as below:

$$a_i = \sqrt{h_{ii}} \qquad (5)$$

Then, the matrix of discrete PSF is:

$$h = ka^T a \qquad (6)$$

where, k is normalized coefficient, and

$$k = 1 \Big/ (\sum_{i=1}^{n} a_i)^2 \qquad (7)$$

For example, define a PSF vector as below:

$$a = (1, 4, 6, 4, 1) \tag{8}$$

Then, the PSF matrix, h, is a 5×5 mask:

$$h = \frac{1}{(\sum\limits_{i=1}^{n} a_i)^2} a^T a = \frac{1}{256} \begin{bmatrix} 1 & 4 & 6 & 4 & 1 \\ 4 & 16 & 24 & 16 & 4 \\ 6 & 24 & 36 & 24 & 6 \\ 4 & 16 & 24 & 16 & 4 \\ 1 & 4 & 6 & 4 & 1 \end{bmatrix} \tag{9}$$

A series of PSF may be used in our work, which is listed in Table 1:

Table 1. PSF vectors employed to fuse multi-focused images ($P_0(r) = 1$)

PSF	P_0	P_1	P_2	P_3	P_4	P_5	P_6	P_7
a	1,3,11,3,1	1,3,10,3,1	1,3,9,3,1	1,3,8,3,1	1,4,8,4,1	1,3,6,3,1	1,4,6,4,1	1,2,3,2,1
k	441	400	289	256	361	196	256	81
$W_H^{(1)}$	0.52	0.61	0.70	0.85	0.91	1.1	1.3	1.4
$Sr^{(2)}$	0.4187	0.3460	0.2803	0.2500	0.2244	0.1837	0.1406	0.1111

(1)$W_H$the width of PSF, i. e. the Full Width at Half Maximum of the PSF, (2)$Sr=P(r)/P_0(r)$, i.e. the central intensity of PSF of a defocused image compared to that of a focused image.

Fig.1 shows the profile of each PSF in Table 1. When we convolve a focused image g(n,m) with a discrete PSF, it is equivalent to getting a new image f(k, l) by low-pass filter. We can set different components in the PSF matrix in Equation (8), and make the blurry image f(k, l) to be blurred with different degree relative to the original sharp image g(n,m). Different PSF provides different results for the same original image.

Clearly, suitable PSF must be selected to simulate the out-focus-blurring by imaging lens, and theoretically, infinite amount of

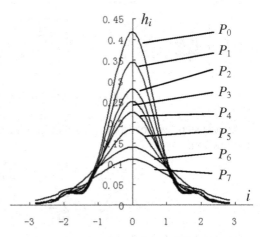

Fig. 1. The profile of different PSF

discrete PSF is required. In most case, the eight preparative functions in Table 1 are enough for depth fusion. The required PSF can may be estimated by inspecting the image.

3 Depth Fusion by PSF

3.1 Identification the In-focus Region

As show in Fig.2, original image g_1 and g_2 are two images respectively with different focused region. For easier to describe our idea, the scenes is simplified that it only consists of two objectives which are respectively imaged within circular and triangular regions in the image, see Fig.2. We describe the focused region with black color, defocused region with dark gray. In image g_1, the triangular region is within the focused range and the circular region is the defocused region. In image g_2, the triangular region is out of focus and the circular region is the in focus.

The blurry image $f_i(k, l)$ may be obtained as below:

$$f_i = g_i * h_i \tag{10}$$

where $i=1,2$, h_i is point spread function, and * represents the convolution.

Convolved by different PSF, the original images will be blurred at different degree.

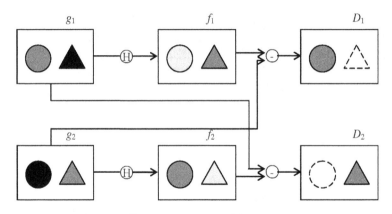

Fig. 2. Forming candidates of focus region

At next step, we obtain the candidate graphics D_i, as shown in Fig. 2, by the following function:

$$D_1 = (\sum_{j \in S} g_{2j} + \sum_{j \in B} g_{2j}) - (\sum_{k \in S} g_{1k} + \sum_{k \in B} g_{1k}) * h_1$$
$$D_2 = (\sum_{k \in S} g_{1k} + \sum_{k \in B} g_{1k}) - (\sum_{j \in S} g_{2j} + \sum_{j \in B} g_{2j}) * h_2 \tag{11}$$

where symbols, S and B, respectively denote the sharp and blur regions in the images.

Equation (11) may be rewritten as:

$$D_1 = (\sum_{j \in S} g_{2j} - \sum_{k \in B} g_{1k} * h_1) + (\sum_{j \in B} g_{2j} - \sum_{k \in S} g_{1k} * h_1)$$
$$D_2 = (\sum_{k \in S} g_{1k} - \sum_{j \in B} g_{2j} * h_2) + (\sum_{k \in B} g_{1k} - \sum_{j \in S} g_{2j} * h_2) \tag{12}$$

In order to enhance the absolute dispersion and avoid the negative value, we use the square number to replace the direct subtraction as below:

$$D_1 = (\sum_{j\in S} g_{2j} - \sum_{k\in B} g_{1k} * h_1)^2 + (\sum_{j\in B} g_{2j} - \sum_{k\in S} g_{1k} * h_1)^2$$

$$D_2 = (\sum_{k\in S} g_{1k} - \sum_{j\in B} g_{2j} * h_2)^2 + (\sum_{k\in B} g_{1k} - \sum_{j\in S} g_{2j} * h_2)^2 \quad (13)$$

The focused region in one image probably present as blurry region out of focus in another image. We simply treat the area of the sharp region in one image as the same as that of the blurry region in another image, which means that:

$$Area(\sum_{j\in S} g_{2j}) \approx Area(\sum_{k\in B} g_{1k})$$

$$Area(\sum_{k\in S} g_{1k}) \approx Area(\sum_{j\in B} g_{2j}) \quad (14)$$

It's apparent that the more desirable the mask coefficients of PSF are, the lower values of the second term in polynomial (13) will be. If set an identification threshold, V_{th}, which is large than the values of the first term in Equation (13) and small than that of the second term. All of the pixels in g_1, g_2, which behave with variant larger than identification threshold in the candidate graphics, D_2, D_1, may be treated as in-focus imaging region. The identification threshold is relative to the scene, such as content, texture and illumination. For idiographic applied field, it may be confirmed via statistic data from experiments or derived from experience. Then a binary feature image, F_1, F_2, may be obtained in which the high gray level pixel is sharp pixel for correspondence original image. The desirable PSF may be chosen from one of the candidate vectors listed in Table 1, according to blurry degree of the image.

In fact, if the discrete PSF is m×m matrix, the area of the high gray level pixel in binary feature images, F_1, F_2, will be little large than that of the sharp region in original images, g_1, g_2. Considering this condition, we use the fusion principle to obtain the fusion image, $I(x, y)$, as below:

$$I(x, y) = \begin{cases} g_1(x, y) & ; F_2(x, y) = 1, F_1(x, y) = 0 \\ g_2(x, y) & ; F_2(x, y) = 0, F_1(x, y) = 1 \\ [g_1(x, y) + g_1(x, y)]/2 & ; other \end{cases} \quad (15)$$

For the case of more than two original multi-focus images, an iterative algorithm may be applied. Assume to there are n images, g_1, g_2, ..., g_n, the steps of iterative algorithm are taken as below:

STEP 0: $g_1 = g_1$, $g_2 = g_2$, i=2;
STEP 1: confirm the PSF mask according to the status of g_1 and g_2;
STEP 2: obtain binary feature images, F_1, F_2;
STEP 3: obtain the fusion image, $I(x, y)$, as below:

$$I(x,y) = \begin{cases} g_1(x,y) & ; F_2(x,y)=1, F_1(x,y)=0 \\ g_2(x,y) & ; F_2(x,y)=0, F_1(x,y)=1 \\ g_1(x,y) & ; F_2(x,y)=0, F_1(x,y)=0 \\ [g_1(x,y)+g_1(x,y)]/2 & ; F_2(x,y)=1, F_1(x,y)=1 \end{cases} \quad (16)$$

STEP 4: $i = i+1$, if $i > n$ go to STEP 6;
STEP 5: $g_1 = I(x, y)$, $g_2 = g_i$, go to STEP 1;
STEP 6: STOP;

3.2 Candidate of PSF

The selection of appropriate PSF has a close connection with blurring degree of the region. So it's important to estimate image blur by some appropriate method.

Suppose that $I_i(x, y)$ and $I_j(x, y)$ are two differently focused images in a series of multi-focus images with different focus settings, a differential image is defined as below:

$$I_{i,j}(x,y) = |I_i(x,y) - I_j(x,y)| \quad (17)$$

where $I_k(x, y)$ is the pixel intensity value at position (x, y) in image k. Generally, set $j=i+1$ correspond to the adjacent image of the ith one.

The value of $I_{i,j}(x, y)$ may indicate the blur degree. But use of solely adjacent pixel-level indicators can make decisions vulnerable to wide fluctuations dependent on specific parameters such as noise, brightness and local contrast. Hence, corroboration from neighboring pixels of decision choices becomes necessary to maintain robustness of the algorithm in the face of the above adverse effect. Adding this corroboration while maintaining pixel-level decisions requires summing the $I_{i,j}(x, y)$s over an $l \times l$ region surrounding each decision-point. This yields an indicator for focus measure:

$$B_{i,j}(x,y) = \sum_{\tau=-l/2}^{l/2} I_{i,j}(x+\tau, y+\tau) \quad (18)$$

where, aggregation size, k, is constant number decided by image resolution. Range of aggregation size is about 20~50 for different resolution.

The larger the relative magnitude of the indicator, the higher the probability that its corresponding blur circle radius is larger, and the larger F_W is of the appropriate PSF in Table 1. The continuous value $B_{i,j}(x, y)$ may be put into several discrete subsections correspond to preset PSF.

4 Experiment Result

In this section, we experimentally demonstrate the effectiveness of the proposed algorithm. Experiments were performed on 256-level original images of size 800×600 by Matlab 6.0 in PC with Pentium 1.4GHz CPU and 628M RAM. Experiments were carried out on two actual scenarios, which contain multiple objects at different distances from the camera one or more objects naturally become out of focus when the image is taken. For example, the focus is on the trunk in Fig 3(a), while that in Fig 3(b) on the shrub and bench.

(a) image A1 of scene A

(b) image A2 of scene A

Fig. 3. Original image pair

Convolved by PSF, the original images will be blurred as shown in Fig.4. We use two PSF to respectively blur the original image, by PSF with vector $a = (1, 4, 6, 4, 1)$ to get blurry image B1 and $a = (1, 2, 3, 2, 1)$ to get blurry image B2 as illustrated in Fig.4.

(a) blurry image B1 from image A2

(b) blurry image B2 from image A1

Fig. 4. Blurry image pair

The fused image, with chosen identification threshold equaling to 152, is shown in Fig. 5. The fused image merges the respective portions that are in focus from the original images and all objects in the image are sharp. It can be seen in the final fused image that any undesired discontinuities at the region boundaries are prevented. The elapsed time is 1.362s to achieve the composite result.

Fig.6 shows some images with another typical content to depict the similar process.

Fig.5. Result image of fusion

The result of image fusion applied to two images having different depth of focus, with the focus in on the scrip in Fig 6(a), while that in Fig.6(b) on the book. Please notice that, in Fig.6(b), the scrip is blurred by Gaussion blur and little moving blur. The fused image in Fig.6(c) is gotten by PSF with vector $a = (1, 4, 6, 4, 1)$ to successively convolve the original images with two times, 178 for identification threshold. It takes 1.322s to finish the conduct.

(a) Original image 1   (b) Original image 2   (c) Fused image

Fig. 6. Second example for the proposed algorithm

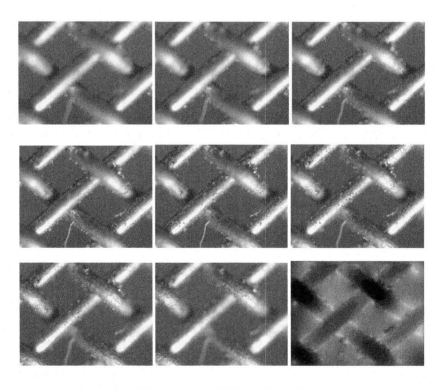

Fig. 7. Image series of different focused range

Fig.7 shows some images with another typical content to depict the similar process. The image series includes eight different focused images captured by a microscope. The depth distribution of the scene is shown at the right-bottom, in which the more dark is the gray level, the more close the object is to the lens of imaging system. Result image with depth extension is shown in the bottom in Fig. 8. It is gotten by PSF with vector $a = (1, 4, 6, 4, 1)$, 178 for identification threshold.

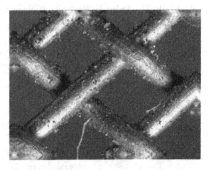

Fig. 8. Depth extension for image series

5 Conclusion

The defocus portions in one image represent dissimilar blurry degree, thus the vector of PSF is different for original image. The "ideal" way to choose the appropriate vector is accurately evaluate the blurry degree of each image's region. The blurry degree may be confirmed by different way such as gray gradient difference, wavelet transform and so on. In mathematics, the local regularity of the image function is often measures with Lipschitz exponent. Our method for blur estimation is based on calculating the Lipschitz exponent in all points where a change in intensity is found either in the horizontal or vertical direction. The strategy for detecting and characterizing singularities of the original images is important for this fusion theorem. The detail method may be described in the other paper. In fact, assisting in correlative technology, the proposed algorithm may provide acceptable results in most cases.

References

1. Zhang, Z., Blum, R.S.: A categorization of multiscale-decomposition-based image fusion schemes with a performance study for a digital camera application. Proc. IEEE 87(8), 1315–1326 (1999)
2. Piella, G.: A general framework for multiresolution image fusion: from pixels to regions. Information Fusion 4, 259–280 (2003)
3. Maik, V., Shin, J., Paik, J.: Pattern Selective Image Fusion for Multi-focus Image Reconstruction. In: Gagalowicz, A., Philips, W. (eds.) CAIP 2005. LNCS, vol. 3691, pp. 677–684. Springer, Heidelberg (2005)
4. Pajares, G., de la Cruz, J.M.: A wavelet-based image fusion tutorial. Pattern Recognition (2004)
5. Forster, B., Van de Vill, D., et al.: Complex Wavelets for Extended Depth-of-Field: A NewMethod for the Fusion of Multichannel Microscopy Images. Microscopy Research and Technique 65, 33–42 (2004)
6. Li, S., Kwok, J.T., Tsang, I.W., et al.: Fusing Images with Different Focuses. Using Support Vector Machines 15(6), 1555–1561 (2004)

7. Maik, V., Shin, J., Paik, J.: Pattern Selective Image Fusion for Multi-focus Image Reconstruction. In: Gagalowicz, A., Philips, W. (eds.) CAIP 2005. LNCS, vol. 3691, pp. 677–684. Springer, Heidelberg (2005)
8. Aizawa, K., Kodama, K., Kubota, A.: Producing object-based special effects by fusing multiple differently focused images. IEEE Transactions on Circuits and Systems for Video Technology 10(2), 323–330 (2000)

Wang Yuanqing graduated from Zhejiang University, Hangzhou, Zhejiang, P.R.C. and received his Master degree from the same University, all in the optical & electrical engineering, received his Doctorial degree from Nanjing University,. Now, he acts as a professor in the department of electronic science & engineering in Nanjing University. He has been involved in research work on auto-stereoscopic display, non-intrusive human-computer interface, . His present research interest includes the opti-electronics and information process, especially in sensor, 3D vision and display. He has compered many research projects such as National Hi-tech Research and Develop Plan (863 Plan) of China, National Natural Science Foundation of China, Jiangsu Hi-tech Plan, the 11th Five Years Research Plan and so on.
yqwang@nju.edu.cn

Image Annotation Using Sub-block Energy of Color Correlograms

Jingsheng Lei

School of Computer and Information Engineering, Shanghai University of Electric Power,
Shanghai 200090, China
jshlei@126.com

Abstract. This paper proposes an algorithm using local energy of color correlograms without any explicit using sub-block energy of color correlograms. The sub-block energy is defined as sub-windows from color correlograms information based on color distribution of original image. The model for image annotation involves computing histogram using color correlograms and analysis its sub-block characteristics. Similarly, sub-block energy is applied to annotate image's class and got a satisfied result. The model is fast and invariant to image's size and rotation. The comparison with SVM is done by experiments demonstrate the model is quite successful in annotation of image.

Keywords: Image Annotation, Color Correlogram, Sub-block Engery.

1 Introduction

With the development of network and multimedia technology, the storage of image information is expanding quickly and image retrieval has become a hotspot of image research field [1-2]. Content-based image retrieval (CBIR) which uses image content such as color and texture to compute the similarity of images have succeeded in fingerprint and logo recognition, etc. However, there is a huge gulf which is called "semantic gap", the lack of coincidence between the low-level features extracted from the visual features and high-level semantics concept, result that CBIR cannot provide meaningful results, and current state-of-art computer vision technology lags far behind the human's ability to assimilate information at a semantic level [3].

Using different models and machine learning methods to find the relation between image visual features and keywords for them from labeled images, and propagating keywords to unlabeled images, this is image annotation. There are three main methods for image annotation: annotation based on image segmentation, fixed size block and annotation based on image classification. Annotation based on image segmentation [4] depends on image visual features and precise results of segmentation [5-7]. The main point is how to correspond features of the regions to keywords. In the ideal condition, every segmented region corresponds to one object. However, the outcome of image segmentation is not satisfied at present. So there exists a big diversity between the expression of image object level part and human vision system. And this problem also exists in fixed size image division, for it may divide one object into

H. Deng et al. (Eds.): AICI 2009, LNAI 5855, pp. 555–562, 2009.

several blocks or put several objects into one block. Compared with the two methods, image annotation based on image classification can avoid the low accuracy caused by wrong image division. Cusano et al. [8] categorized images into various groups with Support Vector Machines (SVMs) and counted the frequencies of co-occurrence between the keywords in each group. They then used them to annotate an input image. Gauld et al. [9] treated features separately and trained the optimal classifier for each feature in advance. But these methods using features extracted from whole images that result in high computing cost which is not suitable for large data set. To avoid the problem mentioned above, this paper realize an annotation procedure by sub-block energy based on color correlograms [10] which indicates the color and spatial distributions of the image directly.

To address the problems mentioned above, in this work, a fast algorithm for image annotation based on color correlograms is proposed. Using sub-block energy of correlograms, the algorithm can find the relative energy blocks in Euclid space. Differ from the traditional annotation approach based on the whole image features, the proposed method analyze the central area which associate with the image semantics and only vision features of the area are extracted, then LIBSVM is used for image classification to get the relationship between image visual features and semantics, lastly image annotation is executed.

The remainder of this paper is organized as follows: The sub-block energy of color correlogram are introduced in section 2 and a fast image annotation approach based on central area analysis and key issues are discussed in detail in section 3. In section 4, experimentation results are presented and the concluding remarks are given in the last section.

2 Sub-block Energy of Color Correlograms

2.1 Color Correlograms

Traditionally the distance of points in an image can be measured by Euclid space. Correlograms are graphs giving the information of auto-correlation changes with distance, it captures the spatial correlation of colors in an image, rectifies the major drawbacks of the classical histogram method

For an image distributed in a discrete domain $I \subset Z^2$ and its gray level is L, $c_i, i = 1...C$, where c_i is the i color index and $C \leq L$ is the gray level for statistic,. h_{c_i} is the histogram value of color correlogram for c_i.

In Fig.1., a window $w_h \times w_v$ is used as a distance to compute histogram of gray value of p_1 with p_2 and p_3. When moving the window in the image, we can get a whole histogram table values which indicates the distance of $c_i \rightarrow c_j$ in C. Therefore, If we want the color auto-correlograms, we can set $i = j$ as the value of histogram.

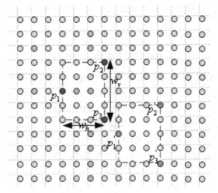

Fig. 1. Array image and its color correlograms window

2.2 Sub-block Energy

The correlograms shows the changes of the color histogram with the spatial distribution of pixels.

The color correlogram is selected to implement the sub-block energy model because it is color histogram with spatial information. It signals the presence of local color distribution. In the sub-block model, a given image is computed with the feature of correlogram. Consequently, the correlogram is pieced into sub-blocks. The sub-blocks are

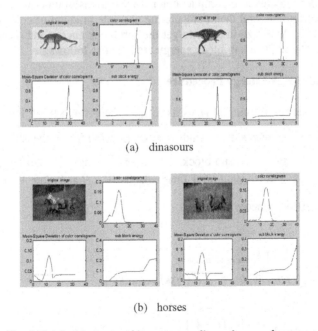

(a)   dinasours

(b)   horses

Fig. 2. Original images and its corresponding color correlograms

computed by absolute gradient of correlograms. Here we define the absolute gradient of h_{c_i} as below

$$Cov(k) = |h_{c_i} - \overline{h_c}| \tag{1}$$

In order to use sub-blocks to annotate the image, we should erase the spatial changes of sub-blocks. In Fig.2., we can note that the color correlograms is different and the corresponding sub-block energy changes dynamically. The curve line of sub-block is similar when they belong to a class while different classes are very different. Covariance values of correlograms give variability of the statistic of gray color. The sub-block energy is computed by summing the squared values of block.

$$E_k = \sum_{i=(k-1)*blocksize}^{(k)*blocksize} COV(i) \tag{2}$$

For the implementation of annotation of image, each type of images has their uniquely identifiable features which can discriminate from others, surely other features can be combined for annotation, but here we only put the sub-block vectors into high-space. When a test sample is filtered into its sub-block energy, we can compute its Euclid distance in high-dimensional space.

3 Sub-block Energy Algorithm

In this paper, the basic principle of sub-block energy algorithm is indicated as: firstly, chose a proper gray level to compute color correlograms in order to produce the spatial color distribution, and then compute the absolute gradient of color correlograms, finally, select a block size in order to get the sub block energy and order them by the energy size. The steps of sub-block energy are given as follows.

3.1 The Sub-block Energy Procedure

Input: The set of training images is $S_{training}$ and the set of testing images is $S_{testing}$. $w_h \times w_v$ is selected as window size. *Levelwidth* is the step size of L for histogram and sub-block size is $Sub - Block$. $L = 256$, $C = \lfloor 256/levelwidth \rfloor$. Image size is $W \times H$ and often $W = H$.

Output: Sub-Block Energy of Images

```
Begin
    For each  I ∈ [S_training, S_testing]
        I ← ⌊I/levelwidth⌋, C ← ⌊256/levelwidth⌋;
        TempC ← [1:C;1:C] ← (0,0);
        For each  i ∈ [1:H - w_v]
            For each  j ∈ [1:W - w_h]
```

$$m \leftarrow I(i, j) + 1;$$
$$n \leftarrow I(i + w_v, j + w_h) + 1;$$
$$TempC(m, n) \leftarrow TempC(m, n) + 1;$$
$$TempC(n, m) \leftarrow TempC(n, m) + 1;$$

   End
  End
  For each $i \in [1 : H - w_v]$
    For each $j \in [w_h + 1 : W]$

$$m \leftarrow I(i, j) + 1;$$
$$n \leftarrow I(i + w_v, j - w_h) + 1;$$
$$TempC(m, n) \leftarrow TempC(m, n) + 1;$$
$$TempC(n, m) \leftarrow TempC(n, m) + 1;$$

    End
  End
  For each $i \in [1 : C]$

$$hc(i) \leftarrow [TempC(i, i) / \sum_{i=1}^{C} TempC(i, i)];$$

  End
  For each $i \in [1 : C]$

$$Cov(i) \leftarrow hc(i) - \overline{hc};$$

  End
  For each $k \in [1 : C]$

$$E_k \leftarrow \sum_{i=(k-1)*blocksize}^{(k)*blocksize} COV(i) \quad ;$$

  End

$$E \leftarrow Sort(E);$$

 End
End

After getting the results of sub-block energy, we can use it for annotating a testing set. When an unknown image is input into the model, firstly, its sub-energy is calculated base on the algorithm. For the reason of the sort of energy, there are is no need to align at location. When we describe energy sub-block as the point of dimensional space $\lfloor C / blocksize \rfloor$. We can use the Euclid distance to measure the class.

3.2 The Annotation Procedure

Input: The sub-block of training images with annotation and the sub-block of testing images without annotation.

Output: The set of testing with annotation.

```
Begin
```

$$\text{For each } e_{testing} \in E_{testing}$$

$$D \leftarrow [];$$

$$\text{For each } e_{training} \in E_{training}$$

$$D \leftarrow [D, Dis \tan ce_{Eculid} (e_{testing} - e_{training})];$$

```
        End
```

$$D \leftarrow Sort(D);$$

$$L(e_{testing}) \leftarrow L(\{e_{training} \mid e_{training} = \min(D)_{e_{training}}\})$$

```
    End
End
```

After the annotation, images are labeled. In order to improve the performance, the training set energy are represented by vectors. In Additional, the algorithm can parallel executed to speed up the efficiency of computing.

4   Experimental Results

In order to verify the correctness and effectiveness of our proposed solution, an annotation system is developed. The scheme is tested on the real images with different semantic. The color correlogram parameter *levelwidth* is set as 8, others(w_h, w_v, *blocksize*) are variable. For the LIBSVM [11], its parameters are s=C-SVC, c=sigmoid, *gama*=1/k, *degree*=3. Firstly, the testing image is transformed into

Fig. 3. Annotation Result (a)image to be annotated; (b)first match image; (c) match image after first; (d) match image after second

gray where range value of pixel is 0 to 255. Using the proposed annotation, we can label the testing image properly and also we can get a list of compared images based on distance from testing image to training one. And also, the list gives the information of similar images. From the Fig.3., we can notice that the testing image(dinosaurs) is labeled correctly and the found list have some similar images(buses are the second and third match image). Therefore, the sub-block energy algorithm is suitable for annotation of images. On the other hand, the annotation results between sub-block algorithm and LIBSVM are compared. The test is performed using 80 items for training and 20 for each category with cross validation. Table.1 shows the result. From the table, we can find the sub-block algorithm is better than LIBSVM. The result indicates that the sub-block energy algorithm uses the characteristics of features, and can achieve some better correct annotation for images with complex background. And the small sub-block energy can gives more high speed for large dataset.

Table 1. Annotation Compare of Sub-Block Energy and SVM (E :Sum-Block Energy; S:SVM)

Train=240			BlockSize				
Test = 40			16	8	6	4	2
w_v = w_h	2	E	67.5	67.5	65.0	70.0	67.5
		S	58.5	50.0	50.0	60.0	50.0
	4	E	70.0	62.5	62.5	65.0	65.0
		S	55.0	55.0	50.0	45.0	50.0
	8	E	67.5	60.0	67.5	62.5	67.5
		S	57.5	50.0	50.0	40.0	50.0
	16	E	70.0	82.5	62.5	70.0	60.0
		S	60.0	50.0	50.0	42.5	47.5

The sub-block energy also has the ability of find similar images from the measure energy list. Therefore, the result of annotation for images is satisfied and the computing cost is low based on the proposed algorithm.

5 Conclusions

In this paper, we present an algorithm for annotation of images. The contributions of this paper can be summarized as follows:

(1) The approach uses an image's histogram and its spatial information with small feature vector, it can reduce computing cost dynamically. The proposed algorithm can applied to large image dataset.

(2) This paper proposes a model using sub-block energy based on color correlograms. The representation of the sub-block is a small vector. The vector can be fast computed and can be regarded as a criterion for image's annotation using spatial information. The algorithm can annotate images with different complex background. The algorithm is robust.

Experimental results indicate our method is superior in terms of accuracy, robustness, and stability when compared to SVM method. The novelty of this solution is to find the local relationship of image for annotation. In addition, the proposed method can reduce computing cost. More studies in the aspect of image's color correlograms and its application should be invested in future.

References

1. Smeulders, A., et al.: Content-based image retrieval at the end of the early years. IEEE Trans. PAMI 22(12), 1349–1380 (2000)
2. Zhang, R., Zhang, Z.: Effective image retrieval based on hidden concept discovery in image database. IEEE Trans. on Image Processing 16(2), 562–572 (2007)
3. Naphade, M., Huang, T.: Extracting semantics from audiovisual content: The final frontier in multimedia retrieval. IEEE Trans. on Neural Networks 13(4), 793–809 (2002)
4. Xuelong, H., Yuhui, Z., Li, Y.: A New Method for Semi-Automatic Image Annotation. In: The Eighth International Conference on Electronic Measurement and Instruments, vol. 2, pp. 866–869 (2007)
5. Mori, Y., Takahashi, H., Oka, R.: Image-to-word transformation based on dividing and vector quantizing images with words. In: Proceedings of the International Workshop on Multimedia Intelligent Storage and Retrieval Management (1999)
6. Duygulu, P., Barnard, K., Freitas, J., Forsyth, D.: Object recognition as machine translation: Learning a lexicon for a fixed image vocabulary. In: Heyden, A., Sparr, G., Nielsen, M., Johansen, P. (eds.) ECCV 2002. LNCS, vol. 2353, pp. 97–112. Springer, Heidelberg (2002)
7. Jeon, J., Lavrenko, V., Manmatha, R.: Automatic image annotation and retrieval using cross media relevance models. In: Proceedings of the ACM SIGIR Conference on Research and Development in Information Retrieval (2003)
8. Cusano, C., Ciocca, G., Scettini, R.: Image annotation using SVM. In: Proceedings of Internet Imaging IV (2004)
9. Gauld, M., Thies, C., Fischer, B., Lehmann, T.: Combining global features for content-based retrieval of medical images. Cross Language Evaluation Forum (2005)
10. Huang, J., Kumar, S.R., Mitra, M., Zhu, W.J., Zabih, R.: Image indexing using color correlograms. In: Proc. 16th IEEE Conf. on Computer Vision and Pattern Recognition, pp. 762–768 (1997)
11. Chang, C., Lin, C.-J.: LIBSVM: - A Library for Support Vector Machines [EB/OL], http://www.csie.ntu.edu.tw/~cjlin/libsvm

Optimal Evolutionary-Based Deployment of Mobile Sensor Networks

Niusha Shafiabady[1] and Mohammad Ali Nekoui[2]

[1] Azad University, Science and Research Branch
[2] K.N.T. University of Technology
nshafiabady@yahoo.com, manekoui@eetd.kntu.ac.ir

Abstract. Sensor deployment is an important issue in designing sensor networks. Here both energy consumption and coverage factors of mobile sensor networks are optimized in two phases using fully informed particle swarm optimization (FIPSO) that has proved to have a good achievement and given good results. The method has been applied on three sets of data to test its efficiency. All results have been perfectly good.

Keywords: Fully Informed Particle Swarm Optimization, Mobile sensor networks.

1 Introduction

Mobile sensor networks consist of the sensor nodes that are deployed in a large area collecting important information from the sensor field. The communication between the nodes is wireless. Since energy is an important resource [4], the nodes' energy consumption must be kept at a minimum rate. Optimum placement of the sensors also includes their coverage that means the sensors should be places in the best position assigned by a coverage function to have the best functionality. Some random deployments [1] do not apply a uniform distribution over a surface that can be important in some cases [3].

This paper uses FIPSO to find the optimal placement and energy usage of the mobile sensor networks. In section 2 FIPSO is discussed. Section 3 describes applying FIPSO for deployment of mobile sensor networks and section 4 describes the results and section 5 denotes the conclusion.

2 Fully Informed Particle Swarm Optimization

The canonical particle swarm algorithm is a new approach to optimization, drawing inspiration from group behavior and establishment of social norms [6]. It is gaining popularity because of its being easy to use and its speed of convergence. This method keep the individuals fully informed that means unlike simple PSO, every single member of the population's experience is taken into account.

H. Deng et al. (Eds.): AICI 2009, LNAI 5855, pp. 563–567, 2009.
© Springer-Verlag Berlin Heidelberg 2009

$$\vec{v}_i(t) = \phi_1 \vec{v}_i(t-1) + \vec{r}_1 \vec{c}_1 \otimes (\vec{x}_{personalbest} - \vec{x}_i(t)) + \vec{r}_2 \vec{c}_2 \otimes (\vec{x}_{globalbest} - \vec{x}_i(t))$$

$$\vec{x}_i(t) = \vec{v}_i(t) + \vec{x}_i(t-1)$$
(1)

$$\phi_1 = 1 - 0.5 \times \frac{1-t}{1-t_{max}}$$

As it is clear in Eq. (1), the experience of every member of the population is taken into consideration and this method has a better performance than the simple PSO.

3 Applying FIPSO for Deployment of Mobile Sensor Networks

3.1 Optimization of Coverage

Coverage is considered to be the first optimization objective. The coverage of each sensor can be defined either by a binary sensor model or a probabilistic sensor model as described in Fig. 1.

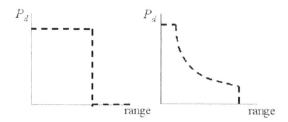

Fig. 1. Sensor coverage models a) binary b) probabilistic

In the binary sensor models, the detection probability of the event of interest is 1 within the sensing range, otherwise, the probability is zero. Although the binary sensor model is simpler, it is not realistic as it assumes that sensor readings have no associated uncertainty. In reality sensor detections are imprecise, hence the coverage has to be described in probabilistic terms. The probabilistic sensor model used in this work is given in Eq. (2).

$$c_{ij}(x,y) = \begin{cases} 0 & , \ if \quad r+r_e \leq d_{ij}(x,y) \\ e^{-\lambda a^\beta} & , \ if \ r-r_e < d_{ij}(x,y) < r+r_e \\ 1 & , \ if \quad r+r_e \geq d_{ij}(x,y) \end{cases}$$
(2)

The sensor field is represented by a $n \times m$ grid where an individual sensor is placed in point S at grid point (x,y). Each sensor has a detection range r. For any grid point P at (i,j), the Euclidean distance between P and the grid point at (x,y) is denoted as $d_{ij}(x,y) = \sqrt{(x-i)^2 + (y-j)^2}$, the equation expresses the coverage $c_{ij}(x,y)$ of a grid point at (i,j) by a sensor S at (x,y). In this equation λ and β are the parameters that measure the detection probability when a target is at distance greater than r_e but within a distance from the sensor and $a = d_{ij}(x, y) - (r - r_e)$.

3.2 Optimization of Energy Consumption

After optimization of coverage, all the deployed sensor nodes move to their own position. Now our goal is to minimize the energy usage in a cluster based sensor network topology by finding the optimal cluster head positions. For this purpose a power consumption model [5] for the radio hardware dissipation where the transmitter dissipates energy to run the radio electronics and the power amplifier and the receiver dissipates energy to run the radio electronics, is used to be optimized. The fitness function to be minimized is given in Eq. (3).

$$f = \sum_{j=1}^{m} \sum_{i=1}^{n} \left(0.01 dis_{ij}^2 + \frac{1.3 \times 10^{-6} Dis_j^4}{n_j} \right) \tag{3}$$

For this approach both the free space (distance^2 power loss) and the multi-path fading (distance^4 power loss) channel models were used. Assume that the sensor nodes inside a cluster have short distance *dis* to the cluster head but each cluster head has long distance *Dis* to the base station. The base station is situated at the position (25,80).

4 The Simulation Results

The program is simulated in MATLAB. The particles are randomly produced at the beginning. Three different results are given in Fig. 2-5.

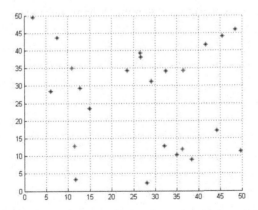

Fig. 2. The result showing the heads and the nodes

The red spots denote the chosen heads for each selected cluster among the four chosen clusters.

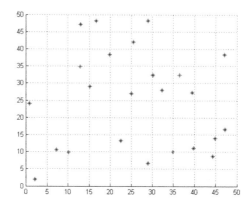

Fig. 3. The result showing the heads and the nodes

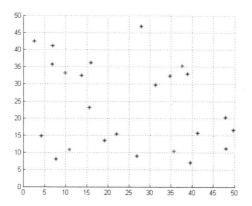

Fig. 4. The result showing the heads and the nodes

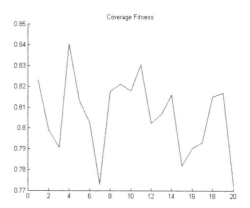

Fig. 5. The error of the first phase

The error of the first phase is given in Fig. 5. It shows that it has efficiently placed the nodes in the grid's positions.

The results are achieved in two phases. At first the particles' coverage cost function is minimized using FIPSO. Then FIPSO is applied to the achieved set of particles to find the head and the best position of the node that will be chosen as the head.

FIPSO has been able to give better results in compare with the previous works [1-5].

5 Conclusion

The results show that the proposed method (FIPSO) has successfully placed the mobile sensors and then chosen the head as an appropriate member of the clusters according to the cost functions. This method is more efficient than the simple PSO as it uses the information from all the members rather than only some of the particles. The results are achieved fast and the coverage and energy functions are optimized simultaneously.

References

1. Chakrabarty, K., Iyengar, S., Qi, H., Cho, E.: Grid coverage for surveillance and target location in distributed sensor networks. IEEE Transactions on Computers 51, 1448–1453 (2002)
2. Jourdan, B., Weck, O.: Layout optimization for a wireless sensor network using multi-objective genetic algorithm. In: IEEE VTC 2004 Conference, vol. 5, pp. 2466–2470 (2004)
3. Howard, A., Mataric, M.J., Sukhatme, G.S.: Mobile sensor network deployment using potential field: a distributed, scalable solution to the area coverage problem. In: Proc. Int. Conf. on Distributed Autonomous Robotics Systems, pp. 299–308 (2002)
4. Heo, N., Varshney, P.K.: Energy-efficient deployment of intelligent mobile sensor networks. IEEE Transaction on Systems, Man and Cybernetics 35(1), 78–92 (2005)
5. Heinzelman, W.B., Chandrakasan, A.P., Balakrishnan, H.: An application specific protocol architecture for wireless microsensor networks. IEEE Transactions on Wireless Communications 1(4), 660–670 (2002)
6. Mendes, R., Kennedy, J., Neves, J.: The fully informed particle swarm. IEEE Transactions on Evolutionary Computation 1(1) (January 2005)

Research of Current Control Strategy of Hybrid Active Power Filter Based on Fuzzy Recursive Integral PI Control Algorithm

Zhong Tang and Daifa Liao

School of Computer and information Engineering, Shanghai University of Electric Power,
Shanghai 200090, China
tangzhong64@163.com, liaodaifa0724@163.com

Abstract. According to the current control characters of hybrid active power filter (HAPF), the current control model of HAPF is designed. The fuzzy recursive integral PI control algorithm is presented when it is compared to conventional PI control method. The control algorithm is applied to auto-regulate the proportional and integral parameters of PI. Thus, the robustness and response speed is enhanced; the dynamic performance of the HAPF device is improved. Under Matlab/Simulink background, a fuzzy recursive integral PI controller is designed and it is applied in a HAPF model in PSCAD/EMTDC. The results prove the feasibility and effectiveness of this fuzzy recursive integral PI control algorithm.

Keywords: Hybrid active power filter, Recursive integral PI, Fuzzy control, PSCAD/EMTDC.

1 Introduction

Combined with the advantages of passive power filter (PPF) and active power filter (APF), hybrid active power filter (HAPF) is a powerful apparatus for reactive power compensation, harmonic suppression; it is also an effective device to resolve power quality problems [1-3] . It is a key link to keep DC voltage stable, ensure compensate current signal timely, track instruction current correctly in active power filtering technology [4].

Related work. Recently, there are many current control methods, mainly including linear-current control, digital-deadbeat control, triangular-wave control, hysteresis control and one-cycle control [5-6]. But because of the influences of detecting precision, phase shift of output filter and time delay of control method, the tracking ability of the system is become non-ideal. Because of its advantages of simple algorithm, good robustness and high reliability, PI control method has been widely used in industry control system [7]. The model parameters are required invariable and proportional integral parameters are hard to determine in conventional PI control. Because the periodic characteristics of error signal in HAPF system, the application of conventional PI control is limited seriously. Fuzzy logical control has ideal dynamic performance and is insensitive of producer parameter; it has high robustness and can overcome the influence of non-linear factors. So, it has been widely used in control system [8-10].

H. Deng et al. (Eds.): AICI 2009, LNAI 5855, pp. 568–578, 2009.

In this paper, fuzzy logical control is applied to adjust the parameters of recursive integral PI control method, the response speed is enhanced and dynamic performance is improved. Based on fuzzy recursive integral PI control strategy, a HAPF current controller is designed and it is applied to auto-regulate the proportional and integral parameters. The simulation results prove the feasibility and effectiveness of this control algorithm.

2 Configuration of HAPF and Control Model

The HAPF circuit configuration suitable for medium voltage and low voltage system is showed in Fig.1, which consists of PPF and APF, and is directly shunted to the 10KV bus. Fundamental resonance circuit (FRC) consists of L_1, C_1 which is tuned at fundamental frequency; an injection capacitance C_g is directly connected with FRC in series which is used to suppress 6th harmonic current when it is combined with FRC circuit. Because of the FRC suppressing fundamental current, the capacitance and initial investment of APF are decreased sharply. In this circuit PPF are mainly used to filter the vast majority of the5th, 7th, 11th, 13th harmonic currents and compensate reactive power of the system.

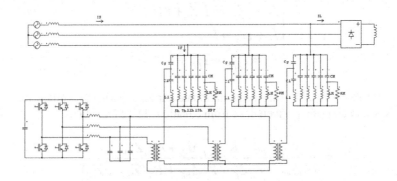

Fig. 1. Shunt hybrid active power filter diagram

The single-phase equivalent circuit of HAPF is showed in Fig.2, in which the APF is controlled as an ideal harmonic voltage resource U_F and the load is controlled as a harmonic current resource i_L. Where I_S and I_L represent source current and load current; I_{PF} and I_{APF} represent the current of PPF circuit and injected current of APF. Z_S and Z_{PF} represent the equivalent impedance of source and PPF branch; Z_C and Z_R represent the equivalent impedance of injected capacitance and FRC branch; Z_{C0}, L_{C0} are capacitive reactance and inductance of output filter branch. The ratio of coupling transformer is n: 1.

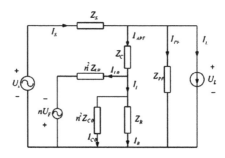

Fig. 2. Single phase equivalent circuit of HAPF

In Fig.2 if we use Z_1 represents the shunt impedance of n^2Z_{C0} and Z_R, according to KVL and KCL we can obtain:

$$
\begin{cases}
U_S = U_L + I_S Z_S \\
U_L = I_{PF} Z_{PF} \\
I_{APF} Z_C + I_1 Z_1 = U_L \\
I_{L0} n^2 Z_{L0} = I_1 Z_1 + n U_F \\
I_1 + I_{L0} = I_{APF} \\
I_S = I_{APF} + I_{PF} + I_L
\end{cases}
\tag{1}
$$

When HAPF solve and compensate harmonic currents, harmonic voltage distortion is very small and can be neglected, so from equation (1) we can obtain

$$
I_S = \frac{(Z_C Z_{PF} + K_2 Z_{PF})I_L + n K_1 Z_{PF} U_F}{Z_C Z_{PF} + Z_S Z_C + Z_S Z_{PF} + K_2(Z_{PF} + Z_S)}
\tag{2}
$$

where

$$
\begin{cases}
K_1 = Z_1 / (Z_1 + n^2 Z_{L0}) \\
K_2 = n^2 Z_1 Z_{L0} / (Z_1 + n^2 Z_{L0})
\end{cases}
$$

If we assume

$$
G_1(S) = \frac{n K_1 Z_{PF}}{Z_C Z_{PF} + Z_S Z_C + Z_S Z_{PF} + K_2(Z_{PF} + Z_S)}
\tag{3}
$$

$$G_2(S) = \frac{Z_C Z_{PF} + K_2 Z_{PF}}{Z_C Z_{PF} + Z_S Z_C + Z_S Z_{PF} + K_2(Z_{PF} + Z_S)} \tag{4}$$

So we can obtain closed-loop control system of source current I_{sh} showed in Fig.3. Where $I_C^*(S)$ represents reference signal of source current, $G_{con}(S)$ and $G_{inv}(S)$ represent controller function and transfer function of active inverter. If $I_C^*(S) = 0$ that the closed-loop transfer function of I_{sh} will be

$$G_S(S) = \frac{I_{Sh}(S)}{I_C^*(S)} = \frac{G_{con}(S)G_{inv}(S)G_1(S)}{1 + G_{con}(S)G_{inv}(S)G_1(S)} \tag{5}$$

The current I_{Lh} is a disturbance to this closed-loop control system of I_{sh}, and the transfer function of I_{sh} to disturbance is

$$G_L(S) = \frac{I_{Sh}(S)}{I_{Lh}(S)} = \frac{G_2(S)}{1 + G_{con}(S)G_{inv}(S)G_1(S)} \tag{6}$$

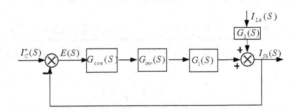

Fig. 3. Closed-loop control of the system current

3 The Design of Fuzzy Recursive Integral PI Controller

Conventional PI control method can reach deadbeat control when it control DC signals and changing slowly signals, but there are steady-state errors when it control sinusoid signals directly. In this paper the system reference signals are combined with all kinds of harmonic currents which are multiple of fundamental current whose period is 20ms. Recursive integral PI control method based on conventional PI control method is used to implement deadbeat control. Every corresponding sampling point of error $e(t)$ is integrated in each cycle when use this kind of control algorithm, and it is equivalent to N parallel PI controller to implement PI control of the system. Equation (7) is recursive integral PI control algorithm, where sampling value of controller output is $U_R(K)$ and sampling value of error is $e(K)$ at the moment of K; N is sampling point of a cycle, K_P, K_I respectively represents proportion and integral coefficient. This kind of algorithm is equivalent to integrate error of each cycle.

$$\begin{cases} U_R(K) = K_P e(K) + \sum_{i=0}^{C} K_I e(K - iN) \\ C = ent \dfrac{K}{N} \end{cases} \tag{7}$$

In order to simplify the calculation, we use incremental form of $U_R(K)$ to calculate, and we can obtain control low of incremental form

$$U_R(K) = U_R(K - N) + K_P e(K) - K_P e(K - N) + K_I e(K) \tag{8}$$

From the equation (8) we can obtain pulse transfer function of recursive integral PI controller

$$G_{con}(S) = \frac{U_R(S)}{E(S)} = K_P + \frac{K_I}{1 - e^{-SNT}} \tag{9}$$

Where N is sampling points in each cycle of controlled object, T is sampling time. In Fig.3 we use recursive integral PI controller and can obtain

$$G_S(S) = \frac{K_P G_{inv}(S)G_1(S)(1 - e^{-SNT}) + K_I G_{inv}(S)G_1(S)}{(1 + K_P G_{inv}(S)G_1(S)) \times (1 - e^{-SNT}) + K_I G_{inv}(S)G_1(S)} \tag{10}$$

$$G_L(S) = \frac{G_2(S)(1 - e^{-SNT})}{(1 + K_P G_{inv}(S)G_1(S)) \times (1 - e^{-SNT}) + K_I G_{inv}(S)G_1(S)} \tag{11}$$

So the frequency characteristic equations of the two functions are

$$\begin{aligned} G_S(j\omega) = &[K_P G_{inv}(j\omega)G_1(j\omega)(1 - \cos \omega NT + j \sin \omega NT) \\ &+ K_I G_{inv}(j\omega)G_1(j\omega)] \Big/ [(1 + K_P G_{inv}(j\omega)G_1(j\omega)) \\ &\times ((1 - \cos \omega NT + j \sin \omega NT)) + K_I G_{inv}(j\omega)G_1(j\omega)] \end{aligned} \tag{12}$$

$$\begin{aligned} G_L(j\omega) = &[G_2(j\omega)(1 - \cos \omega NT + j \sin \omega NT)] \Big/ [(1 + K_P G_{inv}(j\omega)G_1(j\omega)) \\ &\times ((1 - \cos \omega NT + j \sin \omega NT)) + K_I G_{inv}(j\omega)G_1(j\omega)] \end{aligned} \tag{13}$$

When $\omega = 2n\pi f$, from equation (12) (13) we can obtain

$$G_S(jn \cdot 2\pi f) = G_L(jn \cdot 2\pi f) = 1 \qquad (14)$$

Equation (14) illustrates that the amplitude of closed-loop system transfer function and HAPF system to disturbance transfer function is 1, and the phase is 0; output of the system can track the reference signal bitterly; the influence of load current become 0 with time increasing, when it's frequency is power frequency or Integral Multiple power frequency.

The systemic steady-state error can be eliminated using recursive integral PI control algorithm, but its robustness and dynamic performance are not ideal. So in this paper fuzzy logical algorithm is applied to enhance the systemic response speed by auto-regulating K_P and K_I parameters of recursive integral PI control method [9]. Compound control of fuzzy recursive integral PI is showed in figure 4. Where error e and error rate e_c is fuzzed into corresponding fuzzy variable E and E_C. Firstly, it seeks the fuzzy relation of K_P, K_I, E and E_C, continuously detects E and E_C within operating time. Then, according to fuzzy logical control algorithm, K_P and K_I are auto-regulated online to meet various requirement of recursive integral PI controller in various E and E_C. Thus makes the controlled object better robustness and dynamic performance.

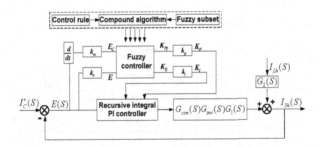

Fig. 4. Configuration of fuzzy recursive integral PI control

Firstly, inputs are fuzzed by fuzzy controller; the linguistic values of input variable and output variable are equally divided into 7 parts: {NB, NM, NS, 0, PS, PM, PB}. According to engineering experience, using triangle membership functions, the universe of e and e_c is [-6 6]. The proportion link is to proportionally reflect the systemic error signal e. The integral link is to improve systemic indiscrimination degree. When $|E|$ is bigger, in order to have better tracking performance K_P should be bigger. At the same time, in order to avoid bigger overshoot of systemic response, the effect of integral should be restricted, usually we set $K_I = 0$. When $|E|$ is medium, in order to have smaller overshoot of systemic response, K_P should be smaller. When $|E|$ is smaller, in order to have better steady-state performance, K_P and K_I both should be bigger. So, according to the above analysis, we can obtain the fuzzy control rules of K_P and K_I in Table 1 and Table 2.

Table 1. Fuzzy control rules of ΔK_P

E	e_c						
	NB	NM	NS	0	PS	PM	PB
NB	PB	PB	NB	PM	PS	PM	0
NM	PB	PB	NM	PM	PS	0	0
NS	PM	PM	NS	PS	0	NS	NM
0	PM	PS	0	0	NS	NM	NM
PS	PS	PS	0	NS	NS	NM	NM
PM	0	0	NS	NM	NM	NM	NB
PB	0	NS	NS	NM	NM	NB	NB

Table 2. Fuzzy control rules of ΔK_I

E	e_c						
	NB	NM	NS	0	PS	PM	PB
NB	0	0	NB	NM	NM	0	0
NM	0	0	NM	NM	NS	0	0
NS	0	0	NS	NS	0	0	0
0	0	0	NS	NM	PS	0	0
PS	0	0	0	PS	PS	0	0
PM	0	0	PS	PM	PM	0	0
PB	0	0	NS	PM	PB	0	0

Fuzzy querying table reflects the final results of fuzzy control algorithm. It is calculated off-line through afore-hand and stored in computer memory. In real-time control system, the auto-regulate modification process of recursive integral PI control parameters is translated into minute calculating querying process of control rule table.

4 Simulation Results and Analysis

In this paper a HAPF model controlled by current tracking strategy is set up, and its simulation is carried out, in which current detection use fuzzy recursive integral PI control algorithm, compared with the results using conventional PI control algorithm. Based on Matlab/Simulink toolbox and corresponding function, a fuzzy recursive integral PI controller is established and it is applied in the HAPF model which is set up based on PSCAD/EMTDC V4.2 of Manitoba HVDC research center.

In this simulation model, line voltage of the source is 10KV; capacitance of the source is 100MVA; capacitance of coupling transformer is 4MVA, and its ratio is 10000:380; where the parameters of FRC are $C_1 = 960uF$ and $L_1 = 15.47mH$, injecting capacitance $C_G = 19.65uF$. A single-tuned passive power filter to suppress 6th harmonic current is consisted of the injecting capacitance and FRC branch. A six pulse rectifier bridge is used as a harmonic source and its power frequency is 50HZ. The parameters of PPF are showed in table 3 and the high-pass filter quality factor $Q = 1.25$. The simulation circuit and configuration of HAPF system is showed in Fig.5.

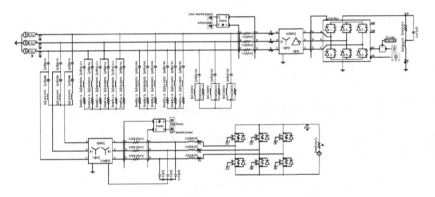

Fig. 5. Hybrid active power filter model under PSCAD/ EMTDC

Table 3. Parameters of the LC passive power filter

Harmonics	C	L	R
5th branch	$C_5 = 117.03uF$	$L_5 = 3.46mH$	$R_5 = 0.11\Omega$
7th branch	$C_7 = 70.46uF$	$L_7 = 2.93mH$	$R_7 = 0.15\Omega$
11th branch	$C_{11} = 94.50uF$	$L_{11} = 0.89mH$	$R_{11} = 0.10\Omega$
13th branch	$C_{13} = 34.30uF$	$L_{13} = 1.70mH$	$R_{13} = 0.13\Omega$
HPF	$C_H = 20.00uF$	$L_H = 3.1mH$	$R_H = 12.06\Omega$

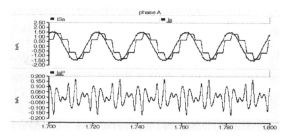

(a) source current and active power filter compensated current waveform using conventional PI control

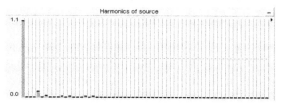

(b) the frequency spectrum of load current using conventional PI control

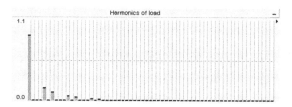

(c) the frequency spectrum of source current using conventional PI control

Fig. 6. The harmonic suppression and compensation results of HAPF using conventional PI control

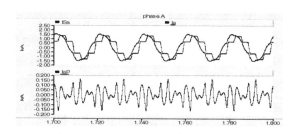

(a) source current and active power filter compensated current waveform using fuzzy recursive integral PI control

Fig. 7. The harmonic suppression and compensation results of HAPF using fuzzy recursive integral PI control

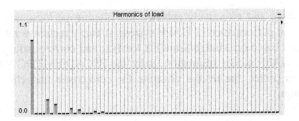

(b) the frequency spectrum of load current using fuzzy recursive integral PI control

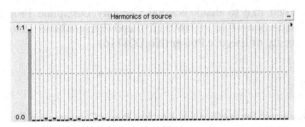

(c) the frequency spectrum of source current using fuzzy recursive integral PI control

Fig. 7. (*Continued*)

Table 4. Harmonic current analysis of two kinds of control methods

Current times	5th	7th	11th	13th
Pre-filtering(A)	116.69	77.79	39.61	28.29
Conventional PI (A)	50.13	30.21	15.55	9.12
Fuzzy recursive integral PI (A)	16.27	12.02	6.36	7.07

Fig.6 shows the harmonic suppression and compensation results of HAPF system using conventional PI control method. Fig.7 shows the harmonic suppression and compensation results when use fuzzy recursive integral PI control algorithm. ISa, Ia and IaP respectively represents the source current of phase A, the load current of phase A and APF compensated current of phase A. Table 4 shows the currents value of 5th, 7th, 11th and 13th harmonics when use conventional PI control method and fuzzy recursive integral PI control algorithm. Analyzed from Fig.6, Fig.7 and Table 4, it is proved that fuzzy recursive integral PI control algorithm has better and more ideal effect in HAPF harmonic suppression and compensation.

5 Conclusions

In this paper aiming at the harmonic detecting and control problems of HAPF system, a fuzzy recursive integral PI control algorithm is been proposed based on the conventional

PI control method, and effectively enhance the filtering performance, robustness and dynamic response performance of HAPF system. A fuzzy recursive integral PI controller is established based on Matlab/Simulink toolbox and corresponding function, and it is applied in a HAPF model simulation under the PSCAD/EMTDC V4.2 of Manitoba HVDC research center, the results illustrate the correctness and effectiveness of this control algorithm.

Acknowledgement

The authors would like to thank Natural Capacity Discipline Project of Shanghai Local High Institutions (No. 071605125) and Postgraduate Innovation Fund of Shanghai University of Electric Power (No.D08116).

References

1. Luo, A.: Harmonic Suppression and Reactive Power Compensation Equipment and Technology. China Electric Power System Press, Beijing (2006)
2. Darwin, R., Luis, M., Juan, W.: Improving passive filter compensation performance with active techniques. IEEE Trans on Industrial Electronics 50(1), 161–170 (2003)
3. He, N., Huang, L.: Multi-objective optimal design for passive part of hybrid active power filter based on particle swarm optimization. Proceeding of the CSEE 28(27), 63–69 (2008)
4. Luo, A., Fu, Q., Wang, L.: High-capacity hybrid power filter for harmonic suppression and reactive power compensation in the power substation. Proceeding of the CSEE 24(9), 115–223 (2004)
5. Buso, s., Malesani, L.: Design and Fully Digital Control of Parallel Active Power Filters for Thyristor Rectifiers to Comply with IEC 1000-3-2. IEEE Trans. on Industry Application 34(2), 508–517 (1998)
6. Tang, X., Luo, A., Tu, C.: Recursive Integral PI for Current Control of Hybrid Active Power Filter. Proceedings of the CSEE 23(10), 38–41 (2003)
7. Sun, M., Huang, B.: Iterative Learning Control. National Defense Industry Press, Beijing (1999)
8. Zhou, K., Luo, A., Tang, J.: PI iterative learning for current-tracking control of active power filter. Power Electronics 40(4), 53–55 (2006)
9. Xu, W.-f., Luo, A., Wang, L.: Development of hybrid active power filter using intelligent controller. Automation of Electric Power Systems 27(10), 49–52 (2003)
10. Fukuda, S., Sugawa, S.: Adaptive signal processing based control of active power filters. In: Proceeding of IEEE IAS Annual Meeting (1996)

A Fuzzy Query Mechanism for Human Resource Websites

Lien-Fu Lai[1], Chao-Chin Wu[1], Liang-Tsung Huang[2], and Jung-Chih Kuo[1]

[1] Department of Computer Science and Information Engineering,
National Changhua University of Education, Taiwan
{lflai,ccwu}@cc.ncue.edu.tw, verygoodtony@hotmail.com
[2] Department of Information Communication, MingDao University, Taiwan
larry@mdu.edu.tw

Abstract. Users' preferences often contain imprecision and uncertainty that are difficult for traditional human resource websites to deal with. In this paper, we apply the fuzzy logic theory to develop a fuzzy query mechanism for human resource websites. First, a storing mechanism is proposed to store fuzzy data into conventional database management systems without modifying DBMS models. Second, a fuzzy query language is proposed for users to make fuzzy queries on fuzzy databases. User's fuzzy requirement can be expressed by a fuzzy query which consists of a set of fuzzy conditions. Third, each fuzzy condition associates with a fuzzy importance to differentiate between fuzzy conditions according to their degrees of importance. Fourth, the fuzzy weighted average is utilized to aggregate all fuzzy conditions based on their degrees of importance and degrees of matching. Through the mutual compensation of all fuzzy conditions, the ordering of query results can be obtained according to user's preference.

Keywords: Fuzzy Query, Fuzzy Weighted Average, Human Resource Websites.

1 Introduction

In traditional human resource websites [1,2,3,4,5], users must state clear and definite conditions to make database queries. Unfortunately, users' preferences often contain imprecision and uncertainty that are difficult for traditional SQL queries to deal with. For example, when a user hopes to find a job which is near Taipei City and pays good salary, he can only make a SQL query like "SELECT * FROM Job WHERE (Location='Taipei City' or Location= 'Taipei County') and Salary $\geq$ 40000". However, both 'near Taipei City' and 'good salary' are fuzzy terms and cannot be expressed appropriately by merely crisp values. A job which locates in 'Taoyuan County' with salary of 50000 may be acceptable in user's original intention, but it would be excluded by the traditional SQL query. SQL queries fail to deal with the compensation between different conditions. Moreover, traditional database queries cannot effectively differentiate between the retrieved jobs according to the degrees of satisfaction. The results to a query are very often a large amount of data, and the problem of the information overload makes it difficult for users to find really useful information.

H. Deng et al. (Eds.): AICI 2009, LNAI 5855, pp. 579–589, 2009.

Hence, it is required to sort results based on the degrees of satisfaction to the retrieved jobs. Computing the degree of satisfaction to a job needs to aggregate all matching degrees on individual conditions (e.g. location, salary, industry type, experience, education etc.). It is insufficient for merely using the ORDER BY clause in SQL to sort results based on some attribute. In addition, traditional database queries do not differentiate between conditions according to the degrees of importance. One condition may be more important than another condition for some user (e.g. salary is more important than location in someone's opinion). Both the degree of importance and the degree of matching to every condition should be considered to compute the degree of satisfaction to a job. We summarize the problems of traditional human resource websites as follows.

- Users' preferences are usually imprecise and uncertain. Traditional database queries are based on total matching which is limited in its ability to come to grips with the issues of fuzziness.
- In users' opinions, different conditions may have different degrees of importance. Traditional database queries treat all conditions as the same importance and can not differentiate the importance of one condition from that of another.
- The problem of information overload makes it difficult for users to find really useful information from a large amount of query results. Traditional database queries do not support the ordering of query results by aggregating the degrees of matching to all conditions (i.e. no compensation between conditions).

To solve the mentioned problems, we apply the fuzzy logic theory [15] to develop a fuzzy query mechanism for human resource websites. First, a storing mechanism is proposed to store fuzzy data into conventional database management systems without modifying DBMS models. Second, a fuzzy query language is proposed for users to make fuzzy queries on fuzzy databases. User's fuzzy requirement can be expressed by a fuzzy query which consists of a set of fuzzy conditions. Third, each fuzzy condition associates with a fuzzy importance to differentiate between fuzzy conditions according to their degrees of importance. Fourth, the fuzzy weighted average is utilized to aggregate all fuzzy conditions based on their degrees of importance and degrees of matching. Through the mutual compensation of all fuzzy conditions, the ordering of query results can be obtained according to user's preference.

2   The Fuzzy Query Mechanism for Human Resource Websites

Applying the fuzzy logic theory to develop fuzzy queries on human resource websites, there are two issues that should be addressed:

- A storing mechanism is required to represent and store fuzzy data. It should be applied directly to existing databases without modifying DBMS models. Moreover, it should not only deal with the continuous and numerical fuzzy data but the discrete and lexical fuzzy data.
- A fuzzy query language is required to make fuzzy queries on fuzzy databases. The degrees of importance should be considered to differentiate between fuzzy conditions. The query results should be sorted by mutual compensation of all fuzzy conditions.

2.1 Storing Fuzzy Data into Databases

Galindo et al. [8] classify fuzzy data into four types: (1) Type 1 contains attributes with precise data. This type of attributes is represented in the same way as crisp data, but can be transformed or manipulated using fuzzy conditions. (2) Type 2 contains attributes that gather imprecise data over an ordered referential. These attributes admit both crisp and fuzzy data, in the form of possibility distributions over an underlying ordered domain (fuzzy sets). (3) Type 3 contains attributes over data of discrete non-ordered dominion with analogy. In these attributes, some labels are defined with a similarity relationship defined over them. The similarity relationship indicates to a degree that each pair of labels resembles each other. (4) Type 4 contains attributes that are defined in the same way as Type 3 attributes, without being necessary for a similarity relationship to exist between the labels.

By analyzing several popular human resource websites [1,2,3,4,5], we sum up that the major fuzzy data needed to be stored are location preference, salary preference, industry preference, job category preference, experience preference, education preference, department preference, job seeker's profile, and hiring company's profile. We adopt the notions of [6] to classify these fuzzy data into three types: (1) Discrete fuzzy data include location preference, industry preference, job category preference, education preference, and department preference. (2) Continuous fuzzy data include salary preference and experience preference. (3) Crisp data include job seeker's profile and hiring company's profile.

2.1.1 Discrete Fuzzy Data

The discrete fuzzy data (i.e. type 3 and type 4 fuzzy data) is represented by a discrete fuzzy set which consists of a set of discrete data items with their degrees of conformity. How to grade numeric degrees of conformity may confuse users. Hence, we use linguistic degrees of conformity (i.e. totally unsatisfactory, unsatisfactory, rather unsatisfactory, moderately satisfactory, rather satisfactory, very satisfactory, and totally satisfactory) to make it easier and clearer for users to grade degrees. 'Totally unsatisfactory' and 'totally satisfactory' stand for 0 and 1 respectively, while the others grade values between 0 and 1. For example, a fuzzy term 'near Taipei City' can be expressed as a discrete fuzzy set like {(Taipei City, totally satisfactory), (Taipei County, very satisfactory), (Taoyuan County, moderately satisfactory)}. We utilize the membership functions in [13] to define the linguistic degree of conformity between a discrete data item and a fuzzy term (see Figure 1).

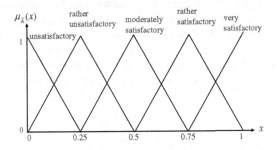

Fig. 1. Membership functions for degrees of conformity

A fuzzy number $\tilde{A}$ can be defined by a triplet (a, b, c) and the membership function $\mu_{\tilde{A}}(x)$ is defined as:

$$\mu_{\tilde{A}}(x) = \begin{cases} 0 & ,x < a \\ \dfrac{x-a}{b-a} & ,a \leq x \leq b \\ \dfrac{c-x}{c-b} & ,b \leq x \leq c \\ 0 & ,x > c \end{cases}$$

Therefore, each linguistic degree of conformity can be mapped to a triangular fuzzy number [11], e.g. 'rather satisfactory' is mapped to (0.5,0.75,1).

To store the discrete fuzzy data in conventional databases, a new table is needed to store the set of discrete data items with their degrees of conformity. For example, job seekers' profiles are stored in a *Resume* table (see Figure 2). If the location preference of a job seeker numbered 10001 is 'near Taipei City', the discrete fuzzy set inputted by users can be stored in a new *Location_Preference* table shown in Figure 3. In the *Location_Preference* table, the *RID* attribute serves as a foreign key to reference the primary key of the *Resume* table. Hence, we can retrieve a job seeker's location preference via joining the two tables on equivalent RIDs.

RID	Name	SSN	Education	Experience	...
10001	C.M. Wang	N123456789	Bachelor	2	...

Fig. 2. The *Resume* table

RID	Location	SatisfactoryDegree
10001	Taipei City	Totally satisfactory
10001	Taipei County	very satisfactory
10001	Taoyuan County	moderately satisfactory

Fig. 3. The *Location_Preference* table

2.1.2 Continuous Fuzzy Data

The continuous fuzzy data (i.e. type 2 fuzzy data) is represented by a continuous fuzzy set which consists of a set of continuous data items with their degrees of conformity. For example, a fuzzy term 'good salary' can be expressed as a continuous fuzzy set like {(50000, totally satisfactory), (45000, very satisfactory), (30000, totally unsatisfactory)}. The degree of conformity can be defuzzified by using mathematical integral to compute the center of the area that is covered by the corresponding triangular fuzzy number [18], e.g. 'very satisfactory' is defuzzified by computing its center of gravity of the triangular fuzzy number (0.75, 1, 1) as follows.

$$\frac{\int_{0.75}^{1} x(4x-3)dx}{\int_{0.75}^{1} (4x-3)dx} = 0.92$$

Therefore, the membership function corresponding to the given 'good salary' can be constructed by {(50000,1), (45000,0.92), (30000,0)} (see Figure 4).

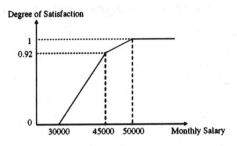

Fig. 4. The membership function of 'good salary'

To store the continuous fuzzy data in conventional databases, a new table is needed to store the set of continuous data items with their degrees of conformity. A *Salary_Preference* table is used to store the corresponding membership function in which the degrees of conformity have been defuzzified (see Figure 5). In the *Salary _Preference* table, the *RID* attribute serves as a foreign key to reference the primary key of the *Resume* table.

RID	Salary	SatisfactoryDegree
10001	50000	1
10001	45000	0.92
10001	30000	0

Fig. 5. The *Salary_Preference* table

2.1.3 Crisp Data
Crisp data (i.e. type 1 fuzzy data) contain precise and certain information and can be directly stored in conventional databases. For example, the *Resume* table in Figure 2 stores job seeker's profile including Name, BirthDate, SSN, Education, Department, Experience, Address, and Phone etc.

2.2 Fuzzy Queries on Web Databases

To deal with fuzzy queries on web databases, three major tasks are required to be accomplished: (1) a fuzzy query language for users to make fuzzy queries on fuzzy databases, (2) matching fuzzy conditions in a fuzzy query with fuzzy data in fuzzy databases, and (3) aggregating all fuzzy conditions based on their degrees of importance and degrees of matching.

2.2.1 Making Fuzzy Queries on Web Databases
A fuzzy query consists of a set of fuzzy conditions. Each fuzzy condition associates with a fuzzy importance to differentiate between fuzzy conditions according to the degrees of importance and uses a fuzzy set to state the degrees of conformity for different

attribute values. We use linguistic degrees of importance (i.e. don't care, unimportant, rather unimportant, moderately important, rather important, very important, and most important) to make it easier for users to grade relative importance. Each linguistic degree of importance can be mapped to a triangular fuzzy number as shown in Figure 7.

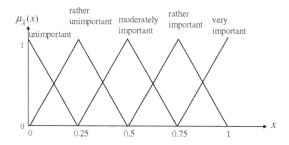

Fig. 7. Membership functions for degrees of importance

The fuzzy set that defines a fuzzy condition could be discrete, continuous, or crisp. In the example shown in Figure 8, a fuzzy query for the resume search consists of 7 fuzzy conditions with their degrees of importance. The hiring company may consider that job seeker's department and salary preference are the most important, experiences and

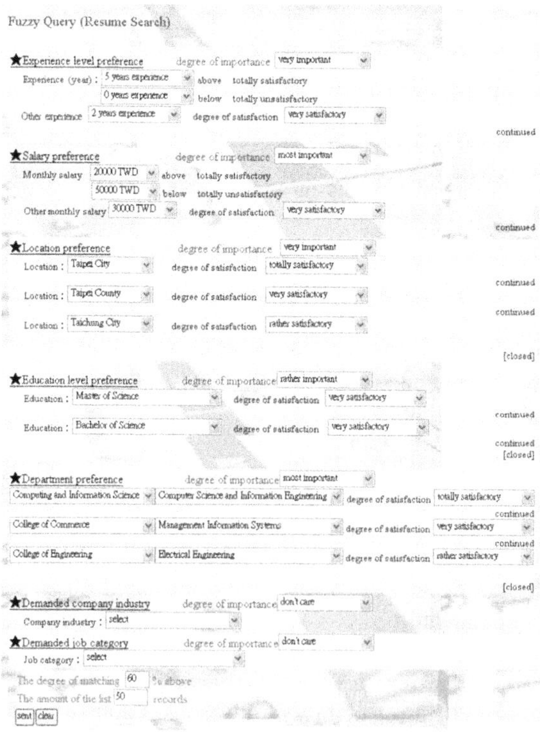

Fig. 8. A fuzzy query for resume search

location preference are very important, the education level is rather important, and the company doesn't care about other fuzzy conditions. In Figure 8, experience preference is defined by a continuous fuzzy set {(5,1), (2,0.92), (0,0)}, while education level preference is defined by a discrete fuzzy set {(Master, very satisfactory), (Bachelor, rather satisfactory)}. Job seekers and hiring companies can make their own fuzzy queries to search jobs and resumes via selecting options on web pages. In addition, users can set the least matching degree and the most amount of data listing to reduce the search space and to avoid the information overload.

2.2.2 Matching Fuzzy Conditions with Fuzzy Data

Once the user makes a fuzzy query, our system would execute a preliminary SQL query to filter out the data that are totally unsatisfactory to some fuzzy condition in the fuzzy query. Reducing the search space could effectively save the execution time of fuzzy matching and fuzzy computation. For example, when a job seeker makes a fuzzy query that the location is near Taipei City with {(Taipei City, totally satisfactory), (Taipei County, very satisfactory), (Taoyuan County, moderately satisfactory)} and the salary is good with {(50000, totally satisfactory), (45000, very satisfactory),(30000, totally unsatisfactory)}, our system would automatically execute a SQL query "CREATE VIEW Job_List AS SELECT * FROM Job WHERE JID in (SELECT DISTINCT JID FROM Job_ Location WHERE Location='Taipei City' or Location= 'Taipei County' or Location='Taoyuan County') and JID in (SELECT JID FROM Job_Salary_Offer WHERE SalaryOffer>30000 and SatisfactoryDegree=0)" to exclude those totally unsatisfactory data.

For matching fuzzy conditions with fuzzy data, we adopt the possibility measure in the fuzzy logic theory [16] to compute the degree of matching between a fuzzy condition $\tilde{A}$ and the fuzzy data $\tilde{B}$:

$$\text{Poss}\{\tilde{B} \text{ is } \tilde{A}\} = \sup_{x \in U}[\min(\mu_{\tilde{A}}(x), \mu_{\tilde{B}}(x))]$$

Where the possibility of $\tilde{B}$ being $\tilde{A}$ is obtained by: for each data item $x \in U$, we get a minimum of two degrees of conformity $\mu_{\tilde{A}}(x)$ and $\mu_{\tilde{B}}(x)$, and the possibility measure is the maximum of these minimums. As fuzzy conditions and fuzzy data may be continues, discrete or crisp, the matching of a fuzzy condition and a fuzzy data could be classified into 9 types according to the possibility measure (see Table 1).

Table 1. Different types of matching

Fuzzy Condition	Fuzzy Data	Result of Matching
continuous	continuous	a crisp value [0,1]
continuous	crisp	a crisp value [0,1]
continuous	discrete	not exist
discrete	continuous	not exist
discrete	crisp	a triangular fuzzy number
discrete	discrete	a triangular fuzzy number
crisp	continuous	a crisp value [0,1]
crisp	crisp	a crisp value 1 or 0
crisp	discrete	a triangular fuzzy number

- In the case of the fuzzy condition and the fuzzy data being two continuous fuzzy sets, the intersection point of two membership functions is the result of matching. For example, when a job seeker's salary preference is {(60000,1), (50000,0.92), (30000,0)} and a job's salary offer is {(20000,1), (30000,0.75), (40000,0)}, we obtain the degree of matching is 0.285 (see Figure 9).
- In the case of one continuous fuzzy set and one crisp value, the point of the crisp value mapping to the continuous fuzzy set is the result of matching. For example, when a hiring company's experience preference is {(5,1), (4,0.92), (1,0)} and a job seeker's experience is 3 years, we obtain the degree of matching is 0.613 (see Figure 10).

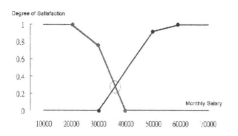 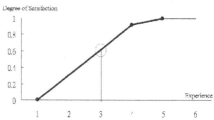

Fig. 9. Matching of two continuous fuzzy sets   **Fig. 10.** Matching of a continuous fuzzy set with a crisp value

- In the case of one continuous fuzzy set and one crisp value, the point of the crisp value mapping to the continuous fuzzy set is the result of matching. For example, when a hiring company's experience preference is {(5,1), (4,0.92), (1,0)} and a job seeker's experience is 3 years, we obtain the degree of matching is 0.613 (see Figure 10).
- The case of one continuous fuzzy set and one discrete fuzzy set does not exist, since an attribute cannot contain both continuous and discrete fuzzy data simultaneously.
- In the case of two discrete fuzzy sets, the possibility measure is a triangular fuzzy number. For example, when a job seeker's location preference is {(Taipei City, totally satisfactory), (Taipei County, very satisfactory), (Taoyuan County, moderately satisfactory)} and a job's location offer is {(Taichung City, very satisfactory), (Taipei city, rather satisfactory), (Taoyuan County, rather satisfactory)}, we can obtain the minimum degree for each data item {(Taipei City, rather satisfactory), (Taipei County, 0), (Taoyuan County, moderately satisfactory), (Taichung City, 0)}. Hence, the result of matching is the maximum 'rather satisfactory' which can be represented by a triangular fuzzy number (0.5,0.75,1).
- In the case of one discrete fuzzy set and one crisp value, the result of matching is the mapping of the crisp value to the discrete fuzzy set. For example, when a hiring company's education preference is {(Master, very satisfactory), (Bachelor, rather satisfactory)} and a job seeker's education is 'Master', we obtain the result of matching is 'very satisfactory' (0.75,1,1).

- In the case of two crisp values, the result of matching is either 1 (the fuzzy data satisfies the fuzzy condition) or 0 (the fuzzy data doesn't satisfy the fuzzy condition).

2.2.3 Aggregating All Fuzzy Conditions

Computing the overall degree of satisfaction between a fuzzy query and the fuzzy data needs to aggregate all fuzzy conditions based on their degrees of importance and degrees of matching. We apply the fuzzy weighted average (FWA) [6,9,10] to calculate the overall degree of satisfaction using triangular fuzzy numbers. In our approach, the matching degrees of all fuzzy conditions are indicators (x_i) that rate the overall degree of satisfaction between a fuzzy query and the fuzzy data. The degrees of importance are weights (w_i) that act upon indicators. Therefore, the fuzzy weighted average y can be defined as:

$$y = f(x_i,...,x_n,w_1,...,w_n) = \frac{\sum_{i=1}^{n} w_i x_i}{\sum_{i=1}^{n} w_i}$$

Where there is n fuzzy conditions, the degree of matching x_i, $1 \leq i \leq n$ is represented by a crisp value or a triangular fuzzy number, and the degree of importance w_i, $1 \leq i \leq n$ is represented by a triangular fuzzy number. Since w_i are no longer crisp numbers, $\sum_{i=1}^{n} w_i = 1$ is not imposed as a requirement. We adopt the approximate expressions on $\oplus$ and $\otimes$ operators for the computation of L-R fuzzy numbers, which is suggested by Dubois and Prade [7]. Consider this example: (1) A job seeker makes a fuzzy query containing a salary preference {(60000,1), (50000,0.92), (30000,0)} with 'rather important', a location preference {(Taipei City, totally satisfactory), (Taipei County, very satisfactory), (Taoyuan County, moderately satisfactory)} with 'most important', his education 'Master' with 'moderately important', and other fuzzy conditions with the default importance 'don't care'. (2) A job stored in databases contains the salary offer {(20000,1), (30000,0.75), (40000,0)}, the location offer {(Taichung City, very satisfactory), (Taipei city, rather satisfactory), (Taoyuan County, rather satisfactory)}, the education preference {(Master, very satisfactory), (Bachelor, rather satisfactory)}, and a set of attributes with fuzzy data. We apply FWA to calculate the overall degree of satisfaction between the fuzzy query and the fuzzy data as follows.

$$y = \frac{(0.5,0.75,1) \otimes 0.285 \oplus (1,1,1) \otimes (0.5,0.75,1) \oplus (0.25,0.5,0.75) \otimes (0.75,1,1)}{(0.5,0.75,1) \oplus (1,1,1) \oplus (0.25,0.5,0.75)}$$

$$= \frac{(0.1425,0.21375,0.285) \oplus (0.5,0.75,1) \oplus (0.125,0.5,0.75)}{(0.5,0.75,1) \oplus (1,1,1) \oplus (0.25,0.5,0.75)}$$

$$= \frac{(0.7675,1.46375,2.035)}{(1.75,2.25,2.75)} \approx \frac{1.431765}{2.25} = 0.63634$$

Applying the mathematical operations on fuzzy numbers [7,12,16], we get two fuzzy numbers (0.7675, 1.46375, 2.035) and (1.75, 2.25, 2.75). The center of gravity is adopted to defuzzify a fuzzy number [14], which is achieved by mathematical integral. Therefore, the overall degree of satisfaction between the fuzzy query and the fuzzy data is 0.63634 (i.e. 63.634%).

By applying FWA to calculate fuzzy data's overall degrees of satisfaction to a fuzzy query, the ordering of all fuzzy data is obtained according to their overall degrees of satisfaction. Figure 11 shows the ordering result of a fuzzy query on the resume search, and users can click the resume code for more detailed information.

Featured Resume Results
1-10 out of 706 Resumes
2 3 4

Resume Code	Experience	Department	Education	Monthly Salary Preference	Matching Degree	View
2643	3 Years	Computer Science and Engineering	Bachelor degree	TWD 30000-45000	71.289%	Detailed
605	2 Years	Computer Science and Engineering	Bachelor degree	TWD 35000-48000	70.81%	Detailed
4657	2 Years	Computer Science and Engineering	Bachelor degree	TWD 40000-60000	70.47%	Detailed
1810	3 Years	Computer Science and Engineering	Master's degree	TWD 45000-55000	69.13%	Detailed
2623	1 Years	Computer Science and Engineering	Bachelor degree	TWD 30000-60000	68.8%	Detailed
4257	1 Years	Computer Science and Engineering	Master's degree	TWD 30000-45000	68.789%	Detailed
2562	4 Years	Computer Science and Engineering	Bachelor degree	TWD 36000-50000	68.69%	Detailed
2494	4 Years	Management Information Systems	Master's degree	TWD 30000-42000	68.36%	Detailed
1835	2 Years	Electrical Engineering	Master's degree	TWD 30000-48000	68.13%	Detailed
1927	4 Years	Management Information Systems	Master's degree	TWD 40000-60000	68.13%	Detailed

Fig. 11. The result of a fuzzy query on resume search

3 Conclusion

In this paper, we apply the fuzzy logic theory to develop a fuzzy query mechanism for human resource websites. The advantages of the proposed approach are as follows.

- Users' preferences often contain imprecision and uncertainty. Our approach provides a mechanism to express fuzzy data in human resource websites and to store fuzzy data into conventional database management systems without modifying DBMS models.
- Traditional SQL queries are based on total matching which is limited in its ability to come to grips with the issues of fuzziness. Our approach provides a mechanism to state fuzzy queries by fuzzy conditions and to differentiate between fuzzy conditions according to their degrees of importance.
- Traditional SQL queries fail to deal with the compensation between different conditions. Our approach provides a mechanism to aggregate all fuzzy conditions based on their degrees of importance and degrees of matching. The ordering of query results via the mutual compensation of all fuzzy conditions is helpful to alleviate the problem of the information overload.

References

1. http://hotjobs.yahoo.com/
2. http://www.104.com.tw/
3. http://www.1111.com.tw/
4. http://www.find-job.net/
5. http://www.monster.com/
6. Chang, P.T., Hung, K.C., Lin, K.P., Chang, C.H.: A Comparison of Discrete Algorithms for Fuzzy Weighted Average. IEEE Transactions on Fuzzy Systems 14(5), 663–675 (2006)
7. Dubois, D., Prade, H.: Fuzzy Sets and Systems: Theory and Applications, New York, London (1980)

8. Galindo, J., Urrutia, A., Piattini, M.: Fuzzy Databases: Modeling, Design and Implementation. Idea Group Publishing, Hershey (2005)

9. Guu, S.M.: Fuzzy Weighted Averages Revisited. Fuzzy Sets and Systems 126, 411–414 (2002)

10. Kao, C., Liu, S.T.: Competitiveness of Manufacturing Firms: An Application of Fuzzy Weighted Average. IEEE Transactions on Systems, Man, and Cybernetics – Part A: Systems and Humans 29(6), 661–667 (1999)

11. Kaufmann, A., Gupta, M.M.: Introduction to Fuzzy Arithmetic: Theory and Applications. Van Nostrand Reinhold, New York (1985)

12. Lai, Y.J., Hwang, C.L.: Fuzzy Mathematical Programming, Methods and Applications. Springer, Heidelberg (1992)

13. Ngai, E.W.T., Wat, F.K.T.: Fuzzy Decision Support System for Risk Analysis in E-Commerce Development. Decision Support Systems 40(2), 235–255 (2005)

14. Tseng, T.Y., Klein, C.M.: A New Algorithm for Fuzzy Multicriteria Decision Making. International Journal of Approximate Reasoning 6, 45–66 (1992)

15. Zadeh, L.A.: Fuzzy Sets. Information and Control 8, 338–353 (1965)

16. Zimmermann, H.J.: Fuzzy Set Theory and Its Applications, 2nd revised edn. Kluwer Academic Publishers, Dordrecht (1991)

Selecting Cooperative Enterprise in Dynamic Enterprise Alliance Based on Fuzzy Comprehensive Evaluation

Yan Li[1], Guoxing Zhang[2], and Jun Liu[1]

[1] College of Mechanical & Electrical Engineering, Henan Agriculture University
[2] Information Center, Henan University of Urban Construction
liyanliulizhi@sina.com, zgxyx@hncj.edu.cn, liujunhenau@163.com

Abstract. Dynamic enterprise alliance is a complex organization system. It has the virtual enterprise establishment, virtual operation and virtual management functions. The suitable cooperative partner is the necessary condition to ensure the efficient functioning of the dynamic enterprises alliance. It is the key technology of selecting the suitable partner reasonably and forming the dynamic enterprises alliance to realize the network manufacturing. In this paper, one reasonable model of fuzzy comprehensive evaluation on the cooperative partner of the dynamic enterprises alliance is established, using the fuzzy multi-grade comprehensive evaluation method based on the theories of fuzzy logic. It can offer a scientific method to select the optimum cooperative partner for the dynamic enterprises alliance, and improve the comparison level for alternatives.

Keywords: cooperative enterprise, dynamic enterprise alliance, fuzzy comprehensive evaluation.

1 Introduction

Dynamic enterprise alliance is a complex organization system. It is one organization of cooperation and competition composed by two or more than two enterprises which have the common strategic interests and share the enterprise resources each other in order to achieve the business strategies and certain objectives, and restricted by the various agreements and contracts in one certain period each other [1-3]. The development of network manufacturing technology is restricted by the network manufacturing environment resource management, and suitable cooperative partner is the necessary condition to ensure the efficient functioning of the dynamic enterprises alliance. It is the key technology to evaluate the cooperative partner synthetically and select the optimum cooperative partner scientifically to realize the network manufacturing.

Selecting the optimum cooperative partner is a typical combinatorial optimization problem, and it also is one key problem in the process of establishing the dynamic enterprise alliance. It is very difficult to select the optimum cooperative partner simply depending on the experience of the influence factors. The fuzzy comprehensive evaluation is one kind of effective multitudinous factors decision method used to evaluate the object synthetically affected by various factors, and the influence factors of selecting the cooperative partner always are uncertain, so the fuzzy comprehensive

H. Deng et al. (Eds.): AICI 2009, LNAI 5855, pp. 590–598, 2009.

evaluation is the optimum method to evaluate the cooperative partner synthetically, and select the optimum cooperative partner [4-7].

2 Decision Making of Fuzzy Comprehensive Evaluation on the Cooperative Partners of the Dynamic Enterprise Alliance

Dynamic enterprise alliance is a generalized business organization. It includes multitudinous cooperative partners, such as design firms, manufacturing enterprises, component manufacturers, assembly plant, wrapping enterprise, advertising company, logistics enterprise, sales company and after-sale service company, and so on. The influence factors on selecting the optimum cooperative partner can be divided into internal factors and external factors. The internal factors mainly include design, innovation, manufacturing, cost, time, quality, sale, serve, informatization, staff, management, culture and credibility, and so on. The external factors mainly include geography condition, cooperation ability, logistical support system and social environment, and so on. Towards the different kind of cooperative enterprise, the thinking emphasis of the influence factors is different. For example, toward the manufacturing enterprise, it mainly thinks of the technological levels of workers, the applicability and advanced degree of the manufacturing equipments, manufacturing technique, the rationality of process and the credibility of the manufacturing enterprise in the entire manufacturing industry. In this paper, the sales enterprise is taken as the example to research the method of selecting cooperative enterprise in dynamic enterprise alliance based on fuzzy comprehensive evaluation.

2.1 The Influence Factor Set Establishment

The influence factor set is one ordinary set comprised by various factors influencing the fuzzy comprehensive evaluation on the cooperative partner. It mainly takes into account six factors while the sales enterprise is evaluated, such as the economic strength, marketing strategy, economic benefit, development prospect, staff quality and sales channels.

The economic strength factor mainly includes registered capital, permanent assets, bank loans and liquid assets. The marketing strategy factor mainly includes marketing purposes, marketing planning, market research and market positioning. The economic benefit factor mainly includes return on assets, sales net profit rate, profit-tax rate of cost and velocity of liquid assets. The development prospect factor mainly includes enterprise culture, customer relationship, sales achievement, service quality and strain capacity. The staff quality factor mainly includes professional dedication, professional skill, insight, psychology bearing capacity and decision-making ability. And the sales channels factor mainly includes the economic benefits of sales channels, the ability of enterprise to control the sales channels and the adaptability of the sales channels to the market environment. Thus, the influence factor set U can be represented as

$$U = \begin{pmatrix} u_1 & u_2 & u_3 & u_4 & u_5 & u_6 \end{pmatrix} \tag{1}$$

In (1), the subset u_1 indicates economic strength, the subset u_2 indicates marketing strategy, the subset u_3 indicates economic benefit, the subset u_4 indicates development prospect, the subset u_5 indicates staff quality and the subset u_6 indicates sales channels.

The economic strength subset u_1 can be represented as

$$u_1 = \begin{pmatrix} u_{11} & u_{12} & u_{13} & u_{14} \end{pmatrix} \tag{2}$$

In (2), u_{11} indicates registered capital, u_{12} indicates permanent assets, u_{13} indicates bank loans and u_{14} indicates liquid assets.

The marketing strategy subset u_2 can be represented as

$$u_2 = \begin{pmatrix} u_{21} & u_{22} & u_{23} & u_{24} \end{pmatrix} \tag{3}$$

In (3), u_{21} indicates marketing purposes, u_{22} indicates marketing planning, u_{23} indicates market research and u_{24} indicates market positioning.

The economic benefit subset u_3 can be represented as

$$u_3 = \begin{pmatrix} u_{31} & u_{32} & u_{33} & u_{34} \end{pmatrix} \tag{4}$$

In (4), u_{31} indicates return on assets, u_{32} indicates sales net profit rate, u_{33} indicates profit-tax rate of cost and u_{34} indicates velocity of liquid assets.

The development prospect subset u_4 can be represented as

$$u_4 = \begin{pmatrix} u_{41} & u_{42} & u_{43} & u_{44} & u_{45} \end{pmatrix} \tag{5}$$

In (5), u_{41} indicates enterprise culture, u_{42} indicates customer relationship, u_{43} indicates sales achievement, u_{44} indicates quality of service and u_{45} indicates strain capacity.

The quality of staff subset u_5 can be represented as

$$u_5 = \begin{pmatrix} u_{51} & u_{52} & u_{53} & u_{54} & u_{55} \end{pmatrix} \tag{6}$$

In (6), u_{51} indicates professional dedication, u_{52} indicates professional skill, u_{53} indicates insight, u_{54} indicates psychology bearing capacity and u_{55} indicates decision-making ability.

The sales channels subset u_6 can be represented as

$$u_6 = \begin{pmatrix} u_{61} & u_{62} & u_{63} \end{pmatrix} \tag{7}$$

In (7), u_{61} indicates economic benefits of sales channels, u_{62} indicates ability of enterprise to control the sales channels and u_{63} indicates adaptability of the sales channels to the market environment.

2.2 The Evaluation Set Establishment

The evaluation set is comprised of various possible evaluate results made by the decision makers. The economic strength subset u_1, marketing strategy subset u_2, economic benefit subset u_3, development prospect subset u_4, staff quality subset u_5 and sales channels subset u_6 can be separately evaluated by the set of {good, relatively good, general, bad, sucks}, {high, relatively high, general, low, very low} and {long, relatively long, general, short, very short}, and so on. In this paper, the evaluation set of economic strength subset u_1, marketing strategy subset u_2, economic benefit subset u_3, development prospect subset u_4, staff quality subset u_5 and sales channels subset u_6 are unified into one evaluation set V, it can be represented as

$$V = \begin{pmatrix} v_1 & v_2 & v_3 & v_4 & v_5 \end{pmatrix} \tag{8}$$

In (8), v_1 means excellent, v_2 means good, v_3 means middle, v_4 means pass muster and v_5 means bad.

2.3 The Weight Set Establishment

The weight set is comprised of every influence factor's weight number. It can reflect every influence factor's importance. Assuming a_i is the weight number of the influence factor u_i, thus, the weight set A can be represented as

$$A = \begin{pmatrix} a_1 & a_2 & \cdots & a_m \end{pmatrix} \tag{9}$$

Usually, every influence factor's weight number should meet polarity and non-negativity constraint. i.e.

$$\begin{cases} \sum_{i=1}^{m} a_i = 1 \\ 0 \le a_i \le 1 \end{cases} \tag{10}$$

Different evaluator maybe has the different attitude toward the same thing, and the weight numbers offered by them also are different. In this paper, the weighted statistics method is adopted to determine the weight number of every influence factor [8]. Firstly, make a weight distribution questionnaire (shown as table 1), then ask some experts or related people fill in the optimum weight number who believe, after taking bake the weight distribution questionnaires, adopt the weighted statistics method to calculate the weight number A.

Table 1. Weight Distribution Questionnaire

influence factor u_i	u_1	u_2	u_3	Σ
weight number a_i	a_1	a_2	a_3	1

The weight number set A can be calculated through the statistical investigation on influence degree of the six subsets u_1, u_2, u_3, u_4, u_5 and u_6 on the making-decision of selecting the cooperative partner decision.

$$A = \begin{pmatrix} 0.2 & 0.23 & 0.12 & 0.08 & 0.15 & 0.22 \end{pmatrix} \qquad (11)$$

In the same way, the weight number sets of the six subsets u_1, u_2, u_3, u_4, u_5 and u_6 also can be calculated as

$$A_1 = \begin{pmatrix} 0.23 & 0.18 & 0.27 & 0.32 \end{pmatrix} \qquad (12)$$

$$A_2 = \begin{pmatrix} 0.22 & 0.28 & 0.19 & 0.31 \end{pmatrix} \qquad (13)$$

$$A_3 = \begin{pmatrix} 0.22 & 0.28 & 0.19 & 0.31 \end{pmatrix} \qquad (14)$$

$$A_4 = \begin{pmatrix} 0.09 & 0.22 & 0.27 & 0.23 & 0.19 \end{pmatrix} \qquad (15)$$

$$A_5 = \begin{pmatrix} 0.29 & 0.22 & 0.18 & 0.12 & 0.19 \end{pmatrix} \qquad (16)$$

$$A_6 = \begin{pmatrix} 0.39 & 0.33 & 0.28 \end{pmatrix} \qquad (17)$$

3 The Membership Function Establishment

It is the key point to determine the membership degree for the fuzzy synthetic evaluation, and it can directly influence the result of evaluation. The membership degree of the j-th evaluating indicator of the influence factor u_i can be calculated through the membership function.

Using the method of fuzzy statistics, to calculate the membership degree of the j-th evaluating indicator of the influence factor u_i offered by every expert, and then the weight of the evaluating indicator a_{ij} or the span of the value a_{ij} of the membership degree μ_{ij} can be calculated using statistical analysis method [9]. The value a_{ij} of the membership degree μ_{ij} is shown as table 2.

Table 2. The Value a_{ij} of the Membership Degree μ_{ij}

factor set	factor subset	v_1	v_2	..	v_m
	u_{i1}	$a_{10} - a_{11}$	$a_{11} - a_{12}$	..	$a_{1m-1} - a_{1m}$
u_i	u_{i2}	$a_{20} - a_{21}$	$a_{21} - a_{22}$	..	$a_{2m-1} - a_{2m}$
	$\vdots$	$\vdots$	$\vdots$	$\vdots$	$\vdots$
	u_{in}	$a_{n0} - a_{n1}$	$a_{n1} - a_{n2}$	..	$a_{nm-1} - a_{nm}$

Based on the table 2, the membership function can be structured and represented as

$$\mu_{i1}(x) = \begin{cases} 1 & (a_{i0} \leq x \leq a_{i1}) \\ (a_{i2} - x)/(a_{i2} - a_{i1}) & (a_{i1} \leq x \leq a_{i2}) \\ 0 & (a_{i2} \leq x \leq a_{im}) \end{cases} \tag{18}$$

In (18), $i = 1, 2, \cdots, n$.

$$\mu_{ij}(x) = \begin{cases} 0 & (a_{i0} \leq x \leq a_{ij-2}) \\ (x - a_{ij-2})/(a_{ij-1} - a_{ij-2}) & (a_{ij-2} \leq x \leq a_{ij-1}) \\ 1 & (a_{ij-1} \leq x \leq a_{ij}) \\ (a_{ij+1} - x)/(a_{ij+1} - a_{ij}) & (a_{ij+1} \leq x \leq a_{im}) \\ 0 & (a_{ij} \leq x \leq a_{ij+1}) \end{cases} \tag{19}$$

In (19), $i = 1, 2, \cdots, n$ and $j = 2, 3, \cdots, m-1$.

$$\mu_{im}(x) = \begin{cases} 0 & (a_{i0} \leq x \leq a_{im-2}) \\ (x - a_{im-2})/(a_{im-1} - a_{im-2}) & (a_{im-2} \leq x \leq a_{im-1}) \\ 1 & (a_{im-1} \leq x \leq a_{im}) \end{cases} \tag{20}$$

In (20), $i = 1, 2, \cdots, n$.

The membership degree μ_{ij} of the j-th evaluating indicator of the influence factor u_i on one certain technological design scheme can be calculated through the membership function.

4 Fuzzy Comprehensive Evaluation

4.1 The Primary Fuzzy Comprehensive Evaluation

Assuming the primary fuzzy comprehensive evaluation is carried out on the influence factor u_i in the influence set U, thus, the membership degree μ_{ij} of the j-th evaluating indicator of the influence factor u_i can be calculated through the membership function, and the evaluation result of the single factor u_i can be represented by the fuzzy set as R_{ij}

$$R_{ij} = \frac{\mu_{i1}}{v_1} + \frac{\mu_{i2}}{v_2} + \cdots + \frac{\mu_{im}}{v_n} \tag{21}$$

In (21), R_{ij} is the single factor evaluation set, it can be simply represented as

$$R_{ij} = \begin{pmatrix} r_{i1} & r_{i2} & \cdots & r_{im} \end{pmatrix} \tag{22}$$

At the same way, the evaluation set corresponding to every influence factor u_i can be obtained. And the single factor evaluation matrix R_i can be represented as

$$R_i = \begin{bmatrix} R_{i1} & R_{i2} & \cdots & R_{i4} \end{bmatrix}^T \tag{23}$$

In (23), R_i is the single factor fuzzy evaluation matrix.

The primary evaluation set B_i can be obtained through the fuzzy comprehensive evaluation on the influence factor u_i.

$$B_i = A_i \circ R_i = \begin{pmatrix} b_{i1} & b_{i2} & \cdots & b_{im} \end{pmatrix} \tag{24}$$

4.2 The Secondary Fuzzy Comprehensive Evaluation

With the primary evaluation set B_1, B_2, $\cdots$, and B_k, the single factor evaluation matrix R of the influence factor set U can be represented as

$$R = \begin{bmatrix} B_1 \\ B_2 \\ \vdots \\ B_K \end{bmatrix} = \begin{bmatrix} b_{11} & b_{12} & \cdots & b_{1m} \\ b_{21} & b_{22} & \cdots & b_{2m} \\ \vdots & \vdots & \vdots & \vdots \\ b_{k1} & b_{k2} & \cdots & b_{km} \end{bmatrix} \tag{25}$$

The fuzzy comprehensive evaluation set B can be obtained through the fuzzy comprehensive evaluation on the influence factor set U, and it can be represented as

$$B = A \circ R = \begin{pmatrix} b_1 & b_2 & \cdots & b_m \end{pmatrix} \tag{26}$$

Through the normalization of the fuzzy comprehensive evaluation set B, the fuzzy comprehensive evaluation on one certain technological design scheme can be carried out with the maximum membership degree method, the weighted average method or the fuzzy distribution method.

5   An Example of Fuzzy Comprehensive Evaluation

The fuzzy comprehensive evaluation on one certain sales enterprise is carried out. Based on the statistical results, the evaluation matrix of the subset u_1 can be obtained and represented as

$$R = \begin{bmatrix} 0.35 & 0.42 & 0.18 & 0.04 & 0.01 \\ 0.20 & 0.53 & 0.23 & 0.02 & 0.02 \\ 0.41 & 0.37 & 0.16 & 0.06 & 0 \\ 0.38 & 0.49 & 0.10 & 0.02 & 0.01 \end{bmatrix} \tag{27}$$

The comprehensive evaluation result of the subset u_1 can be calculated and represented as B_1

$$B_1 = A_1 \circ R_1 = (0.20 \quad 0.37 \quad 0.23 \quad 0.18 \quad 0.18) \tag{28}$$

At the same way, the comprehensive evaluation results of the subset u_2, u_3, u_4, u_5 and u_6 can be calculated and represented as B_2, B_3, B_4, B_5, and B_6. Then, the evaluation matrix R of the factor set U can be calculated and represented as

$$R = \begin{bmatrix} B_1 \\ B_2 \\ B_3 \\ B_4 \\ B_5 \\ B_6 \end{bmatrix} = \begin{bmatrix} 0.20 & 0.37 & 0.23 & 0.18 & 0.18 \\ 0.27 & 0.32 & 0.30 & 0.17 & 0.17 \\ 0.23 & 0.21 & 0.22 & 0.19 & 0.19 \\ 0.22 & 0.29 & 0.25 & 0.09 & 0.09 \\ 0.19 & 0.21 & 0.22 & 0.12 & 0.12 \\ 0.28 & 0.34 & 0.32 & 0.28 & 0.28 \end{bmatrix} \tag{29}$$

The comprehensive evaluation result of the factor set U can be calculated and represented as B

$$B = A \circ R = (0.20 \quad 0.21 \quad 0.22 \quad 0.09 \quad 0.09) \tag{30}$$

The normalized result of the comprehensive evaluation result B can be calculated and represented as B'

$$B' = (b_1' \quad b_2' \quad b_3' \quad b_4' \quad b_5') = (0.247 \quad 0.259 \quad 0.272 \quad 0.111 \quad 0.111) \tag{31}$$

To assign the value to the evaluation set V, assuming the value of the evaluation set V is

$$V = (1 \quad 0.85 \quad 0.75 \quad 0.6 \quad 0.5) \tag{32}$$

Think of b_j is the weight of the evaluation target v_j, and the final evaluation result of the certain sales enterprise can be calculated with the weighted average method.

$$V = \sum_{j=1}^{6} b_j \cdot v_j = \tag{33}$$
$$1 \times 0.247 + 0.85 \times 0.259 + 0.75 \times 0.272 + 0.6 \times 0.111 + 0.5 \times 0.111 = 0.793$$

It means the fuzzy comprehensive evaluation value of the certain sales enterprise is 0.793.

At the same way, the fuzzy comprehensive evaluation value of all the sales enterprises which have the cooperation intent can be obtained. And the sales enterprise it has the maximal fuzzy comprehensive evaluation value can be selected as the optimum cooperative partner.

Similarly, the fuzzy comprehensive evaluation on the other cooperative enterprise in the other business area can be carried out,   and the optimum cooperative partner also can be selected.

6 Conclusion

It can play the role of experts adequately, reduce the harm caused by the personal subjective assume, and provide the scientific basis for evaluating the enterprises which have the cooperation intent based on comprehensive evaluation of the cooperative partners in the dynamic enterprise alliance with the fuzzy comprehensive evaluation method. It can improve the assessment level on the cooperative partners and make the evaluation result more scientific with the fuzzy comprehensive evaluation method.

References

1. Wanshan, W., Yadong, G., Peili, Y.: Networked manufacturing. Northeast University Press, Shenyang (2003) (in Chinese)
2. Jicheng, L., Jianxun, Q.: Dynamic alliance synergistic decision-making model based on business intelligence center. In: International Symposium on Information Processing, ISIP 2008 and International Pacific Workshop on Web Mining and Web-Based Application, WMWA 2008, Moscow, Russia, May 23-25, pp. 219–223 (2008)
3. Congqian, Q., Yi, G.: Partner optimization of enterprises dynamic alliance. Journal of Tongji University 35(12), 1674–1679 (2007) (in Chinese)
4. Lixin, W.: Fuzzy System & Fuzzy Control Tutorial. Tsinghua University Press, Beijing (2003) (in Chinese)
5. Baoqing, H.: The basis of fuzzy theory. Wuhan University Press, Wuhan (2004)
6. Bing, Z., Zhang, R.: Research on fuzzy-grey comprehensive evaluation of software process modeling methods. In: Proceedings - 2008 International Symposium on Knowledge Acquisition and Modeling, KAM 2008, Wuhan, China, December 21-22 (2008)
7. Xulin, L., Baowei, S.: Three level fuzzy comprehensive evaluation based on Grey Relational Analysis and Entropy weights. In: 2008 International Symposium on Information Science and Engineering, ISISE 2008, Shanghai, China, December 20-22, vol. 2, pp. 32–35 (2008)
8. Korotchenko, M.A., Mikhailov, G.A., Rogazinskii, S.V.: Value modifications of weighted statistical modelling for solving nonlinear kinetic equations. Russian Journal of Numerical Analysis and Mathematical Modelling 22(5), 471–486 (2007)
9. Shinguang, C., Yikuei, L.: On performance evaluation of ERP systems with fuzzy mathematics. Expert Systems with Applications 36(3 Part 2), 6362–6367 (2009)

An Evolutionary Solution for Cooperative and Competitive Mobile Agents

Jiancong Fan[1], Jiuhong Ruan[2], and Yongquan Liang[1]

[1] College of Information Science and Engineering, Shandong University of Science and Technology, Qingdao 266510, China
howdoyoudo07@yahoo.com.cn
[2] Scientific Research Department, Shandong Jiaotong University, Jinan 250023, China
jh_ruan@bit.edu.cn

Abstract. The cooperation and competition among mobile agents using evolutionary strategy is an important domain in Agent theory and application. With evolutionary strategy the cooperation process is achieved by training and iterating many times. From evolutionary solution of cooperative and competitive mobile agents (CCMA), a group of mobile agents are partitioned into two populations, cooperative agents group and competitive agent group. Cooperative agents are treated as several pursuers, while a competitive agent is viewed as the pursuers' competitor called evader. The cooperation actions take place among the pursuers in order to capture the evader as rapidly as possible. An agent individual (chromosome) is encoded based on a kind of two-dimensional random moving. The next moving direction is encoded as chromosome. The chromosome can be crossed over and mutated according to designed operators and fitness function. An evolutionary algorithm for cooperation and competition of mobile agents is proposed. The experiments show that the algorithm for this evolutionary solution is effective, and it has better time performances and convergence.

Keywords: evolution, mobile agent, cooperative agents, competitive agents.

1 Introduction

Mobile Agent is a kind of software Agent. Mobile Agent has many characteristics, such as autonomy, sociality, self-learning, and importantly mobility[1]. Mobile Agent can move from one position to another. The moving involves in the states transition and Agent's states change. Under current context and state mobile Agent autonomously determines what time to move and what place it moves to[2,3].

Cooperative mobile agents have the ability of cooperation and adaptability[4,5]. Firstly, agents can cooperate by exchanging data and/or code when they meet on a given condition. Secondly, the behavior of an agent can change based on its current state and the information it has gathered while traveling.

In this paper the cooperation among mobile agents is studied using evolutionary strategy. The determination of cooperation process and form is achieved by training and iterations many times. The evolvement result is that a reasonable cooperation strategy is obtained and stable.

H. Deng et al. (Eds.): AICI 2009, LNAI 5855, pp. 599–607, 2009.
© Springer-Verlag Berlin Heidelberg 2009

2   Relative Work

The combination of evolutionary algorithm and cooperative mobile agents has been studied in some aspects, most of which use agents to study or implement evolutionary computation processes. The literature [6,7] presented an agent-based version of cooperative co-evolutionary algorithm. This type of systems has been already applied to multi-objective optimization. Evolution learning has also been studied for multi-agent with strategic coalition [8], which used the iterated prisoner's dilemma game to model the dynamic system in an evolutionary learning environment. In literature [9], an evolution strategy is introduced on the basis of cooperative behaviors in each group of agents. The evolution strategy helps each agent to be self-defendable and self-maintainable, and agents in same group cooperate with each other. This method use reinforcement learning, enhance neural network and artificial life. In literature [10] a neural network is used for the behavior decision controller. The input of the neural network is decided by the existence of other agents and the distance to the other agents. The output determines the directions in which the agent moves. The connection weight values of this neural network are encoded as genes, and the fitness of individuals is determined using a genetic algorithm. There are also other studies from the perspectives of game theory [11,12,13]. Other studies concentrating on neural computing can be obtained in literatures [14,15,16].

3   Cooperative and Competitive Mobile Agents

In this paper, cooperative agents are treated as several pursuers, while a competitive agent is viewed as the pursuers' competitor called evader. The cooperation actions take place among the pursuers in order to capture the evader as rapidly as possible. However, the competitive agent evader doesn't cooperate with the pursuers so as to avoid the capturing.

The action area of pursuers P=$\{p_1, p_2, ..., p_n\}$ and evaders E=$\{e_1, e_2, ..., e_m\}$ is projected to a two-dimension coordination plane. P and E are represented as particles in the coordinate in our application experiments. The locations of P and E are represented as coordinate variables (x, y). For convenience but not influencing the problem-solving results, the coordinate variables (x, y) are discrete and expanded as square regions. The edge length of each square is ε ($\varepsilon > 0$). And $p_i(i = 1,2,...,n)$ and $e_j(j=1,2,...,m)$ move between two adjacent squares, which is called two-dimensional random moving.

Definition 1. reflection barriers: in two-dimension coordinate plane XOY, (X,0) and (0,Y) are called reflection barriers.

Definition 2. two-dimensional random moving: in two-dimension coordinate plane XOY, Agents move autonomously and accord with the following terms:

(1) Agents can only move to adjacent coordinate cells for each moving;
   At random circumstances,
(2) If one side of Agents current location is reflection barriers, Agents have k ($k =3$ or 5) alternative moving locations. They move to the adjacent areas with probability $1/k$.

(3) If there don't exist reflection barriers around the Agents, the Agents move to the adjacent areas with probability 1/8.

If the Agents have preferences under particular conditions,

(4) The Agents move to the adjacent areas with probability $\lambda_i P(d_i)$, where λ_i is called preference coefficient. λ_i denotes a measure of Agent$_i$'s customs or cognitive styles of moving directions,

d_i denotes moving direction of Agent$_i$ at particular moment, and at this moment there exist reflection barriers adjacent to Agent$_i$, then

$$\sum_{i=1 \ to \ 3} \lambda_i P(d_i) = 1 \ . \quad \text{or} \quad \sum_{i=1 \ to 5} \lambda_i P(d_i) = 1 \ . \tag{1}$$

else,

$$\sum_{i=1 \ to \ 8} \lambda_i P(d_i) = 1 \ . \tag{2}$$

The moving is called two-dimensional random moving if it is accord with the above items.

Example 1. There are three Agents showed in figure 1 (a) (b) (c) respectively. (a) shows an Agent locates at a place where there is no reflection barriers around. There are 8 alternative directions to move towards. It probably likes to move upward vertically or others. (b) shows left side of an Agent's location is reflection barrier. There are 5 directions to choose. (c) shows there are only three directions to choose because there are two reflection barriers around the Agent.

(a) no reflection barriers (b) reflection barriers at one side (c) reflection barriers at two sides

Fig. 1. Agents' two-dimensional random moving

4 Evolutionary Solution for CCMA

There are two populations designed in this paper. One population has k individuals cooperating with each other, that is, population size is k. The other has one individual which is a competitor of the above k cooperative individuals.

4.1 Encoding Schemes for Individual

The chromosome is encoded based on two-dimensional random moving. The next moving direction of individual is encoded as chromosome. One agent locates at one of the two positions, no reflection barriers and reflection barriers. Figure 2 (a) and (b) illustrate respectively. In Figure 2 (a) there are eight selected directions. An agent's next moving scheme can be represented as 8 bits binary code. The codes from left to right correspond to the locations 0 to 7. The bit 1 denotes the agent moves to the direction, and 0 doesn't. On no barriers circumstances the encoding resembles with the above. The only difference is that the directions with barriers are coded with a character neither 1 nor 0, such as an asterisk.

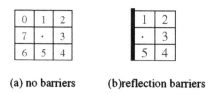

(a) no barriers   (b)reflection barriers

Fig. 2. Numbering of moving directions

For example, the chromosome (0 1 0 0 0 0 0 0) represents an agent moves vertically upward because the bit of location 1 is 1 while the other locations are 0. The chromosome (* 0 0 1 0 0 * *) represents the adjacent left side is reflection barrier and an agent moves right horizontally.

4.2 Evolutionary Operators

Selection. The chromosomes compete against each other and the chromosome with the highest fitness value is selected. In this paper distance-based function is adopted as the fitness function, that is, the Euclid distance between pursuer and evader is the fitness value. The smaller fitness value is better for pursuing agent while the larger fitness value is better for evading agent.

Crossover. The crossover operation exchanges parts of a pair of chromosomes, creating new chromosomes called children or offspring (If crossover does not succeed in the probability test, then the children will be identical copies of their parents). The crossover operation needs two chromosomes, so the population with over two agents can be applied with crossover operator. An agent moves to only one direction. There should be strict limitation that the offspring must have one bit 1 only. Figure 3 illustrates a crossover operation in the solution.

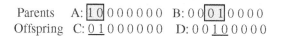

Fig. 3. Crossover operation

Mutation. The mutation operator aims to increase the variability of the population, allowing the evolution to simultaneously search different areas of the solution space. This operator changes at random the value of a chromosome gene, also randomly chosen with a given probability (named mutation rate). Also, an agent moves to only one direction. If one bit 1 becomes 0, some other bit 0 should become 1. Figure 4 shows a mutation operation that changes the value of the forth gene from 0 to 1.

Before mutation   0 0 $\boxed{1}$ 0 0 0 0 0
After mutation   0 0 0 0 $\boxed{1}$ 0 0 0

Fig. 4. Mutation operation

4.3 Evolutionary Algorithm for CCMA

Let k cooperative agents (a cooperative agent abbr. cooAg) and a competitive agent (abbr. comAg) consist of a niche in which k cooAgs pursue a comAg. In the process of all agents' two-dimensional random moving cooAgs pursue comAg and want to capture comAg as rapidly as possible. The comAg tries its best not to be captured. Once one of the cooAgs captures comAg, the algorithm halts. A penalty factor is designed in the algorithm in order to evaluate the quality of cooAgs in pursuing co-mAg. The penalty factor α is defined as formula (3).
Let

$$x' = P(cooAg_i).x - P(cooAg_i).x$$
$$y' = P(cooAg_i).y - P(cooAg_i).y$$
$$D = (x'^2 + y'^2)^{1/2}$$

Then

$$\alpha(cooAg_i) = \begin{cases} \alpha, & D \leq \theta_0 . \\ \alpha+1, & \theta_0 < D \leq \theta_1 . \\ \alpha+2, & \theta_1 < D \leq \theta_2 . \end{cases} \qquad (3)$$

---

(1) Initialize distance threshold value θ_i, time interval T, capture distance d;
      //initialize penalty factor α
(2) for (int i=0; i<=k; i++)
(3) α(cooAg[i]) = 0;
(4) t = 0;   //start from time 0
(5) while (true){
         t = t+T;
(6)   for (int j=0; j<=k; j++)
(7)   Compute D(cooAg[i], comAg);
(8)   if(D<=d){
(9)   Capture succeed and break; }
(10)   else{
(11)   Compute α(cooAg[i]) according to D and formula (1); }
(12)   Select the chromosome with the smallest α to the next generation;
(13)   Crossover the chromosomes with bigger α to produce offspring;
(14)   Mutate the chromosome with the largest α;
(15)   if (t exceeds the longest time threshold) break;
(16) }   //The algorithm halt

Fig. 5. Algorithm description for CCMA

In formula (3), P(·) is agent's position in coordinate system. θ_0, θ_1, θ_2 are threshold values of distance. The distances between cooAgs and comAg are computed at a fixed interval T. From time 0, at time kT (k=1, 2, …) the distance D between each cooAg and comAg is computed and the penalty factor α is computed according to D. The chromosome with smaller α has better fitness. The chromosomes of cooAgs with bigger α are crossed over or mutated. The chromosome with the smallest α is copied to the next generation.

In our algorithm, it is assumed that if the distance D between some cooAg and co-mAg is less than a threshold value d, the cooperative capture succeeds. The threshold d is called capture distance. The algorithm description is given by figure 5.

5 Experimental Analysis

Assume that XOY shown in figure 1 is participants' active area. The whole active area is discrete, and the participants perform two-dimensional random moving in XOY. In our experiments, each cell is represented by coordinate value, that is, a cell as a coordinate point. Although the agents move randomly, they have their own preferences, which are illustrated in figure 6 and 7. In figure 6 two persuading agents and

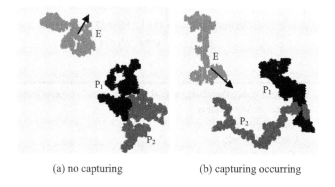

(a) no capturing   (b) capturing occurring

Fig. 6. Area of two cooperative agents and one competitive agent

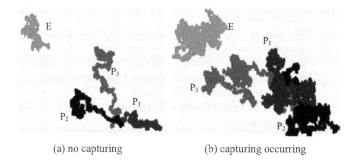

(a) no capturing   (b) capturing occurring

Fig. 7. Area of three cooperative agents and one competitive agent

an evading agent are interacting in some niche. In figure 6(a) the evader prefers to move to the direction on the whole as the arrow shows, but two cooperative persuaders P_1 and P_2 search within the scope of area they like. Under this circumstance, P_1 and P_2 can't capture the evader however they endeavor. On the contrary, figure 6(b) shows that the evader E has preferential moving direction meeting with the pursuer P_1's. Figure 7 shows the similar situation but with three persuading agents.

In this experiments, the evading agent moves at random. The algorithm proposed in this paper is mainly used to solve the problems shown in figure 6(a) and figure 7(a). During evolution the point with the next moving direction towards the evading agent's preferential direction and the nearest distance with the evading agent is selected to continue the persuader's moving. Under other circumstances the agents' chromosome is crossed over or mutated.

Table 1 shows the comparison of random capture times and evolutionary capture times with three cooperative agents. The capture distance d takes values 50, 100 and 200. Maximized moving times are designed with six values. With each d and all designed moving times 50 mobile experiments are implemented with random and evolutionary modes respective. The experiment results show that the times of evolutionary cooperative mobile agents capture successfully are more than random mobile agent obviously.

Table 1. Comparison of random capture times and evolutionary capture times

d	moving times	random mobile		evolutionary mobile	
		succeed	fail	succeed	fail
50	5000	1	49	5	45
	10000	7	43	13	37
	20000	17	33	29	21
	50000	32	18	39	11
	100000	37	13	45	5
	200000	41	9	48	2
100	5000	6	44	18	32
	10000	15	35	31	19
	20000	28	22	40	10
	50000	33	17	44	6
	100000	41	9	50	0
	200000	45	5	50	0
200	5000	25	25	37	13
	10000	31	19	42	8
	20000	33	17	47	3
	50000	37	13	50	0
	100000	42	18	50	0
	200000	45	5	50	0

The average iterative generations and average capture time according to different capture distances 50, 100 and 200 listed in table 2. From the table we can see that at each distance less moving times and more moving times can lead to more average iterative generations and capture time than medium moving times. This is because less moving times make the agents' mobility ceasing much faster, and more moving times make their mobile preferences more diverse.

Table 2. Average iterative generations and average capture time according to different capture distances

Capture distance	Moving times (10^3)	Average Iterative generations	Average capture time (ms)
	5	4301	243
	10	3747	202
50	20	3194	145
	50	7868	798
	100	20075	1892
	5	6862	386
	10	5014	299
100	20	4341	214
	50	9881	1319
	100	17150	3008
	5	4844	312
	10	3892	260
200	20	2368	192
	50	2663	194
	100	6554	702

6 Conclusions

As it can be observed in section 5 analyzing the quality of obtained solutions as the function of iteration and time, evolutionary multi-agent system with co-operation is very attractive alternative since initially solutions proposed by this algorithm were better than solutions proposed by "classical" (non evolution-based) algorithms. Evolution process can assure higher cooperative efficiency and optimal strategy.

The future plans include further investigation of the proposed mechanism. It could be interesting to modify the co-operation proposed in this paper in such a way that agents from different sub-populations (species) would specialize in different criteria and form aggregates (teams) composed of the represents of different species in order to solve the problem, for example, more competitive agents and more cooperative agents, even more populations.

Acknowledgments. This work is supported by Natural Science Foundation of Shandong Province of China (NO G2007Y07) and Spring Bud Planning of SUST(NO 06540040512).

References

1. Rao, A.S., Georgeff, M.P.: BDI agents: From theory to practice. In: Proc. the First International Conference on Multi-Agent Systems, San Franciso, CA, USA, pp. 312–319 (1995)
2. Lange, D.B., Oshima, M.: Seven good reasons for mobile agents. Communications of the ACM 42(3), 88–89 (1999)
3. Kotz, D., Gray, R.S.: Mobile Agents and the Future on the Internet. ACM Operating Systems Review 33(3), 7–13 (1999)
4. Bassett, J., De Jong, K.: Evolving behaviors for cooperating agents. In: Ras, Z. (ed.) Proceedings from the Twelfth International Symposium on Methodologies for Intelligent Systems, Charlotte, NC, pp. 157–165. Springer, Heidelberg (2000)
5. Berenji, H., Vengerov, D.: Learning, cooperation, and coordination in multi-agent systems, Technical Report IIS-00-10, Intelligent Inference Systems Corp., 333 W. Maude Avennue, Suite 107, Sunnyvale, CA 94085–4367 (2000)
6. Dreżewski, R., Siwik, L.: Agent-based co-operative co-evolutionary algorithm for multi-objective optimization. In: Rutkowski, L., Tadeusiewicz, R., Zadeh, L.A., Zurada, J.M. (eds.) ICAISC 2008. LNCS (LNAI), vol. 5097, pp. 388–397. Springer, Heidelberg (2008)
7. Dreżewski, R., Siwik, L.: Multi-objective optimization technique based on co-evolutionary interactions in multi-agent system. In: Giacobini, M. (ed.) EvoWorkshops 2007. LNCS, vol. 4448, pp. 179–188. Springer, Heidelberg (2007)
8. Yang, S.-R., Cho, S.-B.: Co-evolutionary learning with strategic coalition for multiagents. Appl. Soft Comput. 5(2), 193–203 (2005)
9. Lee, M.: A study of evolution strategy based cooperative behavior in collective agents. Artificial Intelligence Review 25(3), 195–209 (2006)
10. Lee, M., Chang, O.-b., Yoo, C.-J., et al.: Behavior evolution of multiple mobile agents under solving a continuous pursuit problem using artificial life concept. Journal of Intelligent and Robotic Systems 39(4), 433–445 (2004)
11. Nitschke, G.: Designing emergent cooperation: a pursuit- evasion game case study. Artificial Life Robotics 9(4), 222–233 (2005)
12. Tanev, I., Shimohara, K.: Effects of learning to interact on the evolution of social behavior of agents in continuous predators-prey pursuit problem. In: Banzhaf, W., Ziegler, J., Christaller, T., Dittrich, P., Kim, J.T. (eds.) ECAL 2003. LNCS (LNAI), vol. 2801, pp. 138–145. Springer, Heidelberg (2003)
13. Tanev, I., Brzozowski, M., Shimohara, K.: Evolution, generality and robustness of emerged surrounding behavior in continuous predators-prey pursuit problem. Genetic Programming and Evolvable Machines 6(3), 301–318 (2005)
14. Lee, M., kang, E.-K.: Learning enabled cooperative agent behavior in an evolutionary and competitive environment. Neural compu. & applic. 15(2), 124–135 (2006)
15. Lee, M.: Evolution of behaviors in autonomous robot using artificial neural network and genetic algorithm. Information Sciences 155(1-2), 43–60 (2003)
16. Caleanu, C.-D., Tiponut, V., et al.: Emergent behavior evolution in collective autonomous mobile robots. In: 12th WSEAS international conference on system, Heraklion, Greece, July 2008, pp. 428–433 (2008)

Towards an Extended Evolutionary Game Theory with Survival Analysis and Agreement Algorithms for Modeling Uncertainty, Vulnerability, and Deception

Zhanshan (Sam) Ma

IBEST (Initiative for Bioinformatics and Evolutionary Studies) & Departments of
Computer Science and Biological Sciences, University of Idaho, Moscow, ID, USA
ma@vandals.uidaho.com

Abstract. *Competition, cooperation* and *communication* are the three fundamental relationships upon which natural selection acts in the evolution of life.
Evolutionary game theory (EGT) is a 'marriage' between game theory and Darwin's evolution theory; it gains additional modeling power and flexibility by
adopting *population dynamics* theory. In EGT, natural selection acts as optimization agents and produces *inherent* strategies, which eliminates some essential
assumptions in traditional game theory such as *rationality* and allows more realistic modeling of many problems. Prisoner's Dilemma (PD) and Sir Philip Sid-
ney (SPS) games are two well-known examples of EGT, which are formulated
to study *cooperation* and *communication*, respectively. Despite its huge
success, EGT exposes a certain degree of weakness in dealing with time-,
space- and covariate-dependent (i.e., *dynamic*) *uncertainty*, *vulnerability* and
deception. In this paper, I propose to extend EGT in two ways to overcome the
weakness. First, I introduce *survival analysis* modeling to describe the lifetime
or fitness of *game players*. This extension allows more flexible and powerful
modeling of the dynamic uncertainty and vulnerability (collectively equivalent
to the dynamic *frailty* in survival analysis). Secondly, I introduce *agreement algorithms*, which can be the Agreement algorithms in distributed computing
(e.g., Byzantine Generals Problem [6][8], Dynamic Hybrid Fault Models [12])
or any algorithms that set and enforce the rules for players to determine their
consensus. The second extension is particularly useful for modeling *dynamic
deception* (e.g., *asymmetric* faults in fault tolerance and *deception* in animal
communication). From a computational perspective, the extended evolutionary
game theory (EEGT) modeling, when implemented in simulation, is equivalent
to an optimization methodology that is similar to evolutionary computing approaches such as Genetic algorithms with dynamic populations [15][17].

Keywords: Evolutionary Game Theory, Extended Evolutionary Game Theory,
Survival Analysis, Agreement Algorithms, Frailty, Deception, Dynamic Hybrid
Fault Models, UUUR (Uncertain, latent, Unobservable or Unobserved Risks),
Strategic Information Warfare, Reliability, Security, Survivability.

1 Background

The endeavor to develop evolutionary game theory (EGT) was initiated in the 1970s
by John Maynard-Smith and Peter Price with the application of game theory to the

H. Deng et al. (Eds.): AICI 2009, LNAI 5855, pp. 608–618, 2009.

modeling of animal conflict resolution [21]. Their pioneering research not only provided a refreshing mathematical theory for modeling animal behavior, but also revealed novel insights to evolution theory and greatly enriched the evolutionary ecology. It also quickly became one of the most important modeling approaches for emerging behavioral ecology in the 1980s. The well-known prisoner's dilemma (PD) game (especially its latter versions), which plays a critical role in the study of cooperation or altruism, is essentially an evolutionary game [1][3]. Today, the applications of EGT have been expanded to fields well beyond behavioral ecology.

Evolutionary game is a hybrid of *continuous static* and *continuous dynamic games* (*differential games*) in the sense that its *fitness* (payoff) functions are continuous, but the strategies are inheritable phenotypes, which usually are discrete. In the study of animal communication or engineering communication (such as wireless sensor networks), the strategy can be the *behavior* of animals or communication nodes. Another feature is that evolutionary game is played by a *population* of players repeatedly. Although the latter is not a unique feature of evolutionary games, the integration with *population dynamics* theory is somewhat unique and makes it particularly powerful for modeling biological systems or any systems where population paradigm is important [4][9][21][23]. One of the most frequently used models is the replicator dynamics model [4][23].

In evolutionary game theory, *replicator dynamics* is described with differential equations. For example, if a population consists of n types $E_1, E_2, ..., E_n$ with frequencies $x_1, x_2, ..., x_n$. The fitness $f_i(x)$ of E_i will be a function of the population structure, or the vector, $x = (x_1, x_2, ..., x_n)$. Following the basic tenet of Darwinism, one may define the success as the difference between the fitness $f_i(x)$ of E_i and the average fitness of the population, which is defined as:

$$f(x) = \sum x_i f_i(x). \tag{1}$$

Then, the simplest replicator model can be defined as:

$$dx_i / dt = x_i[f_i(x) - f(x)] \tag{2}$$

for $i = 1, 2, ..., n$. The population $x(t) \in S_n$, where S_n is a simplex, which is the space for population composition, is similar to mixed strategies in traditional games [4][23]. Another formulation is the *Fitness Generating Function* (*G-function*) and is invented by Vincent and Brown (2005) [23] to specify groups of individuals within a population. Individuals are assigned the same *G-function* if they possess the same set of evolutionarily feasible strategies and experience the same fitness consequences within a given environment [23].

EGT still exposes certain limitations in dealing with dynamic (time-, space-, and/or covariate-dependent) *uncertainty*, *vulnerability*, and *deception* of game players. For example, the so-called UUUR (Unknown, latent, Unobserved, or Unobservable Risks) events exist in many problems such as the study of reliability, security and survivability of computer networks, and UUUR events are largely associated with uncertainty, vulnerability, or their combinations (the frailty) [9][19]. Another challenge is the modeling of *deception* in communication among the players: the players not only communicate, but also need to reach consensus (agreement) or make decisions with the existence of dynamic frailty and *deception* [9][10].

In this paper, I introduce two extensions to EGT to primarily deal with the above mentioned limitations. In the first extension, I introduce *survival analysis* and its 'sister' subjects (*competing risks analysis* and *multivariate survival analysis*), which have some uniquely powerful features in dealing with UUUR events, to deal with dynamic *uncertainty* and *vulnerability*. The second extension is designed to address *deception* of game players: the consequence of *dynamic deception* and *frailty* on the game strategies, as well as the capability for game players to reach an agreement under the influences of deception and frailty. There is a third extension—using hedging principle from mathematical finance for decision-making, which is necessary for some applications such as survivability analysis [9], prognostic and health management in aerospace engineering [19], and strategic information warfare [10] [18], but is skipped here.

The significance of EGT is probably best revealed by a brief analysis of the fundamental elements that Darwin's evolutionary theory addresses. *Competition, cooperation* and *communication* were clearly in Darwin's primary concerns. According to an historical analysis by Dugatkin [3] in his landmark volume "*Origin of Species*" (Darwin 1859), Darwin focused on the *competition*, or the struggle for life, but Darwin was also clearly concerned with the *cooperation* or *altruism* in nature. In his "*Origin of Species*," Darwin's solution to *altruism* was that selection may be applied to the *family*, and individual may get the desired benefit ultimately. It took near a century for scientists to formalize Darwin's idea mathematically with a simple equation, known as Hamilton's Rule, first formulated by William Hamilton (1964). About two decades later, Hamilton's collaborative work with political scientist Robert Axelrod ([1]), which implemented Robert Trivers' (1971) suggestion of using PD game to study altruism, led to the eruption of the study of cooperation in the last three decades [3]. Today, there is hardly a major scientific field to which PD game has not been applied. The PD game, especially its latter versions, is essentially the evolutionary game. Besides *competition* and *cooperation*, Darwin also published a volume on *communication*, titled "*The Expression of the Emotions in Man and Animals*" in 1872. A century later, "*The Handicap principle: a missing piece of Darwin's Puzzle*" by Zahavi (1997) opened a new chapter in the study of *animal communication*. It was also the EGT modeling that led to the wide acceptance of the *Handicap principle*. The simplest EGT model for this principle is the Sir Philip Sydney (SPS) game, which has been used to elucidate the *handicap principle*—that animal communication is honest (reliable) as long as the signaling is costly in a *proper* way. Therefore, the phenomenon that EGT seems to offer the most lucid modeling elucidation for both *cooperation* and *communication* shows its power. Of course, the fact that EGT is a 'marriage' between game theory and Darwin's evolution theory is the best evidence of its importance in the study of *competition* (the struggle for life).

The far reaching cross-disciplinary influences of Darwin's evolution theory and PD game also suggest that the study of *communication*, especially with EGT (e.g., SPS game) may spawn important interdisciplinary research topics and generate significant cross-disciplinary ramifications that are comparable to the study of cooperation and PD games [9][16]. It is hoped that the extensions to EGT in this paper will also be helpful to the cross-disciplinary expansion because these extensions address some critical issues that exist across several fields from reliability and survivability of distributed networks, prognostics and health management in aerospace engineering, machine learning, to animal communication networks [9][10][16]–[20].

In the following, we assume that the players of an evolutionary game form a *population*, similar to a biological population. Population *dynamics* over space and time is influenced by environmental covariates (factors). The covariates often affect the lifetime or fitness of game players by introducing *uncertainty* and/or *frailty*. Players may have stochastically variable *vulnerabilities*. In addition, the individual players may form various spatial structures or topologies, which can be captured with a graph model or *evolutionary games on graphs*. Players can form a complex communication network; in the networks, some act as *eavesdroppers, bystanders* or *audience*. The communication signals may be *honest* or *deceitful* depending on circumstance, or the so-called *dynamic deception*. Finally, the *'failure'* (termination of lifetime or a process) of players in engineering applications is often more complex than biological death in the sense that a failing computer node may send conflicting signals to its neighbor nodes, or the so-called *asymmetric faults*. Indeed, the challenge in dealing with somewhat arbitrary *Byzantine fault* as well as *dynamic deception* is the major motivation why we introduce agreement algorithm extension to EGT.

The remainder of this paper is organized as follows: Section 2 and 3 summarize the first extension with survival analysis, and the second extension with agreement algorithms, to EGT, respectively. Section 4 discusses implementation issues related to the EEGT, with the modeling of animal communication networks as an example. Due to the complexity involved, the EEGT has to be implemented as a simulation environment in the form of software. The page space limitation makes it only feasible to sketch a framework of the EGGT. In the following, I often use the *network* settings for survivability analysis, strategic information warfare, or animal communication as examples, and use the terms *node, individual,* and *player* interchangeably.

2   The Extension to EGT with Survival Analysis

Survival analysis has become a *de facto* standard in much of the biomedical research and also found applications in many other disciplines. There are more than a dozen monographs of survival analysis since the 1980s: e.g., [7] for (univariate) survival analysis; [5] for multivariate survival analysis; [2] for competing risks analysis. A series of reviews on survival analysis in reliability, network survivability and computer science can be found in [11][13][14].

Survival analysis can be considered as the study of *time-to-event* random variables (also known as *failure* or *survival* time) with *censoring*. Time-to-event random variables exist widely, e.g., lifetime of animals or wireless sensor network nodes, duration of eavesdropping in animal communication, duration of aggression displays in conflict resolution. The *survivor function* $S(t)$ is defined as the probability that *survival time* (or *failure* time) T exceeds t; i.e.,

$$S(t) = P(T \geq t), \quad 0 < t < \infty \tag{3}$$

This definition is exactly the same as the traditional *reliability* function, but survival analysis has much rich models and statistical procedures such as Cox's proportional hazard modeling [equ. (4) & (5)] and accelerated failure time modeling. Dedicated volumes have been written to extend the Cox model (see citations in [11]).

Cox proportional hazard model represents *conditional* survivor function:

$$S(t \mid z) = [S_0(t)]^{\exp(z\beta)} \tag{4}$$

where
$$S_0(t) = \exp\left[- \int_0^t \lambda_0(u)\,du \right]. \tag{5}$$

and z is the vector of *covariates*, which can be any factors that influence the survival or lifetime of *game players*.

Competing risks analysis is concerned with the scenario that multiple risks exists but only one of the risks leads to the failure ([2][13]). The single risk that leads to the failure becomes the failure cause aftermath, and the other risks that competed for the 'cause' are just *latent* risks. Competing risks analysis is neither univariate nor truly multivariate given its univariate cause and multivariate latent risks. *Multivariate survival analysis* deal with truly multivariate systems where multiple failure-causes and multiple failures (modes or failure times) may exist/occur simultaneously [5][14]. Furthermore, the multiple failure causes (risks) may be dependent with each other, and so do the multiple failures (modes or times). Indeed, multivariate survival analysis often offers the most effective statistical approaches to study dependent failures [5][9][14].

Observation (or information) *censoring* refers to the incomplete observation of either failure events or failure causes or both. Censoring is often unavoidable in the studies of time-to-event random variables such as reliability analysis. However, traditional reliability analysis does not have a rigorous procedure to handle censored observations; either including or excluding the censored individuals may cause bias in statistical inference.

In network reliability and survivability analysis, Ma (2008) proposed to utilize survival analysis to assess the *consequences* of the UUUR (Unpredictable, latent, Unobserved or Unobservable Risks) events [9]. Mathematically, although the probabilities of UUUR events are unknown, survival analysis does provide procedures to assess their *consequences, e.g.,* in the form of the survivor function variations at various censoring level, the influences of latent risks on failure time and/or modes, or the effects of *shared frailty* (the unobserved or unobservable risks) on failure times, modes, and/or dependencies. From the perspective of application problems such as network reliability and survivability, UUUR covers a class of risks that are particularly difficult to characterize, such as malicious intrusions, virus or worm infections, software vulnerabilities that are poorly understood. The event probabilities associated with those kinds of risks are often impossible to obtain in practice. Furthermore, these risks are time, space and covariate dependent or dynamic, and this further complicates the problem. On the other hand, survival analysis models such as Cox proportional hazard models are designed to deal with covariate dependent dynamic hazards (risks). Therefore, survival analysis provides a set of ideal tools to quantitatively assess the consequences of UUUR events, which is recognized as one of the biggest challenges in network security and survivability research.

In the context of evolutionary games, the concept of UUUR events can be used to describe *'risk'* factors that affect the lifetime or fitness of game players. Of course, those factors may affect the fitness *positively, negatively,* or *variable (nonlinearly)* at different times. Therefore, the term 'risk' may bear a neutral consequence, rather than consistently negative. In general, UUUR captures *uncertainty* or *frailty.* The latter can be considered as the combinations or mixture of *uncertainty* and *vulnerability.* For example, information can dispense uncertainty and consequently eliminate some *perceived* vulnerability [9]–[11]. In addition, *deception* in some applications such as the study of animal communication networks or strategic information warfare can also be described with UUUR risks from the perspective of the *players* [10][16][18].

From above discussion it can be seen that survival analysis models can be harnessed to describe the lifetime and/or survivor probability of individual players, which can often be transformed into the *individual* fitness. An alternative modeling strategy is to utilize survival analysis models for *population* dynamics of the game players, or the dynamics of meta-populations. For example, in the Dove-Hawk game, there are two populations: one is the hawk population and another is the dove population. Both populations can be described with survival analysis models. The population modeling with survival analysis should be similar to the modeling of biological population dynamics [9].

In summary, the advantages from introducing survival analysis include: (*i*) flexible *time-, space-* and *covariate*-dependent *fitness* function; (*ii*) unique approaches to assess the consequences of UUUR events; (*iii*) *deception* modeling, which can be recast as a reliability problem; (*iv*) effective modeling of the *dependency* between '*failure*' events or between the '*risks*' that influence the '*failure*' events.

3   The Extension to EGT with Agreement Algorithms

The second extension to EGT with agreement algorithms is actually built upon the first extension with survival analysis. It was initially conducted with the Agreement algorithms in distributed computing ([6][8]) and the resulting models were termed *Dynamic Hybrid Fault* (DHF) Models ([9][12]). One of the earliest Agreement algorithm problems was formulated as the Byzantine Generals problem (BGP) by Lamport (1982)[6], in which the components of a computer system are abstracted as generals of an army. Loyal generals (nodes or players) need to find a way (algorithm) to reach a consensus (e.g., to attack or retreat) while traitors (or bad nodes) would try to confuse others by sending conflicting messages. Because the focus of Agreement algorithms is to reach a consensus, in the traditional *hybrid fault models* (which are simply the constraints, often in the form of inequalities, used to classify faults based on the Agreement algorithms), the *failure rate* (λ) is often ignored or is implicitly assumed to be constant. In other words, the hybrid fault models only specify whether or not an agreement can be reached, given a certain number of traitors, but they do not keep track of *when* the generals committed treason. This assumption is appropriate in the study of Agreement algorithms because, even if the voting is dynamic, they are abstracted as multiple rounds of voting to study the possibility to reach a consensus. However, when the fault models are applied to analyze a real-world system consisting of multiple components (generals), the *history* of the generals must be considered. Some generals may be loyal for their entire lifetimes; some may quickly become 'corrupted;' still others may be loyal for a long time but ultimately become 'corrupted.'

Each of the generals may have different (inhomogeneous) *time-variant* (not constant) *hazard functions*, $\lambda_i(t)$, $i = 1, 2, ..., g$, where g is the number of generals.

To overcome this limitation of lacking *real time* notion in tradition hybrid fault models, Ma and Krings (2008) extended the traditional hybrid fault models with the so-called *Dynamic Hybrid Fault models* (DHF) [9][12]. Essentially, there are two limitations with the traditional hybrid fault models, when they are applied to reliability analysis. The first is the lack of the notion of *real time*, as explained in the previous paragraph, and the second is the lack of *approaches* to incorporate hybrid fault models into reliability analysis *after* the issues associated with the first limitation is resolved. The latter depends on the integration of the dynamic hybrid fault models with the evolutionary game theory.

The solution to the first aspect, the missing notion of *real time*, or the *Agreement-algorithm* aspect of the problem, is to introduce survival analysis. In particularly, time and covariate dependent survivor functions or hazard functions can be used to describe the survival of the Byzantine generals. For example, the constraint of the BGP under *oral message* assumption is,

$$N \geq 3m + 1 \tag{6}$$

is replaced with the following model in the dynamic hybrid fault models:

$$N(t) \geq 3m(t) + 1 . \tag{7}$$

Further assuming that the survivor function of *generals* is $S(t|z)$, a simplified conceptual model can be:

$$N(t) = N(t-1) * S(t \mid z) \tag{8}$$

$$m(t) = m(t-1) * S_m(t \mid z) \tag{9}$$

where $N(t)$ and $m(t)$ are the number of total *generals* and treacherous *generals* (*traitors*) at time t, respectively. $S(t|z)$ and $S_m(t|z)$ are the corresponding *conditional* survivor functions for the total number of generals and traitors, respectively. They can use any major survival analysis models, such as Cox proportional hazard model, expressed in equation (4). One immediate benefit of the dynamic hybrid fault models is that it is now possible to predict the *real-time fault tolerance level* in a system.

The above extension with survival analysis models is necessary, but not sufficient for applying the dynamic hybrid fault models to reliability analysis, except for extremely simple cases. The difficulty arises when there are multiple types of failure behaviors. This is typical in real world dynamic hybrid fault models. For example, the failure modes could be *symmetric* vs. *asymmetric*, *transmissive* vs. *omissive*, *benign* vs. *malicious*, etc. Besides different failure modes, node behaviors can also include: cooperative vs. non-cooperative, mobile nodes vs. access points (which might be sessile), etc. To model the different behaviors, we will need multiple groups, or system of equations (8) and (9). The challenge is that we lack an approach to synthesize the models to study reliability and survivability. This is the second limitation (aspect) of traditional hybrid fault models. The solution to the second limitation (aspect), is a new notion termed 'Byzantine Generals Playing Evolutionary Games,' first outlined in Ma (2007, unpublished dissertation proposal).

Evolutionary game theory models such as represented with replicator dynamics [(1) and (2)] can be used to synthesize the fitness functions for various failure behaviors in dynamic hybrid fault models [(4)–(9)]. In the case of BGP, there are only two strategies (behaviors): *loyal* (reliable) or *treacherous* (deceitful), i.e., types E_1 and E_2 in (1) and (2). The fitness function $f_i(x)$ can adopt survivor functions such as represented with Cox models [(4)–(5), (8)–(9)]. Since strategies in EGT are discrete, Agreement algorithms are introduced to perform *dynamic 'voting'*. In the case of BGP, the constraint [inequality (7)] to reach an agreement is *dynamically* checked, which is a reason for the notion of '*Byzantine generals playing evolutionary games*' [9][12].

In principle, the similar procedure with dynamic hybrid fault models, briefly summarized above, can be generalized to any evolutionary games. Of course, the constraint for reaching an agreement does not have to be BGP constraints or any of the Agreement algorithms in distributed computing [6][8] at all. The constraints can be any algorithms that set the rules for players to assess their global status or reach an agreement if requested. Therefore, the term *agreement algorithm* in the context of EEGT can refer to either the existing Agreement algorithms (from distributed computing [6][8]) or any user-defined agreement algorithms '*customized*' for a specific problem. Nevertheless, I do suggest that the agreement *algorithms* take similar forms with *dynamic hybrid fault models*, i.e., taking advantages of survival analysis based fitness function. In other words, the extension with agreement algorithms is built upon the first extension with survival analysis.

In the following, I use an even simpler legislature scenario to illustrate the general idea for extending evolutionary game theory with agreement algorithms. Assume a legislature bill is presented to the Congress. Two opposing parties may try to defeat each other. Lobbyists may try to influence the delegates. Also assume the bill has to be voted before the end of the summer recess. It is a *rule* that 2/3 majority is needed to pass the bill. However, there are time-dependent covariates that influence the delegates. The *minds* of the delegates may change from day to day. Each delegate has a different *fitness* function, which determines how long it takes for a delegate to make the decision. Since the time-to-decision is a *time-to-event* (survival or failure time) random variable, survival analysis is ideally applicable. The fitness function can be,

$$\lambda_i(t) = f \ (\textit{effort of lobbyist, media influence, constituent, health, scandal, etc}). \qquad (10)$$

The voting result or the agreement should depend on *time*. At the time of voting, determining the result is simple; for example, a simple 2/3 majority rule. However, assessing probability to reach an agreement *each day* is not trivial at all, e.g.,

$$\lambda(t) = g[\lambda_i(t)] \qquad (11)$$

or
$$\lambda(t) = g[\lambda_R(t), \lambda_D(t), \lambda_I(t)], z] \qquad (12)$$

z could be lobbyists or other factors that influence the *fitness* functions.

Up to this point, we only completed the extension of evolutionary game theory with survival analysis. The real complexity lies in the secondary extension with agreement algorithms. The extension starts with the function g. What is the form of *function g*? The delegates may be 'playing' the game, and we assume that they 'play' evolutionary game. Simple addition or even complex survival function is not sufficient because the *behavior (strategy)* change is not a "continuous" function and furthermore, the players may form complex coalitions. Beyond those complexities, there are *conditional commitments*, which evolve daily. To deal with all the complexity, the *'Byzantine generals playing evolutionary games'* paradigm, which turns the problem into an evolutionary game, is necessary. Different problems may have different function *g*, but the principle should be similar.

In principle, there are four major types of *function g*: (*i*) borrow from traditional games; e.g., Hawk-Dove game; (*ii*) utilize the *replicator dynamics modeling* [4]; (*iii*) *Fitness Generating Function* (G-function) [23]; (*iv*) *'Byzantine Generals Playing EG'*—*the approaches outlined above.*

Actually, there is a third extension to the EGT with the principle of hedging, which is necessary when UUUR events dominantly influence decision-making, e.g., dealing with malicious actions in survivability analysis. The three-layer survivability analysis is essentially the application of the EEGT to survivable network systems at the tactical and strategic levels, and further extended the EGT with the hedging principle at the operational level, which is responsible for the decision-making in managing survivable network systems or planning information warfare [9][10][18][19].

4   Modeling Animal Communication Networks with the EEGT

One of the most important results from EGT is *evolutionary stable strategies* (ESS). Obtaining ESS can be a challenging computational problem for complex evolutionary game models. This is yet another reason that simulation and heuristic algorithms are usually necessary for the EEGT modeling, especially given the added complexity from the extensions. Evolutionary computing is particularly suitable for implementing the EEGT. When implemented in simulation, EEGT itself is equivalent to an optimization methodology [15][17]. The paradigm of EEGT has been applied to the studies of reliability and survivability of wireless sensor networks [9], strategic information warfare research [10][19], and aerospace PHM engineering [18].

Animal communication has been studied with the sender-receiver dyadic paradigm traditionally, and so does the theory of honesty (reliability) in animal communication. In the last decade, the network paradigm for animal communication has emerged, in which the third-party receivers, such as eavesdroppers, bystanders and audience, are part of an animal communication network and their roles are considered [22]. However, much of the current research in animal communication *networks* is still qualitative. Particularly, the modeling of the *reliability (honesty)* of animal communication *networks*, similar to the SPS game for dyadic paradigm, is still largely an uncharted area. As argued previously, both the studies of animal communication *networks* and their reliability may spawn exciting cross-disciplinary research, such as offering a paradigm for resilient and pervasive computing, serving as a *trust* model in computer science, which are further expanded as seven *open problems* in [16].

Here, I suggest that the EEGT, which has been applied to modeling of reliability and survivability of distributed computer networks [9], should provide an effective modeling architecture for studying animal communication networks and their *reliability, resilience*, and *evolution*. The following diagram (Fig.1) shows a simulation modeling architecture for such a study. In Fig. 1, the top rectangle box represents a population of animals (or game players) consisting of senders, receivers, eavesdroppers, bystanders, audience, etc. The second row of five boxes lists the major issues of animal communication. The third row is the major models (algorithms) involved, such as those from the EEGT, models for behavior adaptation and evolution (e.g., the Handicap principle), and other constraints (competition and cooperation). The last row shows the major results obtainable from the EEGT modeling.

In perspective, it is my opinion that the study of animal communication networks and their reliability is on the verge to spawn cross-disciplinary research similar to that generated by the study of cooperation and PD games. Honesty or deception is an issue potentially interested in by researchers from many disciplines, such as psychology, sociology, criminal justice study, political and military sciences, economics, and computer science. Deception can create uncertainty, vulnerability and frailty, but what makes it particularly hard to model is its asymmetric or Byzantine nature. Furthermore, Byzantine deception and frailty are likely to be the major factors that often contribute to the unexpected but catastrophic 'crashes.' This may explain the phenomenon that the fraud and corruption in social and economic systems, as well as deception in information warfare are often among the most illusive and difficult objects to describe in any quantitative modeling research. It is hoped that the extensions to EGT introduced in this paper will relieve some of the difficulties.

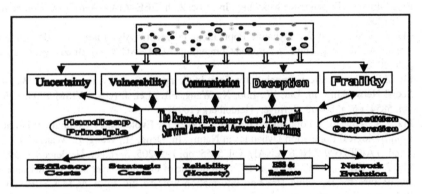

Fig. 1. The major modules of modeling animal communication networks with the EEGT

References

1. Axelrod, R., Hamilton, W.D.: The evolution of cooperation. Science 211, 1390–1396 (1981)
2. Crowder, M.J.: Classical Competing Risks Analysis, p. 200. Chapman & Hall, Boca Raton (2001)
3. Dugatkin, L.A.: The Altruism Equation. Princeton University Press, Princeton (2006)
4. Hofbauer, J., Sigmund, K.: Evolutionary Games and Population Dynamics, p. 323. Cambridge University Press, Cambridge (1998)
5. Hougaard, P.: Analysis of Multivariate Survival Data, p. 560. Springer, Heidelberg (2000)
6. Lamport, L., Shostak, R., Pease, M.: The Byzantine Generals Problem. ACM Transactions on Programming Languages and Systems 4(3), 382–401 (1982)

7. Lawless, J.F.: Statistical models and methods for lifetime data, 2nd edn. Wiley, Chichester (2003)
8. Lynch, N.: Distributed Algorithms. Morgan Kaufmann Press, San Francisco (1997)
9. Ma, Z.S.: New Approaches to Reliability and Survivability with Survival Analysis, Dynamic Hybrid Fault Models, and Evolutionary Game Theory. PhD Dissertation, University of Idaho (2008)
10. Ma, Z.S.: Extended Evolutionary Game Theory Approach to Strategic Information Warfare Research. Journal of Information Warfare 8(2), 25–43 (2009)
11. Ma, Z.S., Krings, A.W.: Survival Analysis Approach to Reliability Analysis and Prognostics and Health Management (PHM). In: Proc. 29th IEEE–AIAA AeroSpace Conference, p. 20 (2008a)
12. Ma, Z.S., Krings, A.W.: Dynamic Hybrid Fault Models and their Applications to Wireless Sensor Networks (WSNs). In: The 11-th ACM/IEEE International Symposium on Modeling, Analysis and Simulation of Wireless and Mobile Systems, ACM MSWiM 2008, Vancouver, Canada, p. 9 (2008)
13. Ma, Z.S., Krings, A.W.: Competing Risks Analysis of Reliability, Survivability, and Prognostics and Health Management (PHM). In: Proc. 29th IEEE–AIAA AeroSpace Conference, p. 21 (2008)
14. Ma, Z.S., Krings, A.W.: Multivariate Survival Analysis (I): Shared Frailty Approaches to Reliability and Dependence Modeling. In: Proc. 29th IEEE–AIAA AeroSpace Conference, p. 21 (2008)
15. Ma, Z.S., Krings, A.W.: Dynamic Populations in Genetic Algorithms, SIGAPP. In: 23rd Annual ACM Symposium on Applied Computing (ACM SAC 2008), Brazil, March 16-20, p. 5 (2008)
16. Ma, Z.S.: The Handicap Principle for Trust in Computer Security, the Semantic Web and Social Networking. In: Liu, W., Luo, X., Wang, F.L., Lei, J. (eds.) WISM 2009. LNCS, vol. 5854, pp. 458–468. Springer, Heidelberg (2009)
17. Ma, Z.S.: Towards a Population Dynamics Theory for Evolutionary Computing: Learning from Biological Population Dynamics in Nature. In: Deng, H., Wang, L., Wang, F.L., Lei, J. (eds.) AICI 2009. LNCS (LNAI), vol. 5855, pp. 195–205. Springer, Heidelberg (2009)
18. Ma, Z.S., Krings, A.W., Sheldon, F.T.: An outline of the three-layer survivability analysis architecture for modeling strategic information warfare. In: Fifth ACM CSIIRW, Oak Ridge National Lab (2009)
19. Ma, Z.S.: A New Life System Approach to the Prognostic and Health Management (PHM) with the Three-Layer Survivability Analysis. In: The 30th IEEE-AIAA AeroSpace Conference, p. 20 (2009)
20. Ma, Z.S., Krings, A.W.: Insect Sensory Systems Inspired Computing and Communications. Ad Hoc Networks 7(4), 742–755 (2009)
21. Maynard Smith, J., Price, G.R.: The logic of animal conflict. Nature 246, 15–18 (1973)
22. McGregor, P.K. (ed.): Animal Communication Networks. Cambridge University Press, Cambridge (2005)
23. Vincent, T.L., Brown, J.L.: Evolutionary Game Theory, Natural Selection and Darwinian Dynamics, p. 382. Cambridge University Press, Cambridge (2005)
24. Zahavi, A., Zahavi, A.: The Handicap Principle: A Missing Piece of Darwin's Puzzle. Oxford University Press, Oxford (1997)

Uncovering Overlap Community Structure in Complex Networks Using Particle Competition

Fabricio Breve, Liang Zhao, and Marcos Quiles

Institute of Mathematics and Computer Science, University of São Paulo,
São Carlos SP 13560-970, Brazil
{fabricio,zhao,quiles}@icmc.usp.br

Abstract. Identification and classification of overlap nodes in communities is an important topic in data mining. In this paper, a new clustering method to uncover overlap nodes in complex networks is proposed. It is based on particles walking and competing with each other, using random-deterministic movement. The new community detection algorithm can output not only hard labels, but also continuous-valued output (soft labels), which corresponds to the levels of membership from the nodes to each of the communities. Computer simulations were performed with synthetic and real data and good results were achieved.

Keywords: complex networks, community detection, overlap community structure, particle competition.

1 Introduction

In the last years, the advances and the convergence of computing and communication has rapidly increased our capacities of generating and collecting data. However, most of this data is in its raw form, and it is not useful until it is discovered and articulated. Data Mining is the process of extracting the implicit potentially useful information from the data. It is a multidisciplinary field, drawing works from areas including statistics, machine learning, artificial intelligence, data management and databases, pattern recognition, information retrieval, neural networks, data visualization, and others [1,2,3,4,5].

Community Detection is one of the data mining problems that arose with the advances in computing and the increasingly interest in complex networks, which studies large scale networks with non-trivial topological structures, such as social networks, computer networks, telecommunication networks, transportation networks, and biological networks [6,7,8]. Many of these networks are found to be divided naturally into communities or modules, therefore discovering of these communities structure became an important topic of study [9,10,11,12,13]. Recently, a particle competition approach was successfully applied to detect communities modeled in networks [14].

The notion of communities in networks is straightforward, they are defined as a subgraph whose nodes are densely connected within itself but sparsely connected with the rest of the network. However, in practice there are common

H. Deng et al. (Eds.): AICI 2009, LNAI 5855, pp. 619–628, 2009.

cases where some nodes in a network can belong to more than one community. For example: in a social network of friendship, individuals often belong to several communities: their families, their colleagues, their classmates, etc. These nodes are often called overlap nodes, and most known community detection algorithms cannot detect them. Therefore, uncovering the overlapping community structure of complex networks becomes an important topic in data mining [15,16,17].

In this paper we present a new clustering technique, based on particle walking and competition. We have extended the model proposed in [14] to output not only hard labels, but also a fuzzy output (soft labels) for each node in the network. The continuous-valued output can be seen as the levels of membership from each node to each community. Therefore, the new model is able to uncover the overlap community structure in complex networks.

The rest of this paper is organized as follows: Section 2 describes the model in details. Section 3 shows some experimental results from computer simulations, and in Section 4 we draw some conclusions.

2   Model Description

The model we propose in this paper is an extension of the particle competition approach proposed by [14], which is used to detect communities in networks. Some particles walk in a network, competing with each other for the possession of nodes, while rejecting intruder particles. After a number of iterations, each particle will be confined within a community of the network, so the communities can be divided by examining the nodes ownership. The new model is not only suitable to detect community structure, but it can also uncover overlap community structure. In order to achieve that, we have changed the nodes and particles dynamics, and introduced a few new variables, among other details that will follow.

Let the network structure be represented as a graph $\mathbf{G} = (\mathbf{V}, \mathbf{E})$, with $\mathbf{V} = \{v_1, v_2, \ldots, v_n\}$, where each node v_i is an element from the network. An adjacency matrix $\mathbf{W}$ defines which network nodes are interconnected:

$$W_{ij} = 1, \quad \text{if there is an edge between nodes } i \text{ and } j, \tag{1}$$
$$W_{ij} = 0, \quad \text{otherwise,} \tag{2}$$

and $W_{ii} = 0$.

Then, we create a set of particles $\mathbf{P} = (\rho_1, \rho_2, \ldots, \rho_c)$, in which each particle corresponds to a different community. Each particle ρ_j has a variable $\rho_j^\omega(t) \in [\omega_{\min} \quad \omega_{\max}]$ is the particle potential characterizing how much the particle can affect a node at time t, in this paper we set the constants $\omega_{\min} = 0$ and $\omega_{\max} = 1$.

Each node v_i has two variables $v_i^\omega(t)$, and $v_i^\lambda(t)$. The first variable is a vector $v_i^\omega(t) = \{v_i^{\omega_1}(t), v_i^{\omega_2}(t), \ldots, v_i^{\omega_c}(t)\}$ of the same size of $\mathbf{P}$, where each element $v_i^{\omega_j}(t) \in [\omega_{\min} \quad \omega_{\max}]$ corresponds to the instantaneous level of ownership by particle ρ_j over node v_i. The sum of the levels of ownership of each node is always a constant, because a particle increases its own ownership level and, at the same time, decreases the other particles ownership levels. Thus, the following equations always holds:

$$\sum_{j=1}^{c} v_i^{\omega_j} = \omega_{\max} + \omega_{\min}(c-1). \tag{3}$$

The second variable is also a vector $v_i^{\lambda}(t) = \{v_i^{\lambda_1}(t), v_i^{\lambda_2}(t), \ldots, v_i^{\lambda_c}(t)\}$ of the same size of $\mathbf{P}$ and it also represents ownership levels, but unlike $v_i^{\omega}(t)$ which denotes the instantaneous ownership levels, $v_i^{\lambda_j}(t) \in [0 \ \ \infty]$ rather denotes long term ownership levels, accumulated through the whole process. The particle with higher ownership level in a given non-overlap node after the last iteration of the algorithm is usually the particle which have visited that node more times, but that does not always apply to overlap nodes, in which sometimes the dominant particle could easily change in the last iterations, and thus it would not correspond to the particle which have dominated that node for more instants of time. Therefore, the new variable $v_i^{\lambda}(t)$ was introduced in order to define the ownership of nodes considering the whole process. Using a simple analogy, we can say that now the champion is not the one who have won the last games, but rather the one who have won more games in the whole championship. Notice that the long term ownership levels only increases and their sum is not constant, they are normalized only at the end of the iterations.

We begin the algorithm by setting the initial level of instantaneous ownership vector v_i^{ω} by each particle ρ_j as follows:

$$v_i^{\omega_j}(0) = \omega_{\min} + \left(\frac{\omega_{\max} - \omega_{\min}}{c}\right), \tag{4}$$

which means that all nodes starts with all particles instantaneous ownership levels equally set. Meanwhile, the long term ownership levels $v_i^{\lambda}(t)$ are all set to zero:

$$v_i^{\lambda_j}(0) = 0. \tag{5}$$

The initial position of each particle $\rho_j^v(0)$ is set randomly, to any node in $\mathbf{V}$, and the initial potential of each particle is set to its minimum value, as follows:

$$\rho_j^{\omega}(0) = \omega_{\min}. \tag{6}$$

Each particle will choose a neighbor to visit based in a random-deterministic rule. At each iteration, each particle will chose between *random walk* or *deterministic walk*, where *random walk* means the particle will try to move to any neighbor randomly chosen, i.e., the particle ρ_j will try to move to any node v_i chosen with the probabilities defined by:

$$p(v_i|\rho_j) = \frac{W_{ki}}{\sum_{q=1}^{n} W_{qi}}, \tag{7}$$

where k is the index of the node node being visited by particle ρ_j, so $W_{ki} = 1$ if there is an edge between the current node and v_i, and $W_{ki} = 0$ otherwise.

The *deterministic walk* means that the particle will try to move to a neighbor with probabilities according to the nodes instantaneous ownership levels, i.e., the particle ρ_j will try to move to any neighbor v_i chosen with probabilities defined by:

$$p(v_i|\rho_j) = \frac{W_{ki}v_i^{\omega_j}}{\sum_{q=1}^{n} W_{qi}v_i^{\omega_j}},\qquad(8)$$

again, k is the index of the node stored being visited by particle ρ_j.

At each iteration, each particle has probability p_{det} of taking deterministic movement and probability $1 - p_{\text{det}}$ of taking random movement, with $0 \le p_{\text{det}} \le 1$. Once the random movement or deterministic movement is chosen, a target neighbor $\rho_j^\tau(t)$ will be randomly chosen with probabilities defined by Eq. 7 or Eq. 8 respectively.

Regarding the node dynamics, at time t, each instantaneous ownership level $v_i^{\omega_k}(t)$ of each node v_i, which was chosen by a particle ρ_j as its target $\rho_j^\tau(t)$, is defined as follows:

$$v_i^{\omega_k}(t+1) = \begin{cases} \max\{\omega_{\min}, v_i^{\omega_k}(t) - \frac{\Delta_v \rho_j^\omega(t)}{c-1}\} & \text{if } k \neq j \\ v_i^{\omega_k}(t) + \sum_{q\neq k} v_i^{\omega_q}(t) - v_i^{\omega_q}(t+1) & \text{if } k = j \end{cases},\qquad(9)$$

where $0 < \Delta_v \le 1$ is a parameter to control the changing rate of the instantaneous ownership levels. If Δ_v takes a low value, the node ownership levels change slowly, while if it takes a high value, the node ownership levels change quickly. Each particle ρ_j will increase their corresponding instantaneous ownership level $v_i^{\omega_j}$ of the node v_i they are targeting, while decreasing the instantaneous ownership levels (of this same node) that corresponds to the other particles, always respecting the conservation law defined by Eq. 3.

Regarding the particle dynamics, at time t, each particle potential $\rho_j^\omega(t)$ is set as:

$$\rho_j^\omega(t+1) = \rho_j^\omega(t) + \Delta_\rho(v_i^{\omega_j}(t+1) - \rho_j^\omega(t))\qquad(10)$$

where $v_i(t+1)$ is the node rho_j is targeting, $0 < \Delta_\rho \le 1$ is a parameter to control the particle potential changing rate. Therefore, every particle ρ_j have their potential ρ_j^ω set to approximate the value of instantaneous ownership level $v_i^{\omega_j}$ from the node it is currently targeting. In this sense, a particle gets stronger when it is visiting a node with higher ownership level of its own, but it will be weakened if it tries to invade a node dominated by other particle.

The long term ownership levels are adjusted only when the particle selects the random movement. This rule is important because although the deterministic movement is useful to prevent particles from abandoning their neighborhood, which would let it susceptible to other particles attack, it is also a mechanism that makes a node gets more visits from the particle that currently dominates it. We consider only when the *random movement* was chosen because, in this case, particles will choose a target node based only in their current neighborhood, and not in their instantaneous ownership levels that are important for community detection, but are too volatile in overlap nodes. Therefore, for each particle selected in *random movement* by a particle ρ_j, the long term ownership levels $v_i^{\lambda_j}$ are update as follows:

$$v_i^{\lambda_j}(t+1) = v_i^{\lambda_j}(t) + \rho_j^{\omega}(t), \tag{11}$$

where v_i is the node ρ_j is targeting. The increase will always be proportional to the current particle potential, which is a desirable feature because the particle will probably have a higher potential when it is arriving from its own neighborhood, while it will have a lower potential when it is arriving from a node from other particles neighborhoods.

It should be noted that a particle really visits a target node only if its ownership level in that node is higher than the others; otherwise, a shock happens and the particle stays at the current node until next iteration.

At the end of the iterations, the degrees of membership $f_i^j \in [0 \quad 1]$ for each node v_i are calculated using the long term ownership levels, as follows:

$$f_i^j = \frac{v_i^{\lambda_j}(\infty)}{\sum_{q=1}^c v_i^{\lambda_q}(\infty)} \tag{12}$$

where f_i^j represents the final membership level of the node v_i to community j.

In summary, our algorithm works as follows:

1. Build the adjacency matrix $\mathbf{W}$ by using Eq. 1,
2. Set nodes ownership levels by using Eq. 4 and Eq. 5,
3. Set particles initial positions randomly and their potentials by using Eq. 6,
4. Repeat steps 5 to 8 until convergence or for a pre-defined number of steps,
5. Select the target node for each particle by using Eq. 8 or Eq. 7 for deterministic movement or random movement respectively,
6. Update nodes ownership levels by using Eq. 9,
7. If the random movement was chosen, update the long term ownership levels by using Eq. 11,
8. Update particles potentials by using Eq. 10,
9. Calculate the membership levels (fuzzy classification) by using Eq. 12.

3   Computer Simulations

In order to test the overlap detection capabilities of our algorithm, we generate a set of networks with community structure using the method proposed by [13]. Here, all the generated networks have $n = 128$ nodes, split into four communities containing 32 nodes each. Pairs of nodes which belongs to the same community are linked with probability p_{in}, whereas pairs belonging to different communities are joined with probability p_{out}. The total average node degree k is constant and set to 16. The value of p_{out} is taken so the average number of links a node has to nodes of any other community, z_{out}, can be controlled. Meanwhile, the value of p_{in} is chosen to keep the average node degree k constant. Therefore, z_{out}/k defines the mixture of the communities, and as z_{out}/k increases from zero, the communities become more diffuse and harder to identify. In each of these generated networks we have added a 129th node and created 16 links between the new node and nodes from the communities, so we could easily determine an expected "fuzzy" classification for this new node based on the count of its links with each community.

The networks were generated with $z_{out}/k = 0.125$, 0.250, and 0.375 and the results are shown in Tables 1, 2, and 3 respectively. The first column of these

Table 1. Fuzzy classification of a node connected to network with 4 communities generated with $z_{out}/k = 0.125$

Connections	Fuzzy Classification			
A-B-C-D	A	B	C	D
16-0-0-0	0.9928	0.0017	0.0010	0.0046
15-1-0-0	0.9210	0.0646	0.0079	0.0065
14-2-0-0	0.8520	0.1150	0.0081	0.0248
13-3-0-0	0.8031	0.1778	0.0107	0.0084
12-4-0-0	0.7498	0.2456	0.0032	0.0014
11-5-0-0	0.6875	0.3101	0.0016	0.0008
10-6-0-0	0.6211	0.3577	0.0111	0.0101
9-7-0-0	0.5584	0.4302	0.0011	0.0103
8-8-0-0	0.4949	0.4944	0.0090	0.0017
8-4-4-0	0.5025	0.2493	0.2461	0.0021
7-4-4-1	0.4397	0.2439	0.2491	0.0672
6-4-4-2	0.3694	0.2501	0.2549	0.1256
5-4-4-3	0.3144	0.2491	0.2537	0.1828
4-4-4-4	0.2512	0.2506	0.2504	0.2478

Table 2. Fuzzy classification of a node connected to network with 4 communities generated with $z_{out}/k = 0.250$

Connections	Fuzzy Classification			
A-B-C-D	A	B	C	D
16-0-0-0	0.9912	0.0027	0.0024	0.0037
15-1-0-0	0.9318	0.0634	0.0026	0.0023
14-2-0-0	0.8715	0.1219	0.0023	0.0044
13-3-0-0	0.8107	0.1827	0.0036	0.0030
12-4-0-0	0.7497	0.2437	0.0044	0.0022
11-5-0-0	0.6901	0.3036	0.0034	0.0029
10-6-0-0	0.6298	0.3654	0.0020	0.0028
9-7-0-0	0.5584	0.4360	0.0026	0.0030
8-8-0-0	0.4952	0.4985	0.0027	0.0036
8-4-4-0	0.5060	0.2485	0.2427	0.0028
7-4-4-1	0.4442	0.2477	0.2429	0.0652
6-4-4-2	0.3762	0.2465	0.2514	0.1259
5-4-4-3	0.3178	0.2500	0.2473	0.1849
4-4-4-4	0.2470	0.2518	0.2489	0.2523

Table 3. Fuzzy classification of a node connected to network with 4 communities generated with $z_{out}/k = 0.375$

Connections	Fuzzy Classification			
A-B-C-D	A	B	C	D
16-0-0-0	0.9709	0.0092	0.0108	0.0091
15-1-0-0	0.9160	0.0647	0.0093	0.0101
14-2-0-0	0.8571	0.1228	0.0104	0.0097
13-3-0-0	0.8008	0.1802	0.0100	0.0090
12-4-0-0	0.7422	0.2385	0.0095	0.0098
11-5-0-0	0.6825	0.2958	0.0123	0.0093
10-6-0-0	0.6200	0.3566	0.0111	0.0123
9-7-0-0	0.5582	0.4181	0.0128	0.0109
8-8-0-0	0.4891	0.4846	0.0130	0.0133
8-4-4-0	0.5045	0.2437	0.2406	0.0113
7-4-4-1	0.4397	0.2461	0.2436	0.0705
6-4-4-2	0.3797	0.2471	0.2445	0.1287
5-4-4-3	0.3175	0.2439	0.2473	0.1913
4-4-4-4	0.2462	0.2494	0.2549	0.2495

tables shows the number of links the 129th node has to communities A, B, C, and D, respectively. Notice that in each configuration the 129th node has different overlap levels, varying from the case where it fully belongs to a single community up to the case where it belongs to the four communities almost equally. From 2nd to 5th column we have the fuzzy degree of membership of the 129th node relative to communities A, B, C, and D respectively, obtained by our algorithm. The presented values are the average of 100 realizations with different networks. For these simulations, the parameters were set as follows: $p_{det} = 0.5$, $\Delta_v = 0.4$ and $\Delta_\rho = 0.9$.

The results shown that the method was able to accurately identify the fuzzy communities of the overlap nodes. The accuracy gets lower as z_{out}/k increases, this was expected, since a higher z_{out}/k means that the communities are more diffuse and the observed node can be connected to nodes that are overlap nodes themselves.

Based on this data, we have created an overlap measure in order to easily illustrate the application of the algorithm in more complex networks with lots of overlap nodes. Therefore, the overlap index o_i for a node v_i is defined as follow:

$$o_i = \frac{f_i^{j**}}{f_i^{j*}} \tag{13}$$

where $j* = \arg\max_j f_i^j$ and $j ** = \arg\max_{j,j\neq j*} f_i^j$, and $o_i \in [0 \quad 1]$, where $o_i = 0$ means completely confidence that the node belongs to a single community, while $o_i = 1$ means the node is completely undefined among two or more communities.

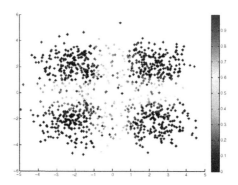

Fig. 1. Problem with 1000 elements split into four communities, colors represent the overlap index from each node, detected by the proposed method

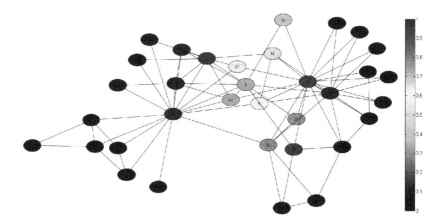

Fig. 2. The karate club network, colors represent the overlap index from each node, detected by the proposed method

Then, we have applied the algorithm to a problem with 1000 elements, split into four communities with 250 elements each. There are four gaussian kernels in a two dimensional plane and the elements are distributed around them. To build the network, each element is transformed into a network node. Two elements i and j are connected if their Euclidean distance $d(i,j) < 1$. The algorithm parameters were set as follows: $p_{\mathrm{det}} = 0.6$, $\Delta_v = 0.4$ and $\Delta_\rho = 0.9$. In Figure 1 the overlap index of each node is indicated by their colors. It is easy to realize that the closer to the communities frontier the nodes are, the higher are their respective overlap indexes.

Finally, the algorithm was applied to the famous Zachary's Karate Club Network [18] and the results are shown in Figure 2. The algorithm parameters were set as follows: $p_{\mathrm{det}} = 0.6$, $\Delta_v = 0.4$ and $\Delta_\rho = 0.9$. Again, the overlap index of each node is indicated by their colors. In Table 4 the fuzzy classification of all the nodes on this network are shown.

Table 4. Fuzzy classification of the Zachary's Karate Club Network achieved by the proposed method

Node	Community A	Community B	Node	Community A	Community B
1	0,8934	0,1066	18	0,9861	0,0139
2	0,8744	0,1256	19	0,0126	0,9874
3	0,5727	0,4273	20	0,6452	0,3548
4	0,9470	0,0530	21	0,0140	0,9860
5	0,9960	0,0040	22	0,9877	0,0123
6	0,9979	0,0021	23	0,0168	0,9832
7	0,9979	0,0021	24	0,0106	0,9894
8	0,9282	0,0718	25	0,0275	0,9725
9	0,3703	0,6297	26	0,0262	0,9738
10	0,2750	0,7250	27	0,0050	0,9950
11	0,9968	0,0032	28	0,1319	0,8681
12	0,9957	0,0043	29	0,2298	0,7702
13	0,9791	0,0209	30	0,0123	0,9877
14	0,7510	0,2490	31	0,3293	0,6707
15	0,0238	0,9762	32	0,2188	0,7812
16	0,0100	0,9900	33	0,0878	0,9122
17	1,0000	0,0000	34	0,1342	0,8658

4   Conclusions

This paper presents a new clustering technique using combined random-deterministic walking and competition among particles, where each particle corresponds to a class of the problem. The algorithm outputs not only hard labels, but also soft labels (fuzzy values) for each node in the network, which corresponds to the levels of membership from that node to each community. Computer simulations were performed in both synthetic and real data, and the results shows that our model is a promising mechanism to uncover overlap community structure in complex networks.

Acknowledgements

This work is supported by the State of São Paulo Research Foundation (FAPESP) and the Brazilian National Council of Technological and Scientific Development (CNPq).

References

1. Han, J., Kamber, M.: Data Mining: Concepts and Techniques, 2nd edn. Morgan Kaufmann, San Francisco (2006)
2. Witten, I.H., Frank, E.: Data Mining: Practical Machine Learning Tools and Techniques, 2nd edn. Morgan Kauffman, San Francisco (2005)

3. Hand, D.J., Mannila, H., Smyth, P.: Principles of Data Mining. MIT Press, Cambridge (2001)
4. Weiss, S.M., Indurkhya, N.: Predictive Data Mining: A Practical Guide. Morgan Kaufmann, San Francisco (1998)
5. Tan, P.N., Steinbach, M., Kumar, V.: Introduction to Data Mining. Pearson/Addison Wesley (2005)
6. Newman, M.E.J.: The structure and function of complex networks. SIAM Review 45, 167–256 (2003)
7. Dorogovtsev, S., Mendes, F.: Evolution of Networks: From Biological Nets to the Internet and WWW. Oxford University Press, Oxford (2003)
8. Bornholdt, S., Schuster, H.: Handbook of Graphs and Networks: From the Genome to the Internet. Wiley-VCH (2006)
9. Newman, M.E.J., Girvan, M.: Finding and evaluating community structure in networks. Physical Review E 69, 026113 (1–15) (2004)
10. Newman, M.: Modularity and community structure in networks. Proceedings of the National Academy of Science of the United States of America 103, 8577–8582 (2006)
11. Duch, J., Arenas, A.: Community detection in complex networks using extremal optimization. Physical Review E 72, 027104 (1–4) (2006)
12. Reichardt, J., Bornholdt, S.: Detecting fuzzy community structures in complex networks with a potts model. Physical Review Letters 93, 218701 (1–4) (2004)
13. Danon, L., Díaz-Guilera, A., Duch, J., Arenas, A.: Comparing community structure identification. Journal of Statistical Mechanics: Theory and Experiment 9, P09008 (1–10) (2005)
14. Quiles, M.G., Zhao, L., Alonso, R.L., Romero, R.A.F.: Particle competition for complex network community detection. Chaos 18, 033107 (1–10) (2008)
15. Zhang, S., Wang, R.S., Zhang, X.S.: Identification of overlapping community structure in complex networks using fuzzy c-means clustering. Physica A Statistical Mechanics and its Applications 374, 483–490 (2007)
16. Palla, G., Derényi, I., Farkas, I., Vicsek, T.: Uncovering the overlapping community structure of complex networks in nature and society. Nature 435, 814–818 (2005)
17. Zhang, S., Wang, R.S., Zhang, X.S.: Uncovering fuzzy community structure in complex networks. Physical Review E 76, 046103 (1–7) (2007)
18. Zachary, W.W.: An information flow model for conflict and fission in small groups. Journal of Anthropological Research 33, 452–473 (1977)

Semi-supervised Classification Based on Clustering Ensembles

Si Chen, Gongde Guo, and Lifei Chen

School of Mathematics and Computer Science,
Fujian Normal University, Fuzhou, China
chensi072@163.com, {ggd,clfei}@fjnu.edu.cn

Abstract. In many real-world applications, there only exist very few labeled samples, while a large number of unlabeled samples are available. Therefore, it is difficult for some traditional semi-supervised algorithms to generate the useful classifiers to evaluate the labeling confidence of unlabeled samples. In this paper, a new semi-supervised classification based on clustering ensembles named SSCCE is proposed. It takes advantages of clustering ensembles to generate multiple partitions for a given dataset, and then uses the clustering consistency index to determine the labeling confidence of unlabeled samples. The algorithm can overcome some defects about the traditional semi-supervised classification algorithms, and enhance the performance of the hypothesis trained on very few labeled samples by exploiting a large number of unlabeled samples. Experiments carried out on ten public data sets from UCI machine learning repository show that this method is effective and feasible.

Keywords: semi-supervised learning, classification, clustering ensembles, clustering consistency.

1 Introduction

In traditional supervised learning, a large number of labeled samples are required to learn a well-performed hypothesis, which is used to predict the class labels of unlabeled samples. In many real-word applications such as toxicity prediction of chemical products, computer-aided medical diagnosis, labeled samples are very difficult to obtain because labeling samples needs to spend much time and effort [1]. However, a large number of unlabeled samples are easily available. If only using very few labeled samples, it is difficult to generate a good classification model, and a lot of available unlabeled samples are wasted. Thus, when the labeled samples are very limited, how to enhance the performance of the hypothesis trained on very few labeled samples, by making use of a large number of unlabeled samples, has become a very hot topic in machine learning and data mining [1].

Semi-supervised learning is one of the main approaches exploiting unlabeled samples [1]. This paper mainly focuses on semi-supervised classification algorithms which aim to improve the performance of the classifier by making fully use of the labeled and unlabeled data. A kind of semi-supervised classification learning approaches are called pre-labelling approaches [2]. The method firstly trains an initial

H. Deng et al. (Eds.): AICI 2009, LNAI 5855, pp. 629–638, 2009.

classifier on a few labeled samples and then the unlabeled samples are utilized to improve the performance of the classifier. For example, self-training [5] is a well-known pre-labelling method. In this method, only a classifier is trained on the labeled training set. In some pre-labelling methods such as co-training [7], tri-training [9], and co-forest [6], more classifiers are used in their training processes. However, when there only exist very few labeled samples, some pre-labelling approaches [5, 6, 7, 9] are difficult to generate the initial useful classifier to estimate the labeling confidence of unlabeled samples, so that a certain number of unlabeled samples may be misclassified. Additionally, some algorithms place constraints to the base learner [6, 8], or require the sufficient and redundant attribute sets [7].

Another kind of semi-supervised classification learning methods are called post-labelling approaches [2, 17]. The method firstly generates a data model on all the available data by applying a data density estimation method or a clustering algorithm, and then labels the samples in every cluster or estimates the class conditional densities. However, using only a single clustering algorithm may generate the bad-performed data model, and thus affect the performance of the final classifier.

In this paper, a new post-labelling algorithm named SSCCE, i.e., Semi-Supervised Classification based on Clustering Ensembles, is proposed. Based on clustering ensembles, SSCCE firstly generates multiple partitions on all the samples, and matches clusters in different partitions. Then by analyzing the multiple clustering results, the unlabeled samples with high labeling confidence are selected to label and added into the labeled training set. Finally, a learner is trained on the enlarged labeled training set. Experiments carried out on ten UCI data sets show that this method is effective and feasible. SSCCE is easier to understand, which uses the clustering consistency index to estimate the labeling confidence of unlabeled samples, and requires neither special classifiers nor the sufficient and redundant attribute sets.

The rest of the paper is organized as follows: Section 2 gives a brief review on semi-supervised learning and clustering ensembles. Section 3 presents the SSCCE algorithm. Section 4 reports the experimental results carried out on ten UCI data sets. Section 5 concludes this paper and raises several issues for future work.

2   Related Work

2.1 Semi-supervised Learning

Semi-supervised learning methods include pre-labelling approaches and post-labelling approaches [2], etc. For example, self-training [5] is a pre-labelling method, in which, however, the unlabeled samples may be misclassified when the initial labeled samples are insufficient for learning a well-performed hypothesis [14]. Some pre-labelling methods [6, 7, 8, 9], such as co-training, more classifiers are used in the training process. However, co-training requires the sufficient and redundant attribute sets. In [8], Goldman et al. proposed a new algorithm extending co-training which requires the two special classifiers, each of which can divide the data space into several equivalence classes. Zhou and Li developed the tri-training algorithm [9], which does not require the sufficient and redundant attribute sets and special classifiers. The algorithm implicitly determines the labeling confidence by prediction consistency for unlabeled samples. But this implicit determination may not accuracy enough, and the initial labeled samples are not large enough to train a high accuracy classifier, which is likely to lead to misclassify a certain number of unlabeled samples. In [6], Li et al. proposed the co-forest algorithm which incorporates Random Forest [18] algorithm to extend the co-training paradigm. But this algorithm tends to under-estimate the error rates of concomitant ensembles, and it places constraints to the base learner and ensemble size.

The post-labelling approach [2, 17] firstly generates a data model on all the available data by applying a data density estimation method or a clustering algorithm, and then labels the samples in every cluster or estimates the class conditional densities. An unlabeled sample is labeled mainly according to its relative position with the labeled data. The kind of algorithm is formally described in [2]. In [17], Parzen windows are used for estimating the class conditional distribution and a genetic algorithm is applied to maximize the posteriori classification of the labeled samples.

2.2 Clustering Ensembles

Recently, a number of clustering algorithms [10] have been proposed. However, it is difficult to find a single clustering algorithm which can handle all types of cluster shapes and sizes, and it is hard to determine what kind of clustering algorithm should be used for a particular dataset [15]. Therefore, many scholars begin to study clustering ensembles methods. In [11], it shows that clustering ensembles can go beyond a single clustering algorithm in robustness, novelty, stability, confidence estimation, parallelization and scalability.

Let $X = \{x_1, x_2, ..., x_n\}$ be a data set with n samples in multidimensional space. Consider H as ensemble size (i.e., the number of partitions), and $\pi = \{\pi_1, \pi_2, ..., \pi_H\}$ as the set of H partitions. $\pi_i = \{C_1^i, C_2^i, ..., C_{n_i}^i\}$ ($i \in \{1, 2, ..., H\}$) denotes a data partition, where C_j^i is the jth cluster in the partition π_i, and n_i is the number of the cluster labels in π_i. Let $\pi_i(x)$ be the cluster label of a sample x in partition π_i. Clustering ensembles methods aim to combine H partitions $\pi_1, \pi_2, ..., \pi_H$ into a better final partition, where the samples in the same cluster are more similar while the samples belonging to different clusters are less similar. Recently, the researches for clustering ensembles mainly focus on how to generate effective multiple partitions [16], and how to combine multiple partitions [12].

The k-means algorithm [10] with different initial cluster centers is often used to generate multiple partitions for a given dataset. The k-means algorithm has low computational complexity, and does not require many user-defined parameters. Because the k-means algorithm is sensitive to the choice of the initial cluster centers, SSCCE generates H different partitions by using the k-means algorithm which randomly selects k different initial cluster centers each time, and then combines the H different partitions.

3 SSCCE Algorithm

3.1 Outline of SSCCE

In this paper, a novel semi-supervised classification based on clustering ensembles named SSCCE is proposed. SSCCE firstly generates H different partitions on all the samples using the k-means algorithm [10], and then matches clusters in different partitions. Secondly, the samples with high labeling confidence are selected and the cluster labels of these selected unlabeled samples are matched with the real class labels of the labeled samples. Then the selected unlabeled samples are labeled and added into the initial labeled training set. Finally, a hypothesis is trained on the enlarged training set. The SSCCE algorithm is summarized as follows.

Algorithm: SSCCE
Input: labeled training set $L=\{(x_1,y_1),(x_2,y_2),...,(x_{|L|},y_{|L|})\}$;
　　　　unlabeled training set $U=\{x_1,x_2,...,x_{|U|}\}$;
　　　　$X=L \cup U$;
　　　　H: number of partitions (i.e., ensemble size);
　　　　π_a: reference partition;
　　　　α: threshold value of clustering consistency index;
　　　　k: number of cluster centers of k-means algorithm;
Output: a hypothesis h.
Step1: for i=1 to H
　　　　1.1. Randomly select k initial cluster centers from the data set X;
　　　　1.2. Use the k-means algorithm to generate partition π_i on the data set X;
　　　　end for
Step2: Match clusters in different partitions according to the matching clusters method described in Section 3.2.
Step3: Compute clustering consistency index (CI) and clustering consistency label (CL) of each sample in the data set X.
Step4: Select the samples with $CI > \alpha$ from the data set X, and these samples are divided into clusters $C_1, C_2,...,C_m$ according to their corresponding clustering consistency labels $CL_1, CL_2,...,CL_m$ (m is the number of labels).
Step5: Using the methods described in Section 3.4, match the corresponding clustering consistency labels $CL_1, CL_2,...,CL_m$ and the real class labels of the labeled training set L.
Step6: Assign the re-matched clustering consistency labels to all the selected unlabeled samples in clusters $C_1, C_2,...,C_m$, and add them into the labeled training set L.
Step7: A hypothesis h is trained on the enlarged labeled training set L.

According to the above outline of SSCCE, the following three main issues are required to address: 1) how to match the clusters in different partitions? , 2) how to define the clustering consistency index (CI) and clustering consistency label (CL)? , and 3) how to match the cluster labels of unlabeled samples and the real class labels? We will discuss these issues in the following sections.

3.2 Matching Clusters

In the multiple different partitions, the similar clusters in different partitions may be assigned with different labels. For example, the cluster labels of five samples are (1, 2, 2, 1, 2) and (2, 1, 1, 2, 1) in two different partitions, respectively. In fact, these two partitions are exactly the same, and are only assigned with different cluster labels.

How to match clusters in different partitions is one of the most important problems needed to be resolved.

Based on the methods proposed in [12, 19], SSCCE uses Jaccard index to determine the best matching pair of clusters in different partitions after converting clusters C^i_j (the jth cluster in partition π_i) into a binary valued vector X^i_j. The value of lth entry of X^i_j, say 0 or 1, denotes absence or presence of lth sample in the cluster C^i_j. Note that any partition can be arbitrarily selected as the reference partition from H partitions, so that the clusters of the other partitions are matched with those of the selected reference partition.

3.3 Clustering Consistency Index and Clustering Consistency Label

After matching the clusters in different partitions, it is easy to find that some samples are steadily assigned to the same cluster, whereas the clusters are frequently changed for some examples in H partitions. In this paper, the samples with the stable cluster assignment are considered as the ones with high labeling confidence, and then these unlabeled samples are selected to label. Therefore, SSCCE introduces clustering consistency index (CI) to measure the labeling confidence. Clustering consistency index (CI) [16] is defined as the radio of the maximal number of times which the sample is assigned in a certain cluster to the total number of partitions H. The clustering consistency index $CI(x)$ of a sample x is defined by Eq. (1).

$$CI(x) = \frac{1}{H} \max \left\{ \sum_{i=1}^{H} \delta(\pi_i(x), C) \right\}_{C \in \text{ cluster labels}} \quad , \text{with } \delta(a,b) = \begin{cases} 1, & \text{if } a = b \\ 0, & \text{otherwise} \end{cases}. \quad (1)$$

By analyzing these partitions $\pi_1, \pi_2, ..., \pi_H$, the samples with the clustering consistency index greater than the given threshold value α are considered to be steadily assigned to a certain cluster. The label of this cluster is defined as clustering consistency label (CL). For a sample x, if $CI(x) > \alpha$, then its corresponding clustering consistency label $CL(x)$ is defined by Eq. (2).

$$CL(x) = \arg \max_{C} \left\{ \sum_{i=1}^{H} \delta(\pi_i(x), C) \right\}_{C \in \text{ cluster labels}}. \quad (2)$$

The greater the clustering consistency index of a sample is, the higher the labeling confidence of the sample is. SSCCE needs to label the samples with high clustering consistency index. However, their corresponding clustering consistency labels do not match with the real class labels of the labeled samples. Therefore, we should match clustering consistency labels and real class labels.

3.4 Matching Clustering Consistency Labels and Real Class Labels

Suppose that the clustering consistency labels of the selected samples with high clustering consistency index are respectively $CL_1, CL_2, ..., CL_m$ (m is the total number of CL). Then the clusters $C_1, C_2, ..., C_m$ only including the selected samples, corresponding to the clustering consistency labels $CL_1, CL_2, ..., CL_m$, may be the following three cases: (1) both labeled samples and unlabeled samples are contained; (2) only the unlabeled samples are contained; (3) only the labeled samples are contained. The third case will not be taken into account because there are not any unlabeled samples in clusters. The first two cases will be analyzed as follows.

In the first case, according to the principle of majority vote, if cluster C_i ($i \in \{1, ..., m\}$) contains both labeled samples and unlabeled samples, then its corresponding label CL_i is re-matched with the majority label of the labeled samples in cluster C_i. In the second case, according to k-nearest-neighbor [13], if cluster C_i contains only the unlabeled samples, then its corresponding label CL_i is re-matched with the majority label of the labeled nearest neighbors of all the unlabeled samples in cluster C_t. Finally, all the unlabeled samples in cluster C_i are labeled with the re-matched clustering consistency labels, and then added into the labeled training set.

4 Experiments

In order to test the performance of SSCCE, we use ten public data sets from the UCI machine learning repository [4], named solar-flare, breast-w, Australian, credit-a, tic-tac-toe, Liver, heart-statlog, heart-c, colic, and auto-mpg, respectively. For each data set, 10-fold cross validation is employed for evaluation. In each fold, training data are randomly partitioned into labeled training set L and unlabeled training set U under the label rate only 1%, i.e., just 1% training data (of the 90% data) are used as labeled samples while the remaining 99% training data are used as unlabeled samples. The entire 10-fold cross validation process is repeated ten times. In each time, randomize the ordering of samples, and then run a tested algorithm to generate an experimental result using 10-fold cross validation. Finally, the ten results of 10-fold cross validation are averaged for each test.

For SSCCE, SimpleKMeans algorithm from the Weka Software package [3] is used to generate multiple partitions, and the number of clusters is set to the number of classes of the labeled training data. When using the matching clusters algorithm, reference partition is set to partition π_1. Use 1-nearest-neighbor to deal with the second case that a cluster only contains unlabeled samples in Section 3.4.

For comparison, we evaluate the performances of SSCCE, the supervised learning algorithm trained on only the labeled training set L, co-forest [6] and self-training [5]. For SSCCE, the ensemble size H of clustering ensembles is set to 100. In co-forest, the number of classifiers is set to 10. For fair comparison, Random Forest [18] in WEKA [3] is used as the classifier in all the above algorithms. In either co-forest or self-training, the threshold value of labeling confidence is set to 0.75. For each algorithm, the average classification accuracy, i.e., the ratio of the number of test samples correctly classified to the total number of test samples, is obtained with 10 times of 10-fold cross validation on each data set. Table 1 illustrates the average classification accuracies of the supervised learning algorithm using Random Forest (denoted by RF), SSCCE using Random Forest denoted by SSCCE (RF), co-forest, and self-training using Random Forest denoted by self-training (RF). The last row Average shows the average results over all the experimental data sets using Random Forest. The highest average accuracy is shown in bold style on each dataset.

Table 1. Average classification accuracies of the compared algorithms using RF (%)

Data set	RF	SSCCE (RF)	co-forest	self-training (RF)
solar-flare	77.92	**80.04**	77.00	77.11
breast-w	83.23	**84.72**	82.58	79.07
Australian	61.43	**65.19**	59.58	56.28
credit-a	59.62	**61.48**	59.52	58.06
tic-tac-toe	61.44	**62.91**	61.47	60.62
Liver	50.19	50.63	**51.71**	49.96
heart-statlog	48.96	**53.44**	47.00	47.04
heart-c	52.97	**59.03**	52.18	52.18
colic	58.42	58.16	**58.64**	55.69
auto-mpg	52.36	55.86	**57.00**	52.36
Average	60.66	**63.15**	60.67	58.84

Table 1 shows that SSCCE benefits much from the unlabeled samples since the performances of SSCCE (RF) are better on 7 data sets than RF, co-forest, and self-training (RF). Furthermore, Table 1 also shows that SSCCE outperforms self-training on all the data sets, and the average result of SSCCE (RF) over all the experimental data sets is highest. Therefore, Table 1 supports that SSCCE is effective and feasible, and can improve the performance of the classifier trained on very few labeled samples by utilizing a large number of unlabeled data under the label rate 1%.

For further analysis on SSCCE, Table 2 tabulates the average threshold value α of clustering consistency index. In the experiments, set the threshold value α to the arithmetical mean of the clustering consistency indices of all the training data on each data set. The column add_ratio shows the ratio of the number of unlabeled samples added into the labeled training set L to the total number of unlabeled samples. In

Table 2. Further analysis of SSCCE using RF

Data set	α	add_ratio (%)	one-k-means (%)	SSCCE (RF) (%)
solar-flare	0.21	42.79	77.92	**80.04**
breast-w	0.54	58.02	83.23	**84.72**
Australian	0.54	49.49	61.43	**65.19**
credit-a	0.54	47.46	59.62	**61.48**
tic-tac-toe	0.58	48.32	61.44	**62.91**
Liver	0.54	42.48	50.19	**50.63**
heart-statlog	0.53	49.32	48.96	**53.44**
heart-c	0.25	45.72	52.97	**59.03**
colic	0.54	46.72	58.42	58.16
auto-mpg	0.3	46.82	52.36	**55.86**
Average	0.46	47.71	60.66	**63.15**

addition, the column one-k-means shows the average classification accuracy of the extreme case of SSCCE using Random Forest when $H = 1$, i.e., the k-means algorithm is run only one time to generate a partition, and then all the unlabeled samples are given the matched class labels and added into the training set L. In Table 2, SSCCE (RF) is the same as those of Table 1. All the parameters of the algorithms still keep unchanged.

From Table 2, it can be observed that the average 47.71% of unlabeled samples are added into the labeled training set L. And when H=100, the average threshold value α is about 0.46. It can be concluded that SSCCE can effectively select a certain number of unlabeled samples with high labeling confidence. Moreover, the average accuracies of SSCCE when H=100 are better than those of one-k-means on 9 data sets. Therefore, it validates that clustering ensembles can outperform the individual clustering, and thus enhance the performance of SSCCE.

In the experiments, we also use J48 [32] from the Weka Software package [3] to observe the impact of different classifiers on the average classification accuracies of these compared algorithms. The results show that the performances of SSCCE (J48) are better than the supervised algorithm J48 and self-training using J48 on 6 data sets among 10 data sets, and the average result of SSCCE using J48 over all the experimental data sets is highest.

Therefore, SSCCE is effective when using either Random Forest or J48, and co-forest and self-training fail to perform well on the most data sets. One possible explanation is when the number of labeled samples is very few, i.e., the label ratio only 1%, it is difficult to generate the initial useful classifier to estimate the labeling confidence of unlabeled samples, and a certain number of noises can be added into the initial labeled training set for co-forest and self-training. However, SSCCE can benefit much from a large number of unlabeled samples based on clustering ensembles, which can select a certain number of unlabeled samples with high labeling confidence to label and enhance the performance of the initial classifier.

Table 3. Average classification accuracies of SSCCE (RF) over different ensemble sizes (%)

Data set	H=50	H=100	H=150
solar-flare	80.20	80.04	**81.16**
breast-w	85.46	84.72	**86.69**
Australian	64.93	65.19	**68.45**
credit-a	61.07	**61.48**	60.25
tic-tac-toe	60.59	**62.91**	60.36
Liver	**51.58**	50.63	50.16
heart-statlog	55.74	53.44	**56.59**
heart-c	57.43	**59.03**	56.06
colic	**58.64**	58.16	57.43
auto-mpg	**57.30**	55.86	52.60
Average	**63.29**	63.15	62.98

Note that in previous experiments, the ensemble size H of SSCCE is fixed to 100. But the different ensemble size might affect the performance of SSCCE. Table 3 tabulates the performances of SSCCE using Random Forest on $H=50$, $H=100$, and $H=150$. From Table 3, it can be seen that the last row average accuracy on $H=50$ is better than that on $H=100$ and $H=150$. Ensemble sizes of the highest average accuracies of SSCCE (RF) are the same on most data sets except heart-statlog.

5 Conclusion

In this paper, a new semi-supervised classification based on clustering ensembles named SSCCE is proposed. It firstly generates multiple different partitions on all the samples using the k-means algorithm which uses the different cluster centers each time, and then matches clusters in different partitions. Moreover, the samples with high clustering consistency index are selected, and their corresponding clustering consistency labels are matched with the real class labels. Then the selected unlabeled samples are labeled with the re-matched clustering consistency labels. Finally, these unlabeled samples are added into the initial labeled training set, and then a hypothesis is trained on the enlarged labeled training set.

Using an easily understandable method of estimating the labeling confidence of unlabeled samples, SSCCE can make full use of the labeled and unlabeled samples, and overcome some defects in some traditional semi-supervised learning algorithms. Experiments carried out on ten UCI data sets show that SSCCE outperforms the supervised learning algorithm only trained on labeled training data. Compared with some traditional semi-supervised learning algorithms such as co-forest and self-training, SSCCE can also perform better on most data sets, when the number of the labeled training data is very few.

The ensemble size H, i.e., the number of partitions, might impact the performance of SSCCE. Therefore, how to overcome the effect of ensemble size and further improve the performance of SSCCE will be studied in the further work.

Acknowledgments. This work was supported by the Natural Science Foundation of Fujian Province of China under Grants No. 2007J0016 and No. 2009J01273, and the Scientific Research Foundation for the Returned Overseas Chinese Scholars, Ministry of Education of China under Grant No. [2008] 890.

References

1. Zhou, Z.H., Li, M.: Semi-Supervised Regression with Co-Training Style Algorithms. IEEE Transactions on Knowledge and Data Engineering 19, 1479–1493 (2007)
2. Gabrys, B., Petrakieva, L.: Combining Labelled and Unlabelled Data in the Design of Pattern Classification Systems. International Journal of Approximate Reasoning, 251–273 (2004)
3. Witten, I.H., Frank, E.: Data Mining: Practical Machine Learning Tools and Techniques, 2nd edn. Morgan Kaufmann, San Francisco (2005)
4. Asuncion, A., Newman, D.J.: UCI Machine Learning Repository. School of Information and Computer Science. University of California, Irvine, CA, http://www.ics.uci.edu/~mlearn/MLRepository.html

5. Nigam, K., Ghani, R.: Analyzing the Effectiveness and Applicability of Co-Training. In: Proceedings of the 9th International Conference on Information and Knowledge Management, pp. 86–93 (2000)

6. Li, M., Zhou, Z.H.: Improve Computer-Aided Diagnosis with Machine Learning Techniques Using Undiagnosed Samples. IEEE Transactions on Systems, Man and Cybernetics – Part A: Systems and Humans 37, 1088–1098 (2007)

7. Blum, A., Mitchell, T.: Combining Labeled and Unlabeled Data with Co-Training. In: Proceedings of the 11th Annual Conference on Computational Learning Theory (COLT 1998), MI, Wisconsin, pp. 92–100 (1998)

8. Goldman, S., Zhou, Y.: Enhancing Supervised Learning with Unlabeled Data. In: Proceedings of the 17th International Conference on Machine Learning (ICML 2000), CA, San Francisco, pp. 327–334 (2000)

9. Zhou, Z.H., Li, M.: Tri-training: Exploiting Unlabeled Data Using Three Classifiers. IEEE Transactions on Knowledge and Data Engineering 17, 1529–1541 (2005)

10. Han, J., Kamber, M.: Data Mining: Concepts and Techniques, 2nd edn. Morgan Kaufmann, San Francisco (2006)

11. Topchy, A., Jain, A.K., Punch, W.: A Mixture Model for Clustering Ensembles. In: Proceeding of the 4th SIAM International Conference on Data Mining, pp. 379–390 (2004)

12. Fred, A.: Finding Consistent Clusters in Data Partitions. In: Kittler, J., Roli, F. (eds.) MCS 2001. LNCS, vol. 2096, pp. 309–318. Springer, Heidelberg (2001)

13. Roussopoulos, N., Kelly, S., Vincent, F.: Nearest Neighbor Queries. In: Proceedings of the 1995 ACM SIGMOD International Conference on Management of Data, pp. 71–79 (1995)

14. Li, M., Zhou, Z.H.: SETRED: Self-Training with Editing. In: Ho, T.-B., Cheung, D., Liu, H. (eds.) PAKDD 2005. LNCS (LNAI), vol. 3518, pp. 611–621. Springer, Heidelberg (2005)

15. Dubes, R., Jain, A.K.: Clustering Techniques: the User's Dilemma. Pattern Recognition 41, 578–588 (1998)

16. Topchy, A., Minaei-Bidgoli, B., Jain, A.K., Punch, W.F.: Adaptive Clustering Ensembles. In: Proceedings of the 17th International Conference on Pattern Recognition (ICPR 2004), vol. 1, pp. 272–275 (2004)

17. Kothari, R., Jain, V.: Learning from Labeled and Unlabeled Data. In: IEEE World Congress on Computational Intelligence, IEEE International Joint Conference on Neural Networks, HI, Honolulu, USA, pp. 1468–1474 (2002)

18. Breiman, L.: Random Forests. Machine Learning 45, 5–32 (2001)

19. Zhou, Z.H., Tang, W.: Clusterer Ensemble. Knowledge-Based Systems 9, 77–83 (2006)

Research on a Novel Data Mining Method Based on the Rough Sets and Neural Network

Weijin Jiang[1], Yusheng Xu[2], Jing He[1], Dejia Shi[1], and Yuhui Xu[1]

[1] School of computer, Hunan University of Commerce, Changsha 410205, China
jwjnudt@163.com
[2] China north optical-electrical technology, Beijing 100000, China
yshxu520@163.com

Abstract. As both rough sets theory and neural network in data mining have special advantages and exiting problems, this paper presented a combined algorithm based rough sets theory and BP neural network. This algorithm deducts data from data warehouse by using rough sets' deduct function, and then moves the deducted data to the BP neural network as training data. By data deduct, the expression of training will become clearer, and the scale of neural network can be simplified. At the same time, neural network can easy up rough set's sensitivity for noise data. This paper presents a cost function to express the relationship between the amount of training data and the precision of neural network, and to supply a standard for the change from rough set deduct to neural network training.

1 Introduction

As a product of the combination of many kinds of knowledge and technology, Data mining (acronym: DM) is a new technology rose in 90's of twentieth century. On August, 1989, Detroit of American, the concept of KDD (vivid form: Knowledge Discovery in Database) is presented firstly on the eleventh international artificial intelligence union meeting[1,2]. 1995, American computer meeting (ACM) proposed Data Mining, and it vividly described large database as a valuable mineral resources, from which useful information can be found by efficient knowledge-finding technology[3,4]. Because Data Mining is the key step of KDD, people usually will not distinguish Knowledge Discovery in Database (KDD) from Data Mining (DM).

Data Mining is a technology which is used to analyze observed data set, usually very large, the intention of DM is to find unknown relation and summarize data under an efficient way, which can be understood by data owner. The common algorithm and theory of Data Mining are Rough Sets[5], Artificial Neural Networks, Decision trees, Genetic algorithm and etc. this paper mainly discussed Rough Sets theory and BP Neural Networks in Artificial Neural Networks[6].

Both Rough sets and BP Neural Networks have classification function in Data Mining. The advantage of Rough Sets is that it is good at parallel execution, description of uncertain information and the strategy dealing of redundant data, and the problem is it is sensitive with object noise[7,8]. BP Neural Networks is the most popular Neural Networks, whose main merits is high precision and non-sensitive with noise,

H. Deng et al. (Eds.): AICI 2009, LNAI 5855, pp. 639–648, 2009.

but its problem is that redundant data can easily cause over-training of neural networks, besides, networks scale and the amount of training sample's influence on the speed of network training and training time are headache problem[9].

As to the merits and the demerits of Rough Set theory and BP Neural Networks, this paper proposed a new Data Mining algorithm that combines Rough Sets theory and BP Neural Network[10]. The algorithm overcame rough sets' sensitive on noise data; meanwhile, it reduced the training time of BP Neural Network, provided constringency, which improved efficiency much.

2 Rough Sets Theory

Rough Sets first proposed by a Poland scholar Z.Pawlak in 1982[3,4]. One of the characteristic is its powerful qualitative analysis ability; it has no need to give Rough Sets quantity description for some character or property in advance, such as probability classification in statistics, subjection degree or subjection function in Blur Sets theory, rough sets can find problem's inner rule directly from the given problem's description set under the way that certain the given problem's approximate area by non-distinguished relation and non-distinguished class.

According to Pawlak's Rough Sets model, the following definition and description are produced.

Definition 1. Decision System regard $S=(U, A\{V_a\}, a)$ as knowledge expression system, and:

S is a non-empty limit set, and it is called topic area;

A is a non-empty limit set, and it is called property set;

V_a is property a∈A 's value area;

$a : U{\to}a : U{\to}V_a$为 is one-to-many reflection, which makes every element of topic area U only has a reflection in V_a.

if A is composed by condition property set C and conclusion property set D, and C, D satisfy $C{\cap}D=\varphi$, then S can be called strategy system.

In a strategy system, every element of U represents a rule; the precondition of rule is decided by C and its value, and the post-condition is decided by D and its value. Under the situation that V_a and $a : U{\to}V_a$ will not rise confuse, people usually use $(U, C{\cup}D)$ to represent strategy system

Definition 2. The non-discriminability relation of DS   as for strategy system $S=(U, C{\cup}D)$, $B0C$ is a subset of condition property set, duality relation IND $(B, D)=\{ (x,y) \in U{\times}U, f(y,a) \forall a{\in}B\}$ is S's non-discriminability relation.

Non-discriminability relation is an equal relation, from which classification of strategy system can be obtained, and here, we called the divided equal class non-discriminability class. Usually, $[x]_{IND}$ (B denotes x 's non-discriminability class, under the unconfused situation, IND (B) is used to present non-discriminability relation $IND(B, D)$.

Definition 3. Upper-approximate and lower-approximate   as for knowledge expression system S (U, A) , we set $B{\subseteq}A$, $X{\subseteq}U$, and:

$$\underline{B}X=\{x\in U[x]_{IND(B)}\in X\}$$

$$\overline{B}X=\{\ x\in U[x]_{IND(B)}\cap X\neq\varphi\}$$

They are X's B-Lower approximation and B-upper approximation. Lower-approximate $\underline{B}X$ is union set of all the atom set of X subset, it is the biggest sets of X; upper-approximate $\overline{B}X$ is union set of atom sets, which has no empty intersect set with X, and it is the smallest set of X.

Definition 4. Positive area, negative area and boundary   the upper-approximate and lower-approximate of set $X\subseteq U$ divided topic area into three uncross area; positive area POS (X) , negative area NEG (X) and boundary area BND (X) :

POS (X) $=\underline{B}$ (X)

NEG (X) $=U\text{-}\overline{B}$ (X)

BMD (X) $=\overline{B}$ (X) $\text{-}\underline{B}$ (X)

Any element x which belongs to POS (X) or NEG (X) certainly not belongs to X, but belongs to X's supplementary set; when an element belongs to BND (X) , it is hard to decide whether it belongs to X or X's supplementary set, as a results, upper boundary area is a kinds of uncertain area to some extent, upper-approximate of one set is the union set of positive area and boundary area, it is $\overline{B}X=POS$ (X) UBND (X). If BND (X) $=\varphi$, then X is precise set, otherwise, X is rough set.

Definition 5. Tthe dependency degree of property   the dependency degree of two sets B, $R\subseteq U$ can be valued by property dependency function (rough membership function), the definition of which is as follows:

$$y_R(B)=\frac{card(POS_R(B))}{card(U)}$$

$$POS_R(B)=\bigcup_{X\in U/IND(B)}\underline{R}\ X$$

And card $(\cdot)$ denotes the base of function, POS_R (B) is property set R's positive area in U/IND (B) .

Definition 6. The importance of property   different property has different influence on dependency relation between condition and decision property. The meaning of Adding property a to R for classification U/IND (B) is:

SGF $(a,\ R,\ B)$ $=y_R(B)\text{-}y_{R\text{-}|a|}(B)$

The importance of property a is relative, and it is dependent on property set B and R. so, the importance of property may different in different situation. If set D as decision property, then the meaning of SGF $(a,\ R,\ D)$ is: After adding property a in property set R, dependency degree between R and D change, and this reflects the importance of property a.

Definition 7. Redundant property   as for property sets D and R, property $a \in R$, if POS_R (D) = $POS_{R-\{a\}}$ (D) ,then a is redundant in property set R, otherwise, a is necessary for Set D.

3 The Design of BP Neural Network

Although BP neural network is the most popular neural network, its design is still dependent on experience.

Input level design: the input level is mainly used to input data, so the design of input level is dependent on input information, for example, object 'person' has two properties: height and weight. As for object 'person', two inputs neural can be defined, inputting height and weight as training information, and one input neural is define will be ok, property 'height' will be used as training information. As for data table in database, input information can be the properties which not belongs to decision property. As results, in data mining, as for data table in database, the number of neural is less than the amount of non-decision property.

Hidden level design: the main task of hidden level design is to confirm the amount of hidden level and neural. This problem is being studied by many people, so the design is mainly dependent on experience. There are two views which are important for the design of hidden level. The first one is: double hidden neural network can solve super-dimensional classification problem, which can overcome almost every classification problem; the second is: the simple hidden level neural network, which contains enough neural, can solve almost every reality classification problem.

The amount of hidden level neural can be made sure from three experience formula:

$$k < \sum_{i=0}^{n} \binom{n_i}{i}$$

(1)

k is the amount of training sample, n is the number of input neural, n_i is the number of hidden level neural, when $i > n_i$, $\binom{n_i}{i} = 0$.

$$n_1 = \sqrt{n - m} + a$$

(2)

m is the number of output neural, n is the number of input neural, a's value is between 1 and 10, n_1 is the number of hidden level neural.

$$n_1 = log_2 n$$

(3)

n is the number of input neural, n_1 is the number of hidden level neural.

When it come to design the number of hidden level neural, these three formulas are always taken in consider, and the most suitable one will be chose.

The design of output level: a neural network with m output neural can provide two kinds of classification. For example, as we mentioned, 'weight' and 'height', 'sex' is needed to output. Here, only one neural can realize the discrimination of man or woman. In order to improve precision, more neural is adopted; such as two outputs neural are adopt to discriminate sex. As a result, the number of output must equal with $\log_2 X$, or bigger than $\log_2 X$, in the formula, X represents the needed classification model. In common situation, the number of output neural always equal to the standard of classification model.

4 The Combined Algorithm of Rough Set and BP Neural Network

A lot of problems will be arisen to access the tremendous data set in data-house: the most conspicuous is problem caused by data gathering, such as, producing redundant data will make the data increased rapidly; secondly, the distortion of big data set, such as noise data. Almost every process of data mining can not avoid these two problems. As for rough set theory, dealing redundant data is its advantage, but it is not sensitive with noise data. As for BP neural network, training speed and training sample is its troublesome problem.

Usually, the clearer the method of data expression, the less the redundant data, then the neural network will be easier to studied, but the number of neural and power value will increase, its training time will be longer, and the extension ability of neural network will be weaker.

One idea is: using rough set to reduce data, then using the reduced data set as the design evidence and training data of BP neural network. And this can make training data clear, and absorbs both advantages of these two methods. While reducing the training time of BP neural network, the BP neural network can reduce the influence of noise. Based on this, the paper proposed a combined algorithm of rough set and neural network.

Data mining based on rough set theory always are property reduction, the steps of property reduction are: first, finding the kernel of property reduction set by discrimination matrix, then using reduction algorithm to calculate reduction set, and deciding the best reduction set by some judge standard. The data mining process based on rough set theory is illustrated in fig. 1.

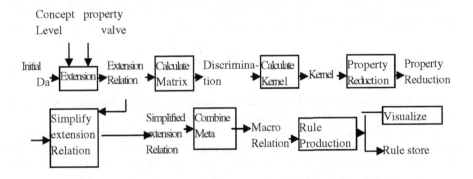

Fig. 1.

The difficulty of the algorithm is the end condition of rough set reduction, in another words, it is hard to decide the selection condition of the amount of BP neural network training data. According to the statement of third part, selecting the amount of training data has big influence for the training time of BP neural network. At present, there is no certain way to decide the amount of training data, just a cursory evaluation method: the amount of training data is twice as many as connect power. For example, if a BP neural network has n input knots, n_1 hidden level knots and m output knots, then it need $2\times$ $(n\times n_1 + n_1 \times m)$ training data[11].

The selection of the amount of training data is concern with the precision of neural network. Usually, error is used to reflect the capability of study. The definition of error is:

$$e\sqrt{\frac{\sum_{i=1}^{m}\sum_{j=1}^{n}(d_{ij}-y_{ij})}{m\cdot n}}$$

m denotes the sample number of training set, n is the amount of the output unit of neural network.

When the amount is increased, the error will be smaller; as a result, adding more training data will help to avoid error. But at the same time, when the training data is increased, the training time also will be longer. Based on this, a cost function is proposed.

Cost function is to describe the relation between the amount of training data and error, the error function can be modified as:

$$e\sqrt{\frac{\sum_{i=1}^{m}\sum_{j=1}^{n}(d_{ij}-y_{ij})}{Xm\cdot n}}$$

A variable X is added into this formula, when X=1, the function will be turned back. The form of cost function is:

$$y = x/(1-\frac{1}{\sqrt{x}})$$

x is coefficient, the value area is >1, y is cost guideline. Table 1 illustrates the relation between cost function, its differential coefficient and the value of x.

From the table, when x's value is 2.25, cost function's one differential coefficient is 0, which is the minimum value of cost function. When cost function's differential coefficient is smaller than -1, or bigger than 1, we can deem that the cost change too fast, and the coefficient is wrong. As a result, the coefficient should be in the area between 1.93 and 4, the optimal selection is 2.25.

Table 1.

x	1.25	1.5	1.75	2	2.25	2.5	2.75	3
y	11.84	8.17	7.17	6.83	6.75	6.2	6.80	7.10
y'	-30	-6.7	-2.24	-0.71	0	0.38	0.61	0.75
x	3.25	3.5	3.75	4	4.25	4.5	4.75	5
y	7.30	7.52	7.75	8	8.25	8.51	8.78	9.05
y'	0.85	0.91	0.96	1	1.02	1.05	1.06	1.08

Cost function can be regarded as the selection guideline or the end rule of rough set reduction. As for the sample with many properties, the optimal cost coefficient 2.25 will be chose to be the selection guideline, as for sample with few properties, cost higher than 2.25 will be chose. Data mining mainly deal with tremendous data, cost 2.25 will be absolutely the best answer, in order to mind special situation, here, the situation with few data has been taken in to consider. The algorithm is written as follows:

Step 1: sampling data, pointing mining condition, and deciding the goal of mining.

Step 2: deleting redundant property by following rough set theory.

Step3: doing property reduction under rough set theory.

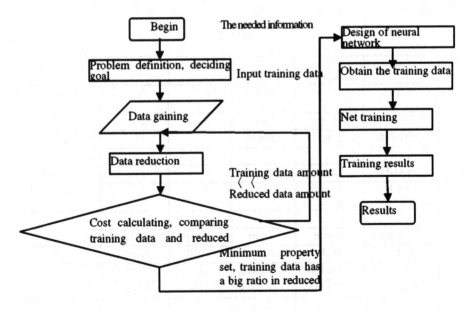

Fig. 2.

Step 4: if the minimum property set has been get, choosing training data set by cost 2.25.otherwise, using the highest cost 4 and reduction property to calculate training sample data, if the results of calculating is smaller than the reduced amount, then turn to step 3, otherwise, choosing training data by the definition and cost function.

Step 5: designing neural network by training data, and training these training data sample.

Step 6: outputting the final results.

The flow chart is illustrated by fig.2[12].

5   Applications

Here, a car data table in reference [2] is used to illustrate the algorithm in table 2.

The decision property is Make-model, others are condition properties. Using rough set to reduce the redundant propertied, table 3 is got.

After reducing the redundant properties, two properties are deleted. Doing data reduction for table 3, taking the user's request that the property reduction set must contain displace and weight, table 4 is obtained.

Then building the neural network, and selecting training sample. The neural net has 4 inputs neural, 3 outputs neural, the hidden level has 4 neural, the structure of the network is illustrated by fig.3.

Following network structure and cost coefficient, the number of training sample is $4 \times 2 \times (4 \times 4 + 4 \times 3) = 224$. Then training these samples, the final results will be output as table 5.

Table 2.

Plate#	Make-model	Color	Cyl	Door	Displace
BCT89C	Ford escoort	Silver	6	2	Medium
RST45W	Dodeg	Green	4	2	Big
IUTY56	Benz	Green	4	2	Big
......	......	......	......	......	......
PLMJH9	Honda	Brown	4	2	Medium
DSA321	Toyota paso	Black	4	2	Small
Compress	Power	Trans	Weight	Mileage	
High	High	Auto	1020	Medium	
High	Medium	Manual	1200	Medium	
Medium	Medium	Manual	1230	High	
......	......	......	......	......	
Medium	Medium	Manual	900	Medium	
Medium	Low	Manual	850	High	

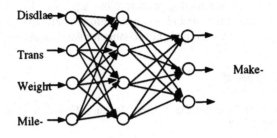

Disdlae

Trans

Weight

Mile-

Make-

Fig. 3.

Table 3.

Obj	Make-model	cyl	Door	Compress	Power	Trans	Weight	Mileage
1	USA	6	2	Medium	High	Auto	Auto	Medium
2	Germany	4	2	Big	Medium	Manual	Heavy	High
3	Japan	4	2	Small	Medium	Low	Light	High
...	...	...	...	...	...	...	...	...

Table 4.

Make-model	Displace	Trans	Weight	Mileage
USA	Medium	Auto	Medium	Medium
USA	Big	Manual	Heavy	Medium
Germany	Big	Manual	Heavy	High
Japan	Small	Manual	Light	High
......	......	......	......	......

Table 5.

Make-model	Displace	Trans	Weight	Mileage
USA	Medium	Auto		Medium
USA	Medium		Medium	
USA	Big		Heary	Medium
Germany	Big		Heary	High
Germany	Big	Auto	Medium	
Japan	Small			High
Japan		Manual	Light	

6 Conclusions

In the mining process of data house which has tremendous data and many properties, this algorithm possesses the advantages of both rough set theory and BP neural network. It can overcome the noise's influence for data sensation, at the same time; it can delete redundant data, provide clearer training data, lessen the scale of network, improves efficiency of mining. The proposal of cost function not only resolved the relation between training data and mining precision, but also provided guideline for the transformation from rough set to neural network. Unfortunately, data mining is aiming to big data warehouse, so the algorithm is not suitable for data mining with small scale.

Acknowledgement

This paper is supported by the Society Science Foundation of Hunan Province of China No. 07YBB239.

References

1. Bazan, J., Son, N.H., Skowron, A., Szczuka, M.S.: A View on Rough Set Concept Approximations. In: RSFDGrC, Chongqing, China (2003)
2. Bazan, J.: Dynamic reducts and statistical inference. In: Sixth International Conference on IPMU, pp. 1147–1152 (1996)
3. Kai, Z., Jue, W.: A Reduction Algorithm Meeting Users' Requirements. Journal Computer Science and Technology 17(15), 578P–P593 (2002)
4. Hu, K., Lu, Y., Shi, C.: Feature Ranking in Rough Sets (2002)
5. Nguyen, H.S., Ślęzak, D.: Approximate Reducts and Association Rules. In: Zhong, N., Skowron, A., Ohsuga, S. (eds.) RSFDGrC 1999. LNCS (LNAI), vol. 1711, pp. 137–145. Springer, Heidelberg (1999)
6. Nguyen, T.T., Skowron, A.: Rough set Approach to Domain Knowledge Approxima-tion. In: RSFDGrC 2003, Chongqing, China (2003)
7. Pawlak, Z.: Rough Sets. Int. J. Comput. In-form. Sci. 11(5), P341–P356 (1982)
8. Pawlak, Z.: Rough sets-Theoretical Aspects of reasoning about data. Kluwer Academic Publishers, Dordrecht (1991)
9. Polkowski, L.: A Rough Set Para-digm for Unifying Rough Set Theory and Fuzzy Set Theory. In: Wang, G., Liu, Q., Yao, Y., Skowron, A. (eds.) RSFDGrC 2003. LNCS (LNAI), vol. 2639, pp. 70–77. Springer, Heidelberg (2003)
10. Polkowski, L., Tsumoto, S., Lin, T.Y.: Rough Set Methods and Applications: New Developments. In: Knowledge Discovery in Information Systems. Springer, Heidelberg (2000)
11. Shaffer, C.A.: A Practical Introduc-tion to Data Structures and algorithm analysis. Prentice Hall, Englewood Cliffs (1998)
12. Weijin, J.: Research on Extracting Medical Diagnosis Rules Based on Rough Sets Theory. Computer Science 31(11), 97–101 (2004)

The Techno-Economic Analysis of Reducing NOx of Industrial Boiler by Recycled Flue Gas

Caiqing Zhang[1], Wenpei Hu[1], Yu Sun[2], and Bo Yang[2]

[1] School of Business Adnimistration, North China Electric Power University,
Baoding 071003, Hebei Province, China
[2] School of Environmental Science and Engineering,
North China Electric Power University,
Baoding 071003, Hebei Province, China
sunyu1031@126.com

Abstract. The quantity of industrial boiler is huge in our country, consuming the coal is about 350Mt/a, so a large number of NOx is emitted every year. It is a complex technology to reduce the emission of NOx. There are various methods according to different control mechanism, from the angel of technology and economy, the feasibility of industrial boilers reducing NOx by recycled flue gas is ananlyzed and get the decrease of NOx is between 30% and 40%. It is a mature, simple technology, lower investment and low cost of runs, what's more offers the reference of the NOx emission control technical reformation for industrial boilers, at last make some suggestions for the emission of NOx.

Keywords: industrial boilers, recycled flue gas, NOx, technicality, economy.

1 Introduction

Coal is the main fuel of industrial boilers in our country, and the primarily boilers are the low-power grate furnace. It is estimated that the number of all kinds of boilers which are being used is more than 50 ten thousand, the quantity is large, and meanwhile, more than 60% is chain boilers that is 0.5~75T/h. The consumption of coal is 350Mt/a [1.2], and 1000 kilogram coal could produce 7.4 kilogram NOx, thus it leads to make 2600 thousand NOx annual. Although our country does not have the standard about the emission of NOx that the industry boilers make, however, the quantity is quite considerable, and its number is still increasing year by year, the capacity is also expanding, and the emissions of nitrogen oxide compound (NOx)is increasing. How to reduce NOx which the industrial boilers emit economically is extremely important to reduce the air pollution.

Reducing the NOx that the industry boilers emit is a complex technology, lots of methods according to different control mechanism. The technology of "selectivity catalytic combustion SCR" is best to reduce NOx, but the investment and operating

H. Deng et al. (Eds.): AICI 2009, LNAI 5855, pp. 649–657, 2009.

Table 1. Comprehensive comparison table of Low N0x control technology

Craft plan	Appraisal index					
	Degree of technical mature	The level of craft difficulty	Site area	Initial investment	Operation cost	Reduction amount of NOx
Fuel graduation	Mature	Complex	A little large	Grading plant equipment (big)	ignored	30−40
Inleting the secondary wind	Mature	Simple	none	Air pipe expend (small)	ignored	20
Recycled flue gas	Mature	Simple	Relatively small	Recycled pipe expend (small)	ignored	30−40

cost are extremely high, a lot of small and medium-sized enterprise can not afford it, therefore it is not realistic. From economic and technological angel, the technology of recycled flue gas was selected to reduce the NOx emissions as much as possible. It will provide the reality basis for government to establish the policy formulation on the NOx emission.

From table 1, which shows that the comparison of low NOx control technology: The technology of recycled flue gas is mature, the craft is simple, at the premise of the NOx which drops largely, and it only adds the recycled pipeline, the cost of the invest at the beginning could be ignored, so it is a feasible method for reducing NOx. The feasibility of the technology was analyzed that uses recycled flue gas to reduce NOx, which is emitted by industrial boiler from the economical and technological points.

2 The Technological Analysis of Recycled Flue Gas

2.1 The Introduction of the Computation Mode

The simulation object is a 35t/h chain boiler in this article. This is a boiler with a single barrel and a natural circulation water-tube boiler, using the lamination coaling installment, the burning way of scale type chain link fire grate level, with the suitable coal of bituminous coal, the chamber size 7500mm×459mm×13500mm .The first inlet sets six wind areas to enter the wind separately, the amount of wind is different, and around the arches are two inlet areas. The functions of the secondary wind is mainly to disturb the above the smoke current, supplying the oxygen promptly, burning the aerosol combustible substance, to enhance the combustion efficiency, thus the design request of the position cannot cause the first wind and the wind pressure fluctuate to influence the combustion. Part flue gas after dust removal is sent to the first wind pipe through the recycling fan from the induced draft fan, then it is sent to the furnace after mixed with the first wind, and the quality of the recycled wind can be adjusted by the both sides of the door of the recycled flue gas.

Because the reaction that occurs in the entrance and combustion mainly concentrates above the chain, the grids of this region are relatively crowded to others in order to increase computation precision; the grid division can not cause the false proliferation in numerical simulation. In order to reduce the false proliferation, the grids that are in the calculate area is supposed to adapt the development of mobile as far as possible, causing the streamline enter the grid in vertical direction to control false proliferation occur. In this article, dividing the grids of the burning area follows this thought. Liking figure 1, firstly, decomposing the entire chamber into 5 parts, because the reaction in above individual is very few, the computation request is not so accurate and the size of the grids is 0.3 meter. The downward has two gusty areas, therefore the request is high, the size of grids is 0.1 meter, the coal entrance, the first wind area as well as the chain are main reaction regions, which are needed to concentrate the encryption, the dense of the grids is dense is 0.05, the diameter of the second wind areas itself only has 0.05 meter, because of its particularity, the grids are 0.02 meter to guarantee continuously and uniformity.

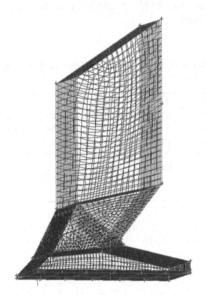

Fig. 1. Grids of the boiler

2.2 The Introduction of Computation Method

The simulation of the combustion process mainly has the mutual couplings of the gas and solid, the combustion reaction also with the turbulence, momentum, energy and mass transfer computation to draw the distribution of the temperature of boiler, the flowing and the component. The solid phase pellet uses the stochastic model, adds to the computation as the separation, the pellet has the independence equations to carry on the computation, considering the mutual functions between the two phase; the coal pellet in the combustion process is thought that the particle size is invariable, along with burning the quality of them reduces. The turbulence model uses the K-ε model, the burning computation uses the simply PDF to compute component score that is in the reaction process, the fuel and the oxidant are thought to react in a rapid rate in this computation, the percent of the fuel and the oxidant is mixed score. The double mixed score is used in this paper, and the experiential form is used in PDF calculation, thinking that the fuel and the oxidant fully contact and react. 90% of thermal transmission comes from radiation; the radiation uses the separate coordinate mode, considering the radiation exchanging. The temperature of surface of the wall is 600K, according to the temperature of saturated water add to revise temperature that takes entire, the revise temperature is considered the general burning situation that the slagging causes heating. According to the function of standard wall, we process the boundary condition of the surface. The entrance of air and the coal pellet entrance are established separately, the inlet area is thought to be the speed entrance. The outlet is established as the free, and meanwhile we suppose that the quality that enters and flows out the chamber is constant. The parameter of the entrance is illustrated in Table 2, the kind of the coal: bituminous coal. The parameters are listed in table 3.

Table 2. Parameters of the simulation of the combustion (kg/s)

The operation of boiler	The quality of the first wind in inlet 1	The quality of the first wind in inlet 2	The quality of the first wind in inlet 3	The quality of the first wind in inlet 4	The quality of the first wind in inlet 5	The quality of the first wind in inlet 6	The quality of the secondary wind in inlet
The combustion of the normal operation of boiler	0.731	1.219	1.920	2.194	1.950	0.731	0.455
The recycled flue gas rate is 30%	0.731	1.700	2.720	3.094	2.650	0.731	0.455

Indication : the temperature of the first and secondary wind is 423K.

Table 3. Analytical table of the characteristic of coal

Name	Mark	Unit	Value
amount of carbon for receiving	Cy	%	55.19
amount of hydrogen for receiving	Hy	%	2.38
amount of oxygen for receiving	Oy	%	1.51
amount of nitrogen for receiving	Ny	%	0.74
amount of sulfur for receiving	Sy	%	2.51
amount of ash for receiving	Ay	%	28.67
amount of water for receiving	Wy	%	9.0
the volatility of combustable part	V	%	26.7
the low calorific capacity for receiving	Qnet,v,ar	kJ/kg	20900

The Fluent software is applied in this computation. Firstly solve the flow field of the uniform temperature when starting calculating, after the momentum equation restraining, then couple pellet field, burning and the radiation heat transfer, secondly carry on the iteration; When the residual error of the equation of continuity and the energy all no longer reduces, each parameter is constant along with the iterative, the computation namely is restraining. At last the numerical simulation of NOx emission is the post-processing of the combustion, after restraining calculate the amount of the NOx, at that time the flow model, the rapids model, the energy equation, the radiation model and the PDF mode of the component do not participate in the computation, the calculate region, the parameter value of other each field are frozen except the component of NOx , that is the NOx reaction and the process of the combustion are not coupled.

2.3   The Numerical Simulation and Discussion of the Result

2.3.1   The Comparison of the Distribution of the Temperature Field with and without Recycled Flue Gas

Figure 2 (a) is the distribution of the temperature field without recycled flue gas, Figure 2 (b) is map that is with recycled flue gas. By contrast we could observe: the average temperature of the boiler with recycled flue gas decreases to some extent, it drops 30~80℃, the central region decreases 90℃.This is because the quantity of the flue which enters to the chamber increases, with it the flue that needs to be heated also increased with recycled flue gas. Moreover, the concentration of the O_2 that is in the recycled flue gas drops obviously, but the concentration of CO_2 rises, this has some influence to burning. Therefore the average temperature of the boiler decreases on the whole, it is an advantage to reducing of NOx, and the temperature that reduces does not affect the normal combustion as well as the movement.

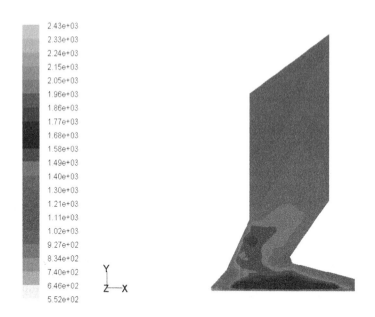

(a)   The distribution of the temperature field without recycled flue gas

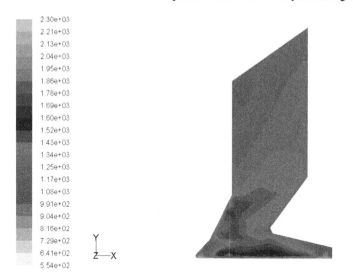

(b)   The distribution of the temperature field with recycled flue gas

Fig. 2. The distribution of the temperature field

2.3.2   The Comparison of the Concentration of NOx with and without Recycled Flue Gas

The change of NOx is illustrated in chart 2-3, the total concentration of NOx is about 394mg/m^3 (192ppm) without recycled flue gas while the concentration reduces to

281mg/m³ (137ppm) with recycled flue gas, in certain region, the emissions of NOx drops from 610mg/m³ to 450mg/m³. Speaking of the total quantity of NOx, the emission of NOx improves greatly.

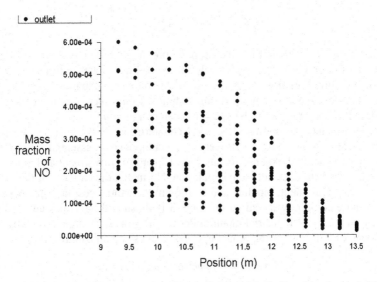

(c)   The quality score of NOx at the outlet along the height of the boiler without recycled flue gas

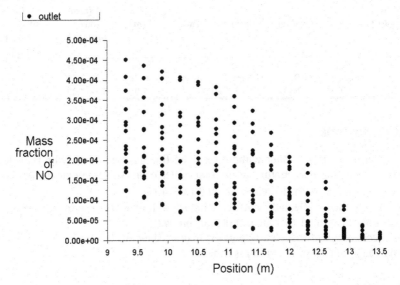

(d)   The quality score of NOx at the outlet along the height of the boiler with recycled flue gas

Fig. 3. Distribution of the quality score of NOx at the outlet along the height of the boiler

3   The Analysis of Economical Property

The pressure of the air blower is obtained from:

$$h_L = \lambda \frac{L}{D} \cdot \frac{\omega^2}{2g} \tag{1}$$

From the Table 4 conclude that: the dropped pressure which the recycled pipe causes is probably ranging 0.71 mm water column to 8.09 mm water column, its respective dropped pressure does not affect the use of air blower comparing with the selected blower, therefore the investment of this technical t only increases the expense of the pipeline and labor, the cost of the operation can be ignored.

In order to study the optimizing control countermeasure, the cost of information on the control technology related nitrogen oxide compound is essential. Under ordinary circumstances, the total expense includes the expense of investment and annual operating, but the cost still does not have stipulation. Normally, the expense of the investment contains react equipment, pipeline and so on; but the annual cost includes depreciation, service, and electric power and so on. table 3 is illustrated Several kinds of low NOx control technology of industrial chain boiler are illustrated in table 3, the beginning of investment, cost of movement, reducing of NOx and so on are compared in it.

Table 4. Parameters of recycled pipe

The quantity (T)	λ	Q(m³/s)	V(m/s)	D(m)	L(m)	hL(mm water column)	The pressure of the air blower (mm water column)
10	0.018	7.29	12-15	0.49-0.61	8-16	3.25-8.09	107
35	0.018	19.7	12-15	1.3-1.45	10-20	1.10-3.81	170
60	0.018	33.33	12-15	2.22-2.78	15-25	0.71-2.32	214

4   Conclusions

1. Comparing with the boiler without recycled flue gas, the boiler with recycled flue gas, the average temperature in internal boiler drops to some extent, the emission concentration of NOx decreases about 30%−40%.
2. Regarding the middle and small scale industrial boilers, the use of recycled flue gas does not affect the use of the original air blower.
3. The technology of recycled flue gas is mature, the craft is simple, the investment is small and the cost of operation is little.

References

1. Commandr, J.-M., Stanmore, B.R., Salvadore, S.: The high temperature reaction of carbon with nitric oxide. Combust Flame, 128 (2005)
2. Glarborg, P., Jensen, A.D., Johnsson, J.E.: Fuel nitrogen conversion in solid fuel fired systems. Progress in Energy and Combustion Science, 29 (2003)

3. Jones, J.M., Patterson, P.M., Pourkashanian, M., Williams, A., Arenillas, F.R., Pis, J.J.: Modeling NOxformation in coal particle combustion at high temperature an investigation of the devolatilisation kinetic factors. Fuel, 88 (1999)
4. Xu, X.C., Chen, C.H., Qi, H.Y., et al.: Development of coal combustion pollution control for SO2 and NOx in China. Fuel Processing Technology 62, 153–160 (2000)
5. Yin, C., Cailat, S., Harion, J.-l.: Bernard Baudo Everest Perez Investigation of the flow utility tangentialy fired pulverized-coal boiler. Fuel, 81 (2002)

Urban Traffic Flow Forecasting Based on Adaptive Hinging Hyperplanes

Yang Lu[1,2], Jianming Hu[1,2], Jun Xu[1,2], and Shuning Wang[1,2]

[1] Department of Automation, Tsinghua University, Beijing, China
[2] Tsinghua National Laboratory for Information Science and Technology
{lu-y08,yun-xu05}@mails.tsinghua.edu.cn,
{hujm,swang}@mail.tsinghua.edu.cn

Abstract. In this paper, after a review of traffic forecasting methods and the development of piecewise linear functions, a new traffic flow forecasting model based on adaptive hinging hyperplanes was proposed. Adaptive hinging hyperplanes (AHH) is a kind of piecewise linear models which can decide its division of the domain and the parameters adaptively. Acceptable results (forecasting error is smaller than 15%) were obtained in the test of the real traffic data in Beijing. After comparison with the results of prediction model base on MARS, the following conclusions can be drawn. First, the two methods have almost the same performance in prediction precision. Second, AHH will be a little more stable and cost less computing time. Thus, AHH model may be more applicable in practical engineering.

Keywords: adaptive hinging hyperplanes, MARS, traffic forecasting.

1 Introduction

Traffic flow forecasting is a very important aspect in Advanced Traffic Information System (ATIS). The quality of the system's service depends on the accuracy of the traffic prediction to a great extent. Over the past 30 years, traffic flow forecasting has been a very active issue but without perfect solutions. There are a variety of forecasting methods[1,2,3,4,5], such as: historical average, time series models (ARIMA), neural network and nonparametric models (k-NN method), which all have their own shortage. Historical average method cannot respond to dynamic changes, ARIMA model is very sensitive to disturbance which cannot be avoided in traffic environment, neural network is possible to overtrain the network and k-NN model has complicated way in looking for neighbors. As we know, the main application of the above methods is in highway traffic flow prediction, but the urban transportation system is more complex. To overcome the drawbacks above, Ye applied MARS in urban traffic flow forecasting[6], and obtained good results.

In another aspect, during the past 30 years, the research about piecewise linear function has been developing gradually and there have been some kinds of representing models, such as canonical piecewise linear representation [7,8,9,10], lattice PWL function model[11] and hinging hyperplanes model[12]. Among them, the HH model is well-known and comparatively practical model which was proposed by Breiman in

H. Deng et al. (Eds.): AICI 2009, LNAI 5855, pp. 658–667, 2009.

1993 with a good capacity in nonlinear function approximation. Due to its special structure, the least square method can be used to make identification. Therefore, the HH model is a very useful tool in black-box modeling. However, in fact, HH model is equivalent with the canonical PWL model. To improve the model's representation ability, Wang introduced generalized hinging hyperplanes (GHH) in 2005, which is actually derived from the lattice PWL function, so the model's general representation ability is proved directly[13].

Adaptive Hinging Hyperplanes model[14], which was proposed by Xu in 2008, based on MARS and GHH and then proved to be a special case of GHH, shares the advantages of the two approaches. The model is adaptive, flexible and its basic function is linear in each subregion, so the acceptable results in traffic flow prediction using AHH can be expected because the model based on MARS has already obtained good results. In this paper, we proposed an urban short-term traffic flow forecasting model based on adaptive hinging hyperplanes, and the real traffic data in Beijing was used to test its performance. Furthermore, a comparison between AHH and MARS is drawn to demonstrate AHH's effectiveness in forecasting.

The paper is organized as follows. Section 2 introduced AHH model and its algorithm after a review of MARS and GHH. How to use AHH model to carry out the traffic flow prediction was described in detail in Section 3. We then compared the results of AHH model and MARS model in Section 4 and made a brief conclusion in Section 5.

2 Adaptive Hinging Hyperplanes

2.1 Review of MARS and GHH

2.1.1 MARS

MARS (Multivariate Adaptive Regression Splines) was first introduced by Friedman in 1991[15]. It is developed from recursive partitioning regression as generally considered. The basic idea of the latter method is to find the most appropriate division of the domain through successive partition. In the beginning of the procedure, there is only one father node and it can be divided into two child nodes, each of them denotes one subregion. One subregion does not intersect with each other and the union of all subregions constructs the whole domain. The above step is repeated until getting the terminal node which cannot be split anymore, while the region it denotes corresponds with one basic function. The partitioning is based on a greedy algorithm where in each step it selects the split which can give the best fit to the data.

Recursive partitioning regression can be written as the following forms of a set of basic functions:

$$\hat{f}(x) = \sum_{m=1}^{M} a_m B_m(x) \tag{1}$$

where the basic function takes the form

$$B_m(x) = I(x \in R_m) \tag{2}$$

in which I is an indicator function having the value one when $x \in R_m$ and zero otherwise.

From the splitting procedure, the disadvantage of recursive partitioning regression is obvious to be known. The subregion is discontinuous and the method cannot represent the functions which have no interactions between their variables. To overcome the problems above, firstly, MARS uses truncated power function $\left[\pm (x-t) \right]_+^q$ to replace the step function, where $q = 1$ and the form $\left[\ \right]_+^q$ denotes the positive part of the expression. The basic function is constructed in the terms of tensor product of the truncated power function. Second, MARS allows the parent basic function which has already been split to be involved in the further splitting, so it has the ability to approximate the additive functions whose variables do not interact with each other.

MARS model can be written as follows:

$$ \hat{f}(x) = a_1 + \sum_{m=2}^{M} a_m \prod_{k=1}^{Km} \left[s_{km} \cdot (x_{v(k,m)} - t_{km}) \right]_+ \tag{3} $$

where K_m is the number of split of the basic function B_m, $s_{km} = \pm 1$ and $x_{v(k,m)}$ is the split variable, t_{km} is the split location in the k-th split of the m-th basic function. $a_1, a_2, ..., a_m$ is the coefficient of each basic function.

2.1.2 GHH

Generalized hinging hyperplanes was first introduced by Wang in 2005 as an extension of the hinging hyperplanes. Although HH model can approximate many nonlinear functions well, it has been proved that the model is lack of the ability to represent CPWL functions in high dimensions. HH model can be written as follows :

$$ \pm \max \left\{ l(x, a), l(x, b) \right\} \tag{4} $$

Wang made some modification to the above form and rewrote it as follows :

$$ \sum_i \sigma_i \max \left\{ l(x, \theta_1(i)), l(x, \theta_2(i)), ..., l(x, \theta_{k_i+1}(i)) \right\} \tag{5} $$

where $l(x, \theta(i))$ is linear function, $k_i \leq n$, $\sigma_i = \pm 1$. This model's representation ability was proved in [13].

2.2 AHH Model

The truncated power splines $\left[s_{km} \cdot (x_{v(k,m)} - t_{km}) \right]_+$ in MARS basic function can be written as $\max\{0, s_{km} \cdot (x_{v(k,m)} - t_{km})\}$, and can be seen as a special HH model. To improve the stability when the dimension of the predictive variable is high, after making a modification to the MARS basic function, replacing the operator "$\prod$" by "min", the AHH model is obtained[14]. AHH model can be written as follows:

$$f(x) = a_1 + \sum_{m=2}^{M} a_m \min_{k \in \{1,2...,Km\}} \left\{ \max\{0, s_{km} \cdot (x_{v(k,m)} - t_{km})\} \right\} \tag{6}$$

where K_m is the number of splits of the basic function B_m, $s_{km} = \pm 1$ and $x_{v(k,m)}$ is the split variable, t_{km} is split location in the k-th split of the m-th basic function. $a_1, a_2, ..., a_m$ is the coefficient of each basic function.

AHH's algorithm was developed analog to the MARS algorithm and the generalized cross validation (GCV) [16] which is shown as following formula was still used as estimating standard of errors.

$$LOF(\hat{f}_M) = GCV(M) = \frac{1}{N} \sum_{i=1}^{N} [y_i - \hat{f}_M(x_i)]^2 \Big/ [1 - \tilde{C}(M)/N]^2 \tag{7}$$

Forward Procedure

Step 1: Initialize $B_1(x)$ as a positive integer which ensures $\min(B_1(x), B_m(x)) = B_m(x)$, set $M = 2$

Step 2: Loop from $B_1(x)$ to $B_{M-1}(x)$, in each step of the loop changes x_v and t, let

$$g = \sum_{i=1}^{M-1} a_i B_i(x) + a_M \min\left\{B_m(x), \max\{(x_v - t), 0\}\right\} + a_{M+1} \min\left\{B_m(x), \max\{-(x_v - t), 0\}\right\} \tag{8}$$

use least square method to decide $a_1, a_2, ..., a_{M+1}$ to get the least $LOF(g)$ of the step. After finishing the loop, set $m^* = m$; $v^* = v$; $t^* = t$, which can lead to the least $LOF(g)$ of all possibilities appeared in Step 2.

Step 3: Set

$$B_M(x) = \min\{B_{m^*}(x), \max\{(x_{v^*} - t^*), 0\}\} \tag{9}$$

$$B_{M+1}(x) = \min\{B_{m^*}(x), \max\{-(x_{v^*} - t^*), 0\}\} \tag{10}$$

Setp 4: $M = M + 2$. If $M < M_{max}$, go to step 2; else finish the algorithm.

The AHH algorithm also needs a backward step just as MARS algorithm. In the backward procedure, the basic functions which do not make contributions to the fitting precision will be removed.

3   AHH Based Traffic Flow Forecasting

3.1   Data Description

In this paper, the real traffic flow and occupancy data collected from 50 detectors are provided by Beijing Traffic Management Bureau, which last for 16 days from Oct.22, 2006 to Nov.6, 2006. The sampling interval is 10mins, which means 144 sample points per day for each detector.

A small part of the network, which is shown in Figure 1, will be used to test the prediction effect of the model. The flow and occupancy data in link1 and its neighbors, such as link2, 3, 4, 5 and 6, will be used to forecast the traffic flow in link1.

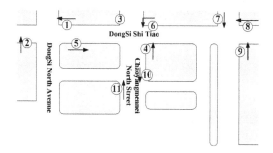

Fig. 1. The topological structure of the test network

3.2   Traffic Flow Forecasting Model and Evaluating Criterions

3.2.1   Forecasting Model

The data lasting for 14 days from Oct.22th, 2006 to Nov.4th, 2006 is used to train the AHH model, and the data in Nov.6th is applied to examine the model's prediction accuracy and obtain the forecasting error at the same time.

The traffic forecasting model which is to predict the flow of link i in t+1 time interval can be written as follows:

$$\hat{flow}_i^{t+1} = f(flow_i^t, flow_i^{t-1}, ..., flow_i^{t-N}, flow_{i+1}^t, flow_{i+2}^t, ..., flow_{i+L}^t, occ_i^t \sigma_i) \tag{11}$$

where $flow_i^t, flow_i^{t-1}, ..., flow_i^{t-N}$ denotes the link's own historic flow data before the prediction interval, in which $t, t-1, ..., t-N$ denote different time intervals; $flow_{i+1}^t, flow_{i+2}^t, ..., flow_{i+L}^t$ denote the neighbors' flow data in t interval, in which L decides how many neighbors should be considered; and $occ_i^t \sigma_i$ is the link's own historic occupancy data in t interval, σ_i can be chosen as 1 or 0.

Some explanation is made here to illustrate why the formula above should contain these factors as predictive variable. First, because the traffic flow can be recognized as a time series which changed gradually and not very fast, the data which is even N time intervals before the prediction time can be considered to be related, not only one interval before. Second, it is obvious that the link to be predicted has correlation with its neighbors, so the neighbor's data should also be taken into account. Third, the situation in the network obtained through prediction with both the flow and occupancy data may be more accurate than with only one of them. However, what value N and L should be is an important issue worth discussing later. If the value is small, the trained network may be incomplete while too large value will cause the overfitting problem in the process of training.

3.2.2 Evaluating Criterions

Two criterions will be used in this paper to evaluate the AHH model's predictive performance. One is mean absolute percentage error (MAPE) which can provide useful information in comparison when the traffic flow changes in a large scale. It can be calculated as following formula:

$$MAPE(\%) = \frac{1}{N}\sum_{i=1}^{N}\frac{predict(i) - obversed(i)}{obversed(i)} \times 100\% \tag{12}$$

Another is called PT criterion which denotes the Percentage of the sample points whose predictive error is smaller than the given Threshold. The criterion can be written as following formula:

$$PT(\%) = \frac{1}{N}\sum_{i=1}^{N}\left| predict(i) \mid \frac{predict(i) - obversed(i)}{obversed(i)} < \alpha \right| \times 100\% \tag{13}$$

where the expression $|\ | = 1$ if the condition in $|\ |$ is satisfied, otherwise $|\ | = 0$ and α is the given threshold.

3.3 Prediction Results and Analysis

3.3.1 The Length of Time Intervals

Table 1 shows the prediction results while only the link's own traffic flow information is considered as predictive variables in the forecasting procedure. N, which represents the length of time intervals being considered before the prediction time, changes from 0 to 9. The threshold in PT criterion is set to 0.2.

Table 1. Results using own flow information in AHH

N	0	1	2	3	4	5	6	7	8	9
MAPE(%)	16.5	14.4	13.9	13.9	14.1	13.3	12.9	13.4	13.5	13.6
PT(%)	69.9	78.9	79.4	80.0	77.7	76.0	77.4	77.9	77.8	77.6

From Table 1, a tendency of MAPE that first becomes smaller and then larger can be observed with the increasing of N. When N is small, the historic information which is needed to make prediction is deficient. But when N is too large, the prediction effect is also worse than the results when N=6. For example, if N=9, the traffic information which is nearly 2 hours before the prediction time will be taken into account.

Actually, the data 2 hours ago does not have much relationship with current situation. However, if these data is seen as predictive variable, the network trained will get accustomed to the situations with no sense and reduce the prediction accuracy.

3.3.2 The Effect of Occupancy

Table 2 can give an answer about whether the occupancy of the link should be a predictive variable. In the procedure of prediction, N also changes from 0 to 9.

After comparing Table 1 and 2, the following conclusions can be drawn. First, when N is small, which means the deficiency of information, the addition of occupancy is useful to improve the precision. However, when N is large enough, which means the information getting from the link itself is sufficient, the results are almost the same. But in practice, considering the computation speed of the computer, the time interval won't be too long. So, in sum, occupancy data should be a predictive variable.

Table 2. Results using own flow and occupancy information in AHH

N	0	1	2	3	4	5	6	7	8	9
MAPE(%)	15.6	14.3	13.8	13.9	13.7	13.3	12.9	13.4	13.5	13.6
PT(%)	73.4	78.8	78.7	80.0	78.4	76.1	77.4	77.9	77.8	77.6

3.3.3 The Number of Neighbor Detectors

Table 3 shows different prediction results with different neighbors' combinations being considered to forecast the flow of link 1 which are link 2, link2+3, link 2+3+4, link 2+3+4+5 and link 2+3+4+5+6, respectively.

Table 3. Mean absolute percentage error results using own flow, occupancy and neighbors' flow information in AHH

Link	N	0	1	2	3	4	5	6	7	8	9
2		16.0	14.3	13.7	13.9	13.0	12.6	13.0	13.4	13.5	13.6
2,3		14.8	13.9	13.5	13.2	13.3	12.8	13.0	13.4	13.8	13.9
2,3,4		13.8	13.4	13.0	13.2	12.7	12.5	13.5	13.5	13.2	13.2
2,3,4,5		14.3	14.2	13.5	13.2	12.7	12.5	13.5	13.5	13.2	13.2
2,3,4,5,6		14.2	14.1	13.8	13.5	13.5	12.8	13.4	13.5	13.2	13.2

In Table 3, there are no obvious advantages or disadvantages among the prediction results of different combinations, but in general the result of link 2+3+4 is relatively best. The reason to explain this phenomenon is just like the reason to decide time intervals. There is no doubt that with the neighbors' information the accuracy will be better. However, the specific relationship between the neighbors' data set is not clear and worth of further research.

3.3.4 Typical Results

In sum, the most precise result is obtained when N=5 and with the contribution of link 1's own occupancy and link 2, 3 and 4's flow information. The typical result is shown in Figure 2.

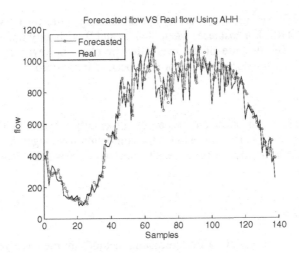

Fig. 2. The typical prediction result

4   Compared with Results Based on MARS

Based on MARS algorithm and adopts the same traffic prediction model form of AHH, the prediction results based on MARS has been obtained and will be compared with the results of AHH in this section.

4.1   General Performance

As seen in Figure 3, in general, the predictive performance of AHH and MARS model using their own flow information are almost the same. In specific terms, when N is small, the precision of AHH model is better with smaller error. But when N becomes large, the MARS model will have some advantages gradually.

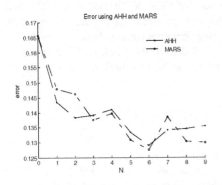
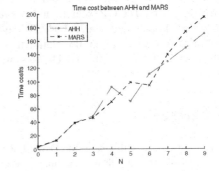

Fig. 3. General Comparison between AHH and MARS

Fig. 4. Time Cost Comparison between AHH and MARS

From Figure 3, we can also observe that there exists more fluctuation in the performance of MARS. For instance, when N=7, 8 and 9, a slightly increase in error is intended to be seen because of the overfitting problem, however, a sharp increase followed with dramatic decrease in error have been found instead.

4.2 Time Cost

As seen in Figure 4, in most circumstances, especially when N is large, the time which MARS model needs to accomplish the computation is longer than AHH model. However, MARS can only get better results with larger N which denotes much more computing time.

5 Conclusions

There have been a variety of traffic forecasting methods during these years, but every approach has its own limitation, so the research about this issue is still very active. We combine the MARS algorithm and the thinking of piecewise linear functions, introducing a new traffic prediction method.

In this paper, after a review of MARS algorithm and the development of piecewise linear functions, a traffic forecasting model based on adaptive hinging hyperplanes was proposed. Acceptable results (error is smaller than 15%) were obtained in the test of the real traffic data in Beijing. After comparison with the results of MARS model, the following conclusions can be drawn. First, the two methods have almost the same performance in precision. Second, AHH will be more stable and cost less computing time so it is more applicable in practice.

However, there are still some problems unsolved. The prediction results based on AHH which is nonparametric model cannot always follow the real data when the traffic flow is becoming very fast. In further research, some other methods may be adopted together with AHH as a combination model.

Acknowledgement. This work described in the paper is partially supported by National Natural Science Foundation of China (NSFC) 50708054, 60674025 and 60534060, Hi-Tech Research and Development Program of China (863Project) 2007AA11Z222 and 2007AA04Z193, National Basic Research Program of China (973Project) 2006CB705506, National Key Technology Research and Development Program 2006BAJ18B02 and The Research Fund for the Doctoral Program of Higher Education 200800030029.

References

1. Smith, B.L., Williams, B.M., Oswald, R.K.: Comparison of parametric and nonparametric models for traffic flow forecasting. Transportation Research Part C 10, 303–321 (2002)
2. Okutani, I., Stephanedes, Y.J.: Dynamic prediction of traffic volume through Kalman filtering theory. Transportation Research Part B 18B, 1–11 (1984)

3. Williams, B.M., Hoel, L.A.: Modeling and Forecasting Vehicular Traffic Flow as a Seasonal ARIMA Process: Theoretical Basis and Empirical Results. ASCE Journal of Transportation Engineering 129(6), 664–672 (2003)
4. Davis, G.A., Nihan, N.L.: Nonparametric Regression and Short-term Freeway Traffic Forecasting. ASCE Journal of Transportation Engineering 117(2), 178–188 (1991)
5. Smith, B.L., Demetsky, M.J.: Traffic Flow Forecasting: Comparison of Modeling Approaches. ASCE Journal of Transportation Engineering 123(4), 261–266 (1997)
6. Ye, S.Q., et al.: Short-Term Traffic Flow Forecasting Based on MARS. In: The 4th International Conference on Natural Computation, pp. 669–675. IEEE Press, Jinan (2008)
7. Chua, L.O., Kang, S.M.: Section-wise piecewise-linear functions: canonical representation, properties, and applications. IEEE Transactions on Circuits and Systems 30(3), 125–140 (1977)
8. Kang, S.M., Chua, L.O.: Global representation of multidimensional piecewise-linear functions with linear partitions. IEEE Transactions on Circuits and Systems 25(11), 938–940 (1978)
9. Kahlert, C., Chua, L.O.: A generalized canonical piecewise linear representation. IEEE Transactions on Circuits and Systems 37(3), 373–382 (1990)
10. Lin, J.N., Unbehauen, R.: Canonical piecewise-linear networks. IEEE Transactions on Neural Networks 6(1), 43–50 (1995)
11. Tarela, J.M., Alonso, E., Martinez, M.V.: A representation method for PWL functions oriented to parallel processing. Mathematical & Computer Modelling 13(10), 75–83 (1990)
12. Breiman, L.: Hinging hyperplanes for regression, classification and function approximation. IEEE Transactions on Information Theory 39(3), 999–1013 (1993)
13. Wang, S.N., Sun, X.S.: Generalization of Hinging Hyperplanes. IEEE Transactions on Information Theory 12(51), 4425–4431 (2005)
14. Xu, J., Huang, X.L., Wang, S.N.: Adaptive Hinging Hyperplanes. In: The 17th World Congress of International Federation of Automatic Control, World Congress, Seoul, Korea, pp. 4036–4041 (2008)
15. Friedman, J.H.: Multivariate adptive regression splines. The Annuals of Statistics 19(1), 1–61 (1991)
16. Friedman, J.H., Silverman, B.W.: Flexible parsimonious smoothing and additive modeling. Technometrics 31, 3–39 (1989)

Turbine Fault Diagnosis Based on
Fuzzy Theory and SVM

Fei Xia[1], Hao Zhang[1], Daogang Peng[1], Hui Li[1], and Yikang Su[2]

[1] College of Electric Power and Automation Engineering,
Shanghai University of Electric Power,
200090 Shanghai, China
[2] Nanchang Power Supply Corporation,
330006 Nanchang, JiangXi, China
{Fei.Xia,Hao.Zhang,Daogang.Peng,Hui.Li,Yikang.Su,
xiafei}@shiep.edu.cn

Abstract. A method based on fuzzy and support vector machine (SVM) is pro-posed to focus on the lack of samples in fault diagnosis of turbine. Typical fault symptoms firstly are normalized by the membership functions perceptively. Then some samples are used to train SVM of fault diagnosis. With the trained SVM, the correct fault type can be recognized. In the application of condenser fault diagnosis, the approach enhances successfully the accuracy of fault diag-nosis with small samples. Compared with the general method of BP neural net-work, the method combining advantages of fuzzy theory and SVM makes the diagnosis results have higher credibility.

Keywords: turbine fault diagnosis, fuzzy theory, SVM.

1 Introduction

The turbine on electric power plant is the important equipment [1]. Its structure is complex and the operation environment is particular. It works under the particular operation environment which is under high temperature, high-pressured, and high speed revolving. Obliviously, the steam turbine has high fault rate and hazardous breakdown. These breakdowns will make the heavy economic loss and the social consequence [2]. Therefore, it needs to use the advanced intelligent technology to monitor and analysis device status parameters in order to judge the corresponding equipment whether it is in a health state.

Condenser is the primarily auxiliary equipment of steam turbine whose quality of work situation affects the safety and economic operation of generator. Therefore, the monitoring and diagnosis of condenser running state are generally concerned by the operation department of the power plant. The research of condenser system and its fault diagnosis has great significance to reduce the downtime of units and improve availability of units.

In the application of modern fault diagnosis, the method of neural network has be-come an important tool with the characteristics of self-learning capability, fault-tolerant capability and parallel calculation capability. Various methods of neural network have

H. Deng et al. (Eds.): AICI 2009, LNAI 5855, pp. 668–676, 2009.

been applied to fault diagnosis of the condenser [3-6]. As the result of nonlinear relationship between the failures and the symptoms, complexity, ambiguity and randomness, it is hard to construct the precise mathematical model for the condenser system. Therefore, the values of symptoms can be transformed by the corresponding membership function with fuzzy theory. After that, more methods of fuzzy neural network are used in the fault diagnosis of condenser [7-8]. Although these methods take into account the merits of the two principles, the neural network is still vulnerable to fall into local minimum point. Even though, the network structure should be determined by the experience.

For solving these problems, the approach combined with fuzzy theory and Support Vector Machine (SVM) is proposed to achieve the fault diagnosis results in the condenser system. SVM is a new machine learning method developed on basis of the statistical learning theory [9]. Based on the principle of structural risk minimization, it can effectively address the learning problem and have better classification accuracy [10]. SVM has been used for face detection, speech recognition and medical diagnosis etc. Through the study of this paper, it can be proved successfully that SVM does very well in the tasks of fault diagnosis of condenser.

2 Fault Analysis of Condenser

Condenser plays an important role in the power plants. However, the condenser often has some faults in the operation process due to the design, the installation, the maintenance ant other reasons, such as low vacuum operation, cooling pipe leakage, super cool condensate water, higher oxygen content in condensate etc.. One of the most common is running low vacuum condenser. When condenser vacuum is too low, the thermal efficiency and output of unit is reduced. Even some security issues are caused by this state. The details are as following.

Due to the increased exhaust pressure, the turbine low pressure rotor blades level chatter and the stress increased which reduce the security of unit. Because of increased exhaust pressure and exhaust steam temperature, the speed ratio increases for the end of lower-level. In the reason of the thermal expansion of the mental increasing, the rotor center line is changed and the unit has the vibration. The increased exhaust temperature results in increase of exhaust hood temperature and condenser temperature. And the relative expansion between the cylinder and the condenser is up to 3 times than the normal value. The relative difference between brass condenser tube and plate rises to bulging population relaxation and destruction of water-tight side which result in that condensation water is contaminated.

Thus the condenser vacuum monitoring is very important in actual operation. In the standard operation of 300MW unit, the value of low vacuum condenser alarm is 87kpa and the value of down trip is 67kpa. According to the survey, the unit availability in thermal power plants over 600 MW declined 3.8% due to the equipment of condenser having the poor reliability.

Fault diagnosis of condenser is mainly discussed in this paper according to previous fault diagnosis methods. In order to carry out fault diagnosis for condenser, fault universe of condenser are divided into three typical fault fuzzy modes through selection combining experience of engineer experts and technicians in this paper, they are

Y1:serious fault of cycle pump; Y2:full water of condenser; Y3:not working of condensate pump. Eight symptoms relating with condenser fault are extracted as follows: X1: vacuum declined significantly and sharply; X2: vacuum declined slowly and slightly; X3: output pressure of condensate pump increased; X4: output pressure of condensate pump decreased; X5: temperature rise of cycle water decreased; X6: terminal difference of condenser increased; X7: coldness of condensation water increased; X8: pressure difference between pumping port and input of extractor decreased.

3 Definition of Membership Function

3.1 Fuzzy Theory

The In the daily life, many concepts are vague. It is far insufficient to describe with belonged absolutely or not belong absolutely. Thus, it has the necessity to break the relations of absolutely subordinates. University of California Professor Zadeh has introduced the fuzzy set theory in 1965. The fuzzy set theory develops the ordinary set theory concept, value scope of characteristic function is expanded from gathers $\{0,1\}$ to $[0,1]$. Some object of the universe is no longer regarded as belongs to this set or not belong to this set, but said that the degree of subordinates in this set is how much [4].

Assuming that a mapping is assigned on universe U:

$$A : U \rightarrow [0,1] \tag{1}$$

$$u \rightarrow A(u) \tag{2}$$

So, A is fuzzy set on the U, all fuzzy set in the U are recorded as (U), that is,

$$(U) = \{A \mid A : U \rightarrow [0,1]\} \tag{3}$$

A fuzzy set definite in the universe, its membership function have many different form. Determines the membership function correctly is the foundation for using fuzzy set to describe fuzzy concept appropriately. One basic step of the application of fuzzy mathematics to solve the practical problem is to find one or a several membership function. If this question has been solved, other questions are easily solved.

3.2 Membership Function

In order to disperse data from sensors, proper membership function should be constructed. Membership function is the foundation of fuzzy sets applied to practical problems, correctly constructing membership function is the key of properly using fuzzy sets. While there is no mature and effective method in the existing membership functions, they are all determined by experience, and then corrected by experiments or feedback information by computer simulating. In the turbine fault diagnosis, if the corresponding membership function is unable to determined, the following three methods can be adopted according to specific conditions:

1) Triangular fuzzy function

$$\mu_A(x) = \begin{cases} (x-a)/(b-a) & a \leq x \leq b \\ (c-x)/(c-b) & b \leq x \leq c \\ 0 & x < a \quad or \quad x > c \end{cases}$$

(4)

2) Trapezoidal fuzzy function

$$\mu_A(x) = \begin{cases} 0 & x < a \\ (x-a)/(b-a) & a \leq x \leq b \\ 1 & b \leq x \leq c \\ (d-x)/(d-c) & c < x \leq d \\ 0 & x > d \end{cases}$$

(5)

3) Normal-shaped fuzzy function

$$\mu_A(x) = e^{-(\frac{x-a}{b})^2}$$

(6)

3.3 Definition

Judging whether fault symptom exist or not directly by values of each thermal parameters of condenser is not accurate at all, because one fault symptom may correspond to different fault, and although the changing trend of same fault symptom of different faults are the same, the degree of changing are different. Therefore, thermal parameters should be fuzzy treated with the concept of fuzzy mathematics. Meanwhile, data from sensors can be dispersed. The changing trends are shown natural and veritable by blurring each input parameters using membership function.

According [11], the following three types of membership functions are applicable to condenser fault symptom:

1) Smaller sized

$$\mu_A(x) = \begin{cases} (1+a(x-c)^b)^{-1} & (x > c) \\ 1 & (x \leq c) \end{cases}$$

(7)

In the formula, $c \in U$ is an arbitrary point,$(a>0,b>0)$.

2) Larger sized

$$\mu_A(x) = \begin{cases} (1+a(x-c)^b)^{-1} & (x > c) \\ 0 & (x \leq c) \end{cases}$$

(8)

In the formula, $c \in U$ is an arbitrary point,$(a>0,b>0)$.

3) Medium sized

$$\mu_A(x) = e^{-k(x-c)^2}$$

(9)

In the formula, $c \in U$ is an arbitrary point; k is a parameter larger than zero $(k>0)$.

According to the upper formulas and combining practical situation, the membership function of operating parameters of 300MW unit condenser corresponding symptom can be obtained and they are shown in TABLE I.

Table 1. Membership function of each fault symptom

Fault symptom	Membership function
vacuum declined significantly and sharply（X1）	$\exp[-0.3(x-87)^2]$
vacuum declined slowly and slightly（X2）	$\exp[-0.3(x-97)^2]$
output pressure of condensate pump increased（X3）	$(x-0.5)/3.7$
output pressure of condensate pump decreased（X4）	$\exp[-4(x-3.7)^2]$
temperature rise of cycle water decreased（X5）	$\exp[-0.2(x-8)^2]$
terminal difference of condenser increased（X6）	$(x-3)/3.56$
coldness of condensation water increased（X7）	$(x-2)/0.5$
pressure difference between pumping port and input of extractor decreased（X8）	$\exp[-0.05(x-8.2)^2]$

4 Support Vector Machine

4.1 Fundamentals

SVM is a class of supervised learning algorithms introduced firstly by Vapnik. SVM has excellent performance in some pattern recognition application fields. Also, it has obtained much success in a variety of domains based on the principle of structural risk minimization. The basic idea of SVM in the application of classification is introduced by the simple example with 2-D space. Considered as shown in Figure 1 corresponds to 2-dimensional space on a simple classification to discuss the function for the linear function f (x) = (w, x) + b. When the classification problem is to be found a suitable straight-line which divided the entire 2-D plane, that is, to determine the direction of w and the intercept method b.

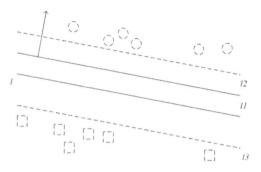

Fig. 1. Schematic Diagram of Classification Plane

Two types of points can be separated from a right line which exists numerously. The classification function is approximated by the following function:

$$f(x,a) = w \cdot x + b \tag{10}$$

Coefficients w and b are estimated by minimizing the regularized risk function:

$$\min \phi(x) = \frac{1}{2}\|w\|^2 \tag{11}$$

$$s.t. \quad y_i[(w \cdot x_i) - b] \geq 1, \quad i = 1, 2, \cdots, n$$

According to Lagrange principle, the above problem can be transformed to solve its antithesis problem. Finally we can get the decision function as follows:

$$f(x) = \text{sgn}\left\{\sum_{i=1}^{n} \alpha_i^* y_i (x_i \cdot x) + b^*\right\} \tag{12}$$

4.2 Method of SVM

With the deep study on support vector machine, many researchers increased and changed the function, the variable coefficient method or formula for the deformation, resulting in advantages that there is a certain application or algorithm with a number of support vector machines deformation algorithm. The main algorithms are the following: C-SVM Series, v-SVM Series, One-class SVM, RSVM (reduced SVM), WSVM (weighted SVM) and LS-SVM (least-square SVM).

C-SVM algorithm for the only adjustable parameter C is not an intuitive explanation, it is very difficult in practice to choose the appropriate value of the defection, Schblkoph proposed v-SVM algorithm. v-SVM with the new parameter v to replace the C, so that we can control the number of support vector or error, but also easy to choose the value than that of C parameter. It has very clear physical meaning for lv which represents the maximum amount of support vector borders and the lower limit of the number of support vector. l is the size of samples. Although the standard v-SVM algorithm is more complex than C-SVM algorithm, it has the effective way to solve the problem of classification with small samples. Chang et al proposed a deformed v-SVM

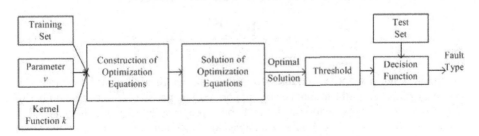

Fig. 2. Processing of v-SVM Algorithm

algorithm which has original objective function with the increment of variable a. In order to distinguish the two v-SVM methods, the later is called Bv-SVM algorithm. For simplification, the v-SVM algorithm is adopted in the fault diagnosis analysis task. A simple process of v-SVM algorithm is displayed in Figure2.

5 Simulation Experiments

5.1 SVM Training

The three classical fault types of condenser mentioned are used in the stimulation of fault diagnosis. The fault set is expressed by Y={Y1, Y2, Y3}. In the set, Y1 represents serious fault of cycle pump, Y2 represents full water of condenser and Y3 represents not working of condensate pump.

For different fault types, 40 samples are collected respectively. Then 20 samples are used in SVM train and the other 20 samples are to be measured the SVM through the previous train. Some examples in rain set and the corresponding fault type are shown in Table 2.

Table 2. Examples of SVM Train Set

	1	2	3	4
X1	0.05	0.06	0.03	0.01
X2	0.65	0.60	0.53	0.75
X3	0.15	0.11	0.24	0.12
X4	0.18	0.17	0.42	0.15
X5	0.56	0.50	0.67	0.86
X6	0.43	0.40	0.58	0.69
X7	0.11	0.09	0.13	0.22
X8	0.37	0.28	0.41	0.73
Fault Type	Y1	Y1	Y2	Y3

There is not a standard to choose the various parameters of SVM in theory. So it is hard to get the correct value in the practical problem. In the problem of condenser fault diagnosis, Gaussian function is used as the kernel function:

$$k(x_i, x_j) = \exp\left\{ \frac{-\left| x_i - x_j \right|^2}{\sigma^2} \right\}, \qquad i = 1, 2, \cdots, n \tag{13}$$

σ, which is the perception variable, determines the width of Gaussian function. The value of σ can control the number of support vector machine. When the number of support vector machine is too much, it can be appropriate by reducing the value of σ. Generally the value is based on data analysis and experiences. To simplify, the value of σ in the work is determined by the follow formula:

$$\sigma^2 = E\left(\left\|x_i - x_j\right\|^2\right) \tag{14}$$

In the above formulation, E represents mathematical expectation. And the value of v is also obtained by the test. In the work of condenser fault diagnosis, v equals 10.

5.2 Stimulation Test

According to the trained SVM, 60 samples of condenser fault are adopted for the verification. And some of the test results are shown in Table 3.

Table 3. Examples of SVM Test Set

	1	2	3	4
X1	0.05	0.03	0.02	0.01
X2	0.62	0.55	0.70	0.72
X3	0.10	0.26	0.11	0.13
X4	0.16	0.41	0.13	0.15
X5	0.52	0.66	0.79	0.84
X6	0.43	0.58	0.71	0.69
X7	0.09	0.14	0.18	0.21
X8	0.30	0.43	0.78	0.71
Fault Type	Y1	Y2	Y3	Y3
Actual Situation	Y1	Y2	Y3	Y3

The results of stimulation test shows that SVM has the ability to classify correctly the fault samples especially for the small samples due to the strong capability of data processing.

In order to compare with the method of SVM proposed in this paper, the same eight fault characteristics and three fault types of condenser are also adopted in the method of BP neural network. For compared the result, the same membership function and three-level BP neural network [12-13] are adopted in tests. Eight fault characteristics are inputs of the neural network and three fault types are the outputs.

Furthermore, the above two methods were tested through 30 simulation data. The results are shown in Table 4.

Table 4. Results of Comparing Experiments

Diagnosis method	True(%)	Fault(%)
SVM	91.3	8.7
BP neural network	83.3	16.7

Diagnostic results show that SVM has a higher diagnostic accuracy than that of BP neural network in the application of small and complex fault samples. Generally the correct classification rate is improved 8%.

6 Conclusion

SVM is a learning algorithm based on the principle of structural risk minimization, which has stronger theoretical basis and better generalization ability than neural network algorithm based on the experience minimization. Steam condenser fault diagnosis is taken as an example to verify the approach of SVM with fuzzy theory in intelligent steam turbine fault diagnosis. Experiment results show that SVM had higher diagnostic accuracy and stronger ability to classify than BP neural network in the complex fault diagnosis with small samples. In the future research, the approach will be applied in the more complicated fault diagnosis to verify the applicability. Furthermore, other intelligent diagnosis methods can be combined with SVM to enhance the accuracy of fault diagnosis.

Acknowledgments. This work is supported by Program of Shanghai Subject Chief Scientist with (09XD1401900) and Natural Science Foundation of Shanghai (No. 09ZR1413300).

References

1. Diao, Y., Passinb, K.M.: Fault diagnosis for a turbine engine. Control Engineering Practice, 1151–1165 (December 2004)
2. Zeng, X., Li, K.K., Chan, W.L., Yin, X., Chen, D., Lin, G.: Discussion on application of information fusion techniques in electric power system fault detection. Electric Power 36, 8–12 (2003)
3. Zhao, H., Li, W., Sheng, D., et al.: Study on Fault Diagnosis of Condenser Based on BP Neural Network. Power System Engineering 20, 32–34 (2004)
4. Chen, Z., Xu, J.: Condenser fault diagnosis based on Elman networks. East China Electric Power 35, 871–874 (2007)
5. Wang, J., Li, L., Tang, G.: Study on Fault Diagnosis of Condenser Based on RBF Neural Network. Electric Power Science and Engineering 23, 27–31 (2007)
6. Ma, Y., Yin, Z., Ma, L.: Study on fault diagnosis of condenser based on SOM neural network. Journal of North China Electric Power University 33, 5–8 (2006)
7. Wang, X., Wang, Q.: A Study on the Diagnosis of the Condenser Faults Based on the Fuzzy Neural Network. Modern Electric Power 18, 12–17 (2001)
8. Wu, Z., Lin, Z.: FNN-based fault diagnosis system for condenser. Gas Turbine Technology 21, 42–45 (2008)
9. Burges, C.J.: A Tutorial on Support Vector Machine for Pattern Recogntion. Data Mining and Knowledge Discovery 2(2) (1998)
10. Vapnik, V.N.: The Nature of Statistical Learning. Tsinghua University Press, Beijing (2004)
11. Jia, X.: Fault Diagnosis in Turbine of Condenser. Harbin Engineering University, Haerbin (2004)
12. FeiSi Center of Technology and Research, Neural Network Theory and MATLAB 7 Application, pp. 44–51. Publishing House of Electric Industry, Beijing (2005)
13. Wu, Z., Lin, Z.: FNN-based fault diagnosis system for condenser. Gas Turbine Technology 21, 42–45 (2008)

Architecture of Multiple Algorithm Integration for Real-Time Image Understanding Application

Bobo Duan, Chunyang Yang, Wei Liu, and Huai Yuan

Software Center, Northeastern University,
110179 Shenyang, China
{duanbb,yangcy,lwei,yuanh}@neusoft.com

Abstract. Robustness and real-time usually are the main challenges when designing image understanding approach for practical application. To achieve robustness, integrating multiple algorithms to make a special hybrid approach are becoming a popular way and there have been a lot of successful hybrid approaches. But this aggravates the difficulties in achieving real-time because of heavy computational workload of multiple algorithms. To design a hybrid approach with real-time constraint more easily, some theoretical researches about multiple algorithm integration are very necessary. This paper presents a common multiple algorithm integration model and architecture for typical image understanding applications. To achieve robustness and real-time in a hybrid approach, the strategies for increasing robustness and speed up are analyzed. Finally a robust hybrid approach for rear vehicle and motorcycle detection and tracking is introduced as a sample.

Keywords: Image Understanding, Robustness, Real-Time, Hybrid, Fusion.

1 Introduction

When developing an image understanding system for practical applications, follow challenges usually are faced: robustness and real-time. For example, an obstacle detection approach for driver assistance system need to provide robust perception in wide variety of outdoor environment, and this procedure need to be finished as quickly as possible to save time for driver reaction [1]. A successful image understanding system should meet the two requirements at the same time, but it is very difficult to design such an approach because the two requirements are usually with strong conflicts.

In image understanding procedure, it is often the case that many algorithms exist to acquire same special information, and each possessing different performance and computing workload characteristics. Since no single algorithm will be robust enough to deal with a wide variety of environment conditions, integrating multiple algorithms to make a special hybrid approach is becoming a popular way to increase robustness. But image understanding algorithms usually have high computational workload, this integration aggravates the difficulty in achieving real-time. So it become very important to do some common research about how to design a hybrid image understanding approach which can achieve robustness and real-time at the same.

H. Deng et al. (Eds.): AICI 2009, LNAI 5855, pp. 677–684, 2009.

Up to now, there have been a lot of researches about multiple algorithms integration in image understanding. Most of them propose their hybrid approaches for special problems, which usually are with a form of multiple kinds of technologies hybrid or multiple visual cues integration. For example, P. Zhang et al. [2] propose a hybrid classifier for handwritten numeral recognition, which uses neural network as coarse classifier and used decision tree as fine one. K. Toyama et al. [3] propose a hybrid visual tracking approach, which includes multiple trackers such as edge of polygon tracker, rectangle tracker and etc. B. Leibe et al. [4] propose a segmentation approach for object recognition that integrates multiple visual cues and combines multiple interest region detectors. Multiple algorithms integration has been proved to be an effective way for increasing robustness, and some theoretical researches have been performed. For example, M. Spengler et al. [5] propose a common framework for visual tracking that can support the integration of multiple cues. These researches can help to design robust approach for some special purposes such as tracking, but deeper researches are still needed to be performed about how to design a robust image understanding hybrid approach for common purpose with real-time constraint.

This paper is organized as follow, firstly a common multiple algorithm integration model is proposed, which can support common purpose such as segmentation, classifying, tracking and etc. Then some strategies are introduced about how to design a robust and real-time hybrid approach base on the model, and design strategies and executing strategies are also analyzed about increasing robustness and speeding up. Then architecture for typical applications is introduced, which can help design a total solution for image understanding task. At last a hybrid approach for rear vehicle detection and tracking is introduced as a sample.

2 Hybrid Model and Architecture

2.1 Hybrid Model

The kernel of multiple algorithm integration is to dynamically select and execute appropriate algorithm sequence with expected performance and executing time. So it is important to analyze those factors that can affect algorithm sequence. Performance and executing time of algorithm usually are related with input data, so input data is the first important factor. Besides, computing power is another important factor. When there is abundance computing power, all algorithms could be executed to get expected information as robust as possible in real-time. But when computing power is limited it is difficult to archive robustness and real-time at the same time, a tradeoff has to been made to determine the weightiness for robustness and real-time. So the third factor is executing strategies which implement the tradeoff. According to above analysis, a multiple algorithm integration model can be given and we call it "HYBRID". As in Fig.1, the model includes follow parts: "Algorithm Pool", "Scheduler", "Evaluator" and "Fuser", and real arrows indicate data flow and dash arrows indicate control flow.

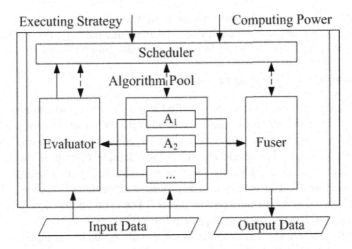

Fig. 1. Multiple Algorithm Integration Model - "HYBRID"

Algorithm Pool includes a set of different algorithms which can provide same output information and each algorithm has a description about its performance and computing workload characteristics. In most cases, quality of output information could be as a usual description of performance characteristic of algorithm, and computing cost could be a usual description of computing workload characteristic. These descriptions may be express in different forms, for example, the average detection rate of segmentation algorithm in some illumination condition may be as the expression of quality of its output information, and average executing time may be as the expression of its computing cost. And this information can be acquired from statistic.

Evaluator is in charge of providing the evaluation values of performance and computing workload characteristics in current input data. For example, to above segmentation algorithms, the evaluation value could be given according to current illumination condition which could be calculated base on current input data.

Fuser is in charge of fusing output information of multiple algorithms to give final output. And most technologies about multi-sensor data fusion and multi-cues integration could be applied very well for this procedure.

Scheduler is in charge of scheduling algorithms according to current available computing power, executing strategy and evaluation values of performance and computing workload characteristics of each algorithm. Usually computing power indicates available executing time of hybrid approach and executing strategy indicates the priority of performance and computing workload requirement, and special executing strategy will be implemented with special scheduling method.

The main advantage of HYBRID model is that it can support common image understanding task. On the one hand, the algorithms of "Algorithm Pool" may be detector, classifier and tracker, so the model can also support different purposes such as detection, classification, tracking and etc. On the other hand, when taking visual cue observation procedures as algorithms of "Algorithm Pool", HYBRID procedure is equivalent to multiple visual cues integration. And when taking information acquiring procedures from sensors as algorithms of "Algorithm Pool", HYBRID procedure is equivalent to multiple sensor data fusion. It shows the model can support multi-cues integration even multi-sensor data fusion. The analyzing indicates that HYBRID model can provide support for current main image understanding purposes.

2.2 Strategies for Hybrid Approach

One important aim of this research is to find some strategies that can help to design a robust and real-time approach base on above HYBRID model. For such an approach, strategies about increasing robustness can be analyzed base on three parts. The first part is "Algorithm Pool". For single algorithm, an adaptive algorithm such as adaptive SOBEL edge detector is helpful for robustness in complexity environment. The second part is "Scheduler". The selection of appropriate algorithms sequence is helpful for increasing robustness. The third part is "Fuser". The fusion of multiple visual cues and multiple sensor data can provide huge help for robustness.

Since the procedure of information fusion usually needs less computing power than image understanding algorithm, the strategies about speeding up hybrid approach will be analyzed base on only "Algorithm Pool" and "Scheduler". For single algorithm in Algorithm Pool, the speeding up can be implemented in two levels. The first level is software implementation of single algorithm. And high efficient implementation in special hardware is an important way to achieve real-time, such as high performance program base on data parallel technology. The second level is base on the reducing of computing complexity of single algorithm, multi-resolution technology is popular way to reduce complexity. For multiple algorithm integration in Scheduler, the determining of appropriate algorithms sequence is helpful for archive real-time.

Above strategies can be implemented in two phases: designing phase and executing phase. Strategy implementations in design phase are mainly focusing on choosing efficient algorithms for "Algorithm Pool". And strategy implementations in executing phase are mainly focusing on scheduling algorithms and fusing their output information. And this paper will mainly provide some analysis about algorithms scheduling method. In actual applications the scheduling strategies usually are various, but they still can be summarized into two patterns in the rough as follow.

2.2.1 Scheduling Pattern – Select-Do
This pattern usually has another name – Adaptive Pattern. This pattern means the scheduler needs to select and execute an algorithm sequence which can provide expected information with enough quality and with least computing cost. To reduce the difficulty of scheduling, this paper will consider only selecting one algorithm from "Algorithm Pool" as a sequence. And an available scheduling procedure is shown as Fig.2. In this procedure, selecting step will choose one algorithm that can provide output information with enough quality for current input and is with lowest computing cost. If there is not an available algorithm for current input, the robustness will not been archived. And when the executing time of selected algorithm is more than available executing time, real-time will not been achieved. A typical example of this pattern has been shown in article [6].

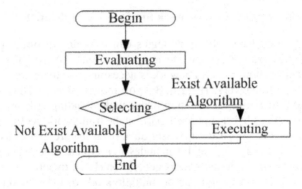

Fig. 2. Scheduling Pattern: Select-Do

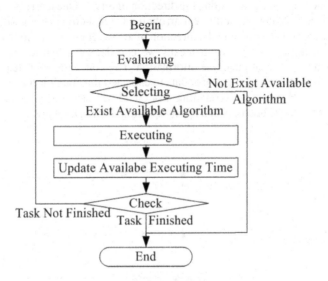

Fig. 3. Scheduling: Do-Till

2.2.2 Scheduling Pattern – Do-Till

This pattern usually means scheduler needs to select and execute a sequence of algorithms which can provide expected information with best quality in a time constraint. And an available scheduling procedure is shown as Fig.3.In this procedure, selecting step will select one algorithm whose expected executing time is less than current available executing time and with highest robustness. And the selecting step will be repeated till there is not available executing time. If there is not any available algorithm, the procedure will also be terminated. After the selected algorithm has been executed, its executing time should been subtract from available executing time. Then check step will judge if the special task has been finished. If the task has been finished, the procedure will be terminated. Two typical samples of this pattern have been shown in article [7] [8].

2.3   Hierarchical Hybrid Framework for Typical Application

The HYBRID model and above strategies can help design one approach for some purpose, but they are not enough for designing a total solution of special image understanding application. So a framework is necessary and there are a lot of researches that provide their frameworks [9] [10]. Base on above researches, this paper customizes a hierarchical framework for common image understanding task and give an analysis about how to design a hybrid approach base on the framework. As Fig.4, the framework includes six kinds of information (sensor data, data, feature, object hypotheses, object observation, object tracking), and it includes five typical processing procedures (preprocess, feature extraction, hypothesis generation (HG), hypothesis verification (HV), tracking), which can transform one or multiple kinds of information to another. One ellipse is a kind of information, and one a dash arrow indicates there is procedure that can implement the transform along the direction of arrow. One circle is a special case of some kind of information, and a real arrow indicates there is an algorithm which can implement the transform along the direction of arrow. If there is a circle towards which there are multiple real arrows, it means we can design a hybrid approach base on HYBRID model for acquiring the information. Then once the expected information is specified, if a data chain from sensor data to expected information could be constructed, a solution of this image understanding task could be determined. And a total solution could be enhanced robustness and speed up using above strategies.

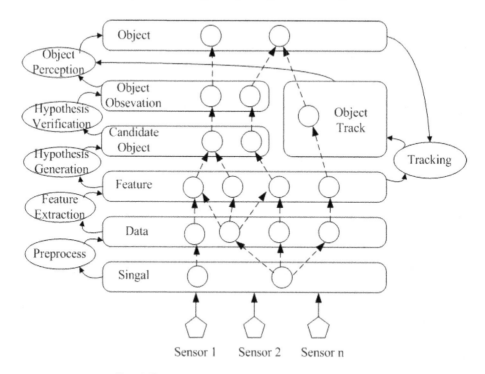

Fig. 4. Framework for Multiple Algorithm Integration

3 Hybrid Approach for Rear Vehicle Detection and Tracking

To validate above model and framework, an approach about rear vehicle and motor-cycle detection and tracking base on image understanding is developed, which need to provide robust result information in real-time (33frame/s) and in different situations such as day or night, sunshine or rain, etc. A part of our system has been shown in Fig 5. Shadow and wheel based hypothesis generation method is applied for daytime and headlight-based method is applied for night, and it is a sample of scheduling strategy pattern 1. And knowledge-based classifier is applied as coarse one and SVM classifier is applied as fine one, and it is a sample of scheduling strategy pattern 2. A real-time prototype of our approach has been finished which can process a frame (640*480) at an average processing time less than 33 ms (with NEC SIMD Computer IMAP-CAR) and the DR (Detection Rate) is more than 95% and FAR (False Detection Rate) is less than 5%. Some detail information about this approach can be fund in [11]. The test result shows the efficiency of multiple algorithms integration and validate our framework and HYBRID Model.

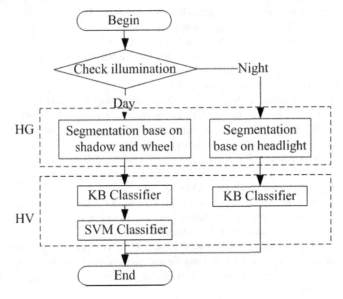

Fig. 5. A Part of the Our Vehicle and Motorcycle Recognition Approach

4 Conclusions

To increase robustness of image understanding approach, integrating multiple algo-rithms to make a special hybrid approach are becoming popular. But this aggravates the difficulties in achieving real-time because of heavy computational workload of multi-ple image understanding algorithms. Some theoretical researches about how to design a robust and real-time hybrid approach are necessary.

This paper presents a common multiple algorithm integration model and analyzes the strategies and scheduling method for increasing robustness and speeding up. Then architecture for typical applications is introduced and it can help to design a total image understanding approach for special application. At last the prototype of rear vehicle and motorcycle detection and tracking validate efficiency of the framework and HYBRID model for achieving robustness and real-time.

Research about multiple algorithms integration can help people to apply image understanding in real applications more easily and it is necessary to do deeper researches. Our next step is to validate them using a multi-sensor system.

References

1. Sun, Z., Bebis, G., Miller, R.: On-road vehicle detection using optical sensors a review. In: IEEE International Conference on Intelligent Transportation Systems, pp. 585–590. IEEE Press, New York (2004)
2. Heidemann, G., Kummert, F., Ritter, H., Sagerer, G.: A Hybrid Object Recognition Architecture. In: Vorbrüggen, J.C., von Seelen, W., Sendhoff, B. (eds.) ICANN 1996. LNCS, vol. 1112, pp. 305–310. Springer, Heidelberg (1996)
3. Tay, Y.H., Khalid, M., Yusof, R., Viard-Gaudin, C.: Offline Cursive Handwriting Recognition System base on Hybrid Markov Pattern and Neural Networks. In: IEEE International Symposium on Computational Intelligence in Robotics and Automation, vol. 3, pp. 1190–1195. IEEE Press, New York (2003)
4. Leibe, B., Mikolajczyk, K., Schiele, B.: Segmentation based multi-cue integration for object detection. In: 17th British Machine Vision Conference, vol. 3, p. 1169. BMVA Press (2006)
5. Spengler, M., Schiele, B.: Towards Robust Multi-Cue Integration fot Visual Tracking. International Journal of Machine Vision and Applications 14, 50–58 (2003)
6. Cao, X., Balakrishnan, R.: Evaluation of an On-line Adaptive Gesture Interface with Command Prediction. In: Graphics Interface Conference, pp. 187–194. Canadian Human-Computer Communications Society (2005)
7. Teng, L., Jin, L.W.: Hybrid Recognition for One Stroke Style Cursive Handwriting Characters. In: 8th International Conference on Document Analysis and Recognition, pp. 232–236. IEEE Computer Press, Los Alamitos (2005)
8. Koerich, A.L., Leydier, Y., Sabourin, R., Suen, C.Y.: A Hybrid Large Vocabulary Handwritten Word Recognition System using Neural Networks with Hidden Markov Patterns. In: 8th International Workshop on Frontiers in Handwriting Recognition, p. 99. IEEE Press, New York (2002)
9. Hall, D.L.: Mathematical Techniques in Multisensor Data Fusion. Artech House, Boston (2002)
10. Scheunert, U., Lindner, P., Richter, E., Tatschke, T., Schestauber, D., Fuchs, E.: Early and Multi Level Fusion for Reliable Automotive Safety Systems. In: IEEE Intelligent Vehicles Symposium, pp. 196–201. IEEE Press, New York (2007)
11. Duan, B., Liu, W., Fu, P., Yang, C., Wen, X., Yuan, H.: Real-Time On-Road Vehicle and Motorcycle Detection Using a Single Camera. In: IEEE International Conference on Industrial Technology, pp. 1–6. IEEE Press, New York (2009)

Formalizing the Modeling Process of Physical Systems in MBD

Nan Wang[1,2], Dantong OuYang[1,3,*], Shanwu Sun[2], and Chengli Zhao[1,2]

[1] Department of Computer Sciences and Technology, Jilin University, Jilin, Changchun 130012
[2] Department of Information, Changchun Taxation College, Jilin, Changchun, 130117
[3] Key Laboratory of Symbol Computation and Knowledge Engineering of the Ministry of Education, Jilin, Changchun 130012

Abstract. Many researchers have proposed several theories to capture the essence of abstraction. The G-KRA model(Genera KRA model), based on the GRA model which offers a framework R to represent the world W where a set of generic abstraction operators allows abstraction to be automated, can represent the world from different abstraction granularity. This paper shows how to model a physical system in model-based diagnosis within the G-KRA model framework using various kinds of knowledge. It investigates, with the generic theory of abstraction, how to automatically generate different knowledge models of the same system. The present work formalizes the process of constructing an abstract model of the considered system (e.g., using functional knowledge) based on the fundamental model and abstract objects database and expects that formalizing the modeling process of physical systems in MBD within the G-KRA framework will open the way to explore richer and better founded kinds of abstraction to apply to the MBD task.

Keywords: G-KRA Model, Knowledge Modeling, Model-Based Diagnosis.

1 Introduction

Abstraction is an essential activity in human perception and reasoning. In the Artificial Intelligence community, abstraction has been investigated mostly in problem solving [1], planning [2], diagnosis [3], and problem reformulation [4]. In general, the proposed approaches differ either in the kind of abstractions they use or in how they formally represent them. They all fail to characterize the practical aspects of the abstraction process. Saitta and Zucker [5] have proposed a model of representation change that includes both syntactic reformulation and abstraction. The model, called KRA (Knowledge Reformulation and Abstraction), is designed to help both the conceptualization phase of a problem and the automatic application of abstraction operators. Compared to the KRA model, the G-KRA model(Genera KRA model) [6] is more general and flexible to represent the world. It can represent the world from different abstraction granularity.

* Corresponding author: ouyangdantong@163.com

H. Deng et al. (Eds.): AICI 2009, LNAI 5855, pp. 685–695, 2009.

The major efforts in model-based reasoning have concentrated on paradigms for qualitative modeling and qualitative reasoning about physical systems [7]. Some approaches have exploited structural and behavioral knowledge in order to support a variety of tasks, such as prediction, diagnosis [8,9], and so on. Some researchers focused on theories of qualitative reasoning based on the use of knowledge about functions and goals [10]. At the same time, the issue of cooperation of multiple models of the same system has been causing increasing attention in order to improve the effectiveness and efficiency of reasoning processes. These approaches failed to investigate the relations between proposed methods and general theories of abstraction.

In this paper, we simply introduce G-KRA model which is based on KRA model and extends it. We show how to model a physical system in model-based diagnosis within the G-KRA model framework using various kinds of knowledge. We investigate with the generic theory of abstraction how to automatically generate different knowledge models of the same system. We expect that formalizing the modeling process of physical systems in MBD within the G-KRA framework will open the way to explore richer and better founded kinds of abstraction to apply to the MBD task.

2   The G-KRA Model Applied to Knowledge Modeling in MBD

The modeling process within the G-KRA model can be divided into two correlative phases: fundamental modeling and abstract modeling. The former one uses the fundamental knowledge, i.e. structural and behavioral knowledge, to derive the fundamental model of the system based on some kind of ontology and representational assumption which is relatively simple and intuitive. While the abstract modeling process automatically constructs a(some) more abstract model(models) for the same system utilizing other knowledge such as functional knowledge, teleological knowledge.

2.1   The General KRA Model of Abstraction

In this section we simply introduce the definition of G-KRA abstraction model, see [6] for details.

Definition 1. A primary perception P is a 5-ple, P = (OBJ,ATT,FUNC,REL,OBS) where OBJ contains the types of objects considered in W, ATT denotes the types of attributes of the objects, FUNC specifies a set of functions and REL is a set of relations among object types.

Definition 2. Let P be a primary perception, A be an agent (who perceives W) and O_a be a database. The objects with some abstract types in O_a are predefined by A. An abstract perception of A is defined as $P^*=\delta_a(P,O_a)$, where δ_a denotes an abstract perception mapping.

Definition 3. Given a primary perception P, an abstract objects database O_a and an abstract perception mapping δ_a, a general representation framework $\mathbf{R}^*$ is a 4-ple (P^*,D^*,L^*,T^*), where $P^*=\delta_a(P,O_a)$ is an abstract perception, D^* is a database, L^* represents a language and T^* specifies a theory.

2.2 Fundamental Modeling

The primary perception can be taken as the fundamental model of W. Firstly we are supposed to define the ontology of the model and identify what to represent of the real system in the model according to the knowledge involved above. There exists representational links and ontological links between the different fundamental models of the same world [11]. Then we can choose some appropriate kinds of structural and behavioral knowledge to build the primary perception P within the framework R and moreover construct the fundamental model of W as shown in figure 1 where L designates the union of representational links L_1 and ontological links L_2.

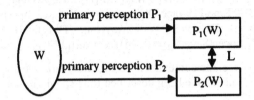

Fig. 1. Fundamental Modeling

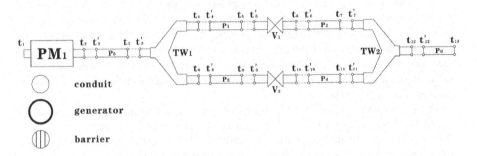

Fig. 2. Hydraulic system used as an example

In this section we take the diagnosis task as an example to show how to build up the fundamental model of a particular physical system. We will consider the hydraulic system reported in Figure 2, which is the same used in [12].

We here perceive four types of components, i.e., pipes, valves, three-way nodes, and pumps. Moreover we treat ports that connect components with each other (internal ports), or connect components to the environment(external ports), as the kind of entities that the model must designate.

We assume the state of the valves has been given that V_1 is open while V_2 is closed without considering the commands set on the valves. The volumetric pump PM_1, that we assume is active, is delivering a constant flow equal to F_k. We choose a qualitative behavioral representation of the hydraulic system used in the example instead of its quantitative version.

The possible behaviors of pipes, valves, pumps and three-ways are reported in [12]. For example, pipes have two behavioral modes: $OK(F_{in}=F_{out})$ and $LK(F_{in}>F_{out})$.

Now we show how to construct the fundamental model of the system in Figure 2. We use the representation in [13] with a few different details.

-OBJ=COMP ∪ {PORT},COMP={PUMP, PIPE, VALVE, THREE-WAY}
-ATT={ObjType: OBJ->{pipe,pump,valve,three-way,port},
Direction: PORT->{in,out},THREE-WAY->{2wayOut,2wayIn},
Observable: PORT->{yes, no}, State : VALVE->{open,closed}}
-FUNC={Bpump : PUMP->{ok,uf,of,lk},Bpipe: PIPE->{ok,lk},
Bvalve: VALVE->{ok,so,sc},Bthree-way: THREE-WAY->{ok}}
-REL'={port-of ⊆ PORT×COMP,connected ⊆ PORT×PORT}
-OBS'={(PM₁,P₁,...,P₆,V₁,V₂,TW₁,TW₂,t₁,...,t₁₃,t₂',...,t₁₂),(ObjType(PM₁)=pump,Ob-jType(P₁)=pipe,...,ObjType(P₆)=pipe,ObjType(V₁)=valve,...,ObjType(TW₁)=three-way, ...,ObjType(t₁)=port,...),(Direction(TW₁)=2wayOut,...,Direction(t₁)=in,...),(Observable(t₁)=yes,...),(State(V₁)=open,State(V₂)=closed),(port-of(t₁,PM₁),...),(connected(t₂,t₂'),...)}

For the sake of space, the contents of the structure/database D, the logical language L and the theory T, which have been described in [13], will not be provided here.

2.3 Abstract Modeling

Two or more objects perceived in the fundamental modeling process may have the same nature from some perspective so that we can take them individually as one more abstract object. It is an automatically matching process, before which the abstract objects data-base must be constructed manually directed by some kind of subject knowledge and at the same time it is important to design abstract mapping in order to automatically op-erate and then generate the more abstract model of the same world appropriate for some particular reasoning tasks.

In the following section we will formalize the abstract modeling process within the G-KRA model based on the fundamental model of the same system built up before. We will give the formal definition of Abstract Objects Database and Abstract Map-ping and take the functional knowledge for example to automatically construct the model of functional roles which is based on an object-centered ontology.

2.4 Abstract Objects Database

We will represent the abstract objects using the concept of the framework R=(P,D,L,T) proposed by Saitta and Zucker. We define P as abstract objects perception, denoted by AOP, where the observation OBS has not been involved as it is not a particular system that we perceive but a set of abstract objects. We will also introduce some assumptive object representation for characterizing the behavior of the abstract objects. The rest of the framework R will be automatically regulated to AOD, AOL, and AOT to form a new framework AOR of different meaning with R.

Definition 4. An abstract objects perception is a 4-ple AOP =(A_OBJ,A_ATT, A_FUNC,A_REL), where A_OBJ contains the types of abstract objects predefined based not on fundamental knowledge, A_ATT denotes the types of attributes of the abstract objects, A_FUNC specifies a set of functions, and A_REL is a set of relations among object types. These sets can be expressed as follows:

-A_OBJ = {A_TYPE$_i$|1 ≤ i ≤ N}
- A_ATT= {A_A$_j$: A_TYPE$_j$ → Λj|1 ≤ j ≤ M}
- A_FUNC= {A_f$_k$: A_TYPE$_{ik}$ × A_TYPE$_{jk}$ × ... → C$_k$|1 ≤ k ≤ S}
- A_REL= {A_r$_h$ ⊆ A_TYPE$_{ih}$ × A_TYPE$_{jh}$ |1 ≤ h ≤ R}

The physical system comprises many individual components which have also been assigned abstract roles in the functional role model. The functional role of a component is an interpretation of the equations describing its behavior, aimed at characterizing how the component contributes to the realization of the physical processes in which it takes part. Here we introduce three kinds abstract functional objects involved in [14]: conduit, which enables a generalized flow (effort, energy or information and so on) from one point to another in the structure of a system; generator, which causes a generalized flow from one point to another in the structure of a system; barrier, which prevents a generalized flow from one point to another in the structure of a system.

Let AOP=(A_OBJ, A_ATT, A_FUNC, A_REL) be the abstract objects perception based on functional knowledge specified as follows:

- A_OBJ= A_COMP ∪ A_PORT ∪ ... where A_COMP=CONDUIT ∪ GENERATOR ∪ BARRIER
    - A_ATT={ A_ObjType: A_OBJ→{conduit,generator,barrier,port},
    A_Direction: A_PORT→{in,out},A_MeasureType: A_PORT→{flow,pressure}}
    - A_FUNC={ A_F$_c$: CONDUIT→{func$_c$},A_F$_g$: GENERATOR→{func$_g$},
        A_F$_b$: BARRIER→{func$_b$}}
    - A_REL={ A_portof ⊆ A_PORT×A_COMP,
    A_connected ⊆ A_PORT×A_PORT }

Going back to the representation framework AOR, we will define three instantiated abstract objects c(CONDUIT), g(GENERATOR) and b(BARRIER) and introduce several ports(A_PORT) $p_1,p_2,...$ to characterize the database AOD which contains the tables described as follows.

-TableAObj=(A_obj, A_objtype, A_direction, A_measuretype), which describes abstract components and their attributes.
-TableAFunc=(A_obj, A_func, A_funcmode), which describes the functional modes of the abstract components.
-TableAPortOf=(A_port, A_comp), which describes what ports is attached to the abstract components.

In the table TableAObj some of the entries can be set to N/A (not applicable), as not all attributes are meaningful for all abstract objects. For the sake of simplification, we will not discuss all of the generalized flow types but assume the measure type of the ports is flow. We list the contents of the three tables as follows.

The definition of AOL and AOT are similar to L and T in the framework R so that we can, for example, describe the structure relating a conduit using AOL:

conduit(c,p_1,p_2) ⟺ A_comp(c)∧A_port(p_1) ∧A_portof(p_1,c) ∧in(p_1) ∧flow(p_1)
∧A_port(p_2) ∧A_portof(p_2,c) ∧out(p_2) ∧flow(p_2)

Table 1. TableAObj

A_obj	A_objtype	A_direction	A_measuretype
c	Conduit	N/A	N/A
g	Generator	N/A	N/A
b	Barrier	N/A	N/A
p1	Ports	in	flow
p2	Ports	out	flow
...	...	...	...

Table 2. TableAFun

A_obj	A_func	A_funmode
c	$func_c$	Pin=Pout
g	$func_g$	Pin=Pout=v≠0
b	$func_b$	Pin=Pout=0

Also the functional modes described with equations in Table 3 can be rewritten in logical terms in the theory. For instance: conduit(c,p_1,p_2) $\wedge$ $func_c$(c)$\rightarrow$ $\triangle$ FlowValue(p_1,p_2)=0.

Table 3. TableAPortOf

A_port	A_comp	A_port	A_comp
p1	c	p4	g
p2	c	p5	b
p3	g	p6	b

Structure Matching Procedure, SMP for short
We say the concrete component is matching the abstract one on structure when they have exactly the same ports, including the number of input ports, the number of output ports and the measure type of each port. Hence before identifying the corresponding abstract object in the abstract objects database for some concrete component perceived in the process of fundamental modeling, we should firstly make structural matching for them. We can describe the matching procedure as follows.

Behavior Matching Procedure, BMP for short
The normal behavior of the considered component exhibits its function to the whole system. For instance, when the pipes behave normally they can transport the flow from one point to another just like the functional role CONDUIT works. So we need only to see if the normal behavior of the component matches the functional mode of some given abstract object. We can describe the matching procedure as follows: (we assume either the behaviors of concrete components or the functional modes of the abstract objects are represented by qualitative equations.

Procedure SMP COMP c_1, A_COMP c_2
    in_1=the set of all in ports of c_1;
    in_2=the set of all in ports of c_2;
    out_1=the set of all out ports of c_1;
    out_2=the set of all out ports of c_2;
    If (|in_1|<>|in_2| or |out_1|<>|out_2|) return FAIL;
    If the types of in_1 and in_2 can't completely match, return FAIL;
    If the types of out_1 and out_2 can't completely match, return FAIL;
    return SUCCEED;

Fig. 3. Structure Matching Procedure

Procedure BMP COMP c_1, A_COMP c_2
    IN={$in_1,in_2,...,in_m$}: the input ports of c1(or c_2) individually;
    OUT={$out_1,out_2,...,out_n$}: the output ports of c_1(or c_2) individually;
    E={$E_1,E_2,...,E_j$}: the equations representing c_1's normal behaviors;
    F={$F_1,F_2,...,F_k$}: the equations representing c_2's functional modes;
    Put the elements of IN and OUT to the corresponding equation of E and F, and receive the
new sets E' and F';
    If E' completely matches F', return SUCCEED;
    return FAIL;

Fig. 4. Behavior Matching Procedure

Procedure CRP ObjType $type_1$, State S, A_ObjType $type_2$
//replace the components having type *type*$_1$ and state *S* with the abstract object having abstract
type *type*$_2$ in the considered system
    IF S is arbitrary {
        OBJ'=COMP' $\cup$ PORT $\cup$ ... where COMP'=COMP-$type_1$ $\cup$ $type_2$;
        ATT'=ATT-{ATTname: type2$\rightarrow$ $\wedge_j$}-{ObjType} $\cup$ { A_ObjType : COMP' $\rightarrow$
$\wedge_{ObjType}$-{$type_1$} $\cup$ {$type_2$}}
        FUNC'=FUNC-{Btype1: type1$\rightarrow c_k$} $\cup$ {A_F_{type2}: type2$\rightarrow func_{type2}$}
        REL'=Keep REL unchanged except for replacing COMP in the relation of REL with
A_COMP.
        Modify OBS to OBS' based on OBJ',ATT', FUNC',and REL';
        P'={OBJ',ATT', FUNC', REL', OBS'};
        Modify the contents of D, L, and T accordingly ,based on P', AOD, AOL,and AOT , to
D', L' and T';
    }

Fig. 5. Components Replacement Procedure

• *Components Replacement Procedure, CRP for short*
After succeeding the matching in structure and behavior we think of both the considered component and all other ones of the same type (sometimes with the same states,

like valves) are functionally equivalent to the given abstract object. So then we can replace the components with the matched abstract type object through the procedure CRP described as follows:

We have only considered in Procedure CRP the situation that the state of components can be ignored (like pipes). While dealing with the components whose state needs to be discussed we must do more works in the procedure. On the one hand, we are not allowed to delete the object type $type_1$ in the more abstract perception, because not all components with type $type_1$ stay in state S. Only, the new type $type_2$ is added. On the other hand, the replacement happens only on the components in state S, so we should check the state of the considered component before we modify it.

Execute the three procedures repeatedly until there is no component matching any abstract object of the abstract objects database, and then we can generate a more abstract model based on functional knowledge with less object types in it. We use the procedure GAMP (from Generate Abstract Model Procedure) to describe the process.

Procedure GAMP

Add a flag ENABLE(initially ENABLE=YES) to each component of the considered system;

//ENABLE=YES represents the component has not been checked or replaced, namely, it can be dealt with. Whereas if ENABLE=NO, then it indicates either the component has been replaced or it does not match any abstract object of the given abstract database.

While there exist components unsettled, do{
    c=any component whose ENABLE value is YES;
    $type_1$=ObjType(c); S=State(c);
    For each object o_j of the given abstract database, do{
      if (SMP(c, o_j)=SUCCEED)
      if(BMP(c, o_j)=SUCCEED) {
         $type_2$=A_ObjType(o_j);
         CRP ($type_1$, State S, $type_2$);
           Set ENABLE=NO to all the components of type $type_1$ and state S;
      }
    Set ENABLE=NO to all the components of type $type_1$ and state S;
    }
  }

Fig. 6. Generate Abstract Model Procedure

In the rest of this section we give an example based on the hydraulic system described in figure 2 to show how to automatically construct the abstract model from the fundamental model using functional knowledge.

According to the procedure GAMP, we compare any component c unsettled to each object of the abstract objects database to find an abstract object completely matching it in both structure and behavior and then replace all components of the same type as c in the considered system with the abstract object. We choose the components in such an order as follows: $PM_1, P_5, TW_1, P_1, V_1, P_2, P_3, V_2, P_4, TW_2, P_6$.

·PM₁ and conduit

·PM, and conduit
PM1 and conduit can be seen as the same one on structure, namely, SMP(PM₁,conduit)=SUCCEED, because they both have two ports (one input port and one output port) and the two ports allow the same measure type. However, when comparing the PM₁'s normal behavior to the conduit's functional mode, we find the equation $F_{in}=F_{out}=F_k\neq0$(the normal mode of pumps) is not completely equivalent to $P_{in}=P_{out}$, i.e., BMP(PM₁,conduit)=FAIL, so we say PM₁ is not completely matching conduit.

Here we stress two completely matching components to insure to avoid choosing the inappropriate abstract object to do the replacing. For example, an active pump requires the input flow value be equal to the output flow value and they both should be a nonzero number which means an active pump delivers a constant flow equal to F_K ($F_K\neq0$). While the function of a conduit of the abstract type CONDUIT is to enable a generalized flow from one point to another, namely, the output flow value should be equal to the input flow value which could be zero. So obviously matching the PUMP type component with the CONDUIT type object, we say, is incomplete that will lead to an incorrect abstract model based on functional knowledge.

·PM₁ and generator

·PM₁ and generator
Similar to the process of comparing PM₁ to conduit, we invoke SMP(PM₁,generator) and BMP(PM₁,generator) individually and they both receive SUCCEED which means PM1 completely matches the abstract object generator. Then we can replace all the active pumps of the system with generators.

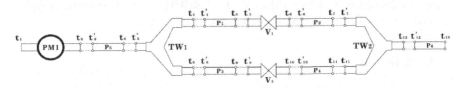

Fig. 7. Replace all the active pumps of the system with generators

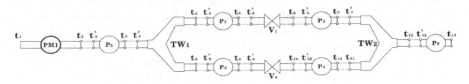

Fig. 8. Replace all pipes with conduits

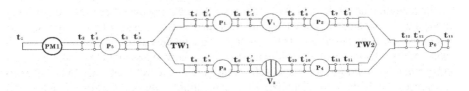

Fig. 9. Replace all open valves with conduits and closed valves with barriers

-P'=(OBJ',ATT',FUNC',REL',OBS') is presented as follows:

-OBJ'=COMP' ∪ {PORT},COMP'={**GENERATOR,   PIPE,   VALVE,** **THREE-WAY**}

-ATT'={ObjType: OBJ->{pipe,**generator**,valve,three-way,port},

Direction: PORT->{in,out},THREE-WAY->{2wayOut,2wayIn},

Observable: PORT->{yes, no},   State : VALVE->{open,closed}}

-FUNC'={ **A_F$_g$: GENERATOR→{func$_g$}**,

Bpipe: PIPE->{ok,lk},Bvalve: VALVE->{ok,so,sc},

Bthree-way: THREE-WAY->{ok}}

-REL'={port-of ⊆ PORT×COMP',connected ⊆ PORT×PORT}

-OBS'={(PM$_1$,P$_1$,...,P$_6$,V$_1$,V$_2$,TW$_1$,TW$_2$,t$_1$,...,t$_{13}$,t$_2$',...,t$_{12}$),(**ObjType(PM$_1$)=gener ator**,ObjType(P$_1$)=pipe,...,ObjType(P$_6$)=pipe,ObjType(V$_1$)=valve,...,ObjType(TW$_1$) =three-way,...,ObjType(t$_1$)=port,...),(Direction(TW$_1$)=2wayOut,...,Direction(t$_1$)=in, ...),(Observable(t$_1$)=yes,...),(State(V$_1$)=open,State(V$_2$)=closed),(port-of(t$_1$,PM$_1$),...), (connected(t$_2$,t$_2$'),...)}

We adopt the representation to the three functional roles of [14] to show the transformed system described as Figure 8.

The next replaced components are pipes(P$_1$∼P$_6$), namely replacing all pipes with the abstract objects of the type CONDUIT, described in Figure 9.

TW$_1$ and TW$_2$ can't match any object of the abstract objects database, so they are not replaced and there exists still the type of three-ways.

The last replaced components are valves(V$_1$ staying open and V$_2$ closed). V$_1$ is replaced with an abstract object with type of CONDUIT while V$_2$ is replaced with an abstract object with type of BARRIER described in Figure 10.

3   Conclusion

Many researchers have explored how to use relevant knowledge to construct the models of physical systems in MBD, for instance, structural and behavioral knowledge(e.g.,[9]), functional and teleological knowledge(e.g.,[14]), multimodeling approach(e.g.,[11]). In recent years, new approaches have been proposed in MBD to exploit the degree of observability for defining useful abstractions (e.g., [12],[15]). While these approaches have developed some interesting solutions to the problem of abstracting models for MBD, they have failed to investigate the relations between proposed methods and general theories of abstraction. The present work represents a step in the direction of filling this gap. In particular, the present paper introduces the extended KRA model, i.e., G-KRA model, which can represent the world more general and flexible. We have shown how to automatically construct an abstract model based on the fundamental model which contributes to formalizing the process of transforming models with fundamental knowledge to other models with abstract knowledge, such as structural and behavioral models to functional role models described above. There still exist some problems we will explore in future, e.g., how to deal with the situation that a component has multiple roles about different domains to automatically build up multiabstract models; how to formalizing the realization of the links of two fundamental models with different ontologies or representation assumption, and so on.

Acknowledgment

The authors are grateful to the support of NSFC Major Research Program under Grant Nos. 60496320 and 60496321, Basic Theory and Core Techniques of Non Canonical Knowledge; NSFC under Grant Nos. 60773097 and 60873148, Program for New Century Excellent Talents in University; Jilin Provine Science and Technology Development Plan under Grant Nos. 20060532 and 20080107, and European Commission under Grant No. TH/Asia Link/010 (111084).

References

1. Holte, R., Mkadmi, T., Zimmer, R., MacDonald, A.: Speeding up problem-solving by abstraction: A graph-oriented approach. J. Art. Intelligence 85, 321–361 (1996)
2. Knoblock, C., Tenenberg, J., Qiang, Y.: A spectrum of abstraction hierarchies for planning. In: Proc. AAAI WS on AGAA, pp. 24–35 (1990)
3. Mozetic, I.: Hierarchical model-based diagnosis. J. Int. Journal of Man-Machine Studies 35(3), 329–362 (1991)
4. Subramanian, D.: Automation of abstractions and approximations: Some challenges. In: Proc. AAAI WS on AGAA, pp. 76–77 (1990)
5. Saitta, L., Zucker, J.: Semantic abstraction for concept representation and learning. In: Proc. SARA, pp. 103–120 (1998)
6. Shan-wu, S., Nan, W., Dan-tong, O.Y.: General KRA Abstraction Model. J. Journal of Jilin University (Science Edition) 47(3), 537–542 (2009)
7. Weld, D., De Kleer, J.: Readings in Qualitative Reasoning about Physical Systems. Morgan Kaufmann, San Mateo (1990)
8. Bobrow, D.G. (ed.): Special Volume on Qualitative Reasoning about Physical Systems. J. Artificial Intell. 24 (1984)
9. Davis, R.: Diagnostic reasoning based on structure and behavior. J. Artificial Intelligence 24, 347–410 (1984)
10. Sticklen, J., Bond, E.: Functional reasoning and functional modeling. IEEE Expert 6(2), 20–21 (1991)
11. Chittaro, L., Guida, G., Tasso, C., Toppano, E.: Functional and teleological knowledge in the multimodeling approach for reasoning about physical system:a case study in diagnosis. IEEE Trans. Syst. Man, Cybern. 23(6), 1718–1751 (1993)
12. Chittaro, L., Ranon, R.: Hierarchical model-based diagnosis based on structural abstraction. Art. Intell. 155(1-2), 147–182 (2004)
13. Saitta, L., Torasso, P., Torta, G.: Formalizing the abstraction process in model-based diagnosis. In: Tr cs, vol. 34, Univ. of Torino, Italy (2006)
14. Chittaro, L., Ranon, R.: Diagnosis of multiple faults with flow-based functional models:the functional diagnosis with efforts and flows approach. Reliability Engineering and System Safety 64, 137–150 (1999)
15. Torta, G., Torasso, P.: A Symbolic Approach for Component Abstraction in Model-Based Diagnosis. In: Proceedings of the Model-Based Diagnosis International Workshop (2008)

Study on Stochastic Programming Methods Based on Synthesizing Effect

FaChao Li[1,2], XianLei Liu[2], and ChenXia Jin[1,2]

[1] School of Economics and Management, Hebei University of Science and Technology,
050018, Shijiazhuang, China
[2] School of Science, Hebei University of Science and Technology,
050018, Shijiazhuang, China
lifachao@tsinghua.org.cn, liuwoolong@163.com,
jinchenxia2005@126.com

Abstract. Stochastic programming is a well-known optimization problem in resource allocation, optimization decision etc. in this paper, by analyzing the essential characteristic of stochastic programming and the deficiencies of the existing methods, we propose the concept of synthesizing effect function for processing the objective function and constraints, and further we give an axiomatic system for synthesizing effect function. Finally, we establish a general solution model (denoted by BSE-SGM for short) based on synthesizing effect function for stochastic programming problem, and analyze the model through an example. All the results indicate that our method not only includes the existing methods for stochastic programming, but also effectively merge the decision preferences into the solution, so it can be widely used in many fields such as complicated system optimization and artificial intelligence etc.

Keywords: Stochastic programming, Stochastic decision-making, Stochastic effect, Synthesizing effect function, Mathematical expectation, Variance, Model.

1 Introduction

Randomness is a widespread phenomenon in the real world and is unavoidable in many practical fields. How to establish an effective and workable method to process random information is a widespread concern on production management, artificial intelligence, optimization of complex systems research, and the stochastic programming is generally involved in these research areas. As a random variable is a family of data that satisfy some laws and dose not have clear order, the current programming methods should not be directly applied to solving the problem of stochastic programming. At present, there are three basic methods to solve the stochastic programming problem: 1) Expectation model, the basic idea is to use mathematical expectation to describe random variables, then turn the stochastic programming into the general programming problem. 2) Chance-constrained model, the basic idea is to convert stochastic constraints and objective functions into ordinary constraints and objective functions through some reliability principles. 3) Dependent-chance programming model, the basic idea is to

H. Deng et al. (Eds.): AICI 2009, LNAI 5855, pp. 696–704, 2009.
© Springer-Verlag Berlin Heidelberg 2009

regard objective functions and constraints as events under random environment and solve the stochastic programming problem by maximizing the chances of all the events happening. These methods have achieved much good applications, such as, in [3], the authors studied measures program of oil field by using expectation model, and paper [4] use chance-constrained model to size batteries for distributed power system, then paper [5] use it to optimize allocation of harmonic filters on a distribution network. But they could not solve the stochastic programming problem under complicated environment effectively, and the deficiencies mostly are: 1) it is hard to use expectations to describe and represent random variables effectively and it is difficult to ensure the reliability of model when the randomness is larger by expectation model; 2) when the stochastic characteristics are complex (that is to say, it is difficult to determine the distribution of the stochastic environment), the computational complexities of the chance-constrained model and dependent-chance programming model are so large to get an analytic form of the program viable solution In response to these problems, many scholars from different perspectives studied stochastic programming problem, for example, papers[6-9] studied to solve chance-constrained model and dependent-chance programming model by using the stochastic simulation technique and genetic algorithm. But all of these methods have strongly points. Thus, so far, there are not systematic and effective stochastic programming methods.

Based on the analyses above, this paper analyses the basic characteristics of stochastic programming problems, combines with the inadequacy of existing methods, and have the following contributions: a) we propose the concept of synthesizing effect function for processing the objective function and constraints, and further we give some commonly-used models of synthesizing effect functions; b) we establish a general solution model (denoted by BSE-SGM for short) based on synthesizing effect function for stochastic programming problem; c) We analyze the characteristic of our model by an example, and the results indicate that our methods are effective.

In this paper, let $(\Omega, \mathcal{B}, P)$ be the probability space, and for every random variable ξ in the $(\Omega, \mathcal{B}, P)$, let $E(\xi)$ and $D(\xi)$ be the mathematical expectation and variation of the random variable ξ respectively.

2   Overview of the Methods to Solve the Stochastic Programming

Stochastic programming problem is the core content in the production management, resource allocation and other practical problems, its general form is:

$$\begin{cases} \max f(x, \xi), \\ \text{s.t. } g_i(x, \xi) \leq 0, i = 1, 2, \cdots, m. \end{cases} \tag{1}$$

Where $x = (x_1, x_2, \cdots, x_n)$ is the decision vector, and $\xi = (\xi_1, \xi_2, \cdots, \xi_n)$ is the given random variable vector in the probability space $(\Omega, \mathcal{B}, P)$, and $f(x, \xi)$, $g_j(x, \xi)$, are random variable functions, $j = 1, 2, \cdots, m$.

As there is no simple order between random variables and $g_j(x, \xi) \leq 0$ mostly could not be completely satisfied, model (1) is just a model and can not be solved directly. So, it is an essential way to solving stochastic programming problems that we can convert the stochastic programming problems to normal programming problems in some strategies.

2.1 Expectation Model

Mathematical expectation is a commonly used tool to describe the values of random variable intently. It has some good theory characters. If we use the mathematical expectation of the random variable to take place of the random variable, the model (1) becomes the following model (2):

$$\begin{cases} \max E(f(x, \xi)), \\ \text{s.t. } E(g_j(x, \xi)) \le 0, j = 1, 2, \cdots, m. \end{cases} \tag{2}$$

Generally, we call the model (2) the Expectation model [3].

When the variation of the random variable is larger, the mathematical expectation could not describe the variable effectively. So we could not get the optimum solution of the stochastic programming by using the model (2).

2.2 Chance-Constrained Mode [1]

For the problem that the constraint of the stochastic programming often can not satisfy absolutely, we can use the reliability to dispose of the constraints and objective functions of the programming. Then the model (1) becomes the following model (3):

$$\begin{cases} \max \bar{f}(x), \\ \text{s.t. } P(f(x, \xi) \ge \bar{f}(x)) \ge \alpha, \\ P(g_j(x, \xi) \le 0) \ge \alpha_j, j = 1, 2, \cdots, m. \end{cases} \tag{3}$$

Generally, we call the model (3) the chance-constrained model. In the model, $\alpha_j, \alpha \in [0, 1]$ are the reliabilities that the solutions satisfy constraints and objective functions, and $P(A)$ is the probability that the event A may happen. Compared to the model (2), this model could control the quality of the decision, but we don't know the range of α_j's value when the model (3) could be soluble.

As stochastic programming is an uncertain decision-making problem, the results of its decision-making in general should not be bound to make the relevant constraint established absolutely. So compared to bounding by the given reliability of constraints and objectives in advance, it is more suitable for the characters of integrated decision-making that considering the constraint satisfaction and the size characteristics of objective function. To establish a general solution model under this concept, we can synthesize the objective function value and constraint satisfaction together through some kind of strategy (called the synthesizing strategy the synthesizing effect function), then discuss the programming based on the synthesized value. Further we give the following random effects multi-attribute axiomatic system of synthesizing effect functions.

3 The Axiomatic System for Synthesizing Effect Function

As the minimum decision-making problems can be through some ways transformation into the maximum decision-making problems, in the following, we mainly discuss the synthesizing strategy of objective function and the satisfaction of constraints on the maximization stochastic programming problem with single objective (real function) and multi-constraints. According to the essential characteristic of the optimal decision, we should obey the following rules in seeking for the decision scheme of stochastic programming.

Principle 1. When the satisfaction degree of constraints is same, the greater the objective value function is, the better the effect is;

Principle 2. When the objective function value is same, the greater the satisfaction degree of constraints is, the better the effect is;

Principle 3. When the constraints are absolutely satisfied, the decision only depends on the objective value function;

Principle 4. When some constraints are absolutely dissatisfied, we can't make a decision.

If we abstractly regard the synthesizing strategy of objective function value and satisfaction degree of constraints as a function $S(u, v) = S(u, v_1, v_2, \cdots, v_m)$ (here, u is the objective function value with the conversion interval Θ; v_i is the satisfaction degree of the i-th constraint, with the conversion interval $[0, 1]$, that is, $S(u, v)$ is a map on $\Theta \times [0, 1]^m \to (-\infty, +\infty)$), then the above principles can be accordingly stated as follows:

Condition 1. For any given $u \in \Theta$, $S(u, v)$ is monotone non-decreasing on each v_i ;

Condition 2. For any given $v = (v_1, v_2, \cdots, v_n) \in [0, 1]^m$, $S(u, v)$ is monotone non-decreasing on each u ;

Condition 3. $S(u, 1, 1, \cdots, 1)$ is strictly monotone increasing;

Condition 4. When $\prod_{j=1}^n v_j = 0$, $S(u_1, v) = S(u_2, v)$ for any $u_1, u_2 \in \Theta$. Obviously, for uncertain decision problem, principles 1~3 are must be obeyed, while principle 4 can be loosed. For convenience, we call $S(u, v)$ satisfying conditions 1~3 synthesizing effect function on Θ, and $S(u, v)$ satisfying conditions 1~4 uniform synthesizing effect function on Θ.

Remark 1. Since the above four principles also must be obeyed for multi-objective decision, we can similarly establish the axiomatic system for synthesizing effect function for multi-objective programming, and for the above symbolic system, we only consider u as $(u_1, u_2, \cdots, u_m)$, change that $S(u, v)$ is monotone non-decreasing on u into that $S(u, v)$ is monotone non-decreasing on each u_i, these changes will not have essential effect for the results.

According to the above definition, we have the following conclusions:

① For any $a, b \in (-\infty, +\infty)$, $a < b$, $S(u, v) = (u-a)(b-a)^{-1} \wedge v_1 \wedge v_2 \wedge \cdots \wedge v_n$ is a uniform synthesizing effect function on $[a, b]$. Here, $\wedge$ is min operation of real numbers.

② For any $k \in (0, +\infty)$, $c \in [0, +\infty)$, $\alpha_j \in (0, +\infty)$, $S(u, v) = k(u+c)\Pi_{j=1}^{n} v_j^{\alpha_j}$ is a uniform synthesizing effect function on $[0, +\infty)$.

③ For any $\alpha \in (0, \infty)$, $\beta_j \in [1, \infty)$, $S(u, v) = u^{\alpha} \Pi_{j=1}^{n} v_j^{\beta_j}$ is a uniform synthesizing effect function on $[0, +\infty)$; $S(u, v) = \exp(\alpha u)\Pi_{j=1}^{n} v_j^{\beta_j}$ is a uniform synthesizing effect function on $(-\infty, +\infty)$.

④ For any $\alpha \in (0, \infty)$, $\lambda_j \in (0, 1]$, $j = 1, 2, \cdots, n$, $S(u, v) = u\Pi_{j=1}^{n} \delta(v_j - \lambda_j)$ is a uniform synthesizing effect function on $[0, +\infty)$; $S(u, v) = \exp(\alpha u) \times \Pi_{j=1}^{n} \delta(v_j - \lambda_j)$ is a uniform synthesizing effect function on $(-\infty, +\infty)$; $S(u, v) = \alpha u + \sum_{j=1}^{n} \eta(v_j - \lambda_j)$ is a uniform synthesizing effect function on $(-\infty, +\infty)$. Here, $\delta(t) = 0$ for each $t < 0$, $\delta(t) = 1$ for each $t \geq 0$, and $\eta(t) = -\infty$ for each $t < 0$, $\eta(t) = 1$ for each $t \geq 0$.

⑤ For any $a, b, a_j \in (0, +\infty)$, $j = 1, 2, \cdots, n$, $S(u, v) = bu + a\sum_{j=1}^{n} a_j v_j$ is a synthesizing effect function on $(-\infty, +\infty)$, but not a uniform synthesizing effect function.

4 The Stochastic Programming Model Based on Synthesizing Effect

Using the synthesizing strategy in section 3, for single-objective (real value function) and multi-attribute stochastic programming problem(1), if we regard u and v_j in $S(u, v_1, v_2, \cdots, v_n)$ as the concentrated quantizing value $C(f(x, \xi))$ of the objective function value $f(x, \xi)$ and the satisfaction degree $\beta_j(x) = \text{Sat}(g_j(x, \xi) \leq 0)$ of the i th constraint for the scheme x, then $S(C(f(x, \xi)), \beta_1, \beta_2, \cdots, \beta_m)$ considering both objective and constraint, is a quantitative descriptive model measuring the quality of the solution, and the model (1) can be converted into the following model (4):

$$\begin{cases} \max S(C(f(x, \xi)), \beta_1(x), \beta_2(x), \cdots, \beta_m(x)), \\ \text{s.t. } x \in X. \end{cases} \tag{4}$$

Here, $(\inf f(x), \sup f(x)) \subset \Theta$, $S(u, v_1, v_2, \cdots, v_n)$ is a (uniform) synthesizing effect function on Θ. For convenience, we call (4) the stochastic programming model based on synthesizing effect, denoted it by BSE-SGM for short.

Remark 2. If we use $E(f(x, \xi))$ to describe the size characteristics of $f(x, \xi)$ in concentrated form, and use $\beta_j(x) = \delta(E(-g_j(x, \xi)))$ to represent the satisfaction of $g_j(x, \xi) \leq 0$, when $S(u, v) = \exp(u) \cdot \eta(\Pi_{j=1}^{m} \delta(v_j - \alpha_j))$, the model (4) is the expectation model (2). Here, $\delta(t)$ satisfies: $\delta(t) = 0$ for each $t < 0$, $\delta(t) = 1$ for each $t \geq 0$; and $\eta(t)$ satisfies $\eta(0) = -\infty, \eta(1) = 1$.

Remark 3. If we use $\bar{f}(x)$ to describe the size characteristics of $f(x, \xi)$ in concentrated form, and use $\beta_j(x) = P(g_j(x, \xi)) \leq 0)$ to represent the satisfaction of $g_j(x, \xi) \leq 0$, when $S(u, v) = \exp(u) \cdot \eta(\prod_{j=1}^{m} \delta(v_j - \alpha_j))$ The model(4) is the chance-constrained model(3). Here, $\bar{f}(x)$ satisfies $P(f(x, \xi)) \geq \bar{f}(x)) \geq \alpha$, and $\delta(t)$ satisfies: when $t < 0$, $\delta(t) = 0$, and when $t \geq 0$, $\delta(t) = 1$; $\eta(t)$ satisfies that when $\eta(0) = -\infty$, $\eta(1) = 1$.

Remark 4. If we use $(E(f(x, \xi)), D(f(x, \xi)))$ to describe the size characteristic of $f(x, \xi)$ in a compound quantification mode, and use $T(E(f(x, \xi)), D(f(x, \xi)))$ to represent $C(f(x, \xi))$ (Here, $T(E(f(x, \xi)), D(f(x, \xi)))$ is one synthesis of $E(f(x, \xi))$ and $D(f(x, \xi))$. Obviously, $T(x, y)$ should satisfy that: 1) $T(s, t)$ monotone non-decreasing on each s; 2) $T(s, t)$ monotone non-increasing on each t), then when $S(u, v) = \exp(u) \cdot \eta(\prod_{j=1}^{m} \delta(v_j - \alpha_j))$, the model (4) is the following model (5) (Here, $\delta(t)$ satisfies that $\delta(t) = 0$ for each $t < 0$, and $\delta(t) = 1$ for each $t \geq 0$; $\eta(t)$ satisfies that $\eta(0) = -\infty$, $\eta(1) = 1$):

$$\begin{cases} \max T(E(f(x, \xi)), D(f(x, \xi))), \\ \text{s.t. } P(g_j(x, \xi) \leq 0) \geq a_j, j = 1, 2, \cdots, m. \end{cases} \tag{5}$$

Remark 5. The above analysis indicate that, model (4) includes the existing stochastic programming model, and also it has better structural characteristic and strong interpretation, therefore model (4) provides a theoretical platform for solving stochastic programming problem. For different problems, we can use different (uniform) synthesizing effect functions to embody and describe different decision consciousness.

5 Example Analysis

In this part, we will further analyze the characteristic of stochastic programming model (4) by an example.

Example 1. Consider the following programming problem.

$$\begin{cases} \max z = -2\xi_1 x_1^2 - \xi_2 x_2 - \xi_3 x_3 - \xi_4 \sin x_4, \\ \text{s.t. } \xi_1 x_1^2 + \xi_2 x_2 + \xi_3 \cos x_3 \leq 5, \\ x_1 x_2 x_3 x_4 \leq 4. \end{cases} \tag{6}$$

Here, ξ_1 is a random variable with the uniform distribution on interval $[2, 12]$; ξ_2 is a random variable with the normal distribution $N(5, 10)$; ξ_3 is a random variable with the normal distribution $N(8, 20)$; ξ_4 is a random variable with the exponential distribution that its parameters is 0.5, and $\xi_1, \xi_2, \xi_3, \xi_4$ are independent with each other.

In order to facilitate the analysis of the characteristics of BSE-SGM, we only solve the problem (6) by using the expectation model (2) and the BSE-SGM as follows.

I. By using the model (2), the problem (6) could be transformed into the general programming problem (7):

$$\begin{cases} \max z^* = -14x_1^2 - 5x_2 - 8x_3 - 2\sin x_4, \\ \text{s.t.} \ \ 7x_1^2 + 5x_2 + 8\cos x_3 - 5 \leq 0, \\ \qquad x_1 x_2 x_3 x_4 \leq 4. \end{cases} \tag{7}$$

II. By using BSE-SGM to solve the problem (6): its parameters settings are: 1) use $(E(z'), D(z'))$ to describe the size characteristic of z' in a compound quantification mode, and use $C(z') = E(z')[1 + k(D(z')^\alpha]^{-1}$ to represent the concentrated quantifying value of z'; 2) use $S(u, v) = uv^b$ to the stochastic synthesizing effect function (that is $S(C(z'), \beta) = E(z')[1 + k(D(z')^\alpha]^{-1}\beta^b)$, then problem (6) could be transformed into the following general programming problem (8):

$$\begin{cases} \max z^* = \dfrac{(-14x_1^2 - 5x_2 - 8x_3 - 2\sin x_4)\beta^b}{1 + k(\dfrac{100}{3}x_1^4 + 10x_2^2 + 20x_3^2 + 4\sin^2 x_4)^\alpha}, \\ \text{s.t.} \ \ \beta = P(\xi_1 x_1^2 + \xi_2 x_2 + \xi_3 \cos x_3 \leq 5), \\ \qquad x_1, x_2, x_3, x_4 \leq 4. \end{cases} \tag{8}$$

Obviously, (7) and (8) are nonlinear programming problems and they could not be solved easily by an analytical way. Genetic Algorithm is used here (its parameters set as follows: Code mode: binary; Mutation probability: 0.001; Crossover probability: 1; Population size: 80; Evolutionary generations: 100). Then the results of the problem under different methods are shown in Table 1 (Here, S.E.V. denotes the Synthesizing Effect Value).

The above analysis and computation results indicate: 1) The variations of the decision results for the same stochastic programming problem by using BSE-SGM are smaller than the one by using expectation model, which shows that BSE-SGM is much closer with the essence of the decision than expectation model; 2) For the different synthesizing effect functions, the decision results are different, and even the difference is great. Therefore, BSE-SGM can effectively merge uncertainty

Table 1. The results of solving the problem (6) under different methods

Solving Model		Optimal Solution	S.E.V.	Expectation	Variation	Reliability
Expectation Model		(-0.0098, -10.000, - 10.000, 4.7214)	——	131.9986	3004.0	0.7355
B S E - S G M	$k=0.1, a=0.5,$ $b=1$	(-0.0098, -9.218, - 8.2796, -7.4585)	14.1813	114.1707	2224.1.	0.7100
	$k=0.1, a=0.5,$ $b=2$	(- 0.0098, -9.6676, - 9.5308, 5.3079)	11.6090	126.2389	2754.1	0.7580
	$k=0.1, a=0.3, b$ $=2$	(- 0.0489, -9.9218, - 9.7458, 4.4282)	36.5984	129.4621	2887.7	0.7690
	$k=0.1, a=0.5, b$ $=3$	(0.0098, -0.2444, - 8.1427, -6.8133)	11.9354	67.3736	1327.7	0.9370
	$k=0.1, a=0.3, b$ $=3$	(-0.0098, -0.1466, - 7.9277, 5.0147)	35.6801	66.0624	1260.8	0.9970
	$k=0.1, a=0.5, b$ $=5$	(-0.0293, -9.9609, - 10.0000, 4.4868)	54.6945	131.7418	2996.0	0.7630

process consciousness into decision process; 3) BSE-SGM has good structural characteristics.

6 Conclusion

In this paper, for the solution of multi-attribute stochastic programming, by analyzing the deficiencies of the existing methods, we propose the concept of multi-attribute synthesizing effect function, give an axiomatic system for multi-attribute synthesizing effect function, and establish a general solution model for stochastic programming problem; we further analyze the characteristic of our model by an example. All the results indicate that the multi-attribute synthesizing effect function is an effective tool for processing decision preference, it can merge the processing thought of stochastic information into the quantitative operation process, and it has theoretical systematization and operational application.

Acknowledgements

This paper is supported by National Natural Science Foundation of China (70671034, 70871036, 70810107018) and the Natural Science Foundation of Hebei Province (F2006000346) and the Ph. D. Foundation of Hebei Province (05547004D-2, B2004509).

References

1. Charnes, A., Cooper, W.W.: Management Models and Industrial Applications of Linear Programming. Johhn Wiley & Sons Inc., New York (1961)
2. Liu, B.: Dependent-chance programming: A class of stochastic programming. Computers & Mathematics with Applications 34(12), 89–104 (1997)
3. Song, J.K., Zhang, Z.X., Zhang, Y.: A Stochastic expected value model for measures program of oil field. Journal of Shandong University of Technology (Sci & Tech) 20(3), 9–12 (2006)
4. Sun, Y.J., Kang, L.Y., Shi, W.X., et al.: tudy on sizing of batteries for distributed power system utilizing chance constrained programming. Journal of System Simulation 17(1), 41–44 (2005)
5. Zhao, Y., Deng, H.Y., Li, J.H., et al.: Chance-constrained programming based on optimal allocation of harmonic filters on a distribution network. Proceedings of the CSEE 21(1), 12–17 (2001)
6. Iwamura, K., Liu, B.: A genetic algorithm for chance constrained programming. Journal of Information & Optimization Sciences 17(2), 40–47 (1996)
7. Zhao, R., Iwamura, K., Liu, B.: Chance constrained interger programming and stochastic simulation based genetic algorithms. Journal of Systems Science and Systems Engineering 7(1), 96–102 (1998)
8. Chen, J.M.: Solving order problems with genetic algorithms based on stochastic simulation. J. Chongqing Technol. Business Univ (Nat. Sci. Ed.) 22(2), 179–181 (2005)
9. Liu, B.: Dependent-chance goal programming and its genetic algorithm based approach. Mathematical and Computer Modelling 24(7), 43–52 (1996)
10. Holland, J.H.: Genetic algorithms and the optimal allocations of trials. SIAM J. of Computing 2, 8–105 (1973)

Rate Control, Routing Algorithm and Scheduling for Multicast with Network Coding in Ad Hoc Networks

Xu Na Miao[*] and Xian Wei Zhou

Department of Communication Engineering, School of Information Engineering,
University of Science and Technology Beijing, Beijing, 100083, P.R. China
miaoxuna1227@126.com

Abstract. In this paper, we developed a distributed rate control and routing algorithm for multicast session in ad hoc networks. We studied the case with dynamic arrivals and departures of the users. With random network coding, the algorithm can be implemented, and work at transport layer to adjust source rates and at network layer to carry out network coding in a distributed manner. The scheduling element of our algorithm is a dynamic scheduling policy. Numerical examples are provided to complement our theoretical analysis. Modeling and solution algorithm can be easily tuned according to a specific networking technology.

Keywords: Cross-layer design, Rate control, Network coding, Stochastic flow, Multicast, Routing, Ad hoc networks.

1 Introduction

As a network layer problem, routing involves simply replicating and forwarding the received packets by intermediate nodes in multi-hop networks. Network coding extends routing by allowing intermediate nodes to combine the information received from multiple links in the subsequent transmissions and enables wired network connections with rates that are higher than those achieved by routing only [1]. Subsequently, important progress has been made regarding the low-complexity construction of network codes. Li et al. [2] showed that the maximum multicast capacity can be achieved by performing linear network coding. Ho et al. [3], Jaggi et al. [4] and Sanders et al. [5] showed that random linear network coding over a sufficiently large finite field can (asymptotically) achieve the multicast capacity. Following these constructive theoretical results about network coding, Chou et al. [6] proposed a practical scheme for performing network coding in real packet networks. Network coding has been extended to wireless environments with distributed implementation ([6], [7], [8]).

In order to achieve high end-to-end throughout and efficient resource utilization, rate control, routing and scheduling need to be jointly designed in ad hoc networks. Cross-layer design is becoming increasingly important for improving the performance of multihop wireless networks ([11], [12], [13], [14], [15]).

In this paper, we consider the problem of rate control and resource allocation (through routing and scheduling) for multicast with network coding over a multi-hop

---

[*] Corresponding author.

H. Deng et al. (Eds.): AICI 2009, LNAI 5855, pp. 705–714, 2009.

wireless ad hoc network. Rate control is a key functionality in modern communication networks to avoid congestion and to ensure fairness among the users ([9],[10]). Although rate control has been studied extensively for wireline networks where links are reliable and link capacities are usually well provisioned. However, these results cannot be applied directly to multihop wireless networks where link is a shared medium and interference-limited. Unlike in the wireline network where flows complete for transmission resources only when they share the same link, here, network layer flows that do not even share a wireless link in their paths can compete. Thus, in ad hoc wireless networks the contention relations between link-layer flows provide fundamental constraints for resource allocation.

2 Related Work

There are several works on rate control of cross-layer design over wireless ad hoc networks and works on rate control of multicast flows. In this section, we will only briefly discuss several references that are directly relevant to this paper. To the best of our knowledge, this paper is the first effort to study cross-layer design for network coding based multicasting in ad hoc networks.

The work in [13], [15], [16], [17] provides a utility–based optimization framework to study rate control over ad hoc networks, the authors study joint rate control and media access control for ad hoc wireless network, and formulate rate allocation as a utility maximization problem with the constraints that arise from contention for channel access. This paper studies the case with dynamic arrivals and departures of the users, and extends this work to include routing and to study cross-layer design for network coding based multicast.

With network coding, the work that are most similar to our work are [18], [19], [20]. What differentiates our work from others are the following: First, we extend this model to ad hoc wireless networks. Second, our rate control algorithm is a dual subgradient algorithm whose dual variables admit concrete and meaningful interpretation as congestion prices. Third, the session scheduling in our cross-layer design is a dynamic scheduling over ad hoc networks.

The rest of the paper is structured as follows. The system model and problem formulation is presented in Section 3. We present a fully distributed cross-layer rate control algorithm in Section 4.In section 5, we describe a scheduling policy over ad hoc networks, and the simulation results are presented in Section 6, and the conclusion is given in Section 7.

3 The System Model

3.1 System Model

We consider a set N of mobile nodes in an ad hoc network, labeled $1, 2, \cdots |N|$. We say that link $l = ij$ joining node i to node j exists if node i can successfully transmit to node j. The set of links so formed is denoted by L. Let $\vec{P} = [P_l, l \in L]$ denote the vector of global power assignments and let $\vec{c} = [c_l, l \in L]$ denote the vector of data rates. We assume that $\vec{c} = u(\vec{P})$ the data rates are completely determined by the global power assignment. The function $u(.)$ is called the rate-power function of the system.

Note that the global power assignment $\vec{P}$ and the rate-power function $u(.)$ summarize the cross-layer control capability of the network at both the physical layer and the MAC layer. Precisely, the global power assignment determines the Signal- to- Interference-Ratio (SIR) at each link. Let M denote the number of multicast sessions. Each session has one source $s^m \in N$ and a set $D_m \subset N$ of destinations. Network coding allows flows for different destinations of a multicast session to share network capacity by being coded together: for a single multicast session m of rate x^m, information must flow at rate x^m to each destination. The arrival process of any multicast session m according to a Poisson process with rate λ_m and that each node brings with it a file for transfer whose size is exponentially distributed with mean $1/\mu_m$. Thus, the traffic intensity brought by sessions of class m is $\rho_m = \lambda_m / \mu_m$. Consider a graph $G = (V, E)$ with capacity set c and a collection of multicast sessions $S_m = (s^m, d, x^m)$, $m = 1, \cdots M$, $d \in D_m$ as the end-to-end traffic demands.

Let $n_m(t)$ ($m = 1, 2, \cdots, M$) denote the number of multicast sessions that are present in the system, In the rate assignment model that follows, the evolution of $n_m(t)$ will be governed by a Markov process. Its transition rates are given by:

$$n_m(t) \to n_m(t) + 1, \text{ with rate } \lambda_m, \text{ and}$$

$$n_m(t) \to n_m(t) - 1, \text{ with rate } \mu_m x_m(t) n_m(t), \text{ if } n_m(t) > 0.$$

Let $p_l^m(t)$ denote congestion cost at link l in the network at time t. As in [17], we say that the network is stable if

$$\limsup_{t \to \infty} \frac{1}{t} \int_0^t 1_{\left\{ \left(\sum_{m=1}^{M} n_m(t) + \sum_{l=1}^{L} p_l^m(t) \right) > H \right\}} dt \to 0,$$

as $H \to \infty$. This means that the fraction of time that the amount of "unfinished work" in the system exceeds a certain level H can be made arbitrary small as $H \to \infty$. In other words, the number of sessions at each source node and the queues at each link must be finite. The capacity region of the network is defined to be the set of user arrival rate vectors for which the network can be stabilized by some scheduling policy. The capacity region of a constrained queuing system, such as a wireless network, is well characterized in [27]. Let $A = [A_{lr}^m]$ denote the multicast matrix, i.e., $A_l^m = 1$ if $l \in L$ and $A_l^m = 0$ otherwise. For our model, the capacity region is given by the set

$$\Lambda = \left\{ \lambda : \left[\sum_{m=1}^{M} \frac{A_l^m \lambda_m}{\mu_m c_l^m} \right]_{l \in L} \in Co(S) \right\}$$

Where $Co(S)$ represents the convex hull of all link schedules S.

With coding the actual physical flow on each link need only be the maximum of the individual destination's flows [1]. For the case of multiple sessions sharing a network, achieving optimal throughout requires in some cases coding across sessions. However, designing such codes is a complex and largely open problem. In this paper, we use intra-session coding approach to solve the multiple sessions problem similar to [26].

In this case, these constrains of the set of feasible flow vectors can be expressed as

$$\sum_{j:(ij)\in L} f^{md}(ij) - \sum_{j:(ji)\in L} f^{md}(ji) = \begin{cases} x^m & if \ i = s^m; \\ -x^m & if \ i = d; \\ 0 & otherwise. \end{cases} \tag{1}$$

$$0 \le f^{md}(ij) \le g^{md}(ij), \ \forall d \in D_m, \forall ij \in L \tag{2}$$

$$\sum_{m=1}^{M} g^{md}(ij) \le c(ij), \ \forall d \in D_m, \ \forall ij \in L \tag{3}$$

Where $f^{md}(ij)$ denote the information flow for destination d of session m, and $g^{md}(ij)$ gives the physical flow for session m. The inequality (2) reflects the network coding condition relating physical rate and information rate:

$$g^{md}(ij) = \max_d f^{md}(ij) \tag{4}$$

In practice, the network codes can be designed using the approach of distributed random linear network coding ([21]). If (1)-(2) holds, each sink receives with high probability a set of packets with linearly independent coefficient vectors, allowing it or decode.

In this paper we consider the problem of rate control (congestion control) over a multihop wireless ad hoc network. We take into among signals transmitted simultaneously by multiple nodes and the fact that a single node's transmission can be received by multiple nodes, referred to as wireless multicast advantage. We model the contention relation between subflows as the interference set of the link, i.e., the links in the interference set contend each other. Now we let $S(l) = \{ m \in S_m \mid l \in L(m), \forall l \in L \}$ be the set of multicast sessions that use link l. Note that $ij \in L(m)$ if and only if $m \in S(l)$. We denote by $I_{s(l)}$ the interference set of link $l = ij$, including the link l itself. This set indicates which groups of subflows interfere with the subflows which go through the link l. Because links included in the interference set $I_{s(l)}$ share the same channel resource $c(l)$ of the link l, only one of the subflows going through link $k \left(k \in I_{s(l)} \right)$ may transmit at any given time. The accurate interference set of the link can be constructed based on the SIR model as proposed in [22].

3.2 Multicast with Network Coding

We assume each session m attains a utility function of the rate vector $U_m(x^m)$. We assume $U_m(x^m)$ is continuously differentiable, increasing, and strictly concave. For example, a utility that leads to proportional fairness is $U_m(x^m) = w_m \log x^m$, where $w_m, m = 1, \cdots, M$ are the weights. With the advantages of network coding [1], we can model the routing problem at the network layer as follows:

$$\max_{x^m, y^m} \sum_m U_m(x^m)$$

$$s.t. \quad A_l^m x^m \le y_l^m, \quad \forall m \in M \tag{5}$$

$$\sum_{k \in I_{S(l)}} \sum_{m \in S(k)} y_l^m \le c_l^m, \quad \forall l \in L$$

The physical flow rate y_l^m for each multicast session m though link l is $y_l^m = \max_r \{ A_{lr}^m x_r^m \}$.

Note that network coding comes into action through the constraint (4). The utility function is a concave function, and it is easy to prove that the network coding region constraint (linear constraint) is a convex set. Therefore, solving the problem (5) is a convex optimization problem. For a data with multiple multicast sessions, the maximum utility and its corresponding optimal routing strategy can be computed efficiently and in a distributed fashion.

4 Cross-Layer Design with Network Coding

4.1 Distributed Rate Control Algorithm

At time t, given congestion price $p(t)$, the source s^m adjusts its sending rate according to aggregate congestion price over the multicast tree T_r^m. This rate control mechanism has the desired price structure and is an end-to-end congestion control mechanism.

The solution to the cross-layered rate control problem is of the following form similar to [28].

$$x^m(t) = x^m(kT) = \min\left(\frac{w_m}{\sum_{l=1}^{L} A_l^m p_l^m(kT)}, V_m \right) \tag{6}$$

For $kT \le t < (k+1)T$, where V_m is the maximum data rate for users of class m. The congestion costs are updated as

$$p_l^m((k+1)T) = [p_l^m(kT) +$$

$$\alpha_l \left(\sum_{m=1}^{M} A_l^m \int_{kT}^{(k+1)T} n_m(t) x^m(kT) dt - c_l(t)T \right)]^+ \tag{7}$$

where α_l is a positive scalar stepsize. $[.]^+$ is a projection on $[0, \infty)$.

4.2 Cross-Layer Design

- For each link l, in the beginning of period t allocates a capacity $\vec{\phi}(t)$ over link l such that

$$\vec{\phi}(t) = \arg\max_{\vec{\phi} \in R} \sum_{l=1}^{L} p_l^m(t) c_l \tag{8}$$

The scheduling (8) is a difficult problem for ad hoc network. Our algorithm is similar to a distributed variant of the sequential greedy algorithm presented in [24].

- Over link l, send an amount of coded packets for the session $m_l(t) = \arg\max_m \sum_l p_l^{m(t)}$ at rate $\vec{\phi}(t)$.

- Routing: over link l, a random linear combination of data of multicast session m_l to all destinations d is sent at an amount of bits according to the rate determined by the above scheduling.

- Coding scheme: In wireless networks, nodes encode and transmit packets or receive and decode packets at different time instants. This coding scheme is a random distributed coding scheme as the schemes in [21].

Directly applying the convergence results [17], we have the following result regarding the convergence property of the distributed rate control algorithm for multicast.

5 Numerical Examples

In this section, we provide numerical examples to complement the analysis in the previous sections. We consider a simple ad hoc network with two multicast sessions shown in Fig. 1. The network is assumed to undirected and a link has equal capacities in both directions. And assume that session one with source node s^1 and destination x^1 and x^2, and session two with source node s^2 and destination d^1 and d^2 with the same utility $U_m(x_m) = \log(x_m)$. We have chosen such a small, simple topology to facilitate detailed discussion of the results.

We now simulate the distributed algorithm with dynamic scheduling and network coding developed in IV. We assume the following link capacities: links (s^1, s^2) and (x^2, d^1) have 2 units of capacity, links $(s^2, y), (y, x^1), (y, d^2)$ and (y, x^2) have 3 units of capacity and all other links have 1 units of capacity when active.

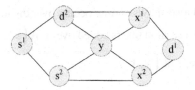

Fig. 1. A network topology with two multicast sessions

Fig. 2 and Fig. 3 show the evolution of source rate and congestion price of each session with stepsize $\alpha_i = 0.01$ in (7). We see that the source rates converge quickly to a neighborhood of the corresponding stable values .

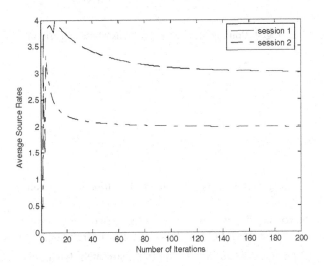

Fig. 2. The average source rates with distributed algorithm

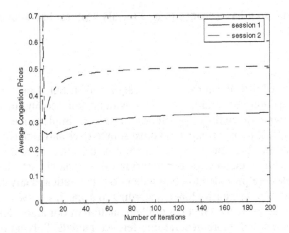

Fig. 3. The congestion prices with distributed algorithm

The simulation result also shows coding occurs over the multicast trees in Fig. 4 and Fig. 5: 2 units of traffic of session one is coded over link $\left(s^2, y\right)$ and 2 units of traffic of session two is coded over link $\left(y, x^1\right)$.

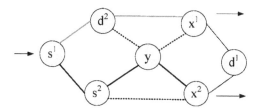

Fig. 4. Routing for session 1with distributed algorithm

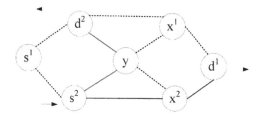

Fig. 5. Routing for session 2 with distributed algorithm

Note that the routing depends not only on the network topology, which determines the interference among links, but also link capacity configurations. So we can see that the link $\left(s^2, x^2\right)$ is not used in Fig. 3. But the distributed dynamic scheduling always picks the link with the locally heaviest weight, this feature such that the link $\left(s^2, y\right)$ is always active due to it has a chance to be a locally heaviest link a lot of time.

6 Conclusions

We have presented a model for the joint design of distributed rate control algorithm, routing and scheduling for multicast session with network coding in ad hoc networks. With random network coding, the algorithm can be implemented in a distributed manner, and work at transport layer to adjust source rates and at network layer to carry out network coding. We study the case with dynamic arrivals and departures of the users. The scheduling element of our algorithm is a dynamic scheduling policy. Numerical examples are provided to complement our theoretical analysis. Modeling and solution algorithm can be easily tuned according to a specific networking technology.

We will further study the practical implementation of our algorithm and extend the results to networks with more general interference models. Solving this problem will further facilitate the practical deployment of network coding in real networks.

Acknowledgement

This work is supported in part by the National Science Foundation of P.R. China under Grant no.60773074 and by the National High Technology Research and Development Program of P. R. China under Agreement no.2009AA01Z209.

References

[1] Ahlswede, R., Cai, N., Li, S.Y.R., Yeung, R.W.: Network information flow. IEEE Trans. Inform. Theory 46, 1204–1216 (2000)

[2] Li, S.Y.R., Yeung, R.W., Cai, N.: Linear network coding. IEEE Trans. Inform. Theory 49, 371–381 (2003)

[3] Ho, T., Koetter, R., Karger, M.D.R., Effros, M.: The benefits of coding over routing in a randomized setting. In: Proc. Int'l Symp. Information Theory, Yokohama, Japan. IEEE, Los Alamitos (2003)

[4] Jaggi, S., Chou, P.A., Jain, K.: Low complexity optimal algebraic multicast codes. In: Proc. Int'l Symp. Information Theory, Yokohama, Japan. IEEE, Los Alamitos (2003)

[5] Sander, P., Egner, S., Tolhuizen, L.: Polynomial time algorithms for network information flow. In: Symposium on Parallel Algorithms and Architectures (SPAA), San Diego, CA, pp. 286–294. ACM, New York (2003)

[6] Lun, D.S., Ratnakar, N., Médard, M., Koetter, R., Karger, D.R.: Minimum-cost multicast over coded packet networks. IEEE Trans. Inform. Theory (2006)

[7] Sagduyu, Y.E., Ephremides, A.: Joint scheduling and wireless network coding. In: Proc. WINMEE, RAWNET and NETCOD 2005 Workshops (2005)

[8] Wu, Y., Chou, P.A., Kung, S.-Y.: Minimum-energy multicast in mobile ad hoc networks using network coding. IEEE Trans. Commun., 1906–1918 (2005)

[9] Deb, S., Srikant, R.: Congestion control for fair resource allocation in networks with ulticast flows. IEEE Trans. on Networking, 274–285 (2004)

[10] Kelly, F.P., Maulloo, A., Tan, D.: Rate control in communication networks: Shadow prices, proportional fairness and stability. Journal of the Operational Research Society, 37–252 (1998)

[11] Huang, X., Bensaou, B.: On Max-min Fairness and Scheduling in Wireless Ad-Hoc Networks: Analytical Framework and Implementation. In: Proceedings of IEEE/ACM MobiHoc, Long Beach, CA, October 2001, pp. 221–231 (2001)

[12] Sarkar, S., Tassiulas, L.: End-to-end Bandwidth Guarantees Through Fair Local Spectrum Share in Wireless Ad-hoc Networks. In: Proceedings of the IEEE Conference on Decision and Control, Hawaii (2003)

[13] Yi, Y., Shakkottai, S.: Hop-by-hop Congestion Control over a Wireless Multi-hop Network. In: Proceedings of IEEE INFOCOM, Hong Kong (2004)

[14] Qiu, Y., Marbach, P.: Bandwith Allocation in Ad-Hoc Networks: A Price-Based Approach. In: Proceedings of IEEE INFOCOM, San Francisco, CA (2003)

[15] Xue, Y., Li, B., Nahrstedt, K.: Price-based Resource Allocation in Wireless Ad hoc Networks. In: Jeffay, K., Stoica, I., Wehrle, K. (eds.) IWQoS 2003. LNCS, vol. 2707, pp. 79–96. Springer, Heidelberg (2003)

[16] Lin, X., Shroff, N.B.: Joint Rate Control and Scheduling in Multihop Wireless networks. Technical Report, Purdue University (2004),
http://min.ecn.purdue.edu/_linx/papers.html

[17] Lin, X., Shroff, N.: The impact of imperfect scheduling on cross-layer rate control in multihop wireless networks. In: Proc. IEEE Infocom (2005)

[18] Wu, Y., Chiang, M., Kung, S.Y.: Distributed utility maximization for network coding based multicasting: A critical cut approach. In: Proc. IEEE NetCod (2006)

[19] Wu, Y., Kung, S.Y.: Distributed utility maximization for network coding based multicasting: A shortest path approach. IEEE Journal on Selected Areas in Communications (2006)

[20] Ho, T., Viswanathan, H.: Dynamic algorithms for multicast with intrasession network coding. In: Proc. Allerton Conference on Communication, Control and Computing (2005)

[21] Chou, P.A., Wu, Y., Jain, K.: Practical network coding. In: Proc. Allerton Conference on Communication. Control and Computing (2003)

[22] Gupta, P., Kumar, P.R.: The capacity of wireless network. IEEE Trans. on Information Theory 46(2), 388–404 (2000)

[23] Shor, N.Z.: Monimization Methods for Non-Differentiable Functions. Springer, Heidelberg (1985)

[24] Preis, R.: Linear time 1/2-approximation algorithm for maximum weighted matching in general graphs. In: Meinel, C., Tison, S. (eds.) STACS 1999. LNCS, vol. 1563, p. 259. Springer, Heidelberg (1999)

[25] Jain, K., Padhye, J., Padmanabhan, V.N., Qiu, L.: Impact of interference on multi-hop wireless network performance. In: Proc. ACM Mobicom (2003)

[26] Ho, T., Viswanathan, H.: Dynamic algorithms for multicast with intra-session network coding. In: Proc. 43rd Annual Allerton Conference on Communication. Control and Computing (2005)

[27] Tassiulas, L., Ephremides, A.: Stability properties of constrained queuing systems and scheduling policies for maximum throughout in multihop radio networks. IEEE Trans. on Automatic Control 37(12), 1936–1948 (1992)

[28] Lin, X., Shroff, N.B.: On the stability region of congestion control. In: Proceedings of the 42nd Annual Allerton Conference on Communication. Control and Computing (2004)

Design and Experimental Study on Spinning Solid Rocket Motor

Heng Xue, Chunlan Jiang, and Zaicheng Wang

State Key Laboratory of Explosion Science and Technology, Beijing Institute of Technology,
Beijing, China
xh930@163.com, {jiangchuwh,wangskyshark}@bit.edu.cn

Abstract. The study on spinning solid rocket motor (SRM) which used as power plant of twice throwing structure of aerial submunition was introduced. This kind of SRM which with the structure of tangential multi-nozzle consists of a combustion chamber, propellant charge, 4 tangential nozzles, ignition device, etc. Grain design, structure design and prediction of interior ballistic performance were described, and problem which need mainly considered in design were analyzed comprehensively. Finally, in order to research working performance of the SRM, measure pressure-time curve and its speed, static test and dynamic test were conducted respectively. And then calculated values and experimental data were compared and analyzed. The results indicate that the designed motor operates normally, and the stable performance of interior ballistic meet demands. And experimental results have the guidance meaning for the pre-research design of SRM.

Keywords: Solid Rocket Motor（SRM）, Interior Ballistic Performance, Static Test, Dynamic Test.

1 Introduction

Solid rocket motor (SRM) which has characteristics of small size, reliable operation, simple structure, low cost and long-term preservation, is widely used in all kinds of small-sized, short range military rocket and missile power plant. In recent years, a series of studies on SRM have been made, and these studies are important for the construction of structure design, internal flow field simulation and improvement of interior ballistic performance [1], [2], [3], [4].

In this paper, spinning SRM is used as the power plant of twice throwing structure of aerial submunition. Its structural characteristic is 4 tangential nozzles distributing uniform on the circumference of the motor case, the thrust generated by nozzle acting along tangential direction of the rocket body and then the motor rotates. Depending on the centrifugal force generated by the rotation, bullets were dispersed in a large area. Meanwhile, rotation ensures the flight stability of submunition.

Firstly, according to technical parameters the general scheme of SRM was determined, and three-dimensional overall structure design was achieved by using Inventor software. Prediction of interior ballistic performance was carried out after the determination of overall structure design and grain design. Finally, static test and dynamic

H. Deng et al. (Eds.): AICI 2009, LNAI 5855, pp. 715–723, 2009.
© Springer-Verlag Berlin Heidelberg 2009

test were conducted, and the test results which were compared and analyzed under different conditions, providing reference for similar SRM design.

2 Spinning Solid Rocket Motor Design

Unlike conventional tail nozzle, this article aims to present a motor with the structure form of 4 tangential nozzles, the key of design is to determine a reasonable and feasible design scheme. According to tactical and technical indexes of the SRM, general parameters were determined, and the main structural design was completed such as the design of propellant charge, combustion chamber, nozzles, etc.

2.1 Grain Design

The size and geometry of grain determine gas producing rate and its change law of SRM, thus determine thrust and the rule pressure with time. At the same time, volume and weight of combustion chamber were determined by the volume of propellant charge.

Considering motor thrust is relatively large, while working time is not long, and meanwhile in order to obtain a higher charge coefficient, tubular grain with seven columns were used. As shown in Fig. 1. The grains with ends coated burn on the internal and external surface simultaneously. The greatest advantage of this program is simple structure, easy to make, neutral burning and constant thrust can be obtained. Diameter and length of the tubular grain were determined by parameters such as combustion chamber pressure, working time, combustion rate of propellant etc.

Fig. 1. Tubular grain configuration

2.2 Structure Design

Combustion chamber is one of the most important components of SRM, and it provides the place of storage and combustion for propellant charge. Furthermore, the combustion chamber bears high temperature, high pressure and under the action of various loads. At the same time, the shell of combustion chamber is opened holes to install nozzles. In order to ensure combustion chamber still work reliably under the above bad working conditions, the combustion chamber should have characteristics of sufficient strength and rigidity, reliable connection and seal, light structural weight etc. Based on the above considerations, shell material uses 35CrMnSi, connection of shell and nozzle, shell and closure head adopts screw connection.

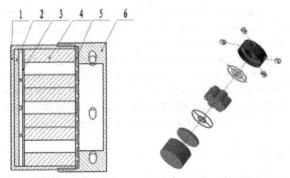

1. Combustion chamber 2. Igniter cartridge 3. Front charge
baffle 4.Tubular charge 5. Back charge baffle plate 6. Nozzle

Fig. 2. Structure of the spinning solid rocket motor

According to requirements of the general parameters, 4 tangential nozzles distributed uniform near the bottom of SRM, and its profile uses laval nozzle. In order to convenient for processing and installation, the axial line of nozzle and the normal line of rocket body inclined to one another at an angle of 45 degrees. Fig. 2 depicts the architecture of the SRM.

3   Interior Ballistic Calculation

Interior ballistic calculation is an important step in the design of SRM, and combustion chamber pressure is an important parameter. To begin with, pressure influences not only the size of SRM thrust but also the combustion time of propellant grain, thus affecting the motor's working time. Secondly, from the perspective of structure design, pressure directly determines the load bear by combustion chamber, thereby affects the structural weight of SRM. From the above we can see that calculation of combustion chamber pressure has vital significance to the prediction of interior ballistic performance, determination of SRM thrust as well as verification of structural strength.

3.1   Basic Assumptions

Through simplifying the working process of SRM, mass conservation equation and state equation of the combustion chamber were established. Simplification is as follows: (1) Combustion chamber gas is ideal gas and heat loss is constant in combustion process; (2) Propellant charge is complete combustion and the burning temperature is unchanged; (3) The charge is ignited under the pressure of ignition pressure; (4) Do not consider effects of erosive burning or the change of burning surface caused by scouring. The charge obeys geometric burning law in combustion process.

Seriously, interior ballistic calculation is relevant to varying with time of flow parameters which distributing along the length of combustion chamber. Then one-dimensional unsteady flow equations seemed to be used to determine the functional relation between time and flow parameters such as gas pressure, temperature, density. However, a series of difficulties will be encountered in mathematics when using this method. Therefore, methods of simplifying problem are usually adopted. Calculation of average pressure and average temperature change with time, which without considering flow parameters change along the length of combustion chamber, is called zero-dimensional ballistic calculation. In addition, ignition pressure is regarded as the starting point of pressure calculation and ignition process is not considered. At this moment, motor working process can be divided into three stages: pressure rising stage, pressure balance stage and after-effect stage.

3.2 Mathematical Model

Generally speaking, pressure rising stage is an unsteady state and it reflects motor's igniting process. The pressure rapidly increases to close to equilibrium state in a very short time. Equation of this stage is as follows [5]:

$$p_c = p_{eq}\left\{1-\left[1-\left(\frac{p_{ig}}{p_{eq}}\right)^{1-n}\right]e^{-\frac{(1-n)\phi_2 A\Gamma\sqrt{RT_{eq}}}{V_c}t}\right\}^{\frac{1}{1-n}} \tag{1}$$

Where p_c is combustion chamber pressure, pa; p_{eq} is average pressure, pa; p_{ig} is ignition pressure, pa; e is grain web thickness, mm; n is pressure index; ϕ_2 is correction coefficient; Γ is specific heat ratio function; t is working time, s; V_C is initial chamber volume, m^3; R is molar gas constant, $J \cdot mol^{-1} \cdot K^{-1}$; T_{eq} is equilibrium temperature, K; A_t is nozzle exit area, m^2.

Because of the designed tubular grain with ends coated burns on the internal and external surface simultaneously, the burning surface is neutral burning. The stage of pressure balance can be calculated as:

$$p_c = p_{eq}^{(1-n)} - \frac{V_c a}{\sqrt{x\phi_2\Gamma^2 A_t kC^*}}\frac{dp_c}{de} \tag{2}$$

Where x is heat release coefficient; a is velocity coefficient; k is specific heat ratio; C^* is characteristic velocity, $m \cdot s^{-1}$.

After the end of combustion, combustion products which no longer producing are continue to be exhausted from the nozzle. The combustion chamber pressure decreases rapidly until reaching balance with the external pressure. The process of gas expansion is after-effect stage and the pressure can be calculated as:

$$p_c = p_{eq}\left[\frac{2V_{cf}}{2V_{cf}+(k-1)\sqrt{RT_{eq}}\phi_2 A_t\Gamma t}\right]^{\frac{2k}{k-1}} \tag{3}$$

Where V_{cf} is combustion chamber volume, m^3.

$$\Gamma = \sqrt{k}\left(\frac{2}{k+1}\right)^{\frac{k+1}{2(k-1)}} \qquad (4)$$

3.3 Calculation and Analysis

Basic parameters are as follows: combustion chamber diameter is $D_i = 132mm$; combustion chamber length is $L = 79mm$; density is $\rho = 1.61g \cdot cm^{-3}$; specific heat ratio is $k = 1.25$; ignition pressure is $p_{ig} = 4MPa$; characteristic velocity is $C^* = 1341m \cdot s^{-1}$; initial chamber volume is $V_c = 0.53 \times 10^{-3} m^3$.

The above differential equations can be solved by using fourth-order runge-kutta, and then the law of combustion chamber pressure changes with time will be obtained. Table 2 presents calculation results, and the calculated pressure-time curve is shown in Fig. 3.

From the results and the pressure-time curve, we can see that the interior ballistic performances of SRM meets design requirements. However, because of grains burn on the internal and external surface simultaneously, the main problem existed in practical combustion is erosive burning, and the actual curve will vary. Therefore, in order to verify the correctness of theoretical model, experimental research should be carried out.

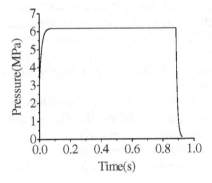

Fig. 3. Pressure-time calculating curve

4 Experimental Studies

In order to study working performance of the spinning SRM and measure pressure-time curve and speed, static test and dynamic test were conducted respectively.

4.1 Static Test

4.1.1 Test Equipment
The static test of spinning SRM is designed to check working reliability of components, rationality of charge structure and interior ballistic performance. As shown in Fig. 4,

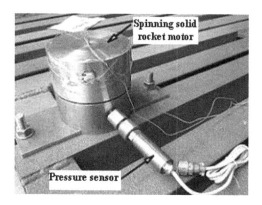

Fig. 4. Static test

the assembled spinning SRM is fixed with bolts in a test stand. By using pressure sensor, pressure of the combustion process was converted into electrical signals. And combustion chamber pressure and working time were measured by data acquisition system.

The test motor consists of a combustion chamber, propellant charge, 4 tangential nozzles, charge baffle plates, igniter cartridge, etc. Combustion chamber shell material uses 35CrMnSi, connection of shell and nozzle, shell and closure head adopt screw connection. Charge baffle plate is made of low carbon steel. As the motor's working time is short, no additional thermal protection measures are need. The motor uses the ways of electric ignition, igniter charge bag is placed in front of the grain, and igniting charge is 2# small powder grain. The charge is double base propellant grain, tubular grain with seven columns are used.

The grains which two ends coated by 2 millimeter thick nitrocellulose lacquer, burn on internal and external surface simultaneously. The ignition charge mass of first and second test are 15gram and 20gram respectively.

Pressure sensor (model ZQ-Y1) was used, and the main technical specifications are as follows, range: 20MPa, accuracy: 0.3%FS, power: 5-15VDC, output: 1.3672mV/V.

4.1.2 Test Results and Analyses

Pressure-time test curves of the first and second test are shown in Fig. 5 and Fig. 6 respectively. Because of the size of spinning SRM is relatively small, ignition charge mass has a relatively high impact on interior ballistic performance. A comparison of Fig. 5 and Fig. 6 indicates that less ignition charge mass make ignition delay too long, and could not meet the needs. However, considering the restriction on motor size and the affordability of solid rocket motor case, too much ignition charge mass is unsuitability. Appropriate amount of ignition charge mass should be selected according to the design requirements. And the ignition pressure can be improved and the problem of ignition delay at the preliminary stage was solved by adjusting ignition charge mass.

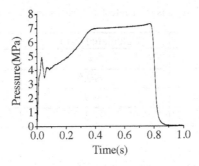

Fig. 5. Pressure-time test curve of the first set

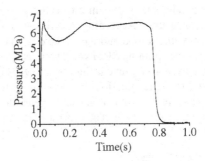

Fig. 6. Pressure-time test curve of the second set

The results also indicate that the designed motor work well, structure is reliable, the thread connections have no loosening, combustion chamber and nozzles have no deformation, and the technical requirements are met.

Table 1 presents the first and second experimental data, and Table 2 presents a comparison of calculated values and experimental data. As can be seen in Table 2, the designed average thrust F_n and average pressure P_n is lower than experimental data, and the burning time T_b is longer than experimental data. The main reason of this discrepancy is that the actual burning rate is higher than theoretical value. And this causes burning time shortened and pressure reduced.

Table 1. Experimental data of spinning SRM

Parameters	First set	Second set
p_m (Mpa)	7.32	6.77
p_n (Mpa)	7.09	6.55
T_b (s)	0.78	0.74
T_k (s)	0.85	0.81

Table 2. Comparison of calculated values and experimental data

Parameters	Calculated values	Experimental data
p_n (Mpa)	6.2	6.55
F_n (N)	1940.5	2058.2
T_b (s)	0.89	0.74
T_k (s)	0.93	0.81

4.2 Dynamic Test

In order to check dynamic performance of the spinning SRM, dynamic test was conducted. As shown in Fig. 7, with bearings installed on the top, the motor was suspended. By using high speed camera, dynamic working process of the motor was recorded and the speed was measured. According to high speed photographic records, the speed is up to 15000r/min without load and meets the design requirements.

The results indicate that the spinning SRM work well, and have no abnormal sounds. However, high-speed rotation has great influence on working performance of the spinning SRM. Compared to static situation, the ablation of nozzle is more serious in rotation conditions. Meanwhile, the centrifugal force generated by the rotation caused metal oxides deposited on the motor case. And this may lead to thermal damage of head structure.

Fig. 7. Dynamic test of the spinning SRM

5 Conclusions

Design and experimental methods of spinning SRM were preliminarily studied in the paper. By comparing with calculated values and experimental data, the following results were obtained:

(1) The ignition charge mass can be of significant effect on the pressure-time curves and internal ballistic performances of the spinning SRM. And in order to obtain satisfied internal ballistic performances, the ignition charge mass must be strictly controlled.

(2) By comparing Fig. 3 with Fig. 6, and from Table.2 we can see that calculated values accord with experimental data, and the calculation method presented in the paper is simple and practicable.

(3) The experimental results indicate that the designed motor operated normally, and the stable performance of interior ballistic meet demands.

(4) High-speed rotation has great influence on working performance of the spinning SRM.

References

1. Kamm, Y., Gany, A.: Solid Rocket Motor Optimization. In: 44th AAIA/ASME/SAE/ASEE Joint Propulsion Conference and Exhibit, AIAA-2008-4695, Hartford, CT (2008)
2. French, J., Flandro, G.: Linked Solid Rocket Motor Combustion Stability and Internal Ballistic Analysis. In: 41st AAIA/ASME/SAE/ASEE Joint Propulsion Conference and Exhibit, AIAA-2005-3998, Tucson, Arizona (2005)
3. Shimada, T., Hanzawa, M.: Stability Analysis of Solid Rocket Motor Combustion by Computational Fluid Dynamics. AIAA Journal 46(4), 947–957 (2008)
4. Willcox, M.A., Brewster, Q., Tang, K.C.: Solid Rocket Motor Internal Ballistics Simulation Using Three-Dimensional Grain Burnback. Journal of Propulsion and Power 23(3), 575–584 (2007)
5. Zhang, P., Zhang, W.S., Gui, Y.: Principle of Solid Rocket Motor. Beijing Institute of Science and Technology Press, Beijing (1992) (in Chinese)

An Energy-Based Method for Computing Radial Stiffness of Single Archimedes Spiral Plane Supporting Spring

Enlai Zheng[1], Fang Jia[1], Changhui Lu[1], He Chen[1], and Xin Ji[2]

[1] School of Mechanical Engineering, Southeast University, Nanjing, 211189, China
[2] Nanjing Jienuo Environment Technology Co.Ltd, Nanjing, 210014, China
xx_xx1111@163.com, fangjia1988@yahoo.com, magiclu2007@126.com,
sjzyzch2222@yahoo.com.cn, njjn@Njjinuo.com

Abstract. With space-based adaptive performance of lower stiffness and greater deformation energy, the plane supporting spring finds its wide application in fields like aeronautics, astronautics, etc. In the current study, the radial stiffness formula of a single Archimedes spiral plane supporting spring is derived by means of energy approach, with three key parameters of the supporting spring as independent variables. A series of the supporting spring FEA models are established via APDL speedy modeling. According to the isolation requirements of electronic equipment for a fighter, an example is presented in the form of finite element analysis. The theoretical calculation and analysis data are studied and fitted by MATLAB using the least-square method to obtain the discipline of the radial stiffness of single spiral plane supporting spring with the changes of its three key parameters. The validity of energy-based radial stiffness formula of the spring is confirmed by the comparison between the theoretical calculation and finite element analysis results.

Keywords: single archimedes spiral, plane supporting spring, energy approach, radial stiffness.

1 Introduction

Precision instrument and system for aeronautics and astronautics usually endure stronger impact or vibration under hostile environments. In order to obtain longer life and higher reliability, it is essential to introduce effective measures for vibration reduction and isolation. Due to certain factors like constraints of assembly space, etc, common isolation components usually fail to meet fully the requirements of vibration isolation. Therefore, with its space-based adaptive performance of smaller volume, lighter weight, lower stiffness, greater deformation energy and better static and dynamic characteristics, the plane supporting spring begins to attract attention recently.

The performance of plane supporting spring is mainly determined by its radial stiffness, and environmental vibration varies from case to case in its requirements for radial stiffness of springs, the study of which thus becomes an important issue. Currently, the general way of obtaining the radial stiffness of plane supporting spring is to calculate by the finite element simulation, or to measure by experiment. In our previous studies, Jia Fang et al [1] give the relationship between turn number and radial stiffness by FEA;

H. Deng et al. (Eds.): AICI 2009, LNAI 5855, pp. 724–733, 2009.

Chen Nan et al [2][3] carry out the FEA and experimental research on the stiffness of the spring; and Gao Weili et al [4] study the stiffness and stress performance of involutes of circle applied to linear compressor by ANSYS. Besides, other researchers, like A. S. Gaunekar et al[5], also make a study on a kind of structural supporting spring by FEA. Nonetheless, finite element analysis is constrained by the need of hardware and software support, and experimental method tends to have the deficiencies of long cycle, high cost and complicated process, etc.

As far as we know, no attempts have ever been reported to have taken an energy-based approach to calculating the radial stiffness of plane supporting spring until our current study, in which the X dimensional radial stiffness formula of single spiral plane supporting spring is derived by means of energy approach and its validity is verified by finite element simulation.

2 Design of Plane Supporting Spring

The structure of a plane supporting spring studied in this paper is shown in Fig. 1. Table 1 shows the independent parameters of the spring.

It uses Archimedes spiral as flexure spiral spring to connect the inner and outer supports. The equation of Archimedes spiral in polar coordinates is

$$\rho = a + t\theta \tag{1}$$

where ρ stands for the polar radius corresponding to θ, mm; a for the polar radius corresponding to $\theta = 0$, mm; t for the Archimedes spiral coefficient, mm/°; and θ for the polar angle. And we have

$$t = (D_{oi} - D_{io}) / 4\pi n_2 \tag{2}$$

where D_{oi} is the outer support inner diameter; D_{io} the inner support outer diameter; n_2 the turn number of flexible spiral spring.

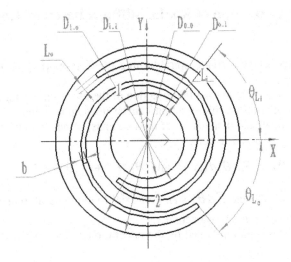

Fig. 1. The geometric model of plane supporting spring

Table 1. The independent parameters of plane supporting spring

No.	Parameter	Symbol
1	Thickness	h
2	Outer support outer diameter	D_{oo}
3	Outer support inner diameter	D_{oi}
4	Inner support outer diameter	D_{io}
5	Inner support inner diameter	D_{ii}
6	Length of inner connecting line	L_i
7	Length of outer connecting line	L_o
8	Width of flexible spiral spring line	b
9	Number of flexible spiral spring	n_1
10	Turn number of flexible spiral spring line	n_2

The flexible spiral spring line is composed of two Archimedes spirals with different starting points. Width of flexible spiral spring line can be adjusted by regulating the starting points.

The inner and outer connecting lines are designed to prevent stress concentration. The outer support is used for spring location; the electronic instrument is tightened on the inner support. The key parameters for radial stiffness include thickness, outer support inner radius, inner support outer radius, width of flexible spiral spring line, and number of flexible spiral spring. The influence of each parameter on radial stiffness is analyzed by control variate method.

3   Formula Derivation of Radial Stiffness Based on Energy Approach[6]

The analysis model of a single spiral/turning plane supporting spring is shown in Fig.2.

As shown in Fig.3, taking the microsection of the plane supporting spring as focus, the visual deformation of microsection can be decomposed into: the axial relative displacement, angle, and dislocation of the two ends, of which only the internal force at both ends works, its value being

$$dW = F_N(\theta)d(\Delta l)* + M(\theta)d\theta* + F_S(\theta)d\lambda* \tag{3}$$

where $F_N(\theta)$, $M(\theta)$ and $F_S(\theta)$ stand for axial force, flexural moment and shearing force under external force F respectively.

The total visual works obtained by the integration of Eq.3 is shown in the following equation.

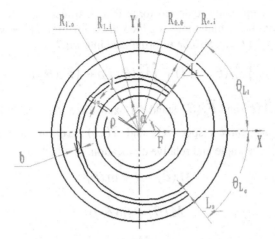

Fig. 2. The analysis model of single spiral/turning plane supporting spring

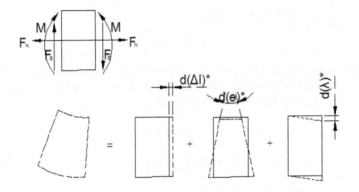

Fig. 3. The visual deformation of microsection

$$W = \int F_N(\theta)d(\Delta l)^* + \int M(\theta)d\theta^* + \int F_s(\theta)d\lambda^* \qquad (4)$$

Suppose the visual displacement of external force F is $\upsilon^*(x)$, as a result of which the visual displacement remain unchanged under external force F and the total visual works can be expressed in another form as follows:

$$W = F\upsilon^*(x) \qquad (5)$$

According to the energy principle, the visual works under external force are equal to those under internal force, as expressed below.

$$F\upsilon^*(x) = \int F_N(\theta)d(\Delta l)^* + \int M(\theta)d\theta^* + \int F_s(\theta)d\lambda^* \qquad (6)$$

With the displacement under external force F in X direction as the visual displacement on condition that a unit force is added in X direction of the single helix plane supporting spring, Eq.6 can also be reformulated into Eq.7.

$$\Delta = \int \overline{F}_N(\theta)d(\Delta l) + \int \overline{M}(\theta)d\varphi + \int \overline{F}_S(\theta)d\lambda \tag{7}$$

where $\overline{F}_N(\theta)$, $\overline{M}(\theta)$ and $\overline{F}_S(\theta)$ stand for axial force, flexural moment and shearing force under unit force respectively and $d(\Delta l)$, $d\varphi$ and $d\lambda$ for relative axial displacement, relative angle and relative dislocation, of which

$$M(\theta) = F \cdot \sin\theta \cdot \rho$$

$$\overline{M}(\theta) = \sin\theta \cdot \rho$$

$$d\varphi = \frac{M(\theta)}{EI} \cdot \rho \cdot \theta$$

Being of Archimedes spiral, we have

$$\rho = R_{io} + t\theta \tag{8}$$

As the plane supporting spring mainly endures bending deformation, the strain energy generated by axial tension and compression and shear can be ignored, unlike that generated by bending deformation. Therefore,

$$\int \overline{F}_N(\theta)d(\Delta l) = 0$$

$$\int \overline{F}_S(\theta)d\lambda = 0$$

The displacement under F in X direction is

$$\Delta = \int_{\theta_{L_i}}^{2\pi - \theta_{L_o}} \sin\alpha \cdot \rho \frac{F \cdot \sin\alpha \cdot \rho}{EI} \cdot \rho \cdot d\alpha$$

$$= \frac{F}{2EI}\left(\frac{(R_{io}+t\theta)^4}{4t} - \frac{1}{2}\sin(2\theta)(R_{io}+t\theta)^3 - \right. \tag{9}$$

$$\left. \frac{3t}{4}\cos(2\theta)(R_{io}+t\theta)^2 + \frac{3t^2}{4}\sin(2\theta)(R_{io}+t\theta) + \frac{3t^3}{8}\cos(2\theta)\right)\Big|_{\theta_{L_i}}^{2\pi-\theta_{L_o}}$$

Of which

$$t = (R_{oi} - R_{io} - b)/2\pi$$

$$\theta_{L_i} = L_i/t$$

$$\theta_{L_o} = b/t$$

The stiffness of X direction is expressed in Eq.10.

$$K_X = \frac{F}{\Delta} = 2EI / \{(\frac{(R_{io}+t\theta)^4}{4t} - \frac{1}{2}\sin(2\theta)(R_{io}+t\theta)^3 -$$

$$\frac{3t}{4}\cos(2\theta)(R_{io}+t\theta)^2 + \frac{3t^2}{4}\sin(2\theta)(R_{io}+t\theta) + \frac{3t^3}{8}\cos(2\theta))|_{\theta_L}^{2\pi-\theta_L} \}$$

(10)

4 Verification of Radial Stiffness Formula

The spring material used for finite element simulation is 65Mn, whose properties are as follows:

$$E = 2.06 \times 10^{11} Pa, u = 0.3, [\sigma] = 475MP,$$

$$\sigma_s = 800MP, \sigma_b = 980MP$$

A group of independent parameters of plane supporting spring for a fighter are shown in table 2.

Table 2. Lists independent parameter of a spring for a fighter

No.	Parameter	Symbol	Value (mm)
1	Thickness	h	1
2	Outer support outer diameter	D_{oo}	34
3	Outer support inner diameter	D_{oi}	30
4	Inner support outer diameter	D_{io}	10
5	Inner support inner diameter	D_{ii}	8
6	Length of inner connecting line	L_i	0.5
7	Length of outer connecting line	L_o	0.5
8	Width of flexible spiral spring line	b	0.5
9	Number of flexible spiral spring	n_1	1
10	Turn number of flexible spiral spring	n_2	1

A. Modeling in APDL language[7] and analyzing

ANSYS creates the geometric model of plane supporting spring using APDL language. And then the model will be divided into meshes. Automatic analysis will begin after adding the load and boundary conditions.

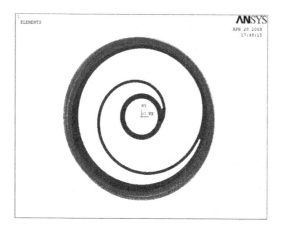

Fig. 4. The finite element analysis model

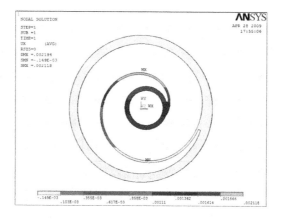

Fig. 5. The displacement of finite element simulation

The finite element model and the displacement chart are demonstrated in Figures 4 and 5.

It can be seen from the Fig.5 that under the action of unit force in the X direction, the value of displacement of finite element simulation is 1.4971mm, while the theoretical displacement by radial stiffness formula is 1.61334mm.

B. Analysis and Comparison between Theoretical Calculation and Finite Element Simulation by Control Variate Method

1) Changing width of Flexible Spiral Spring Line from 0.4mm to1mm by the Step of 0.1mm

The radial stiffness curves of theoretical calculation and finite element simulation with the changes of the width of flexible spiral spring line are shown in Fig.6.

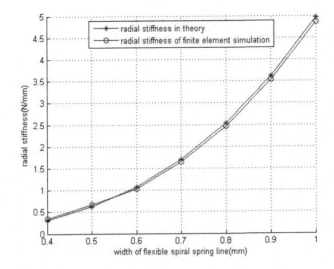

Fig. 6. The radial stiffness of theoretical calculation and finite element simulation varying with changing width of flexible spiral spring line

It can be seen from the graph that under the action of radial unit force, the stiffness goes up with the increasing width of flexible spiral spring line. Radial stiffness in theory increases from 0.3125 N/mm to 0.4958 8 N/mm, and radial stiffness of finite element simulation from 0.3368 N/mm to 0.48652 N/mm. The curves are cubic. The results from theoretical calculation match those from finite element simulation.

2) Changing inner Support Outer Radius from 5mm to 11mm by the Step of 1mm

The radial stiffness curves of theoretical calculation and finite element simulation with the changes of the inner support outer radius of plane supporting spring are shown in Fig.7.

As revealed in Fig.7, the stiffness declines with the increasing inner support outer radius of plane supporting spring under the action of X direction force. Radial stiffness in theory decreases from 0.61985N/mm to 0.3269 N/mm, and radial stiffness of finite element simulation from 0.6680N/mm to 0.4178 N/mm. The deviation between the results of theoretical calculation and simulation turns narrower in that Eq.10 ignores the impact of inner support energy in the calculation.

3) Changing thickness from 0.2mm to 0.8mm by the step of 0.1mm

The thickness-radial stiffness curves of theoretical calculation and finite element simulation are shown in Fig.8.

This graph indicates that the radial stiffness ascends linearly with the increment of thickness under the action of X direction force. Radial stiffness in theory increases from 0.5N/mm to 0.8671 N/mm, and radial stiffness of finite element simulation from 0.5343N/mm to 0.9696 N/mm. The deviation between theoretical calculation and simulation turns narrower due to the same reason mentioned above.

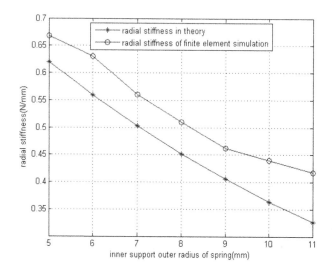

Fig. 7. The radial stiffness of theoretical calculation and finite element simulation varying with changing inner support outer radius of plane supporting spring

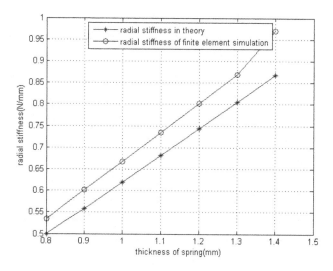

Fig. 8. The radial stiffness of theoretical calculation and finite element simulation varying with changing thickness of plane supporting spring

5 Conclusions

With wider application of the plane supporting spring, an ideal theoretical formula for calculating radial stiffness grows more and more important in the process of its design. This study has proposed a brand-new energy-based method for computing the radial stiffness of single spiral plane supporting spring, with the radial stiffness formula well established theoretically. The results generated by the formula are basically consistent with those from finite element simulation. Comparing both results makes it possible to derive the general discipline of the radial stiffness of single spiral plane supporting spring in relation to changes of the spring's key parameters. And it may fairly be assumed that our study has provided a solid theoretical basis for further research on the multi-spiral/turn plane supporting spring as well.

Acknowledgement

We would like to acknowledge the contributions of a number of staff with Jienuo Environment Technology. Special thanks go to their valuable work and, especially, to their company, Nanjing Jienuo Environment Technology Co., Ltd., which provided us with whatever we needed for the research and experiments. We are also greatly indebted to Professor Li Lingzhen for her successful organization and great support for the project.

References

[1] Jia, F., Zhang, D., Zhang, Z.: Speedy stiffness modeling and designing of plane supporting spring. In: The 15th International Conference on Mechatronics and Machine Vision in Practice (M2VIP 2008), Auckland, New Zealand, pp. 209–214 (2009)

[2] Chen, N., Chen, X., Wu, Y.N., Yang, C.G., Xu, L.: Spiral profile design and parameter analysis of flexure spring. Cryogenics (46), 409–419 (2006)

[3] Nan, C., Xi, C., Yinong, W., Lie, X., Chunguang, Y.: Performance analysis of spiral flexure bearing. Cryogenics and Superconductivity 33(4), 5–8 (2005)

[4] Weili, G., Pengda, Y., Guobang, C.: Influence of geometry parameters on the performance of flexure bearing. Cryogenics (6), 8–11 (2007)

[5] Gaunekar, A.S., Goddenhenrich, T., Heiden, C.: FiniteElement nalysis and Testing of Flexure Bearing Elements. Cryogenics 36(5), 359–364 (1996)

[6] Liu, H.-w.: Mechanics of Materials (II). Higher Education Press, Beijing (2004)

[7] Jiang, S., Jia, F., Zhang, D., Wang, X.: Parameterized Modeling Technology of Plane Supporting Spring Based on APDL. Equipment for Electronic Products Manufacturing 37(12), 46–49 (2008)

Author Index